Frank Clark

| TERM | SYMBOLIC QUANTITY | ABBREVIATION |
|---|---|---|
| Output power | $P_o$ | |
| Reactive power | $VAR$ | var |
| Power factor | | PF |
| Push-pull | | pp |
| Quality | $Q$ | |
| Radius | $R, r$ | rad |
| Reactance | $X$ | |
| Capacitive reactance | $X_C$ | |
| Inductive reactance | $X_L$ | |
| Resistance | $R$ | |
| Root mean square | | rms |
| Revolutions per minute | | rpm |
| Second | $T$ | s |
| Microsecond | | μs |
| Sigma | $\Sigma$ | |
| Sine | | sin |
| Specific resistance | $K$ | |
| Tangent | | tan |
| Television | | TV |
| Time | $T$ | |
| Transistor | $Q$ | |
| Dc base current | $I_B$ | |
| Dc collector current | $I_C$ | |
| Dc emitter current | $I_E$ | |
| Alpha—the ratio of $\Delta I_C$ to $\Delta I_E$ | $\alpha$ | |
| Beta—the ratio of $\Delta I_C$ to $\Delta I_B$ | $\beta$ | |
| Base bias resistor | $R_B$ | |
| Collector-to-base resistance | $R_{CB}$ | |
| Resistance in the collector leg; the load resistor | $R_L$ | |
| Base-to-ground voltage | $V_B$ | |
| Base supply voltage | $V_{BB}$ | |
| Base-to-emitter voltage drop | $V_{BE}$ | |
| Collector-to-ground voltage | $V_C$ | |
| Collector-to-base voltage | $V_{CB}$ | |
| Collector supply voltage | $V_{CC}$ | |
| Collector-to-emitter voltage | $V_{CE}$ | |
| Emitter-to-ground voltage | $V_E$ | |
| Emitter supply voltage | $V_{EE}$ | |
| Turns ratio | | TR |
| Vacuum tube | | VT |
| Plate | $P$ | |
| Plate current | $I_p$ | |
| Plate resistance | $R_p$ | |
| Plate voltage | $E_p$ | |
| Screen current | $I_s$ | |
| Volt | $V$ | V |
| Kilovolt | | kV |
| Millivolt | | mV |
| Microvolt | | μV |
| Voltage | $V$ | |
| Average voltage | $V_{av}$ | |
| Effective value | $V$ | |
| Instantaneous value | $v$ | |
| Maximum value | $V_{\max}$ | |
| Voltampere | $VA$ | VA |
| Kilovoltampere | $KVA$ | kVA |
| Voltage ratio | $VR$ | |
| Watt (unit of electric power) | $P$ | W |
| Kilowatt | | kW |
| Kilowatthour | | kWh |
| Milliwatt | | mW |
| Microwatt | | μW |
| Watthour | | Wh |
| Wavelength | $\lambda$ | |

# BASIC MATHEMATICS FOR ELECTRICITY AND ELECTRONICS

**Fifth Edition**

**BERTRAND B. SINGER**
Former Assistant Chairman
Mathematics Department
Samuel Gompers High School
New York

**HARRY FORSTER**
Assistant Professor
Electronics Department
Miami-Dade Community College
Miami, Florida

**McGraw-Hill Book Company**

New York Atlanta Dallas St. Louis San Francisco Auckland Bogotá Guatemala Hamburg
Lisbon London Madrid Mexico Montreal New Delhi Panama Paris
San Juan São Paulo Singapore Sydney Tokyo Toronto

**Sponsoring Editor:** George Z. Kuredjian
**Editing Supervisor:** Iris Wheat
**Design Supervisor/Cover Designer:** Judith Yourman
**Production Supervisor:** Priscilla Taguer

**Cover Photographer:** Richard Megna/Fundamental Photographs

**Library of Congress Cataloging in Publication Data**

Singer, Bertrand B.
Basic mathematics for electricity and electronics.
Bibliography: p.
Includes index.
1. Electric engineering—Mathematics. 2. Electronics
—Mathematics. I. Forster, Harry, (date). II. Title.
TK153.S54 1984 621.3′01′51 83-929
ISBN 0-07-057477-4

**BASIC MATHEMATICS FOR ELECTRICITY AND ELECTRONICS, FIFTH EDITION**



9 0 DOCDOC 9 0

ISBN 0-07-057477-4

# CONTENTS

# PREFACE

The success of the first four editions of this book has been credited to the presentation of the principles of mathematics as a direct result of a need encountered in the development of the electrical theory. This objective has been emphasized to an even greater degree in this fifth edition.

The following new subjects have been added in an effort to keep up with our modern, rapidly changing electronics technology.

- Job 9-9 on Norton's equivalent is a natural extension of Thevenin's equivalent and a useful tool for conceptualizing circuit operation with current-controlled devices such as transistors. It is also very useful in reducing the complexity of circuit analysis for many network configurations.
- An individual cannot read a manufacturer's transistor data sheets without a knowledge of H parameters. For this reason, a brief introduction to this subject has been included.
- Reading oscilloscope faces is a practical extension of graphs that is intended to take the pure mathematics of graphing and relate it to the real needs and applications of taking data from a CRT display.
- Job 10-13 on the sizing of power resistors is another effort to relate mathematics to the real-world limitations of standard power sizes for resistors. This will give the student an introduction to the problems of component selection and specification.
- The section in the Appendix on standard values of resistors reinforces the job on power sizing. This material will permit the instructor to provide more practical presentations.
- A new chapter, Chap. 23 on mathematics for logic control, introduces concepts that are becoming increasingly important in the advancing age of electronic control.

Please note that all these additions have the objective of bringing the student's electrical education into closer contact with the real world. For example, the usual approach to H parameters has been through abstract mathematics. Our new technique draws only on Thevenin and Norton equivalent circuits. Rather than being a new complicated concept, H parameters has become nothing more than an application of already-learned skills.

Simple practicality is again the keynote in Chap. 23. This chapter avoids the postulates and axioms of boolean albegra. The emphasis is on the fact that mathematics has no value in everyday life unless it relates to our daily experiences and describes and predicts them. For this reason, all of the boolean concepts are developed in terms of both conventional concepts of logic and experience with common switches. Also included is a strong development of the fact that switching logic and digital logic have common analysis concepts but different forms of hardware realization.

As in all previous editions, very little has been taken for granted. Whenever a new mathematical operation is needed—no matter how simple—a checkup and brushup on the operation are provided. Each checkup is a mathematical diagnostic test and motivation for the job that follows. The problems are electrical in nature but can be solved solely on the basis of the mathematical principle involved, without any knowledge of the electrical theory to follow. Each brushup reviews the mathematical concepts immediately necessary for the continuance of the electrical development and application. For example, Jobs 3-1 to 3-5 (checkups and brushups on formulas) precede the algebraic manipulation of the Ohm's law formula for finding current or resistance. The theory is developed in slow, simple stages and is directly and immediately applied to the solution of realistic industry-related problems and actual industrial applications.

Each new concept, electrical or mathematical, is developed from fundamental principles and then applied to a variety of situations. This is accomplished by breaking each chapter down into a series of short "jobs" that concentrate on a small portion of the large concept involved in the chapter. Each job develops either a new mathematical operation or a new electrical concept and is illustrated by a series of examples that increase in both arithmetical difficulty and electrical scope. The many users of this book have indicated that this illustrative method makes the book particularly useful for individuals in the trade as well as for students in basic courses in electricity and electronics.

The illustration method has been further enhanced by programming the last example of each job and the summaries and reviews. These self-test examples serve as a final check on the students' understanding, forcing them to stop, think, and check *each* step of the problem, thus reinforcing the concepts learned in the previous illustrations. A tremendous advantage of these self-tests is that students have a chance to reach their own conclusions *before* they get the answer. Each self-test thus becomes a diagnostic test to pinpoint the precise error in the students' thinking before they tackle the problems of the jobs. The problems in each job are very carefully graded. Answers to the odd-numbered problems are provided to augment the self-help features of the book. Continued review and drill in both mathematical and electrical concepts are obtained by extending the electrical theory and application.

Diagrams (many in color) are still used extensively, with particular attention being paid to the process of changing the words of a problem into a simple sketch, properly labeled with all the information given and to be found. A programmed summary and set of review problems as well as many tests are included in each chapter. The review jobs attempt to attack the problems in the chapter from a slightly different point of view to round up any loose ends in the student's mind. Modern standards as recommended by the Institute of Electrical and Electronics Engineers have been used throughout this book.

The authors wish to extend appreciation to Mr. William Nanes for his review of, and suggestions for, the new chapter, Chap. 23. Finally, the authors wish to acknowledge their indebtedness to the many users of this book for their valuable suggestions for this revision.

*Bertrand B. Singer*
*Harry Forster*

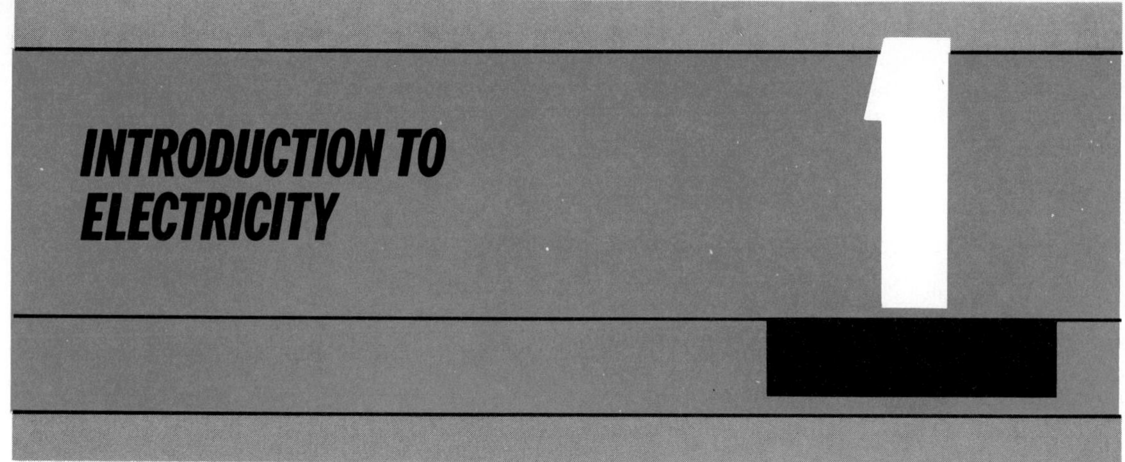

# INTRODUCTION TO ELECTRICITY

## JOB 1-1 BASIC THEORY OF ELECTRICITY

We shall start our study of electricity with an examination of the materials from which electric energy is produced.

**Elements.** Science has discovered more than 100 different kinds of material called *elements.* These cannot be made from other materials and cannot be broken up to form other materials by ordinary methods. Gold, iron, copper, oxygen, and carbon are some examples of elements.

**Atoms.** An atom is the smallest amount of an element which has all the properties of the element. An atom may be broken down into smaller pieces, but these pieces have none of the properties of the element. It is now believed that all material is electrical in nature. All matter is made of combinations of elements, which, in turn, are made of atoms. The atoms themselves are merely combinations of different kinds of electric energy. The theory presently accepted is that an atom, because it is electrically neutral, is made of positive charges of electricity called *protons* and an equal number of negative charges called *electrons.* There are also a number of electrically neutral particles called *neutrons.* Similarly charged particles will repel each other, while particles of opposite charge will attract each other. For example, two electrons will repel each other, and two protons will repel each other, but an electron will be attracted to a proton as shown in Fig. 1-1.

**Structure of the atom.** An atom is believed to resemble our solar system with the sun at the center and the planets revolving around it, as

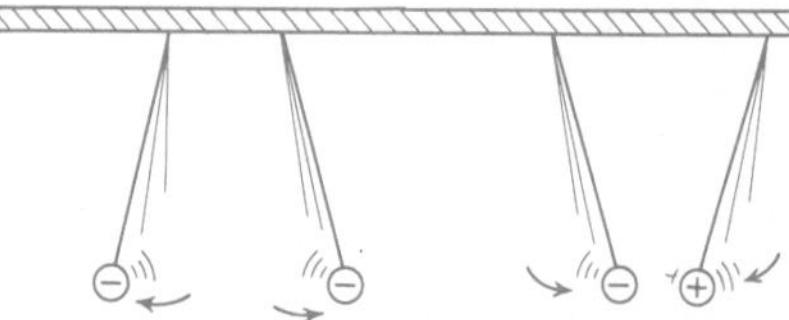

**FIGURE 1-1**
Opposite charges attract; like charges repel each other.

shown in Fig. 1-2*a*. An atom consists of electrons revolving around a nucleus, which contains the protons and neutrons, as shown in Fig. 1-2*b*. The number of revolving, or *planetary,* electrons is equal to the number of positively charged protons in the nucleus.

**Free electrons.** The electrons farthest from the nucleus are called *free electrons* because they are bound very loosely to the nucleus. If you shuffle your feet over a rug on a dry day, some of these free electrons will be rubbed off the rug and will accumulate on your body. Touching a metal doorknob or some other conductor will produce a slight shock as the electrons *flow* through your body to the ground.

**Electron flow.** Notice that no shock will be experienced while the body is accumulating the electrons. The shock will occur only when the electrons *flow* through your body in one concerted surge. This directed flow, or movement, of the electrons is one of the most important ideas in electricity. In order to understand this better, let us compare the particles of matter with a brick wall, as in Fig. 1-3. A brick in a wall is like an atom, since it is the smallest part of the wall that has the characteristics of the wall. If an individual brick were to be powdered into dust, we could compare a single grain of dust with an electron. The grains of brick cannot do any useful work by themselves, but if a powerful pressure like a blast of compressed air were allowed to hit the particles, as would occur in a sandblasting machine, then enormous energies would be available.

We can also consider the drops of water in a stream to be like the electrons in a piece of matter. If the drops of water move aimlessly, as in a water sprinkler, then their energy is small. But if all the drops are forced to move in the *same direction,* as in a high-pressure fire hose, then their energy is large. We can see, then, that if electrons are to do useful work, they must be *moving under pressure,* like the grains of sand or drops of water.

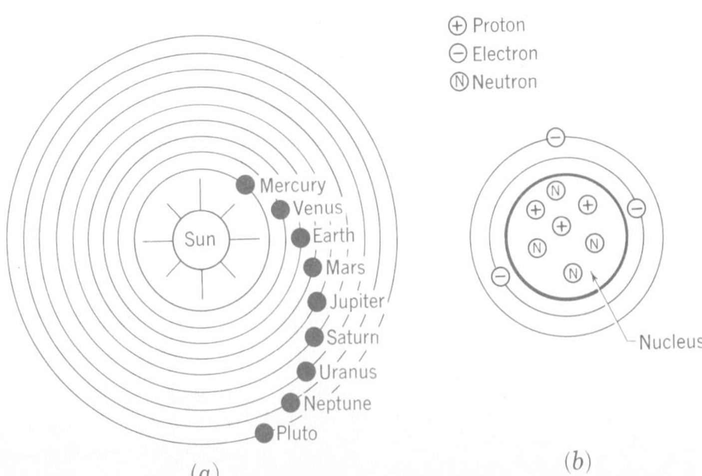


**FIGURE 1-2**
The resemblance between the solar system (*a*) and the theoretical diagram of an atom of lithium (*b*).

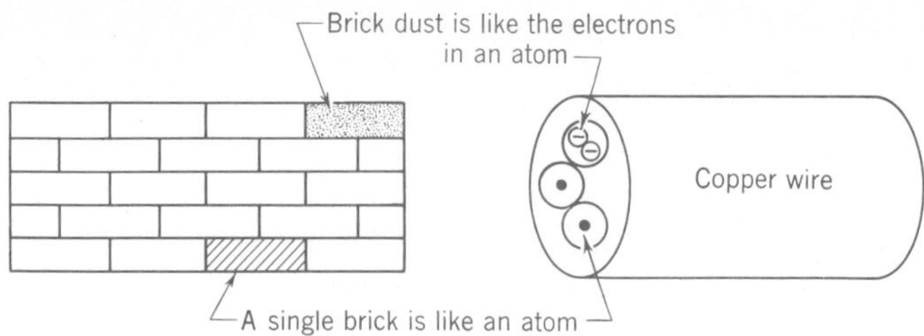


**FIGURE 1-3**
Comparison of brick dust with electrons.

**Electrical pressure: voltage.** In order to move the drops of water or the grains of sand, a mechanical pressure supplied by a water pump or an air compressor is required. Similarly, an electron pressure is required to move the electrons along a wire. This electrical pressure is called the *voltage* and is measured in units of *volts* (V). This unit of measurement was named after Count Alessandro Volta, an Italian physicist (1745–1827).

**Producing electrical pressures.** A working pressure is considered to exist when there is a *difference* in energies. Water in a high tank will exert a pressure because of the *difference* in the height of the water levels, as shown in Fig. 1-4. Elements differ not only in the number of electrons that make up their individual "solar systems" but also in the energies of their electrons. Therefore, if two different materials are brought together under suitable conditions, there will be a *difference* of electron energies. This difference produces the electrical pressure that we call the voltage.

Different combinations of materials will produce different voltages. A dry cell of carbon and zinc produces 1.5 V, a lead and sulfuric acid cell produces 2.1 V, and an Edison storage cell of nickel and iron produces 1.2 V. These combinations are called *batteries.* A voltage pressure may be produced in several other ways, which we shall learn about later. For example, the 110 to 120 V supplied by an ordinary house outlet is produced by a machine called a *generator.* An automobile spark coil delivers about 1500 V. The purpose of all voltages, however produced, is to provide a force to *move* the electrons. It is appropriately termed an

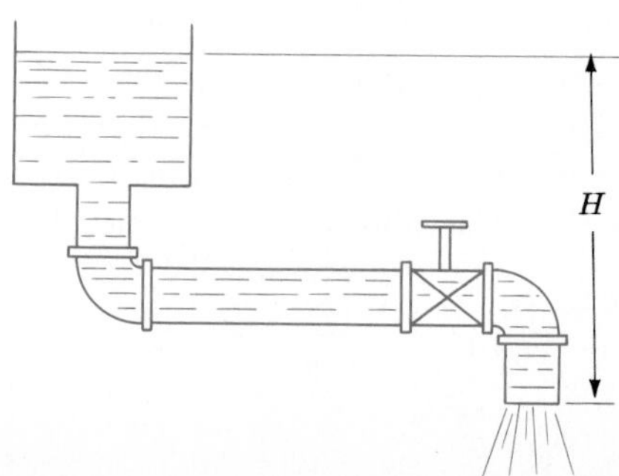


**FIGURE 1-4**
The difference in height between the water levels causes pressure.

electron-moving, or *electromotive, force,* which is abbreviated emf. Modern practice uses the symbol $V$ to indicate emf.

**Quantity of electricity.** The amount of electricity represented by a single electron is very small—much too small to be used as a measure of quantity in practical electrical work. The electron has already been compared with a drop of water. It is obviously ridiculous to measure quantities of water by the number of drops. Instead, we use quantities like a gallon or a liter, each of which represents a certain number of drops. Similarly, the practical unit of electrical quantity represents a certain number of electrons and is called a *coulomb* (C). This unit was named after C. A. Coulomb, a French physicist (1736–1806). The coulomb is equal to about 6 billion billion electrons (6,000,000,000,000,000,000). The symbol for electrical quantity is $Q$.

**Current.** When a voltage is applied across the ends of a conductor, the electrons, which up to now have been moving in many different directions (Fig. 1-5*a*), are forced to move in the *same direction* along the wire (Fig. 1-5*b*). Individual electrons all along the path are forced to leave the atoms to which they are attached. They travel only a short distance until they find an atom that needs an electron. This motion is transmitted along the path from atom to atom, as motion in a whip is transmitted from one end to the other. This *flow* of electrons is called an *electric current* (Fig. 1-6). The speed of this flow is very nearly equal to the speed of light, which is about 186,000 miles per second (mi/s). The actual speed of the individual electrons is much slower, but the effect of a pressure at one end of a wire is felt almost instantaneously at the other end.

The flow of water is measured as the number of gallons per minute, barrels per hour, etc. Similarly, the flow of electricity is measured by the number of electrons that pass a point in a wire in 1 s. We do not have special names for the flow of water, but we do have a special name for the flow of electrons. This name is *ampere* (A) of current. A current of 1 A represents a flow of 1 C of electricity (6 billion billion electrons) past a point in a wire in 1 s, that is, a flow of 1 C/s.

The symbol for current is $I$. The ampere was named after André Marie Ampère, a French physicist and chemist (1775–1836). Since an electron is so small (about 25 trillion to an inch) and since there are so many of them, it is impossible actually to count them as they go by. However, when electrons are moving as an electric current, they can do useful work, such as lighting lamps, running motors, producing heat, and plating metals. We can make use of this last ability of an electric current

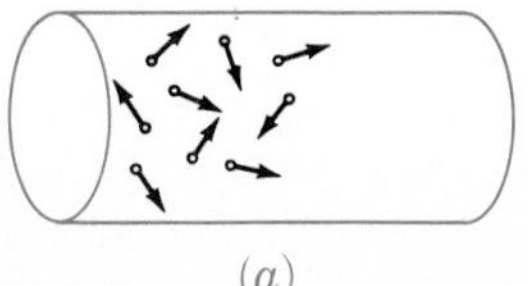
(*a*)

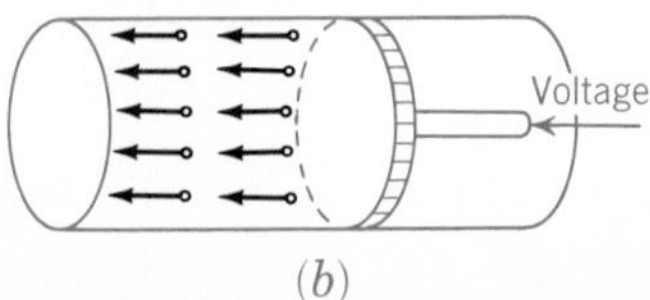

(*b*)

**FIGURE 1-5**
(*a*) The electrons in the wire move in many different directions. (*b*) When a voltage pushes the electrons, they all move in the *same* direction and make an electric current.

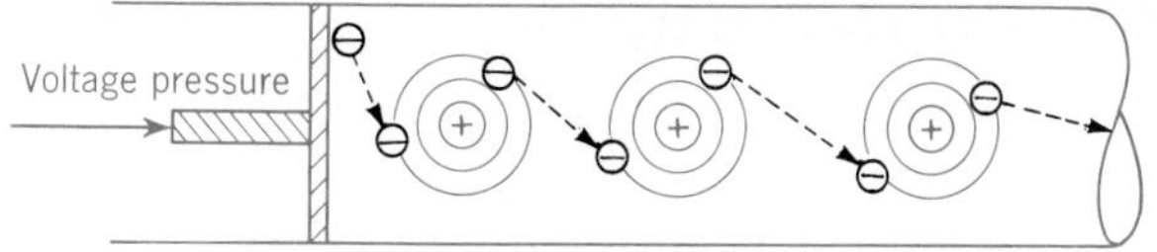


**FIGURE 1-6**
An electric current is produced by billions of electrons moving through a wire.

to measure and define it accurately. An international commission has thus defined an ampere as the number of electrons that can deposit a definite amount of silver [0.001 118 gram (g)] from a silver solution in 1 s. A 100-W 110-V lamp uses about 1 A. A 600-W 110-V electric iron uses about 5.5 A, and the current required by a television transistor may be as low as 0.001 A.

**Resistance.** We have learned that "free" electrons may be forced to move from atom to atom when a voltage pressure is applied. Different materials vary in their number of free electrons and in the ease with which electrons may be transferred between atoms. A *conductor* (Fig. 1-7*a*) is a material through which electrons may travel freely. Most metals are good conductors. An *insulator* (Fig. 1-7*b*) is a material that prevents the electrons from traveling through it easily. Nonmetallic materials like glass, mica, porcelain, rubber, and textiles are good insulators. No material is a perfect insulator or a perfect conductor.

The ability of a material to resist the flow of electrons is called its *resistance* and is measured in units called *ohms* (Ω). This unit of measurement was named after Georg Simon Ohm, a German physicist (1787–1854). The symbol for resistance is $R$. An international agreement defines the ohm as the resistance offered by a column of mercury of uniform cross section, 106.3 cm long (about 41.8 in), and weighing 14.45 g (about ½ oz). For example, 1000 ft of No. 10 copper wire has a resistance of almost exactly 1 Ω; the resistance of a 40-W electric light is 300 Ω when hot; and a 150-V voltmeter has a resistance of 15,000 Ω.

There are many electric devices which make use of the fact that materials offer resistance to the flow of an electric current. Soldering irons, electric heaters, and electric light bulbs contain conductors which have a high resistance compared with the resistance of the connecting wires. Radio and television circuits contain many resistance elements, called *resistors*. They are not always made of special resistance wire. Carbon and mixtures of carbon and insulating materials are molded to form resistors.

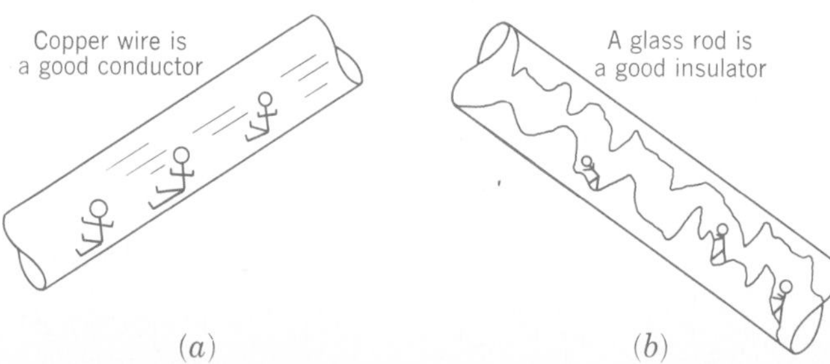


**FIGURE 1-7**
(*a*) Electrons travel freely through conductors. (*b*) Insulators prevent electrons from flowing freely.

## JOB 1-2 ELECTRICAL MEASUREMENTS AND CIRCUITS

The voltage, current, and resistance of an electric circuit are measured with special instruments. An *ammeter* measures the current $I$ in units of amperes. A *voltmeter* measures the voltage $V$ in units of volts. An *ohmmeter* measures the resistance $R$ in units of ohms.

**How to use the ammeter.** In order to measure the flow of water in a pipe, a flowmeter is inserted into the pipe, as shown in Fig. 1-8*a*. The pump supplies the pressure to force the water through the pipe and flowmeter against the resistance of the faucet valve. Since all the water in the system must flow through the flowmeter, it must indicate the number of gallons per minute that flow through the pipe. In the same way, in order to measure the flow of current in a wire, an ammeter must be inserted *directly into the circuit* so that all the electrons will be forced through it. The ammeter has a very low resistance and thus does not stop the flow of current. When large currents are measured, certain adjustments must be made in order to prevent the ammeter from burning out (Job 10-1). In Fig. 1-8*b*, the battery supplies the electrical pressure that forces the electrons through the lamp against the resistance of the lamp. Since all the electrons in the circuit must pass through the ammeter, it will indicate the number of electrons per second, or amperes, passing through it.

**How to use the voltmeter.** As we have already learned, the amount of water that will flow in a pipe depends on the *difference* in the pressure between any two points. In the same way, the electric current that will flow through a resistance depends on the *difference* in electrical pressure (voltage) between the two ends, or *terminals,* of the resistance. In order to measure this difference, a voltmeter must be connected across the ends of the resistance, as in Fig. 1-9. Note that the current that flows through the resistance does *not* flow through the voltmeter. Since a voltmeter is designed to measure the electrical pressure, it should be placed in the circuit so that the pressure across the resistance is also across the voltmeter. An ammeter, on the other hand, is inserted *into* the circuit and receives the full current of the circuit. A voltmeter is merely attached to the ends of the part across which the voltage is to be measured.

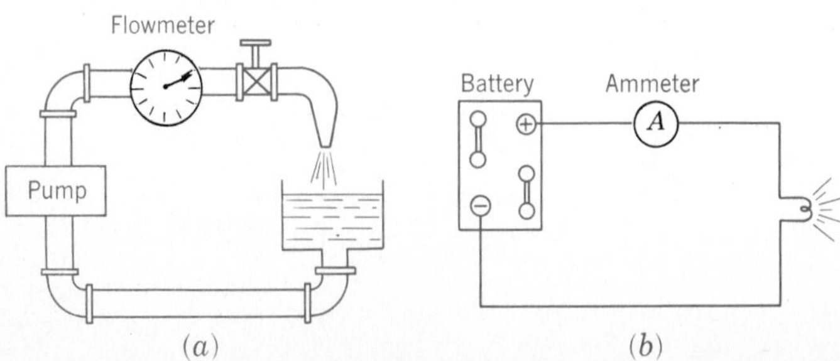


**FIGURE 1-8**
(*a*) A flowmeter is inserted in the line of flow of the water. (*b*) An ammeter is inserted in the line of flow of the electrons.

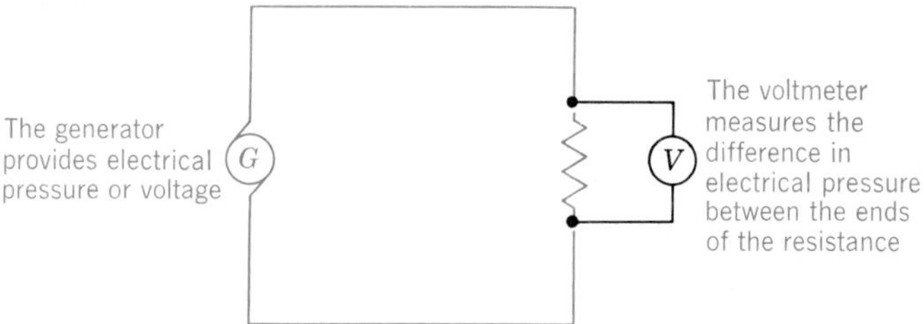
The generator provides electrical pressure or voltage
G
V
The voltmeter measures the difference in electrical pressure between the ends of the resistance

**FIGURE 1-9**
A voltmeter is always connected across the ends of a part when the voltage across it is being measured.

**How to use the ohmmeter.** An ohmmeter is really an ammeter whose dial has been marked to read the resistance in ohms instead of the current in amperes according to a very simple relationship called *Ohm's law,* which we shall study soon. An ohmmeter will measure the resistance of an individual part of a circuit when the leads of the instrument are connected across the ends of the part, as is done with a voltmeter. *Warning:* All power must be *off* when an ohmmeter is used.

**Electric circuits.** A *circuit* is simply a *complete path* along which the electrons can move. A complete circuit must have an *unbroken* path, as shown in Fig. 1-10*a,* with the source of energy acting as an electron pump to force the electrons through the conductor (usually a copper wire) against the resistance of the device to be operated. When the switch is opened as in Fig. 1-10*b,* the electrons cannot leave one side of the switch to enter the other side and so return to their source because of the very high resistance of the air gap. This is an *open circuit,* and no current will flow.

**Fuses.** One of the effects of the passage of an electric current is the production of heat. The larger the current, the more heat is produced. In order to prevent large currents from accidentally flowing through our expensive apparatus and burning it up, a device called a *fuse* is placed directly into the circuit, as in Fig. 1-11, so as to form a part of the circuit through which all the electrons must flow. To protect a circuit, a fuse must be a device which will *open* the circuit whenever a dangerously large current starts to flow. Accordingly, a fuse completes a circuit with a piece of special metal designed to melt quickly when heated excessively, as would occur when a large current flows. The fuse will permit currents smaller than the fuse value to flow but will melt and therefore break the

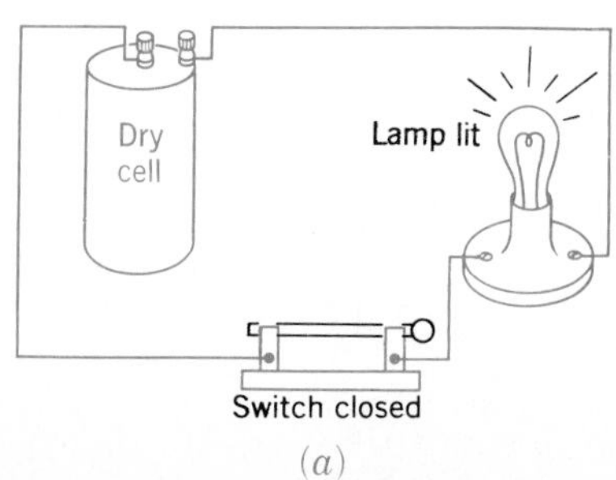
Dry cell
Lamp lit
Switch closed
(a)

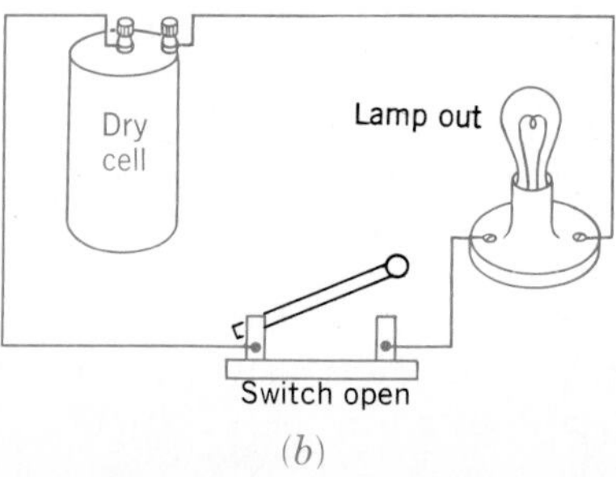
Dry cell
Lamp out
Switch open
(b)

**FIGURE 1-10**
(*a*) With the switch closed, current flows through this complete circuit. (*b*) With the switch open, no current flows through this broken or open circuit.

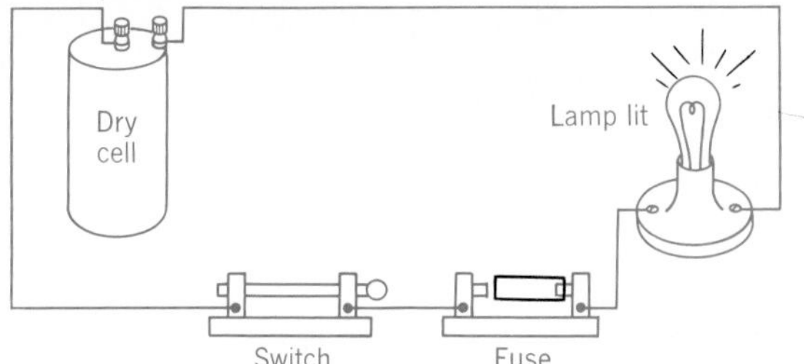


**FIGURE 1-11**
A fuse completes an electric circuit and protects it against the flow of dangerously large currents.

circuit if a larger, dangerous current ever appears. A dangerously large current will flow, for instance, when a "short circuit" occurs. A short circuit is usually caused by an accidental connection between two points in a circuit which offer very little resistance to the flow of electrons. If the resistance is small, there will be nothing to stop the flow of the electrons, and the current will increase enormously. The resulting heat generated might cause a fire. However, if the circuit is protected by a fuse, the heat caused by the short-circuit current will melt the fuse wire, thus breaking the circuit and reducing the current to zero.

**Fuse ratings.** Fuses are rated by the number of amperes of current that can flow through them before they melt and break the circuit. Thus we have 10-, 15-, 30-A, etc., fuses. We must be careful that any fuse inserted in a circuit be rated low enough to melt, or "blow," before the apparatus is damaged. For example, in a house wired to carry a current of 15 A it is best to use a fuse no larger than 15 A so that a current larger than 15 A could never flow.

**Delayed fuses.** In ordinary house wiring, a 15-A fuse is used to protect the line wires from overheating and producing a fire hazard. However, if a current larger than 15 A were to flow for only a few seconds, it would not harm the wires. There are many times when a wire must carry a current larger than it was designed to carry—but only for a short time. For example, a normal 5 A may be flowing in a line. If the motor of a washing machine is suddenly turned on, the current drawn may go up to 35 A for a few seconds but then drop to a normal 10 A after the motor is running. The ordinary fuse would blow, since the current is larger than the rated 15 A. However, since there is no danger, as the current drops again very quickly, we would like a fuse that could carry larger currents for a short time. This type of fuse is called a *time-delay* fuse. It is designed to carry about twice its rated current for 20 to 30 s but to blow if the rated current is exceeded for 1 min. The net effect is that the fuse will hold for *temporary* overloads but will blow on continuous, small overloads or short circuits.

**Symbols.** Most of our diagrams up to this point have used pictures of the various electric devices. However, not everybody can draw pictures quickly and accurately, and so we shall substitute special simple diagrams

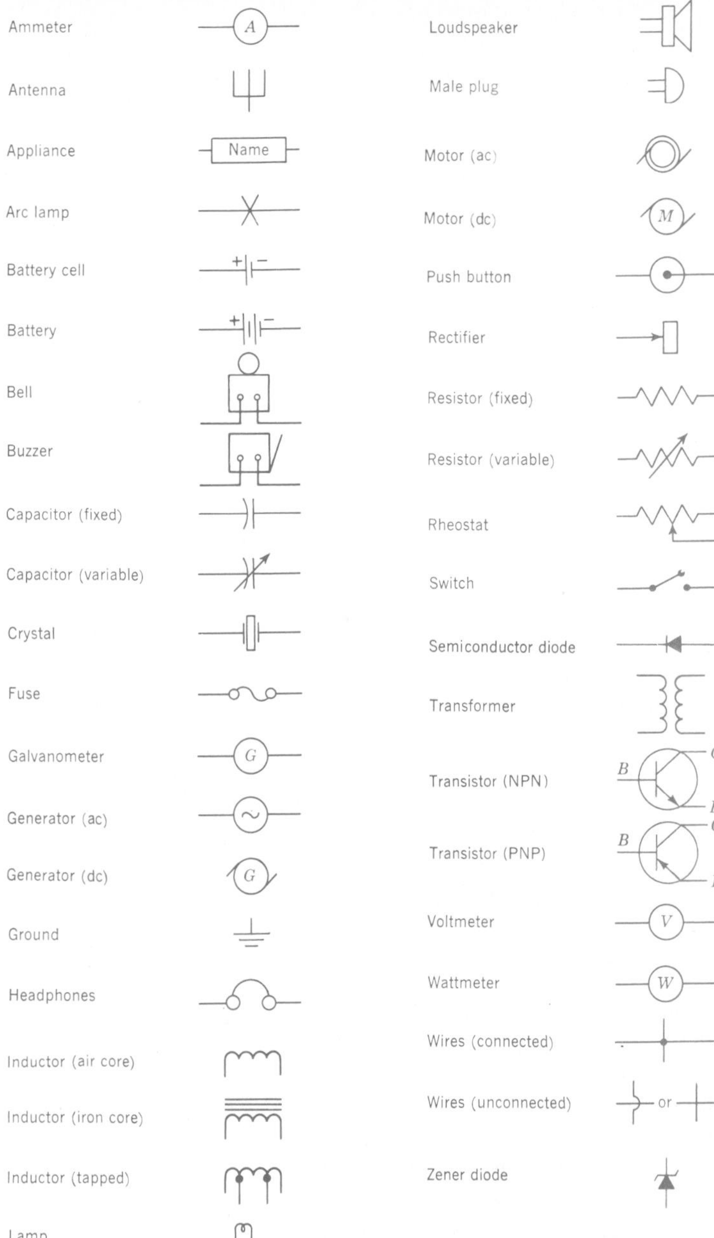


**FIGURE 1-12**
Standard circuit symbols.

to illustrate the various parts of any circuit. Each circuit element is represented by a simple diagram called the *symbol* for the part. The standard symbols for the commonly used electrical and electronic components are given in Fig. 1-12.

**Circuit diagrams.** Every electric circuit must contain the following:

1. A source of electrical pressure or voltage $V$, measured in volts.
2. An unbroken conductor through which the electrons may flow easily. The amount of electron flow per unit of time is the current $I$, measured in amperes.
3. A load, or resistance $R$ to this flow of current, measured in ohms.
4. A device to control the flow of current, or switch.

EXAMPLE 1-1. Draw a circuit containing two battery cells, an ammeter, a fuse, a single-pole switch, a resistor, and a lamp. Label each part, using $V$ for voltage, $R$ for resistance, and $I$ for current.

SOLUTION. See Fig. 1-13.

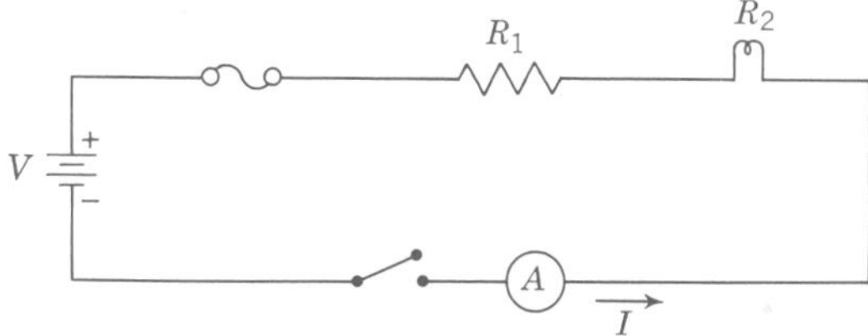


FIGURE 1-13

*Note:* The numbers 1 and 2 under the letter $R$ are a convenient way to indicate the first resistance $R_1$ and the second resistance $R_2$.

**PROBLEMS**

Using the circuit symbols shown in Fig. 1-12, draw a circuit for each problem to include the elements indicated. Label the diagram completely, using the subscripts 1, 2, 3, etc., to indicate the parts of the circuit.

1. A battery of two cells, a fuse, a switch, and two resistors
2. An ac generator, a switch, an ammeter, a bell, and a buzzer
3. A battery, a switch, an ammeter, two resistors, and a lamp with a voltmeter across it
4. A battery, a switch, a bell, and a push button
5. A dc generator, a switch, a rheostat, and a dc motor
6. A dc generator, a switch, a fuse, an arc lamp, and a resistor
7. A male plug, an electric iron, a rheostat, and a switch
8. A male plug, an ac motor, a switch, and a variable resistor

SELF-HELP FEATURES. The summaries and the various steps in some of the illustrative problems that follow are incomplete. The correct answers will be found at the right side of the page. To achieve the maximum benefit from this type of instruction, please follow these instructions carefully.

1. Place the response shield over the answers at the right of the page.

2. Read the statement on the left side of the page, noting the blank space where you are to respond.
3. Write the correct response in the blank space.
4. Slide the response shield down to uncover the correct answer.
5. If you have responded correctly, continue to the next line.
6. If your response is incorrect, *stop!* Review the preceding material to discover the reason for your error.
7. When you are satisfied that you understand, draw a line through the incorrect answer and write the correct answer above it.
8. Repeat steps 2 to 5 above.

## SUMMARY

| Statement | Answer |
|---|---|
| 1. All material is made of ______ charges. | electric |
| 2. Negative charges are called ______. | electrons |
| 3. Positive charges are called ______. | protons |
| 4. Similar charges ______ each other, and ______ charges attract each other. | repel opposite |
| 5. A ______ represents a quantity of 6 billion billion electrons. | coulomb |
| 6. An *ampere* is a current of 1 ____/s. The symbol for current is ____. | C *I* |
| 7. Current is measured by an ______, which is always inserted directly into the circuit and reads ______. | ammeter<br>amperes |
| 8. Electrons move because of the pressure of an electromotive force, or ______. The symbol for voltage is ______. | voltage *V* |
| 9. Voltage is measured by a ______, which is always connected across the ______ of the circuit element and reads ______. | voltmeter<br>ends (or terminals) volts |
| 10. The resistance of a circuit is the resistance offered by the circuit to the flow of ______. The symbol for resistance is ______. | electrons *R* |
| 11. Resistance is measured by an ______, which is always connected across the ______ of the circuit element and reads ______. | ohmmeter<br>ends (or terminals) ohms |
| 12. A *conductor* is a material that allows ______ to flow. | current (or electrons) |
| 13. An ______ is a material that prevents current from flowing easily. | insulator |
| 14. A fuse is a thin strip of easily ______ material. It protects a circuit from large currents by melting quickly, thereby ______ the circuit. | melted<br>breaking |

(See Instructor's Manual for Test 1-1)

# 2 SIMPLE ELECTRIC CIRCUITS

## JOB 2-1 CHECKUP ON FRACTIONS (DIAGNOSTIC TEST)

Before we go any further into our study of electric circuits, let's stop to check up on our knowledge of fractions. The following are some problems often met by electricians and electronic technicians in their daily work. They are all solved by multiplying the numbers involved in the problem. If you have difficulty solving any of these problems, see Job 2-2, which follows.

### PROBLEMS

1. What length of two-conductor BX cable is needed to obtain six pieces each $4^1/_4$ ft long?
2. What is the total horsepower delivered by five $^3/_4$-hp motors?
3. A voltage divider develops an emf of $^1/_{20}$ V for each ohm of resistance. What voltage would be measured across 1800 Ω of resistance?
4. If the resistance of one turn of a variable wirewound resistor is $2^1/_3$ Ω, what is the resistance of three turns?
5. What is the cost of $2^1/_4$ lb of magnet wire at $1.25/lb?
6. An industrial shop uses $9^1/_2$ kWh of electricity per day. How many kilowatthours are used in a month of 24 working days?
7. How many hours of work were spent on an electrical installation if five people each worked $7^1/_2$ hours (h)?
8. A neon electric sign uses $4^1/_2$ W of power per foot of tubing. How many watts are used for a sign which uses 21 ft of tubing?
9. Number 9191 Fiberduct conduit adapters weigh $1^1/_2$ lb each. Find the weight of 24 such fittings.
10. How many feet of antenna wire are there in a $1^1/_8$-lb coil if the wire runs 18 ft to the pound?
11. $2^1/_2 \times 3^1/_5$ 12. $^2/_5 \times 20$ 13. $^{22}/_7 \times 21$ 14. $^1/_2 \times {}^1/_3$
15. $20 \times {}^1/_{1,000}$ 16. $^2/_3 \times {}^3/_4$ 17. $5^1/_3 \times 8^1/_2$ 18. $3^1/_7 \times 5^1/_{11}$

## JOB 2-2 BRUSHUP ON FRACTIONS

**Meaning of a fraction.** A fraction is a shorthand way of describing some part of a total amount. Figure 2-1*a* shows a whole pie. If the pie is cut into four equal pieces, as shown in Fig. 2-1*b*, then each piece is just

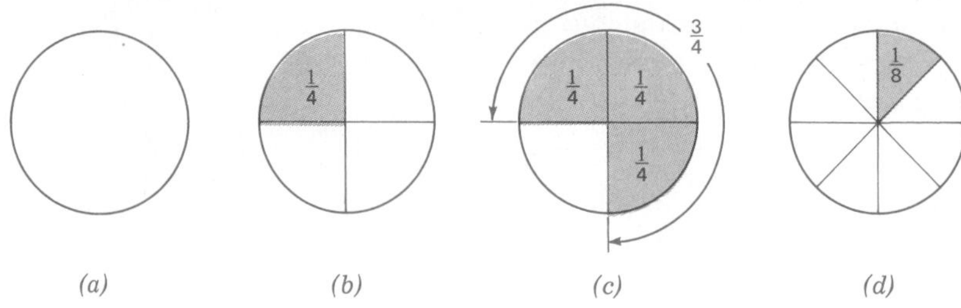


**FIGURE 2-1** Fractional parts of a pie.

one part out of the total of four parts. This is written as the *fraction* $^1/_4$ and means 1 part out of 4 equal parts. The fraction $^3/_4$, as shown by the shaded portion in Fig. 2-1*c*, says that a whole was divided into 4 equal parts and that 3 of these parts were used. The fraction $^1/_8$, as shown in Fig 2-1*d*, says that a whole was divided into 8 equal parts and that 1 of these parts was used. It is evident that we can divide the pie into any number of equal parts. Each part will be 1 part out of the total number of parts. In Fig. 2-2*a* the pie is cut into 8 equal parts. Each part is $^1/_8$ of the pie. The shaded portion is 3 of these, or $^3/_8$ of the pie. In Fig. 2-2*b* the pie is cut into 5 equal parts. Each part is $^1/_5$ of the pie. The shaded portion is 2 of these, or $^2/_5$ of the pie. In Fig. 2-2*c* the pie is again cut into 8 equal parts. Each part is $^1/_8$ of the pie. The shaded portion is 2 of these, or $^2/_8$ of the pie. If the pie had been cut into 4 equal parts, our shaded portion would then have been $^1/_4$. Thus, $^2/_8 = {}^1/_4$.

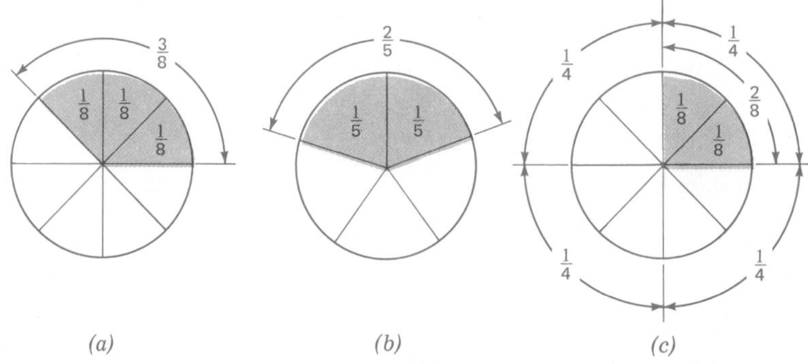


**FIGURE 2-2**

**Simplifying fractions.** We have seen that a portion of a total may be expressed in two different ways. That is, $^1/_4$ is the same as $^2/_8$. The fraction $^1/_4$ is called the *simplified,* or *reduced,* form of $^2/_8$. To simplify a fraction means to change it into another *equivalent* fraction. When the numerator and denominator of a fraction have no common divisor except 1, the fraction is said to be simplified, or reduced, to lowest terms.

**BASIC PRINCIPLE** If a fraction is multiplied or divided by 1, the value of the fraction is unchanged.

The number 1 may be expressed in any number of ways, such as $^{3}/_{3}$, $^{4}/_{4}$, $^{5}/_{5}$, $^{7}/_{7}$, $^{11}/_{11}$, or $^{25}/_{25}$. All that is required for a fraction to be equal to 1 is that the numerator and the denominator be the *same* number. Therefore, to simplify a fraction, divide it by 1, expressed as a fraction with equal numerator and denominator.

To simplify a fraction:

1. Find any number (the largest is best) that will divide evenly into both the numerator and denominator of the fraction.
2. Divide both the numerator and the denominator of the fraction by this number. (This is the same as dividing the fraction by 1.)

EXAMPLE 2-1. Four inches is what fractional part of a foot?

SOLUTION. Since there are 12 in in 1 ft, 4 in is $^{4}/_{12}$ of a foot. Reduce to lowest terms.

$$\frac{4}{12} \div \frac{4}{4} = \frac{4 \div 4}{12 \div 4} = \frac{1}{3} \quad \textit{Ans.}$$

EXAMPLE 2-2. A dropping resistor used 24 V of the total of 96 V supplied to a circuit. Express the fraction $^{24}/_{96}$ in lowest terms.

SOLUTION. We can divide both the 24 and the 96 by 24 to get $^{1}/_{4}$ in one step.

$$\frac{24}{96} \div \frac{24}{24} = \frac{24 \div 24}{96 \div 24} = \frac{1}{4} \quad \textit{Ans.}$$

Or we can simplify in several steps. Dividing both numerator and denominator by 8 will give

$$\frac{24}{96} \div \frac{8}{8} = \frac{24 \div 8}{96 \div 8} = \frac{3}{12}$$

and then dividing both numerator and denominator by 3 will give

$$\frac{3}{12} \div \frac{3}{3} = \frac{3 \div 3}{12 \div 3} = \frac{1}{4} \quad \textit{Ans.}$$

The next example is a self-test type. Please cover the right side of the page with your response card.

SELF-TEST 2-3. The turns ratio of the step-up transformer shown in Fig. 2-3 is $^{36}/_{60}$. Express the ratio in lowest terms.

SOLUTION. To simplify a fraction, we must ______ (multiply/divide) both the | divide

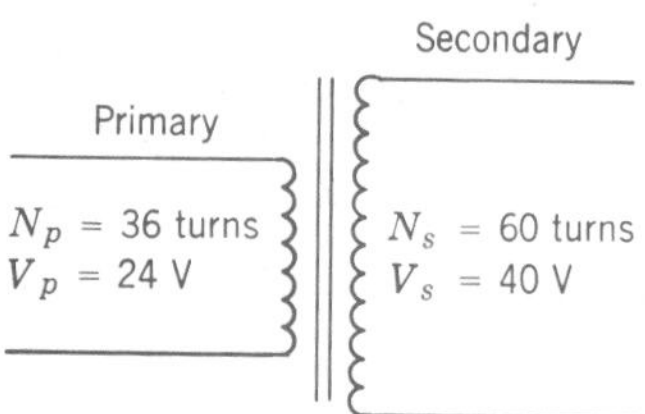


**FIGURE 2-3**
A step-up transformer.

numerator and the denominator of the fraction by the ______ number. The numbers that will divide exactly into both 36 and 60 are 2, 3, ______, ______, and ______. Suppose that we divide the 36 and the 60 by 6. Then,

$$\frac{36 \div 6}{60 \div 6} = \frac{6}{?}$$

Now, to simplify further, divide both the 6 and the 10 by ______.

$$\frac{6 \div 2}{10 \div 2} = \frac{?}{?}$$

Can the fraction $^3/_5$ be simplified any further? ______ (yes/no)
The answer is ______.

same
4
6 12
10
2
$\frac{3}{5}$
no
$^3/_5$

**PROBLEMS**

Express the following fractions in lowest terms:

| | | | |
|---|---|---|---|
| 1. $^4/_8$ | 2. $^3/_9$ | 3. $^4/_{16}$ | 4. $^4/_{80}$ |
| 5. $^{12}/_{16}$ | 6. $^{20}/_{32}$ | 7. $^5/_{20}$ | 8. $^3/_{12}$ |
| 9. $^{24}/_{72}$ | 10. $^{12}/_{36}$ | 11. $^{30}/_{1000}$ | 12. $^{12}/_{48}$ |
| 13. $^{24}/_{64}$ | 14. $^{40}/_{100}$ | 15. $^{14}/_{32}$ | 16. $^{15}/_{60}$ |
| 17. $^{15}/_{45}$ | 18. $^{15}/_{75}$ | 19. $^{60}/_{100}$ | 20. $^{42}/_{72}$ |

21. The ratio of the diameter (in mils) of No. 6 wire compared with No. 12 wire is about $^{162}/_{81}$. Express the ratio in lowest terms.
22. The electric resistance of a piece of iron wire compared with the resistance of an equal length of Nichrome wire is $^{90}/_{100}$. Express the ratio in lowest terms.
23. The current gain $\beta$ of a transistor is the ratio of the change in the output current compared with the change in the input current. If $\beta = {}^{196}/_4$, express it in lowest terms.
24. The turns ratio of a step-up transformer is $^6/_{120}$. Express it in lowest terms.
25. The current in the parallel circuit shown in Fig. 2-4 divides in the ratio of $^{12}/_{30}$. Express it in lowest terms.
26. The National Electrical Code specifies that No. 14 rubber-covered wire can safely carry only 20 A. In lowest terms, what part of the wire's capacity is used by a broiler drawing 12 A?
27. The ratio of 1 horsepower (hp) to 1 kilowatt (kW) is $^{750}/_{1000}$. Express the ratio in lowest terms.

**Reducing mixed numbers to improper fractions.** In a *proper fraction*

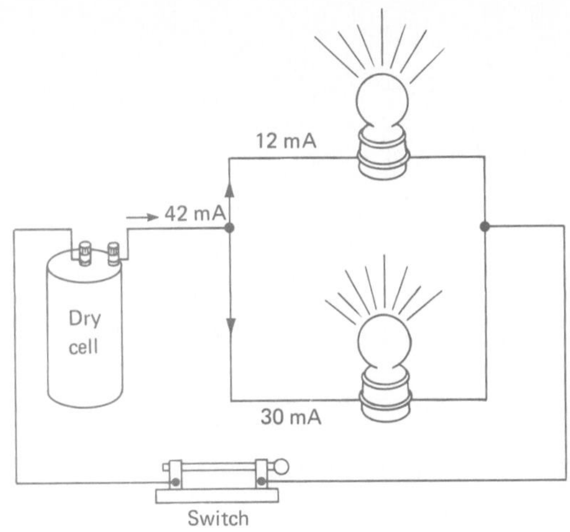


**FIGURE 2-4**
A parallel circuit provides different paths for the electrons to follow.

the top number (the numerator) is always *smaller* than the bottom number (the denominator). Some examples of proper fractions are $^1/_2$, $^3/_4$, and $^8/_{11}$. A *mixed number* is made of two numbers—a whole number and a proper fraction. Some examples are $2^1/_2$, $3^1/_4$, and $5^3/_{16}$. In an *improper fraction* the top number is always *larger* than the bottom number. Some examples are $^5/_3$, $^8/_5$, and $^{22}/_7$.

EXAMPLE 2-4. Change $3^3/_4$ to an improper fraction.

SOLUTION. The improper fraction will be a certain number of fourths. Each whole equals $^4/_4$. In the number 3, there are $3 \times 4$, or 12, fourths—or $^{12}/_4$. But in addition to the 3 wholes, we have $^3/_4$ more. Therefore,

$$3^3/_4 = {}^{12}/_4 + {}^3/_4 = {}^{15}/_4 \qquad \textit{Ans.}$$

**RULE** To change a mixed number to an improper fraction, multiply the denominator of the fraction by the whole number and add the numerator of the fraction. Place this answer over the denominator to make the improper fraction.

EXAMPLE 2-5. Change $2^3/_8$ to an improper fraction.

SOLUTION. The numerator is equal to $2 \times 8$, or 16, plus the 3, which equals 19. Placing this 19 over the denominator 8 gives

$$^{19}/_8 \qquad \textit{Ans.}$$

**PROBLEMS**

Change each of the following mixed numbers to improper fractions:

| | | | |
|---|---|---|---|
| 1. $2^1/_4$ | 2. $3^3/_8$ | 3. $2^2/_9$ | 4. $5^3/_8$ |
| 5. $1^3/_{10}$ | 6. $4^3/_4$ | 7. $4^1/_6$ | 8. $3^7/_{10}$ |
| 9. $4^5/_8$ | 10. $10^1/_2$ | 11. $3^1/_3$ | 12. $2^1/_{12}$ |

## CHANGING IMPROPER FRACTIONS TO MIXED NUMBERS

EXAMPLE 2-6. Change $^{15}/_2$ to a mixed number

SOLUTION. $^{15}/_2$ means 15 divided by 2, or $15 \div 2$. This means that 2 is contained in 15 only 7 times, which is 14, leaving a remainder of 1. This is written as

$$^{15}/_2 = 7^1/_2 \qquad Ans.$$

**RULE** To change an improper fraction to a mixed number, divide the numerator by the denominator. Any remainder is placed over the denominator. The resulting whole number and proper fraction form the required mixed number.

EXAMPLE 2-7. Change $^{25}/_{10}$ to a mixed number.

SOLUTION. $25 \div 10 = 2$ with a remainder of 5, or $2^5/_{10}$. Any remaining fraction like $^5/_{10}$ should be reduced to lowest terms. Therefore,

$$2^5/_{10} = 2^1/_2 \qquad Ans.$$

### PROBLEMS

Change the following improper fractions to mixed numbers or whole numbers:

| | | | | |
|---|---|---|---|---|
| 1. $^7/_4$ | 2. $^9/_2$ | 3. $^{13}/_3$ | 4. $^8/_3$ | 5. $^{15}/_4$ |
| 6. $^{23}/_5$ | 7. $^8/_5$ | 8. $^{23}/_4$ | 9. $^{33}/_{10}$ | 10. $^{14}/_5$ |
| 11. $^{20}/_2$ | 12. $^{43}/_{10}$ | 13. $^{29}/_3$ | 14. $^{26}/_8$ | 15. $^{64}/_4$ |
| 16. $^{16}/_9$ | 17. $^{18}/_5$ | 18. $^{52}/_4$ | 19. $^{96}/_8$ | 20. $^{31}/_4$ |
| 21. $^{70}/_{10}$ | 22. $^{38}/_3$ | 23. $^{17}/_{10}$ | 24. $^{59}/_8$ | 25. $^{47}/_9$ |
| 26. $^{19}/_6$ | 27. $^{108}/_9$ | 28. $^{47}/_{11}$ | 29. $^{147}/_{12}$ | 30. $^{66}/_{10}$ |

## MULTIPLICATION OF FRACTIONS

**RULE** To multiply fractions, place the multiplication of the numerators over the multiplication of the denominators and reduce to lowest terms.

EXAMPLE 2-8. The jaws of the vise shown in Fig. 2-5 will close a distance of $^3/_8$ in for each turn of the handle. How far will the jaws close in only one-half turn?

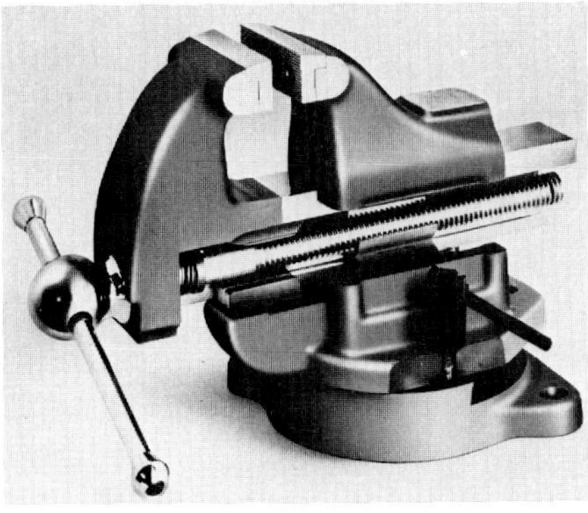

**FIGURE 2-5**
*A vise. (Courtesy The L.S. Starrett Company)*

SOLUTION. One-half turn will close the jaws $^1/_2$ of $^3/_8$ in. In this sense, the word "of" means "multiply." Thus,

$$\frac{1}{2}\text{ of }\frac{3}{8}\qquad\text{means}\qquad\frac{1}{2}\times\frac{3}{8}\qquad\text{or}\qquad\frac{1\times 3}{2\times 8}=\frac{3}{16}\text{ in}\qquad \textit{Ans.}$$

**Cancellation.** To cancel numerators and denominators means to divide both a numerator and a denominator by the same number.

EXAMPLE 2-9. Multiply $^3/_8 \times {}^4/_9$.

SOLUTION. Notice that the 3 on the top and the 9 on the bottom can both be divided by 3. Cross out the 3 and write 1 above it. Cross out the 9 and write 3 under it. Also, the 4 on the top and the 8 on the bottom can both be divided by 4. Cross out the 4 on the top and write 1 above it. Cross out the 8 and write 2 under it as shown below.

$$\frac{3}{8}\times\frac{4}{9}=\frac{\overset{1}{\cancel{3}}}{\underset{2}{\cancel{8}}}\times\frac{\overset{1}{\cancel{4}}}{\underset{3}{\cancel{9}}}=\frac{1\times 1}{2\times 3}=\frac{1}{6}\qquad \textit{Ans.}$$

SELF-TEST 2-10. Multiply

$$\frac{7}{15}\times\frac{25}{14}$$

SOLUTION. The numerator 7 and the denominator ______ may both be divided by ______. Cross out the 7 and write 1 above it. Cross out the 14 and write ______ under it. The numerator 25 and the denominator ______ may be divided by ______. Cross out the 25 and write ______ above it. Cross out the 15 and write ______ under it.

14
7
2 15
5 5
3

$$\frac{\overset{1}{\cancel{7}}}{\underset{3}{\cancel{15}}}\times\frac{\overset{5}{\cancel{25}}}{\underset{2}{\cancel{14}}}=\frac{1\times 5}{3\times 2}=\frac{?}{?}\qquad \textit{Ans.}$$

$\frac{5}{6}$

## PROBLEMS

Multiply the following fractions and reduce to lowest terms:

| | | |
|---|---|---|
| 1. $8 \times {}^{1}/_{2}$ | 2. ${}^{2}/_{5} \times {}^{1}/_{3}$ | 3. ${}^{1}/_{10} \times 3$ |
| 4. ${}^{1}{}_{4} \times 3$ | 5. $5 \times {}^{1}/_{2}$ | 6. ${}^{1}/_{4} \times 7$ |
| 7. ${}^{1}/_{3} \times {}^{1}/_{5}$ | 8. ${}^{3}/_{8} \times 4$ | 9. ${}^{3}/_{8} \times 16$ |

10. A 40-W lamp uses about ${}^{3}/_{8}$ A. How many amperes would four lamps use when connected in parallel?
11. A motor delivers only seven-eighths of the power it receives. Find the power delivered if it receives 6 hp.
12. The actual value of the resistance of many radio resistors may be off as much as one-fifth of the rated value. What would be the possible error in a 230-Ω resistor of this type?
13. What total horsepower is required for the operation of three ${}^{1}/_{6}$-hp single-phase capacitor motors?
14. About two-fifths of the electrolyte in a storage battery is acid. If the battery contains 5 oz of electrolyte, how many ounces of acid are in the cells of the battery?
15. Only four-fifths of the wattage of an electric range is used in computing the size of entrance wires. Find the wattage to use if the range is rated at 4000 W.
16. The load placed on a service line is 375 kilovoltamperes (kVA). If one-third of this load is for motors, three-fifths for lighting, and one-fifteenth for heating, find the kilovoltamperes used for each.

| | | | |
|---|---|---|---|
| 17. ${}^{2}/_{3} \times {}^{1}/_{2}$ | 18. ${}^{7}/_{10} \times {}^{5}/_{14}$ | 19. ${}^{1}/_{7} \times 5$ | 20. ${}^{3}/_{5} \times {}^{5}/_{9}$ |

(See Instructor's Manual for Test 2-1)

## MULTIPLICATION OF MIXED NUMBERS

EXAMPLE 2-11. What length of two-conductor BX cable (Fig. 2-6) is needed to obtain six pieces, each $4^{1}/_{4}$ ft long?

**FIGURE 2-6**
A piece of two-conductor BX cable.

SOLUTION

$$6 \times 4\frac{1}{4} = 6 \times \frac{17}{4}$$

$$\frac{\overset{3}{\cancel{6}}}{1} \times \frac{17}{\underset{2}{\cancel{4}}} = \frac{51}{2} = 25\frac{1}{2} \text{ ft} \qquad \textit{Ans.}$$

EXAMPLE 2-12. If electric energy costs $6^7/_8$ cents per kilowatthour (kWh), find the cost for $5^3/_5$ kWh.

SOLUTION

$$6\frac{7}{8} \times 5\frac{3}{5} = \frac{55}{8} \times \frac{28}{5}$$

$$\frac{\overset{11}{\cancel{55}}}{\underset{2}{\cancel{8}}} \times \frac{\overset{7}{\cancel{28}}}{\underset{1}{\cancel{5}}} = \frac{11 \times 7}{2 \times 1} = \frac{77}{2} = 38\frac{1}{2} \text{ cents} \qquad \textit{Ans.}$$

## PROBLEMS

Multiply the following fractions and mixed numbers. Express the answer in lowest terms.

| | | |
|---|---|---|
| 1. $1^7/_9 \times {}^3/_8$ | 2. ${}^1/_5 \times 4^3/_8$ | 3. $3^1/_6 \times 3$ |
| 4. $1^1/_4 \times 3^2/_5$ | 5. $2^1/_3 \times {}^1/_7$ | 6. $3^1/_3 \times 1^1/_5$ |
| 7. $1^1/_3 \times 3^3/_4$ | 8. $2^1/_4 \times 3^3/_5$ | 9. ${}^1/_5 \times 3^1/_3$ |

10. A neon electric sign uses $11^1/_2$ W of power per meter of tubing. How many watts are used for a sign which uses 20 m of tubing?
11. Find the total voltage supplied by three $1^1/_2$-V dry cells connected in series.
12. UTC power transformers are given a surge test for insulation breakdown of $2^1/_2$ times the normally developed voltage. What is the test voltage for a transformer which delivers 510 V?
13. An electrician gets time and one-half for overtime. If he worked $4^3/_4$ h overtime, for how many hours will he be paid for this overtime?
14. If one-twentieth of the energy put into a motor is lost by friction, copper, and iron losses, how many kilowatts are lost if the motor receives ${}^2/_5$ kW?
15. If one electric outlet requires $13^1/_2$ ft of flexible conduit, how many feet of conduit are required for five such outlets?
16. A capacitor should be checked for dielectric breakdown at $1^1/_2$ times its rated working voltage. If the capacitor to be tested has a working voltage of 600 V, what test voltage should be applied?
17. A factory uses four ${}^3/_4$-hp motors and five ${}^1/_4$-hp motors. What is the total horsepower used when all motors are operating?

SUMMARY

1. A fraction is made of two parts, a ______ and a ______. — numerator denominator
2. The ______ is the top portion of the fraction. — numerator
3. The ______ is the bottom portion of the fraction. — denominator

4. A fraction is a mathematical way to describe some ______ of a total amount. | part

5. If you divide both the numerator and denominator of the fraction $^{10}/_{15}$ by 5, the form of the fraction will be different but the ______ of the fraction will not be changed. | value

6. Consider the fraction $^{12}/_{18}$. The largest whole number which divides evenly into both parts of the fraction is ______. | 6

7. When this number is divided into both parts of the fraction $^{12}/_{18}$, the value of the fraction will not change, and the new fraction will be ______. | $^{2}/_{3}$

8. Can you divide the parts of $^{2}/_{3}$ evenly by any other number except 1? (yes/no) | no

9. When you write $^{12}/_{18} = {}^{2}/_{3}$, you have expressed the fraction in ______ terms. | lowest

10. To express a fraction in lowest terms, we must (multiply/divide) both the numerator and the denominator by the largest number that divides ______ into both of them. | divide | evenly

11. An improper fraction is one in which the numerator is (larger/smaller) than the denominator. | larger

12. Mixed numbers should be changed into ______ fractions before multiplying. | improper

13. To change a mixed number into an improper fraction, multiply the ______ of the fraction by the whole number and add the numerator of the fraction. Place this answer over the ______ to form the improper fraction. | denominator | denominator

14. To change an improper fraction into a mixed number, (multiply/divide) the numerator by the denominator. Any remainder is placed over the ______. | divide | denominator

15. To multiply fractions, place the (product/quotient) of the numerators over the (product/quotient) of the denominators and express in ______ terms. | product | product | lowest

16. To cancel means to ______ *any* one of the numerators and *any* one of the denominators by the ______ number. | divide | same

(See Instructor's Manual for Test 2-2, Fractions)

## JOB 2-3 OHM'S LAW

In Job 1-2 we learned that the following three factors must be present in every electric circuit:

1. The *electromotive force* $V$ expressed in *volts,* which causes the current to flow
2. The *resistance* of the circuit $R$ expressed in *ohms,* which attempts to stop the flow of current
3. The *current* $I$ expressed in *amperes,* which flows as a result of the voltage pressure exceeding the resistance

A definite relationship exists among these three factors, which is known as *Ohm's law.* This relationship is very important, since it is the basis for most of the calculations in electrical and electronic work. In the three circuits shown in Fig. 2-7, the voltage $V$ of the battery is measured by a voltmeter. The current $I$ that flows is measured by an ammeter. The resistance $R$ of the resistor is indicated by the manufacturer by distinctive markings on the resistor. Let us put the information from Fig. 2-7 into a table (see Table 2-1).

TABLE 2-1

| FIGURE | $V$ | $I$ | $R$ | $I \times R$ |
|---|---|---|---|---|
| *a* | $1\frac{1}{2}$ | $\frac{1}{2}$ | 3 | $\frac{1}{2} \times 3 = 1\frac{1}{2}$ |
| *b* | 6 | 2 | 3 | $2 \times 3 = 6$ |
| *c* | 45 | 3 | 15 | $3 \times 15 = 45$ |

In the last column of Table 2-1 we have multiplied the current $I$ by the resistance $R$ for each circuit. Evidently, the result of this multiplication is always equal to the voltage $V$ of the circuit. This is true for all circuits and was first dixcovered by George S. Ohm. This simple relationship is called *Ohm's law*.

**RULE** Voltage equals current multiplied by resistance.

**FORMULA**

$$V = I \times R \quad \boxed{2\text{-}1}$$

In this formula,

$V$ must always be expressed in volts (V).
$I$ must always be expressed in amperes (A).
$R$ must always be expressed in ohms (Ω).

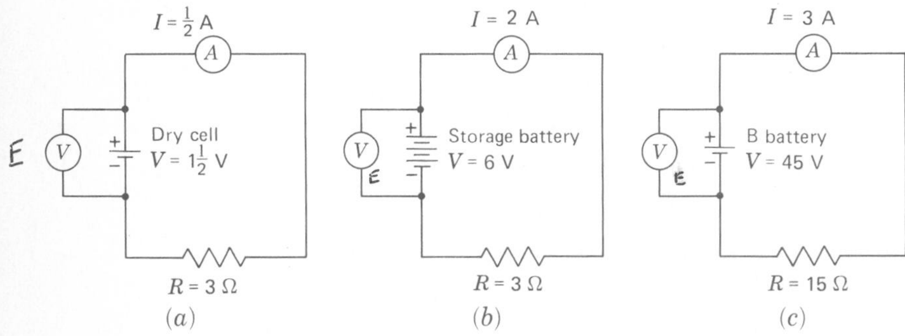


**FIGURE 2-7**
The voltage equals the current multiplied by the resistance in a simple circuit.

## SOLVING PROBLEMS

1. Read the problem carefully.
2. Draw a simple diagram of the circuit.

3. Record the given information directly on the diagram. Indicate the values to be found by question marks.
4. Write the formula.
5. Substitute the given numbers for the letters in the formula. If the number for the letter is unknown, merely write the letter again. Be sure to include all mathematical signs like × or =.
6. Do the indicated arithmetic at the side so as not to interrupt the continued progress of the solution.
7. In the answer, indicate the letter, its numerical value, and the units of measurement.

EXAMPLE 2-13. A doorbell requires ¼ A in order to ring. The resistance of the coils in the bell is 24 Ω. What voltage must be supplied in order to ring the bell?

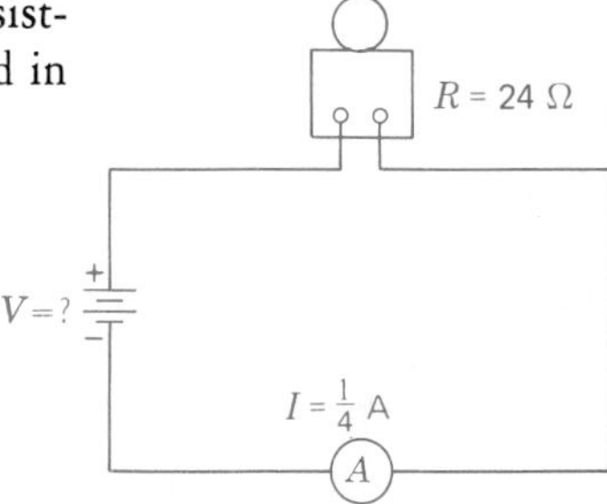


FIGURE 2-8

SOLUTION. The diagram for the circuit is shown in Fig. 2-8.

1. Write the formula. $V = I \times R$
2. Substitute numbers. $V = \frac{1}{4} \times 24$
3. Multiply the numbers. $V = 6$ V *Ans.*

EXAMPLE 2-14. A relay used to control the large current to a motor is rated at 28 Ω resistance. What voltage is required to operate the relay if it requires a current of 0.05 A?

SOLUTION. The diagram for the circuit is shown in Fig. 2-9.

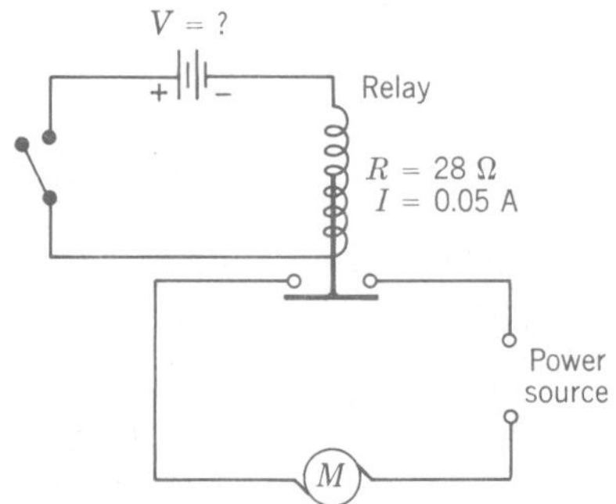


**FIGURE 2-9**
A relay controls the large current drawn by the motor.

1. Write the formula. $V = I \times R$
2. Substitute numbers. $V = 0.05 \times 28$
3. Multiply the numbers. $V = 1.4$ V *Ans.*

SELF-TEST 2-15. An automobile battery supplies a current of 7.5 A to a headlamp with a resistance of 0.84 Ω. Find the voltage delivered by the battery.

SOLUTION. The diagram for the circuit is shown in Fig. 2-10.

1. Write the formula. $V =$ ________ × ________
2. Substitute numbers. $V =$ ________ × 0.84
3. Multiply the numbers. $V =$ ________ V *Ans.*

| $I$ | $R$ |
|---|---|
| 7.5 | |
| 6.3 | |

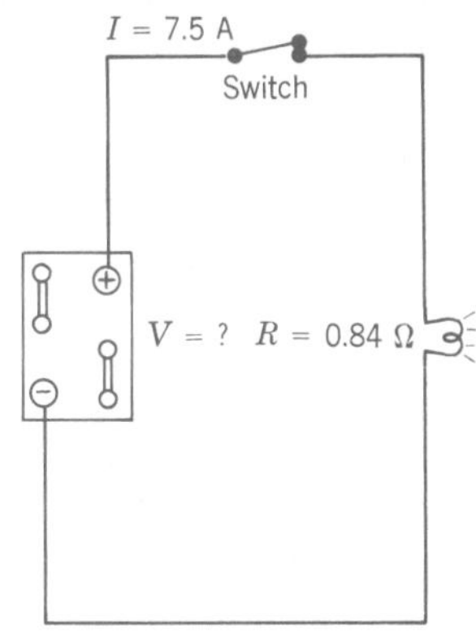


FIGURE 2-10

**PROBLEMS**

1. An arc lamp with a hot resistance of 9 Ω draws 6.2 A. What voltage is required?
2. A 52-Ω electric toaster uses $2^1/_4$ A. Find the required voltage.
3. What voltage is needed to energize the field coil of a loudspeaker if its resistance is 1100 Ω and it uses 0.04 A?
4. The coils of a washing-machine motor have a resistance of 21 Ω. What voltage does it require if the motor draws 5.3 A?
5. What is the voltage required to operate the electric impact wrench shown in Fig. 2-11 if its resistance is 24.5 Ω and it draws 4.8 A?
6. The line from an automobile battery to a distant transmitter must

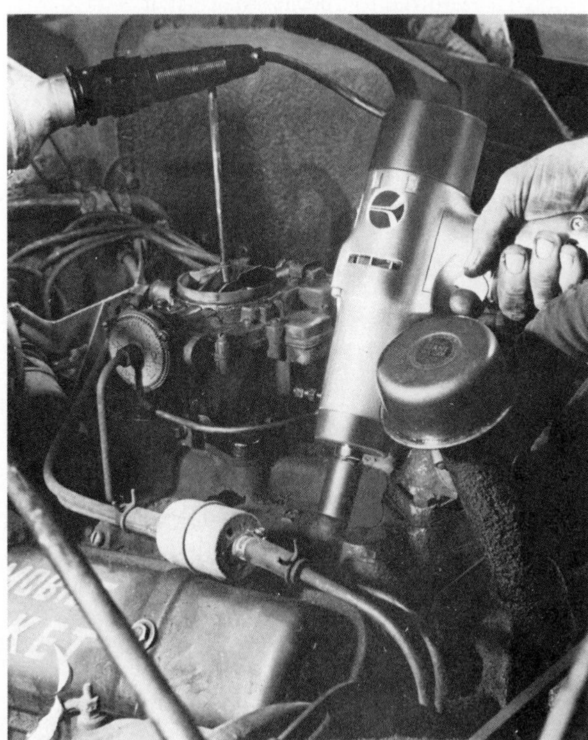

**FIGURE 2-11**
Electric impact wrench. *(Courtesy Rockwell International)*

not develop more than 0.25 V when the transmitter is operating. If the current from the battery is 18 A, will a 0.015-Ω line be satisfactory?

7. The winding of a transformer has a resistance of 63 Ω. What is the voltage drop in the winding when it carries a current of 0.059 A?
8. An electric bell has a resistance of 25 Ω and will not operate on a current of less than 0.25 A. What is the smallest voltage that will ring the bell?
9. A 125-Ω relay coil needs 0.15 A to operate. What is the lowest voltage needed to operate the relay?
10. In the self-bias transistor circuit shown in Fig. 2-12, the load resistor $R_L = 20{,}000$ Ω and carries a current $I_L = 0.000\ 32$ A. Find the voltage drop across the load.

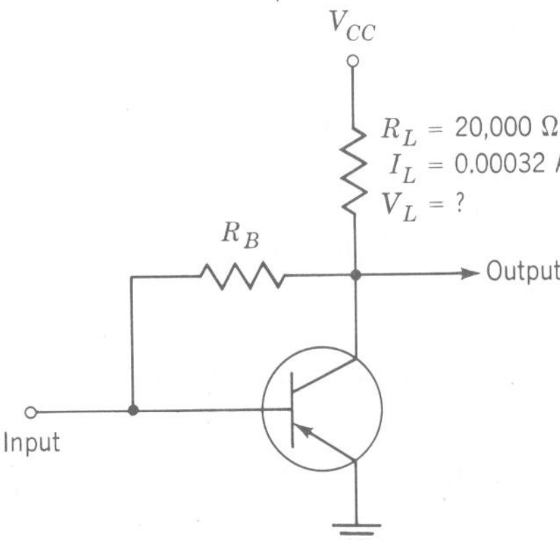


**FIGURE 2-12**
Self-bias transistor circuit.

11. What is the voltage across the shunt of an ammeter if the shunt has a resistance of 0.005 Ω and carries 9.99 A?
12. What voltage is registered by a voltmeter with an internal resistance of 150,000 Ω when 0.001 A flows through it?
13. In the voltage-divider circuit shown in Fig. 2-13, find the voltage across the resistors $R_1$, $R_2$, and $R_3$.

(See Instructor's Manual for Test 2-3)

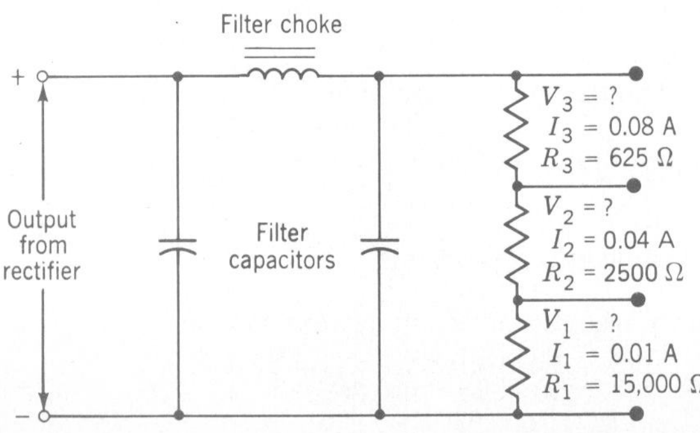


**FIGURE 2-13**
A voltage divider provides different voltages from a single source.

## JOB 2-4 CHECKUP ON DECIMALS (DIAGNOSTIC TEST)

Did you have any difficulty with the decimals in the last job? The following problems occur in the everyday work of the electrician and electronic technician. They all involve decimals and will help you to check up on their use. If you have difficulty with any of these problems, see Job 2-5 which follows.

### PROBLEMS

1. Find the total current drawn by the following appliances by adding the currents: electric iron, 4.12 A; electric clock, 0.02 A; 100-W lamp, 0.91 A; and radio, 0.5 A.
2. Find the total capacitance of a parallel group of capacitors by adding these values: 0.000 25, 0.01, and 0.005 $\mu$F (microfarad).
3. How much larger in diameter is No. 10 copper wire (0.1019 in) than No. 14 wire (0.0641 in)?
4. The current drawn by a motor is 1.21 A at no load and 1.56 A at full load. What is the increase in the current?
5. A motor receives only 117.4 V when connected to a distant generator delivering 120 V. Find the voltage lost in the line wires.
6. The laminated core of a power transformer is made of 15 sheets of steel, each 0.079 in thick. Find the total thickness of the core.
7. The maximum value of an ac voltage wave is 1.414 times its ac meter reading of 46.5 V. Find the maximum value of the voltage wave.
8. If 65 ft of BX cable costs $38.95, what is the cost of 1 ft of this cable?
9. The resistance of 24.5 ft of No. 16 copper wire is 0.0982 Ω. Find the resistance of 1 ft of this wire.
10. Add 7.05, 2, and 3.5.
11. Which current is larger: 0.4 or 0.25 A?
12. Write the following decimals in words: (*a*) 4.3, (*b*) 0.359, and (*c*) 0.41.
13. Multiply $^3/_{10}$ by 0.05.
14. Write as a decimal (*a*) $^1/_4$, (*b*) $^4/_7$, (*c*) $^5/_8$, and (*d*) $^1/_{16}$.
15. Subtract 12 from 18.24.
16. Arrange the following numbers starting with the largest: 0.050, 0.30, 0.0070, 1.1.
17. Find the difference between three-tenths and twenty-five hundredths.
18. What is the excess of 5 over 2.75?

## JOB 2-5 INTRODUCTION TO DECIMAL FRACTIONS

The decimal system is an extremely fast and accurate system to use for most mathematical calculations. The modern electronic calculator is a marvel of speed and capability, and gives answers in extremely accurate decimal figures. Many calculators contain keys for conversions to the metric system of measurement—a decimal system which the United States is expected to adopt completely within the next few years. Also, the accuracy required by modern industry demands the use of the decimal system.

A decimal fraction is a fraction in which the denominator is not written, but with the denominator's value being indicated by the position of a decimal point (.) in the numerator. The denominator of a decimal fraction is always a number like 10, 100, 1000, etc.

The number of digits to the right of the decimal point tells us whether the decimal is to be read as tenths, hundredths, thousandths, etc.

When a decimal has *one* digit to the right of the decimal point, the decimal is read as that many *tenths.* Thus,

$0.3 = {}^{3}/_{10}$ and is read as "3 *tenths*"
$0.9 = {}^{9}/_{10}$ and is read as "9 *tenths*"

When a decimal has *two* digits to the right of the decimal point, the decimal is read as that many *hundredths.* Thus,

$0.23 = {}^{23}/_{100}$ and is read as "23 *hundredths*"
$0.47 = {}^{47}/_{100}$ and is read as "47 *hundredths*"

Now, if we wish to write ${}^{3}/_{100}$ as a decimal fraction, the decimal point must be placed so that there will be *two* digits following it in order to represent *hundredths.* To make up these two digits, we must place a zero between the decimal point and the digit 3. The zero will now push the 3 into the second place, which means hundredths. Thus,

${}^{3}/_{100} = 0.03$ and is read as "3 *hundredths*"

*Note:* The zero *must* be placed between the decimal point and the digit. It is wrong to place the zero after the digit 3, as in 0.30, since this number would be read as 30 hundredths, *not* 3 hundredths.

Similarly,

${}^{7}/_{100} = 0.07$ and is read as "7 *hundredths*"
${}^{9}/_{100} = 0.09$ and is read as "9 *hundredths*"

When a decimal has *three* digits to the right of the decimal point, the decimal is read as that many *thousandths.* Thus,

$0.123 = {}^{123}/_{1000}$ and is read as "123 *thousandths*"
$0.457 = {}^{457}/_{1000}$ and is read as "457 *thousandths*"

If we wish to write ${}^{43}/_{1000}$ as a decimal fraction, the decimal point must be placed so that there will be *three* digits following it in order to represent *thousandths.* To make up these three digits, we must place a zero between the decimal point and the 43. Thus,

$^{43}/_{1000}$ = 0.043 and is read as "43 *thousandths*"
$^{87}/_{1000}$ = 0.087 and is read as "87 *thousandths*"

If we wish to write $^{9}/_{1000}$ as a decimal fraction, the decimal point must still be placed so that there will be *three* digits following it. To make up these three digits, we must now place *two* zeros between the decimal point and the 9.

0.009 = $^{9}/_{1000}$ and is read as "9 *thousandths*"
0.005 = $^{5}/_{1000}$ and is read as "5 *thousandths*"

Following this system, four places means ten-thousandths, five places means hundred-thousandths, six places means millionths, etc. Part of the system is shown in picture form in Fig. 2-14.

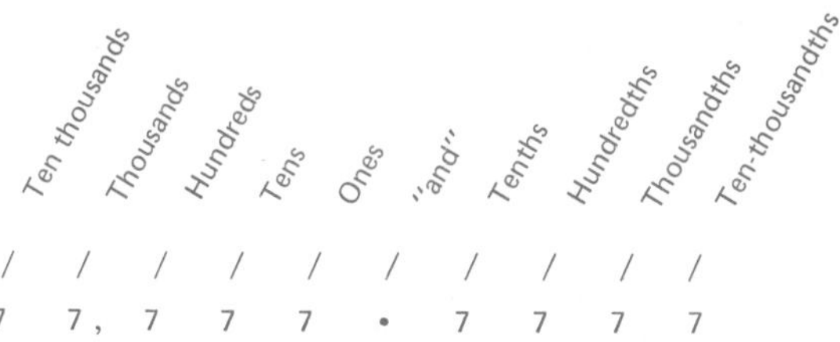


**FIGURE 2-14**
Names of the placeholders in the decimal system.

*Note:* Zeros placed at the *end* of a decimal do *not* change the value of the decimal. They merely describe the decimal in another way. For example: 0.5 (5 tenths) = 0.50 (50 hundredths) = 0.500 (500 thousandths).

A number such as $3^{13}/_{100}$ is written as 3.13 and is read as "3 *and* 13 hundredths." In this type of number, the decimal point is read as the word "and."

**Comparing the value of decimals.** When comparing the value of various decimal fractions, we must first be certain that they have the same denominators. This means that the decimals must be written so that they have the same number of decimal places.

EXAMPLE 2-16. Which is larger: 0.3 or 0.25?

SOLUTION. Since 0.3 has only one decimal place and 0.25 has two decimal places, we must change 0.3 into a two-place decimal by adding a zero. This does *not* change the value but merely expresses it differently. Therefore, 0.3 or 0.30 (30 hundredths) is larger than 0.25 (25 hundredths). *Ans.*

**PROBLEMS**

Write the following fractions as decimal fractions:

| | | | |
|---|---|---|---|
| 1. $^{7}/_{10}$ | 2. $^{29}/_{100}$ | 3. $^{114}/_{1000}$ | 4. $^{3}/_{10}$ |
| 5. $^{6}/_{100}$ | 6. $^{9}/_{1000}$ | 7. $^{18}/_{1000}$ | 8. $^{3}/_{1000}$ |
| 9. $^{11}/_{100}$ | 10. $^{4}/_{10}$ | 11. $^{13}/_{1000}$ | 12. $^{74}/_{100}$ |
| 13. $^{45}/_{1000}$ | 14. $^{316}/_{1000}$ | 15. $^{6}/_{10}$ | 16. $^{23}/_{100}$ |

Arrange the following decimals in order starting with the largest:

| | | |
|---|---|---|
| 17. 0.007, 0.16, 0.4 | 18. 0.2, 0.107, 0.28 | 19. 0.8, 0.06, 0.040 |
| 20. 0.496, 0.8, 0.02 | 21. 0.5, 0.051, 0.18 | 22. 0.90, 0.018, 0.06 |
| 23. 0.1228, 0.236, 0.4 | 24. 0.006, 0.05, 0.3 | 25. 0.19, 0.004, 0.08 |

**Changing mixed numbers to decimals.** When a mixed number is read as a decimal, the word "and" appears as a decimal point. For example, $2\,^{7}/_{10}$ is read as "two *and* seven-tenths" and is written as 2.7. A whole number may be written as a decimal if a decimal point is placed at the *end* of the number. For example, the number 4 means 4.0 or 4.00 or 4.000.

**PROBLEMS**

Change the following mixed numbers to decimals:

| | | | |
|---|---|---|---|
| 1. $2\,^{3}/_{10}$ | 2. $18\,^{5}/_{100}$ | 3. $3\,^{144}/_{1000}$ | 4. $1\,^{17}/_{100}$ |
| 5. $2\,^{25}/_{1000}$ | 6. $7\,^{35}/_{100}$ | 7. $2\,^{20}/_{1000}$ | 8. $3\,^{9}/_{100}$ |
| 9. $1\,^{2}/_{1000}$ | 10. $62\,^{90}/_{100}$ | 11. $9\,^{145}/_{1000}$ | 12. $3\,^{27}/_{1000}$ |

**Changing fractions to decimals.** We shall discover that many of the answers to our electrical problems will be fractions like $^{1}/_{8}$ A, $^{7}/_{40}$ $\mu$F, and $^{3}/_{13}$ $\Omega$. These will be perfectly correct mathematical answers, but they will be completely worthless to a practical electrician or television mechanic. Electric measuring instruments give values expressed as decimals and *not* as fractions. In addition, the manufacturers of electric components give the values of the parts in terms of decimals.

Suppose that we worked out a problem and found that the current in the circuit should be $^{1}/_{8}$ A. Then, using an ammeter, we tested the circuit and found that 0.125 A flowed. Is our circuit correct? How would we know? How can we compare $^{1}/_{8}$ and 0.125? The easiest way is to change the fraction $^{1}/_{8}$ to its equivalent decimal and then to compare the decimals.

**RULE** To change a fraction to a decimal, divide the numerator by the denominator.

EXAMPLE 2-17. Change $^{1}/_{8}$ to an equivalent decimal.

SOLUTION. $^1/_8$ means $1 \div 8$. To write this as a long-division example, place the numerator inside the long-division sign and the denominator outside the sign as shown below.

$$8\overline{)1}$$

We can't divide 8 into 1; but remember that every whole number may be written with a decimal point at the *end* of the number. As many zeros may be added as we desire without changing the value. Our problem now looks like this:

$$8\overline{)1.000}$$

1. Put the decimal point in the answer directly above its position in 1.000.

$$8\overline{)1.000}$$

2. Try to divide the 8 into the first digit. 8 does not divide into 1. Then try to divide the 8 into the first two digits. 8 divides into 10 one time. Place this number 1 in the answer directly above the last digit of the number into which the 8 was divided.

$$\begin{array}{r} 0.1\phantom{00} \\ 8\overline{)1.000} \end{array}$$

3. Multiply this 1 by the divisor 8 and place as shown below. Draw a line and subtract.

$$\begin{array}{r} 0.1\phantom{00} \\ 8\overline{)1.000} \\ \underline{8\phantom{00}} \\ 2\phantom{00} \end{array}$$

4. Bring down the next digit 0 and divide this new number 20 by the 8. The 8 will divide into 20 two times.

$$\begin{array}{r} 0.1\phantom{00} \\ 8\overline{)1.000} \\ \underline{8\downarrow\phantom{0}} \\ 20\phantom{0} \end{array}$$

5. Place this 2 in the answer directly above the last digit brought down.

$$\begin{array}{r} 0.12\phantom{0} \\ 8\overline{)1.000} \\ \underline{8\downarrow\phantom{0}} \\ 20\phantom{0} \end{array}$$

6. Multiply the 2 by the divisor 8, and continue steps 3 to 5. The answer comes out even as 0.125. This means that $^1/_8 = 0.125$ *Ans.*

$$\begin{array}{r} 0.125 \\ 8\overline{)1.000} \\ \underline{8\phantom{00}} \\ 20\phantom{0} \\ \underline{16\phantom{0}} \\ 40 \\ \underline{40} \\ 0 \end{array}$$

EXAMPLE 2-18. The current in a parallel electric circuit divides in the ratio of $^{6}/_{13}$. Express the ratio as a decimal.

SOLUTION. The answer does not come out even:

$$^{6}/_{13} = \begin{array}{r} 0.461 \\ 13\overline{)6.000} \\ \underline{5\,2\phantom{00}} \\ 80\phantom{0} \\ \underline{78\phantom{0}} \\ 20 \\ \underline{13} \\ 7 \end{array}$$

We see that there is a remainder. If the remainder is more than half the divisor, we drop it and add an extra unit to the last place of the answer. Since 7 is more than half of 13,

$$0.461\,^{7}/_{13} \text{ becomes } \begin{array}{r} 0.461 \\ \underline{+1} \\ 0.462 \end{array} \qquad \textit{Ans.}$$

If any remainder is less than half the divisor, drop it completely and leave the answer unchanged. For example,

$$0.236\,^{5}/_{12} = 0.236 \qquad \text{(since 5 is less than half of 12)}$$
$$0.483\,^{1}/_{4} = 0.483 \qquad \text{(since 1 is less than half of 4)}$$

An important question may have occurred to you by now: "If it doesn't come out even, how long should I continue to divide?" The answer to this depends on the use to which the answer is to be put. Some jobs require five or six decimal places, while others need only one place or none at all. For example, the capacitor in the tuning circuit of a radio receiver should be worked out to an answer like 0.000 25 microfarad ($\mu$F). The pitch diameter of a screw thread requires a measurement like 0.498 in, while a bias resistor of 203.4 Ω is just as well written as 203 Ω or even 200 Ω, because precision is not needed in this application.

**Degree of accuracy.** A very general rule for the number of decimal places required in an answer is given in Table 2-2.

TABLE 2-2

| ANSWER | NUMBER OF DECIMAL PLACES | |
|---|---|---|
| Less than 1 | 3 | 0.132, 0.008 |
| From 1 to 10 | 2 | 3.48, 6.07 |
| From 10 to 100 | 1 | 28.3, 52.9 |
| From 100 up | None | 425, 659 |

**PROBLEMS**

Change the following fractions to equivalent decimals:

1. $^1/_4$  2. $^1/_5$  3. $^5/_8$  4. $^1/_3$  5. $^2/_5$
6. $^3/_{10}$  7. $^3/_{20}$  8. $^2/_7$  9. $^7/_8$  10. $^3/_{16}$
11. $^4/_9$  12. $^{13}/_{15}$  13. $^3/_{32}$  14. $^{21}/_{25}$  15. $^{25}/_{40}$
16. $^9/_{16}$  17. $^1/_{50}$  18. $^1/_{200}$  19. $^5/_{26}$  20. $^9/_{64}$

21. A 40-W lamp uses about $^3/_8$ A. Express the current used as a decimal correct to three places.
22. A bell transformer reduced the primary voltage of 120 V to the 18 V delivered by the secondary. Express the voltage ratio ($^{120}/_{18}$) as a decimal.

(See Instructor's Manual for Test 2-4)

**Using the decimal equivalent chart.** There are some fractions that are used very often. These are the fractions which represent the parts of an inch on a ruler, like $^1/_{16}$, $^3/_8$, $^5/_{32}$, $^9/_{64}$, etc. Since they are so widely used, a table of decimal equivalents has been prepared. In order to find the decimal equivalent of a fraction of this type, refer to Table 2-3 on page 33.

**PROBLEMS**

Use the table to find the decimal equivalent of each of the following fractions and mixed numbers:

1. $^5/_8$  2. $^7/_{32}$  3. $^9/_{16}$  4. $^{41}/_{64}$
5. $2^1/_4$  6. $3^{11}/_{32}$  7. $1^9/_{64}$  8. $4^3/_4$

Use the table to find the fraction nearest in value to each of the following decimals:

9. 0.37  10. 0.785  11. 0.449  12. 0.88
13. 0.035  14. 0.525  15. 0.41  16. 0.615

Write each of the following decimals as a mixed number or fraction:

17. 0.75  18. 0.625  19. 0.141  20. 0.5625
21. 2.8125  22. 3.641  23. 1.844  24. 0.9375

TABLE 2-3
TABLE OF DECIMAL EQUIVALENTS

| FRACTION | 1/32DS | 1/64THS | DECIMAL | FRACTION | 1/32DS | 1/64THS | DECIMAL |
|---|---|---|---|---|---|---|---|
| | | 1 | 0.015625 | | | 33 | 0.515625 |
| | 1 | 2 | 0.03125 | | 17 | 34 | 0.53125 |
| | | 3 | 0.046875 | | | 35 | 0.546875 |
| 1/16 | 2 | 4 | 0.0625 | 9/16 | 18 | 36 | 0.5625 |
| | | 5 | 0.078125 | | | 37 | 0.578125 |
| | 3 | 6 | 0.09375 | | 19 | 38 | 0.59375 |
| | | 7 | 0.109375 | | | 39 | 0.609375 |
| 1/8 | 4 | 8 | 0.125 | 5/8 | 20 | 40 | 0.625 |
| | | 9 | 0.140625 | | | 41 | 0.640625 |
| | 5 | 10 | 0.15625 | | 21 | 42 | 0.65625 |
| | | 11 | 0.171875 | | | 43 | 0.671875 |
| 3/16 | 6 | 12 | 0.1875 | 11/16 | 22 | 44 | 0.6875 |
| | | 13 | 0.203125 | | | 45 | 0.703125 |
| | 7 | 14 | 0.21875 | | 23 | 46 | 0.71875 |
| | | 15 | 0.234375 | | | 47 | 0.734375 |
| 1/4 | 8 | 16 | 0.25 | 3/4 | 24 | 48 | 0.75 |
| | | 17 | 0.265625 | | | 49 | 0.765625 |
| | 9 | 18 | 0.28125 | | 25 | 50 | 0.78125 |
| | | 19 | 0.296875 | | | 51 | 0.796875 |
| 5/16 | 10 | 20 | 0.3125 | 13/16 | 26 | 52 | 0.8125 |
| | | 21 | 0.328125 | | | 53 | 0.828125 |
| | 11 | 22 | 0.34375 | | 27 | 54 | 0.84375 |
| | | 23 | 0.359375 | | | 55 | 0.859375 |
| 3/8 | 12 | 24 | 0.375 | 7/8 | 28 | 56 | 0.875 |
| | | 25 | 0.390625 | | | 57 | 0.890625 |
| | 13 | 26 | 0.40625 | | 29 | 58 | 0.90625 |
| | | 27 | 0.421875 | | | 59 | 0.921875 |
| 7/16 | 14 | 28 | 0.4375 | 15/16 | 30 | 60 | 0.9375 |
| | | 29 | 0.453125 | | | 61 | 0.953125 |
| | 15 | 30 | 0.46875 | | 31 | 62 | 0.96875 |
| | | 31 | 0.484375 | | | 63 | 0.984375 |
| 1/2 | 16 | 32 | 0.5 | | 32 | 64 | 1. |

25. A 100-W lamp requires 0.91 A, and a table radio requires 7/8 A. Which item uses more current?
26. An adjustment screw on a carburetor must be at least 0.56 in long. Will a 15/32-in-long screw be acceptable?

(See Instructor's Manual for Test 2-5, Ohm's Law and Decimals)

## JOB 2-6 INTRODUCTION TO THE METRIC SYSTEM OF MEASUREMENT

The metric system is a decimal system of measurement that is the practical standard throughout most of the world. It is expected that this system will be totally adopted by the United States within the next few years. Many industries in the United States have already converted to metric measurements. Signs giving distances in both miles and kilometers are being erected all over the country.

The basic unit of length in the metric system is the meter (m), which is equal to about 39.37 in.

If we divide a meter into 1000 parts, each part is called a *millimeter* (mm), since the prefix "milli" means one one-thousandth of a quantity. A millimeter is approximately equal to the thickness of a dime.

If we divide a meter into 100 parts, each part is called a *centimeter* (cm), since the prefix "centi" means one one-hundredth of a quantity. A centimeter is about $^3/_8$ in, approximately equal to the diameter of an ordinary aspirin tablet.

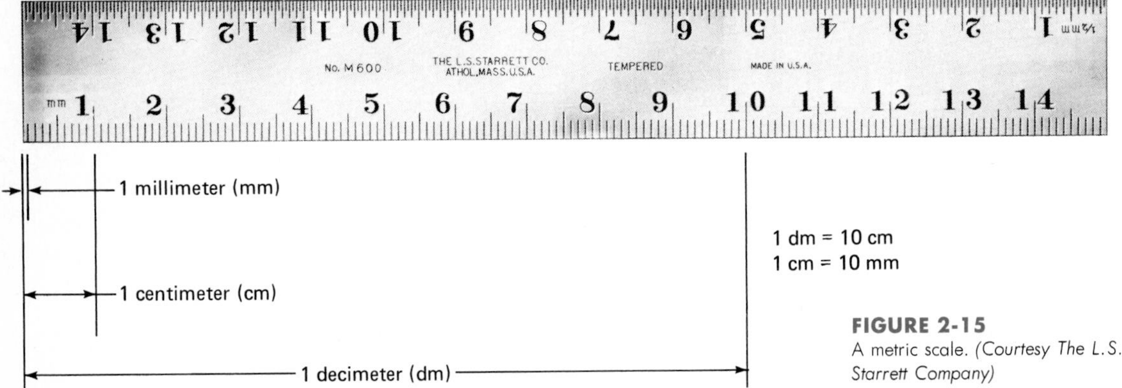


**FIGURE 2-15**
A metric scale. *(Courtesy The L.S. Starrett Company)*

If we divide a meter into 10 parts, each part is called a *decimeter* (dm), since the prefix "deci" means one-tenth of a quantity. A decimeter is almost 4 in long.

The relationship among these three quantities is shown in Fig. 2-15.

For measurements larger than a meter,

1 dekameter (dam) = 10 meters, since the prefix "deka" means 10 times as large.

1 hectometer (hm) = 100 meters, since the prefix "hecto" means 100 times as large.

TABLE 2-4
LENGTHS IN THE METRIC SYSTEM

| | | | | | |
|---|---|---|---|---|---|
| 1 cm | = 10 mm | or | 1 mm | = $^1/_{10}$ cm | = 0.1 cm |
| 1 dm | = 10 cm | or | 1 cm | = $^1/_{10}$ dm | = 0.1 dm |
| 1 m | = 10 dm | or | 1 dm | = $^1/_{10}$ m | = 0.1 m |
| 1 m | = 100 cm | or | 1 cm | = $^1/_{100}$ m | = 0.01 m |
| 1 m | = 1000 mm | or | 1 mm | = $^1/_{1000}$ m | = 0.001 m |
| 1 dam | = 10 m | or | 1 m | = $^1/_{10}$ dam | = 0.1 dam |
| 1 hm | = 100 m | or | 1 m | = $^1/_{100}$ hm | = 0.01 hm |
| 1 km | = 1000 m | or | 1 m | = $^1/_{1000}$ km | = 0.001 km |

1 kilometer (km) = 1000 meters, since the prefix "kilo" means 1000 times as large.

The set of lengths in the metric system is given in Table 2-4. This table is described in picture form in Fig. 2-16.

Before we try to change any of these measurements from one unit to another, let us look at a simple method for multiplying and dividing by numbers like 10, 100, 1000, etc.

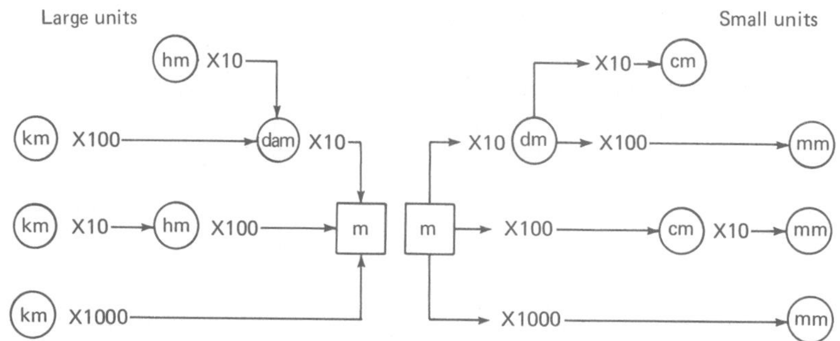


**FIGURE 2-16**
Rules for changing units in the metric system.

## MULTIPLYING BY 10, 100, 1000, ETC.

**RULE** To multiply values by numbers like 100, 1000, 1,000,000 etc., move the decimal point one place to the right for every zero in the multiplier.

EXAMPLE 2-19

1.59 × 10 = 1⊙5.9 = 15.9 (move point 1 place to the right)
4.76 × 100 = 4⊙76.= 476. or 476 (move point 2 places to the right)

A decimal point is ordinarily not written at the end of a whole number. However, it may be written if desired. In the following example, the point must be written in together with two zeros so that we can move the decimal point past the required number of places.

34 × 100 = 34⊙00. = 3400. or 3400

If the decimal point must be moved past more places than are available, zeros are added at the end of the number to make up the required number of places. This is shown in the following example.

0.0259 × 1,000,000 = ⊙025900. = 25,900

SELF-TEST 2-20. Multiply 25.2 × 100.

| | |
|---|---|
| SOLUTION. To multiply, the decimal point is moved to the __(left/right)__. | right |
| The number 100 contains ______ zeros. The point will be moved ______ | 2 2 |
| places. Zeros __(are/are not)__ needed to account for the correct number of | are |
| places. The answer is 25.2 × 100 = ______. | 2520 |

**PROBLEMS**

1. 0.0072 × 1000
2. 45.76 × 100
3. 3.09 × 1000
4. 0.0045 × 100
5. 37 × 100
6. 0.08 × 1000
7. 0.0006 × 1000
8. 0.000 56 × 1,000,000
9. 27 × 100
10. 15 × 1000
11. 0.000 78 × 100
12. 7.8 × 1,000,000
13. 15.4 × 1000
14. 4.5 × 10
15. 0.005 × 100
16. 6 × 10,000
17. 0.008 × 10,000
18. 0.009 × 100,000
19. 2.34 × 1000
20. 0.0008 × 1,000,000

## DIVIDING BY 10, 100, 1000, ETC.

**RULE** To divide values by numbers like 10, 100, 1000, etc., move the decimal point one place to the left for every zero in the divisor.

EXAMPLE 2-21

17.4 ÷ 10 = 1.7⊙4 = 1.74 (move point one place to the left)
45 ÷ 100 = .45⊙ = 0.45 (move point two places to the left)
6.5 ÷ 1000 = .006⊙5 = 0.0065 (move point three places to the left)

SELF-TEST 2-22. Divide 8.5 by 1000.

| | |
|---|---|
| SOLUTION. To divide, the decimal point is moved to the __(left/right)__. | left |
| The number 1000 contains ______ zeros. The point will be moved ______ | 3 3 |
| places. Zeros __(are/are not)__ needed to account for the correct number of places. | are |
| The answer is 8.5 ÷ 1000 = ______. | 0.0085 |

**PROBLEMS**

1. 6500 ÷ 1000
2. 7500 ÷ 1000
3. 880,000 ÷ 1000
4. 32 ÷ 100
5. 6 ÷ 100
6. 17.8 ÷ 10
7. 835 ÷ 1000
8. 550 ÷ 1,000,000
9. 653.8 ÷ 1000
10. 100,000,000 ÷ 1,000,000
11. 0.45 ÷ 1000
12. 0.08 ÷ 10
13. 8.5 ÷ 1000
14. 7 ÷ 1,000,000
15. 28.6 ÷ 1000
16. 15.6 ÷ 100

17. $\frac{2}{1000}$

18. $\frac{0.08}{10}$

19. $\frac{398}{10{,}000}$

20. $\frac{600}{1000}$

(See Instructor's Manual for Test 2-6)

**Changing units of measurement.** There are two factors to be considered when describing any measurement: (1) how many of the measurements and (2) what *kind* of measurements. For example, \$1 may be described as 2 half-dollars, 4 quarters, 10 dimes, 20 nickels, or 100 pennies. When the \$1 was changed into each of the new measurements, *both* the unit of measurement as well as the number of them were changed. For example:

Since 3 ft = 1 yd, 2 yd = 2 × 3 = 6 ft
Since 2000 lb = 1 ton, 3 tons = 3 × 2000 = 6000 lb
Since 100¢ = 1 dollar, 4 dollars = 4 × 100 = 400¢

Large unit → Small unit
Small number → Large number

Notice that a *small number* of *large units* is always changed into a *large number* of *small units* by *multiplying* by the number showing the relationship between the units.

**RULE** To change from a large unit to a small unit, multiply by the number showing the relationship between the units.

Now let us reverse the process and change small units into large units. For example:

Since 3 ft = 1 yd, 6 ft = 6 ÷ 3 = 2 yd
Since 2000 lb = 1 ton, 6000 lb = 6000 ÷ 2000 = 3 tons
Since 100¢ = 1 dollar, 400 ¢ = 400 ÷ 100 = 4 dollars

Small unit → Large unit
Large number → Small number

Notice that a *large number* of *small units* is always changed into a *small number* of *large units* by *dividing* by the number showing the relationship between the units.

**RULE** To change from a small unit to a large unit, divide by the number showing the relationship between the units.

Figure 2-16 illustrates the use of these last two rules for ordinary metric units of measurement.

EXAMPLE 2-23. Change the following units of measurement:

*a.* Change 8 cm to millimeters.

There are 10 mm in each centimeter, and since we are changing large units (cm) to small units (mm), we *multiply* by 10.

$$8 \text{ cm} = 8 \times 10 = 80 \text{ mm} \qquad \textit{Ans.}$$

*b.* Change 42 mm to centimeters.

There are 10 mm in each centimeter, and since we are changing small units (mm) to large units (cm), we *divide* by 10.

$$42 \text{ mm} = 42 \div 10 = 4.2 \text{ cm} \qquad \textit{Ans.}$$

*c.* Change 2 m to decimeters.

There are 10 dm in each meter, and since we are changing large units (m) to small units (dm), we *multiply* by 10.

$$2 \text{ m} = 2 \times 10 = 20 \text{ dm} \qquad \textit{Ans.}$$

*d.* Change 300 cm to meters.

There are 100 cm in each meter, and since we are changing small units (cm) to large units (m), we *divide* by 100.

$$300 \text{ cm} = 300 \div 100 = 3 \text{ m} \qquad \textit{Ans.}$$

*e.* Change 5 m to millimeters.

There are 1000 mm in each meter, and since we are changing large units (m) to small units (mm), we *multiply* by 1000.

$$5 \text{ m} = 5 \times 1000 = 5000 \text{ mm} \qquad \textit{Ans.}$$

*f.* Change 125 m to dekameters.

There are 10 m in each dekameter, and since we are changing small units (m) to large units (dam), we *divide* by 10.

$$125 \text{ m} = 125 \div 10 = 12.5 \text{ dam} \qquad \textit{Ans.}$$

*g.* Change 6450 m to kilometers.

There are 1000 m in each kilometer, and since we are changing small units (m) to large units (km), we *divide* by 1000.

$$6450 \text{ m} = 6450 \div 1000 = 6.45 \text{ km} \qquad \textit{Ans.}$$

SELF-TEST 2-24. Change the following units of measurement.

1. Change 85 mm to centimeters.

Since there are ______ mm in each centimeter,

85 mm = 85 __(×/÷)__ 10 = ______ cm

2. Change 2 m to centimeters.

Since there are ______ cm in each meter,

2 m = 2 __(×/÷)__ 100 = ______ cm

| | |
|---|---|
| 10 | |
| ÷ | 8.5 |
| 100 | |
| × | 200 |

3. Change 4.5 km to meters.
Since there are ______ m in each kilometer,

4.5 km = 4.5 ___(×/÷)___ 1000 = ______ m

1000
× 4500

**PROBLEMS**

Change the following units of measurement:

1. 35 cm to millimeters
2. 48 mm to centimeters
3. 2.4 dm to centimeters
4. 80 cm to decimeters
5. 36 m to decimeters
6. 3 dm to meters
7. 5 m to centimeters
8. 325 cm to meters
9. 2.45 m to millimeters
10. 6,500 mm to meters
11. 125 mm to centimeters
12. 6.8 cm to millimeters
13. 4.2 hm to dekameters
14. 36 dam to hectometers
15. 75 km to hectometers
16. 8 hm to kilometers
17. 3.9 hm to meters
18. 100 m to hectometers
19. 20 km to meters
20. 3498 m to kilometers
21. 1.5 dam to decimeters
22. Arrange in order of size, starting with the largest: 0.15 m, 20 cm, and 180 mm.
23. In a vacuum test on an engine operating at 3000 ft, the vacuum gage read 34.56 cm. Express this measurement as millimeters.

(See Instructor's Manual for Test 2-7)

## JOB 2-7 BRUSHUP ON DECIMAL FRACTIONS

**Addition of decimals.** When adding decimals, be sure to write the numbers so that the decimal points are in a straight vertical line.

EXAMPLE 2-25. Add 2.52 + 0.007 + 13.03 + 5 + 0.4.

SOLUTION. The number 5 is written as 5.000 to locate the decimal point.

```
  2|52            2.520
  0|007           0.007
 13|03     or    13.030
  5|              5.000
  0|4             0.400
 ------          ------
   |             20.957   Ans.
   |<------ Lined-up decimal points
```

The empty spaces are filled in with zeros, as shown in the column at the right, to aid in keeping the numbers in the correct columns.

EXAMPLE 2-26. Find the total thickness of insulation on shielded radio wire covered with resin (0.55 mm), lacquered cotton braid (0.49 mm), and copper shielding (0.225 cm).

SOLUTION

1. Change 0.225 cm to millimeters

$$0.225 \text{ cm} = 0.225 \times 10 = 2.25 \text{ mm}$$

2. The total thickness is the sum of the individual thicknesses.

$$\begin{array}{r} 0.55 \text{ mm} \\ 0.49 \text{ mm} \\ 2.25 \text{ mm} \\ \hline 3.29 \text{ mm} \end{array}$$

*Ans.*

Lined-up decimal points

**PROBLEMS**

Add the following decimals:

1. 3.28 + 9.5 + 0.634 + 0.078
2. 56.09 + 14 + 4.876 + 49.007
3. 13 + 3.072 + 0.7 + 6.06
4. 54 + 0.033 + 0.713 + 8.05
5. 0.087 + 6.18 + 4 + 1.7
6. What is the cost to rewind a motor if the following materials and labor were used: top sticks at $0.42, No. 17 wire at $3.25, No. 26 wire at $2.29, armature lacquer at $1.09, and labor at $32.50?
7. Find the total drop in voltage in a distribution system if the voltage drops across the sections are 1.06, 36.4, and 8 V.
8. Find the total thickness of insulation on shielded radio wire covered with resin (0.56 mm), lacquered cotton braid (0.47 mm), and copper shielding (0.76 mm).
9. Number 14 copper wire has a diameter of 0.0641 in; No. 10 wire is 0.0378 in larger. Find the diameter of No. 10 copper wire.
10. Find the total resistance of the leads of an installation if the resistances are 0.054, 1.004, 1.2, and 1.2 Ω.
11. The emitter current of a transistor is always equal to the sum of the collector current and the base current. Find the emitter current if the collector current is 0.03 A and the base current is 0.0015 A.
12. What is the cost to repair three electric outlets if each outlet requires one loom box at $2.50, one toggle switch at $2.95, one hickey at $0.32, and 1 h of labor at $22.75?
13. A carbon brush 2.46 cm thick has a copper plating of 0.39 mm on each side. What is the total thickness?
14. The insulation to be used in a slot in a motor is as follows: fish paper, 0.0156 in; Tufflex, 0.0313 in; varnished cambric, 0.0156 in;

and top stick, 0.125 in. What is the total thickness of all the insulation?

15. The following materials were charged to an electrical wiring job: conduit, \$4.25; No. 8 wire, \$1.75; BX cable, \$18.50; conduit fittings, \$3.85; outlet boxes, \$6.58; switches, \$5.72; and \$4.25 for tape, solder, and pipe clips. What was the total amount charged for materials?

**Subtracting decimals.** When subtracting decimals, write the numbers in columns as for addition, lining up the decimal points in a straight vertical column.

EXAMPLE 2-27. The electric-meter readings for successive months were 70.08 and 76.49. Find the difference.

SOLUTION

$$\begin{array}{r} 76.49 \\ -70.08 \\ \hline 6.41 \end{array} \quad \textit{Ans.}$$

EXAMPLE 2-28. Subtract 1.04 from 3.

SOLUTION. The number after the word "from" is written on top. The number after the word "subtract" is written underneath. The number 3 is written as 3.00 to locate the decimal point correctly.

$$\begin{array}{r} 3.00 \\ -1.04 \\ \hline 1.96 \end{array} \quad \textit{Ans.}$$

**PROBLEMS**

| | | |
|---|---|---|
| 1. 0.26 − 0.03 | 2. 1.36 − 0.18 | 3. 0.4 − 0.06 |
| 4. 0.05 − 0.004 | 5. 18.92 − 11.36 | 6. $^5/_8$ − 0.002 |
| 7. 0.627 − 0.31 | 8. 0.827 − 0.31 | 9. 3 − 0.08 |
| 10. 0.5 − 0.02 | 11. 6 − 0.1 | 12. 0.83 − $^1/_2$ |
| 13. 2.89 − 0.5 | 14. 12.6 − 7 | 15. 0.316 − 0.054 |
| 16. 14 − 8.06 | 17. 5$^1/_4$ − 2.63 | 18. 3.125 − $^1/_8$ |

19. Subtract $^1/_4$ from 0.765.
20. Find the difference between 110 and 54.9.
21. Find the difference between (*a*) 0.316 and 0.012; (*b*) 3.006 and 1.9; (*c*) 0.5 and 0.11; (*d*) 7.07 and 1.32; and (*e*) 2 and 0.02.
22. Subtract (*a*) 0.008 from 0.80; and (*b*) 0.216 from 2.16.
23. What is the difference in the diameters of No. 1 wire (0.2893 in) and No. 7 wire (0.1447 in)?
24. The resistance of the windings of an electromagnet at room tem-

perature is 28.69 Ω. If its resistance at 175°F (degrees Fahrenheit) is 36.98 Ω, find the increase in the resistance.

25. A locating pin which should be 0.875 cm in diameter measures 9.18 mm in diameter. How much is it oversize?
26. An electric generator delivers 223.8 V. If 4.95 V is lost in the line wires, find the voltage delivered at the end of the line.
27. A circuit in a television receiver calls for a 0.0005-μF capacitor. A capacitor valued at 0.00035 μF is available. How much extra capacitance is needed if connected in parallel?
28. The intermediate frequency at the output of a converter stage is found by obtaining the difference between the oscillator frequency and the radio frequency. If the radio frequency is 1.1 MHz (megahertz) and the oscillator frequency is 1.555 MHz, what will be the intermediate frequency?

(See Instructor's Manual for Test 2-8)

**Multiplication of decimals.** Decimals are multiplied in exactly the same way as ordinary numbers are multiplied. However, in addition to the normal multiplication, the decimal point must be correctly set in the answer.

| **RULE** | The number of decimal places in a product is equal to the sum of the number of decimal places in the numbers being multiplied. |
|---|---|

EXAMPLE 2-29. Multiply 0.62 by 0.3.

SOLUTION
```
  0.62 (multiplicand, 2 places)
 ×0.3  (multiplier, 1 place)
 0.186 (product, 2 + 1 = 3 places)     Ans.
```

EXAMPLE 2-30. Multiply 0.35 by 0.004.

SOLUTION
```
    0.35 (multiplicand, 2 places)
  ×0.004 (multiplier, 3 places)
0.001 40 (product, 2 + 3 = 5 places)    Ans.
```

In this problem, extra zeros must be inserted between the decimal point and the digits of the answer to make up the required number of decimal places.

**PROBLEMS**

Find the product of

| | | |
|---|---|---|
| 1. 0.005 × 82 | 2. 1.732 × 40 | 3. 1.13 × 0.41 |
| 4. 0.9 × 0.09 | 5. 0.866 × 35 | 6. 44.6 × 805 |

7. $7.63 \times 0.029$  8. $0.354 \times 0.008$  9. $6.2 \times 0.003$
10. $0.033 \times 0.0025$  11. $106 \times 0.045$  12. $73.8 \times 1.09$

13. If a 100-W lamp uses 0.91 A, how much current would be used by five such lamps in parallel?
14. If BX cable costs $0.875/ft, what would be the cost of 52.5 ft?
15. The number of milliamperes is found by multiplying the number of amperes by 1000. Find the number of milliamperes equal to (*a*) 0.25 A; (*b*) 0.025 A; and (*c*) 2.5 A.
16. An electrician worked 1.5 h to install a junction box. If her rate of pay is $14.25/h, how much did she earn?
17. The standard unit of resistance is measured by the resistance of a column of mercury 106.3 cm high. If 1 cm equals 0.3937 in, what is the height of the mercury column correct to the nearest thousandth of an inch?
18. The inductive reactance of a coil is found by multiplying the constant 6.28 by the frequency and the inductance. Find the inductive reactance of a coil if the frequency is 60 Hz and the inductance is 0.15 H (henry).
19. What is the capacity of a battery (expressed in ampere-hours) if it discharges at the rate of 9.6 A for 7.25 h?

## Division of decimals

EXAMPLE 2-31. A contractor must locate nine equally spaced electric outlets in a school corridor. The total distance from the first outlet to the last is 117.2 ft. Find the distance between two adjacent outlets.

SOLUTION. The number of spaces between outlets is one less than the number of outlets. Therefore, the distance between any two adjacent outlets is the distance 117.2 divided by 8. This division is accomplished in the same manner as in changing fractions to decimals. See Examples 2-17 and 2-18.

```
   14.65 ft     Ans.
8)117.20
  8
  37
  32
   5 2
   4 8
    40
    40
     0
```

EXAMPLE 2-32. Divide 1.38 by 0.06.

SOLUTION. When dividing a decimal, it is best to move the decimal point all the way over to the right so as to bring it to the end of the divisor.

```
⊙06.)1.38
```

If this is done, the decimal point in the dividend must also be moved to the right *for the same number of places.* This is the equivalent of multiplying both dividend and divisor by 100. Now we can divide as before.

```
            23.     Ans.
    ⊙06.)1⊙38.
          1 2
          ---
           18
           18
           --
            0
```

EXAMPLE 2-33. Divide 3.6 by 0.08.

SOLUTION. A zero must be added after the 6 to provide the two places that the decimal point must be moved to the right.

```
            45.     Ans.
    ⊙08.)3⊙60.
          3 2
          ---
           40
           40
           --
            0
```

EXAMPLE 2-34. Divide 0.0007 by 0.125.

SOLUTION

```
               0.0056     Ans.
    ⊙125.)⊙000.7000
                 625
                 ---
                 750
                 750
                 ---
                   0
```

**PROBLEMS**

1. $3.9 \div 0.3$
2. $12.56 \div 0.4$
3. $80.5 \div 0.5$
4. $51 \div 0.06$
5. $38.54 \div 8.2$
6. $1591 \div 0.43$
7. $2.8296 \div 0.0036$
8. $140.7 \div 0.021$
9. $9.1408 \div 3.94$
10. Using the formula $I = V/R$, find $I$ if $V = 79.5$ V and $R = 265\ \Omega$.
11. Shielded rubber-jacketed microphone cable weighs 0.075 lb/ft. How many feet of cable are there in a coil weighing 15 lb?
12. What is the smallest number of insulators, each rated at 12,000 V, that should be used to safeguard a 220,000-V transmission line?

13. The $Q$, or "quality," of a coil is a measure of its worth in a tuned circuit. It is found by dividing the reactance of the coil by its effective resistance (see Job 16-3). Find the $Q$ of a coil if its reactance is 1820 Ω and its effective resistance is 30 Ω.
14. A 50-ft-long wire has a resistance of 10.35 Ω. What is the resistance of 1 ft of this wire?
15. The current-amplifying ability $\beta$ of a transistor is obtained by dividing the collector current $I_C$ by the base current $I_B$. Find $\beta$ if $I_C$ = 0.0004 A and $I_B$ = 0.000 01 A.

(See Instructor's Manual for Test 2-9)

## SUMMARY—WORKING WITH DECIMALS

| | |
|---|---|
| 1. Decimal fractions are fractions whose ________ are numbers like 10, ________, 1000, etc. | denominators<br>100 |
| 2. The denominator is shown by the number of digits to the ________ of the decimal point. Thus, 0.6 represents six-________. | right<br>tenths |
| 0.57 represents fifty-seven ________.<br>0.123 represents one hundred twenty-three ________.<br>3.09 represents three and nine-________. | hundredths<br>thousandths<br>hundredths |
| 3. Decimals can be compared only when they have the ________ number of decimal places. | same |
| 4. The word "and" in a mixed number such as five and three-hundredths is written as a ________ point. This number would be written as ________. | decimal 5.03 |
| 5. Fractions are changed to decimals by ________ the ________ by the ________. | dividing numerator<br>denominator |
| 6. A whole number always has a decimal point understood to be at the (beginning/end) of the number. | end |
| 7. When dividing decimals, if a remainder is more than ________ of the divisor, drop it and add a full unit to the last ________ of the answer. If the remainder is ________ than half, drop it completely. | half<br>digit<br>less |
| 8. When adding or subtracting decimals, line up the decimal points in a ________ column. | vertical |
| 9. The product of two decimals has as many decimal places as the ________ of the places in the numbers being multiplied. | sum |
| 10. When you divide decimals, move the decimal point in the divisor to the ________ as many places as is necessary to bring the point behind the last digit. Then move the point in the dividend to the right for the ________ number of places. | right<br>same |
| 11. 18.5 × 100 = ________<br>6.28 ÷ 1000 = ________ | 1850<br>0.006 28 |
| 12. To change large units into small units, we (multiply/divide). To change small units into large units, we (multiply/divide). | multiply<br>divide |
| 13. 1 cm = ________ mm<br>1 km = ________ m<br>1 m = ________ cm | 10<br>1000<br>100 |

1 m = ________ mm | 1000
1 m = ________ dm | 10

## PROBLEMS

1. Find the dimensions $A$ and $B$ in the spindle shaft shown in Fig. 2-17.

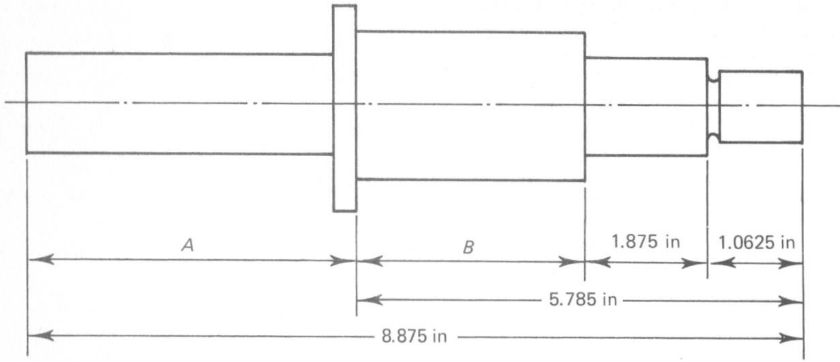


FIGURE 2-17

2. In an ac voltage wave, the maximum voltage is equal to 1.414 times the effective voltage as read on a meter. Find the maximum voltage of a wave whose meter reading is 117 V.
3. The power factor (pf) of an ac circuit is found by dividing its resistance $R$ by its impedance $Z$. Find the power factor of a circuit if $R = 2.46\ \Omega$ and $Z = 30\ \Omega$.
4. The laminated core of a transformer is made of 32 thicknesses of metal. If each lamination is 7.5 mm thick, find the total thickness of the core in centimeters.
5. Multiply: (*a*) $0.735 \times 1000$; (*b*) $81.7 \times 1{,}000{,}000$; (*c*) $0.0035 \times 1000$; and (*d*) $0.004 \times 100$. Divide: (*e*) $0.07 \div 100$; (*f*) $6.28 \div 1000$; (*g*) $4750 \div 100$; and (*h*) $397 \div 1{,}000{,}000$.
6. A rebuilt alternator that cost $95.50 was sold for $147.95. (*a*) Find the profit. (*b*) What decimal part of the cost was the profit?
7. If BX cable costs $0.825/ft, what would be the cost of 52.5 ft?
8. What is the total current supplied to a parallel circuit consisting of a broiler drawing 9.25 A, a radio drawing 0.93 A, and two lamps, each drawing 0.91 A?
9. The diameter of a motor shaft bearing is 0.12 mm larger than the shaft of the motor. What is the diameter of the bearing if the shaft has a diameter of 2.25 cm?
10. Ms. Graham received an increase from $12.40 to $13.10 per hour for her work as an electrician. Find the amount of increase for a 40-h week.
11. A 60-W lamp uses about 0.625 A when connected in an ordinary house line. How many amperes would three such lamps use when connected in parallel?
12. An electrician figured the cost to rewind a motor as follows: No. 17

wire at $2.29, No. 28 wire at $1.65, No. 00 top sticks at $0.60, armalac at $1.25, and 2 h of labor at $14.25/h. Find the total cost.

13. A neon electric sign uses 4.75 W of power for each foot of tubing. Find the power consumption for a sign which uses 34.5 ft of tubing.
14. A 10.5-cm-diameter piece of work is turned down in a lathe in three cuts. The first cut takes off 40 mm, the second takes off 2.4 mm, and the third removes 1.7 mm. What is the finished diameter of the work?
15. An electrician is to wire a light switch to operate the following light bulbs on a truck: two marker lamps at 0.3 A each; three clearance lamps at 0.4 A each; two stop lights at 2.4 A each; two front parking lamps at 1.5 A each; and two headlamps at 7.6 A each. The maximum current-carrying capacities of wires are: 16 gage, 6 A; 14 gage, 15 A; 12 gage, 20 A; and 10 gage, 30 A. With all lights operating, find the smallest gage wire that may be safely used.

(See Instructor's Manual for Test 2-10, Decimals)

# FORMULAS

# 3

## JOB 3-1 CHECKUP ON FORMULAS IN ELECTRICAL WORK (DIAGNOSTIC TEST)

In Job 2-3 we used our first *formula.* When Ohm's law is written using only the letters which represent the words of Ohm's law, it is called a *formula.* As we continue with our study of electricity, we shall meet many new formulas. Some are simple like Ohm's law, but others are more complicated. Let us check up on our knowledge of how to use formulas. The following problems involve the use of formulas. The electrician and the electronic technician find it necessary to solve problems like these in their daily work. If you have any difficulty with these problems, see Job 3-2 which follows.

### PROBLEMS

1. Using Ohm's law ($V = I \times R$), find the current $I$ drawn by a 10-Ω automobile horn $R$ from a 6-V battery $V$.
2. Using Ohm's law, find the number of ohms of resistance $R$ needed to obtain a bias voltage $V$ of 6 V if the current $I$ is 0.02 A.
3. Using the formula for electric power, $P = V \times I$, find the voltage $V$ necessary to operate a 500-W electric percolator $P$ if it draws a current $I$ of 4.5 A.
4. Using the formula $P = V \times I$, find the current $I$ drawn by a 440-W vacuum cleaner $P$ from a 110-V line $V$.
5. Write the formula for the following rule: kilowatts (kW) equals current $I$ multiplied by voltage $V$ and divided by 1000.
6. Using the series-circuit fomula $I = V/(R_1 \times R_2)$, find $I$ if $V = 100$ V, $R_1 = 20$ Ω, and $R_2 = 30$ Ω.
7. Using the ac formula $I = V/Z$, find the impedance $Z$ of an ac circuit if the voltage $V$ is 50 V and the current $I$ is 2 A.
8. Using the formula mA = A × 1000, find the number of amperes which is the equivalent of 125 mA (milliamperes).
9. $I_T = I_1 + I_2 + I_3$ is the formula for the total current in a parallel circuit. Find the total current $I_T$ if $I_1 = 2$ A, $I_2 = 5$ A, and $I_3 = 4$ A.

10. The formula for the number of coulombs of electricity which can be placed on the plates of a capacitor is $Q = C \times V$. Find the voltage $V$ which is necessary to place a charge $Q$ of 0.000 000 2 C on the plates of a capacitor whose capacitance $C$ is 0.000 000 002 F (farad).

## JOB 3-2 BRUSHUP ON FORMULAS

**Meaning.** A formula is a convenient shorthand method for expressing and writing a rule or relationship among several quantities.

**Signs of operation.** The quantities involved in any simple relationship are held together by one or more of the following operations:

1. Multiplication (×)
2. Division (÷)
3. Addition (+)
4. Subtraction (−)
5. Equality (=)

Each of these operations may be written in several ways.

**Multiplication.** The multiplication of two quantities is often expressed as the "product of" the two quantities. This may be written as follows:

1. Using a multiplication sign (×) between the numbers or letters
2. Using a dot (·) between the numbers or letters
3. Writing nothing at all between the numbers or letters

For example, the product of 3 and 4 may be written as (1) $3 \times 4$ or (2) $3 \cdot 4$. The third method cannot be used when only numbers are involved because the meaning would not be clear. For example, 34 would mean the number thirty-four and *not* $3 \times 4$. This method of indicating multiplication by omitting all signs between the quantities can be used only for combinations of numbers and letters or combinations of letters.

The product of 6 and $R$ may be written as (1) $6 \times R$, (2) $6 \cdot R$, or (3) $6R$. All three forms indicate that 6 is to be multiplied by the quantity called $R$.

The product of $P$, $R$, and $T$ may be written as (1) $P \times R \times T$, (2) $P \cdot R \cdot T$, or (3) $PRT$. All three forms indicate that the quantity $P$ is to be multiplied by the quantity $R$ and then multiplied by the quantity $T$.

**Division.** The division of two quantities is often expressed as the "quotient of" the two quantities. This may be written as follows:

1. Using a division sign (÷) between the numbers or letters
2. Using a fraction bar to indicate division

For example, the quotient of 8 divided by 2 may be written as (1) $8 \div 2$ or (2) $^8/_2$.

The quotient of 12 divided by $I$ may be written as (1) $12 \div I$ or (2) $12/I$.

The quotient of $V$ divided by $R$ may be written as (1) $V \div R$ or (2) $V/R$.

**Addition.** The addition of two or more quantities is often expressed as the "sum of" the quantities and is indicated by a plus sign (+) between the quantities. For example,

The sum of 6 and 4 is written as $6 + 4$.
The sum of 3 and $R$ is written as $3 + R$.
The sum of $V_1$ and $V_2$ is written as $V_1 + V_2$.

**Subtraction.** The subtraction of two quantities is often expressed as the "difference between" the quantities and is indicated by a minus sign (−) between them. For example,

The difference between 9 and 4 is written as $9 - 4$. This is read as (1) 9 minus 4 or (2) 4 subtracted from 9.

The difference between 20 and $R$ is written as $20 - R$. This is read as (1) 20 minus $R$ or (2) $R$ subtracted from 20.

The difference between $V_T$ and $V_1$ is written as $V_T - V_1$. This is read as (1) $V_T$ minus $V_1$ or (2) $V_1$ subtracted from $V_T$.

**Equality.** An equality sign (=) is used to indicate that the combination of numbers and letters on one side of the equality sign has the same value as the combination of numbers and letters on the other side. For example,

$$3 \times 4 = 12$$
$$a \cdot b = ab$$
$$2R = 10$$

**Changing rules into formulas.** To change a rule into a formula:

1. Replace each quantity with a convenient letter.
2. Rewrite the rule. Substitute these letters for the words of the rule. Include the symbols for the signs of operation.

*Note:* The letter used to replace a word is usually the first letter of the word representing the quantity. However, any letter may be used. For example, if a letter has already been used to represent some quantity, it cannot be used again in the same formula to represent *another* quantity. In this event a letter which is *not* the first letter of the word would be used.

EXAMPLE 3-1. Change the following rule into a formula: The area of a rectangle is equal to its length multiplied by its width.

SOLUTION. The length is replaced by the letter $L$. The width is replaced by the letter $W$. The area is replaced by the letter $A$.

$$\underbrace{\text{The area}}_{A}\ \underbrace{\text{is equal to}}_{=}\ \underbrace{\text{the length}}_{L}\ \underbrace{\text{multiplied by}}_{\times}\ \underbrace{\text{the width.}}_{W}$$

Thus,

$$A = L \times W \qquad \text{or} \qquad A = L \cdot W \qquad \text{or} \qquad A = LW \qquad \textit{Ans.}$$

EXAMPLE 3-2. Change the following rule into a formula: The voltage is equal to the current multiplied by the resistance.

SOLUTION. The voltage is replaced by the letter $V$. The current is replaced by the letter $I$. The resistance is replaced by the letter $R$.

$$\underbrace{\text{The voltage}}_{V}\ \underbrace{\text{is equal to}}_{=}\ \underbrace{\text{the current}}_{I}\ \underbrace{\text{multiplied by}}_{\times}\ \underbrace{\text{the resistance.}}_{R}$$

Thus,

$$V = I \times R \qquad \text{or} \qquad V = I \cdot R \qquad \text{or} \qquad V = IR \qquad \textit{Ans.}$$

EXAMPLE 3-3. Change the following rule into a formula: The voltage $V$ of a series circuit of two resistors is equal to the current $I$ multiplied by the sum of the resistances $R_1$ and $R_2$.

SOLUTION. The word "sum" is indicated by a plus sign. This sum is actually a *single* quantity obtained by adding $R_1$ and $R_2$. This must be shown by enclosing $R_1$ and $R_2$ in a pair of parentheses. The current $I$ will then be multiplied by the parentheses. Thus,

$$V = I \times (R_1 + R_2) \qquad \text{or} \qquad V = I(R_1 + R_2) \qquad \textit{Ans.}$$

SELF-TEST 3-4. Change the following rule into a formula: The total resistance $R_T$ of two resistors in parallel is equal to the product of the resistances $R_1$ and $R_2$ divided by the sum of the resistances.

SOLUTION. The diagram for the circuit is shown in Fig. 3-1.

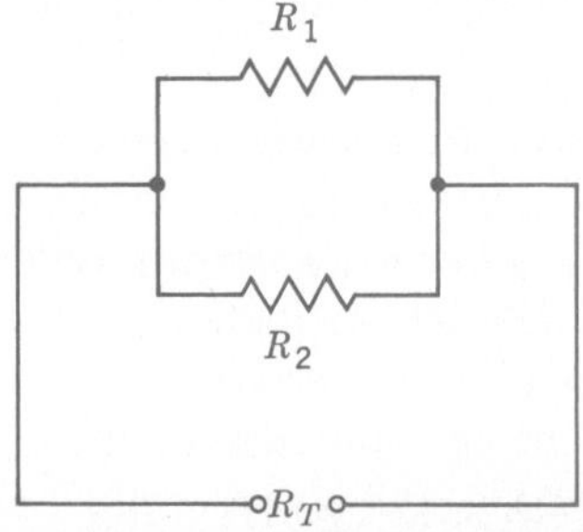


**FIGURE 3-1**

Resistances in parallel.

1. The word "product" means to ___(add/multiply/divide)___. multiply
2. The product of $R_1$ and $R_2$ is written as $R_1$ ______ $R_2$. $\times$
3. The word "sum" means to ___(add/subtract)___. add
4. The sum of $R_1$ and $R_2$ is written as $R_1$ ______ $R_2$. $+$
5. The division in our formula may be written as a fraction. The numerator of the fraction will be the ___(sum/product)___ of the resistances. product
6. The denominator will be the ___(sum/product)___ of the resistances. sum
7. The formula will be read as:

$$R_T = \frac{R_1 \text{ ______ } R_2}{R_1 \text{ ______ } R_2} \quad \textit{Ans.}$$

$\times$
$+$

**PROBLEMS**

Write the formula for each of the rules given. Use the italic letters and abbreviations in parentheses to indicate each word.

1. The electric power $P$ is equal to the current $I$ multiplied by the voltage $V$.
2. The effective voltage $V$ of an ac voltage wave is equal to 0.707 times the maximum value $V_{\text{max}}$.
3. The efficiency (eff) of a motor is equal to the power output $P_o$ divided by the power input $P_i$.
4. The total resistance $R_T$ of a series circuit is equal to the sum of the individual resistances $R_1$, $R_2$, and $R_3$.
5. The capacitive reactance $X_C$ of a capacitor is equal to 159,000 divided by the product of the frequency $f$ and the capacitance $C$.
6. The power factor (PF) of an ac circuit is equal to the total resistance $R_T$ divided by the impedance Z.
7. The perimeter $P$ of a rectangle is equal to the sum of twice the length $L$ and twice the width $W$.
8. The current $I_m$ in an ammeter is the difference between the line current $I$ and the shunt current $I_s$.
9. The total current $I_T$ in a series circuit of two resistors is equal to the total voltage $V_T$ divided by the sum of the resistances $R_1$ and $R_2$.
10. The resistance $R_s$ of an ammeter shunt is equal to the meter resistance $R_m$ divided by 1 less than the multiplying factor $N$.

**Substitution in formulas.** As we have learned, we can change a rule into a formula by substituting letters for words. Since each word or letter actually represents some number in a specific problem, we can go one step further and substitute specific numbers for the letters in any formula or expression. This is called *substitution in a formula.* The numbers are then combined according to the signs of operation shown by the formula.

EXAMPLE 3-5. Find the value of $10 - 2 \times 4$.

SOLUTION. If we do this arithmetic in order from left to right—subtract first and then multiply—we get

$$10 - 2 \times 4$$

or
$$8 \times 4$$

or
$$32 \qquad \textit{Ans.}$$

But if we multiply first and then subtract, we get

$$10 - 2 \times 4$$

or
$$10 - 8$$

or
$$2 \qquad \textit{Ans.}$$

Both answers are correct, and depend on which mathematical operation is performed first. However, if we are all expected to get the *same* answer in a particular problem, we must agree on which operation is to be done first, second, etc. Over the entire world, mathematicians have agreed on a definite *order* in which the operations of arithmetic should be performed. Failure to follow this order often gives results which do *not* represent what was actually meant.

## ORDER OF OPERATIONS FOR EVALUATING MATHEMATICAL EXPRESSIONS

1. Substitute the given numbers for the letters.
2. Find the value of all expressions in parentheses.
3. Do all multiplications in order from left to right.
4. Do all divisions in order from left to right.
5. Do all additions and subtractions in order from left to right.

EXAMPLE 3-6. Find the value of $12 - 2 \times 4 + 3$.

SOLUTION

| | |
|---|---|
| 1. Write the expression. | $12 - 2 \times 4 + 3$ |
| 2. Multiply first. | $12 - 8 + 3$ |
| 3. Add or subtract. | $4 + 3$ |
| 4. Answer. | $7$ |

EXAMPLE 3-7. Find the value of $10 - 2 \times 3 + {}^{8}/_{2}$.

| | |
|---|---|
| 1. Write the expression. | $10 - 2 \times 3 + {}^{8}/_{2}$ |
| 2. Multiply first. | $10 - 6 + {}^{8}/_{2}$ |
| 3. Divide next. | $10 - 6 + 4$ |
| 4. Add or subtract. | $4 + 4$ |
| 5. Answer. | $8$ |

EXAMPLE 3-8. Using the formula $P = VI$, find the power $P$ needed to operate an electric iron using a voltage $V$ of 110 V and a current $I$ of 6 A. Refer to Fig. 3-2.

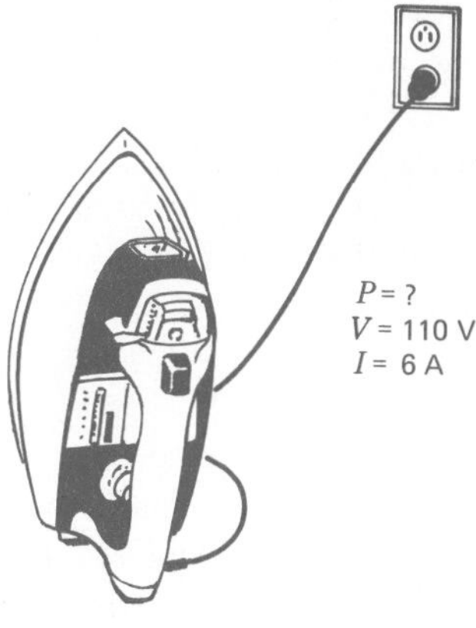


FIGURE 3-2

SOLUTION

1. Write the formula. $P = VI$
2. Substitute numbers. $P = 110 \times 6$

Note that no sign of operation between the letters means multiply.

3. Multiply the numbers. $P = 660$ W *Ans.*

SELF-TEST 3-9. Using the formula $V_T = I(R_1 + R_2)$, find the total voltage $V_T$ in the series circuit shown in Fig. 3-3.

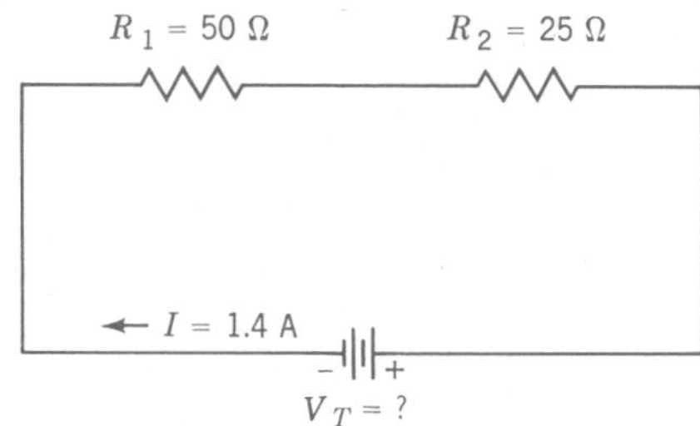


FIGURE 3-3

SOLUTION

1. Write the formula. $V_T = I(R_1 + R_2)$
2. Substitute numbers. $V_T = 1.4(50 + 25)$
3. The arithmetic to be done first is the ________ (multiplication/addition) of ________ and ________.
4. Simplifying the parentheses gives $V_T = 1.4 \times$ ________.
5. Multiply. $V_T =$ ________ V *Ans.*

addition 50 25
75
105

**PROBLEMS**

Evaluate the following expressions:

1. $9 + 2 \times 4 - 3$
2. $12 - 2 \times 5$
3. $2 \times 7 + 2 \times 3$
4. $5(8 - 3) + 7$
5. $5 \times 6 - 4 \times 3$
6. $15 - 2(8 - 3)$
7. $6 + {}^{2}/_{8}$
8. $\dfrac{6 \times 7}{2 + 5}$
9. $3(10 - 7) - 2$
10. $21 - 2(15 - 5)$

Solve the following problems using the formulas that are given in each problem. If there is no diagram that applies to the problem, set down the given information in the space ordinarily used for the diagram.

11. Using the formula $A = LW$, find the number of square feet of area $A$ in a rectangle if the length $L$ is 15 ft and the width $W$ is 9 ft.
12. Using the formula $I = V/Z$, find the number of amperes $I$ if $V$ is 110 V and $Z$ is 22 Ω.
13. Using the formula $I_T = I_1 + I_2 + I_3$, find the total number of amperes $I_T$ if $I_1 = 2$ A, $I_2 = 3$ A, and $I_3 = 5$ A.
14. Using the formula kW $= VI/1000$, find the kilowatts (kW) of power if the voltage $V$ is 200 V and the current $I$ is 2 A.
15. Using the formula $R_T = R_1 + R_2 + R_3$, find the total resistance $R_T$ of a series circuit if $R_1$ is 18.2 Ω, $R_2$ is 45.8 Ω, and $R_3$ is 76.4 Ω.
16. The formula for the shunt current in an ammeter hookup is $I_s = I - I_m$. Find the current $I_s$ through the shunt if the line current $I$ is 0.045 A and the meter current $I_m$ is 0.009 A.
17. Using the formula $R_T = (R_1 \times R_2)/(R_1 + R_2)$, find the total resistance $R_T$ of a parallel circuit if $R_1$ is 30 Ω and $R_2$ is 60 Ω.
18. Using the formula $X_L = 6.28fL$, find the inductive reactance $X_L$ of a coil to a frequency $f$ of 60 Hz if the coil has an inductance $L$ of 0.15 H.
19. Using the series-circuit formula $I = V/(R + r)$, find the current $I$ if $V = 6$, $R = 16$, and $r = 8$.
20. Using the formula $F = {}^{9}/_{5}C + 32$ to change Celsius temperature to Fahrenheit temperature, find $F$ if $C = 20°$.
21. Using the formula $R_t = R_o(1 + 0.0042t)$ for the effect of temperature on resistance, find the resistance $R_t$ at a temperature $t = 50°$, if the resistance at 0°C ($R_o$) is 60 Ω.

(See Instructor's Manual for Test 3-1, Substitution in Formulas)

## JOB 3-3 POSITIVE AND NEGATIVE NUMBERS

Many of the calculations, graphs, and tables used to solve problems in the jobs to follow require an understanding of positive and negative numbers. These numbers, commonly called *signed* numbers, are used to indicate *opposite* amounts, such as a *gain* or a *loss* in voltage, an *increase* or *decrease* in loudness, or currents that flow in *opposite directions.*

**Writing positive and negative numbers.** In our daily conversations we often indicate opposite quantities by pairs of words such as north or south, up or down, gain or loss, and win or lose. In our electrical work, it is much easier to indicate opposite quantities by the use of the plus sign (+) or the minus sign (−). For example, 5° above zero may be written as +5°, and 5° below zero may be written as −5°. A current in one direction is written as + 10 A, but as − 10 A if the flow is in the opposite direction. Numbers preceded by a minus sign are called *negative numbers.* The minus sign must always be written before a negative number. If no sign is written before a number, the quantity is understood to be a positive number. Signed numbers are usually written to agree with the following system. Gains, increases, or directions to the right or upward are written as positive (+). Losses, decreases, or directions to the left or downward are written as negative (−).

### PROBLEMS

Write the following quantities as signed numbers:

1. A loss of $3.
2. A temperature of 8° below zero.
3. 23° north latitude.
4. 70° east longitude.
5. An increase in loudness of 2 dB (decibels).
6. A drop of 10° in temperature.
7. Ten miles per hour slower.
8. Eight paces to the right.
9. Five blocks downtown.
10. Twenty feet below sea level.
11. If the voltage of the ground is considered to be 0 V, indicate a voltage of 4 V below ground.
12. Indicate a grid bias of 2 V below ground.

## ADDITION OF SIGNED NUMBERS

**RULE** To add two positive numbers, add the numbers and place the plus sign before the sum.

EXAMPLE 3-10. "A gain of 7 plus a gain of 3 equals a gain of 10" may be indicated as

$$(+7) + (+3) = +10 \qquad \textit{Ans.}$$

**RULE** To add two negative numbers, add the numbers and place the minus sign before the sum.

EXAMPLE 3-11. "A loss of 5 plus a loss of 2 equals a loss of 7" may be indicated as

$$(-5) + (-2) = -7 \qquad \textit{Ans.}$$

**RULE** To add two numbers of *different* sign, *subtract* the numbers and place the sign of the larger before the answer.

EXAMPLE 3-12. "A gain of 7 plus a loss of 3 equals a total gain of 4" may be indicated as

$$(+7) + (-3) = +4 \qquad \textit{Ans.}$$

EXAMPLE 3-13. "A loss of 9 plus a gain of 4 equals a total loss of 5" may be indicated as

$$(-9) + (+4) = -5 \qquad \textit{Ans.}$$

EXAMPLE 3-14. Combine the following signed numbers:

$$7 - 2 - 3 + 5 + 4 - 2$$

SOLUTION. The signs in problems of this type are *never* meant to represent addition or subtraction. These signs are the signs of the numbers. To combine signed numbers *always* means to *add* the signed numbers using the rules for algebraic addition. Therefore, reading from left to right, we shall consider this problem to mean

+7 *added to* −2 is +5 (7 means +7)
+5 *added to* −3 is +2
+2 *added to* +5 is +7
+7 *added to* +4 is +11
+11 *added to* −2 is +9 *Ans.*

SELF-TEST 3-15. Fill in the blank spaces with the correct responses.

1. (−4) + (−6) means to ____(add/subtract)____ the numbers and prefix the sum with a ________ sign. The answer is ________. | add; minus −10

2. (+4) + (+8) means to ________ the numbers and prefix the sum with a ________ sign. The answer is ________. | add; plus +12

3. (7) + (+4) means to ________ the numbers and prefix the sum with a ________ sign. The answer is ________. | add; plus +11

4. (+6) + (−4) means to ________ the numbers and prefix the difference with a ________ sign because the sign of the larger number is +. The answer is ________. | subtract; plus; +2

5. (−12) + (−3) means to ________ the numbers and prefix the answer with a ________ sign. The answer is ________. | add; minus −15

| | |
|---|---|
| 6. (−10) + (+4) means to _______ the numbers and prefix the answer with a _______ sign because the sign of the larger is _______. The answer is _______. | subtract<br>minus minus<br>−6 |
| 7. 7 − 4 means (+7) + (_______). The answer is _______. | −4 +3 |
| 8. (23) + (−5) means to _______ the numbers and prefix the answer with a _______ sign. The answer is _______. | subtract<br>plus +18 |
| 9. −5 − 7 means (−5) + (_______). The answer is _______. | −7 −12 |
| 10. 7 − 3 − 9 − 2 means (+7) + (−3) + (_______) + (−2): | −9 |
| +7 added to −3 = _______<br>+4 added to −9 = _______<br>−5 added to −2 = _______ *Ans.* | +4<br>−5<br>−7 |

**PROBLEMS**

Add the signed numbers indicated in each problem.

1. (+3) + (+9)
2. (−6) + (−5)
3. (+8) + (−2)
4. (−10) + (+3)
5. (−12) + (−5)
6. (+16) + (+3)
7. (+18) + (−11)
8. (−21) + (+9)
9. (−8) + (−9)

10. $\begin{array}{r} +18 \\ \underline{+14} \end{array}$ 11. $\begin{array}{r} -26 \\ \underline{+12} \end{array}$ 12. $\begin{array}{r} +36 \\ \underline{-14} \end{array}$

13. $\begin{array}{r} -6 \\ \underline{+22} \end{array}$ 14. $\begin{array}{r} -47 \\ \underline{-23} \end{array}$ 15. $\begin{array}{r} -75 \\ \underline{+23} \end{array}$

16. $\begin{array}{r} -8.2 \\ \underline{+11.6} \end{array}$ 17. $\begin{array}{r} -10.5 \\ \underline{+12.4} \end{array}$ 18. $\begin{array}{r} -16.8 \\ \underline{+7} \end{array}$

19. $(-1/4) + (+1/2)$
20. $(+5/8) + (-1/4)$
21. $(-1/2) + (-1/3)$

22. $\begin{array}{r} -16\,7/8 \\ \underline{+8\,3/4} \end{array}$ 23. $\begin{array}{r} +3\,9/16 \\ \underline{-9\,1/4} \end{array}$ 24. $\begin{array}{r} -15\,3/4 \\ \underline{+28\,1/2} \end{array}$

25. +8 − 2 − 9 + 6
26. −6 − 3 + 11 + 4 − 5
27. 13 − 2 − 3 + 4 − 5
28. −9 + 3 − 2 + 4 − 6
29. −3 − 5 + 2 − 26 + 7
30. 6 − 2 − 9 − 1 + 3
31. 1.4 − 0.2 − 0.7
32. −16 + 4.8 + 3.4
33. 6.4 − 8.5 + 3.2 − 4
34. 3.2 − 8 + 2.5 + 5.7

(See Instructor's Manual for Test 3-2)

## MULTIPLYING AND DIVIDING SIGNED QUANTITIES

The following rules for signs apply to *both* multiplication and division.

**RULE** When multiplying or dividing quantities with the same sign, the answer is plus.

**RULE** When multiplying or dividing quantities with different signs, the answer is minus.

EXAMPLE 3-16. Perform the indicated operation.

$$(+4)\times(+2)=+8 \qquad (-6)\times(-2)=+12$$
$$(+4)\times(-2)=-8 \qquad (-3)\times(+4)=-12$$
$$(+2)\times(+3R)=+6R \qquad (-4)\times(-2I)=+8I$$
$$(-4)\times(0)=0 \qquad (+3)\times(0)=0$$

$$\frac{+20}{+5}=+4 \qquad \frac{-12}{-4}=+3 \qquad \frac{+3}{+6}=+\frac{1}{2} \qquad \frac{-2}{-8}=+\frac{1}{4}$$
$$\frac{+20}{-5}=-4 \qquad \frac{-12}{+4}=-3 \qquad \frac{+3}{-6}=-\frac{1}{2} \qquad \frac{-2}{+8}=-\frac{1}{4}$$
$$\frac{-8R}{-2}=+4R \qquad \frac{-12I}{+6}=-2I \qquad \frac{+6R}{-2}=-3R \qquad \frac{-3R}{-6}=+\frac{1}{2}R$$

**PROBLEMS**

Multiply:

1. (+6) by (+3)
2. (−3) by (−4)
3. (+4) by (−2)
4. (−3) by (6)
5. (−5) by (−6)
6. (+2) by (3$R$)
7. (6$R$) by (+3)
8. (−3) by (4$I$)
9. (5$R$) by (−2)
10. (−3) by (−5$R$)
11. (−2$R$) by (−5)
12. (−12) by (−6)
13. (−6) by (0)
14. (−8$T$) by (0)
15. (8) by (−13)
16. (−25) by (−6)
17. (−5$R$) by (17)
18. (2) by (−13$I$)
19. (−4$R$) by (36)
20. (0.2) by (−10)
21. (−0.3) by (20)
22. (−0.2) by (5$R$)
23. (−1.2) by (2$I$)
24. (−1/2) by (−20)
25. (−1/4) by (12$R$)
26. (−1.2) by (−18)
27. (4.6) by (−3.2)
28. (−16$I$) by (−0.2)
29. (3.4$R$) by (−1.5)
30. (−0.6) by (8.2$R$)

Divide:

1. (+12) by (+3)
2. (−24) by (−4)
3. (+8) by (−2)
4. (−16) by (+2)
5. (−84) by (−7)
6. (24$R$) by (6)
7. (24$R$) by (−4)
8. (−60) ÷ (15)
9. (20) ÷ (−40)
10. (−48) ÷ (−8)
11. (−16) ÷ (−32)
12. (+9) ÷ (−27)
13. (0) by (−6)
14. (0) ÷ (+7)
15. (63) ÷ (−9)
16. (−100) ÷ (+5)
17. (−60) ÷ (+8)
18. (−4) ÷ (20)
19. $\frac{-3.6}{1.2}$
20. $\frac{4}{-7}$
21. $\frac{-1.6}{-0.5}$
22. $\frac{4.2}{-3}$
23. $\frac{-65}{-0.5}$
24. $\frac{0.4}{-0.1}$
25. $\frac{-2.6}{1.3}$
26. $\frac{0.4}{-3.2}$
27. $\frac{9.6}{-3}$
28. $\frac{-2.4}{-0.6}$
29. $\frac{-50}{0.8}$
30. $\frac{-10.8}{-0.3}$

(See Instructor's Manual for Test 3-3)

## JOB 3-4 CHECKUP ON USING FORMULAS WITH EXPONENTS (DIAGNOSTIC TEST)

Many problems in electrical work are solved with formulas which use special mathematical symbols called *exponents.* The following problems involve the use of formulas which contain exponents. If you have any difficulty with these problems, see Job 3-5 which follows.

### PROBLEMS

1. Using the formula $P = V^2/R$, find the value of $P$ if $V = 100$ and $R = 50$.
2. Using the formula $P = I^2R$, find the value of $P$ if $I = 3$ and $R = 40$.
3. Using the formula $\text{emf} = (L \times n)/10^8$, find the value of emf if $L = 50 \times 10^6$ and $n = 40$.
4. Using the formula $R = (k \times l)/D^2$, find the value of $R$ if $k = 10.4$, $l = 100$, and $D = 4$.
5. Using the formula $\text{F} = \mu\text{F}/10^6$, find the number of farads equal to 0.5 $\mu$F.

## JOB 3-5 BRUSHUP ON FORMULAS CONTAINING EXPONENTS

**Meaning of an exponent.** Exponents provide a convenient shorthand method for writing and expressing many mathematical operations. For example, $2 \times 2 \times 2$ may be written as $2^3$. The number 3 above and to the right of the 2 is called an *exponent.* The exponent says, "Write the number beneath it as many times as the exponent indicates, and then multiply." An exponent may be used with letters as well as numbers. Thus, $R^3$ means $R \times R \times R$. The exponent 2 is read as the word "square." The exponent 3 is read as the word "cube." When the exponent is any other number, it is read as "to the fourth power," "to the seventh power," etc. For example:

$$3^2 \text{ (3 square)} = 3 \times 3 = 9$$
$$5^3 \text{ (5 cube)} = 5 \times 5 \times 5 = 125$$
$$2^4 \text{ (2 to the fourth power)} = 2 \times 2 \times 2 \times 2 = 16$$
$$10^1 \text{ (10 to the first power)} = 10$$
$$10^2 \text{ (10 square)} = 10 \times 10 = 100$$
$$10^3 \text{ (10 cube)} = 10 \times 10 \times 10 = 1000$$
$$10^6 \text{ (10 to the sixth power)} = 10 \times 10 \times 10 \times 10 \times 10 \times 10 = 1{,}000{,}000$$

EXAMPLE 3-17. The formula for the area of a circle is $A = \pi R^2$. Find the area of a circle if the radius $R$ equals 5 in and $\pi = 3.14$.

SOLUTION. Given: $A = \pi R^2$ Find: $A = ?$

$R = 5$ in
$\pi = 3.14$

1. Write the formula. $A = \pi R^2$
2. Substitute numbers. $A = 3.14 \times 5^2$
3. Simplify the exponent. $A = 3.14 \times 25$
4. Multiply. $A = 78.5\ \text{in}^2$ *Ans.*

SELF-TEST 3-18. The resistance $R$ of a copper wire is found by the formula $R = 10.4 \times L/D^2$, in which $D$ = the diameter of the wire in mils (1 mil = 0.001 in) and $L$ = the length of the wire in feet. Find the resistance of 5000 ft of wire whose diameter is 50 mil.

SOLUTION. Given: $L = 5000$ ft Find: ______ = ? | $R$
$D$ = ______ mil | 50

1. Write the formula. $R = \dfrac{10.4 \times L}{D^2}$
2. Substitute numbers. $R = \dfrac{10.4 \times ?}{(?)^2}$ | 5000 | 50
3. Simplify the exponent. $R = \dfrac{10.4 \times 5000}{?}$ | 2500
4. Cancel. $R = 10.4 \times$ ______ | 2
5. Multiply. $R$ = ______ Ω *Ans.* | 20.8

SELF-TEST 3-19. The formula for the power gain of a common-base transistor circuit is

$$PG = \alpha^2 \times \frac{R_L}{R_i}$$

Find the power gain of a transistor if the current gain $\alpha = 0.96$, the load resistance $R_L = 15{,}000\ \Omega$, and the input resistance $R_i = 300\ \Omega$.

SOLUTION. Given: $\alpha = 0.96$ Find: PG = ?
$R_L = 15{,}000\ \Omega$
$R_i = 300\ \Omega$

1. Write the formula. $PG = \alpha^2 \times \dfrac{R_L}{R_i}$
2. Substitute numbers. $PG = (0.96)^2 \times \dfrac{15{,}000}{?}$ | 300
3. Simplify the exponent. PG = ______ $\times \dfrac{15{,}000}{300}$ | 0.92
4. Simplify the fraction. $PG = 0.92 \times$ ______ | 50
5. Multiply. PG = ______ *Ans.* | 46

## PROBLEMS

1. Using the formula $A = D^2$, find the area $A$, in circular mils, of a wire whose diameter $D = 60$ mil.
2. The volume of a sphere is given by the formula $V = 4.2R^3$, in which $V$ is the volume in cubic inches and $R$ is the radius of the sphere. Find the volume of a sphere whose radius is 4 in.
3. Using the formula $S = \frac{1}{2}gt^2$, find the distance $S$ that a body will fall if the acceleration of gravity $g = 32$ feet per second per second (ft/s²) and the time $t = 8$ s.
4. Using the formula for the volume of a cylinder, $V = \pi R^2 H$, find the volume $V$ if $\pi = {}^{22}/_7$, $R = 3.5$, and $H = 14$.
5. Using the formula A = mA/$10^3$, change 450 mA into amperes.
6. Using the formula for the horsepower rating of a gasoline engine, $H = (nD^2)/2.5$, find the horsepower rating $H$ of an engine if the number of cylinders $n = 6$ and the diameter of each cylinder $D = 3$ in.
7. Using the formula $V = (L \times n)/10^8$, find the value of $V$ if $L = 40{,}000$ and $n = 80$.
8. Using the formula $P = I^2R$, find the power used by an electric appliance if the current $I = 6$ A and the resistance $R = 20\ \Omega$.
9. Another formula for electric power is $P = V^2/R$. Find the power consumed by a 100-Ω electric iron when it is operating on a 115-V line.

(See Instructor's Manual for Test 3-4)

## JOB 3-6 POWERS OF 10

In many problems in electricity and electronics, the usual units of amperes, volts, and ohms are either too large or too small. It has been found more convenient to use new units of measurement. These new units are formed by placing a special word or prefix in front of the unit. These range from exa E (1 billion billion) to atto a (1 billionth of a billionth). As we learned in Job 3-4, these numbers may be written as powers of 10.

$$10 = 10^1 = 10 \text{ to the } \textit{first} \text{ power}$$
$$100 = 10^2 = 10 \text{ to the } \textit{second} \text{ power}$$
$$1000 = 10^3 = 10 \text{ to the } \textit{third} \text{ power}$$
$$10{,}000 = 10^4 = 10 \text{ to the } \textit{fourth} \text{ power}$$
$$100{,}000 = 10^5 = 10 \text{ to the } \textit{fifth} \text{ power}$$
$$1{,}000{,}000 = 10^6 = 10 \text{ to the } \textit{sixth} \text{ power}$$

The use of these units will clearly involve multiplying and dividing by numbers of this size and notation. Our work will be considerably simplified if we learn to apply the following rules.

## MULTIPLYING BY POWERS OF 10

**RULE** To multiply values by numbers expressed as 10 raised to some power, move the decimal point to the right as many places as the exponent indicates.

EXAMPLE 3-20

$0.345 \times 10^2 = \odot 34.5$ or 34.5 (move point two places right)

$0.345 \times 10^3 = \odot 345.$ or 345 (move point three places right)

$0.345 \times 10^6 = \odot 345000.$ or 345,000 (move point six places right)

$0.0065 \times 10^3 = \odot 006.5$ or 6.5 (move point three places right)

### PROBLEMS

1. $0.0072 \times 1000$
2. $45.76 \times 100$
3. $3.09 \times 1000$
4. $0.0045 \times 100$
5. $37 \times 100$
6. $0.08 \times 10^2$
7. $0.0006 \times 10^3$
8. $0.000\,56 \times 10^6$
9. $27 \times 10^2$
10. $15 \times 10^3$
11. $0.000\,78 \times 10^2$
12. $7.8 \times 10^6$
13. $15.4 \times 10^3$
14. $3 \times 15 \times 10^2$
15. $0.005 \times 2 \times 10^2$
16. $6 \times 10^4$
17. $0.008 \times 10^4$
18. $0.009 \times 10^5$
19. $2.34 \times 10^5$
20. $0.000\,000\,5 \times 10^8$

## DIVIDING BY POWERS OF 10

**RULE** To divide values by numbers expressed as 10 raised to some power, move the decimal point to the left as many places as the exponent indicates.

EXAMPLE 3-21

$467.9 \div 10^2 = 4.67\odot 9 = 4.679$ (move point two places left)

$59 \div 10^2 = .59\odot = 0.59$ (move point two places left)

$8.5 \div 10^3 = .008\odot 5 = 0.0085$ (move point three places left)

$5{,}500{,}000 \div 10^8 = .05500000\odot = 0.055$ (move point eight places left)

### PROBLEMS

1. $6500 \div 100$
2. $7500 \div 10^3$
3. $880{,}000 \div 1000$
4. $32 \div 10^2$
5. $6 \div 10^2$
6. $17.8 \div 10$
7. $835 \div 10^3$
8. $550 \div 10^6$
9. $653.8 \div 10^3$
10. $100{,}000{,}000 \div 10^6$

11. $0.45 \div 10^2$
12. $0.08 \div 10$
13. $8.5 \div 10^3$
14. $7 \div 10^6$
15. $\dfrac{28.6}{10^3}$
16. $\dfrac{2 \times 1000}{10^3}$
17. $0.02 \div 10^3$
18. $\dfrac{180}{2 \times 10^2}$

## NEGATIVE POWERS OF 10

If $x^5$ means $x \cdot x \cdot x \cdot x \cdot x$, and $x^3$ means $x \cdot x \cdot x$, then

$$\frac{x^5}{x^3} = \frac{\overset{1}{\cancel{x}} \cdot \overset{1}{\cancel{x}} \cdot \overset{1}{\cancel{x}} \cdot x \cdot x}{\underset{1}{\cancel{x}} \cdot \underset{1}{\cancel{x}} \cdot \underset{1}{\cancel{x}}} = x^2 \text{ or } x^{(5-3)}$$

In general,

$$\frac{a^m}{a^n} = a^{(m-n)} \qquad \text{when } a \neq 0$$

It follows, then, that

$$\frac{a^2}{a^5} = \frac{\overset{1}{\cancel{a}} \cdot \overset{1}{\cancel{a}}}{\underset{1}{\cancel{a}} \cdot \underset{1}{\cancel{a}} \cdot a \cdot a \cdot a} = \frac{1}{a^3}$$

or $$\frac{a^2}{a^5} = a^{(2-5)} = a^{-3}$$

Thus, $$a^{-3} = \frac{1}{a^3}$$

or $$a^{-n} = \frac{1}{a^n}$$

or $$10^{-n} = \frac{1}{10^n} \quad \text{and} \quad \frac{1}{10^n} = 10^{-n}$$

Thus, $50 \times \dfrac{1}{10^2}$ is the same as $50 \times 10^{-2}$

or $\dfrac{50}{10^2}$ is the same as $50 \times 10^{-2}$

Therefore, if dividing by $10^2$ means to move the decimal point to the left for two places, then the $-$ sign in the exponent $-2$ must mean to do the same thing.

## MULTIPLYING BY NEGATIVE POWERS OF 10

**RULE** To multiply values by numbers expressed as 10 raised to some negative power, move the decimal point to the *left* as many places as the exponent indicates.

EXAMPLE 3-22

$50 \times 10^{-2} = .50\odot$ or 0.5 (move point two places left)

$64.9 \times 10^{-2} = .64\odot 9$ or 0.649 (move point two places left)

$50{,}000 \times 10^{-6} = .050000\odot$ or 0.05 (move point six places left)

$0.2 \times 10^{-3} = .000\odot 2$ or 0.0002 (move point three places left)

**PROBLEMS**

1. $25{,}000 \times 10^{-3}$
2. $250 \times 10^{-2}$
3. $1.5 \times 10^{-1}$
4. $0.5 \times 10^{-2}$
5. $6250 \times 10^{-3}$
6. $100{,}000 \times 10^{-5}$
7. $250{,}000 \times 10^{-8}$
8. $6 \times 10^{-3}$
9. $75.4 \times 10^{-4}$
10. $16.5 \times 10^{-2}$

DIVIDING BY NEGATIVE POWERS OF 10

Since by definition,

$$\frac{1}{10^n} = 1 \times 10^{-n}$$

and

$$1 \times 10^{-n} = \frac{1}{10^n}$$

we can transfer any power of 10 from numerator to denominator, or vice versa, by simply changing the sign of the exponent.

EXAMPLE 3-23

$$\frac{15}{10^{-2}} = 15 \times 10^2 = 1500$$

$$\frac{0.05}{10^{-3}} = 0.05 \times 10^3 = 50$$

$$\frac{15{,}000}{10^3} = 15{,}000 \times 10^{-3} = 15$$

$$\frac{5 \times 0.2}{10^{-4}} = 1.0 \times 10^4 = 10{,}000$$

**PROBLEMS**

1. $\frac{19.2}{10^{-2}}$
2. $\frac{25.6}{10^2}$
3. $\frac{0.85}{10^{-3}}$
4. $\frac{85}{10^3}$
5. $\frac{0.0072}{10^{-6}}$
6. $\frac{0.9}{10^{-4}}$
7. $\frac{0.0045}{10^2}$
8. $\frac{0.96}{10^{-3}}$
9. $\frac{880{,}000}{10^{-2}}$
10. $\frac{0.005}{10^{-2} \times 10^{-3}}$

11. $\dfrac{6 \times 5}{10^{-2}}$

12. $\dfrac{0.5 \times 0.04}{10^{-3}}$

13. $\dfrac{30 \times 10^{3}}{10^{-2}}$

14. $\dfrac{50 \times 10^{2}}{10^{-3}}$

15. $\dfrac{60 \times 10^{-2}}{10^{3}}$

16. $\dfrac{64.9 \times 10^{3}}{10^{5}}$

## EXPRESSING NUMBERS AS POWERS OF 10

As you have seen, multiplying and dividing numbers by powers of 10 is just a set of simple mental problems. It certainly will be to our advantage then if we can express the numbers in any problem as powers of 10 before we multiply or divide.

## EXPRESSING NUMBERS LARGER THAN 1 AS A SMALL NUMBER TIMES A POWER OF 10

**RULE** To express a large number as a smaller number times a power of 10, move the decimal point to the *left* as many places as desired. Then multiply the number obtained by 10 to a power which is equal to the number of places moved.

EXAMPLE 3-24

*a.* 3000 = 3.000⊙ (moved three places left)
3000 = $3 \times 10^{3}$

*b.* 4500 = 4.500⊙ (moved three places left)
4500 = $4.5 \times 10^{3}$

*c.* 770,000 = 77.0000⊙ (moved four places left)
770,000 = $77 \times 10^{4}$

*d.* 800,000 = 8.00000⊙ (moved five places left)
800,000 = $8 \times 10^{5}$

*e.* Express the following number to three significant figures and express it as a number between 1 and 10 times the proper power of 10.
7831 = 7830 (since only three significant figures are wanted)
7830 = 7.830⊙ (moved three places left)
7830 = $7.83 \times 10^{3}$

*f.* Express 62,495 using three significant figures as a number between 1 and 10 times the proper power of 10.
62,495 = 62,500 (written as three significant figures)
62,500 = 6.2500⊙ (moved four places left)
62,500 = $6.25 \times 10^{4}$

## PROBLEMS

Express the following numbers to three significant figures and write them as numbers between 1 and 10 times the proper power of 10.

| | | |
|---|---|---|
| 1. 6000 | 2. 5700 | 3. 150,000 |
| 4. 500,000 | 5. 235,000 | 6. 7,350,000 |
| 7. 4960 | 8. 62,500 | 9. 980 |
| 10. 175 | 11. 12.5 | 12. 7303 |
| 13. 48.2 | 14. 12,600 | 15. 880,000,000 |
| 16. 54,009 | 17. 38,270 | 18. 8019.7 |
| 19. 1,754,300 | 20. 2,395,000 | 21. 482,715 |

## EXPRESSING NUMBERS LESS THAN 1 AS A WHOLE NUMBER TIMES A POWER OF 10

**RULE** To express a decimal as a whole number times a power of 10, move the decimal point to the right as many places as desired. Then multiply the number obtained by 10 to a *negative* power which is equal to the number of places moved.

EXAMPLE 3-25

*a.* 0.005 = 0⊙005. (moved three places right)
$0.005 = 5 \times 10^{-3}$

*b.* 0.006 72 = 0⊙006.72 (moved three places right)
$0.006\,72 = 6.72 \times 10^{-3}$

*c.* 0.0758 = 0⊙07.58 (moved two places right)
$0.0758 = 7.58 \times 10^{-2}$

*d.* 0.000 008 9 = 0⊙000008.9 (moved six places right)
$0.000\,008\,9 = 8.9 \times 10^{-6}$

*e.* Express the number 0.000 357 8 to three significant figures and then write it as a number between 1 and 10 times the proper power of 10.

0.000 357 8 = 0.000 358 (written as three significant figures)
0.000 358 = 0⊙0003.58 (moved four places right)
$0.000\,358 = 3.58 \times 10^{-4}$

## PROBLEMS

Express the following numbers to three significant figures and write them as numbers between 1 and 10 times the proper power of 10.

| | |
|---|---|
| 1. 0.006 | 2. 0.0075 |
| 3. 0.0035 | 4. 0.08 |

5. 0.456
6. 0.0357
7. $785 \times 10^{-2}$
8. 0.000 000 12
9. 0.0965
10. 0.004 82
11. 0.5
12. 0.000 374 3
13. 0.008 147
14. 0.000 007 949
15. $0.000\ 725 \times 10^{5}$
16. 0.013 33
17. $0.0006 \times 10^{3}$
18. $3200 \times 10^{-5}$
19. $360 \times 10^{-4}$
20. $0.000\ 008 \times 10^{4}$

## MULTIPLYING WITH POWERS OF 10

If $x^3$ means $x \cdot x \cdot x$, and $x^2$ means $x \cdot x$, then $x^3 \cdot x^2$ means $x \cdot x \cdot x \cdot x \cdot x = x^5$ or, $x^3 \cdot x^2 = x^{(3+2)} = x^5$, which gives us the following rule.

**RULE** The multiplication of two or more powers using the *same* base is equal to that base raised to the sum of the powers.

EXAMPLE 3-26

*a.* $a^4 \times a^5 = a^{(4+5)} = a^9$

*b.* $10^2 \times 10^3 = 10^5$

*c.* Multiply $10{,}000 \times 1000$.
If $10{,}000 = 10^4$ and $1000 = 10^3$, then

$$10{,}000 \times 1000 = 10^4 \times 10^3 = 10^{(4+3)} = 10^7 \quad \textit{Ans.}$$

*d.* Multiply $25{,}000 \times 4000$.
If $25{,}000 = 25 \times 10^3$ and $4000 = 4 \times 10^3$, then

$$\begin{aligned} 25{,}000 \times 4000 &= 25 \times 10^3 \times 4 \times 10^3 \\ &= 25 \times 4 \times 10^3 \times 10^3 \\ &= 100 \times 10^6 \\ &= 10^2 \times 10^6 = 10^8 \quad \textit{Ans.} \end{aligned}$$

*e.* Multiply $0.000\ 05 \times 0.003$.
If $0.000\ 05 = 5 \times 10^{-5}$ and $0.003 = 3 \times 10^{-3}$, then

$$\begin{aligned} 0.000\ 05 \times 0.003 &= 5 \times 10^{-5} \times 3 \times 10^{-3} \\ &= 5 \times 3 \times 10^{-5} \times 10^{-3} \\ &= 15 \times 10^{[-5+(-3)]} \\ &= 15 \times 10^{-8} \quad \textit{Ans.} \end{aligned}$$

*f.* Multiply $7000 \times 0.000\ 91$.
If $7000 = 7 \times 10^3$ and $0.000\ 91 = 9.1 \times 10^{-4}$, then

$$\begin{aligned} 7000 \times 0.000\ 91 &= 7 \times 10^3 \times 9.1 \times 10^{-4} \\ &= 7 \times 9.1 \times 10^3 \times 10^{-4} \\ &= 63.7 \times 10^{[3+(-4)]} \\ &= 63.7 \times 10^{-1} \\ &= 6.37 \quad \textit{Ans.} \end{aligned}$$

*g.* Multiply 0.000 05 × 20,000 × 1500.
If 0.000 05 = $5 \times 10^{-5}$, and 20,000 = $2 \times 10^{4}$, and 1500 = $1.5 \times 10^{3}$, then

$$\begin{aligned} 0.000\,05 \times 20{,}000 \times 1500 &= 5 \times 10^{-5} \times 2 \times 10^{4} \times 1.5 \times 10^{3} \\ &= 5 \times 2 \times 1.5 \times 10^{-5} \times 10^{4} \times 10^{3} \\ &= 15 \times 10^{2} \\ &= 1500 \quad \textit{Ans.} \end{aligned}$$

**PROBLEMS**

Multiply the following numbers.

1. $5000 \times 0.001$
2. $850 \times 2000$
3. $16 \times 10^{2} \times 4 \times 10^{3}$
4. $0.0004 \times 5 \times 10^{2}$
5. $250 \times 4000 \times 3 \times 10^{-2}$
6. $1000 \times 10^{-4} \times 0.02$
7. $3 \times 10^{-5} \times 4 \times 10^{6}$
8. $15 \times 10^{-4} \times 20{,}000 \times 0.04$
9. $200{,}000 \times 0.000\,005 \times 3 \times 10^{-2}$
10. $0.004 \times 0.0005 \times 5000$
11. $0.005 \times 5 \times 10^{-3} \times 0.02$
12. $6{,}000{,}000 \times 0.000\,25 \times 0.3 \times 10^{-2}$
13. $0.3 \times 10^{-2} \times 800{,}000 \times 400 \times 10^{-3}$
14. $(500)^{2} \times 0.0002 \times 4000$
15. $500{,}000{,}000 \times 0.000\,004 \times 3.14 \times 10^{2}$
16. As we shall see in Job 16-2, the inductive reactance of a coil is given by the formula

$$X_L = 6.28\,fL$$

where $f$ = frequency, Hz
$L$ = inductance of the circuit, H
$X_L$ = reactance, Ω

Find the inductive reactance when:
*a.* $f = 60$ Hz and $L = 0.025$ H
*b.* $f = 1{,}000{,}000$ Hz and $L = 0.25$ H
*c.* $f = 10{,}000$ Hz and $L = 0.000\,025$ H

## DIVISION WITH POWERS OF 10

As noted in the section on Dividing by Negative Powers of 10 of this job, we can transfer any power of 10 from numerator to denominator, or vice versa, by simply changing the sign of the exponent. This will permit us to change *all* division problems into multiplications, which are generally easier to do.

EXAMPLE 3-27

*a.* $10^{6} \div 10^{2} = \dfrac{10^{6}}{10^{2}} = 10^{6} \times 10^{-2} = 10^{4}$ *Ans.*

*b.* $\dfrac{4000}{10^{2}} = 4 \times 10^{3} \times 10^{-2} = 4 \times 10^{1} = 40$ *Ans.*

*c.* $\dfrac{35{,}000}{0.005} = \dfrac{35 \times 10^3}{5 \times 10^{-3}} = 7 \times 10^3 \times 10^3$

$= 7 \times 10^6$ *Ans.*

*d.* $\dfrac{144{,}000}{12 \times 10^3} = \dfrac{144 \times \overset{1}{\cancel{10^3}}}{12 \times \underset{1}{\cancel{10^3}}} = 12$ *Ans.*

Note that *any* factor divided by itself cancels out to 1, and that it is not necessary to transfer any powers. That is, $10^3/10^3 = 10^{(3-3)} = 10^0 = 1$.

*e.* $\dfrac{0.000\,75}{500} = \dfrac{75 \times 10^{-5}}{5 \times 10^2} = 15 \times 10^{-5} \times 10^{-2}$

$= 15 \times 10^{-7}$ *Ans.*

*f.* $\dfrac{60}{0.0003 \times 40{,}000} = \dfrac{60}{3 \times 10^{-4} \times 4 \times 10^4} = \dfrac{5}{10^0}$

Now, since $10^0 = 1$, $\dfrac{5}{10^0} = \dfrac{5}{1} = 5$ *Ans.*

**PROBLEMS**

Divide the following numbers.

1. $\dfrac{10^8}{10^3}$
2. $\dfrac{10^3}{10^5}$
3. $\dfrac{60{,}000}{5 \times 10^2}$
4. $\dfrac{50{,}000}{0.05}$
5. $\dfrac{10}{50{,}000}$
6. $\dfrac{20}{0.0005}$
7. $\dfrac{0.0001}{500}$
8. $\dfrac{20}{4000 \times 0.005}$
9. $\dfrac{1000 \times 0.008}{0.002 \times 500}$
10. $\dfrac{150{,}000}{3 \times 10^5}$
11. $\dfrac{0.000\,15}{3 \times 10^{-2}}$
12. $\dfrac{1}{4 \times 100{,}000 \times 0.000\,05}$
13. As we shall see in Job 17-4, the capacitive reactance of a capacitor is given by the formula

$$X_C = \frac{1}{2\pi f C}$$

where $f$ = frequency, Hz
$C$ = capacitance, F
$\pi = 3.14$
$X_C$ = reactance, Ω

Find the capacitive reactance when:
*a.* $f = 60$ Hz and $C = 0.000\,05$ F

*b.* $f = 1000$ Hz and $C = 0.000\ 002\ 5$ F
*c.* $f = 1{,}000{,}000$ Hz and $C = 0.000\ 000\ 05$ F

(See Instructor's Manual for Test 3-5, Powers of 10)

## JOB 3-7 UNITS OF MEASUREMENT IN ELECTRONICS

Ohm's law and other electrical formulas use the simple electrical units of volts, amperes, and ohms. However, if the measurements given or obtained in a problem were stated in kilovolts, milliamperes, or megohms, it would be necessary to change these units of measurement to the units required by the formula (Tables 3-1 to 3-3).

TABLE 3-1
METRIC PREFIXES

| PREFIX | SYMBOL | MULTIPLIER | PREFIX | SYMBOL | MULTIPLIER |
|---|---|---|---|---|---|
| Exa | E | $10^{18}$ | Milli | m | $10^{-3}$ |
| Peta | P | $10^{15}$ | Micro | $\mu$ | $10^{-6}$ |
| Tera | T | $10^{12}$ | Nano | n | $10^{-9}$ |
| Giga | G | $10^{9}$ | Pico | p | $10^{-12}$ |
| Mega | M | $10^{6}$ | Femto | f | $10^{-15}$ |
| Kilo | k | $10^{3}$ | Atto | a | $10^{-18}$ |

## CHANGING UNITS OF MEASUREMENT

TABLE 3-2
CHANGING LARGE UNITS TO SMALL UNITS

| MULTIPLY | BY | TO OBTAIN |
|---|---|---|
| Tera units | $10^{12}$ | Units |
| Tera units | $10^{9}$ | Kilo units |
| Tera units | $10^{6}$ | Mega units |
| Tera units | $10^{3}$ | Giga units |
| Giga units | $10^{9}$ | Units |
| Giga units | $10^{6}$ | Kilo units |
| Giga units | $10^{3}$ | Mega units |
| Mega units | $10^{6}$ | Units |
| Mega units | $10^{3}$ | Kilo units |
| Kilo units | $10^{3}$ | Units |
| Units | $10^{3}$ | Milli units |
| Units | $10^{6}$ | Micro units |
| Units | $10^{9}$ | Nano units |
| Units | $10^{12}$ | Pico units |
| Milli units | $10^{3}$ | Micro units |
| Milli units | $10^{6}$ | Nano units |
| Milli units | $10^{9}$ | Pico units |
| Micro units | $10^{3}$ | Nano units |
| Micro units | $10^{6}$ | Pico units |
| Nano units | $10^{3}$ | Pico units |

EXAMPLE 3-28. Change:

*a.* 2 GV to volts: 2 GV $= 2 \times 10^9 =$ 2,000,000,000 V
*b.* 0.25 MΩ to ohms: 0.25 MΩ $= 0.25 \times 10^6 =$ 250,000 Ω
*c.* 0.3 kV to volts: 0.3 kV $= 0.3 \times 10^3 =$ 300 V
*d.* 880 kHz to hertz: 880 kHz $= 880 \times 10^3 =$ 880,000 Hz
*e.* 2.4 V to millivolts: 2.4 V $= 2.4 \times 10^3 =$ 2400 mV
*f.* 0.0004 A to microamperes: 0.0004 A $= 0.0004 \times 10^6 =$ 400 μA
*g.* 0.000 05 F to microfarads: 0.000 05 F $= 0.000\,05 \times 10^6 =$ 50 μF
*h.* 0.000 000 005 F to picofarads: 0.000 000 005 F $= 0.000\,000\,005 \times 10^{12} =$ 5000 pF
*i.* 20 mA to microamperes: 20 mA $= 20 \times 10^3 =$ 20,000 μA
*j.* 0.000 35 μF to picofarads: 0.000 35 μF $= 0.000\,35 \times 10^6 =$ 350 pF
*k.* 0.007 μF to nanofarads: 0.007 μF $= 0.007 \times 10^3 =$ 7 nF
*l.* 0.000 005 TV to megavolts: 0.000 005 TV $= 0.000\,005 \times 10^6 =$ 5 MV

TABLE 3-3
CHANGING SMALL UNITS TO LARGE UNITS

| MULTIPLY | BY | TO OBTAIN |
|---|---|---|
| Pico units | $10^{-3}$ | Nano units |
| Pico units | $10^{-6}$ | Micro units |
| Pico units | $10^{-9}$ | Milli units |
| Pico units | $10^{-12}$ | Units |
| Nano units | $10^{-3}$ | Micro units |
| Nano units | $10^{-6}$ | Milli units |
| Nano units | $10^{-9}$ | Units |
| Micro units | $10^{-3}$ | Milli units |
| Micro units | $10^{-6}$ | Units |
| Milli units | $10^{-3}$ | Units |
| Units | $10^{-3}$ | Kilo units |
| Units | $10^{-6}$ | Mega units |
| Units | $10^{-9}$ | Giga units |
| Units | $10^{-12}$ | Tera units |
| Kilo units | $10^{-3}$ | Mega units |
| Kilo units | $10^{-6}$ | Giga units |
| Kilo units | $10^{-9}$ | Tera units |
| Mega units | $10^{-3}$ | Giga units |
| Mega units | $10^{-6}$ | Tera units |
| Giga units | $10^{-3}$ | Tera units |

EXAMPLE 3-29. Change:

*a.* 500,000 Ω to megohms: 500,000 Ω $= 500{,}000 \times 10^{-6} =$ 0.5 MΩ

*b.* 660 kHz to megahertz: $660\ \text{kHz} = 660 \times 10^{-3} = 0.66\ \text{MHz}$
*c.* 600 V to kilovolts: $600\ \text{V} = 600 \times 10^{-3} = 0.6\ \text{kV}$
*d.* 14.5 mA to amperes: $14.5\ \text{mA} = 14.5 \times 10^{-3} = 0.0145\ \text{A}$
*e.* 2.5 μF to farads: $2.5\ \mu\text{F} = 2.5 \times 10^{-6} = 0.000\ 002\ 5\ \text{F}$
*f.* 30,000,000 pF to farads: $30{,}000{,}000\ \text{pF} = 30{,}000{,}000 \times 10^{-12} = 0.000\ 03\ \text{F}$
*g.* 400 μV to millivolts: $400\ \mu\text{V} = 400 \times 10^{-3} = 0.4\ \text{mV}$
*h.* 350 pF to microfarads: $350\ \text{pF} = 350 \times 10^{-6} = 0.000\ 35\ \mu\text{F}$
*i.* 4000 W to kilowatts: $4000\ \text{W} = 4000 \times 10^{-3} = 4\ \text{kW}$
*j.* 1,010,000 Hz to kilohertz: $1{,}010{,}000\ \text{Hz} = 1{,}010{,}000 \times 10^{-3} = 1010\ \text{kHz}$
*k.* 356 mV to volts: $356\ \text{mV} = 356 \times 10^{-3} = 0.356\ \text{V}$
*l.* 15,000 μV to volts: $15{,}000\ \mu\text{V} = 15{,}000 \times 10^{-6} = 0.015\ \text{V}$
*m.* 0.005 TV to kilovolts: $0.005\ \text{TV} = 0.005 \times 10^{9} = 5 \times 10^{6}\ \text{kV}$
*n.* 100,000 kV to gigavolts: $100{,}000\ \text{kV} = 100{,}000 \times 10^{-6} = 0.1\ \text{GV}$
*o.* 500 nF to microfarads: $500\ \text{nF} = 500 \times 10^{-3} = 0.5\ \mu\text{F}$

Tables 3-2 and 3-3 are combined into Table 3-4. To use the table:

1. Locate the original unit in the left-hand column.
2. Read to the right until you get to the column headed by the unit desired.
3. The number and arrow at this point indicate the number of places and the direction that the decimal point is to be moved.

TABLE 3-4
METRIC CONVERSION TABLE

| ORIGINAL UNIT | DESIRED UNIT | | | | | | | | |
|---|---|---|---|---|---|---|---|---|---|
| | TERA | GIGA | MEGA | KILO | UNITS | MILLI | MICRO | NANO | PICO |
| Tera | | 3→ | 6→ | 9→ | 12→ | 15→ | 18→ | 21→ | 24→ |
| Giga | ← 3 | | 3→ | 6→ | 9→ | 12→ | 15→ | 18→ | 21→ |
| Mega | ← 6 | ← 3 | | 3→ | 6→ | 9→ | 12→ | 15→ | 18→ |
| Kilo | ← 9 | ← 6 | ← 3 | | 3→ | 6→ | 9→ | 12→ | 15→ |
| Units | ←12 | ← 9 | ← 6 | ← 3 | | 3→ | 6→ | 9→ | 12→ |
| Milli | ←15 | ←12 | ← 9 | ← 6 | ← 3 | | 3→ | 6→ | 9→ |
| Micro | ←18 | ←15 | ←12 | ← 9 | ← 6 | ← 3 | | 3→ | 6→ |
| Nano | ←21 | ←18 | ←15 | ←12 | ← 9 | ← 6 | ← 3 | | 3→ |
| Pico | ←24 | ←21 | ←18 | ←15 | ←12 | ← 9 | ← 6 | ← 3 | |

Check the problems in Examples 3-28 and 3-29 using Table 3-4.

## PROBLEMS

Change the following units of measurement.

1. 225 mA to amperes
2. 0.076 V to millivolts
3. 3.5 MΩ to ohms
4. 5 kW to watts

5. 550 kHz to hertz
6. 700,000 Hz to kilohertz
7. 70,000 Ω to megohms
8. 0.000 08 F to microfarads
9. 0.065 A to milliamperes
10. 6500 W to kilowatts
11. 75 mV to volts
12. 2.3 MHz to hertz
13. 6000 μA to amperes
14. 0.007 F to microfarads
15. 3.9 mA to amperes
16. 75,000 W to kilowatts
17. 0.005 μF to picofarads
18. 1/4 A to milliamperes
19. 1000 kHz to hertz
20. 0.5 MΩ to ohms
21. 0.008 V to millivolts
22. 0.0045 W to milliwatts
23. 0.000 06 μF to picofarads
24. 0.15 A to milliamperes
25. 0.15 μF to farads
26. 125 mV to volts
27. 8000 W to kilowatts
28. 4.16 kW to watts
29. 0.000 004 A to microamperes
30. 0.6 MHz to hertz
31. 0.5 THz to megahertz
32. 0.000 05 nF to picofarads
33. 15 MV to gigavolts
34. 0.000 25 μF to nanofarads

(See Instructor's Manual for Test 3-6)

## JOB 3-8 USING ELECTRONIC UNITS OF MEASUREMENT IN SIMPLE CIRCUITS

All the formulas for Ohm's law, series circuits, parallel circuits, and power demand that the measurements be given in units of amperes, volts, and ohms only. If a certain problem gives the measurements in units other than these, we must change all the measurements to amperes, volts, and ohms before we can use any of these formulas.

EXAMPLE 3-30. Find the voltage that will force 28.6 μA of current through a 70-kΩ resistor in the base circuit of a transistor.

SOLUTION. Given: $I = 28.6\ \mu\text{A}$ Find: $V = ?$
$R = 70\ \text{k}\Omega$

1. Change μA to A. $28.6\ \mu\text{A} = 28.6 \times 10^{-6}\ \text{A}$
2. Change 70 kΩ to Ω. $70\ \text{k}\Omega = 70 \times 10^{3}\ \Omega$
3. Find the voltage $V$. $V = I \times R$ (2-1)

$$V = 28.6 \times 10^{-6} \times 70 \times 10^{3}$$
$$V = 28.6 \times 7 \times 10^{-2}$$
$$V = 200.2 \times 10^{-2} = 2\ \text{V} \quad \textit{Ans.}$$

SELF-TEST 3-31. A bias voltage of 300 mV is developed across a 2-MΩ resistor. Find the current flowing using the formula $I = V/R$.

SOLUTION. Given: $V = 300\ \text{mV}$ Find: $I = ?$
$R =$ _______

1. Change 300 mV to volts. $300\ \text{mV} = 300 \times$ _______ V
2. Change 2 MΩ to ohms. $2\ \text{M}\Omega = 2 \times$ _______ Ω
3. Find the current $I$. $I = \dfrac{V}{R}$

2 MΩ
$10^{-3}$
$10^{6}$

$$= \frac{300 \times 10^{-3}}{2 \times 10^{6}}$$

$$= 150 \times 10^{-?} \text{ A}$$ 9

4. Change amperes to microamperes.

$$150 \times 10^{-9} \text{ A} = 150 \times 10^{-9} \times 10^{6}\ \mu\text{A}$$

$$= 150 \times 10^{-?}\ \mu\text{A}$$ 3

$$= \underline{\hspace{2cm}}\ \mu\text{A}$$ 0.15

## PROBLEMS

1. If 2 $\mu$A of current flow in an antenna whose resistance is 50 Ω, find the voltage drop in the antenna.
2. Find the voltage drop across the 5-kΩ load of a transistor if the collector current is 2 mA.
3. Using the formula Watts $= VI$, find the number of kilowatts used by a 24-Ω device drawing 5 A from a 120-V source.
4. Using the formula $I = V/R$, find how many milliamperes of current will flow through a 100-Ω resistor if the voltage across its ends is 20 mV.
5. Using the formula $I = V/R$, find the number of microamperes flowing through a 2-MΩ grid leak if the voltage drop across it is 1000 mV.
6. Using the formula for total current in a parallel circuit, $I_T = I_1 + I_2 + I_3$, find $I_T$ if $I_1 = 40$ mA, $I_2 = 6000\ \mu$A, and $I_3 = 0.013$ A.
7. Using the formula for total resistance in series, $R_T = R_1 + R_2 + R_3$, find the total resistance $R_T$ if $R_1 = 0.2$ MΩ, $R_2 = 5$ kΩ, and $R_3 =$ 10,000 Ω.
8. Using the formula $R = V/I$, find the resistance $R$ of a coil if an emf of 200 $\mu$V sends 10 mA of current through it.
9. Using the formula for capacitances in parallel, $C_T = C_1 + C_2 + C_3$, find the total capacity $C_T$ of a 0.0025-$\mu$F and a 125-pF capacitor in parallel.
10. Using the formula $R_T = (R_1 \times R_2)/(R_1 + R_2)$, find the total resistance $R_T$ of a 1000-Ω and a 4-kΩ resistor in parallel.
11. The time constant of an $RC$ circuit is equal to the product of the resistance (ohms) and the capacitance (farads). Find the time constant of a circuit if $R = 10$ kΩ and $C = 0.004\ \mu$F.
12. The radiation resistance of a shortwave antenna is 100 Ω. Using the formula Watts $= I^2R$, find the number of watts radiated if the transmitter delivers 900 mA to the antenna.
13. A 470-kΩ resistor in the base circuit of a 2N2924 phase-shift oscillator circuit carries a current of 30 $\mu$A. Find the voltage drop across the resistor.
14. A photoelectric-cell circuit contains a resistance of 0.12 MΩ and carries a current of 50 $\mu$A. Find the voltage drop in the resistor.

(See Instructor's Manual for Test 3-7, Electronic Units of Measurement)

## JOB 3-9 SOLVING THE OHM'S-LAW FORMULA FOR CURRENT OR RESISTANCE

**Formulas are equations.** As you may have noticed, every formula contains the sign of equality. The statement that a combination of quantities is *equal* to another combination of quantities is called an *equation.* In this sense, every formula is an equation. Examples of some equations are

$$3 \times 4 = 12$$
$$2 \times I = 10$$
$$12 = 3 \times R$$
$$V = IR$$

In order that a statement be termed an equation, it is necessary only that the value on the left side of the equality sign be *truly equal* to the value on the right side.

**Working with equations.** Many mathematical operations may be performed on an equation. Whatever is done, however, the basic equality of the statements on each side of the equality sign must not be destroyed. This equality must be maintained if the equation is to remain an equation. This is accomplished by applying the following basic principle.

| | |
|---|---|
| **BASIC PRINCIPLE** | Any mathematical operation performed on one side of an equality sign must also be performed on the other side. |

For example, let us subject the equation $3 \times 4 = 12$ to different mathematical operations.

| | |
|---|---|
| **RULE 1** | The same number may be added to both sides of an equality sign without destroying the equality. |

1. Write the equation. $3 \times 4 = 12$
2. Add 3 to both sides. $(3 \times 4) + 3 = 12 + 3$
3. Do the arithmetic. $12 + 3 = 12 + 3$

or $15 = 15$

which is a true equation, since the left side is still equal to the right side.

| | |
|---|---|
| **RULE 2** | The same number may be subtracted from both sides of an equality sign without destroying the equality. |

1. Write the equation. $3 \times 4 = 12$
2. Subtract 3 from both sides. $(3 \times 4) - 3 = 12 - 3$

3. Do the arithmetic. $12 - 3 = 12 - 3$

or $9 = 9$

which is a true equation, since the left side is still equal to the right side.

**RULE 3** Both sides of an equality sign may be multiplied by the same number without destroying the equality.

1. Write the equation. $3 \times 4 = 12$
2. Multiply both sides by 3. $(3 \times 4) \times 3 = 12 \times 3$
3. Do the arithmetic. $12 \times 3 = 12 \times 3$

or $36 = 36$

which is a true equation, since the left side is still equal to the right side.

**RULE 4** Both sides of an equality sign may be divided by the same number without destroying the equality.

1. Write the equation. $3 \times 4 = 12$
2. Divide both sides by 3. $\frac{3 \times 4}{3} = \frac{12}{3}$
3. Do the arithmetic. $\frac{12}{3} = \frac{12}{3}$

or $4 = 4$

which is a true equation, since the left side is still equal to the right side.

These examples show that if the same mathematical operation (other than division by zero) is performed on *both* sides of an equation, it does not destroy the basic equality of the equation.

**Solving equations.** Consider the equation $2R = 10$. To solve this equation means to find the value of the unknown letter $R$ in the equation. This value is found *when the letter stands all alone on one side of the equality sign.* When this occurs, the equation has the form

$$R = \text{some number}$$

This number will obviously be the value of the letter $R$, and the equation will be solved.

**How do we get the letter all alone?** In the equation $2R = 10$, the letter will be alone on the left side of the equality sign if we can somehow eliminate the number 2 on that side. The number 2 will actually be eliminated if we can change it to a 1, since $1R$ means the same as $R$. This can be done by applying the basic principle that permits us to perform the same operation on both sides of the equality sign. But which mathematical operation will eliminate the number 2? In general, the operation to be performed on both sides of the equality sign will be the *opposite* of that used in the equation. Since $2R$ means 2 *multiplied* by $R$, we shall use rule 4 above, and *divide both sides* of the equation by that same number 2.

1. Write the equation. $2R = 10$
2. Divide both sides by 2 $\frac{2R}{2} = \frac{10}{2}$
3. Simplify each side separately. $\frac{\overset{1}{\cancel{2}}R}{\underset{1}{\cancel{2}}} = \frac{\overset{5}{\cancel{10}}}{\underset{1}{\cancel{2}}}$

or $1R = 5$

or $R = 5$

Only those numbers that appear on the same side of the equality sign may be canceled. Do *not* cancel across the equality sign.

Suppose the unknown letter appears on the right side of the equality sign as in the equation $12 = 3R$. In this situation, we proceed exactly as before. To solve the equation means to get the letter *all alone* on one side of the equality sign. It is not important which side is chosen, as long as the letter is *all alone* on that side. We can get the letter $R$ all alone on the right side by changing the $3R$ to $1R$ by applying the *opposite* operation of *dividing both sides* of the equation by the number 3.

1. Write the equation. $12 = 3R$
2. Divide both sides by 3. $\frac{12}{3} = \frac{3R}{3}$
3. Simplify each side separately. $\frac{\overset{4}{\cancel{12}}}{\underset{1}{\cancel{3}}} = \frac{\overset{1}{\cancel{3}}R}{\underset{1}{\cancel{3}}}$

or $4 = 1R$ or $4 = R$ or $R = 4$ *Ans.*

Notice that the number that is multiplied by the unknown letter will be canceled out *only* if we divide both sides of the equality sign *by that same number.* Dividing both sides by any other number will *not* eliminate this number.

| **RULE 5** | To eliminate the number which is multiplied by the unknown letter, divide both sides of the equality sign by the multiplier of the letter. |
|---|---|

EXAMPLE 3-32. Solve the following equations for the values of the unknown letters.

$$2R = 10 \quad \text{and} \quad 14 = 7V$$

SOLUTION

| | | |
|---|---|---|
| 1. Write the equations. | $2R = 10$ | $14 = 7V$ |
| 2. Divide both sides of each equation by the multiplier of the letter. | $\frac{2R}{2} = \frac{10}{2}$ | $\frac{14}{7} = \frac{7V}{7}$ |
| 3. Cancel out the multiplier of the letter. | $R = \frac{10}{2}$ | $\frac{14}{7} = V$ |
| 4. Divide | $R = 5$ *Ans.* | $2 = V$ *Ans.* |

We are now in a position to shorten our work. Notice that in each example, the effect of dividing both sides of the equality sign by the multiplier of the letter has been to *move* the multiplier *across the equality sign* into the position shown in step 3. Since this will always occur, we can eliminate step 2 and proceed as shown in Example 3-33 below.

EXAMPLE 3-33. Solve the following equations for the values of the unknown letters:

$$3R = 12 \qquad \text{and} \qquad 15 = 5V$$

SOLUTION

| | | |
|---|---|---|
| 1. Write the equations. | $3R = 12$ | $15 = 5V$ |
| 2. Divide the quantity all alone on one side of the equality sign by the multiplier of the letter. | $R = \frac{12}{3}$ | $\frac{15}{5} = V$ |
| 3. Divide. | $R = 4$ *Ans.* | $3 = V$ *Ans.* |

> **RULE 6** To solve a simple equation of the form "a number multiplied by a letter equals a number," divide the number all alone on one side of the equality sign by the multiplier of the letter.

EXAMPLE 3-34. Solve the equation $18 = 0.3Z$ for the value of Z.

SOLUTION

$$18 = 0.3Z$$

$$\frac{18}{0.3} = Z$$

$$60 = Z$$

or $Z = 60$ *Ans.*

SELF-TEST 3-35. In a common-emitter transistor circuit, the relationship between the base current $I_B$ and the collector current $I_C$ is given as $\beta I_B = I_C$. Find the value of $I_B$ if $\beta = 50$ and $I_C = 0.002$ A.

SOLUTION

1. Write the formula. $\beta I_B = I_C$
2. Substitute numbers. $50I_B =$ ______
3. Solve for $I_B$. $I_B = \dfrac{?}{?}$
4. Divide the numbers. $I_B =$ ______ A = ______ mA

0.002
$\dfrac{0.002}{50}$
0.000 04 0.04

## PROBLEMS

Solve the following equations for the value of the unknown letter:

1. $3I = 15$
2. $5R = 20$
3. $2V = 12$
4. $48 = 8R$
5. $7I = 63$
6. $4L = 21$
7. $3R = 41$
8. $16R = 4$
9. $20I = 117$
10. $19 = 2V$
11. $\frac{1}{2}W = 20$
12. $20 = 100R$
13. $0.3R = 120$
14. $0.04Z = 60$
15. $40 = 0.2Z$
16. $8 = 0.4Z$
17. $0.15R = 120$
18. $0.003R = 78$
19. $117 = 0.3Z$
20. $\frac{3}{5}T = 12$
21. $8V = \frac{1}{2}$

**Solving the formula for Ohm's law.** The formula for Ohm's law is actually an equation. By applying rule 6, we can solve Ohm's law for any unknown value of current or resistance.

EXAMPLE 3-36. The total resistance of a relay coil is 50 Ω. What current will it draw from a 20-V source?

SOLUTION. The diagram for the circuit is shown in Fig. 3-4.

1. Write the formula. $V = IR$
2. Substitute numbers. $20 = I \times 50$
3. Solve for $I$. $\dfrac{20}{50} = I$
4. Divide the numbers. $0.4 = I$

or $I = 0.4$ A *Ans.*

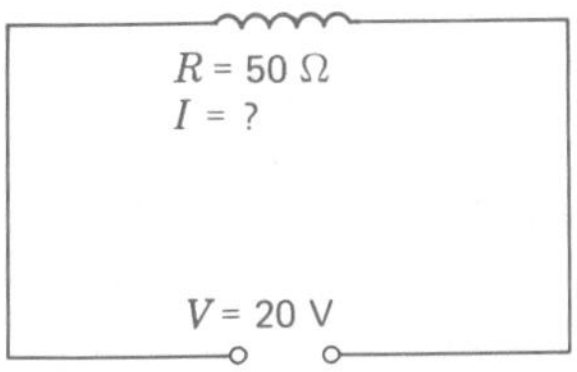

$R = 50\ \Omega$
$I = ?$
$V = 20$ V

FIGURE 3-4

SELF-TEST 3-37. Find the total resistance of a telegraph coil if it draws 0.015 A from a 6.6-V source

SOLUTION. The diagram is shown in Fig. 3-5.

1. Write the formula. $V = I \times$ ______
2. Substitute numbers. ______ $= 0.015 \times R$
3. Solve for $R$. $\dfrac{?}{?} = R$
4. Divide the numbers. ______ $= R$

or $R =$ ______ Ω *Ans.*

R
6.6
$\dfrac{6.6}{0.015}$
440
440

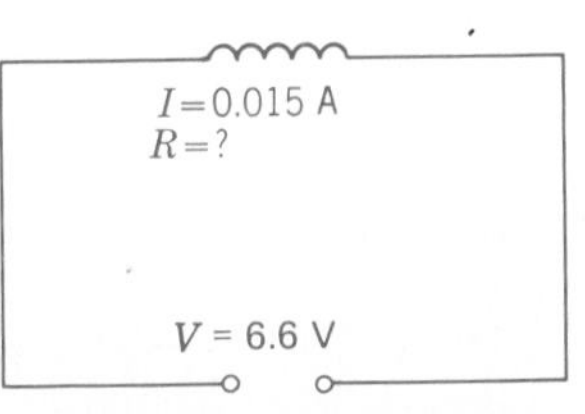


FIGURE 3-5

## PROBLEMS

1. What is the hot resistance of an arc lamp if it draws 15 A from a 30-V line?
2. The resistance of the motor windings of an electric vacuum cleaner is 20 Ω. If the voltage is 120 V, find the current drawn.
3. An electric enameling kiln draws 9 A from a 117-V line. Find the resistance of the coils.
4. The field magnet of a loudspeaker carries 40 mA when connected to a 40-V supply. Find its resistance.
5. How much current is drawn from a 12-V battery when operating an automobile horn of 8-Ω resistance?
6. What is the hot resistance of a tungsten lamp if it draws 250 mA from a 110-V line?
7. What current would flow in a 0.3-Ω short circuit of a 6-V automobile ignition system?
8. A 2N525 transistor is used as a transistor switch to control a 20-V source across a 100-Ω load. Find the current that flows when the switch is conducting.
9. Find the resistance of an electric iron if it draws 4.8 A from a 120-V line.
10. A dry cell indicates a terminal voltage of 1.2 V when a wire of 0.2-Ω resistance is connected across it. What current flows in the wire?
11. Find the resistance of an automobile starting motor if it draws 90 A from the 12-V battery.
12. A meter registers 0.2 mA when the voltage across it is 3 V. Find the total resistance of the meter circuit.
13. What is the resistance of a telephone receiver if there is a voltage drop of 24 V across it when the current is 20 mA?
14. What current is drawn by a 5-kΩ electric clock when operated from a 110-V line?
15. Find the current drawn by a 52-Ω toaster from a 117-V line.
16. Find the resistance of an electric furnace drawing 41 A from a 230-V line.
17. The resistance of the field coils of a shunt motor is 60 Ω. What is the field current when the voltage across the coils is 220 V?
18. The resistance of a common Christmas-tree lamp is about 50 Ω. What is the current through it if the voltage across the lamp is 14 V?
19. The large copper leads on switchboards are called *busbars*. What is the resistance of a busbar carrying 400 A if the voltage across its ends is 0.6 V?
20. If a radio receiver draws 0.85 A from a 110-V line, what is the total resistance (impedance) of the receiver?
21. A 32-candela (cd) lamp in a truck headlight draws 3.4 A from the 6-V battery. What is the resistance of the lamp?

22. If the resistance of the air gap in an automobile spark plug is 2.5 kΩ, what voltage is needed to force 0.16 A through it?
23. A voltage of 5 V appears across the 2.5-kΩ load resistor in a self-biased transistor circuit. Find the current in the resistor.
24. What current flows through an automobile headlight lamp of 1.2-Ω resistance if it is operated from the 6-V battery?
25. What is the resistance of a buzzer if it draws 0.14 A from a 3-V source?

## JOB 3-10. ESTIMATING ANSWERS

In many applications it is important only to estimate the answer. It may only be necessary to know the order of magnitude of the answer and perhaps one digit of the answer. For example, you must know whether the voltage is 10, 100, or 1000 V in order to adjust a meter to the proper scale. Or you may be measuring resistance and need to know whether the range is GΩ, kΩ, or MΩ.

The scientific and exponential notations that have been used in this chapter will make approximations very easy. In scientific notation, all numbers are converted to a number between 1 and 10 with the appropriate power of 10. By looking only at the powers of 10, you usually can determine the power of 10 in the answer. Since all the other numbers are between 1 and 10, the actual arithmetic is easy. This method is particularly useful for problems with which you are unfamiliar.

EXAMPLE 3-38. A large launching rocket was moved 0.0301 mi in 2.2 h. Find the speed in ft/min. Is the speed closer to 1, 10, or 100 ft/min?

SOLUTION

1. **Change units.**
   $0.0301 \text{ m} \times 5280 = \text{ft}$
   $2.2 \text{ h} \times 60 \text{ min/h} = \text{min}$
2. Divide feet by minutes.
   $$\frac{0.0301 \times 5280}{2.2 \times 60}$$
3. Write the problem in scientific notation.
   $$\frac{3.0 \times 10^{-2} \times 5.28 \times 10^{3}}{2.2 \times 6 \times 10}$$
4. Look only at the powers of 10.
   $$\frac{10^{-2} \times 10^{3}}{10} = 10^{0} = 1$$
5. Now estimate the numbers.
   $$\frac{3 \times 5}{2 \times 6} = \frac{15}{12} = 1.25$$

Therefore, the answer is approximately $1.25 \times 1 = 1.25$ ft/min. Since the actual answer is 1.204 ft/min, the error is only 0.046 ft/min, or about 3.8 percent.

Use the following procedure to estimate an answer.

1. Change the measurements to the proper units.
2. Write out the problem.
3. Convert the numbers to scientific notation.

4. Solve the powers of 10.
5. Estimate the numbers.

SELF-TEST 3-39. Find the voltage drop across the 2000-Ω load of a transistor if the collector current is 0.015 A.

SOLUTION.
Given: $R = 2000$ Find: $V = ?$
$I = 0.105$ A

1. $V = I \times R$
2. $V =$ ________ $\times 2000$ — 0.015
3. $V = 1.5 \times$ ________ $\times 2 \times 10^3$ — $10^{-2}$
4. $V \times 1.5 \times 2 \times$ ________ — 10
5. $V -$ ________ $\times 10 = 30$ V *Ans.* — 3

## PROBLEMS

Solve the problems in Job 3-8 by estimation and compare the answers with the actual answers.

## JOB 3-11 REVIEW OF OHM'S LAW

In any electric circuit,

1. The voltage forces the ________ through a conductor against its resistance. — current
2. The ________ tries to stop the current from flowing. — resistance
3. The current that flows in a circuit depends on the ________ and the resistance. — voltage

The relationship among these three quantities is described by Ohm's law. Ohm's law applies to an entire circuit or to any component part of a circuit.

**FORMULA** $$V = IR$$ 2-1

where $V$ = voltage, V
$I$ = current, A
$R$ = resistance, Ω

The formula for Ohm's law may be used to find the value of any one of the quantities in the formula. It is of equal importance that the student be able to determine the relative values of each quantity as one of the other quantities is changed in amount.

In Fig. 3-6, if the resistance $R = 10$ Ω remains unchanged,

When the voltage $V = 10$ V, the current $I =$ ________. — 1 A
When the voltage $V = 20$ V, the current $I =$ ________. — 2 A
When the voltage $V = 50$ V, the current $I =$ ________. — 5 A
When the voltage $V = 100$ V, the current $I =$ ________. — 10 A

As you can see, when the resistance remains constant,

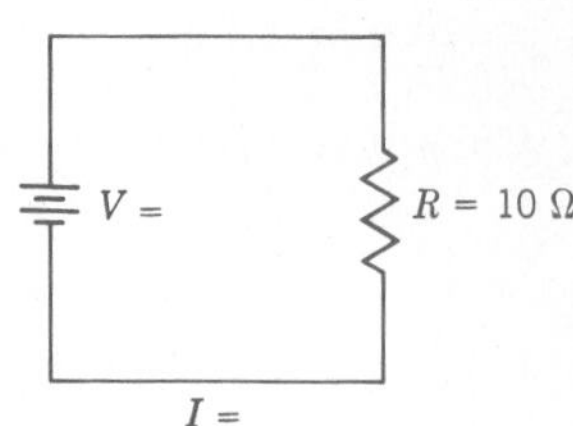


**FIGURE 3-6**
When the resistance remains constant, the larger the voltage the larger the current.

The larger the voltage, the larger the current.
The smaller the voltage, the ________ the current. | smaller

In Fig. 3-7, if the voltage $V = 100$ V remains unchanged,

When the resistance $R = 1\ \Omega$, the current $I =$ ________. | 100 A
When the resistance $R = 10\ \Omega$, the current $I =$ ________. | 10 A
When the resistance $R = 50\ \Omega$, the current $I =$ ________. | 2 A
When the resistance $R = 100\ \Omega$, the current $I =$ ________. | 1 A

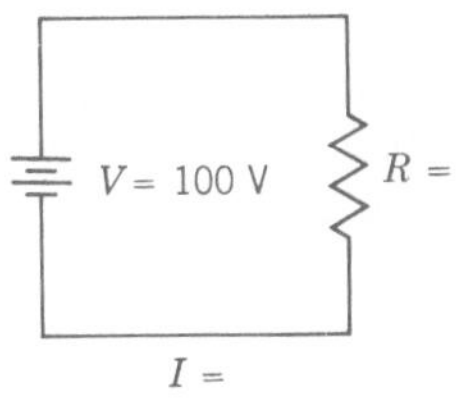

As you can see, when the voltage remains constant,

The larger the resistance, the smaller the current.
The smaller the resistance, the ________ the current. | larger

**FIGURE 3-7**
When the voltage remains constant, the larger the resistance the smaller the current.

**Formulas in electrical work.** A formula is a shorthand method for writing a rule. Each letter in a formula represents a number which may be substituted for it. The signs of operation tells us what to do with these numbers.

## Steps in solving problems

1. Read the problem carefully.
2. Draw a simple diagram of the circuit.
3. Record the given information directly on the diagram. Indicate the values to be found by question marks.
4. Write the formula.
5. Substitute the given numbers for the letters in the formula. If the number for the letter is unknown, merely write the letter again. Include all mathematical signs.
6. Do the indicated arithmetic.
   *a.* If after substitution the unknown letter is multiplied by some number, divide the number all alone on one side of the equality sign by the multiplier of the unknown letter.
7. In the answer, indicate the letter, its numerical value, and the units of measurement.

SELF-TEST 3-40. In Fig. 3-8, the bleeder resistance draws 4 mA from the 300-V power supply when no loads are connected to the various voltage-divider terminals. Find the total bleeder resistance.

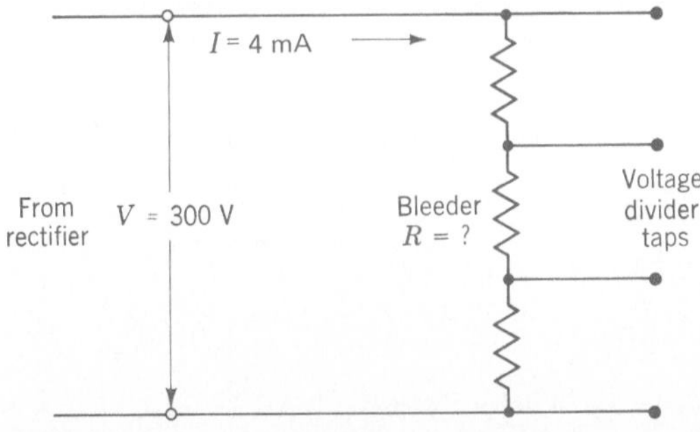


**FIGURE 3-8**
The bleeder resistor is made of three resistors in series.

SOLUTION

| | | | Answers |
|---|---|---|---|
| 1. | The diagram for the circuit is shown in Fig. 3-8. | | |
| 2. | Write the formula. | $V = I \times$ ______ | $R$ |
| 3. | Substitute numbers. | $300 =$ ______ $\times R$ | 0.004 |
| 4. | Solve the equation. | $\frac{300}{?} = R$ | 0.004 |
| 5. | Divide the numbers. | ______ $= R$ | 75,000 |
| or | | $R = 75{,}000\ \Omega = 75$ ______ *Ans.* | kΩ |

## PROBLEMS

1. The resistance of an electric percolator is 22 Ω. If it draws 5 A, what is the operating voltage?
2. An electric heater whose coil is wound with No. 18 iron wire is connected across 110 V. If it draws a current of 10 A, what is the value of its resistance?
3. According to the National Electrical Code, No. 14 asbestos-covered type A wire should never carry more than 30 A. Is this wire safe to use to carry power to a 10-Ω 230-V motor?
4. A washing-machine motor has a total resistance of 39 Ω and operates on 117 V. Find the current taken by the motor.
5. What is the voltage drop across an Allied model BK relay of 12 kΩ resistance if it carries 1.5 mA?
6. What is the voltage across a telephone receiver of 800 Ω resistance if the current flowing is 30 mA?
7. A spot welder delivers 7000 A when the voltage is 4.9 V. What is the resistance of the piece being welded?
8. A 600-W soldering iron is used on a 120-V line and has a resistance of 24 Ω. Find the current drawn.
9. A volume control similar to that used in the Panasonic RF 738 transistor radio is shown in Fig. 3-9. How many ohms of resistance are engaged in the potentiometer if the voltage drop is 7 V and it passes a current of 0.7 mA?

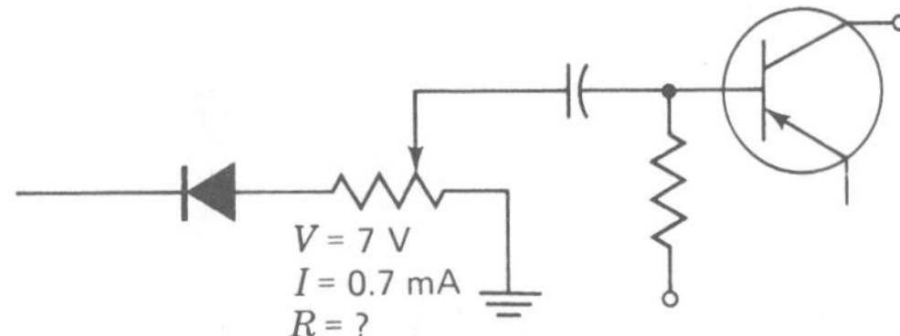


**FIGURE 3-9**
Volume control in a simple transistor radio.

10. If the full-scale reading of an ammeter is 10 A, what is its resistance if this current causes a voltage drop of 0.05 V?
11. What is the resistance of a busbar carrying 300 A if the voltage drop across it is 1.2 V?

12. A sensitive dc meter takes 9 mA from a line when the voltage is 108 V. What is the resistance of the meter?
13. An electromagnet draws 5 A from a 110-V line. What current will it draw from a 220-V line?
14. The resistance of the series field coils of a compound motor is 0.24 Ω, and they carry a current of 72 A. Find the voltage drop across these coils.
15. A 5-kΩ resistor in a voltage divider reduces the voltage across it by 150 V. What current flows through the resistor?
16. A series resistor is used to reduce the voltage to a motor by 45 V. What must be the resistance of the resistor if the motor draws 0.52 A?
17. A voltmeter has a resistance of 27 kΩ. What current will flow through the meter when it is placed across a 220-V line?
18. A series of insulators leak 30 μA at 9 kV. Find the resistance of the insulator string.

(See Instructor's Manual for Test 3-8, Ohm's Law)

# SERIES CIRCUITS

# 4

## JOB 4-1 VOLTAGE, CURRENT, AND RESISTANCE IN SERIES CIRCUITS

**Wiring a series circuit.** A series circuit is one in which all the component parts are connected in succession from plus (+) to minus (−), as shown in Fig. 4-1. In a series circuit there is *only one path* through which the electrons may flow. The flow of current in a series circuit may be compared with the flow of water in a series-connected water system. In Fig. 4-2*a,* the pump forces the water through the three valves in succession. In Fig. 4-2*b*, the electron-moving pump—the battery—forces the electrons through the three resistors in succession. In both series circuits there is *only one path* that the water or the electrons may travel. If the water circuit is broken at any point, by either closing a valve or breaking a pipe, the flow of water around the system will stop. If the electric series circuit is broken at any point, no energy will be available to any part of the circuit, since there will be no return path for the electrons to follow. For example, in Fig. 4-3, a push button is usually placed in series with the bell it controls. As long as the button is held up by the force of the spring inside it, the circuit is broken. Since no current can flow in a broken, or "open," circuit, the bell will not ring. When the button is depressed, the wires make contact, completing the circuit. The current then flows through the bell, and the bell rings.

**Symbols for series circuits.** Numbers or letters written underneath

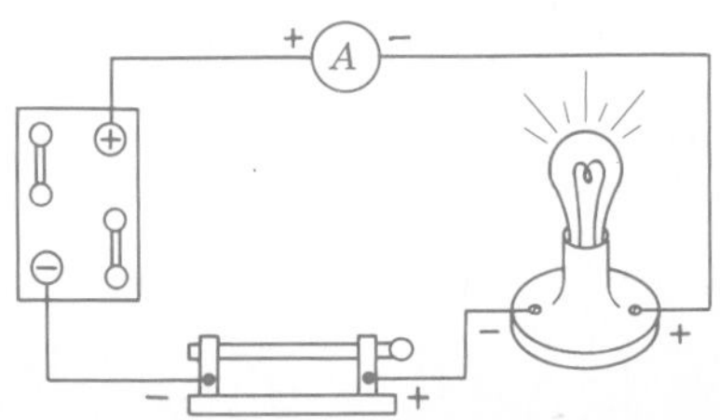


**FIGURE 4-1**
Simple series circuit. The parts are connected so that the current can flow in only one path.

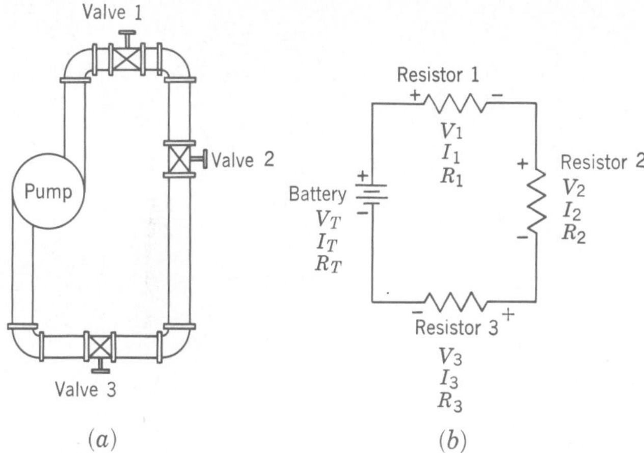


**FIGURE 4-2**
(*a*) Valves in series in a water-supply system. (*b*) Resistors in series in an electric circuit.

other numbers or letters are called *subscripts.* Numbers like 1, 2, or 3 are written under the letters *V, I,* or *R* to indicate these quantities in the first, second, or third part of the circuit. For example, in Fig. 4-2*b*:

$V_1$ represents the voltage across the first resistor.
$I_2$ represents the current through the second resistor.
$R_3$ represents the resistance of the third resistor.
$V_T$ represents the total voltage in the circuit.
$I_T$ represents the total current in the circuit.
$R_T$ represents the total resistance in the circuit.

**Total current in a series circuit.** In Fig. 4-2*a,* the water was forced through each valve in turn because it had no other place to go. Whatever quantity of water flowed through the first valve had to flow through the second and third valve also. In Fig. 4-2*b,* the electrons that were forced through the first resistor $R_1$ also had to flow through the second resistor $R_2$ and through the third resistor $R_3$ because there was no other place for them to go. It follows, then, that any current entering the circuit must flow *unchanged* through all the other parts of the circuit.

**RULE** The total current in a series circuit is equal to the current in any other part of the circuit.

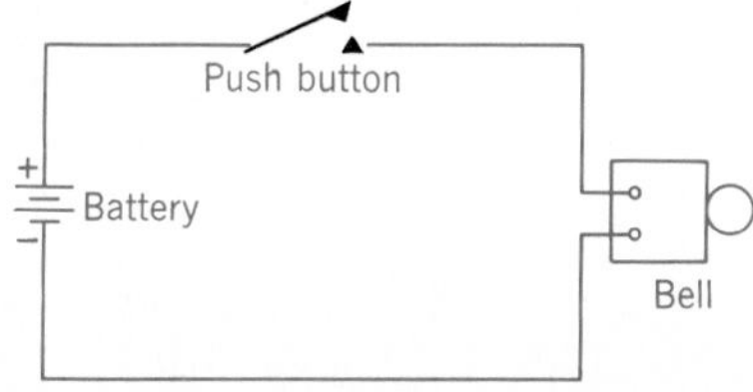


**FIGURE 4-3**
A push button in series controls the flow of current from the battery to the bell by opening or closing the circuit.

**FORMULA** $I_T = I_1 = I_2 = I_3 = \text{etc.}$ **4-1**

where $I_T$ = total current
$I_1$ = current in first part
$I_2$ = current in second part
$I_3$ = current in third part,
etc.

**Total voltage in a series circuit.** In Fig. 4-4*a,* the total force required to lift the weights must be equal to the *sum* of the forces required to lift the individual weights. In Fig. 4-4*b,* the total electrical pressure supplied by the battery must be equal to the *sum* of the pressures required by each lamp. Since the available voltage pressure decreases, or "drops" as the current is forced through each resistance, the voltages $V_1$ and $V_2$ are called voltage *drops*.

**RULE** The total voltage in a series circuit is equal to the sum of the voltage drops across all the parts of the circuit.

**FORMULA** $V_T = V_1 + V_2 + V_3 + \text{etc.}$ **4-2**

where $V_T$ = total voltage
$V_1$ = voltage across first part
$V_2$ = voltage across second part
$V_3$ = voltage across third part
etc.

**Total resistance in a series circuit.** In Fig. 4-4*a,* the resistance that must be overcome by the body is equal to the *sum* of the weights. In Fig. 4-4*b,* the total electric resistance of the circuit is equal to the *sum* of the resistances of all the lamps.

**RULE** The total resistance of a series circuit is equal to the sum of the resistances of all the parts of the circuit.

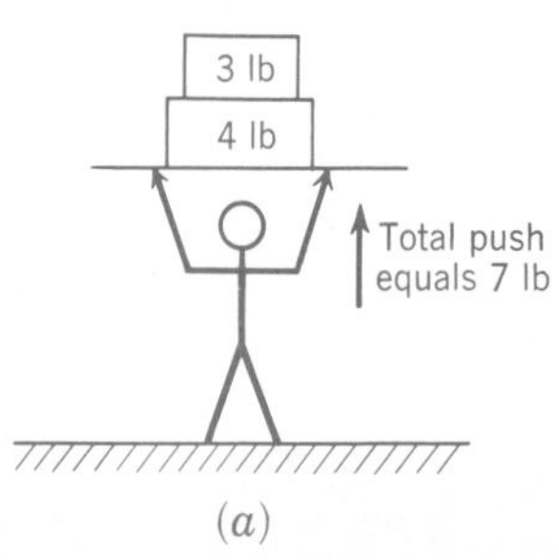


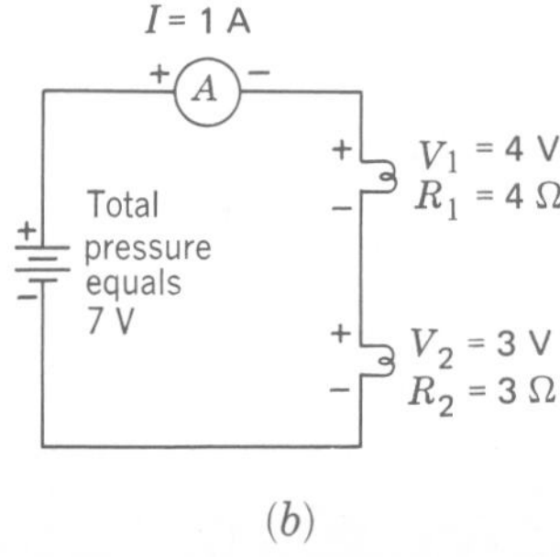


**FIGURE 4-4**
Similarity between (*a*) a force system and (*b*) an electric circuit.

**FORMULA** $R_T = R_1 + R_2 + R_3 + \text{etc.}$ 4-3

where $R_T$ = total resistance
$R_1$ = resistance of first part
$R_2$ = resistance of second part
$R_3$ = resistance of third part
etc.

EXAMPLE 4-1. A 6-V 20-Ω filament and a 12-V 40-Ω filament are connected in series with a 20-Ω limiting resistor using 6 V and 0.3 A. Find (*a*) the total voltage, (*b*) the total current, and (*c*) the total resistance.

SOLUTION. Draw the circuit diagram as shown in Fig. 4-5.

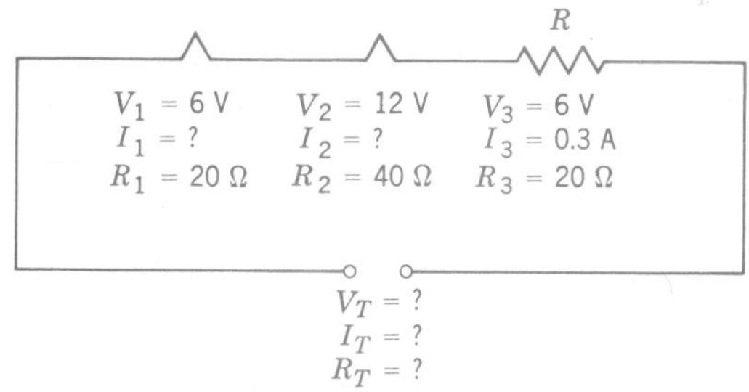


FIGURE 4-5

*a.* Find the total voltage.

1. Write the formula. $V_T = V_1 + V_2 + V_3$ (4-2)
2. Substitute numbers. $V_T = 6 + 12 + 6 = 24\ \text{V}$ *Ans.*

*b.* Find the total current.

1. Write the formula. $I_T = I_1 = I_2 = I_3$ (4-1)
2. Substitute numbers. $I_T = I_1 = I_2 = 0.3\ \text{A}$ *Ans.*

*c.* Find the total resistance.

1. Write the formula. $R_T = R_1 + R_2 + R_3$ (4-3)
2. Substitute numbers. $R_T = 20 + 40 + 20 = 80\ \Omega$ *Ans.*

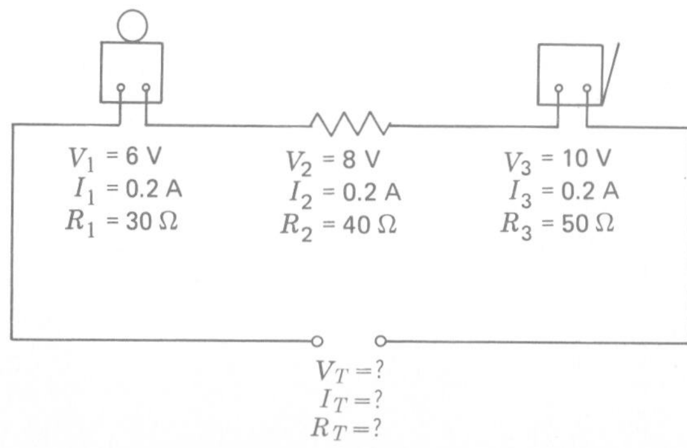


FIGURE 4-6

## PROBLEMS

1. In the circuit shown in Fig. 4-6, find (*a*) the total voltage, (*b*) the total current, and (*c*) the total resistance.
2. In an antique car, a 3-V 1.5-Ω dash light and a 3-V 1.5-Ω taillight are connected in series to a battery delivering 2 A as shown in Fig. 4-7. Find (*a*) the total voltage and (*b*) the total resistance.

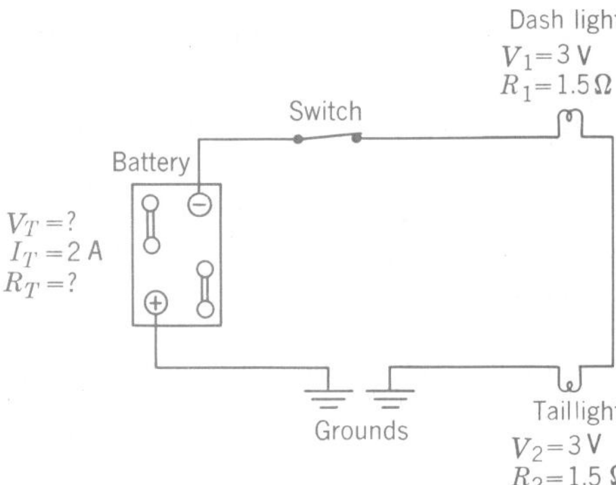


**FIGURE 4-7** Series-connected automobile dash- and taillight for Prob. 2.

3. Three resistances are connected in series. $V_1 = 6.3$ V, $I_1 = 0.3$ A, $R_1 = 21\ \Omega$, $V_2 = 12.6$ V, $R_2 = 42\ \Omega$, $V_3 = 24$ V, and $R_3 = 80\ \Omega$. Find (*a*) the total voltage, (*b*) the total current, and (*c*) the total resistance.
4. The receiver, transmitter, and line coil of a telephone circuit are connected in series. For the receiver: $V = 2.5$ V, $I = ?$, and $R = 12.5\ \Omega$. For the transmitter: $V = 18.6$ V, $I = 0.2$ A, and $R = 93\ \Omega$. For the line coil: $V = 6.7$ V, $I = ?$, and $R = 33.5\ \Omega$. Find (*a*) the total voltage, (*b*) the total current, and (*c*) the total resistance.

## JOB 4-2 USING OHM'S LAW IN SERIES CIRCUITS

Ohm's law may be used for the individual parts of a series circuit. When it is used on a particular part of a circuit, great care must be taken to use *only* the voltage, current, and resistance of that particular part. That is, the *voltage of a part* is equal to the *current in that part* multiplied by the *resistance of that part.* This may be easily remembered by using the correct subscripts when writing the Ohm's-law formula for a particular part.

For the first part:

$$V_1 = I_1 \times R_1 \tag{4-4}$$

For the second part:

$$V_2 = I_2 \times R_2 \tag{4-5}$$

For the third part:

$$V_3 = I_3 \times R_3 \qquad \boxed{4\text{-}6}$$

EXAMPLE 4-2. Solve the circuit shown in Fig. 4-8 for all missing values of (*a*) current, (*b*) voltage, and (*c*) resistance.

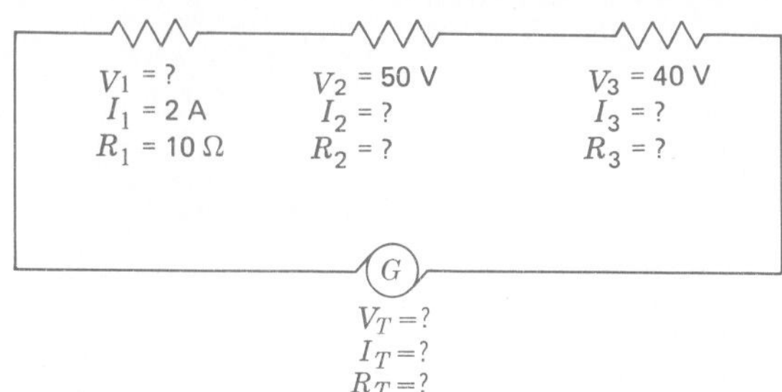


FIGURE 4-8

SOLUTION. We can find the total values by the following formulas:

$$I_T = I_1 = I_2 = I_3 \qquad (4\text{-}1)$$
$$V_T = V_1 + V_2 + V_3 \qquad (4\text{-}2)$$
$$R_T = R_1 + R_2 + R_3 \qquad (4\text{-}3)$$

However, in order to use these formulas, we must know the individual values for each part of the circuit. These values may be found by using the Ohm's-law formulas for each part.

$$V_1 = I_1 \times R_1 \qquad (4\text{-}4)$$
$$V_2 = I_2 \times R_2 \qquad (4\text{-}5)$$
$$V_3 = I_3 \times R_3 \qquad (4\text{-}6)$$

*a.* Since the current has the same value at every point in a series circuit, it is easiest to find the current first.

1. Write the formula. $I_T = I_1 = I_2 = I_3$ (4-1)
2. Substitute numbers. $I_T = 2 = I_2 = I_3$
3. The current value is $I_T = I_1 = I_2 = I_3 = 2$ A *Ans.*

*b.* Find $V_1$. Use $I_1 = 2$ A from step *a*.

1. Write the formula. $V_1 = I_1 \times R_1$ (4-4)
2. Substitute numbers. $V_1 = 2 \times 10 = 20$ V *Ans.*

Find $V_T$. Use $V_1 = 20$ V from step *b*.

1. Write the formula. $V_T = V_1 + V_2 + V_3$ (4-2)
2. Substitute numbers. $V_T = 20 + 50 + 40 = 110$ V *Ans.*

*c.* Find $R_2$ and $R_3$. Use $I_2 = I_3 = 2$ A from step *a*.

1. Write the formula. $V_2 = I_2 \times R_2$ $\quad V_3 = I_3 \times R_3$
2. Substitute numbers. $50 = 2 \times R_2$ $\quad 40 = 2 \times R_3$
3. Solve. $50/2 = R_2$ $\quad 40/2 = R_3$
   $R_2 = 25\ \Omega$ *Ans.* $\quad R_3 = 20\ \Omega$ *Ans.*

Find $R_T$. Use $R_2 = 25\ \Omega$ and $R_3 = 20\ \Omega$ from step *c*.

1. Write the formula. $R_T = R_1 + R_2 + R_3$ (4-3)
2. Substitute numbers. $R_T = 10 + 25 + 20 = 55\ \Omega$ *Ans.*

SELF-TEST 4-3. Part of the first stage of a two-stage transistorized amplifier is shown in Fig. 4-9. Solve the circuit for all missing values of (*a*) current, (*b*) voltage, and (*c*) resistance.

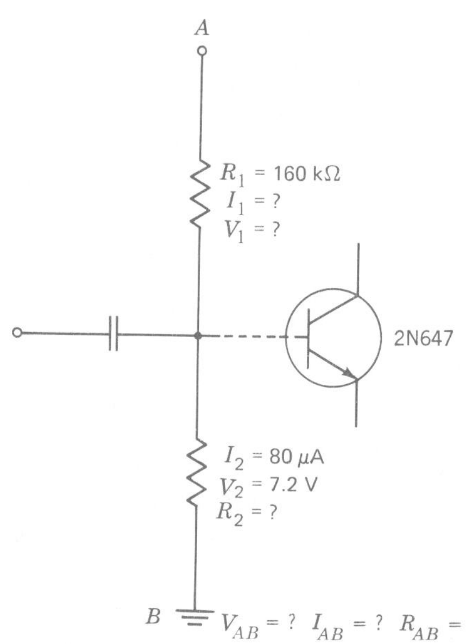


**FIGURE 4-9**
Base circuit of a transistor amplifier.

SOLUTION

*a.* Find the total current.

1. Write the formula. $I_{AB} = I_1 =$ ______ $=$ ______ μA *Ans.* — $I_2$ 80

*b.* Find the voltages. Find $V_1$. Use $I_1 =$ ______ μA from step *a.* — 80

1. Write the formula. $V_1 = I_1 \times$ ______ (4-4) — $R_1$
2. Substitute numbers. $V_1 = 80 \times 10^{-?} \times 160 \times 10^{?}$ — 6 3

   $V_1 = 12{,}800 \times 10^{-?} =$ ______ V *Ans.* — 3 12.8

Find $E_{AB}$. Use $V_1 = 12.8$ V from step *b*.

1. Write the formula. $V_{AB} = V_1 +$ ______ (4-2) — $V_2$
2. Substitute numbers. $V_{AB} = 12.8 +$ ______ $=$ ______ V *Ans.* — 7.2 20

*c.* Find the resistances. Find $R_2$. Use $I_2 =$ ______ μA — 80

1. Write the formula. $V_2 =$ ______ $\times R_2$ (4-5) — $I_2$
2. Substitute numbers. $7.2 =$ ______ $\times R_2$ — $80 \times 10^{-6}$
3. Solve. $R_2 = \dfrac{7.2}{80 \times 10^{-6}} = 0.09 \times 10^{?}$ — 6

   $R_2 =$ ______ $\Omega =$ ______ kΩ — 90,000 90

Find $R_{AB}$. Use $R_2 =$ ______ kΩ from step *c*. — 90

1. Write the formula. $R_{AB} = R_1 +$ ______ — $R_2$
2. Substitute numbers. $R_{AB} = 160\ \text{k}\Omega + 90\ \text{k}\Omega =$ ______ kΩ *Ans.* — 250

## PROBLEMS

1. A lamp using 10 V, a 10-Ω resistor drawing 4 A, and a 24-V motor are connected in series. Find (*a*) the total current, (*b*) the total voltage, and (*c*) the total resistance.
2. A small arc lamp designed to operate on a current of 6 A has a resistance of 14 Ω. It is used in series with a limiting resistor of 6 Ω. Find (*a*) the total current, (*b*) the total voltage, and (*c*) the total resistance.
3. A series-connected automobile dash light and taillight circuit similar to that shown in Fig. 4-7 operates from a 6-V battery. The dash light operates on 2 V and 0.8 A. The taillight requires 4 V. Find (*a*) the resistance of each light, (*b*) the total resistance, and (*c*) the total current.
4. A lamp, a resistor, and a soldering iron are connected in series. The lamp has a voltage of 16 V across it. The voltage across the resistor is 12.8 V. The iron has a resistance of 6 Ω and carries a current of 3.2 A. Find (*a*) the total current, (*b*) the total voltage, and (*c*) the total resistance.
5. A 17,000-Ω 150-V scale voltmeter is combined in series with a 51-kΩ resistor in order to increase the range of the meter to read 600 V. Find the current in the meter when it reads 600 V.
6. Using the circuit shown in Fig. 4-10, find the total voltage between points *A* and *B*.

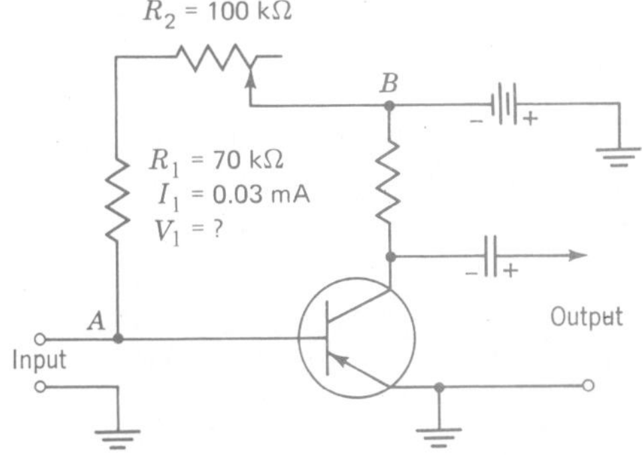


**FIGURE 4-10**
Class A audio amplifier.

7. The first video IF amplifier in the RCA TV chassis KCS 176 uses a base circuit similar to that shown in Fig. 4-11. Find (*a*) $V_1$, (*b*) $V_2$, and (*c*) $V_T$.
8. The Magnavox color TV chassis T940 uses a high-voltage regulator circuit essentially the same as that shown in Fig. 4-12. Find all missing values of (*a*) current, (*b*) voltage, and (*c*) resistance.

(See Instructor's Manual for Test 4-1, Ohm's Law in Series Circuits)

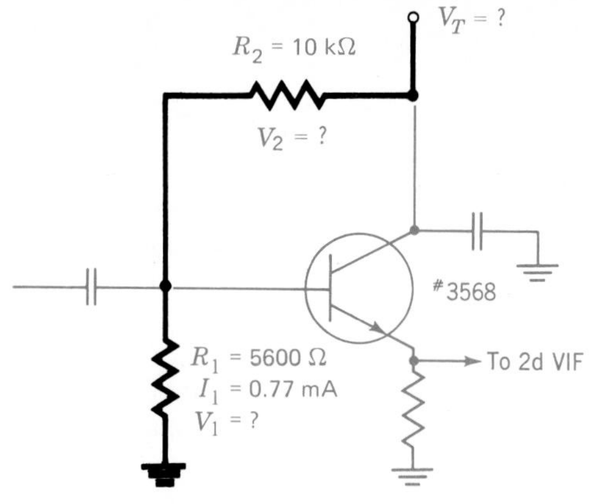


**FIGURE 4-11**
First video amplifier circuit in a modern TV chassis.

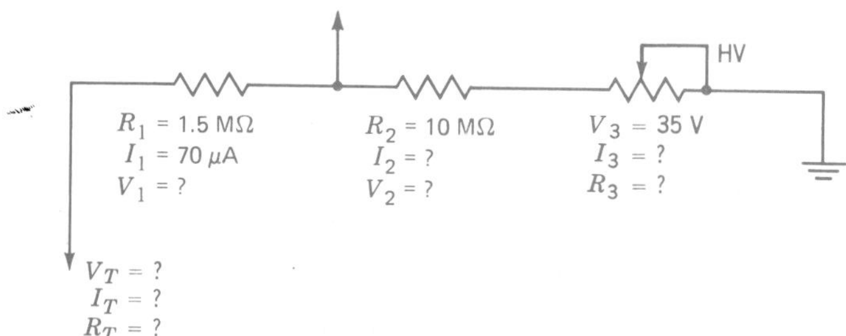


**FIGURE 4-12**
High-voltage regulator circuit in a color television receiver.

## JOB 4-3 USING OHM'S LAW FOR TOTAL VALUES IN SERIES CIRCUITS

Ohm's law was used in the last job to find the voltage, current, and resistance of the individual parts of a series circuit. Ohm's law may also be used to find the *total values* of voltage, current, and resistance in a series circuit.

**RULE** In a series circuit, the total voltage is equal to the total current multiplied by the total resistance.

**FORMULA**

$$V_T = I_T \times R_T \qquad (4\text{-}7)$$

where $V_T$ = total voltage, V
$I_T$ = total current, A
$R_T$ = total resistance, Ω

### FINDING THE TOTAL VOLTAGE

EXAMPLE 4-4. A resistor of 45 Ω, a bell of 60 Ω, and a buzzer of 50 Ω are connected in series as shown in Fig. 4-13. The current in each is 0.2 A. What is the total voltage?

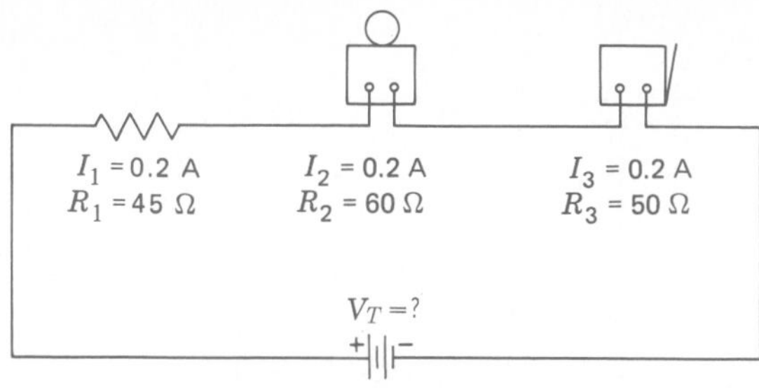


FIGURE 4-13

SOLUTION. In order to use Ohm's law to find the total voltage, we must first obtain the values for the total current and the total resistance. These values may be found by the following formulas:

$$I_T = I_1 = I_2 = I_3 \qquad (4\text{-}1)$$

$$R_T = R_1 + R_2 + R_3 \qquad (4\text{-}3)$$

1. Find the total current.

$$I_T = I_1 = I_2 = I_3 = 0.2 \text{ A} \quad \textit{Ans.} \qquad (4\text{-}1)$$

2. Find the total resistance.

$$R_T = R_1 + R_2 + R_3 \qquad (4\text{-}3)$$

$$R_T = 45 + 60 + 50 = 155\ \Omega \quad \textit{Ans.}$$

3. Find the total voltage.

$$V_T = I_T \times R_T \qquad (4\text{-}7)$$

$$V_T = 0.2 \times 155 = 31 \text{ V} \quad \textit{Ans.}$$

## FINDING THE TOTAL CURRENT

EXAMPLE 4-5. A portable spotlight of 3 Ω resistance is connected to a 6-V power pack with two wires, each of 0.4 Ω resistance. Find the total current drawn from the power pack.

SOLUTION. The diagram for the circuit is shown in Fig. 4-14. In order to use Ohm's law to find the total current, we must first obtain the values for the total voltage and the total resistance.

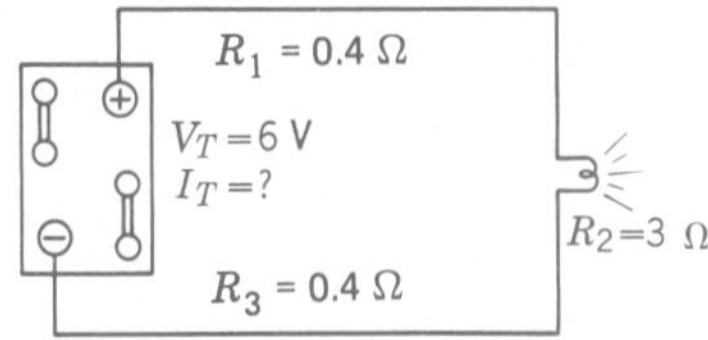


FIGURE 4-14

1. The total voltage is known: $V_T = 6$ V.

2. Find the total resistance.

$$R_T = R_1 + R_2 + R_3 \qquad (4\text{-}3)$$

$$R_T = 0.4 + 3 + 0.4 = 3.8\ \Omega \quad \textit{Ans.}$$

3. Find the total current.

$$V_T = I_T \times R_T \qquad (4\text{-}7)$$

$$6 = I_T \times 3.8$$

$$\frac{6}{3.8} = I_T$$

or $$I_T = 1.58\ \text{A} \qquad \textit{Ans.}$$

## FINDING THE TOTAL RESISTANCE

EXAMPLE 4-6. A motor, a lamp, and a rheostat are connected in series. The motor uses 80 V, the lamp takes 10 V and 2 A, and the rheostat uses 30 V. Find the total resistance of the circuit.

SOLUTION. The diagram for the circuit is shown in Fig. 4-15. In order to use Ohm's law to find the total resistance, we must first obtain the values for the total voltage and the total current.

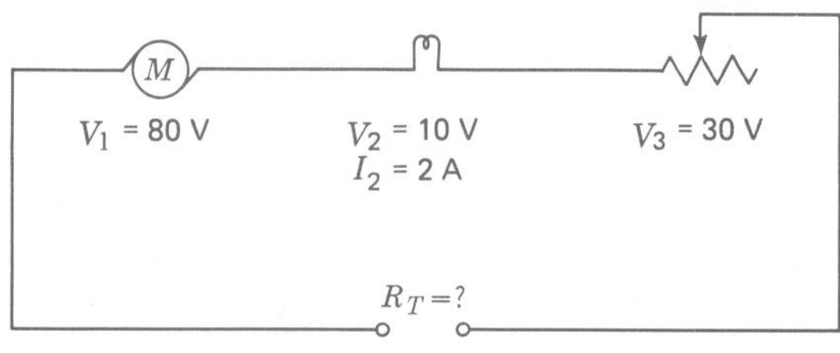


FIGURE 4-15

1. Find the total voltage. $V_T = V_1 + V_2 + V_3$ (4-2)

   $V_T = 80 + 10 + 30 = 120\ \text{V}$ *Ans.*

2. Find the total current. $I_T = I_1 = I_2 = I_3 = 2\ \text{A}$ *Ans.* (4-1)

3. Find the total resistance. $V_T = I_T \times R_T$ (4-7)

$$120 = 2 \times R_T$$

$$\frac{120}{2} = R_T$$

or $$R_T = 60\ \Omega \qquad \textit{Ans.}$$

SELF-TEST 4-7. A Philco-Ford TV chassis 20L23 uses a brightness-control circuit similar to that shown in Fig. 4-16. Find the total voltage between ground and point *B*.

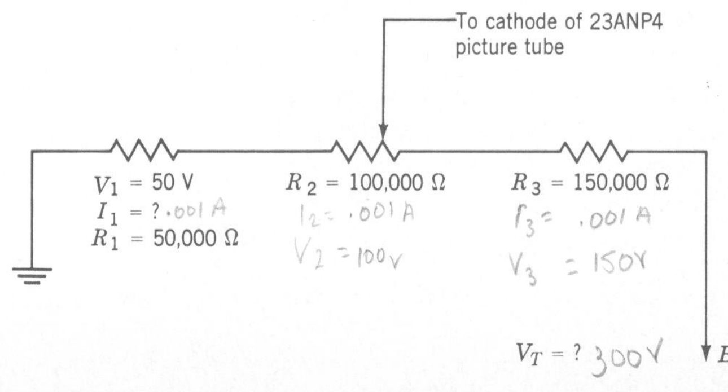


FIGURE 4-16
Brightness-control circuit in a television receiver.

SOLUTION. In order to use Ohm's law to find the total voltage, we must first obtain the values for the total ______ and the total ______. | current resistance

1. Find the total current.

We can find the total current if we can find the current in any resistance. The only resistor about which there is sufficient information is ______. | $R_1$

Find $I_1$.

$$V_1 = \_\_\_\_ \times R_1 \qquad (4\text{-}4)$$ | $I_1$

$$50 = I_1 \times \_\_\_\_$$ | 50,000

$$\frac{50}{?} = I_1$$ | 50,000

$$I_1 = \_\_\_\_ \text{ A}$$ | 0.001

Now, in a series circuit,

$$I_T = I_1 = I_2 = I_3 = \_\_\_\_ \text{ A} \quad \textit{Ans.} \qquad (4\text{-}1)$$ | 0.001

2. Find the total resistance.

$$R_T = R_1 + R_2 + \_\_\_\_ \qquad (4\text{-}3)$$ | $R_3$

$$R_T = 50{,}000 + \_\_\_\_ + 150{,}000$$ | 100,000

$$R_T = \_\_\_\_ \ \Omega \quad \textit{Ans.}$$ | 300,000

3. Find the total voltage.

$$V_T = I_T \times \_\_\_\_ \qquad (4\text{-}7)$$ | $R_T$

$$V_T = \_\_\_\_ \times 300{,}000$$ | 0.001

$$V_T = \_\_\_\_ \text{ V} \quad \textit{Ans.}$$ | 300

## PROBLEMS

1. Five lamps are connected in series for use in a subway lamp bank. Each lamp has a resistance of 110 Ω. What is the subway-circuit total voltage if the total current drawn is 1 A?
2. A railroad signal lamp and the coil of a semaphore are connected in series. The resistance of the lamp is 16 Ω, and that of the coil is 5 Ω. The current is 2 A. What is the total voltage?
3. In a telephone circuit, the receiver, transmitter, and line coil are connected in series. The receiver resistance is 2.2 Ω; the transmitter resistance is 6.2 Ω; the coil resistance is 1.2 Ω. If the current in the receiver is 0.5 A, what is the total voltage of the circuit?
4. A motor, a lamp, and a rheostat are connected in series. The voltage across the motor is 96 V, across the lamp 24 V, and across the rheostat 40 V. If the current in the motor is 2 A, find (*a*) the total current, (*b*) the total voltage, and (*c*) the total resistance.
5. A current of 0.003 A flows through a resistor that is connected to a 1.5-V dry cell. If three additional 1.5-V cells are connected in series to the first cell, find the current now flowing through the resistor.

6. A series circuit consists of a heating coil, an ultraviolet lamp, and a motor for an electric clothes drier. The current in the motor is 5 A. The voltages are 50 V for the lamp, 80 V for the motor, and 100 V for the heating coil. Find (*a*) the total current, (*b*) the total voltage, and (*c*) the total resistance.
7. Four lamps are wired in series. The resistance of each lamp is 0.5 Ω. The voltage across each lamp is 0.4 V. Find (*a*) the total current, (*b*) the total voltage, and (*c*) the total resistance.
8. Three arc lights are connected in series with a resistance of 21 Ω. The resistance of each arc light when hot is 10 Ω. The voltage across each arc light is 40 V, and the voltage across the resistance is 84 V. Find (*a*) the total voltage, (*b*) the total current, and (*c*) the total resistance.
9. A voltage divider in a television receiver consists of a 6-kΩ, a 3-kΩ, and a 1500-Ω resistor in series. If the total current is 15 mA, find the total voltage drop.

## JOB 4-4 INTERMEDIATE REVIEW OF SERIES CIRCUITS

A series circuit is one in which the electrons may flow in only ______ path. | one

**Finding the total voltage.** The total voltage in a series circuit may be found by use of the following formulas:

$$V_T = V_1 + V_2 + \_\_\_\_\_\_ \quad (4\text{-}2)$$ | $V_3$

$$V_T = \_\_\_\_\_\_ \times R_T \quad (4\text{-}7)$$ | $I_T$

**Finding the total resistance.** The total resistance in a series circuit may be found by use of the following formulas:

$$R_T = R_1 + R_2 + \_\_\_\_\_\_ \quad (4\text{-}3)$$ | $R_3$

$$V_T = I_T \times \_\_\_\_\_\_ \quad (4\text{-}7)$$ | $R_T$

**Finding the total current.** The total current in a series circuit may be found by use of the following formulas:

$$I_T = I_1 = \_\_\_\_\_\_ = I_3 \quad (4\text{-}1)$$ | $I_2$

$$\_\_\_\_\_\_ = I_T \times R_T \quad (4\text{-}7)$$ | $V_T$

SELF-TEST 4-8. A 12-Ω spotlight in a theater is connected in series with a dimming resistor of 31 Ω. If the voltage drop across the light is 31.2 V, find (*a*) the current in the light, (*b*) the current in the dimmer, (*c*) the total current, (*d*) the voltage drop across the dimmer, and (*e*) the total voltage and total resistance.

SOLUTION. The diagram for the circuit is shown in Fig. 4-17.

*a*. Find the light current.

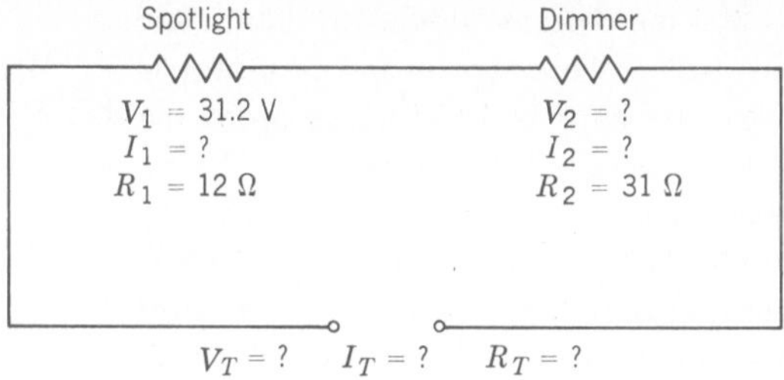


**FIGURE 4-17**

| | | |
|---|---|---|
| $V_1 = I_1 \times$ ______ | (4-4) | $R_1$ |
| $31.2 =$ ______ $\times 12$ | | $I_1$ |
| $\frac{31.2}{?} = I_1$ | | 12 |
| $I_1 =$ ______ A *Ans.* | | 2.6 |

*b.* Find the current in the dimmer.

| | | |
|---|---|---|
| $I_T = I_1 =$ ______ | (4-1) | $I_2$ |
| $I_T =$ ______ $= I_2$ | | 2.6 |
| $I_2 =$ ______ A *Ans.* | | 2.6 |

*c.* The total current $= I_T =$ ______ A *Ans.* 2.6

*d.* Find the voltage drop across the dimmer.

| | | |
|---|---|---|
| $V_2 =$ ______ $\times R_2$ | (4-5) | $I_2$ |
| $V_2 =$ ______ $\times 31 =$ ______ V *Ans.* | | 2.6 80.6 |

*e.* Find the total voltage.

| | | |
|---|---|---|
| $V_T = V_1 +$ ______ | (4-2) | $V_2$ |
| $V_T =$ ______ $+ 80.6$ | | 31.2 |
| $V_T =$ ______ V *Ans.* | | 111.8 |

Find the total resistance.

| | | |
|---|---|---|
| $R_T = R_1 +$ ______ | (4-3) | $R_2$ |
| $R_T =$ ______ $+ 31$ | | 12 |
| $R_T =$ ______ Ω *Ans.* | | 43 |

*Check:* If we have been correct in our calculations, our values for the total voltage, total current, and total resistance should satisfy the Ohm's-law formula for total values.

| | | |
|---|---|---|
| $V_T =$ ______ $\times R_T$ | (4-7) | $I_T$ |
| ______ $= 2.6 \times 43$ | | 111.8 |
| $111.8 =$ ______ | | 111.8 |

The equation __(does/does not)__ check. does

**PROBLEMS**

1. Three resistors are connected in series. $R_1$ has a resistance of 40 Ω. The second resistor causes a 6-V drop, and the third resistor

has a resistance of 120 Ω. Find the total voltage across the circuit if the series current is 0.3 A.

2. A 40-Ω resistor is in series with a 25-Ω resistor. Find the total voltage if the series current is 0.5 A.
3. Three resistors are in series, $I_1 = 0.5$ A, $R_1 = 2$ Ω, $R_2 = 3$ Ω, and $V_3 = 3.5$ V. Find the total voltage.
4. Two resistors and a motor are connected in series. The motor current is 3 A, and its resistance is 25 Ω. The voltage across each resistor is 37.5 V. Find (*a*) the total current, (*b*) the total voltage, and (*c*) the total resistance.
5. A voltage divider consists of a 3-, a 5-, and a 10-kΩ resistor in series. The series current is 0.015 A. Find (*a*) the voltage drop across each resistance, (*b*) the total voltage, and (*c*) the total resistance.
6. A lamp and two bells are connected in series in a burglar-alarm circuit. The lamp draws a current of 0.25 A and has a resistance of 160 Ω. Each bell uses 35 V. Find (*a*) the total current, (*b*) the total voltage, and (*c*) the total resistance.
7. Three resistors are in series: $V_1 = 31.2$ V, $V_2 = 48$ V, $I_2 = 1.2$ A, and $R_3 = 44$ Ω. Find (*a*) the total current, (*b*) the total voltage, and (*c*) the total resistance of the circuit.
8. In a series circuit, $I_1 = 0.5$ A, $R_1 = 35$ Ω, $V_2 = 91$ V, $R_3 = 23$ Ω, and $V_4 = 40$ V. Find the total voltage.
9. Three resistances are in series. They use 8, 10, and 14 V, respectively. The total resistance of the circuit is 80 Ω. Find all missing values of current, voltage, and resistance, including the total values.
10. Three resistances of 30, 40, and 50 Ω are in series. The total voltage impressed across the circuit is 24 V. Find all missing values of current, voltage, and resistance, including the total values.

(See Instructor's Manual for Test 4-2, Series Circuits)

## JOB 4-5 CHECKUP ON FORMULAS INVOLVING ADDITION AND SUBTRACTION (DIAGNOSTIC TEST)

In our next electrical job we shall be required to solve some formulas which are slightly different from any we have solved up to this point. The following 10 problems are of this type. If you have any difficulty with them, see Job 4-6 which follows.

### PROBLEMS

1. Using the formula $V_T = V_1 + V_2 + V_3$, find $V_2$ if $V_T = 78$, $V_1 = 17$, and $V_3 = 32$.
2. Using the formula $P = 2L + 2W$, find $L$ if $P = 80$ and $W = 14$.
3. Using the formula $I_T = I_1 + I_2 + I_3$, find $I_3$ if $I_T = 5$ A, $I_1 = 1.6$ A, and $I_2 = 2.3$ A.
4. Find $R$ in the equation $7 + R = 5.4 + 19$.
5. Find $V$ in the equation $3V + 8 = 29$.

6. Find $I$ in the equation $I - 5 = 40$.
7. Find $I$ in the equation $3I - 4 = 32$.
8. Find $R$ in the formula $R + r = V/I$ if $V = 60$, $I = 5$, and $r = 9$.
9. What resistance must be placed in series with six 15-V Christmas-tree lights in order to operate them on a 110-V circuit? Each light requires 0.8 A.
10. What series resistor is necessary to operate a circuit requiring 79 V at 0.3 A if it is to be operated from a 110-V line?

## JOB 4-6 BRUSHUP ON FORMULAS INVOLVING ADDITION AND SUBTRACTION

Solving a formula means to find the value of the unknown letter in the formula. In Job 3-8, after substituting the numbers for the letters, we obtained statements of equality such as $2 \times R = 10$ or $12 = 3 \times R$. In each instance, a number *multiplied* by a letter was equal to another number. We obtained the value of the unknown letter by eliminating the number multiplied by it. This was accomplished by *dividing both sides* of the equality sign by that *same* number. In general, to solve *any* formula or equation, we must eliminate all numbers and letters which appear on the same side of the equality sign as the *unknown* letter. This is done by applying the basic principle given in Job 3-8.

| **BASIC PRINCIPLE** | Any mathematical operation performed on one side of an equality sign must also be performed on the other side. |
|---|---|

In simple language this says, "whatever we do to one side of an equality sign must also be done to the other side."

### FORMULAS INVOLVING ADDITION

| **RULE** | The same number may be subtracted from both sides of an equality sign without destroying the equality. |
|---|---|

SELF-TEST 4-9. Solve the equation

$$x + 3 = 7$$

SOLUTION. Solving an equation means to get the letter all alone on one side of the equality sign, with its value on the other side. Therefore, all we have to do to solve the equation is to get the $x$ alone. This will be accomplished if we can get rid of the number _______. We can do this by applying the *opposite* operation of _______ 3 *from both sides* of the equality sign.

+3
subtracting

$$\begin{array}{rr} x + 3 = & 7 \\ -3 & -3 \\ \hline x + 0 = & \_\_\_\_ \\ x = 4 & \textit{Ans.} \end{array}$$ 4

SELF-TEST 4-10. Solve the equation $y + 4 = 9$.

SOLUTION. We can get $y$ alone on the left side of the equality sign if we can eliminate the number +4 from that side. This can be done by applying the *opposite* operation of *subtracting* ______ *from both sides.* 4

$$\begin{array}{rcr} y + 4 & = & 9 \\ -\ 4 & & -4 \\ \hline y + \_\_\_\_ & = & 5 \\ y = \_\_\_\_ & & \textit{Ans.} \end{array}$$ 0 5

EXAMPLE 4-11. Using the formula $R_T = R_1 + R_2 + R_3$, find the resistance $R_1$ if $R_T = 100\ \Omega$, $R_2 = 20\ \Omega$, and $R_3 = 40\ \Omega$.

SOLUTION

1. Write the formula. $R_T = R_1 + R_2 + R_3$ (4-3)
2. Substitute numbers. $100 = R_1 + 20 + 40$
3. Simplify (add 20 + 40). $100 = R_1 + 60$

We can get $R_1$ alone on the right side of the equality sign if we can eliminate the number +60 from that side. This can be accomplished by applying the *opposite* operation of *subtracting* 60 *from both sides* of the equality sign.

4. Subtract 60 from both sides. $$\begin{array}{rr} 100 = R_1 + & 60 \\ -\ 60 \quad & -\ 60 \\ \hline 40 = R_1 + & 0 \end{array}$$
5. Subtract.
6. The resistance is $R_1 = 40\ \Omega$ *Ans.*

## PROBLEMS

Solve each of the following equations for the value of the unknown letter:

1. $V + 3 = 9$
2. $2 + R = 10$
3. $10 + P = 80$
4. $110 = V + 60$
5. $V + 18 = 70$
6. $I + 2^1/_2 = 3^1/_2$
7. $V + 40 + 30 + 25 = 120$
8. $I + ^1/_2 = 5$
9. $I + 3.5 = 10.8$
10. $20 + 10 + V = 110$
11. $V + 6.8 = 35$
12. $I + 2^1/_2 = 15^1/_2$
13. $R + ^3/_4 = 6$
14. $V + 12.9 = 75.6$
15. $I + 0.2 + 1.3 = 7.6$
16. $12.6 + 6.3 + V = 120$
17. $110 = 35 + 12.6 + 6.3 + V$
18. $I + 2^1/_2 = 8.4$

19. Using the formula $R_T = R_1 + R_2 + R_3$, find the resistance $R_1$ if $R_T = 175$, $R_2 = 40$, and $R_3 = 15$.
20. Using the formula $P_T = P_1 + P_2 + P_3$, find the power $P_3$ if $P_T = 1100$, $P_1 = 300$, and $P_2 = 275$.
21. Using the formula $I_T = I_1 + I_2$, find the current $I_2$ if $I_T = 99$ and $I_1 = 26$.
22. Using the formula $I_s + I_m = I$, find the shunt current $I_s$ if the line current $I = 1.64$ and the meter current $I_m = 0.014$.
23. Using the formula $C_1 + C_2 + C_3 = C_T$, find $C_1$ if $C_T = 0.000\,25$, $C_2 = 0.000\,12$, and $C_3 = 0.000\,05$.
24. Using the formula $V_1 + (I_2 \times R_2) = V_T$, find $V_1$ if $I_2 = 2$, $R_2 = 25$, and $V_T = 117$.
25. Using the formula $I_T = I_1 + (V_2/R_2)$, find $I_1$ if $I_T = 1.5$, $V_2 = 6.3$, and $R_2 = 126$.

## FORMULAS INVOLVING SUBTRACTION

**RULE** The same number may be added to both sides of an equality sign without destroying the equality.

SELF-TEST 4-12. Solve the equation $x - 3 = 7$.

SOLUTION. Solving an equation means to get the letter all alone on one side of the equality sign and its value on the other side. We can get $x$ alone on the left side of the equality sign if we can get rid of the number _____. We can do this by applying the *opposite* operation of _____ 3 to both sides of the equality sign. | −3 | adding

$$x - 3 = 7$$
$$+3 \quad +3$$
$$x + \_\_\_ = 10$$
$$x = \_\_\_ \quad Ans.$$

| 0 | 10

SELF-TEST 4-13. Using the series-circuit formula $V - Ir = IR$, find the voltage $V$ in the circuit shown in Fig. 4-18.

SOLUTION

1. Write the formula. $V - Ir = IR$
2. Substitute numbers. $V - 2 \times 20 = 2 \times$ _____ | 40
3. Multiply numbers. $V -$ _____ $= 80$ | 40

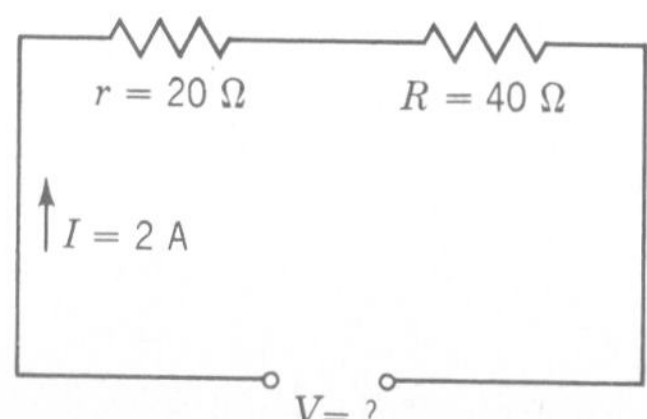


FIGURE 4-18

| | | |
|---|---|---|
| 4. Add 40 to both sides. | $V - 40 = \quad 80$ <br> $+40 = +40$ | |
| 5. Add. | $V + \ 0 =$ ______ | 120 |
| 6. The voltage is | $V =$ ______ V *Ans.* | 120 |

## PROBLEMS

Solve each of the following equations for the value of the unknown letter:

1. $R - 5 = 12$
2. $12 = V - 3$
3. $I - 4 = 18$
4. $30 + R - 10 = 50$
5. $I - 1/2 = 4\,1/2$
6. $P - 3.2 = 8.3$
7. $9 = I - 3.5$
8. $V - 2.2 = 6.3$
9. $V - 17 = 62$
10. $82 = R - 14$
11. $I - 7\,3/4 = 3$
12. $I - 1\,1/2 = 6\,3/4$
13. $12 = R - 3\,1/4$
14. $72\,1/2 = R - 5\,1/4$
15. $120 = R + 70 - 30$
16. $I - 0.045 = 0.85$
17. $C - 0.08 = 0.019$
18. $0.000\,25 = C - 0.000\,05$
19. Using the formula $C = S - P$, find $S$ if $C = \$18.23$ and $P = \$3.60$.
20. Using the formula $I_1 = I_T - I_2$, find $I_T$ if $I_1 = 8$ and $I_2 = 6$.
21. Using the formula $V_T - V_h = V_R$, find $V_T$ if $V_R = 110$ and $V_h = 62$.
22. Using the formula $V_T - (I_1 \times R_1) = V_2$, find $V_T$ if $I_1 = 2$, $R_1 = 30$, and $V_2 = 57$.
23. Using the formula $S - 5L = A$, find $S$ if $L = 3.2$ and $A = 1.05$.
24. Using the formula $V - R = RM$, find $V$ if $R = 8$ and $M = 2.5$.
25. Using the formula $R_1 = R_2 - V/I$, find $R_2$ if $R_1 = 9$, $V = 2.5$, and $I = 0.025$.

(See Instructor's Manual for Test 4-3)

**Transposition.** The method of eliminating a number from one side of an equality sign by adding or subtracting the same number may be shortened by the method known as *transposition.* Let us investigate the following four examples in an effort to determine the rule for the transposition of quantities.

EXAMPLE 4-14. Find $R$ in the equation $R + 2 = 10$.

SOLUTION

| | |
|---|---|
| 1. Write the equation. | $R + 2 = 10$ |
| 2. Subtract 2 from both sides. | $R + 2 - 2 = 10 - 2$ |
| 3. Subtract. | $R + 0 = 10 - 2$ |
| | $R = 8$ *Ans.* |

EXAMPLE 4-15. Find $V$ in the equation $8 = V + 3$.

SOLUTION

1. Write the equation. $8 = V + 3$
2. Subtract 3 from both sides. $8 - 3 = V + 3 - 3$
3. Subtract. $8 - 3 = V + 0$

$5 = V$

or $V = 5$ *Ans.*

EXAMPLE 4-16. Find $S$ in the equation $S - 4 = 10$.

SOLUTION

1. Write the equation. $S - 4 = 10$
2. Add 4 to both sides. $S - 4 + 4 = 10 + 4$
3. Add. $S + 0 = 10 + 4$

$S = 14$ *Ans.*

EXAMPLE 4-17. Find $I$ in the equation $9 = I - 5$.

SOLUTION

1. Write the equation. $9 = I - 5$
2. Add 5 to both sides. $9 + 5 = I - 5 + 5$
3. Add. $9 + 5 = I + 0$

$14 = I$

or $I = 14$ *Ans.*

Notice that in each of the last four examples, the number to be eliminated—the number in color—seems to have *moved* from one side of the equality sign to the *other side.* Also, the sign in front of the number changed from + to − or from − to +. This gives us the following rule for transposing.

**RULE** Plus or minus quantities may be moved from one side of an equality sign to the other if the sign of the quantity is changed from + to − or from − to +.

EXAMPLE 4-18. Find $V$ in the equation $V + 8 = 12$.

SOLUTION

1. Write the equation. $V + 8 = 12$
2. Transpose the 8. $V = 12 - 8$
3. Subtract. $V = 4$ *Ans.*

EXAMPLE 4-19. Find $R$ in the equation $9 = R + 3$.

SOLUTION

1. Write the equation. $9 = R + 3$
2. Transpose the 3. $9 - 3 = R$

3. Subtract. $6 = R$

or $R = 6$ *Ans.*

EXAMPLE 4-20. Find $S$ in the equation $S - 5 = 16$.

SOLUTION

1. Write the equation. $S - 5 = 16$
2. Transpose the 5. $S = 16 + 5$
3. Add. $S = 21$ *Ans.*

EXAMPLE 4-21. Find $I$ in the equation $4 = I - 9$.

SOLUTION

1. Write the equation. $4 = I - 9$
2. Transpose the 9. $4 + 9 = I$
3. Add. $13 = I$

or $I = 13$ *Ans.*

SELF-TEST 4-22. Using the series-circuit formula $R + r = V/R$, find the resistance $r$ in the circuit shown in Fig. 4-19.

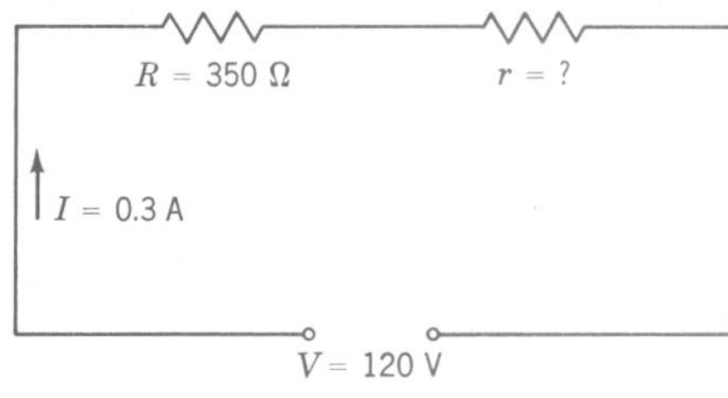


FIGURE 4-19

SOLUTION

1. Write the formula.

$$R + r = \frac{V}{I}$$

2. Substitute numbers.

$$350 + \_\_\_\_\_\_ = \frac{120}{0.3}$$

3. Divide numbers.

$$350 + r = \_\_\_\_\_\_$$

r | 400

4. Transpose the 350. (A + sign is understood to be present in front of the 350.)

$$r = \_\_\_\_\_\_ -350$$

5. Subtract.

$$r = \_\_\_\_\_\_ \ \Omega \quad \textit{Ans.}$$

400 | 50

## PROBLEMS

Solve the following equations by transposition:

1. $V + 3 = 14$
2. $16 = V + 3$
3. $2 + R = 8$
4. $13 = I + 2$
5. $6 + 4 + V = 20$
6. $3 + R + 2 = 15$
7. $21 = 4 + 5 + I$
8. $6 + I = 15$
9. $P - 2 = 6$
10. $14 = R - 4$
11. $0.06 + I = 1.6$
12. $20 = V - 2$
13. $18 - 3 + R = 25$
14. $16 = R - 4 - 5$
15. $R + 6.7 = 18.2$
16. $I + 0.045 = 0.09$
17. $0.025 = C + 0.004$
18. $X + 2.6 + 1.05 + 3.0 = 9.4$
19. $R + 7 - 2.40 + 3.60 = 19.68$
20. $I + 0.07 = 3.4$
21. $X + 3^1/_2 = 5$
22. $I + 0.03 = 6.3$
23. Using the formula $P_T = P_1 + P_2 + P_3$, find $P_1$ if $P_T = 600, P_2 = 120$, and $P_3 = 300$.
24. Using the formula $I_1 = I_T - I_2$, find $I_T$ if $I_1 = 8$ and $I_2 = 3$.
25. Using the formula $V_T = V_1 + V_2 + V_3$, find $V_3$ if $V_T = 110$, $V_1 = 30$, and $V_2 = 50$.
26. Using the formula $R_T = R_1 + R_2 + R_3$, find $R_2$ if $R_T = 245, R_1 = 85$, and $R_3 = 90$.
27. Using the formula $I_s = I_L - I_m$, find $I_L$ if $I_s = 75$ and $I_m = 1.5$.
28. Using the formula $V_T - V_h = V_R$, find $V_T$ if $V_R = 97.2$ and $V_h = 6.3$.
29. For $C_T = C_1 + C_2$, find $C_1$ if $C_T = 0.000\ 35$ and $C_2 = 0.0001$.
30. For $C_T = C_1 + C_2$, find $C_2$ if $C_T = 0.004$ and $C_1 = 0.000\ 15$.

(See Instructor's Manual for Test 4-4)

## JOB 4-7 CONTROL OF CURRENT IN A SERIES CIRCUIT

In Job 3-9 we learned that resistance affects the current in an electric circuit. The greater the resistance, the smaller the current; the smaller the resistance, the greater the current. There are many instances where a certain current must be maintained in a circuit. We can accomplish this by adjusting the resistance of the circuit so as to obtain any required current.

EXAMPLE 4-23. How much resistance must be added to the circuit shown in Fig. 4-20 in order to allow only the rated current of 2 A to flow?

SOLUTION. The diagram for the circuit is shown in Fig. 4-20. If we can find the total resistance required to limit the current to 2 A, then we can use formula (4-3) to find the missing extra resistance.

Find $R_T$.

1. Write the formula. $V_T = I_T \times R_T$ (4-7)
2. Substitute numbers. $110 = 2 \times R_T$
3. Solve for $R_T$. $\dfrac{110}{2} = R_T$
4. Divide. $R_T = 55\ \Omega$ *Ans.*

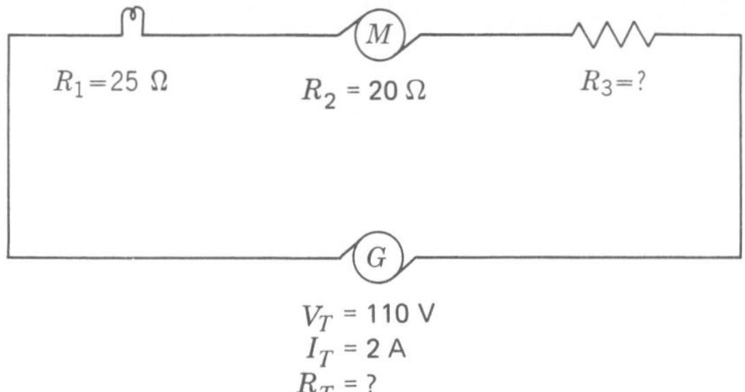


FIGURE 4-20

Find $R_3$, the resistance to be added.

1. Write the formula. $R_T = R_1 + R_2 + R_3$ (4-3)
2. Substitute numbers. $55 = 25 + 20 + R_3$
3. Simplify (add 25 + 20). $55 = 45 + R_3$
4. Transpose the 45. $55 - 45 = R_3$
5. Subtract. $10 = R_3$

or $R_3 = 10\ \Omega$ *Ans.*

EXAMPLE 4-24. What resistance must be added in series with a lamp rated at 12 V and 0.3 A in order to operate it from a 24-V source?

SOLUTION. The diagram for the circuit is shown in Fig. 4-21. If we can discover how many volts are available in excess of that required by the lamp, then we can find the resistance which will use up that extra voltage at the rated 0.3 A.

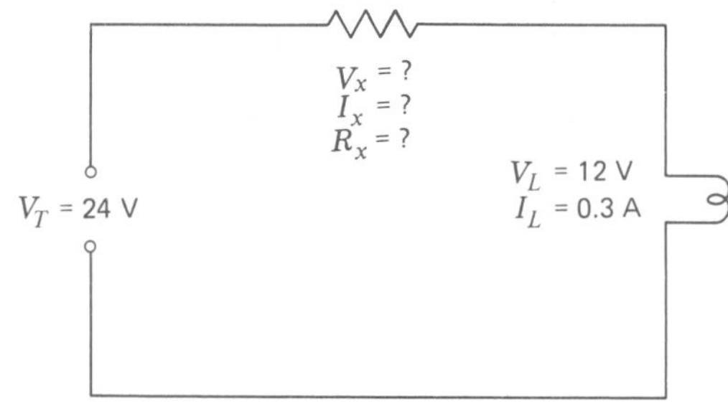


FIGURE 4-21

Find the extra voltage supplied $V_x$.

1. Write the formula. $V_T = V_x + V_L$ (4-2)
2. Substitute numbers. $24 = V_x + 12$
3. Transpose the 12. $24 - 12 = V_x$
4. Subtract. $12 = V_x$

or $V_x = 12$ V *Ans.*

Find the current $I_x$ in the unknown resistor.

1. Write the formula. $I_T = I_x = I_L$ (4-1)
2. Substitute numbers. $I_T = I_x = 0.3$

$I_x = 0.3$ A *Ans.*

Find the resistance $R_x$ which will use up the extra 12 V at the rated series current of 0.3 A.

1. Write the formula. $V_x = I_x \times R_x$ (2-1)
2. Substitute numbers. $12 = 0.3 \times R_x$
3. Solve for $R_x$. $\frac{12}{0.3} = R_x$
4. Divide. $40 = R_x$

or $R_x = 40\ \Omega$ *Ans.*

SELF-TEST 4-25. In the voltage-divider circuit shown in Fig. 4-22, the total load resistance between points $A$ and $B$ also serves to "bleed off" any charge on the filter capacitors after the rectifier is turned off. If the bleeder current $I_b$ is to be limited to 0.015 A, find the resistance of $R_1$.

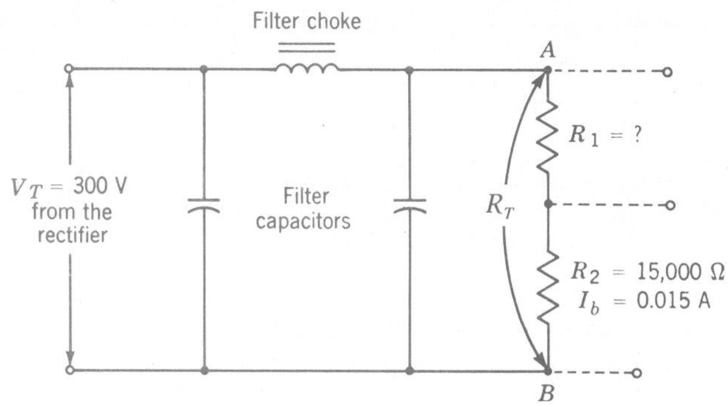


**FIGURE 4-22** Voltage divider with no loads attached.

SOLUTION. If we can find the total resistance of the load which will limit the bleeder current to 0.015 A, then we can use the formula for total ________ to find $R_1$. — resistance

Find $R_T$.

1. Write the formula. $V_T =$ ________ $\times R_T$ (4-7) — $I_T$
2. Substitute numbers. ________ $= 0.015 \times R_T$ — 300
3. Solve for $R_T$. $\frac{300}{?} = R_T$ — 0.015
4. Divide. $R_T =$ ________ Ω *Ans.* — 20,000

Find $R_1$, the resistance to be added to $R_2$ to make a total of ________ Ω. — 20,000

1. Write the formula. $R_T = R_1 + R_2$ (4-3)
2. Substitute numbers. $20{,}000 = R_1 +$ ________ — 15,000
3. Transpose the 15,000. $20{,}000 - 15{,}000 =$ ________ — $R_1$
4. Subtract. ________ $= R_1$ — 5000

or $R_1 =$ ________ Ω *Ans.* — 5000

## PROBLEMS

1. How much resistance must be connected in series with a 30-Ω lamp rated at 2 A if it is to be used on a 110-V line?
2. A 10-Ω lamp rated at 5 A is to be operated from a 120-V line.

How much resistance must be added in series to reduce the current to the desired value?

3. A motor of 22 Ω resistance and a signal lamp of 25 Ω resistance are connected in series across 110 V. Find the series resistor that must be added to limit the current to 2 A.
4. A boy wants to illuminate five houses on his model railroad, using five 20-Ω lamps in series. What resistance must be connected in series in order to limit the current to the 0.5 A needed by the lamps if the circuit is to be operated from the ordinary 110-V house line?
5. What resistance must be placed in series with a 12-Ω bell if it is to draw exactly 1/4 A from a 24-V source?
6. Five ordinary 0.90-A 110-V lamps must be used in series when operated from the 550-V subway system. If only one of these lamps were used, what resistance would have to be placed in series with it to operate it from the 550-V line?
7. A 3-V airplane instrument lamp is to be operated from the 12-V electrical system. What resistance should be inserted in series so that the lamp will receive its rated current of 0.1 A?
8. The Panasonic AM-FM RE 7329 receiver uses an output stage similar to that shown in Fig. 4-23. If the effective $R = 100\ \Omega$, the collector current $I_C = 5$ mA, and the collector supply voltage $V_{CC} =$ 9 V, find the collector voltage $V_C$.

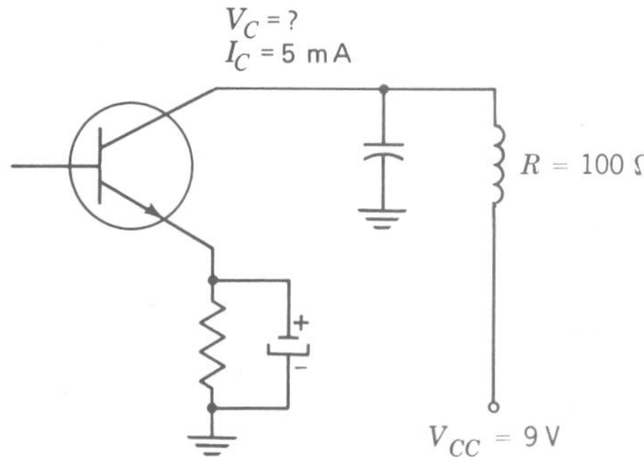


**FIGURE 4-23**
Output stage of a transistor radio.

9. An 80-A motor is connected to a 250-V generator through leads which have a resistance of 0.3 Ω for each lead. What is the voltage available at the motor?
10. The wiring in a house has a resistance of 0.4 Ω. What is the voltage available at an electric range using 12 A if the voltage at the meter is 117 V?
11. A floodlamp of 10 Ω resistance is connected in series with a variable dimming resistor of 0 to 45 Ω and is operated from a 110-V line. What are the maximum and minimum currents that may be supplied to the lamp?
12. The voltage drop in the line cord of a portable transmitter must not exceed 0.25 V. If the transmitter draws 20 A from a 12-V battery, find the maximum resistance of the line cord.

13. A portable transmitter of 0.575 Ω resistance is to operate at 11.5 V. What is the maximum resistance of the line cord connecting the transmitter to a 12-V battery?
14. A relay has a coil resistance of 550 Ω and is designed to operate on 150 mA. If the relay is used on a 120-V source, what value of resistance should be connected in series with the coil?

## JOB 4-8 REVIEW OF SERIES CIRCUITS

**Definition.** A series circuit is a circuit in which the electrons can flow in only one path.

**RULE 1** The total current is equal to the current in any part of the circuit.

$$I_T = I_1 = I_2 = I_3 \tag{4-1}$$

**RULE 2** The total voltage is equal to the sum of the voltages across all the parts of the circuit.

$$V_T = V_1 + V_2 + V_3 \tag{4-2}$$

**RULE 3** The total resistance is equal to the sum of the resistances of all the parts of the circuit.

$$R_T = R_1 + R_2 + R_3 \tag{4-3}$$

**RULE 4** Ohm's law may be used for any part of a series circuit.

$$V_1 = I_1 \times R_1 \tag{4-4}$$
$$V_2 = I_2 \times R_2 \tag{4-5}$$
$$V_3 = I_3 \times R_3 \tag{4-6}$$

**RULE 5** Ohm's law may be used for total values in a series circuit.

$$V_T = I_T \times R_T \tag{4-7}$$

SELF-TEST 4-26. In Fig. 4-24 the type 2N247 transistor has a collector current $I_C$ of 0.01 A and a collector voltage $V_C$ of 9.2 V. Find the total supply voltage $V_{CC}$.

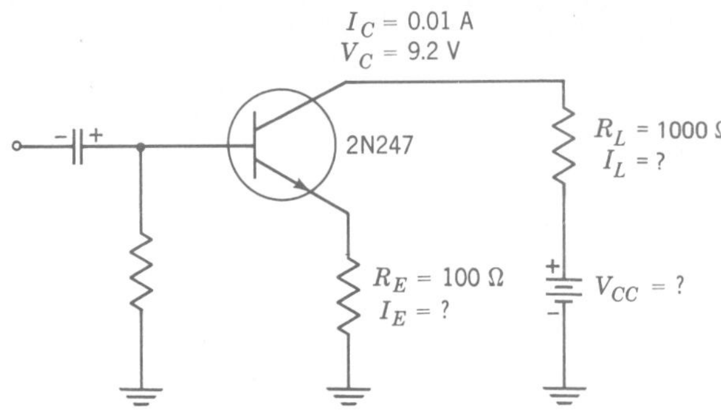


**FIGURE 4-24**
The total supply voltage is equal to the sum of the voltages around the circuit.

SOLUTION. Since the electrons complete the circuit flowing from the emitter to the collector of the transistor, it is a ______(series/parallel)______ circuit.

series

Find the current in each part of the circuit.

$$I_T = I_E = \_\_\_\_\_\_ = I_L \quad (4\text{-}1)$$

$I_C$

$$I_T = I_E = 0.01 = I_L$$

Find the voltage across each part of the circuit.

$$V_{RE} = I_E \times \_\_\_\_\_\_ \quad (4\text{-}4)$$

$R_E$

$$V_{RE} = 0.01 \times \_\_\_\_\_\_ = \_\_\_\_\_\_ \text{ V}$$

100 1

$$V_L = \_\_\_\_\_\_ \times R_L \quad (4\text{-}5)$$

$I_L$

$$V_L = 0.01 \times \_\_\_\_\_\_ = \_\_\_\_\_\_ \text{ V}$$

1000 10

Find the total supply voltage $V_{CC}$.

$$V_{CC} = V_{RE} + V_C + \_\_\_\_\_\_ \quad (4\text{-}2)$$

$V_L$

$$V_{CC} = \_\_\_\_\_\_ + 9.2 + 10 = \_\_\_\_\_\_ \text{ V} \quad \textit{Ans.}$$

1 20.2

## PROBLEMS

1. Three resistors are connected in series. $V_1 = 24$ V, $I_2 = 2$ A, $R_2 = 30\ \Omega$, and $R_3 = 13\ \Omega$. Find all missing values of voltage, current, and resistance, including the total values.
2. The two field coils of a generator have a resistance of 55 Ω each and are connected in series across the brushes, which deliver 110 V. What is the current in the field coils?
3. An electric heater whose resistance is 10 Ω is to be used on a 220-V line. The maximum current permitted through it is 10 A. What resistance must be added in series with the heater in order to hold the current to 10A?
4. In the universal transistor circuit shown in Fig. 4-25, the resistances between points $A$ and $B$ are in series. Find ($a$) the total current and ($b$) the voltage drop across each resistance.
5. The field coils of a motor draw 4 A from a 112-V line. What is the resistance of the coils? If a 14-Ω resistor is added in series with the coils, find the current in the coils and the voltage across the coils.

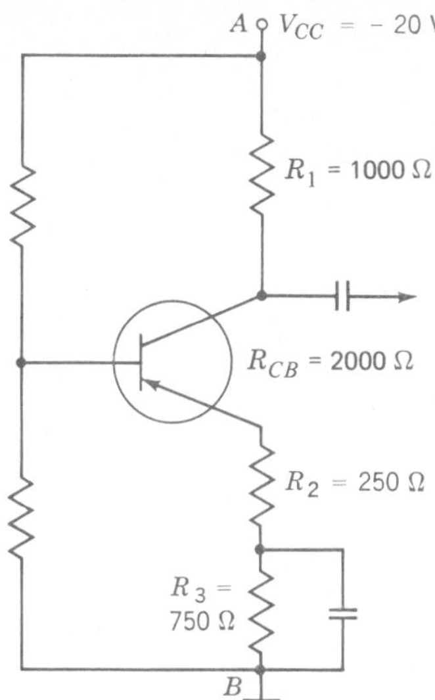


FIGURE 4-25
Universal transistor circuit.

6. What value of resistance must be placed in series with two 50-Ω lamps, each taking 50 V, if they are to be operated from a 220-V line?
7. Three resistors are connected in series across 220 V. $R_1 = 15\ \Omega$, $R_2 = 25\Omega$, and $R_3 = 60\ \Omega$. (*a*) Find the total current. (*b*) If the maximum permissible current is 2 A, how much resistance must be added in series to keep the current at this value?
8. In order to dim a bank of stage lights, a rheostat may be connected in series with the lights to reduce the current and therefore the brightness of the lights. What value of resistance must be connected in series with a lamp bank drawing 20 A from a 120-V line in order to reduce the total current drawn to 5 A?
9. Two 25-V 25-Ω incandescent lamps are connected in series with a demonstration motor requiring 30 V at 1 A. What extra resistance should be added in series in order to draw the required current from a 110-V line?
10. When a voltmeter indicates its maximum rated voltage of 150 V, it is drawing a current of 0.01 A. What value of resistance is in series with the moving coil of the voltmeter if its resistance is 20 Ω?

(See Instructor's Manual for Test 4-5, Series Circuits)

# PARALLEL CIRCUITS

# 5

## JOB 5-1 TOTAL VOLTAGE, TOTAL CURRENT, AND TOTAL RESISTANCE IN PARALLEL CIRCUITS

**Recognizing a parallel circuit.** A parallel circuit is a circuit connected in such a manner that the current flowing into it may *divide* and flow in *more than one path.* In the parallel circuit shown in Fig. 5-1, the current divides at point $A$, part of the current flowing through the lamp in path 1 and the rest of the current flowing through the motor in path 2.

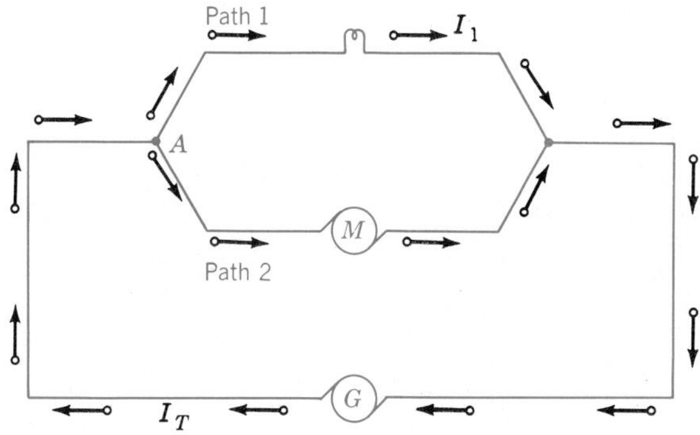


**FIGURE 5-1**
The current in a parallel circuit flows in more than one path.

The various paths of a parallel circuit are called the *branches* of the circuit. Notice how different this is from a series circuit shown in Fig. 5-2, in which the current can flow in *only one path.*

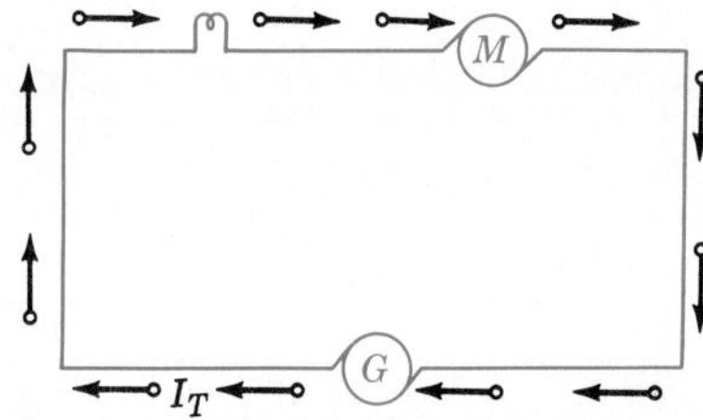


**FIGURE 5-2**
The current in a series circuit can flow in only one path.

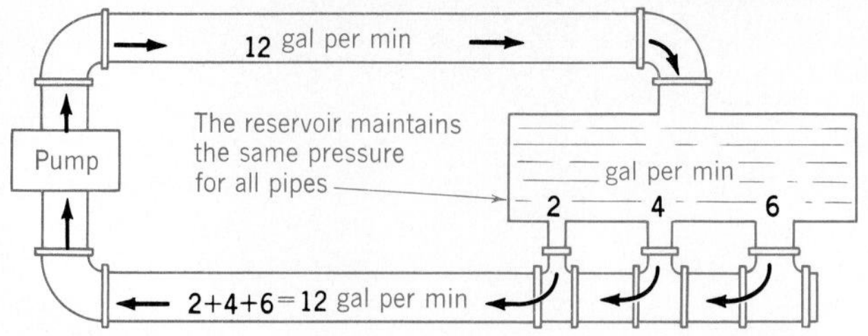


**FIGURE 5-3** Water-distribution system in "parallel."

**Total voltage in a parallel circuit.** Figure 5-3 shows a water-distribution system which might be called a parallel system because the water can flow in more than one pipe. A similar electrical system is shown in Fig. 5-4. In the circuit, the 110-V outlet acts as an electric pump and supplies an *equal* pressure to all the branches in parallel.

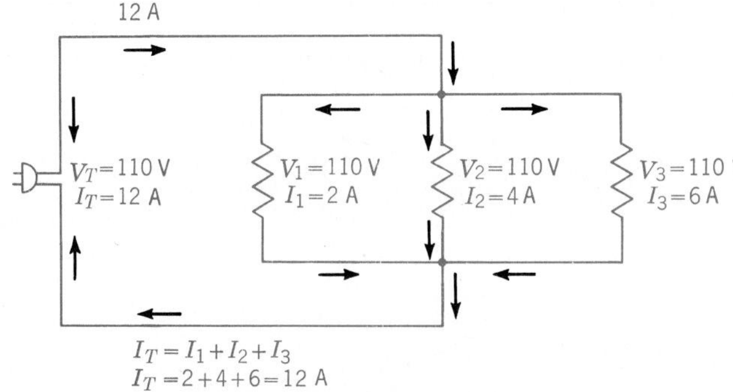


**FIGURE 5-4** Three resistors in parallel. The voltages across all the branches are equal.

**RULE** The voltage across any branch in parallel is equal to the voltage across any other branch and is also equal to the total voltage.

**FORMULA**

$$V_T = V_1 = V_2 = V_3 = \text{etc.} \tag{5-1}$$

**Total current in a parallel circuit.** In Fig. 5-3, we can see that the total number of gallons per minute flowing in the main pipe equals the number of gallons per minute discharged by all the pipes together, or 12 gal/min. In the same way, the total current that flows in the circuit of Fig. 5-4 is equal to 2 + 4 + 6 = 12 A.

**RULE** The total current in a parallel circuit is equal to the sum of the currents in all the branches of the circuit.

**FORMULA**

$$I_T = I_i + I_2 + I_3 + \text{etc.} \tag{5-2}$$

**Distribution of current in parallel.** The current in a parallel circuit

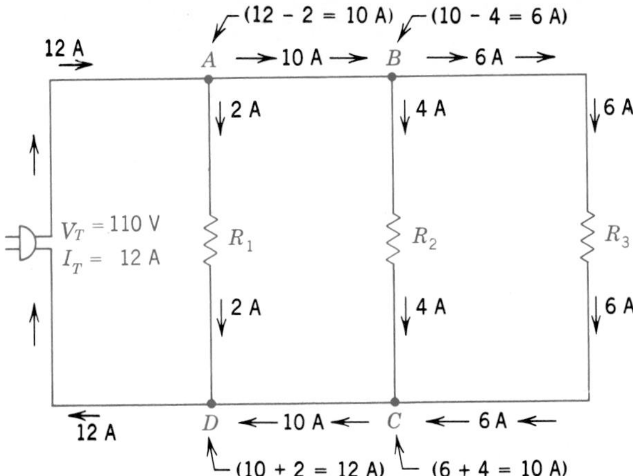


**FIGURE 5-5**
Distribution of current in a simple parallel circuit.

might be distributed as shown in Fig. 5-5. A total of 12 A is drawn from the line. At *A*, 2 A is drawn off through $R_1$, leaving only 10 A to flow along to point *B*. At *B*, 4 A is drawn off through $R_2$, leaving 6 A to flow through the rest of the circuit $R_3$ around to *C*. At *C*, this 6 A combines with 4 A from $R_2$ to give 10 A, which flows along to *D*. Here it combines with the 2 A from $R_1$ to make up the total of 12 A available originally.

**Line drop.** The distribution of the current as indicated above is based on the assumption that the connecting wires have no resistance, which is never true. In most instances, however, the resistance of the connecting wires is so small that we may neglect it completely. If the line wires are so long that their resistance is large, its effect must be included. This case will be taken up later in Job 6-4 under Line Drop.

## TOTAL RESISTANCE IN A PARALLEL CIRCUIT

**RULE** The total resistance in a parallel circuit is found by applying Ohm's law to the total values of the circuit.

**FORMULA** $$V_T = I_T \times R_T \qquad (4\text{-}7)$$

EXAMPLE 5-1. A toaster, a waffle iron, and a hot plate are connected in parallel across a house line delivering 110 V. The current through the toaster is 2 A, through the waffle iron 6 A, and through the hot plate 3 A. Find (*a*) the total current drawn from the line, (*b*) the voltage across each device, and (*c*) the total resistance of the circuit.

SOLUTION. The diagram for the circuit is shown in Fig. 5-6.

*a.* Find the total current $I_T$.

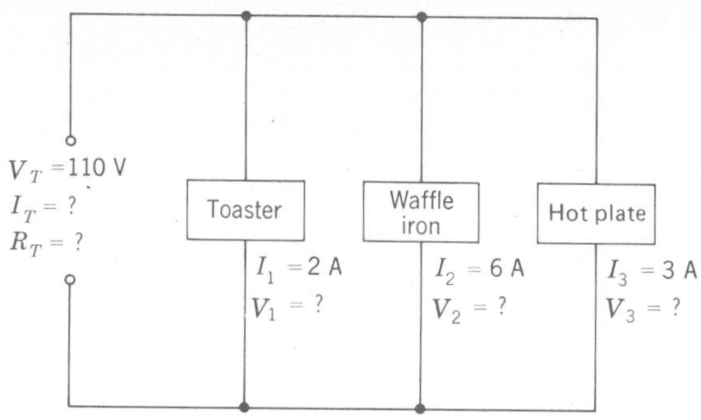


FIGURE 5-6

1. Write the formula. $I_T = I_1 + I_2 + I_3$ (5-2)
2. Substitute numbers. $I_T = 2 + 6 + 3 = 11$ A *Ans.*

*b.* Find the voltage across each device.

1. Write the formula. $V_T = V_1 = V_2 = V_3$ (5-1)
2. Substitute numbers. $V_T = V_1 = V_2 = V_3 = 110$ V *Ans.*

*c.* Find the total resistance $R_T$.

1. Write the formula. $V_T = I_T \times R_T$ (4-7)
2. Substitute numbers. $110 = 11 \times R_T$
3. Solve for $R_T$. $\frac{110}{11} = R_T$
4. Divide. $R_T = 10\ \Omega$ *Ans.*

SELF-TEST 5-2. An ammeter carrying 0.06 A is in parallel with a shunt resistor carrying 1.84 A, as shown in Fig. 5-7. If the voltage drop across the combination is 3.8 V, find (*a*) the total current and (*b*) the total resistance of the combination.

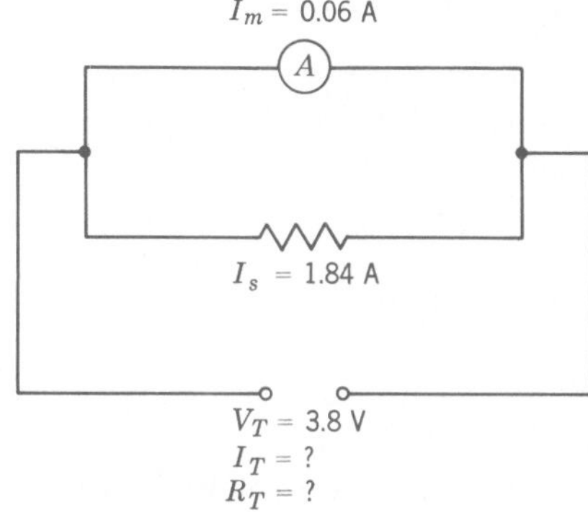


FIGURE 5-7
The shunt prevents large currents from damaging the meter.

SOLUTION

*a.* Find the total current $I_T$.

1. Write the formula. $I_T = I_m +$ ______
2. Substitute numbers. $I_T =$ ______ $+ 1.84 =$ ______ A *Ans.*

$I_s$

0.06 1.9

*b.* Find the total resistance $R_T$.

| | | | |
|---|---|---|---|
| 1. Write the formula. | $\_\_\_\_\_ = I_T \times R_T$ | (4-7) | $V_T$ |
| 2. Substitute numbers. | $3.8 = \_\_\_\_\_ \times R_T$ | | 1.9 |
| 3. Solve for $R_T$. | $\frac{3.8}{?} = R_T$ | | 1.9 |
| 4. Divide. | $R_T = \_\_\_\_\_ \Omega$ *Ans.* | | 2 |

## PROBLEMS

1. Two lamps each drawing 2 A and a third lamp drawing 1 A are connected in parallel across a 110-V line. Find (*a*) the total current drawn from the line, (*b*) the voltage across each lamp, and (*c*) the total resistance of the circuit.
2. The ignition coil and the starting motor of an automobile are connected in parallel across a 12-V battery through the ignition switch as shown in Fig. 5-8. Find (*a*) the total current drawn from the battery, (*b*) the voltage across the coil and the motor, and (*c*) the total resistance of the circuit.

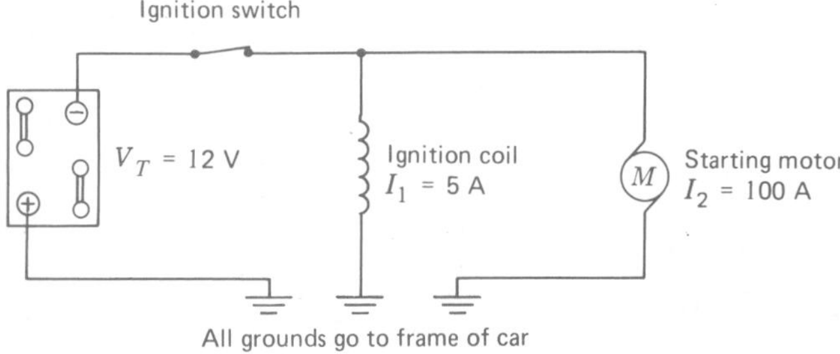


**FIGURE 5-8**
The ignition coil and the starting motor of an automobile are in parallel.

3. Find the total current drawn by eight trailer-truck warning lights in parallel if each takes 0.5 A.
4. A motor, a heating coil, and an ultraviolet lamp of a modern clothes drier are connected in parallel across 117 V. The lamp current is 1 A, the coil current is 4 A, and the motor current is 3 A. Find (*a*) the total current drawn and (*b*) the total resistance of the circuit.
5. In Table 13-1 the National Electrical Code specifies that No. 14 rubber-covered wire can safely carry only 20A. How many 0.5-A lamps could be safely operated at a time on a line using this wire? How many lamps could be safely operated at one time if a 5-A electric iron were connected in the circuit?
6. A toaster drawing 2 A, a coffee percolator drawing 3.5 A, and a refrigerator motor drawing 4.5 A are connected in parallel across a 110-V line. Find (*a*) the total current drawn, (*b*) the voltage across each device, and (*c*) the total resistance.
7. A bank of ten 110-V 100-W lamps is connected in parallel in a stage lighting circuit. Each lamp uses 0.91 A. Find the total current drawn. Will a fuse rated at 10 A safely carry this load?

8. A 40-W lamp drawing 0.36 A, a 60-W lamp drawing 0.54 A, and a 100-W lamp drawing 0.9 A are connected in parallel across 110 V. Find (*a*) the total current, (*b*) the voltage across each lamp, and (*c*) the total resistance of the circuit.
9. Two 32-cd headlight lamps each drawing 3.9 A and two taillight lamps (4 cd, 0.85 A each) are wired in parallel to the 12-V storage battery. Find the total current drawn and the total resistance of the circuit.
10. A washing machine drawing 7.5 A, an electric fan drawing 0.85 A, and an electric clock drawing 0.02 A are in parallel with a 110-V line. Find the total current and the total resistance.

## JOB 5-2 USING OHM'S LAW IN PARALLEL CIRCUITS

In some instances, it may be impossible to find the total current by adding the individual currents because the individual currents may be unknown. Therefore, the current in each branch must be found before we can find the total current.

EXAMPLE 5-3. The circuit of an electric clothes drier is shown in Fig. 5-9. The ultraviolet lamp $R_1$ is 120 Ω and draws 1 A. The heating coil $R_2$ is 30 Ω, and the motor $R_3$ draws a current of 4 A. Find (*a*) the voltage for each part and the total voltage, (*b*) the current in each part and the total current, and (*c*) the resistance of each part and the total resistance.

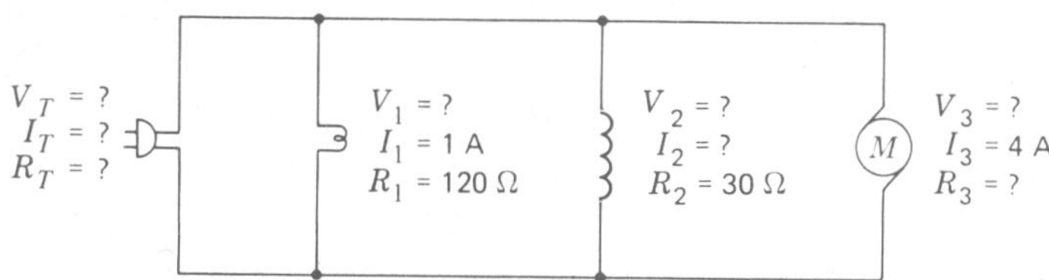


FIGURE 5-9

SOLUTION

*a.* Find $V_1$.

1. Write the formula. $V_1 = I_1 \times R_1$ (4-4)
2. Substitute numbers. $V_1 = 1 \times 120 = 120$ V *Ans.*

Find $V_T$ and the voltage across each part of the circuit.

1. Write the formula. $V_T = V_1 = V_2 = V_3$ (5-1)
2. Substitute numbers. $V_T = V_1 = V_2 = V_3 = 120$ V

*b.* Find the current $I_2$.

1. Write the formula. $V_2 = I_2 \times R_2$ (4-5)
2. Substitute numbers. $120 = I_2 \times 30$
3. Solve for $I_2$. $I_2 = \frac{120}{30} = 4$ A *Ans.*

Find the total current $I_T$.

1. Write the formula. $I_T = I_1 + I_2 + I_3$ (5-2)
2. Substitute numbers. $I_T = 1 + 4 + 4 = 9$ A *Ans.*

*c.* Find the resistance $R_3$.

1. Write the formula. $V_3 = I_3 \times R_3$ (4-6)
2. Substitute numbers. $120 = 4 \times R_3$
3. Solve for $R_3$. $R_3 = \frac{120}{4} = 30\ \Omega$ *Ans.*

Find the total resistance $R_T$.

1. Write the formula. $V_T = I_T \times R_T$ (4-7)
2. Substitute numbers. $120 = 9 \times R_T$
3. Solve for $R_T$. $R_T = \frac{120}{9} = 13.3\ \Omega$ *Ans.*

In the last example, $R_2$ and $R_3$ are both equal to 30 Ω. They both draw 4 A of current. This gives us the following rules.

**RULE** If the branches of a parallel circuit have the same resistance, then each will draw the same current.

**RULE** If the branches of a parallel circuit have different resistances, then each will draw a different current. The larger the resistance, the smaller the current drawn.

**PROBLEMS**

1. Two resistors of 3 and 6 Ω are connected in parallel across 18 V. Find (*a*) the voltage across each resistor, (*b*) the current in each resistor and the total current, and (*c*) the total resistance.
2. Find all missing values of voltage, current, and resistance in the circuit shown in Fig. 5-10.

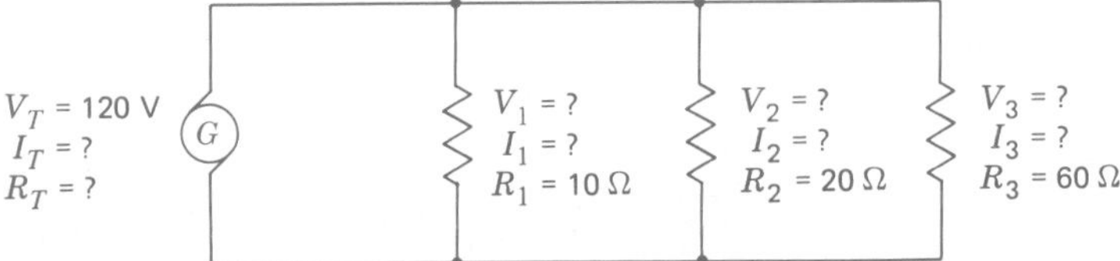


**FIGURE 5-10**

3. A parallel circuit has three branches of 12, 6, and 4 Ω resistance. If the current in the 6-Ω branch is 4 A, what current will flow in each of the other branches? What is the total current?
4. The secondary of a power transformer is connected across the motors of three toy trains in parallel. Motor 1 has a resistance of

50 Ω and draws 0.4 A. Motor 2 has a resistance of 40 Ω. Motor 3 draws a current of 0.3 A. Find (*a*) the total voltage, (*b*) the total current, and (*c*) the total resistance.

5. Three resistors are in parallel. $I_1 = 12$ A, $V_2 = 114$ V, $R_2 = 19\ \Omega$, and $R_3 = 57\ \Omega$. Find (*a*) the voltage across each resistor and the total voltage, (*b*) the current in each resistor and the total current, and (*c*) the resistance of each resistor and the total resistance.
6. Three buzzers are wired in parallel. They draw currents of 0.2, 0.4, and 0.6 A, respectively. If their total resistance is 6 Ω, find (*a*) the resistance of each buzzer and (*b*) the voltage across each buzzer.
7. Four $1\frac{1}{2}$-V lamps are wired in parallel. Three of these lamps draw a current of 0.05 A each. The fourth draws a current of 0.1 A. (*a*) What is the resistance of each filament? (*b*) What is the total current drawn? (*c*) What is the total voltage required? (*d*) What is the total resistance of the four filaments in parallel?
8. Solve the circuit shown in Fig. 5-11 for the values indicated.

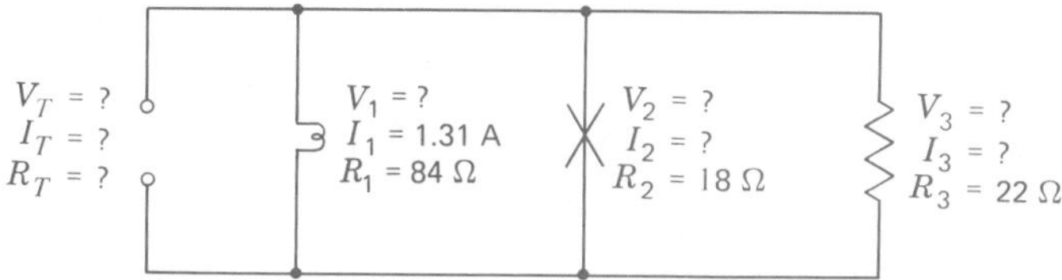


FIGURE 5-11

9. The following GE panel lights are to be wired in parallel from a 6.8-V source: two No. 40 lamps (0.15 A) and three No. 44 lamps (0.25 A). Find the total current, total voltage, and total resistance of the circuit.
10. What is the total current drawn from the 12-V automobile battery by two 4-Ω headlights and a 12-Ω taillight if they are all connected in parallel?

(See Instructor's Manual for Test 5-1)

## JOB 5-3 CHECKUP ON ADDITION AND SUBTRACTION OF FRACTIONS (DIAGNOSTIC TEST)

Our next electrical job considers problems which are solved by the addition of fractions. In order to prepare for them, let us try the following problems which are often met by electricians and electronic technicians in their work. If you have any difficulty with any of these, see Job 5-4 which follows.

### PROBLEMS

1. What is the total horsepower delivered by a $\frac{1}{3}$-, a $\frac{1}{4}$-, and a $\frac{1}{8}$-hp motor in parallel?
2. Add the following conductances in a circuit to obtain the total conductance: $\frac{1}{60}$, $\frac{1}{20}$, and $\frac{1}{10}$ S (siemens).

3. In rewinding the armature of a motor, the following thicknesses of insulation were used: $^1/_{16}$, $^1/_{32}$, and $^3/_{64}$ in. What was the total thickness of the insulation?
4. What length of a bolt is covered by a lock washer ($^1/_{32}$ in), a washer ($^1/_8$ in), and a nut ($^5/_{16}$ in)?
5. What is the total weight of three coils of wire weighing $16^1/_4$, $4^1/_8$, and $2^1/_2$ lb?
6. Add $^1/_5$, $^1/_3$, and $^1/_{10}$.
7. Add $^3/_4$, $^7/_8$, and $^3/_{16}$.
8. Add $^3/_4$, $^5/_6$, and $^7/_8$.
9. Add $^5/_6$, $^1/_9$, and $^2/_3$.
10. Add $^1/_2$, $^5/_7$, and $^3/_4$.
11. Subtract $1^1/_2$ from $3^5/_8$.
12. What is the difference between $9^3/_8$ and $4^7/_{16}$?
13. Subtract: $8 - 3^3/_4$.
14. Subtract: $5^1/_4 - 2^9/_{16}$.
15. Subtract: $^1/_{24} - ^1/_{40}$.

## JOB 5-4 BRUSHUP ON ADDITION OF FRACTIONS

### ADDING FRACTIONS WITH THE SAME DENOMINATOR

EXAMPLE 5-4

1 apple + 3 apples + 2 apples = ______ apples. | 6

1 Ω + 3 Ω + 2 Ω = ______ Ω. | 6

1 A + 3 A + 2 A = 6 ______. | A

1 V + 3 V + 2 V = 6 ______. | V

Apparently, in order to add the same *kind* of thing, it is only necessary to add the numbers involved. Thus

One-*eighth* + three-*eighths* + two-*eighths* = six-______ | eighths

or

$$\frac{1}{8} + \frac{3}{8} + \frac{2}{8} = \_\_\_\_ \qquad \Big| \; \frac{6}{8}$$

And $^6/_8$ may be simplified to $^3/_4$. *Ans.*

**RULE** To add fractions with the same denominator, add the numerators and place the sum over the same denominator.

EXAMPLE 5-5. Add $^3/_{16}$, $^5/_{16}$, and $^7/_{16}$.

SOLUTION

$$\frac{3}{16} + \frac{5}{16} + \frac{7}{16} = \frac{3+5+7}{16} = \frac{15}{?} \qquad \Big| \; 16$$

## ADDING FRACTIONS WITH DIFFERENT DENOMINATORS

### Procedure

1. Find the *least common denominator.* (See Example 5-8 below.)
2. Change the fractions to *equivalent* fractions using this new denominator.
3. Add these fractions with the same denominator.

EXAMPLE 5-6. Add $^1/_4 + {}^3/_8$.

SOLUTION. The least common denominator (LCD) is a number *into which* all the denominators will evenly divide. Both the 4 and the 8 will divide into 8 evenly. Therefore, 8 is the least common denominator. In Fig. 5-12, since $^1/_4 = {}^2/_8$, we have

$$\frac{2}{8} + \frac{3}{8} = \frac{5}{8} \qquad \textit{Ans.}$$

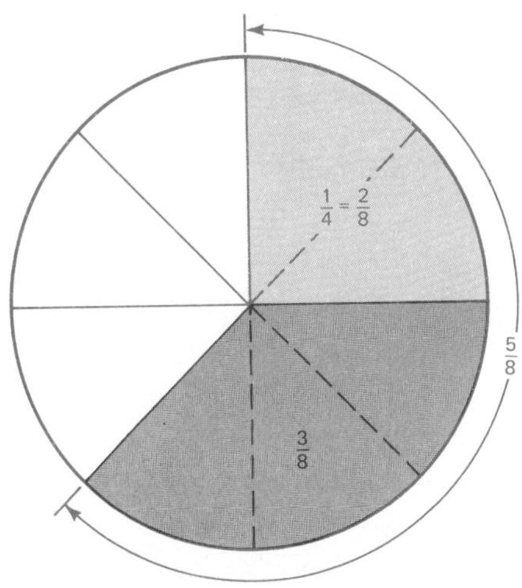


FIGURE 5-12

In this problem, it was very easy to change all the fractions into equivalent fractions with the same denominator. Not all problems are so simple. One of the difficulties will be to decide on what the new denominator will be.

## HOW TO FIND THE LEAST COMMON DENOMINATOR

EXAMPLE 5-7. Find the LCD for $^1/_2 + {}^1/_3 + {}^1/_8$.

SOLUTION

1. Start with the largest denominator—the 8.
2. Try to divide the other denominators into the 8. They must divide exactly. If not,

3. Multiply the 8 by 2 to get 16.

4. Try to divide the other denominators into the 16. If they all divide evenly, then 16 is the LCD. If not, multiply the 8 by 3 and then 4, etc., until you find a number into which all the denominators will evenly divide. This last number will be the LCD.

5. Following this method, we multiply the 8 by 2 and get 16. The 2 and the 8 will divide evenly into the 16, but the 3 will not. We multiply the 8 by 3 and get 24. The 2, the 3, and the 8 all divide into 24 evenly. Therefore,

24 is the LCD *Ans.*

EXAMPLE 5-8. Add: $^1/_7 + {}^1/_2 + {}^1/_4$.

SOLUTION

1. Find the LCD. Start with the 7. This is not the LCD because the 2 and the 4 do not divide evenly into 7. Multiply the 7 by 2 to get 14. This is not the LCD because the 4 does not divide evenly into 14. Multiply the 7 by 3 to get 21. This is not the LCD because neither the 2 nor the 4 will divide evenly into 21. Multiply the 7 by 4 to get 28. This *is* the LCD because the numbers 2, 4, and 7 *all divide evenly into* 28.

2. Set up the problem like this:

| | 28 (LCD) |
|---|---|
| $^1/_7$ | |
| $^1/_2$ | |
| $^1/_4$ | |

3. Change each fraction into an equivalent fraction whose denominator is 28. We can do this by multiplying the fraction by 1, written as a fraction with identical numerator and denominator. This will not change the value of the fraction, but will merely express it as an equivalent fraction with a different denominator. We get the multiplier by dividing the denominator into the LCD (28).

*a.* For the $^1/_7$: 7 into 28 is 4. Therefore,

$$\frac{1}{7} \times \frac{4}{4} = \frac{4}{28}$$

*b.* For the $^1/_2$: 2 into 28 is 14. Therefore,

$$\frac{1}{2} \times \frac{14}{14} = \frac{14}{28}$$

*c.* For the $^1/_4$: 4 into 28 is 7. Therefore,

$$\frac{1}{4} \times \frac{7}{7} = \frac{7}{28}$$

4. The problem now looks as shown on page 126. Add the numerators of the equivalent fractions and place the result over the LCD. If necessary, reduce the fraction to lowest terms.

$$\begin{array}{r|l} & 28 \text{ (LCD)} \\ \hline \frac{1}{7} & \frac{4}{28} \\ \frac{1}{2} & \frac{14}{28} \\ +\ \frac{1}{4} & \frac{7}{28} \\ \hline & \frac{25}{28} \quad \textit{Ans.} \end{array}$$

SELF-TEST 5-9. What length of cable is needed to install the lighting outlets shown in Fig. 5-13?

SOLUTION. The total length required is equal to the ______ of all the lengths. The total length = $9^3/_4 + 8^1/_2 + 14^5/_8 + 6^5/_{12}$. | sum or addition

1. Find the LCD. Start with the number ______. This is not the LCD because the denominator ______ will not divide evenly into it. Multiply the | 12 | 8

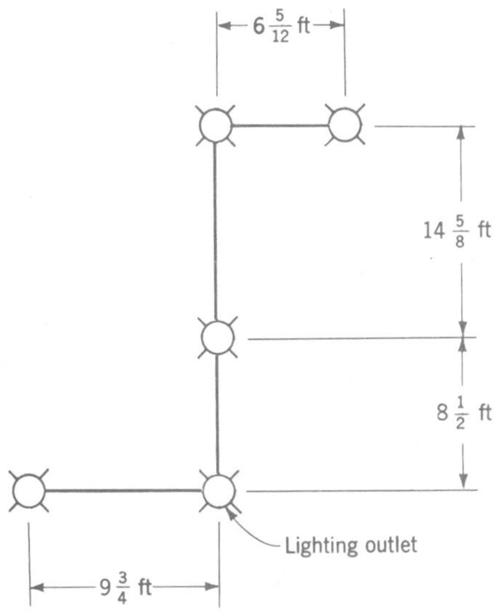


FIGURE 5-13

12 by ______ to get 24. This number *is* the LCD because ______ the denominators divide ______ into 24. | 2 all | evenly

2. Set up the problem like this:

$$\begin{array}{r|l} & 24 \text{ (LCD)} \\ \hline 9^3/_4 & \\ 8^1/_2 & \\ 14^5/_8 & \\ 6^5/_{12} & \\ \hline \end{array}$$

3. Change each fraction to an equivalent fraction whose denominator is 24. Multiply each fraction by the proper common multiplier.

*a.* For the $^3/_4$: 4 into 24 is 6. Therefore,

$$\frac{3}{4} \times \frac{6}{6} = \frac{?}{?}$$ $\frac{18}{24}$

*b.* For the $^1/_2$: 2 into 24 is 12. Therefore,

$$\frac{1}{2} \times \frac{12}{12} = \frac{?}{?}$$ $\frac{12}{24}$

*c.* For the $^5/_8$: 8 into 24 is 3. Therefore,

$$\frac{5}{8} \times \frac{3}{3} = \frac{?}{?}$$ $\frac{15}{24}$

*d.* For the $^5/_{12}$: 12 into 24 is 2. Therefore,

$$\frac{5}{12} \times \frac{2}{2} = \frac{?}{?}$$ $\frac{10}{24}$

4. The problem now looks as shown below. Add the equivalent fractions and place the sum over the ________. Then add the whole numbers. LCD

| | 24 |
|---|---|
| $9\frac{3}{4}$ | $\frac{18}{24}$ |
| $8\frac{1}{2}$ | $\frac{12}{24}$ |
| $14\frac{5}{8}$ | $\frac{15}{24}$ |
| $6\frac{5}{12}$ | $\frac{10}{24}$ |
| 37 | $\frac{?}{24}$ |

55

5. Change $^{55}/_{24}$ to a mixed number by ________ 24 into 55. This will give 2 and ________. Add the whole number 2 to the 37 to give the final answer which is ________$^7/_{24}$ ft. *Ans.*

dividing
$^7/_{24}$
39

## PROBLEMS

Add the following fractions and express in lowest terms.

1. $^1/_8 + {}^3/_8 + {}^7/_8$
2. $^3/_4 + {}^5/_6 + {}^9/_{16}$
3. $^1/_4 + {}^5/_6 + {}^3/_8$
4. $^5/_6 + {}^1/_9 + {}^2/_3$
5. $^1/_{60} + {}^1/_{20} + {}^1/_{10}$
6. $^1/_3 + {}^1/_6 + {}^1/_9$
7. $^1/_4 + {}^1/_{12} + {}^1/_{15}$
8. $^1/_5 + {}^1/_7 + {}^1/_9$
9. $^1/_{200} + {}^1/_{300}$
10. $^1/_{50} + {}^1/_{100}$
11. $^3/_{16} + {}^5/_{64} + {}^5/_8$
12. $^5/_{32} + {}^3/_8 + {}^9/_{16}$
13. $^1/_{10} + {}^1/_{30} + {}^1/_{60}$
14. $^1/_{500} + {}^1/_{200} + {}^1/_{400}$
15. $^3/_4 + {}^9/_{32} + {}^5/_{16}$
16. $1^3/_8 + 3^1/_2 + 2^3/_{16}$
17. $1^3/_4 + 8^5/_{16} + 5^3/_8$
18. $1^5/_8 + 3^2/_3$
19. A carbon brush $^{29}/_{32}$ in thick is coated with copper to a thickness of $^1/_{64}$ in on each side. Find the total thickness.

20. What is the total thickness of insulation made by $^1/_{64}$-in fish paper, $^1/_{32}$-in tufflex, $^3/_{64}$-in varnished cambric, and $^1/_8$-in top stick?
21. In Fig. 5-14, the reciprocal of the resistance is known as the conductance, whose symbol is $G$. If the total conductance of a parallel circuit is the sum of the conductances, find the total conductance $G_T$, expressed in siemens (S).

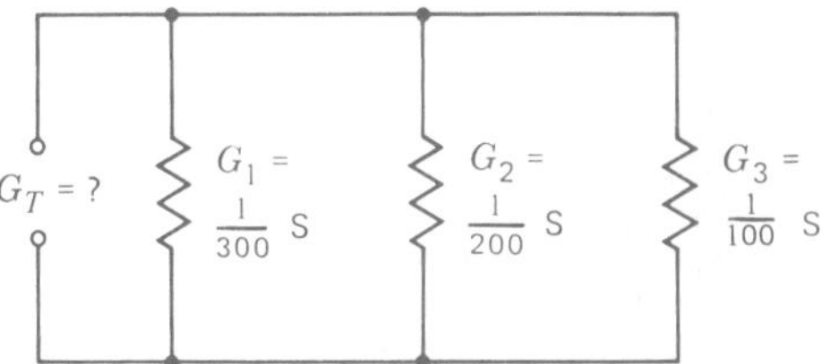

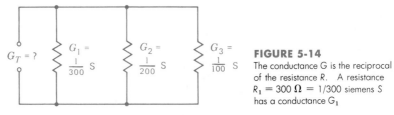

**FIGURE 5-14** The conductance G is the reciprocal of the resistance $R$. A resistance $R_1 = 300\ \Omega = 1/300$ siemens S has a conductance $G_1$

22. Find the total resistance of a series circuit containing a $2^1/_2$-, a $1^3/_{10}$-, and a $^9/_{100}$-MΩ resistor.
23. In a layout similar to that shown in Fig. 5-13, find the total length of cable needed if the measurements between outlets are: $6^1/_4$, $8^7/_{12}$, $10^5/_8$, and $4^5/_{10}$ ft.

(See Instructor's Manual for Test 5-2)

## JOB 5-5 BRUSHUP ON SUBTRACTION OF FRACTIONS

### SUBTRACTING FRACTIONS WITH THE SAME DENOMINATOR

**RULE** To subtract fractions with the same denominator, subtract the numerators and place the difference over the same denominator.

EXAMPLE 5-10. Subtract $^3/_8$ from $^5/_8$.

SOLUTION

$$\frac{5}{8} - \frac{3}{8} = \frac{5-3}{8} = \frac{2}{8} = \frac{1}{4} \quad \textit{Ans.}$$

### SUBTRACTING FRACTIONS WITH DIFFERENT DENOMINATORS

#### Procedure

1. Find the *least common denominator.*
2. Change the fractions to *equivalent* fractions using this new denominator.
3. Subtract these fractions with the same denominator.

EXAMPLE 5-11. Subtract $^3/_{16}$ from $^3/_4$.

SOLUTION

1. Find the LCD, which is 16.
2. Set up the problem as shown below and change the fractions to equivalent fractions as described in Examples 5-8 and 5-9.

| | 16 |
|---|---|
| $^3/_4$ | $^{12}/_{16}$ |
| $-^3/_{16}$ | $^3/_{16}$ |

3. Subtract these fractions with the same denominator and reduce to lowest terms.

$$\frac{12}{16} - \frac{3}{16} = \frac{9}{16} \quad \textit{Ans.}$$

EXAMPLE 5-12. Subtract $3^1/_4$ from 9.

SOLUTION

1. Set up the problem as shown below.

$$\begin{array}{r} 9 \\ -3\ ^1/_4 \\ \hline \end{array}$$

Since $^1/_4$ cannot be subtracted from nothing, *borrow* one unit from the 9, leaving 8, and replace this unit as the fraction $^4/_4$. The unit that is borrowed is always replaced as a fraction with identical numerator and denominator, each being equal to the denominator of the fraction being subtracted.

$$\begin{array}{r} 9 \\ -3\ ^1/_4 \\ \hline \end{array} = \begin{array}{r} 8 \\ \not{9}\ ^4/_4 \\ -3\ ^1/_4 \\ \hline \end{array}$$

2. Subtract the fractions ($^4/_4 - ^1/_4 = ^3/_4$). Subtract the whole numbers ($8 - 3 = 5$).
3. Write the answer.

$5^3/_4$ *Ans.*

EXAMPLE 5-13. Subtract $2^5/_8$ from $8^3/_{16}$.

SOLUTION

1. Set up the problem as shown below.

$$\begin{array}{r} 8\ ^3/_{16} \\ -2\ ^5/_8 \\ \hline \end{array}$$

2. Find the LCD, which is 16.
3. Change the fractions to equivalent fractions using this new denominator.

$$\begin{array}{r} 8\ ^3/_{16} \\ -2\ ^5/_8 \\ \hline \end{array} = \begin{array}{r|r} & 16 \\ 8 & ^3/_{16} \\ -2 & ^{10}/_{16} \\ \hline \end{array}$$

4. As you can see, $^{10}/_{16}$ cannot be subtracted from only $^3/_{16}$. Therefore, we

shall *borrow* one unit from the 8 as in Example 5-12, and replace it as $^{16}/_{16}$ as shown below.

$$\begin{array}{r} 8\ ^{3}/_{16} \\ -2\ ^{5}/_{8} \\ \hline \end{array} = \begin{array}{r|l} \overset{7}{\cancel{8}} & \overset{16}{^{3}/_{16} + ^{16}/_{16}} \\ -2 & ^{10}/_{16} \\ \hline \end{array}$$

5. Combine the $^{3}/_{16} + ^{16}/_{16}$ into $^{19}/_{16}$ and subtract the $^{10}/_{16}$ from the $^{19}/_{16}$ as shown below.

$$\begin{array}{r|l} 7 & \overset{16}{^{3}/_{16} + ^{16}/_{16} = ^{19}/_{16}} \\ -2 & ^{10}/_{16} \\ \hline 5 & ^{9}/_{16} \quad \textit{Ans.} \end{array}$$

SELF-TEST 5-14. What resistance must be placed in series with a $12^{3}/_{4}$-Ω bell in order to provide a total resistance of $84^{3}/_{5}$ Ω?

SOLUTION. The required resistance may be found by ______ the $12^{3}/_{4}$ Ω from the total of $84^{3}/_{5}$ Ω. The number after the word "from" is written on the (top/bottom) of the subtraction problem. The LCD for the denominators 4 and 5 is ______. | subtracting, top, 20

1. Set up the problem as shown below.

$$\begin{array}{r|l} & 20 \\ 84\ ^{3}/_{5} & \\ -12\ ^{3}/_{4} & \\ \hline \end{array}$$

2. Change the fractions to equivalent fractions using ______ as the new denominator. | 20

$$\begin{array}{r} 84\ ^{3}/_{5} \\ -12\ ^{3}/_{4} \\ \hline \end{array} = \begin{array}{r|l} & 20 \\ 84\ ^{3}/_{5} & = \_\_\_\_ \\ -12\ ^{3}/_{4} & = \_\_\_\_ \\ \hline \end{array}$$

$^{12}/_{20}$

$^{15}/_{20}$

3. As you can see, $^{15}/_{20}$ cannot be subtracted from only $^{12}/_{20}$. Therefore, we must ______ one unit from the 84 and replace it as ______ as shown below. | borrow $^{20}/_{20}$

$$\begin{array}{r} 84\frac{3}{5} \\ -12\frac{3}{4} \\ \hline \end{array} = \begin{array}{r|l} \overset{3}{8\cancel{4}}\frac{3}{5} & \overset{20}{\frac{12}{20} + \frac{?}{20}} \\ -12\frac{3}{4} & \frac{15}{20} \\ \hline \end{array}$$

20

4. Combine the $^{12}/_{20}$ and the $^{20}/_{20}$ into ______ and subtract the $^{15}/_{20}$ from it. Complete the problem as shown below. | $^{32}/_{20}$

$$\begin{array}{r|l} & 20 \\ 83 & \frac{12}{20} + \frac{20}{20} = \frac{32}{20} \\ -12 & \frac{15}{20} \\ \hline 71 & \frac{?}{20}\ \Omega \quad \textit{Ans.} \end{array}$$

17

## PROBLEMS

Subtract the following fractions and express in lowest terms.

| | | |
|---|---|---|
| 1. $3/4 - 1/4$ | 2. $7/8 - 3/8$ | 3. $11/16 - 5/16$ |
| 4. $5\,5/8 - 1\,3/8$ | 5. $9/16 - 1/4$ | 6. $5/6 - 1/3$ |
| 7. $3/4 - 2/3$ | 8. $5\,5/6 - 2\,2/3$ | 9. $1/2 - 1/3$ |
| 10. $1/50 - 1/100$ | 11. $1/10 - 1/60$ | 12. $1/90 - 1/100$ |
| 13. $1/40 - 1/80$ | 14. $1/12 - 1/60$ | 15. $1/250 - 1/500$ |
| 16. $1/480 - 1/2400$ | 17. $7 - 2\,1/4$ | 18. $13 - 5\,3/8$ |
| 19. $8 - 2/3$ | 20. $5\,1/4 - 2\,5/8$ | 21. $8\,3/16 - 2\,3/4$ |
| 22. $5\,2/3 - 3\,5/8$ | 23. $1/200 - 1/600$ | 24. $1/1000 - 1/2500$ |

25. How many feet of push-back wire are left from a 100-ft roll if a radio mechanic used $23\,1/4$ ft?
26. An electrician had a $16\,1/4$-ft-long piece of conduit. If she used $11\,5/12$ ft, how many feet were left?
27. The length of threading on a BX box connector is $7/16$ in. If $7/32$ in of thread is covered with washers, lock nuts, and metal, how much thread is left for attaching a threaded bushing?
28. Find the inside diameter of an electric conduit shown in Fig. 5-15 if its outside diameter is $2\,1/4$ in and the thickness of the conduit is $3/16$ in.

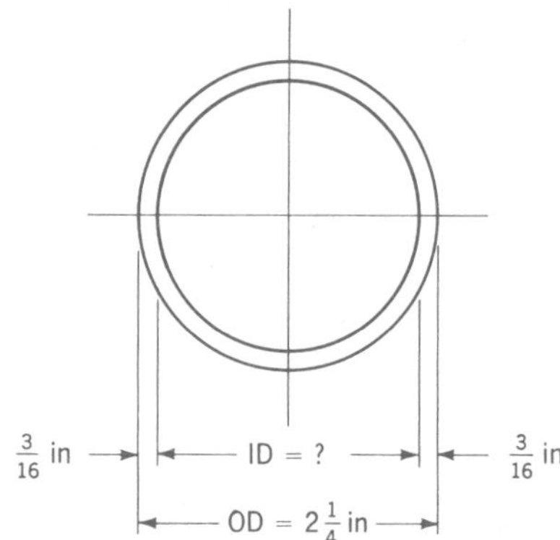


**FIGURE 5-15**
Cross section of an electric conduit.

29. Kirchhoff's law states that the current that enters a point in a circuit is equal to the sum of the currents that leave the point. In the circuit shown in Fig. 5-16, find the current $I_2$.

(See Instructor's Manual for Test 5-3)

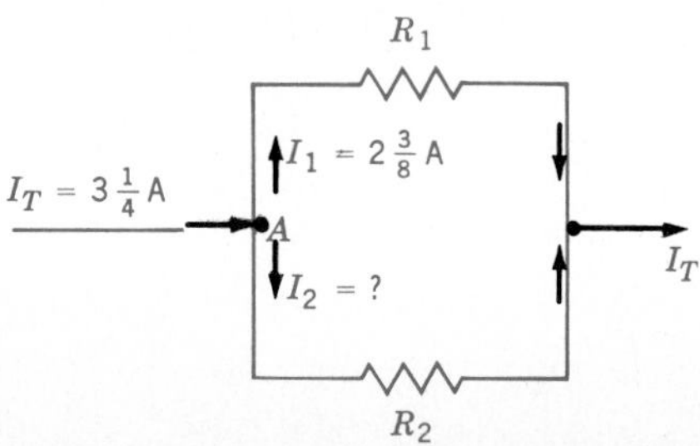


**FIGURE 5-16**
The current that enters point *A* is equal to the sum of the currents that leave the point.

## JOB 5-6 CHECKUP ON SOLUTION OF FORMULAS INVOLVING FRACTIONS (DIAGNOSTIC TEST)

In addition to adding fractions, our next electrical job will require that we be able to solve formulas which contain fractions. Following are some problems of this type. If you have any difficulty solving any of them, see Job 5-7 which follows.

### PROBLEMS

1. Using the formula $f = PS/120$, find the speed $S$ of an alternator if the frequency $f$ is 60 Hz and the number of poles $P$ is 4.
2. $R_1/R_2 = L_1/L_2$ is a formula used to compare the resistances of different lengths of wire. Find $R_1$ if $R_2 = 100$, $L_1 = 500$, $L_2 = 800$.
3. $V_1/V_2 = I_2/I_1$ is the formula for the relation between the voltages and the currents in the primary and secondary windings of a transformer. Find $I_1$ if $V_1 = 110$ V, $V_2 = 22$ V, and $I_2 = 10$ A.
4. The fundamental equation for the Wheatstone bridge (a resistance-measuring device) is $R_1/R_2 = R_3/R_x$. Find $R_x$ if $R_1 = 1000$, $R_2 = 10{,}000$, and $R_3 = 84.3$.
5. In series circuits, the voltage drops are proportional to the resistances. This is stated mathematically as $V_1/V_2 = R_1/R_2$. Find $V_1$ if $V_2 = 8$ V, $R_1 = 3\ \Omega$, and $R_2 = 4\ \Omega$.

## JOB 5-7 BRUSHUP ON SOLUTION OF FRACTIONAL EQUATIONS

**Solving simple fractional equations.** Many electrical formulas are stated in the form of a fraction. After substituting the given numbers for the letters of the formula, we might get an equation like $V/4 = 5$. To find the value of $V$, we must get the letter $V$ all by itself on one side of the equality sign. In other words, we must eliminate the number 4. We can do this by applying the general rule (Job 3-8) which says that we may do anything to one side of an equality sign provided we do the same thing to the other side. In general, the operation to be performed on both sides of the equality sign will be the *opposite* of that used in the equation. Since our equation involves division, the opposite operation of multiplication should be performed on both sides of the equality sign.

**RULE** Both sides of an equality sign may be multiplied by the same number without destroying the equality.

SELF-TEST 5-15. Find the value of $V$ in the equation

$$\frac{V}{4} = 5$$

SOLUTION. Solving an equation means to get the letter all alone on one side of the equality sign. We can get $V$ alone on the left side of the equality sign if we can get rid of the denominator 4. Since the equation says that $V$ is *divided* by 4, we can eliminate the denominator 4 by doing the opposite operation of ______ both sides of the equality sign by that same number 4. | multiplying

1. Write the equation. $\frac{V}{4} = 5$

2. Multiply both sides by 4. $\overset{1}{\not{4}} \times \frac{V}{\underset{1}{\not{4}}} = 5 \times$ ______ | 4

3. Multiply each side separately. $1 \times V =$ ______ | 20

or $V =$ ______ *Ans.* | 20

SELF-TEST 5-16. Using the formula $I = \frac{V}{R}$, find $V$ if $I = 2$ and $R = 6$.

SOLUTION

1. Write the formula. $I = \frac{V}{R}$

2. Substitute numbers. $2 = \frac{V}{?}$ | 6

Step 2 asks the question, "What number divided by 6 gives 2 as an answer?" We can find $E$ if we can eliminate the denominator 6. This may easily be done by ______ both sides of the equality sign by ______. | multiplying 6

3. Multiply both sides by 6. $6 \times 2 = \frac{V}{\underset{1}{\not{6}}} \times \overset{1}{\not{6}}$

4. Multiply both sides separately. ______ $= V \times 1$ | 12

or $V =$ ______ *Ans.* | 12

Notice that in each of the last three examples we eliminated the number in the denominator by *multiplying both sides by that same number.* If we were to multiply both sides by any other number, we would still be left with a number in the denominator and the letter would not stand alone. Therefore, to eliminate a number in the denominator of a fraction, we use the following rule:

**RULE** To eliminate a number divided into a letter, multiply both sides of the equality sign by that same number.

**PROBLEMS**

Find the value of the unknown letter in each problem.

1. $\frac{V}{3} = 8$ 2. $4 = \frac{V}{10}$ 3. $\frac{P}{6} = 2$

4. $0.5 = \frac{V}{3}$ 5. $\frac{M}{0.2} = 5$ 6. $\frac{A}{10} = 0.35$

7. $\frac{V}{3} = \frac{2}{3}$ 8. $0.4 = \frac{M}{0.8}$ 9. $\frac{P}{1/2} = 16$

10. $\frac{V}{4} = 2^1/_2$ 11. $150 = \frac{x}{45}$ 12. $117 = \frac{P}{4.7}$

13. In the formula $I = V/R$, find $V$ if $I = 3$ and $R = 18$.
14. In the formula $V = P/I$, find $P$ if $V = 110$ and $I = 5$.
15. In the formula $Q = X_L/R$, find $X_L$ if $Q = 100$ and $R = 40$.
16. In the formula Eff $= O/I$, find $O$ if $I = 36$ and Eff $= 0.85$.
17. In the formula $Z = V/I$, find $V$ if $Z = 2000$ and $I = 0.015$.
18. In the formula $\cos \theta = I_R/I_T$, find $I_R$ if $I_T = 10$ and $\cos \theta = 0.866$.
19. In the formula $\sin \theta = i/I_{max}$, find $i$ if $I_{max} = 100$ and $\sin \theta = 0.9397$.
20. In the formula $R_1 + R_2 = V_T/I_T$, find $V_T$ if $R_1 = 25$, $R_2 = 85$, and $I_T = 2$.

## USING THE LEAST COMMON DENOMINATOR TO SOLVE FRACTIONAL EQUATIONS

EXAMPLE 5-17. Find $V$ in the equation

$$\frac{2V}{9} = \frac{4}{3}$$

SOLUTION. When fractions appear on both sides of the equality sign, the solution will be simplified if we eliminate *all* denominators first. We could do this one denominator at a time by the method used in Examples 5-15 and 5-16, but it is faster if we eliminate both denominators at the same time. In order to eliminate *both* the 9 and the 3 at the same time, we must multiply both sides of the equality sign by some number so that *both* denominators will be canceled out. This must be a number into which *both* denominators will evenly divide—which is the *least common denominator*. In this problem, the LCD of 9 and 3 is 9.

1. Write the equation. $\frac{2V}{9} = \frac{4}{3}$

2. Multiply both sides by the LCD (9). $\overset{1}{\cancel{9}} \times \frac{2V}{\underset{1}{\cancel{9}}} = \frac{4}{\underset{1}{\cancel{3}}} \times \overset{3}{\cancel{9}}$

3. Multiply each side separately. $2V = 12$

4. Solve for $V$. $V = \frac{12}{2} = 6$ *Ans.*

EXAMPLE 5-18. Find $L$ in the equation

$$\frac{6}{L} = \frac{2}{3}$$

SOLUTION

1. Write the equation.

$$\frac{6}{L} = \frac{2}{3}$$

2. Multiply both sides by the LCD. In general, although it may not be the *least* common denominator, a workable common denominator will be the value obtained by multiplying all the denominators. Thus, in this problem, the LCD is $3L$.

$$\overset{1}{3\cancel{L}} \times \frac{6}{\underset{1}{\cancel{L}}} = \frac{2}{\underset{1}{\cancel{3}}} \times \overset{1}{\cancel{3}L}$$

Note that $L$ cancels $L$ on the left side, and 3 cancels 3 on the right side.

3. Multiply each side separately. $18 = 2L$

4. Solve for $L$. $L = \frac{18}{2} = 9$ *Ans.*

**RULE** When fractions appear on both sides of the equality sign, eliminate the denominators by multiplying both sides by the least common denominator.

SELF-TEST 5-19. $\frac{R}{K} = \frac{L}{A}$ is a formula that may be used to calculate the length of a wire of particular material and cross section that will provide a definite resistance. Find the length $L$ if $K = 60$ for pure iron wire, $A = 25$ cmil, and $R = 24\ \Omega$.

SOLUTION

1. Write the formula. $\frac{R}{K} = \frac{L}{A}$
2. Substitute numbers. $\frac{24}{60} = \frac{L}{25}$
3. The LCD for the numbers 60 and 25 is ______. | 300
4. Multiply both sides by ______. | 300

$$\overset{5}{\cancel{300}} \times \frac{24}{\underset{1}{\cancel{60}}} = \frac{L}{\underset{1}{\cancel{25}}} \times \overset{12}{\cancel{300}}$$

5. Multiply each side separately. $120 =$ ______ | $12L$
6. Solve for $L$. $\frac{120}{?} = L$ | 12
7. Divide. $L =$ ______ ft *Ans.* | 10

**PROBLEMS**

Solve each equation for the value of the unknown letter.

1. $\dfrac{R}{3} = \dfrac{2}{3}$
2. $\dfrac{V}{8} = \dfrac{3}{4}$
3. $\dfrac{2}{3} = \dfrac{P}{6}$
4. $\dfrac{2V}{3} = \dfrac{4}{9}$
5. $\dfrac{M}{5} = \dfrac{2}{3}$
6. $\dfrac{4}{5} = \dfrac{T}{2}$
7. $\dfrac{2V}{5} = \dfrac{3}{0.5}$
8. $\dfrac{2B}{5} = \dfrac{7}{10}$
9. $\dfrac{N}{20} = \dfrac{3}{4}$
10. $\dfrac{4}{7} = \dfrac{S}{2}$
11. $\dfrac{0.3}{V} = \dfrac{1}{4}$
12. $\dfrac{3}{5} = \dfrac{2}{R}$
13. $\dfrac{2}{L} = \dfrac{6}{30}$
14. $\dfrac{10}{C} = \dfrac{150}{200}$
15. $\dfrac{22}{7} = \dfrac{11}{R}$
16. $\dfrac{100}{250} = \dfrac{0.1}{R}$
17. $\dfrac{N}{0.4} = \dfrac{3.9}{0.2}$
18. $\dfrac{9}{L} = \dfrac{0.3}{8.9}$
19. $\dfrac{21}{0.3} = \dfrac{2V}{9}$
20. $\dfrac{37}{250} = \dfrac{74}{R}$
21. $N_1/N_2 = V_1/V_2$ is a formula used in transformer calculations. Find $N_2$ if $N_1 = 40$ turns, $V_1 = 6$ V, and $V_2 = 18$ V.
22. Using the formula of Prob. 21, find $V_1$ if $N_1 = 30$ turns, $N_2 = 70$ turns, and $V_2 = 21$ V.
23. $A_2/A_1 = R_1/R_2$ is a formula used to calculate the sizes of wires in electrical installations. Find $A_2$ if $A_1 = 100$, $R_1 = 1000$, $R_2 = 3000$.
24. $I_1/I_2 = R_2/R_1$ is a formula used for calculating the way the current divides in a parallel circuit. Find $I_1$ if $I_2 = 2$ A, $R_2 = 100\ \Omega$, and $R_1 = 25\ \Omega$.
25. Using the same formula as in Prob. 24, find $R_2$ if $I_1 = 3$ A, $I_2 = 5$ A, and $R_1 = 100\ \Omega$.
26. $R_1/R_2 = R_3/R_x$ is the formula used for calculations in the Wheatstone-bridge method for measuring resistance. Find $R_x$ if $R_1 = 1000$, $R_3 = 26.9$, and $R_2 = 10{,}000$.
27. $E/R = A_1/A_2$ is a formula used to determine the effort required to raise a large weight in a hydraulic lift. Find the effort $E$ required to raise a weight $R$ of 200,000 lb if the area of the small piston $A_1$ is 4 in$^2$ and the area of the large piston $A_2$ is 1600 in$^2$.
28. $R_1/R_2 = L_2/L_1$ is a formula used to locate the position of a break in

an underground cable. Find the length to the break in the cable $L_2$ if the length of the unbroken cable $L_1$ is 4000 ft, $R_1$ is 10 Ω, and $R_2$ is 250 Ω.

29. $R_s/R_m = I_m/I_s$ is the formula for finding the shunt resistor needed to extend the range of an ammeter. Find $R_s$ if $I_m = 0.001$, $R_m = 50$, and $I_s = 0.049$.
30. $R_1/R_2 = C_x/C_1$ is a formula used to measure the capacitance of an unknown capacitor. Find $C_x$ if $R_1 = 100\ \Omega$, $R_2 = 425\ \Omega$, and $C_1 = 0.5\ \mu F$.

(See Instructor's Manual for Test 5-4)

**Cross multiplication.** When a fractional equation contains *only two fractions* equal to each other, the equation may be solved by a very simple method known as cross multiplication. This method automatically multiplies both sides of the equality sign by the LCD and therefore eliminates one step in the solution. This method is most useful when the unknown letter is in the denominator of one of the fractions.

**RULE** To cross-multiply, the product of the numerator of the first fraction and the denominator of the second is set equal to the product of the numerator of the second fraction and the denominator of the first.

It is easier to understand this rule by putting it in picture form. The multiplication of the numbers along one diagonal line is equal to the multiplication of the numbers along the other diagonal line.

$$\frac{2}{3} = \frac{4}{6} \qquad\qquad \frac{A}{B} = \frac{C}{D}$$

$$2 \times 6 = 4 \times 3 \qquad\qquad A \times D = C \times B$$

or $\qquad\qquad$ or

$$3 \times 4 = 6 \times 2 \qquad\qquad B \times C = D \times A$$

EXAMPLE 5-20. Find the value of $R$ in the equation

$$\frac{3}{R} = \frac{2}{5}$$

SOLUTION

1. Write the equation. $\quad \dfrac{3}{R} = \dfrac{2}{5}$
2. Cross-multiply. $\quad 2 \times R = 3 \times 5$
3. Simplify each side. $\quad 2R = 15$
4. Solve for $R$. $\quad R = \dfrac{15}{2} = 7\tfrac{1}{2}$ *Ans.*

## PROBLEMS

Find the value of the unknown letter in each equation.

1. $\frac{V}{10} = \frac{3}{5}$
2. $\frac{8}{3} = \frac{R}{6}$
3. $\frac{60}{A} = \frac{3}{4}$
4. $\frac{9}{V} = \frac{15}{40}$
5. $\frac{84}{V} = \frac{28}{17}$
6. $\frac{84}{28} = \frac{66}{R}$
7. $\frac{V}{18} = \frac{3}{2}$
8. $9 = \frac{54}{R}$
9. $\frac{50}{T} = 2$
10. $\frac{2R}{3} = \frac{10}{5}$
11. $\frac{24}{2} = \frac{6M}{5}$
12. $\frac{3P}{0.4} = 6$
13. $\frac{3.6}{5} = \frac{A}{2}$
14. $0.88 = \frac{R}{3}$
15. $\frac{3}{12} = \frac{4}{3T}$
16. $5 = \frac{1}{0.2V}$
17. $\frac{1}{R} = 10$
18. $80 = \frac{1}{R}$
19. $\frac{1}{R} = \frac{13}{24}$
20. $\frac{10}{R} = 50$

(See Instructor's Manual for Test 5-5, Parallel Circuits)

## JOB 5-8 TOTAL RESISTANCE IN A PARALLEL CIRCUIT

When resistances are connected in parallel, the total resistance is always *less* than the resistance of any branch. When resistances are added to a circuit in parallel, they merely provide extra paths for the current to follow. Each extra path will draw its own current from the voltage source, *increasing* the total current drawn. But resistance is the quality of a circuit which attempts to stop the flow of current. If the addition of a resistance *increases* the current, then this resistance must have *reduced* the total resistance. For example, in the circuit of Fig. 5-17*a*, the current that flows in the resistor is 1 A and the total resistance $R_T$ is 50 Ω. Now let us add an extra 50-Ω resistor in parallel as shown in Fig. 5-17*b*. Since equal resistors carry equal currents, the total current equals $I_1 + I_2 = 2$ A. The total resistance will be

$$V_T = I_T \times R_T \qquad (4\text{-}7)$$

$$50 = 2 \times R_T$$

$$R_T = \frac{50}{2} = 25\ \Omega$$

Thus we see that the net effect of adding another resistance in parallel *reduced* the total resistance from 50 to 25 Ω. Also, the current drawn by the new circuit *increased* from 1 to 2 A.

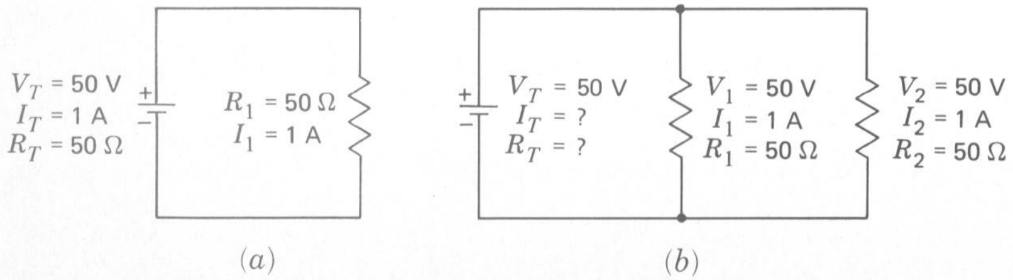

**FIGURE 5-17**
Adding resistances in parallel reduces the total resistance.

The total resistance in parallel is given by

**FORMULA** $$\frac{1}{R_T} = \frac{1}{R_1} + \frac{1}{R_2} + \frac{1}{R_3} + \text{etc.} \qquad \boxed{5\text{-}3}$$

where $R_T$ is the total resistance in parallel, and $R_1$, $R_2$, and $R_3$ are the branch resistances.

EXAMPLE 5-21. Find the total resistance of a 3-, a 4-, and an 8-Ω resistor in parallel.

SOLUTION. The diagram for the circuit is shown in Fig. 5-18.

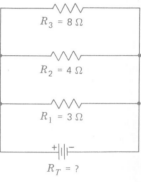

1. Write the formula. $\frac{1}{R_T} = \frac{1}{R_1} + \frac{1}{R_2} + \frac{1}{R_3}$ (5-3)
2. Substitute numbers. $\frac{1}{R_T} = \frac{1}{3} + \frac{1}{4} + \frac{1}{8}$
3. Add fractions. $\frac{1}{R_T} = \frac{17}{24}$
4. Cross-multiply. $17 \times R_T = 1 \times 24$
5. Simplify $17R_T = 24$
6. Solve for $R_T$. $R_T = \frac{24}{17} = 1.4\ \Omega$ *Ans.*

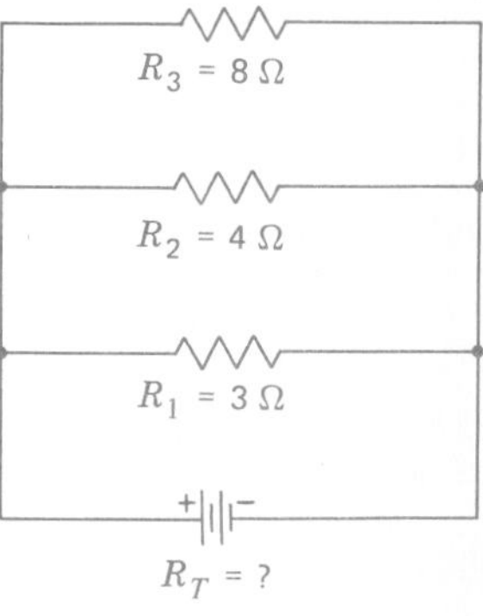


**FIGURE 5-18**

EXAMPLE 5-22. What resistance must be connected in parallel with a 40-Ω resistance in order to provide a total resistance of 24 Ω?

SOLUTION. Given: $R_T = 24\ \Omega$ Find: $R_2 = ?$
$R_1 = 40\ \Omega$

1. Write the formula. $\frac{1}{R_T} = \frac{1}{R_1} + \frac{1}{R_2}$ (5-3)
2. Substitute numbers. $\frac{1}{24} = \frac{1}{40} + \frac{1}{R_2}$
3. Transpose the $^1/_{40}$ to the left side of the equality sign.
$$\frac{1}{24} - \frac{1}{40} = \frac{1}{R_2}$$
4. Subtract fractions. $\frac{1}{60} = \frac{1}{R_2}$
5. Cross-multiply. $R_2 = 60\ \Omega$ *Ans.*

SELF-TEST 5-23. A 40-, a 70-, and a 150-Ω resistor are connected in parallel. Find the total resistance.

SOLUTION. Given: $R_1 = 40\ \Omega$ Find: $R_T = ?$
$R_2 = 70\ \Omega$
$R_3 = 150\ \Omega$

1. Write the formula. $\dfrac{1}{?} = \dfrac{1}{R_1} + \dfrac{1}{R_2} + \dfrac{1}{R_3}$ | $R_T$

2. Substitute numbers. $\dfrac{1}{R_T} = \dfrac{1}{40} + \dfrac{1}{?} + \dfrac{1}{150}$ | 70

The next step would be to add these fractions. However, it is very difficult to find the LCD when the denominators are as large as these. In this situation, it is best to find the decimal equivalent for each fraction and use these decimals instead of the fractions.

$$^1/_{40} = 40\overline{)1.000} = 0.025 \qquad \begin{array}{r} 80 \\ \hline 200 \\ 200 \\ \hline \end{array}$$

$$^1/_{70} = 70\overline{)1.000} = 0.014 \qquad \begin{array}{r} 70 \\ \hline 300 \\ 280 \\ \hline 20 \end{array}$$

$$^1/_{150} = 150\overline{)1.000} = 0.006^2/_3 = 0.007 \qquad \begin{array}{r} 900 \\ \hline 100 \end{array}$$

3. Substitute the decimals for the fractions.

$$\frac{1}{R_T} = 0.025 + 0.014 + \underline{\qquad\quad}$$ | 0.007

4. Add the decimals. $\dfrac{1}{R_T} = \underline{\qquad\quad}$ | 0.046

5. Cross-multiply. $0.046R_T = \underline{\qquad\quad}$ | 1

6. Solve for $R_T$. $R_T = \dfrac{1}{?}$ | 0.046

7. Divide. $R_T = \underline{\qquad\quad}\ \Omega$ *Ans.* | 21.7

## PROBLEMS

1. Find the total resistance of a 3-, a 4-, and a 12-Ω resistor in parallel.
2. Find the total resistance of the circuit shown in Fig. 5-19.

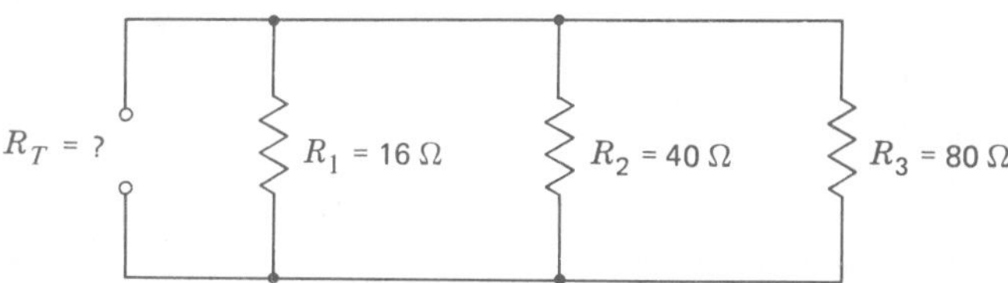


**FIGURE 5-19**

3. Three resistances of 3, 6, and 9 kΩ are in parallel. What is the combined resistance?
4. Find the total resistance of a 25-Ω coffee percolator and a 30-Ω toaster in parallel.
5. Find the total resistance of 4, 8, 12, and 15 Ω in parallel.
6. Find the total resistance of 6, 8, and 4.8 Ω in parallel.
7. Find the total resistance of 15, 7.5, and 5 Ω in parallel.
8. Find the total resistance of 100, 250, and 500 Ω when connected in parallel.

9. Find the resistance of each group of resistors in the circuit shown in Fig. 5-20.

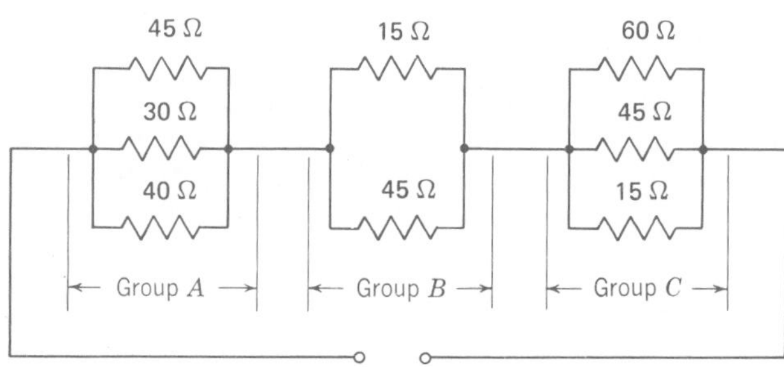


FIGURE 5-20

10. Find the total resistance of 1, 1.5, and 2 kΩ when connected in parallel.
11. What resistance must be connected in parallel with a 20-Ω resistance in order to provide a total resistance of 15 Ω?
12. What resistance must be connected in parallel with a 100-Ω resistance in order to provide a total resistance of 90 Ω?
13. What resistance must be connected in parallel with a 600-Ω resistance in order to provide a total resistance of 400 Ω?
14. A voltage of 120 V is applied across a parallel combination of a 100-Ω resistor and an unknown resistor. If the total current is 1.5 A, find the value of the unknown resistor.
15. A spotlight of unknown resistance is placed in parallel with an automobile cigarette lighter of 80 Ω resistance. If a current of 0.75 A flows when a voltage of 12 V is applied, find the resistance of the spotlight.

## TOTAL RESISTANCE OF A NUMBER OF EQUAL BRANCHES

**RULE** The total resistance of a number of equal resistors in parallel is equal to the resistance of one resistor divided by the number of resistors.

**FORMULA**

$$R_T = \frac{R}{N} \qquad \boxed{5\text{-}4}$$

where $R_T$ = total resistance of equal resistors in parallel
$R$ = resistance of one of the equal resistors
$N$ = number of equal resistors

EXAMPLE 5-24. Three lamps, each having a resistance of 60 Ω, are connected in parallel. Find the total resistance of the combination.

SOLUTION. Given: $R_1 = R_2 = R_3 = 60\ \Omega$ Find: $R_T = ?$

$$R_T = \frac{R}{N} \tag{5-4}$$

$$R_T = \frac{60}{3} = 20\ \Omega \qquad Ans.$$

**PROBLEMS**

1. Find the total resistance of two 100-Ω lamps in parallel.
2. Find the total resistance of three 48-Ω bells in parallel.
3. Find the total resistance of two 1.5-Ω headlight lamps in parallel.
4. If the total resistance of two identical bells in parallel is 20 Ω, what is the resistance of each bell?
5. A toaster, an electric iron, and a coffee percolator, all of 22 Ω resistance, are connected in parallel across a 110-V line. Find (*a*) the total resistance and (*b*) the total current.

**Two resistors in parallel.** When only two resistors are in parallel, the total resistance may be calculated by a simple rule.

**RULE** To find the total resistance of only two resistors in parallel, multiply the resistances and then divide the product by the sum of the resistors.

**FORMULA**

$$R_T = \frac{R_1 \times R_2}{R_1 + R_2} \tag{5-5}$$

where $R_T$ is the total resistance in parallel, and $R_1$ and $R_2$ are the two resistors in parallel.

EXAMPLE 5-25. Find the total resistance of a 4- and a 12-Ω resistor in parallel.

SOLUTION. Given: $R_1 = 4\ \Omega$ Find: $R_T = ?$
$R_2 = 12\ \Omega$

$$R_T = \frac{R_1 \times R_2}{R_1 + R_2} \tag{5-5}$$

$$R_T = \frac{4 \times 12}{4 + 12}$$

$$R_T = \frac{48}{16} = 3\ \Omega \qquad Ans.$$

**PROBLEMS**

1. Find the total resistance of two 40-Ω coils in parallel.
2. Find the total resistance of a 20- and a 60-Ω motor in parallel.
3. An ammeter has a coil whose resistance is 56 Ω and is shunted by a 42-Ω resistor in parallel. Find the equivalent resistance of the combination.

4. Find the total resistance of 90-Ω galvanometer in parallel with a 10-Ω shunt resistor.
5. A section of the picture-control circuit of a television receiver uses a 10,000- and a 25,000-Ω resistor in parallel. Find the total resistance of the combination.
6. A 24- and a 48-Ω solenoid used in semaphore signals of an H-O electric train set are connected in parallel. Find the total resistance.
7. Find the total resistance of a 5- and a 12-kΩ resistor in parallel.
8. Find the total resistance of 9 kΩ and 2000 Ω when connected in parallel.
9. A 40-Ω soldering iron and a 100-Ω lamp are connected in parallel. Find the total resistance.
10. Find the total resistance of 20 and 27 kΩ when connected in parallel.

(See Instructor's Manual for Test 5-6)

## JOB 5-9 TOTAL VOLTAGE IN A PARALLEL CIRCUIT

EXAMPLE 5-26. What voltage is needed to send 3 A through a parallel combination of a 3-, a 4-, and a 12-Ω resistance?

SOLUTION. The diagram for the circuit is shown in Fig. 5-21. In order to find the total voltage $V_T$, we must know the value of the total current $I_T$ and the total resistance $R_T$. If either value is unknown, it must be found first.

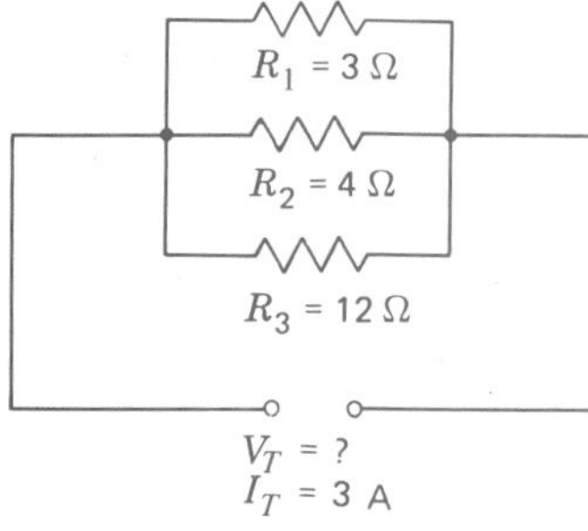


FIGURE 5-21

1. Find the total resistance $R_T$.

$$\frac{1}{R_T} = \frac{1}{R_1} + \frac{1}{R_2} + \frac{1}{R_3} \tag{5-3}$$

$$\frac{1}{R_T} = \frac{1}{3} + \frac{1}{4} + \frac{1}{12}$$

$$\frac{1}{R_T} = \frac{2}{3}$$

$$2R_T = 3$$

$$R_T = \frac{3}{2} = 1.5\ \Omega \qquad \textit{Ans.}$$

2. Find the total voltage $V_T$.

$$V_T = I_T \times R_T \qquad (4\text{-}7)$$

$$V_T = 3 \times 1.5 = 4.5 \text{ V} \qquad \textit{Ans.}$$

## PROBLEMS

1. Find the voltage needed to send 2 A through a parallel combination of three 60-Ω resistors.
2. Find the total voltage needed to send 9 A through a parallel circuit consisting of a 25-Ω percolator and a 30-Ω refrigerator motor.
3. Find the voltage needed to send 3 A through a parallel combination of a 2- and an 8-Ω resistor.
4. Find the voltage needed to send 2 A through a parallel combination of a 3- and a 6-Ω resistor.
5. Find the voltage required to send a current of 2.4 A through a 10-Ω coil, a 20-Ω coil, and a 60-Ω motor if they are wired in parallel.
6. In Fig. 5-22, a 25-Ω galvanometer is shunted with a 4-Ω resistor when indicating a current of 3 mA. What is the voltage across the galvanometer?

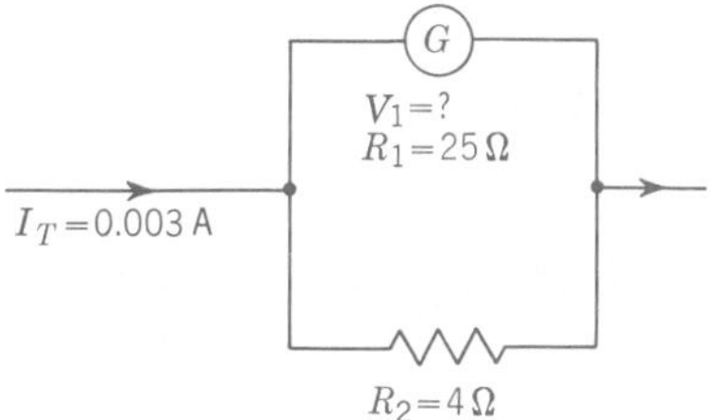


**FIGURE 5-22**
A shunt across the galvanometer carries most of the line current.

7. Find the voltage required to send 2 A through a parallel combination of a 20-, a 30-, and a 40-Ω resistance.
8. Find the total voltage required to send 0.15 A through three coils in parallel if their effective resistances are 200, 400, and 800 Ω.
9. A faulty resistor $R_L$ in the self-bias circuit shown in Fig. 5-23*a* was replaced with a 6000- and a 4000-Ω resistor in parallel as shown in Fig. 5-23*b*. If $I_L$ = 2 mA, find the voltage across the parallel combination.

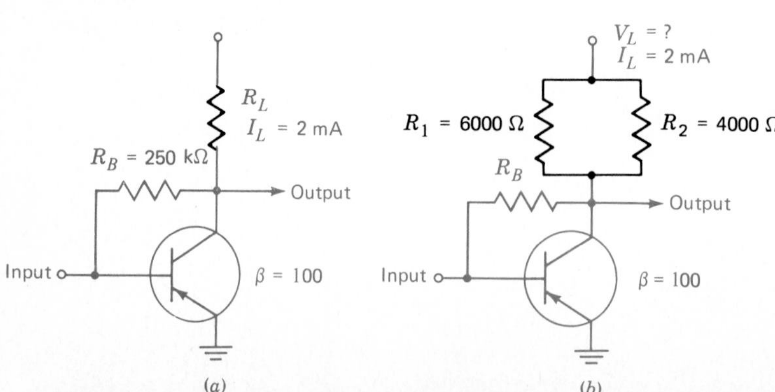


**FIGURE 5-23**
(*a*) Self-bias circuit. (*b*) $R_L$ is replaced with an equivalent parallel combination.

## JOB 5-10 DIVISION OF CURRENT IN A PARALLEL CIRCUIT

In our study of Ohm's law we learned that the resistance of a circuit affected the current in the circuit. The greater the resistance, the smaller the current. The smaller the resistance, the greater the current. When the current in a parallel circuit reaches a point at which it may divide and flow in more than one path, the largest current will naturally flow in that portion of the circuit which offers the smallest resistance, and the smallest current will flow through the largest resistance. The exact manner in which a current will divide in a parallel circuit is shown in the following example.

EXAMPLE 5-27. Find the current that flows in each branch of the parallel circuit shown in Fig. 5-24.

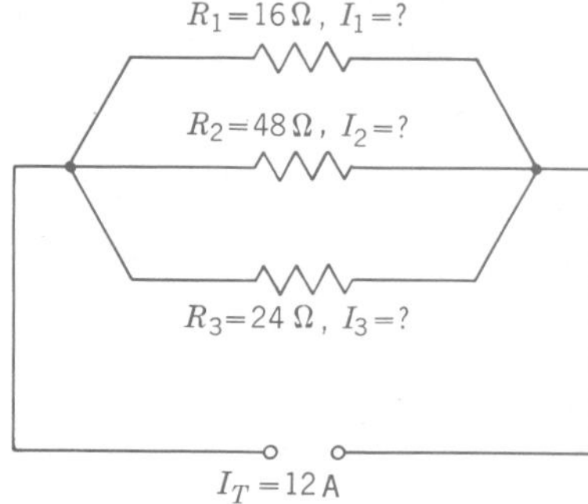


FIGURE 5-24

SOLUTION

1. Find the total resistance $R_T$.

$$\frac{1}{R_T} = \frac{1}{R_1} + \frac{1}{R_2} + \frac{1}{R_3} \qquad (5\text{-}3)$$

$$\frac{1}{R_T} = \frac{1}{16} + \frac{1}{48} + \frac{1}{24}$$

$$\frac{1}{R_T} = \frac{1}{8}$$

$$R_T = 8\ \Omega$$

2. Find the total voltage $V_T$.

$$V_T = I_T \times R_T \qquad (4\text{-}7)$$

$$V_T = 12 \times 8 = 96\text{ V}$$

3. Find the branch voltages.

$$V_T = V_1 = V_2 = V_3 = 96\text{ V} \qquad (5\text{-}1)$$

4. Find the branch currents.

$$V_1 = I_1 \times R_1 \qquad\qquad V_2 = I_2 \times R_2$$

$$96 = I_1 \times 16 \qquad\qquad 96 = I_2 \times 48$$

$$I_1 = \frac{96}{16} = 6\text{ A} \quad \textit{Ans.} \qquad\qquad I_2 = \frac{96}{48} = 2\text{ A} \quad \textit{Ans.}$$

$$V_3 = I_3 \times R_3$$
$$96 = I_3 \times 24$$
$$I_3 = \frac{96}{24} = 4 \text{ A} \qquad \textit{Ans.}$$

5. *Check:*

$$I_T = I_1 + I_2 + I_3 \qquad (5\text{-}2)$$
$$12 = 6 + 2 + 4$$
$$12 = 12 \qquad \textit{Check}$$

EXAMPLE 5-28. This example will illustrate the problem of attempting to use discrete transistors in parallel to handle higher current loads. Because each transistor is manufactured separately, the parameter variations cause an unequal distribution of the load current. Two 2N2652 transistors with input impedances of $R_1 = 2\ \text{k}\Omega$ and $R_2 = 8\ \text{k}\Omega$ are used in the circuit shown in Fig. 5-25*a*. The equivalent input circuit is shown in Fig. 5-25*b*. Find the current into the base of each transistor.

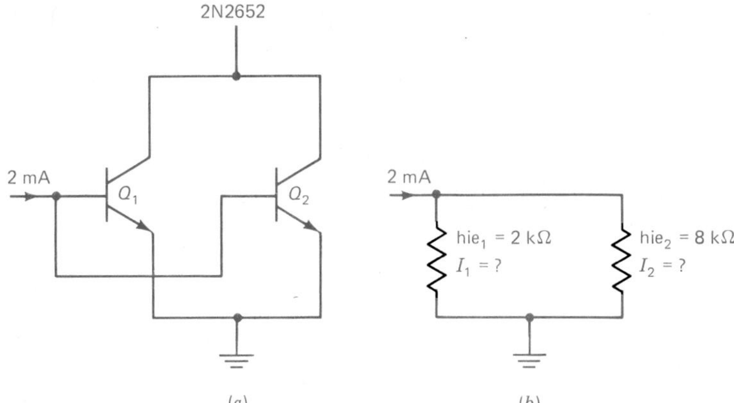


FIGURE 5-25

SOLUTION

1. Find the resistance of the parallel combination.

$$R_T = \frac{R_1 \times R_2}{R_1 + R_2} \qquad (5\text{-}5)$$
$$R_T = \frac{2\ \text{k}\Omega \times 8\ \text{k}\Omega}{2K + 8K}$$
$$R_T = \frac{16}{10}\text{k}\Omega = 1.6\ \text{k}\Omega$$

2. Find the total voltage.

$$V_T = I_T \times R_T \qquad (4\text{-}7)$$
$$V_T = 2\ \text{mA} \times 1.6\ \text{k}\Omega = 3.2\ \text{V}$$

3. Find the branch voltages.

$$V_T = V_1 = V_2 = 3.2\ \text{V} \qquad (5\text{-}1)$$

4. Find the branch currents.

Transistor No. 1

$V_1 = I_1 \times R_1$

$3.2 = I_1 \times 2\ \text{k}\Omega$

$I_1 = \dfrac{3.2}{2\ \text{k}\Omega}$

$I_1 = 1.6\ \text{mA}$ *Ans.*

Transistor No. 2

$V_2 = I_2 \times R_2$

$3.2 = I_2 \times 8\ \text{k}\Omega$

$I_2 = \dfrac{3.2}{8\ \text{k}\Omega}$

$I_2 = 0.4\ \text{mA}$ *Ans.*

## DIVISION OF CURRENT IN TWO BRANCHES IN PARALLEL

**RULE** When only two branches are involved, the current in one branch will be only some fraction of the total current. This fraction is the quotient of the second resistance divided by the sum of the resistances.

**FORMULA** $$\boldsymbol{I_1 = \frac{R_2}{R_1 + R_2} \times I_T} \qquad \boxed{5\text{-}6}$$

**FORMULA** $$\boldsymbol{I_2 = \frac{R_1}{R_1 + R_2} \times I_T} \qquad \boxed{5\text{-}7}$$

EXAMPLE 5-29. We can now solve Example 5-28 by this method.

SOLUTION. Given: $R_1 = 2\ \text{k}\Omega$, $R_2 = 8\ \text{k}\Omega$, $I_T = 2\ \text{mA}$ Find: $I_1 = ?$, $I_2 = ?$

1. Find the current $I_1$.

$$I_1 = \frac{R_2}{R_1 + R_2} \times I_T \qquad (5\text{-}6)$$

$$I_1 = \frac{8\ \text{k}\Omega}{2\ \text{k}\Omega + 8\ \text{k}\Omega} \times 2\ \text{mA} = \frac{8}{10} \times 2\ \text{mA}$$

$I_1 = 0.8 \times 2\ \text{mA} = 1.6\ \text{mA}$ *Ans.*

2. Find the current $I_2$.

$$I_2 = \frac{R_1}{R_1 + R_2} \times I_T \qquad (5\text{-}7)$$

$$I_2 = \frac{2\ \text{k}\Omega}{2\ \text{k}\Omega + 8\ \text{k}\Omega} \times 2\ \text{mA} = \frac{2}{10} \times 2\ \text{mA}$$

$I_2 = 0.2 \times 2\ \text{mA} = 0.4\ \text{mA}$ *Ans.*

SELF-TEST 5-30. A 2N405 in a two-transistor receiver receives 36 mA from a parallel combination of a 40-kΩ resistor and a 5-kΩ headphone, as shown in Fig. 5-26. Find the current flowing in each part.

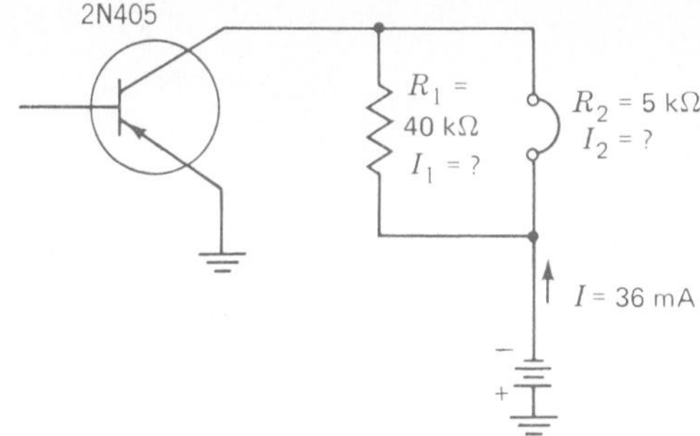

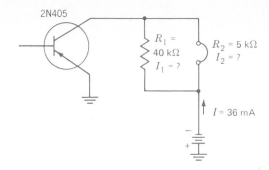

**FIGURE 5-26**
Division of current in a parallel circuit.

SOLUTION

1. Find the current $I_2$ in the headphone.
Write the formula.

$$I_2 = \frac{?}{R_1 + R_2} \times I_T$$ $R_1$

$$= \frac{40\ \text{k}\Omega}{40\ \text{k}\Omega + 5\ \text{k}\Omega} \times 36\ \text{mA}$$

$$= \frac{40 \times 10^3}{45 \times 10^3} \times 36 \times \_\_\_\_\_\_$$ $10^{-3}$

$$= 32 \times 10^{-3}\ \ \underline{\quad(\text{A/mA})\quad}$$ A

$$= \_\_\_\_\_\_\ \text{mA} \quad \textit{Ans.}$$ 32

2. Find the current $I_1$ in the resistor.
Write the formula.

$$I_1 = \frac{?}{R_1 + R_2} \times I_T$$ $R_2$

$$= \frac{5\ \text{k}\Omega}{40\ \text{k}\Omega + 5\ \text{k}\Omega} \times 36\ \text{mA}$$

$$= \frac{5 \times 10^3}{45 \times 10^3} \times 36 \times 10^{-3}$$

$$= \_\_\_\_\_\_ \times 10^{-3}\ \ \underline{\quad(\text{mA/A})\quad}$$ 4 A

$$= \_\_\_\_\_\_\ \text{mA} \quad \textit{Ans.}$$ 4

**PROBLEMS**

1. Two resistances are connected in parallel. $R_1 = 48\ \Omega$, $R_2 = 48\ \Omega$, and $I_T = 8$ A. Find the current flowing in each branch. On the basis of your answers, state a rule about the division of current between equal resistors in parallel.
2. Find the current in each branch of a parallel circuit consisting of a 20-Ω percolator and a 30-Ω toaster if the total current is 9 A.
3. Two 1.5-Ω automobile headlight lamps in parallel draw a total of 8 A. Find the total voltage supplied and the current drawn by each lamp.
4. A galvanometer with a resistance of 48 Ω and a parallel shunt of 2 Ω

draw a total current of 0.2 A. Find the current through the galvanometer and through the shunt.

5. A generator supplies a current of 19.5 A to three small electroplating tanks arranged in parallel. The resistances of the tanks are 8, 12, and 16 Ω. What current does each tank draw?
6. A generator supplies a current of 26 A to three motors arranged in parallel. The resistances of the motors are 24, 36, and 48 Ω. What current does each motor draw?
7. Two resistances are arranged in parallel as shown in Fig. 5-27. Find the current in each resistance.

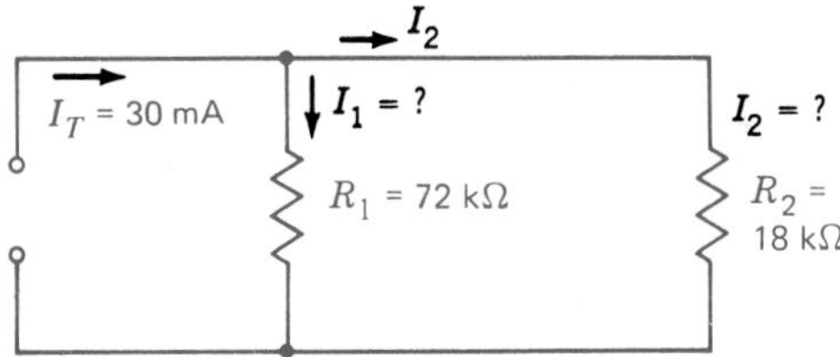


FIGURE 5-27

8. The resistance of an ammeter is 2.8 Ω. A shunt of 0.02 Ω is connected in parallel with it. If the combination is inserted into a line carrying 10 A, how much current actually flows through the ammeter?
9. In a circuit similar to that shown in Fig. 5-21, $R_1 = 5\ \Omega$, $R_2 = 7\ \Omega$, $R_3 = 8\ \Omega$, and $I_T = 13.1$ A. Find (*a*) the total resistance of the circuit, (*b*) the total voltage of the circuit, (*c*) the voltage across each branch, and (*d*) the current in each branch.
10. A low-power flasher circuit contains a resistor $R_1 = 16$ kΩ in parallel with another resistor $R_2 = 4$ kΩ. If the current entering the combination is 3 mA, find the current in each resistor.

## JOB 5-11 REVIEW OF PARALLEL CIRCUITS

**Definition.** A parallel circuit is a circuit in which the current may divide so as to flow in ______ than one path. | more

### Rules and formulas

1. The total voltage across a parallel circuit is ______ to the voltage across any branch of the circuit. | equal

$$V_T = V_1 = \_\_\_\_\_\_ = \_\_\_\_\_\_ \qquad \boxed{5\text{-}1}$$ | $V_2$ $V_3$

2. The total current in a parallel circuit is equal to the ______ of the currents in all the branches of the circuit. | sum

$$I_T = I_1 + \_\_\_\_\_\_ + \_\_\_\_\_\_ \qquad \boxed{5\text{-}2}$$ | $I_2$ $I_3$

3. The total resistance in a parallel circuit may be found by applying Ohm's law to the ______ values of the circuit. | total

$$V_T = I_T \times \_\_\_\_\_\_ \quad \boxed{4\text{-}7}$$ $R_T$

4. The total resistance may also be found as follows:
For any number of resistors:

$$\frac{1}{?} = \frac{1}{R_1} + \frac{1}{R_2} + \frac{1}{R_3} \quad \boxed{5\text{-}3}$$ $R_T$

For just two resistors:

$$R_T = \frac{R_1 ? R_2}{R_1 ? R_2} \quad \boxed{5\text{-}5}$$ $\times$ $+$

For any number $N$ of equal resistors of $R$ Ω each:

$$R_T = \frac{R}{?} \quad \boxed{5\text{-}4}$$ $N$

5. Ohm's law may be used on any branch of a parallel circuit.

$$V_1 = I_1 \times \_\_\_\_\_\_ \quad \boxed{4\text{-}4}$$ $R_1$

$$V_2 = \_\_\_\_\_\_ \times R_2 \quad \boxed{4\text{-}5}$$ $I_2$

$$\_\_\_\_\_\_ = I_3 \times R_3 \quad \boxed{4\text{-}6}$$ $V_3$

6. The total resistance in parallel is always ___(more/less)___ than the resistance of any branch. less

7. Division of current between two branches in parallel:

$$I_1 = \frac{?}{R_1 + R_2} \times I_T \quad \boxed{5\text{-}6}$$ $R_2$

$$I_2 = \frac{?}{R_1 + R_2} \times I_T \quad \boxed{5\text{-}7}$$ $R_1$

8. If the branches of a parallel circuit have the same resistance, then each will draw the ______ current. If the branches of a parallel circuit have different resistances, then each will draw a ______ current. The larger the resistance, the ______ the current drawn. same different smaller

9. Adding or subtracting fractions:
In order to add or subtract fractions, all the fractions must have the ______ denominator. same
The LCD is the ______ number into which ______ the denominators will evenly divide. smallest *all*

10. Solving fractional equations:
Simplify fractions whenever possible.

When a fractional equation contains only ______ fractions equal to each other, the equation may be solved by cross ______. two multiplication

To cross-multiply, the product of the numerator of the first fraction and the denominator of the second is set ______ to the product of the numerator of the second fraction and the denominator of the first. For example, if equal

$$\frac{C}{D} = \frac{E}{F}$$

then $C \times$ ________ $= E \times D$ | $F$

or $E \times D =$ ________ $\times$ F | $C$

## PROBLEMS

1. An electric iron, a radio, and an electric clock are connected to a three-way 110-V kitchen outlet which puts the appliances in parallel. The iron draws 5 A, the radio draws 0.5 A, and the clock draws 0.25 A. Find (*a*) the total current drawn from the line, (*b*) the voltage across each device, and (*c*) the total resistance of the circuit.
2. A semaphore signal, a floodlight tower, and a coal loader of a model railroad are connected in parallel across the 12-V winding of the power transformer. The signal draws 0.1 A, and the coal loader draws 0.2 A. The floodlight tower has a resistance of 48 Ω. Find (*a*) the voltage across each device, (*b*) the resistances of the semaphore and the coal loader, (*c*) the total current, and (*d*) the total resistance of the circuit.
3. Three motors are wired in parallel across 440 V. Motor 1 draws a current of 10 A, and motor 3 draws 15 A. Motor 2 has a resistance of 20 Ω. Solve the circuit for all missing values of current, voltage, and resistance.
4. Find all missing values of voltage, current, and resistance in the circuit shown in Fig. 5-28.

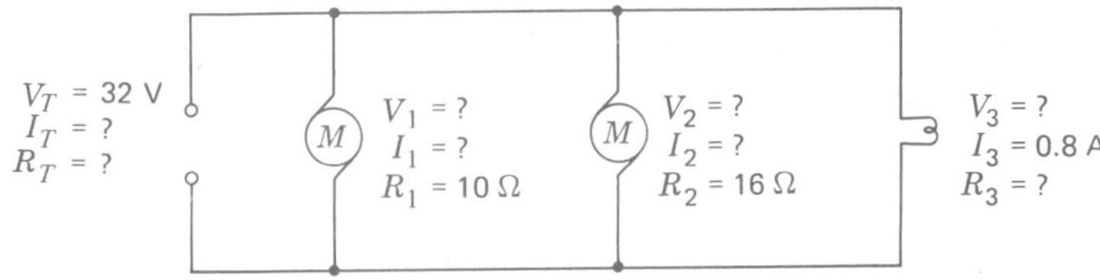


**FIGURE 5-28**

5. Find the total resistance of a 60-, an 80-, and a 120-Ω resistor in parallel.
6. A 2000-and a 5000-Ω resistor are connected in parallel as shown in Fig. 5-29. Find the total resistance of the combination.

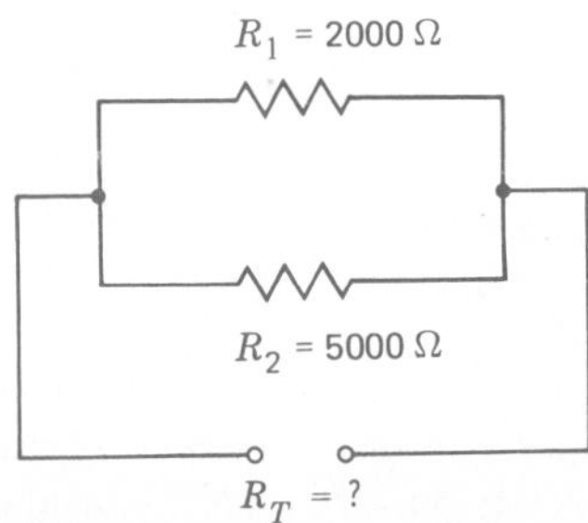


**FIGURE 5-29**

7. In the circuit shown in Fig. 5-30, find (*a*) the total resistance $R_A$ of group *A* and (*b*) the total resistance $R_B$ of group *B*. (*c*) Draw a new circuit using a single resistor ($R_A$ and $R_B$) in place of the groups they represent. (*d*) Is the new circuit a series or a parallel circuit? (*e*) What is the total resistance of the new circuit? (*f*) Find the total current in the new circuit.

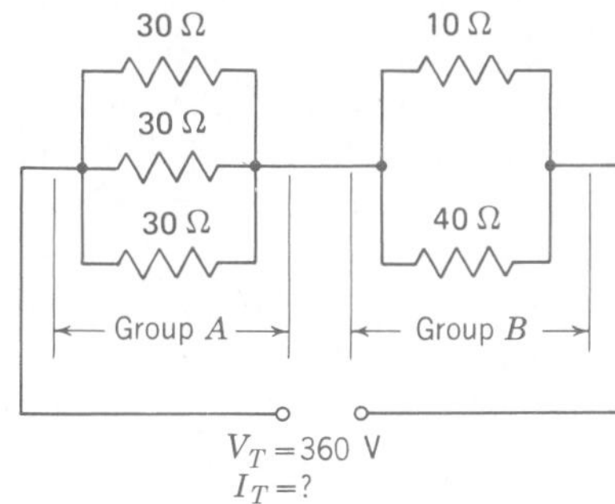


FIGURE 5-30

8. What is the combined resistance of a 480-Ω galvanometer and its parallel 20-Ω shunt?
9. What resistance must be connected in parallel with a 2400-Ω resistor in order to provide a total resistance of 480 Ω?
10. What resistance must be connected in parallel with a 20- and a 60-Ω resistor in parallel in order to provide a total resistance of 10 Ω?
11. Find the voltage needed to send 2 A through a parallel combination of a 75- and a 100-Ω resistor.
12. A 30-Ω resistor is connected in parallel with an unknown resistor across a 120-V source. If the total current is 5 A, find the value of the unknown resistor.
13. In a circuit similar to that shown in Fig. 5-27, $R_2$ equals 3 kΩ, $R_1$ equals 17 kΩ, and $I_T$ equals 40 mA. Find (*a*) the total resistance and (*b*) the total voltage.
14. A generator supplies 26 A to three motors in parallel. The resistances of the motors are 12, 18, and 24 Ω. What current does each motor draw?

(See Instructor's Manual for Test 5-7, Parallel Circuits)

# COMBINATION CIRCUITS

## JOB 6-1 INTRODUCTION TO COMBINATION CIRCUITS

Each simple circuit has its own advantages and disadvantages.

**Series circuits.** An advantage of a series circuit is that it may be used to connect small voltages to obtain high voltages. Also, high voltages may be reduced by connecting resistances in series. Series circuits provide a means for reducing and controlling the current by connecting resistances in series. However, this series current remains unchanged throughout the circuit. This is a serious disadvantage. Since the current is constant, we are forced to use only those devices which require the same current. Thus it would be impossible to use any two household appliances at the same time because their current requirements range from 0.25 to 10 A. In addition, if any part of a series circuit should burn out, it would cause an open circuit and put the entire circuit out of operation. Series circuits are used where different voltage drops and a constant current are needed.

**Parallel circuits.** If a break should occur in any branch of a parallel circuit, it would not affect the other branch circuits. Houses are wired in parallel so that any device may be operated independently of any other device. This is both an advantage and a disadvantage. In a parallel circuit, more branches may be added at any time. Each new load draws current from the line. So much current may be drawn that the original line wires may not be able to carry the new current and the fuse will "blow." In this event, it is necessary to rewire the circuits completely, using wire capable of carrying the larger currents, or to put in extra independent circuits to supply the installation. Parallel circuits are used wherever a constant voltage and a large supply of current are required.

**Combination circuits.** If we combine series circuits with parallel circuits, we produce a combination circuit which makes use of the best features of each. A combination circuit makes it possible to obtain the different voltages of a series circuit and the different currents of a parallel circuit. This is the condition most generally required, particularly when the different voltages and currents must be supplied from the same source of power. The different voltage and current needs of the different circuits of a radio receiver are all obtained from combination circuits drawing power from a single power supply.

Simple combination circuits are of two types:

1. A *parallel-series* circuit (Fig. 6-1) is a circuit in which one or more *groups* of resistances in series are connected in parallel.
2. A *series-parallel* circuit (Fig. 6-12) is a circuit in which one or more *groups* of resistances in parallel are connected in series.

## General method for solving combination circuits

1. A *group* of resistances is a simple combination of two or more resistances which are arranged in either a *simple* series or a *simple* parallel circuit. Locate these groups.
2. Every group must be removed from the circuit as a unit and *replaced* by a single resistor which offers the identical resistance. This equivalent resistance is the total resistance of the group.
3. Redraw the circuit, using the equivalent resistance in place of each group.
4. Solve the resulting simple circuit for all missing values.
5. Go back to the *original* circuit to find the voltage, current, and resistance for each resistance in the circuit.

## JOB 6-2 SOLVING PARALLEL-SERIES CIRCUITS

EXAMPLE 6-1. Solve the circuit shown in Fig. 6-1 for all missing values of voltage, current, and resistance.

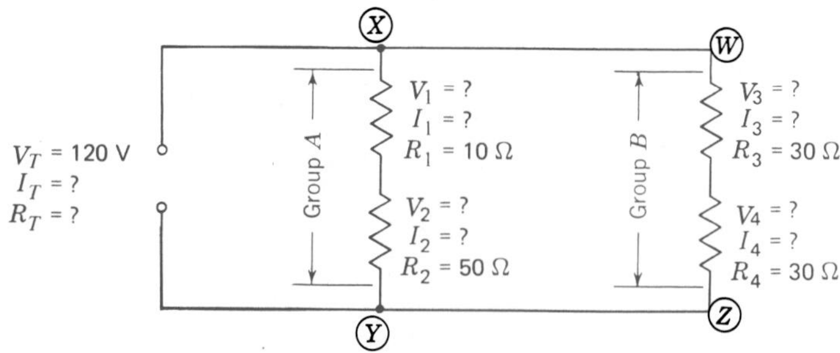


**FIGURE 6-1**
Parallel-series circuit.

SOLUTION

1. Locate groups $A$ and $B$ as simple series circuits.
2. Find the equivalent resistance of each group. This means that the circuit is broken at points $X$ and $Y$; resistors $R_1$ and $R_2$ are removed and replaced by a

single resistance $R_A$. This single resistance will do the work of the combination of $R_1$ and $R_2$. Similarly, break the circuit at points $W$ and $Z$; resistors $R_3$ and $R_4$ are removed and replaced by a *single* resistance $R_B$. This single resistance will do the work of the combination of $R_3$ and $R_4$.

3. Since the resistors of groups $A$ and $B$ are in *series*,

$$R_A = R_1 + R_2 \qquad (4\text{-}3)$$
$$R_A = 10 + 50 = 60\ \Omega \qquad \textit{Ans.}$$
$$R_B = R_3 + R_4 \qquad (4\text{-}3)$$
$$R_B = 30 + 30 = 60\ \Omega \qquad \textit{Ans.}$$

4. Redraw the circuit, using these 60-Ω resistors in place of the series groups as shown in Fig. 6-2.

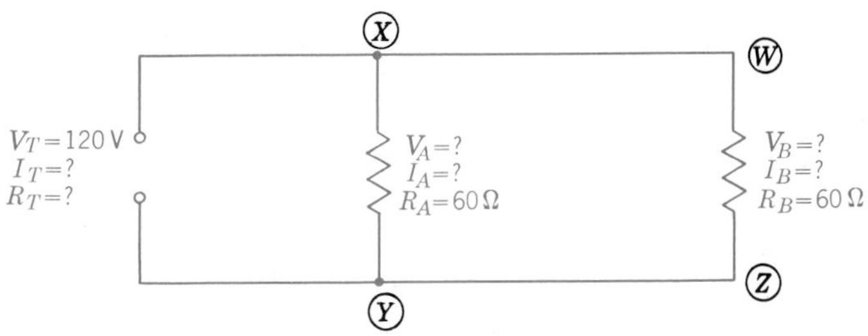


**FIGURE 6-2**
The resistance $R_A$ replaces the series combination of $R_1$ and $R_2$. The resistance $R_B$ replaces the series combination of $R_3$ and $R_4$.

5. Solve the new *parallel* circuit.

Find the voltage for each group.

$$V_T = V_A = V_B = 120\text{ V} \qquad \textit{Ans.} \qquad (5\text{-}1)$$

Find the current in each group.

$$V_A = I_A \times R_A \qquad\qquad V_B = I_B \times R_B$$
$$120 = I_A \times 60 \qquad\qquad 120 = I_B \times 60$$
$$I_A = \frac{120}{60} = 2\text{ A} \quad \textit{Ans.} \qquad\qquad I_B = \frac{120}{60} = 2\text{ A} \quad \textit{Ans.}$$

Find the total current $I_T$.

$$I_T = I_A + I_B \qquad (5\text{-}2)$$
$$I_T = 2 + 2 = 4\text{ A} \qquad \textit{Ans.}$$

Find the total resistance $R_T$.

$$V_T = I_T \times R_T \qquad (4\text{-}7)$$
$$120 = 4 \times R_T$$
$$R_T = \frac{120}{4} = 30\ \Omega \qquad \textit{Ans.}$$

6. Go back to the original circuit to find the voltage and current for each resistor.

Find the current in each resistor.

$$I_A = I_1 = I_2 = 2 \text{ A} \quad Ans. \qquad (4\text{-}1)$$

$$I_B = I_3 = I_4 = 2 \text{ A} \quad Ans. \qquad (4\text{-}1)$$

Find the voltage drop across each resistor.

$V_1 = I_1 \times R_1$
$V_1 = 2 \times 10$
$V_1 = 20$ V *Ans.*

$V_3 = I_3 \times R_3$
$V_3 = 2 \times 30$
$V_3 = 60$ V *Ans.*

$V_2 = I_2 \times R_2$
$V_2 = 2 \times 50$
$V_2 = 100$ V *Ans.*

$V_4 = I_4 \times R_4$
$V_4 = 2 \times 30$
$V_4 = 60$ V *Ans.*

SELF-TEST 6-2. In Fig. 6-3, find the resistance $R_3$ that must be connected in parallel with the resistances of group $A$ in order to obtain a total resistance of 15 Ω.

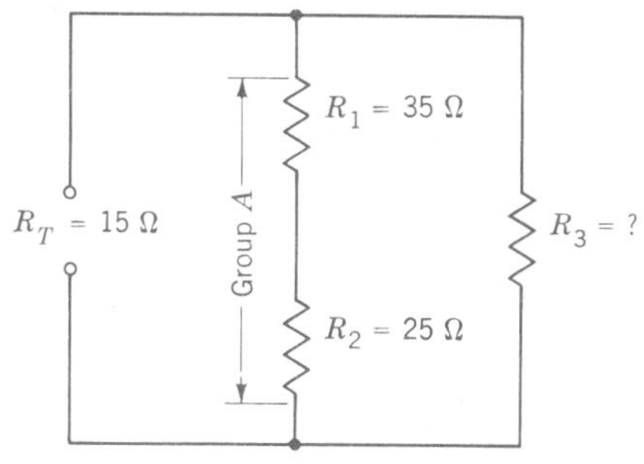


**FIGURE 6-3**
The series combination of $R_1$ and $R_2$ is in parallel with $R_3$.

SOLUTION

1. Locate group $A$ as a simple ______ circuit. — series
2. Find the resistance of group $A$.

$$R_A = R_1 + \underline{\qquad} \qquad (4\text{-}3)$$
$$R_A = 35 + \underline{\qquad} = \underline{\qquad}\ \Omega$$

$R_2$
25 60

3. Redraw the circuit, using $R_A = 60\ \Omega$ in place of the series group as shown in Fig. 6-4.

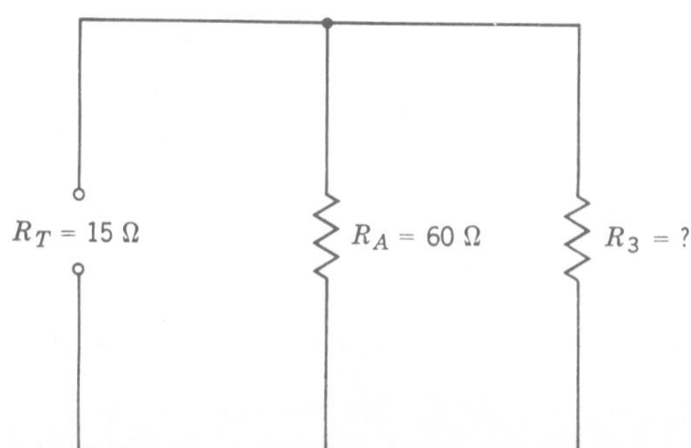


**FIGURE 6-4**
The resistance $R_A$ replaces the series combination of $R_1$ and $R_2$ in Fig. 6-3.

4. Find $R_3$.

$$\frac{1}{R_T} = \frac{1}{?} + \frac{1}{R_3} \qquad (5\text{-}3)$$

$$\frac{1}{15} = \frac{1}{60} + \frac{1}{R_3}$$

$$\frac{1}{15} - \_\_\_\_\_ = \frac{1}{R_3}$$

$$\_\_\_\_\_ = \frac{1}{R_3}$$

$R_3 =$ _____ Ω *Ans.*

| $R_A$ |
|---|
| $\frac{1}{60}$ |
| $\frac{1}{20}$ |
| 20 |

**PROBLEMS**

1. Find all missing values in the circuit shown in Fig. 6-5.

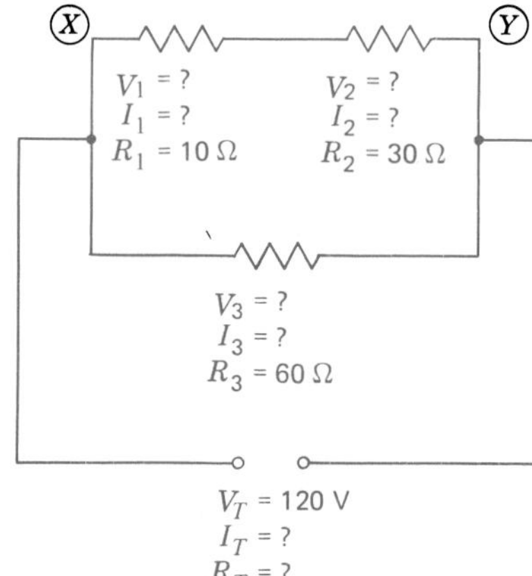


FIGURE 6-5

2. Find the total resistance and the total current for the circuit shown in Fig. 6-6.

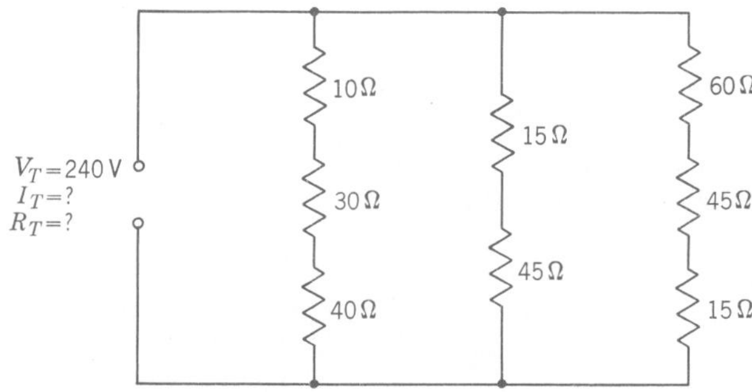


FIGURE 6-6

3. Find the total resistance and the total current for the circuit shown in Fig. 6-7.

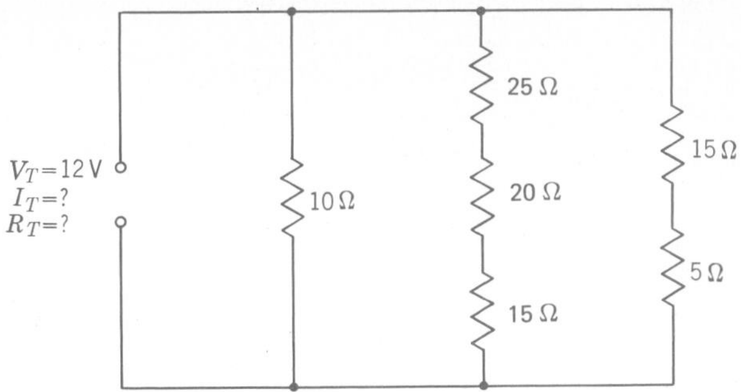


FIGURE 6-7

4. Find all missing values in the circuit shown in Fig. 6-8.

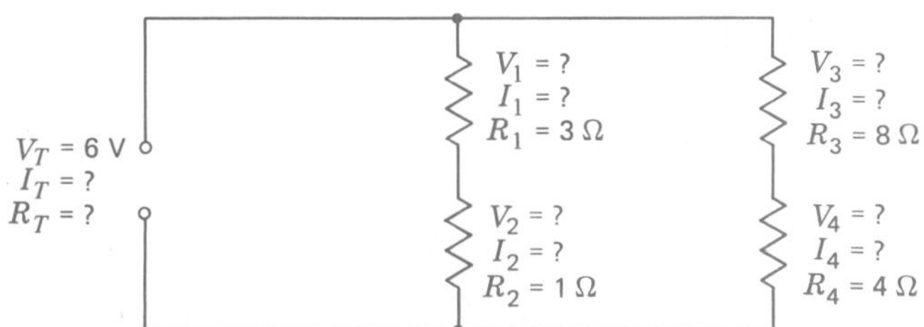


FIGURE 6-8

5. In the circuit shown in Fig. 6-9, find (*a*) the total resistance, (*b*) the total current, (*c*) the voltage $V_1$, (*d*) the current $I_1$, (*e*) the current $I_3$, (*f*) the voltage $V_2$, and (*g*) the voltage $V_3$.

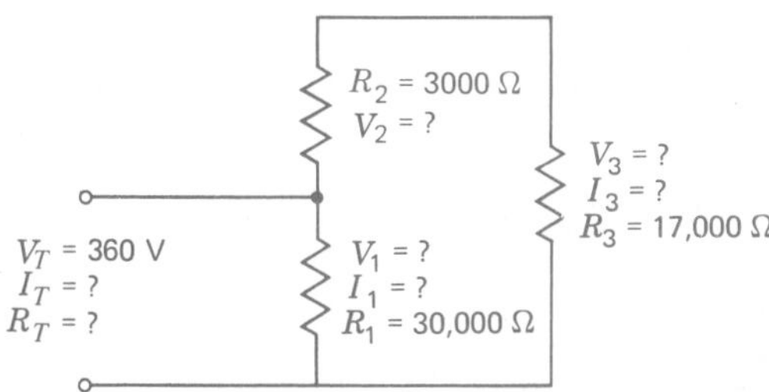


FIGURE 6-9

6. Part of the picture-control circuit of a television receiver is shown in Fig. 6-10. Find the total resistance of the circuit when the variable resistor $R_3$ has engaged (*a*) 6000 Ω, (*b*) 10,000 Ω, and (*c*) 5000 Ω.

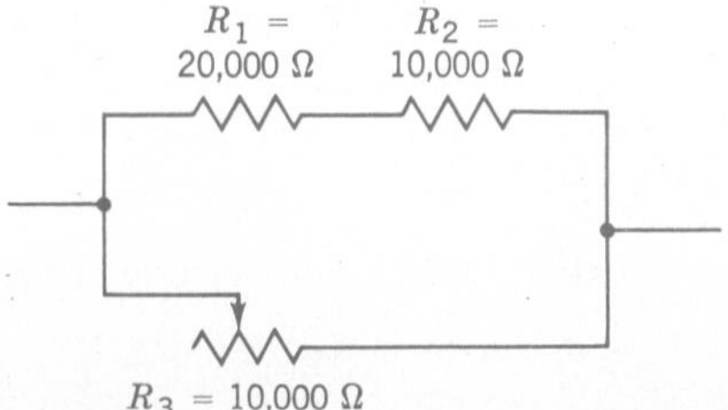


FIGURE 6-10

7. In a circuit similar to that shown in Fig. 6-3, find $R_3$ if $R_1 = 6\ \Omega$, $R_2 = 18\ \Omega$, and $R_T = 6\ \Omega$.
8. Find $R_T$ in the circuit shown in Fig. 6-11. *Hint:* Find the equivalent resistance of $R_3$ and $R_4$ in parallel. Then proceed as in a normal parallel-series circuit. See also Example 6-3 below.

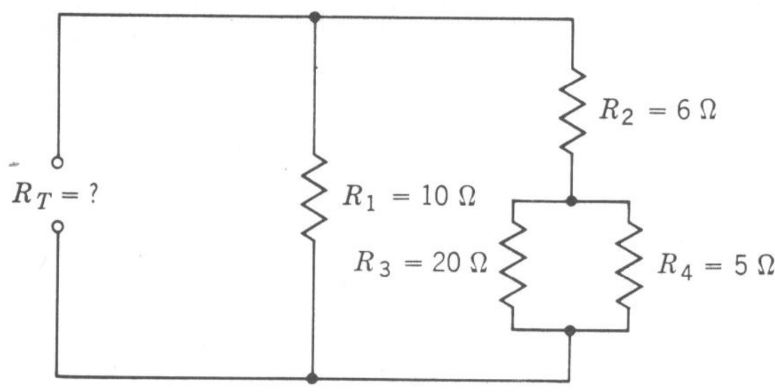


**FIGURE 6-11**

(See Instructor's Manual for Test 6-1)

## JOB 6-3 SOLVING SERIES-PARALLEL CIRCUITS

EXAMPLE 6-3. Solve the circuit shown in Fig. 6-12 for all values of voltage, current, and resistance.

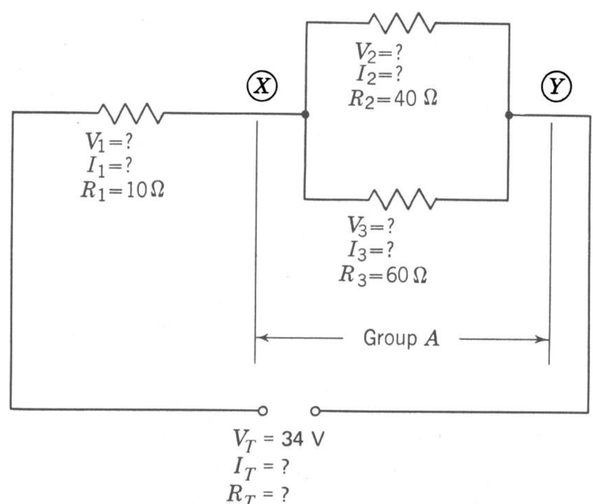


**FIGURE 6-12**
Series-parallel circuit.

SOLUTION

1. In Fig. 6-12, the resistors $R_2$ and $R_3$ form the *parallel* group $A$. Find the total resistance $R_A$ of group $A$. This means that the circuit is broken at points $X$ and $Y$; resistors $R_2$ and $R_3$ are removed and replaced by a *single* resistance $R_A$. This single resistance will do the work of the combination of $R_2$ and $R_3$.
2. Since $R_2$ and $R_3$ are in parallel,

$$R_A = \frac{R_2 \times R_3}{R_2 + R_3} \qquad (5\text{-}5)$$

$$R_A = \frac{40 \times 60}{40 + 60} = \frac{2400}{100} = 24\ \Omega \qquad Ans.$$

3. Redraw the circuit using this 24-Ω resistor $R_A$ in place of the combination as shown in Fig. 6-13.

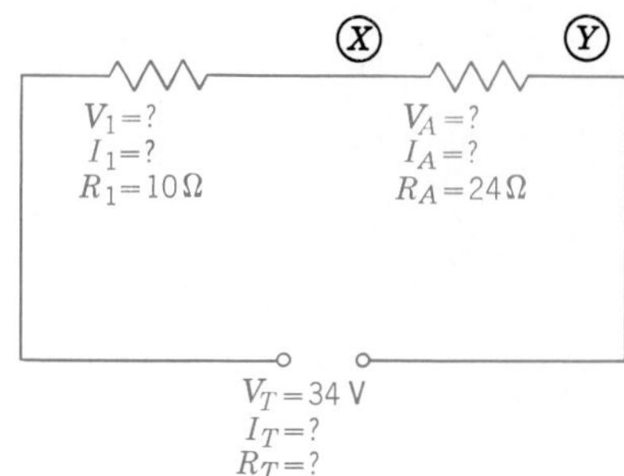


**FIGURE 6-13**
The resistance $R_A$ replaces the parallel combination of $R_2$ and $R_3$ of Fig. 6-12.

4. Solve the new circuit. Notice that we have a simple *series* circuit. Find the total resistance $R_T$.

$$R_T = R_1 + R_A \qquad (4\text{-}3)$$
$$R_T = 10 + 24 = 34\ \Omega \qquad Ans.$$

Find the total current $I_T$.

$$V_T = I_T \times R_T \qquad (4\text{-}7)$$
$$34 = I_T \times 34$$
$$I_T = \frac{34}{34} = 1\text{ A} \qquad Ans.$$

Find the current in each part of the series circuit.

$$I_T = I_1 = I_A = 1\text{ A} \qquad Ans. \qquad (4\text{-}1)$$

Find the voltage in each part of the series circuit.

$$V_1 = I_1 \times R_1 \qquad\qquad V_A = I_A \times R_A$$
$$V_1 = 1 \times 10 = 10\text{ V} \qquad Ans. \qquad V_A = 1 \times 24 = 24\text{ V} \qquad Ans.$$

5. Go back to group $A$ in Fig. 6-12 to find $V$, $I$, and $R$ for each resistance in group $A$ as shown in Fig. 6-14.

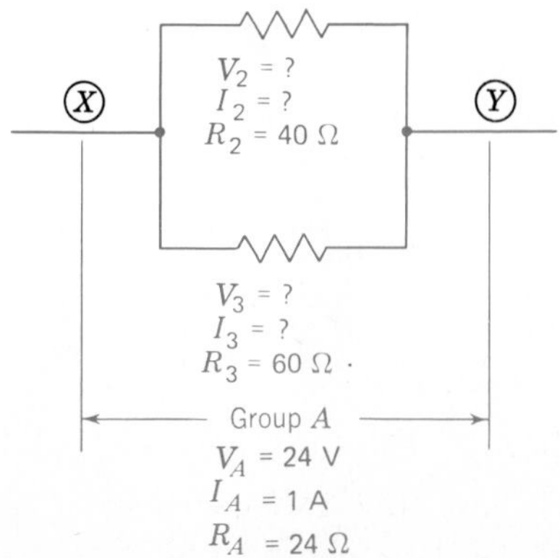


**FIGURE 6-14**

Since $V_A$ represents the total voltage of the parallel group $A$,

$$V_A = V_2 = V_3 = 24\text{ V} \qquad \textit{Ans.} \qquad (5\text{-}1)$$

Find the current in each resistor of group $A$.

$$V_2 = I_2 \times R_2 \qquad V_3 = I_3 \times R_3$$
$$24 = I_2 \times 40 \qquad 24 = I_3 \times 60$$
$$I_2 = \frac{24}{40} = 0.6\text{ A} \qquad I_3 = \frac{24}{60} = 0.4\text{ A} \qquad \textit{Ans.}$$

6. *Check:*

$$I_T = I_2 + I_3 \qquad (5\text{-}2)$$
$$I_T = 0.6 + 0.4$$
$$1 = 1 \qquad \textit{Check}$$

EXAMPLE 6-4. Solve the circuit shown in Fig. 6-15 for all values of voltage, current, and resistance.

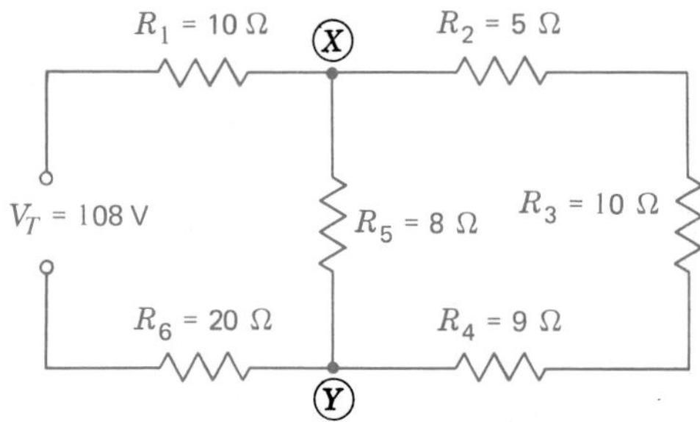


**FIGURE 6-15**

SOLUTION

1. Find the total resistance. It is usually very helpful if the circuit is redrawn so as to put it in standard form with easily recognizable series or parallel subcircuits. Therefore, redraw Fig. 6-15 to appear as shown in Fig. 6-16.

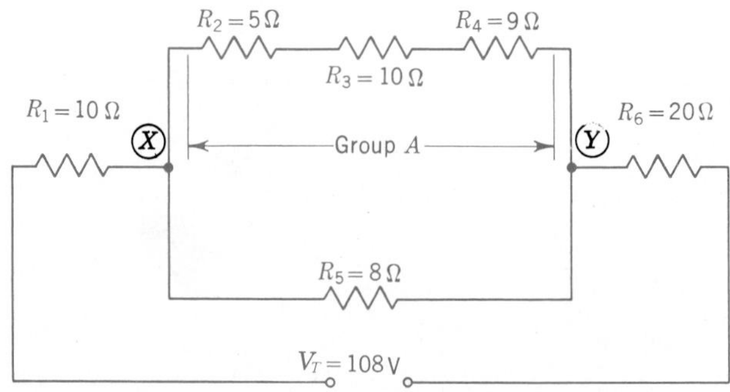


**FIGURE 6-16**
The circuit of Fig. 6-15 is redrawn in standard form.

*a.* In Fig. 6-16, the resistors $R_2$, $R_3$, and $R_4$ form the *series* group $A$. Find the total resistance of this group and replace the group with the equivalent resistance $R_A$ as shown in Fig. 6-17.

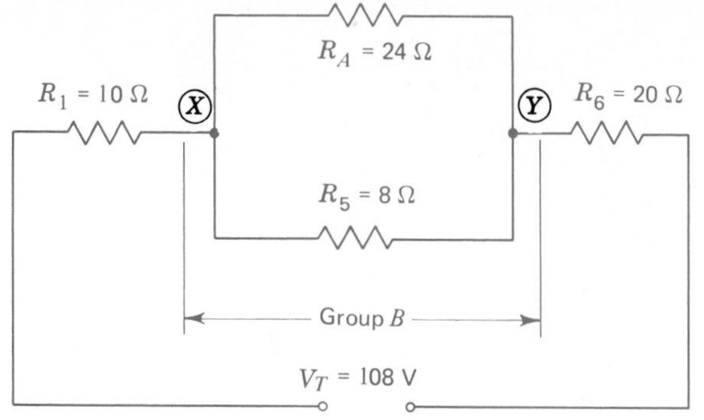


**FIGURE 6-17**
The resistance $R_A$ replaces the series combination of $R_2$, $R_3$, and $R_4$ of Fig. 6-16.

$$R_A = R_2 + R_3 + R_4 \qquad (4\text{-}3)$$
$$R_A = 5 + 10 + 9 = 24\ \Omega \qquad \textit{Ans.}$$

*b.* In Fig. 6-17, the resistances $R_A$ and $R_5$ form the *parallel* group $B$. Find the total resistance of group $B$ and replace the group with the equivalent resistance $R_B$ as shown in Fig. 6-18,

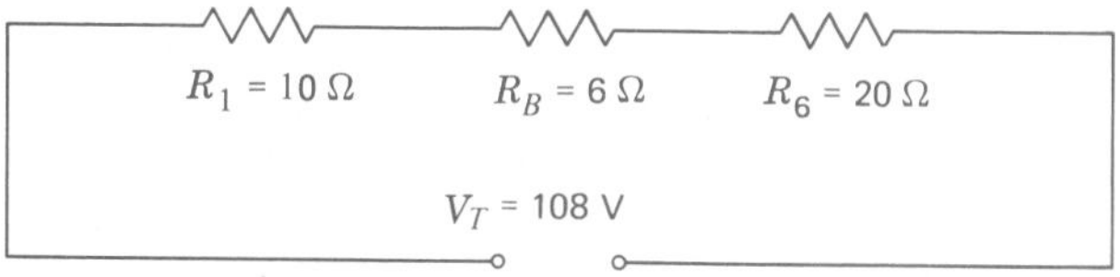

$R_1$ = 10 Ω $R_B$ = 6 Ω $R_6$ = 20 Ω

$V_T$ = 108 V

**FIGURE 6-18**
The resistance $R_B$ replaces the parallel combination of $R_A$ and $R_5$ of Fig. 6-17.

$$R_B = \frac{R_A \times R_5}{R_A + R_5} = \frac{24 \times 8}{24 + 8} = \frac{192}{32} = 6\ \Omega \qquad \textit{Ans.}$$

*c.* In Fig. 6-18, the resistances $R_1$, $R_B$, and $R_6$ form a simple *series* circuit. Find the total resistance of the entire circuit.

$$R_T = R_1 + R_B + R_6 \qquad (4\text{-}3)$$
$$R_T = 10 + 6 + 20 = 36\ \Omega \qquad \textit{Ans.}$$

2. Find the total current $I_T$.

$$V_T = I_T \times R_T \qquad (4\text{-}7)$$
$$108 = I_T \times 36$$
$$I_T = \frac{108}{36} = 3\ \text{A} \qquad \textit{Ans.}$$

3. Find the currents and voltages in each part. Start with the *simplest* circuit obtained in the calculation of the total resistance. This would be Fig. 6-18. In this figure, $R_1$, $R_B$, and $R_6$ are in *series*. Therefore,

$$I_T = I_1 = I_B = I_6 = 3\ \text{A} \qquad (4\text{-}1)$$

Find the voltage across each part by Ohm's law.

$$V_1 = I_1 \times R_1 \qquad V_B = I_B \times R_B \qquad V_6 = I_6 \times R_6$$
$$V_1 = 3 \times 10 \qquad V_B = 3 \times 6 \qquad V_6 = 3 \times 20$$
$$V_1 = 30\ \text{V} \qquad V_B = 18\ \text{V} \qquad V_6 = 60\ \text{V}$$

These values should be entered on the figure so that it will appear as in Fig. 6-19.

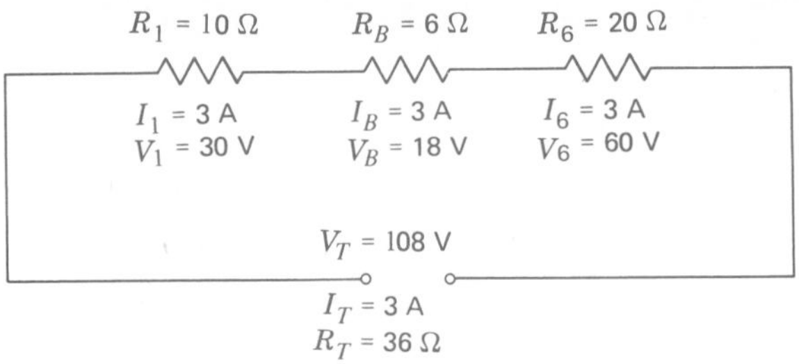

$R_1 = 10\ \Omega$ $R_B = 6\ \Omega$ $R_6 = 20\ \Omega$

$I_1 = 3$ A $I_B = 3$ A $I_6 = 3$ A

$V_1 = 30$ V $V_B = 18$ V $V_6 = 60$ V

$V_T = 108$ V

$I_T = 3$ A

$R_T = 36\ \Omega$

**FIGURE 6-19**
Values of voltage and current are entered on the circuit of Fig. 6-18.

4. We are now ready to find the voltages and currents for group $A$ and resistor $R_5$. In Fig. 6-17, $R_A$ and $R_5$ are in parallel. Therefore,

$$V_B = V_A = V_5 = 18 \text{ V} \qquad (5\text{-}1)$$

Find the current in $R_A$ and $R_5$ by Ohm's law.

$$V_A = I_A \times R_A \qquad V_5 = I_5 \times R_5$$
$$18 = I_A \times 24 \qquad 18 = I_5 \times 8$$
$$I_A = \frac{18}{24} = 0.75 \text{ A} \qquad I_5 = \frac{18}{8} = 2.25 \text{ A}$$

Enter these values on Fig. 6-17 so that it will appear as shown in Fig. 6-20.

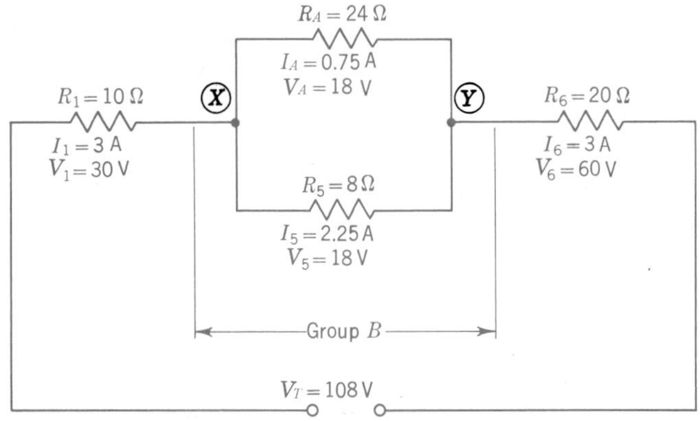


**FIGURE 6-20**
Values of voltage and current are entered on the circuit of Fig. 6-17.

5. We are now ready to find the currents and voltages for the individual resistors of group $A$. In Fig. 6-16, $R_2$, $R_3$, and $R_4$ are in series. Therefore,

$$I_A = I_2 = I_3 = I_4 = 0.75 \text{ A} \qquad (4\text{-}1)$$

Find the voltage across these resistors by Ohm's law.

$$V_2 = I_2 \times R_2 \qquad V_3 = I_3 \times R_3 \qquad V_4 = I_4 \times R_4$$
$$V_2 = 0.75 \times 5 \qquad V_3 = 0.75 \times 10 \qquad V_4 = 0.75 \times 9$$
$$V_2 = 3.75 \text{ V} \qquad V_3 = 7.5 \text{ V} \qquad V_4 = 6.75 \text{ V}$$

6. *Check:* The voltage across $V_2$, $V_3$, and $V_4$ should equal $V_A$.

$$V_A = V_2 + V_3 + V_4 = 18 \text{ V} \qquad (4\text{-}2)$$
$$18 = 3.75 + 7.50 + 6.75$$
$$18 = 18 \quad \textit{Check}$$

EXAMPLE 6-5. Simplify the circuit shown in Fig. 6-21 and find the total resistance of the circuit.

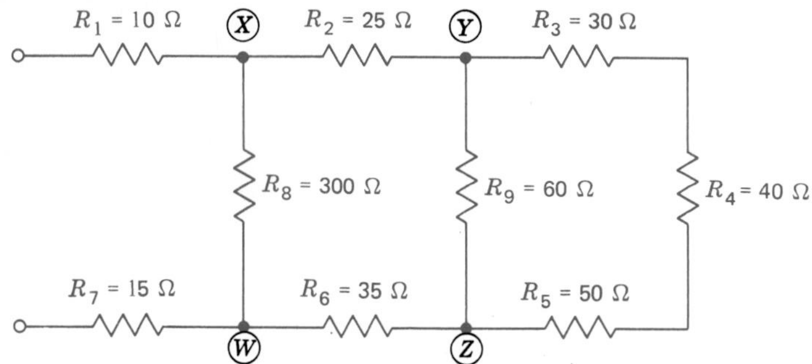


**FIGURE 6-21**

SOLUTION

1. Redraw the circuit in standard form as shown in Fig. 6-22.

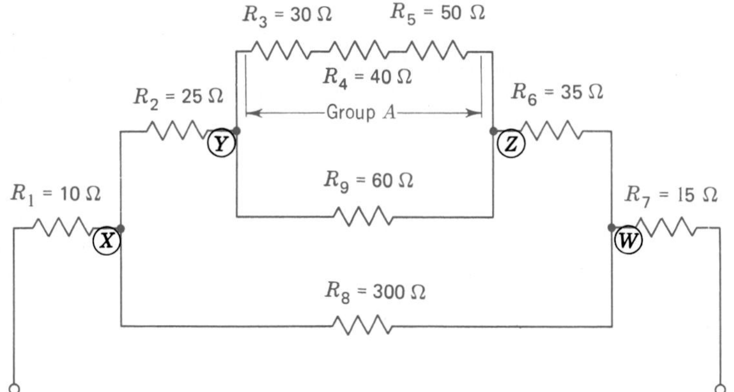


**FIGURE 6-22**
The circuit of Fig. 6-21 is redrawn in standard form.

2. In Fig. 6-22 the resistors $R_3$, $R_4$, and $R_5$ form the *series* group $A$. Find the total resistance of the group and replace the group with the equivalent resistance $R_A$ as shown in Fig. 6-23.

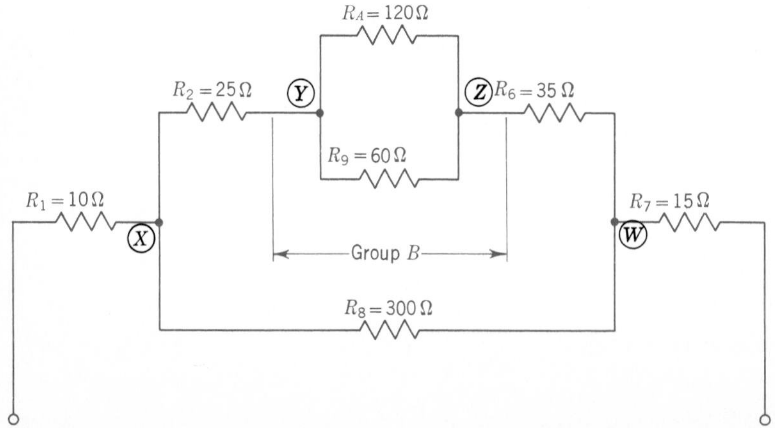


**FIGURE 6-23**
The resistance $R_A$ replaces the series combination of $R_3$, $R_4$, and $R_5$ of Fig. 6-22

$$R_A = R_3 + R_4 + R_5 \qquad (4\text{-}3)$$
$$R_A = 30 + 40 + 50 = 120\ \Omega \qquad \textit{Ans.}$$

3. In Fig. 6-23 the resistances $R_A$ and $R_9$ form the *parallel* group $B$. Find the total resistance of the group and replace the group with the equivalent resistance $R_B$ as shown in Fig. 6-24.

$$R_B = \frac{R_A \times R_9}{R_A + R_9} \qquad (5\text{-}5)$$
$$R_B = \frac{120 \times 60}{120 + 60} = \frac{7200}{180} = 40\ \Omega \qquad \textit{Ans.}$$

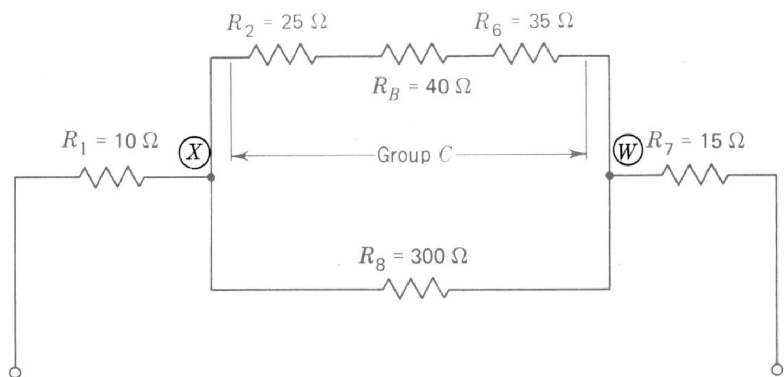


**FIGURE 6-24**
The resistance $R_B$ replaces the parallel combination of $R_A$ and $R_9$ of Fig. 6-23.

4. In Fig. 6-24 the resistances $R_2$, $R_B$, and $R_6$ form the *series* group $C$. Find the total resistance of the group and replace the group with the equivalent resistance $R_C$ as shown in Fig. 6-25.

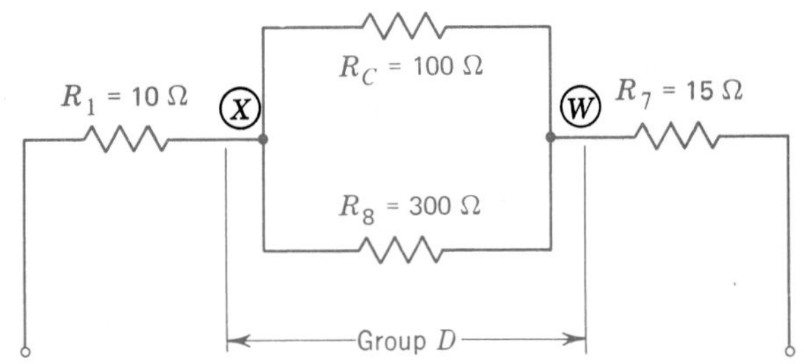


**FIGURE 6-25**
The resistance $R_C$ replaces the series combination of $R_2$, $R_B$, and $R_6$ of Fig. 6-24.

$$R_C = R_2 + R_B + R_6 \qquad (4\text{-}3)$$
$$R_C = 25 + 40 + 35 = 100\ \Omega \qquad \textit{Ans.}$$

5. In Fig. 6-25 the resistances $R_C$ and $R_8$ form the *parallel* group $D$. Find the total resistance of the group and replace the group with the equivalent resistance $R_D$ as shown in Fig. 6-26.

$$R_D = \frac{R_C \times R_8}{R_C + R_8} \qquad (5\text{-}5)$$
$$R_D = \frac{100 \times 300}{100 + 300} = \frac{30{,}000}{400} = 75\ \Omega \qquad \textit{Ans.}$$

6. In Fig. 6-26 the resistances $R_1$, $R_D$, and $R_7$ form a *series* circuit. Find the resistance of the entire circuit.

$$R_T = R_1 + R_D + R_7 \qquad (4\text{-}3)$$

$$R_T = 10 + 75 + 15 = 100\ \Omega \qquad Ans.$$

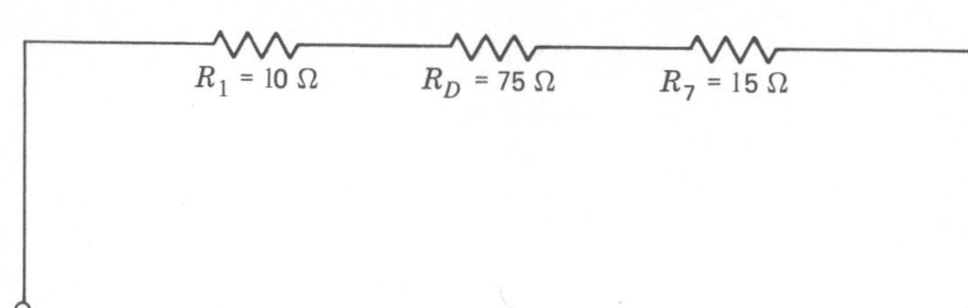


**FIGURE 6-26** The resistance $R_D$ replaces the parallel combination of $R_C$ and $R_8$ of Fig. 6-25.

SELF-TEST 6-6. Find the total resistance of the circuit shown in Fig. 6-27.

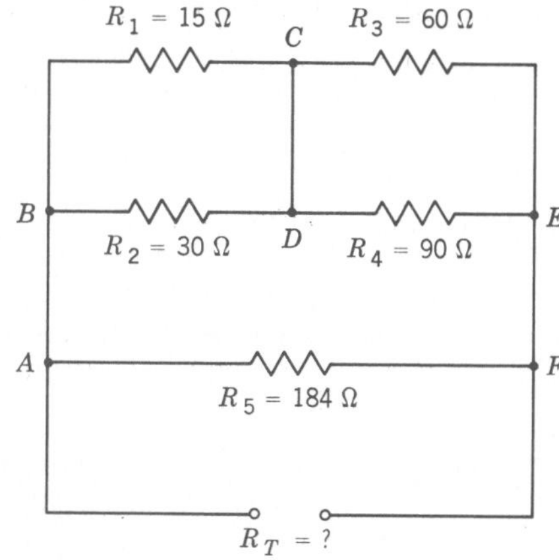


**FIGURE 6-27**

SOLUTION

1. Redraw the circuit in standard form as shown in Fig. 6-28.

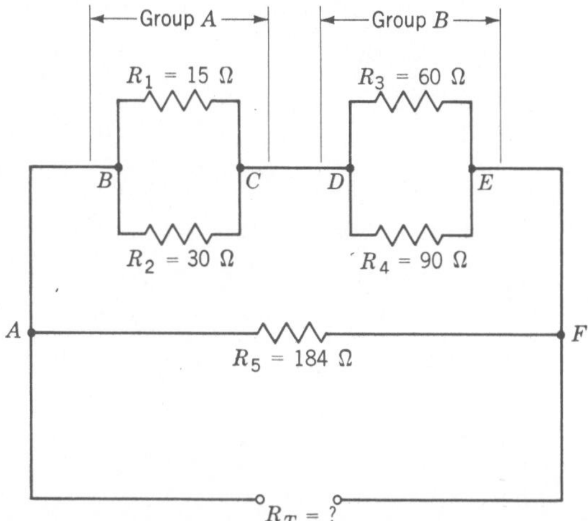


**FIGURE 6-28** Figure 6-27 redrawn in standard form.

2. In Fig. 6-28, the resistors $R_1$ and $R_2$ are in ___(series/parallel)___, forming group *A*. In Fig. 6-28, the resistors $R_3$ and $R_4$ are in ___(series/parallel)___, forming group *B*.

parallel
parallel

3. Find the total resistance of each group.

$$R_A = \frac{R_1 \times R_2}{?}$$

$$R_A = \frac{15 \times 30}{? + 30}$$

$$R_A = \frac{450}{45} = \_\_\_\_\_\_ \Omega \quad \textit{Ans.}$$

$$R_B = \frac{60 \times ?}{60 + 90}$$

$$R_B = \frac{5400}{150} = \_\_\_\_\_\_ \Omega \quad \textit{Ans.}$$

| $R_1 + R_2$
| 15
| 10
| 90
| 36

4. Redraw the circuit, replacing each group with its equivalent resistance $R_A$ and $R_B$ as shown in Fig. 6-29.

5. In Fig. 6-29, the resistances $R_A$ and $R_B$ form the ____ (series/parallel group $C$.

| series

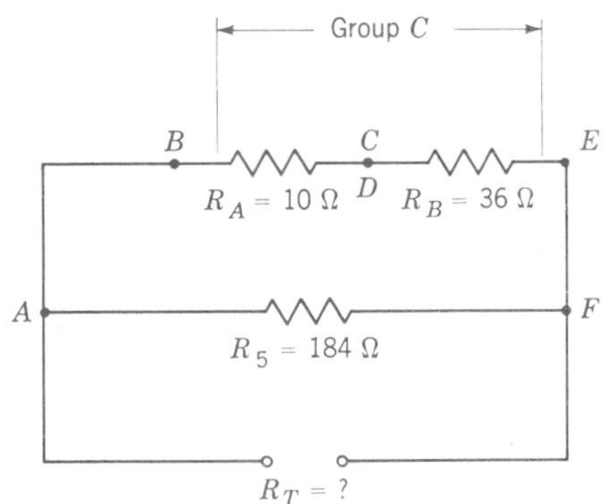


**FIGURE 6-29**
The resistance $R_A$ replaces the parallel combination of $R_1$ and $R_2$; $R_B$ replaces the parallel combination of $R_3$ and $R_4$ in Fig. 6-28.

Find the total resistance of the group $C$.

$$R_C = \_\_\_\_\_\_ \quad (4\text{-}3)$$

$$R_C = 10 + \_\_\_\_\_\_ = \_\_\_\_\_\_ \Omega \quad \textit{Ans.}$$

| $R_A + R_B$
| 36 46

6. Redraw the circuit, replacing $R_A$ and $R_B$ with their equivalent resistance $R_C$ as shown in Fig. 6-30.

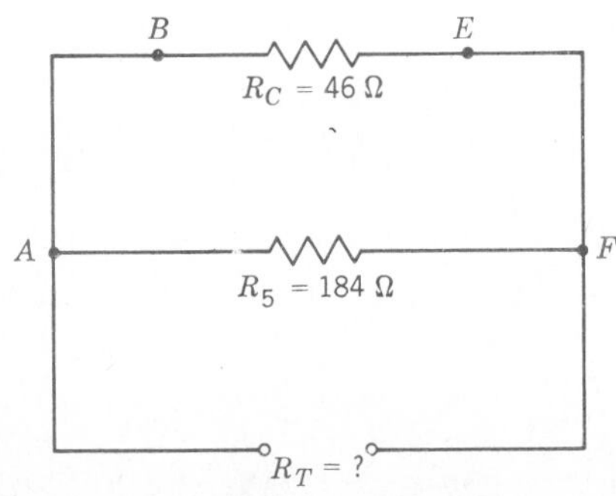


**FIGURE 6-30**
The resistance $R_C$ replaces the series combination of $R_A$ and $R_B$ in Fig. 6-29.

7. In Fig. 6-30, the resistances $R_C$ and $R_5$ are in ______ (series/parallel). Find $R_T$. | parallel

$$R_T = \frac{?}{?}$$ | $\frac{R_C \times R_5}{R_C + R_5}$

$$R_T = \frac{46 \times ?}{46 + ?}$$ | 184, 184

$$R_T = \frac{8464}{230} = \underline{\qquad}\ \Omega \quad Ans.$$ | 36.8

**PROBLEMS**

1. In a circuit similar to that shown in Fig. 6-12, $V_T = 130$ V, $R_1 = 10$ Ω, $R_2 = 4$ Ω, and $R_3 = 12$ Ω. Find all missing values of voltage, current, and resistance.
2. Find all missing values in the circuit shown in Fig. 6-31.

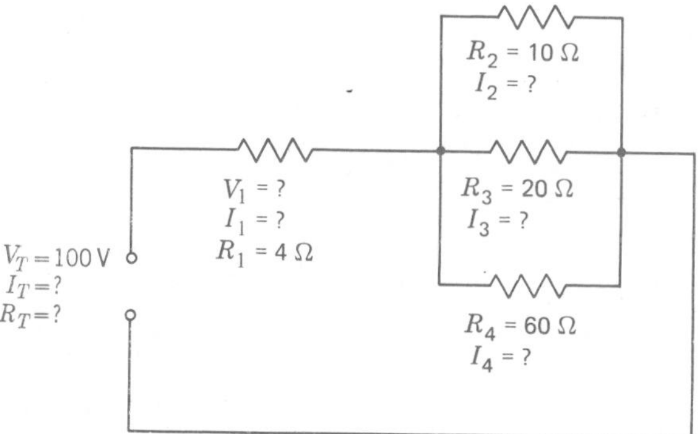


FIGURE 6-31

3. In a circuit similar to that shown in Fig. 6-31, find the total current if $V_T = 8.5$ V, $R_1 = 12$ Ω, $R_2 = 10$ Ω, $R_3 = 15$ Ω, and $R_4 = 30$ Ω.

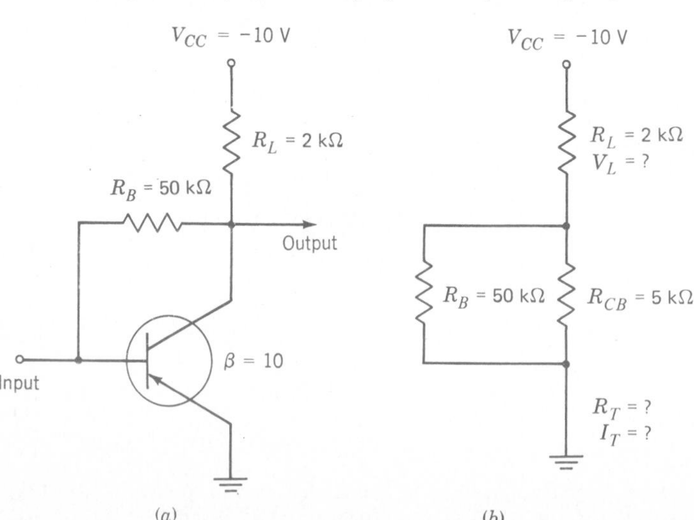


FIGURE 6-32
(a) Self-bias transistor circuit. (b) The dc equivalent circuit.

4. Use a circuit similar to that shown in Fig. 6-15. $V_T = 120$ V, $R_1 = 5\ \Omega$, $R_2 = 1\ \Omega$, $R_3 = 3\ \Omega$, $R_4 = 6\ \Omega$, $R_5 = 10\ \Omega$, and $R_6 = 10\ \Omega$. Find (*a*) the total resistance, (*b*) the total current, (*c*) the current in each resistor, and (*d*) the voltage drop across each resistor.
5. A self-bias transistor circuit is shown in Fig. 6-32*a*, and its dc equivalent circuit in Fig. 6-32*b*. Find (*a*) the total resistance of the circuit, (*b*) the total current in the circuit, and (*c*) the voltage across $R_L$.
6. Use a circuit similar to that shown in Fig. 6-21. $V_T = 400$ V, $R_1 = 50\ \Omega$, $R_2 = 25\ \Omega$, $R_3 = 8\ \Omega$, $R_4 = 2\ \Omega$, $R_5 = 10\ \Omega$, $R_6 = 15\ \Omega$, $R_7 = 105\ \Omega$, $R_8 = 450\ \Omega$, and $R_9 = 20\ \Omega$. Find (*a*) the total resistance, (*b*) the total current, (*c*) the current in each resistor, and (*d*) the voltage drop across each resistor.
7. Find the total resistance of the circuit shown in Fig. 6-33.

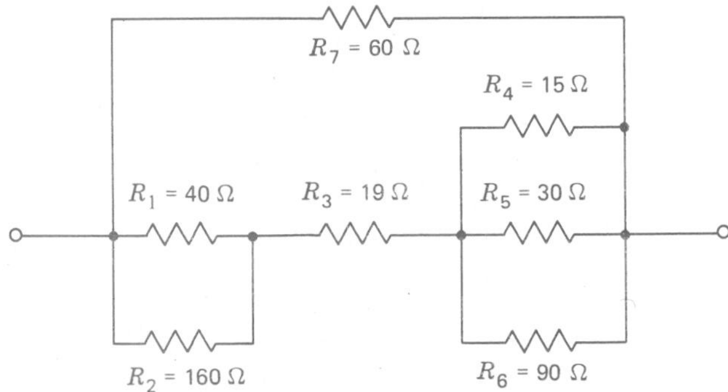


**FIGURE 6-33**

8. Find the total resistance of the circuit shown in Fig. 6-27 if the 60-Ω resistor were to burn out and open.
9. In Fig. 6-34, (*a*) draw the equivalent resistance network, (*b*) find the total resistance between the point *P* and the ground *G*, and (*c*) find the resistance that must be connected between *P* and *G* (in

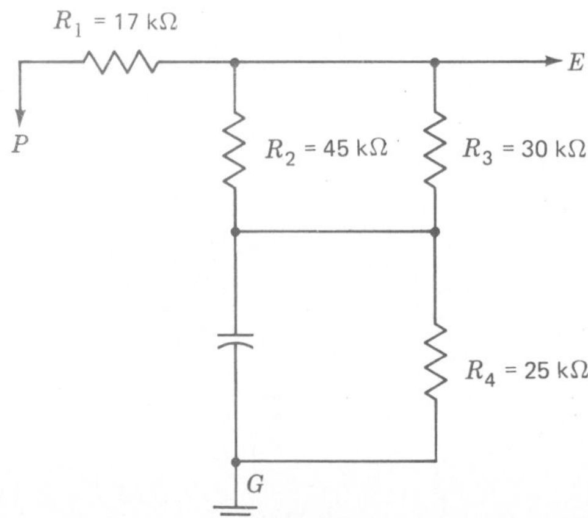


**FIGURE 6-34**

parallel with the entire circuit) to reduce the total resistance to 36 kΩ.

10. In Fig. 6-35, find the total resistance from $A$ to $B$.

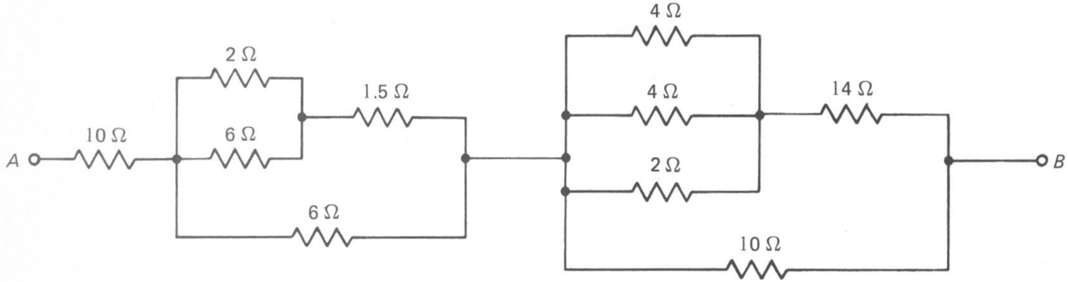


FIGURE 6-35

11. Find the total resistance of the circuit shown in Fig. 6-36.

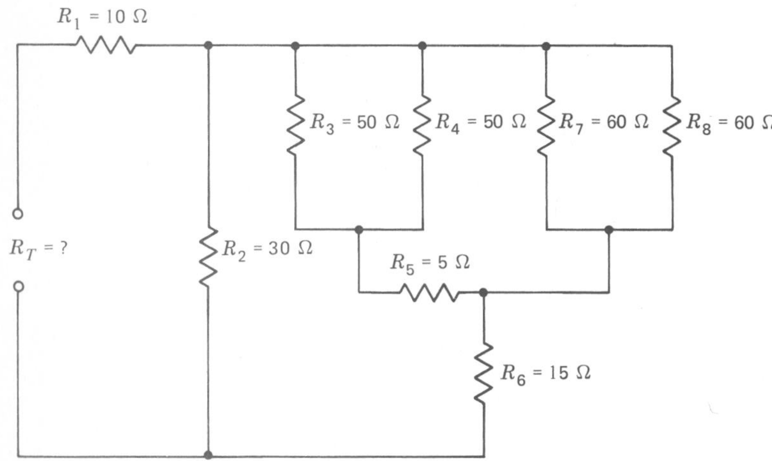


FIGURE 6-36

12. Find the total resistance of the circuit shown in Fig. 6-37.

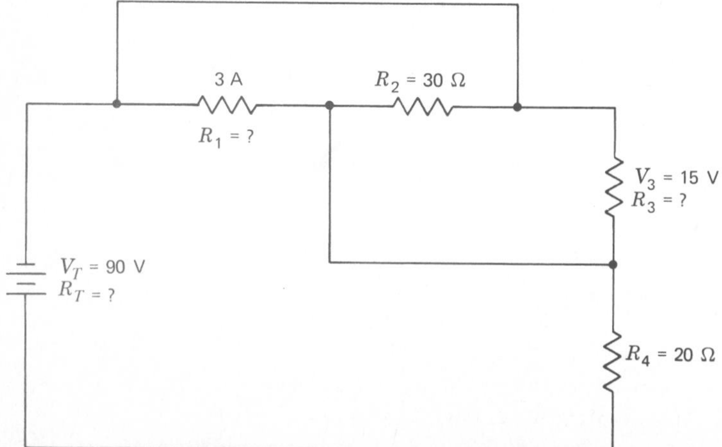


FIGURE 6-37

## SUMMARY—COMBINATION CIRCUITS

### Finding the total resistance

1. Redraw the circuit in ________ form if necessary. | standard
2. First simplification.
   *a.* Locate the combination of resistors which forms a simple ________ or parallel circuit. | series
   *b.* Indicate the beginning and ________ of the group with a dimension line and name it "group ________." | end | $A$
   *c.* Calculate the resistance of this group and label it ________. | $R_A$
3. Redraw the circuit, substituting ________ in place of the group. | $R_A$
4. Second simplification.
   *a.* Repeat steps 2*a*, 2*b*, and 2*c*, using the name "group ________" and $R_B$ in the circuit of step 3. | $B$
5. Repeat steps 2 and 3 using the names "group *C*," "group *D*," etc., until the circuit is reduced to the total resistance named ________. | $R_T$

### Solving combination circuits

1. Find the total resistance $R_T$ by repeated simplification of the original circuit.
2. Find the total current $I_T$ using the formula

$$V_T = \_\_\_\_\_\_\_\_ \times R_T \qquad (4\text{-}7)$$ | $I_T$

3. Find the current and voltage in each resistor.
   *a.* Start with the ______(simplest/most complicated)______ circuit obtained in the calculations for $R_T$. | simplest
   *b.* The value of current in this circuit will be the ________ current. Enter this value of $I_T$ on this circuit diagram. | total
   *c.* Calculate the voltage drops across all resistors of this circuit by the formula

________ | $V = I \times R$

   *d.* Enter these values on the next simplest circuit obtained in the original calculation for ________. | $R_T$
   *e.* Solve this circuit for all missing values of current and ________. | voltage
   *f.* Repeat steps *d* and *e* on successive circuits found in the ________ simplification until all parts have been found. | resistance

(See Instructor's Manual for Test 6-2, Combination Circuits)

## JOB 6-4 LINE DROP

**Meaning of line drop.** In our last two jobs, we worked with circuits in which the connecting wires were very short. The resistances of these short lengths were so very small that we did not bother to include them in our calculations. In home and factory installations, however, where long lines of wire (feeders) are used, the resistance of these long lengths must be included in all calculations.

In Fig. 6-38, the voltage that is needed to force the current through the resistance of the line wires is called the *line drop.* For example, if a generator delivers 120 V but the voltage available at a motor some distance away is only 116 V, then there has been a "drop" in voltage of 4 V. The connecting wires apparently had enough resistance to use up 4 V of electrical pressure.

We must be very careful about the kind and size of wires used in any installation. If the wires are poorly chosen, then the *line drop* may be very large and the voltage available to the electric apparatus will be too low for proper operation. The national and city electrical codes permit definite amounts of line drop for specific installations. This limitation of the line drop is accomplished by specifying definite sizes of wire to be used in specific installations. We shall study this in detail in Chap. 13.

You may have noticed the effect of line drop in your home. The lights may suddenly get dim when the refrigerator motor starts. The large current drain required to start the motor increases the line drop in the house wiring so that the voltage left over for the lights is less than normal. Since the running current is much less than the starting current, as soon as the motor has started the current drain decreases—the line drop decreases—and the lights once again come up to full brilliance.

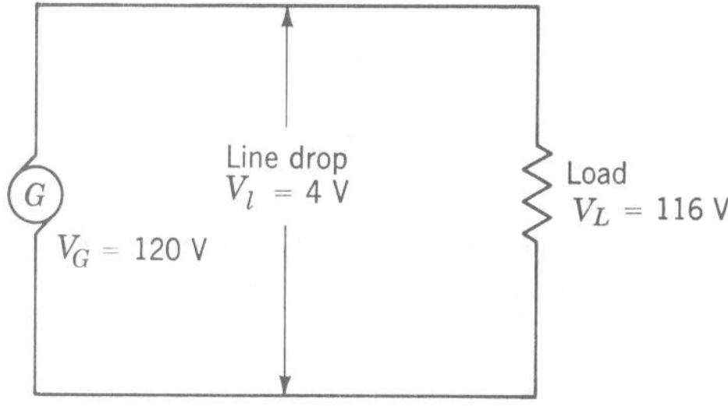


**FIGURE 6-38** Voltage is lost in the line wires between a generator and a load.

## DEFINITIONS

1. Generator voltage ($V_G$): The total voltage supplied to the circuit from the source of voltage
2. Load voltage ($V_L$): The voltage which is available to operate the devices or loads
3. Line drop ($V_l$): The voltage which is lost in sending the current through the line wires

## RULES AND FORMULAS

1. By Ohm's law, the total line drop is equal to the line current multiplied by the total line resistance.

$$V_{\text{line}} = I_{\text{line}} \times R_{\text{line}}$$

or

**FORMULA** $$V_l = I_l \times R_l \qquad \boxed{6\text{-}1}$$

2. The generator voltage is the total voltage of the series circuit, made up of the line drop and the load voltage.

$$V_T = V_1 + V_2 \tag{4-2}$$

or

$$V_{\text{generator}} = V_{\text{line}} + V_{\text{load}}$$

**FORMULA** $$V_G = V_l + V_L \qquad \boxed{6\text{-}2}$$

3. The load voltage is equal to the generator voltage minus the line drop. From

$$V_l + V_L = V_G \tag{6-2}$$

we get, by transposing the $V_l$,

**FORMULA** $$V_L = V_G - V_l \qquad \boxed{6\text{-}3}$$

and, by transposing the $V_L$,

**FORMULA** $$V_l = V_G - V_L \qquad \boxed{6\text{-}4}$$

When using these formulas, be sure that

1. The line current $I_l$ is the current flowing in the line wires and *not* the current in the load.
2. The line drop $V_l$ is the voltage lost in the line wires only.
3. The line resistance is the resistance of the connecting wires only. The resistances of *both* lead and return wires must be considered when finding the total resistance of the line.

EXAMPLE 6-7. A lamp bank consisting of three lamps, each drawing 2 A, is connected to a 120-V source. Each line wire has a resistance of 0.2 Ω. Find the drop in voltage in the line and the voltage available at the load.

SOLUTION. The diagram for the circuit is shown in Fig. 6-39.

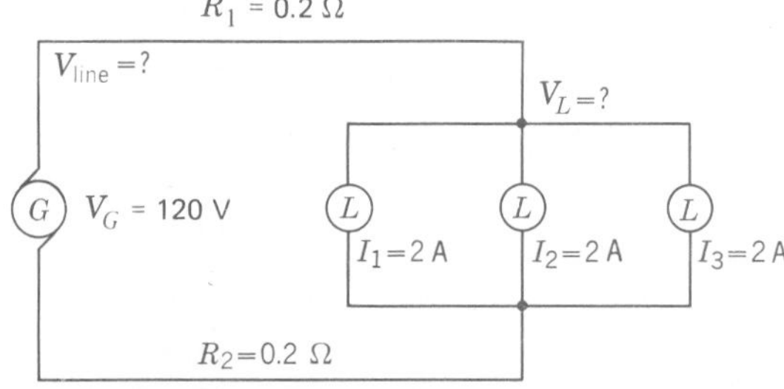


FIGURE 6-39

1. Find the current drawn by the load.

$$I_L = I_1 + I_2 + I_3 \quad (5\text{-}2)$$
$$I_L = 2 + 2 + 2 = 6 \text{ A} \quad \textit{Ans.}$$

2. Find the line current. Since the line wires are in series with the load,

$$I_l = I_L = 6 \text{ A} \quad \textit{Ans.} \quad (4\text{-}1)$$

3. Find the resistance of the line wires. Since the line wires are in series,

$$R_l = R_1 + R_2 \quad (4\text{-}3)$$
$$R_l = 0.2 + 0.2 = 0.4\ \Omega \quad \textit{Ans.}$$

4. Find the total line drop.

$$V_l = I_l \times R_l \quad (6\text{-}1)$$
$$V_l = 6 \times 0.4 = 2.4 \text{ V} \quad \textit{Ans.}$$

5. Find the voltage available at the load.

$$V_L = V_G - V_l \quad (6\text{-}3)$$
$$V_L = 120 - 2.4 = 117.6 \text{ V} \quad \textit{Ans.}$$

EXAMPLE 6-8. Find the generator voltage required for the circuit shown in Fig. 6-40.

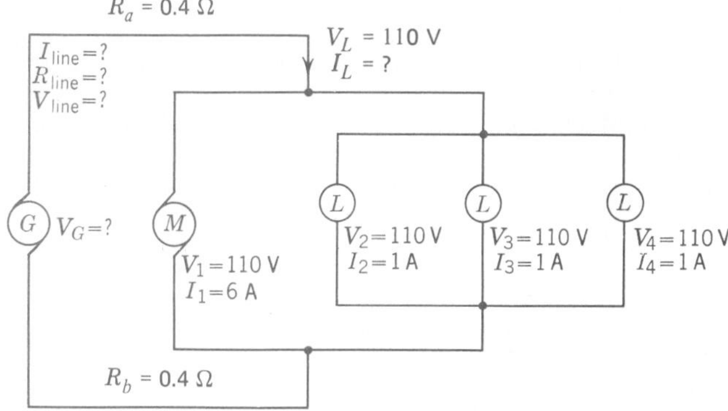


**FIGURE 6-40**

SOLUTION

1. Find the current drawn by the load.

$$I_L = I_1 + I_2 + I_3 + I_4 \quad (5\text{-}2)$$
$$I_L = 6 + 1 + 1 + 1 = 9 \text{ A} \quad \textit{Ans.}$$

2. Find the line current.

$$I_l = I_L = 9 \text{ A} \quad \textit{Ans.} \quad (4\text{-}1)$$

3. Find the resistance of the line wires.

$$R_l = R_a + R_b \quad (4\text{-}3)$$
$$R_l = 0.4 + 0.4 = 0.8\ \Omega \quad \textit{Ans.}$$

4. Find the total line drop.

$$V_l = I_l \times R_l \qquad (6\text{-}1)$$
$$V_l = 9 \times 0.8 = 7.2 \text{ V} \qquad \textit{Ans.}$$

5. Find the generator voltage.

$$V_G = V_l + V_L \qquad (6\text{-}2)$$
$$V_G = 7.2 + 110 = 117.2 \text{ V} \qquad \textit{Ans.}$$

SELF-TEST 6-9. Find the resistance of each line wire in the circuit shown in Fig. 6-41.

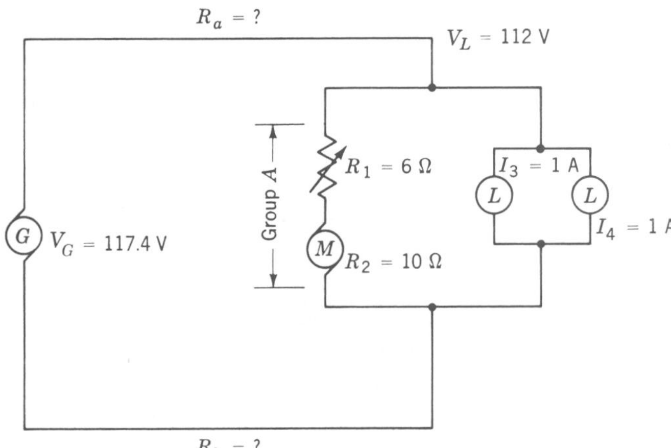


**FIGURE 6-41**

SOLUTION. In order to find the line resistance we must know the voltage drop in the line and the current in the line. Find the current distribution by investigating group $A$ first.

1. Find the total resistance of group $A$.

$$R_A = R_1 + R_2 \qquad (4\text{-}3)$$
$$R_A = 6 + 10 = \underline{\qquad} \ \Omega$$ 16

2. Since $V_A$ is in parallel with the lamp load and the load voltage $V_L$,

$$V_A = V_L = \underline{\qquad} \text{ V} \qquad (5\text{-}1)$$ 112

3. Find the current in group $A$.

$$V_A = I_A \times R_A \qquad (4\text{-}7)$$
$$112 = I_A \times \underline{\qquad}$$ 16
$$I_A = \frac{112}{16} = \underline{\qquad} \text{ A}$$ 7

4. Find the total load current $I_L$.

$$I_L = I_A + I_3 + I_4 \qquad (5\text{-}2)$$
$$I_L = \underline{\qquad} + 1 + 1 = \underline{\qquad} \text{ A}$$ 7 9

5. Since the line wires are in series with the load,

$$I_l = I_L = \_\_\_\_\_\_ \text{ A} \qquad (4\text{-}1)$$ 9

6. Find the voltage drop in the line wires.

$$V_l = V_G - V_L \qquad (6\text{-}4)$$

$$V_l = 117.4 - 112 = \_\_\_\_\_\_ \text{ V}$$ 5.4

7. Find the resistance of the line wires.

$$V_l = I_l \times R_l \qquad (6\text{-}1)$$

$$5.4 = \_\_\_\_\_\_ \times R_l$$ 9

$$R_l = \frac{5.4}{9} = \_\_\_\_\_\_ \ \Omega$$ 0.6

8. Find the resistance of each line wire.

$$R_a = R_b = \frac{1}{2} \times R_l$$

$$R_a = R_b = \frac{1}{2} \times 0.6 = \_\_\_\_\_\_ \ \Omega \qquad \textit{Ans.}$$ 0.3

**PROBLEMS**

1. In a circuit similar to that shown in Fig. 6-39, the generator voltage is 117 V. Each line wire has a resistance of 0.4 Ω, and each lamp draws 1 A. Find the line drop and the voltage available at the lamps.
2. A motor is connected by two wires of 0.15 Ω each to a generator. The motor takes 30 A at 211 V. What must be the generator voltage?
3. If the voltage at a load drawing 6 A is 117 V while the generator voltage is 120 V, what is the resistance of each line wire? *Hint:* Find the line drop, then the line resistance, and finally the resistance of each wire.
4. Home wiring is often done with No. 16 wire, which has a resistance of 0.401 Ω for a 100-ft length. What is the loss in voltage from the house meter to an electric broiler using 12 A and located 100 ft from the meter?
5. What would be the voltage drop if No. 14 wire (0.252 Ω/100 ft) were used in Prob. 4? Which size of wire is better for wiring homes?
6. In a circuit similar to that shown in Fig. 6-39, the generator voltage is 117 V. Each line wire has a resistance of 0.45 Ω, and the lamps draw currents of 0.9, 1.4, and 1.8 A. Find the line drop and the voltage available at the lamps.
7. In a circuit similar to that shown in Fig. 6-40, the motor draws 8.2 A and each of the three lamps draws 0.92 A. Each line wire has a resistance of 0.15 Ω. Find the generator voltage if the load voltage must be 110 V.
8. In a circuit similar to that shown in Fig. 6-39, the generator voltage is

117 V, the resistance of each line wire is 0.2 Ω, and the total resistance of the lamp bank is 16.1 Ω. Find (*a*) the total resistance of the circuit, (*b*) the current delivered to the lamp bank, and (*c*) the voltage across the lamp bank.

## JOB 6-5 DISTRIBUTION SYSTEMS

In order to distribute current throughout an installation, the various loads are connected in parallel across the feeder lines. The feeder lines form various combination circuits with the loads. To solve circuits like these, we must first break down the combination into simple series or parallel circuits. It is best to follow a definite system like the following:

1. Find the current distribution. Start with the section *farthest* from the generator. Find the current in this section and work backward toward the generator, finding the current in the different parts of the circuit.
2. Name the sections. Call the section nearest to the generator section *A*, the next section *B*, etc.
3. Find the resistance of each *pair* of line wires for each section using Eq. (4-3).
4. Find the line drop for each section of the circuit using Eq. (6-1). The line wires connecting the generator to section *A* are called line 1, the next line-wire pair is called line 2, etc.
5. Find the voltage across each section. Start where the voltage is known, and apply Eqs. (6-2) and (6-3).

EXAMPLE 6-10. Find the voltage across the motor and across the lamp bank of the circuit shown in Fig. 6-42.

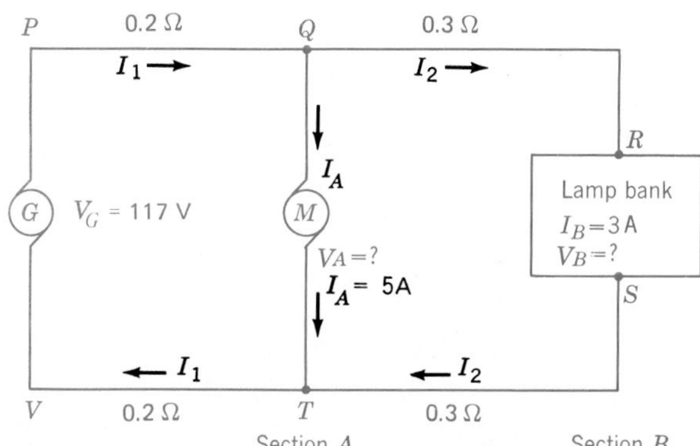


FIGURE 6-42

1. Find the current distribution. Start with the section *B* farthest from the generator. $I_B = 3$ A. Line 2, from *Q* to *R* and from *S* to *T*, must carry this 3 A. Therefore, $I_2 = 3$ A. Since $I_A = 5$ A, the wire from *Q* to *T* must carry this 5 A. At point *Q*, we have the beginning of a parallel circuit made of the motor (section *A*) and the lamp bank (section *B*).

$$I_1 = I_A + I_2 \tag{5-2}$$
$$I_1 = 5 + 3 = 8 \text{ A}$$

Thus, the current in line 1 from $P$ to $Q$ equals 8 A. Similarly, at $T$, the current $I_A$ from the motor and the current $I_2$ from the lamp bank will combine.

$$I_1 = I_A + I_2 \tag{5-2}$$
$$I_1 = 5 + 3 = 8 \text{ A}$$

Thus, the current in line 1 from $T$ to $V$ equals 8 A.

2. Find the resistance of the *pairs* of line wires.

$$R_{l_1} = 0.2 + 0.2 = 0.4\ \Omega \qquad R_{l_2} = 0.3 + 0.3 = 0.6\ \Omega \tag{4-3}$$

3. Find the line drop for each section.

$$V_{l_1} = I_1 \times R_{l_1} \qquad V_{l_2} = I_2 \times R_{l_2} \tag{6-1}$$
$$V_{l_1} = 8 \times 0.4 = 3.2 \text{ V} \qquad V_{l_2} = 3 \times 0.6 = 1.8 \text{ V}$$

4. Find the load voltages.

$$V_A = V_G - V_{l_1} \tag{6-3}$$
$$V_A = 117 - 3.2 = 113.8 \text{ V} \quad \textit{Ans.}$$
$$V_B = V_A - V_{l_2} \tag{6-3}$$
$$V_B = 113.8 - 1.8 = 112 \text{ V} \quad \textit{Ans.}$$

EXAMPLE 6-11. Find the generator voltage and the voltage across the motor in the circuit shown in Fig. 6-43.

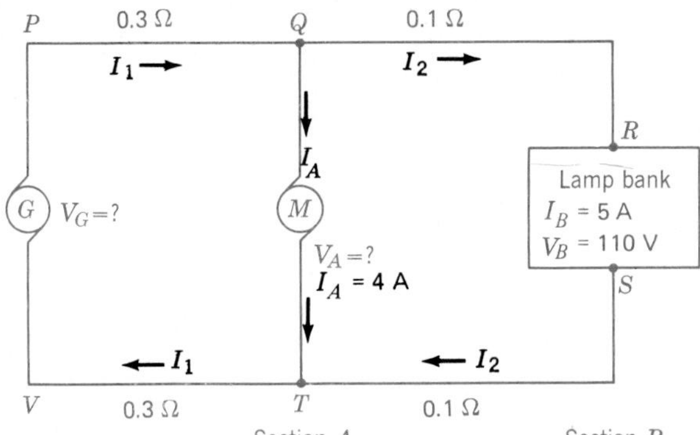

FIGURE 6-43

SOLUTION

1. Find the current distribution. Start with the section $B$ farthest from the generator. $I_B = 5$ A. Line 2, from $Q$ to $R$ and from $S$ to $T$, must carry this 5 A. Therefore, $I_2 = 5$ A.

Since $I_A = 4$ A, the wire from $Q$ to $T$ must carry this 4 A.

At point $Q$: At point $T$:

$$I_1 = I_A + I_2 \qquad I_1 = I_A + I_2 \tag{5-2}$$
$$I_1 = 4 + 5 = 9 \text{ A} \qquad I_1 = 4 + 5 = 9 \text{ A}$$

2. Find the resistance of the *pairs* of line wires.

$$R_{l_1} = 0.3 + 0.3 = 0.6\ \Omega \qquad R_{l_2} = 0.1 + 0.1 = 0.2\ \Omega \tag{4-3}$$

3. Find the line drop for each section.

$$V_{l_1} = I_1 \times R_{l_1} \qquad V_{l_2} = I_2 \times R_{l_2} \qquad (6\text{-}1)$$
$$V_{l_1} = 9 \times 0.6 = 5.4 \text{ V} \qquad V_{l_2} = 5 \times 0.2 = 1 \text{ V}$$

4. Find the load and generator voltages.

$$V_A = V_{l_2} + V_B \qquad (6\text{-}2)$$
$$V_A = 1 + 110 = 111 \text{ V} \quad \textit{Ans.}$$
$$V_G = V_{l_1} + V_A$$
$$V_G = 5.4 + 111 = 116.4 \text{ V} \quad \textit{Ans.}$$

SELF-TEST 6-12. Find the voltage across (*a*) the lamp bank, (*b*) motor 1, and (*c*) motor 2 in the circuit shown in Fig. 6-44.

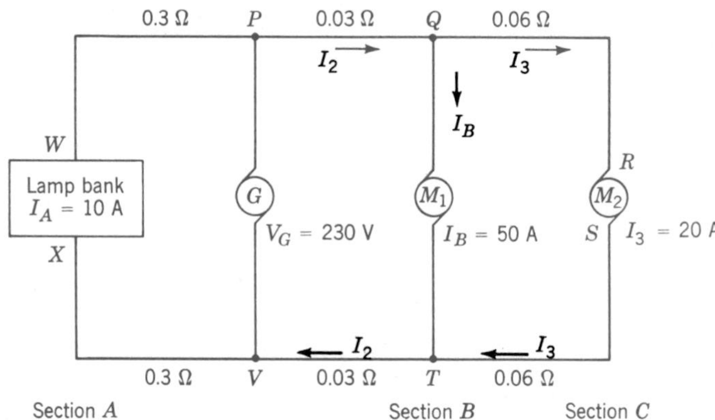


FIGURE 6-44

SOLUTION

1. Find the current distribution. Start with section ________. $I_3$ from $Q$ to $R$ and from $S$ to $T$ = ________ A. At point $Q$, the incoming current $I_2$ divides into two branches, ________ and ________.

*C* | 20 | $I_3$ $I_B$

$$I_2 = I_3 + \_\_\_\_\_\_$$
$$I_2 = 20 + \_\_\_\_\_\_ = \_\_\_\_\_\_ \text{ A}$$

$I_B$ | 50 70

Thus, the current from $P$ to $Q$ = ________ A. At point $T$, $I_3$ and $I_B$ combine to form ________. Thus, the current from $T$ to $V$ = ________ A. The current $I_A$ from $W$ to $P$ and from $V$ to $X$ equals ________ A.

70 | $I_2$ 70 | 10

2. Find the resistance of the *pairs* of line wires.

$$R_{l_A} = 0.3 + 0.3 = \_\_\_\_\_\_ \ \Omega$$
$$R_{l_2} = 0.03 + \_\_\_\_\_\_ = \_\_\_\_\_\_ \ \Omega$$
$$R_{l_3} = \_\_\_\_\_\_ + \_\_\_\_\_\_ = 0.12 \ \Omega$$

0.6 | 0.03 0.06 | 0.06 0.06

3. Find the line drop for each section.

$$V_{l_A} = I_A \times \_\_\_\_\_\_$$
$$V_{l_A} = 10 \times \_\_\_\_\_\_ = \_\_\_\_\_\_ \text{ V}$$
$$V_{l_2} = \_\_\_\_\_\_ \times R_{l_2}$$

$R_{l_A}$ | 0.6 6 | $I_2$

$V_{l_2} = 70 \times$ ________ $=$ ________ V  0.06  4.2

$V_{l_3} = I_3 \times$ ________  $R_{l_3}$

$V_{l_3} =$ ________ $\times 0.12 =$ ________ V  20  2.4

4. Find the load voltages. The voltage across the lamp bank is called $V_A$.

$V_A = V_G -$ ________ (6-3)  $V_{l_A}$

$V_A =$ ________ $- 6 =$ ________ V *Ans.*  230  224

The voltage across motor 1 is called $V_B$.

$V_B =$ ________ $- V_{l_2}$ (6-3)  $V_G$

$V_B = 230 -$ ________ $= 225.8$ V *Ans.*  4.2

The voltage across motor 2 is called ________.  $V_C$

$V_C = V_B -$ ________ (6-3)  $V_{l_3}$

$V_C = 225.8 - 2.4 =$ ________ V *Ans.*  223.4

## PROBLEMS

1. In the circuit shown in Fig. 6-45, the motor takes 30 A and the lamp bank takes 8 A. The generator voltage is 117 V. Find (*a*) the line drop in each section, (*b*) the voltage across the motor, and (*c*) the voltage across the lamp bank.

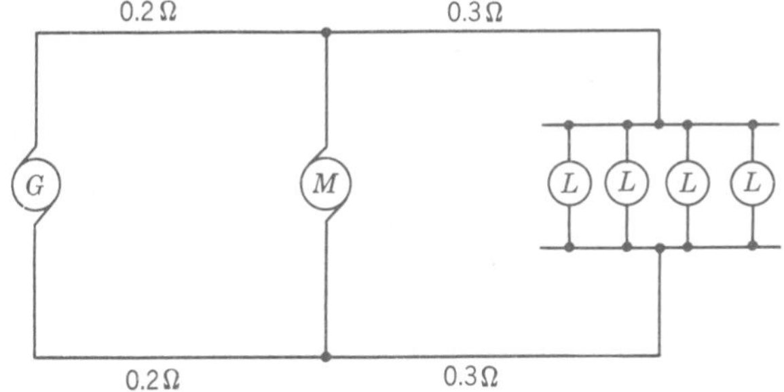


FIGURE 6-45

2. Each lamp in the circuit shown in Fig. 6-45 takes 1.5 A. The motor takes 12 A. The generator voltage is 115 V. Find (*a*) the line drop in each section, (*b*) the voltage across the motor, and (*c*) the voltage across the lamp bank.
3. Each lamp in the diagram shown in Fig. 6-46 takes 0.5 A. Find $V_A$ and $V_B$.

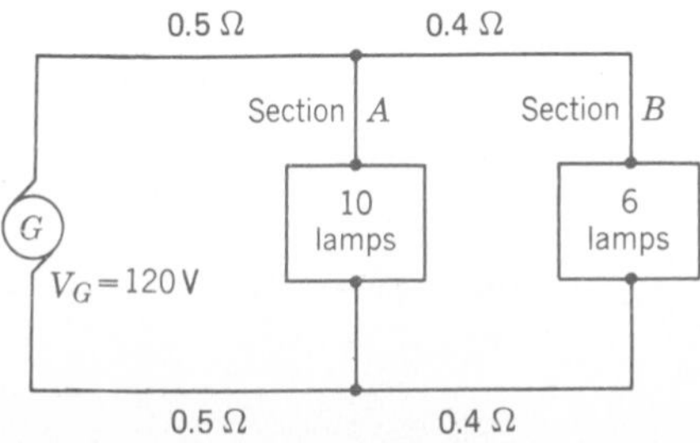


FIGURE 6-46

4. In the circuit shown in Fig. 6-47, the resistance of the wires $AB$ is 0.3 Ω, $BC$ is 0.5 Ω, $EF$ is 0.5 Ω, and $DE$ is 0.3 Ω. If each lamp takes 1 A, what is the terminal voltage of the generator? The voltage across $CF$ is 105 V.

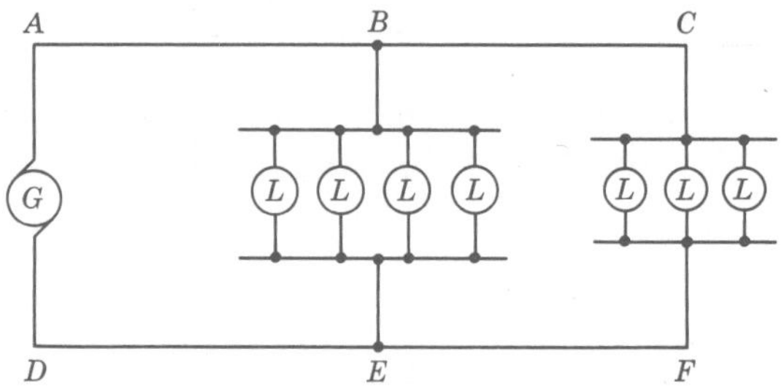


FIGURE 6-47

5. In the circuit shown in Fig. 6-48, each lamp takes 1.2 A. $V_A = 113$ V, and $I_A = 12$ A. Find $V_G$ and $V_B$.

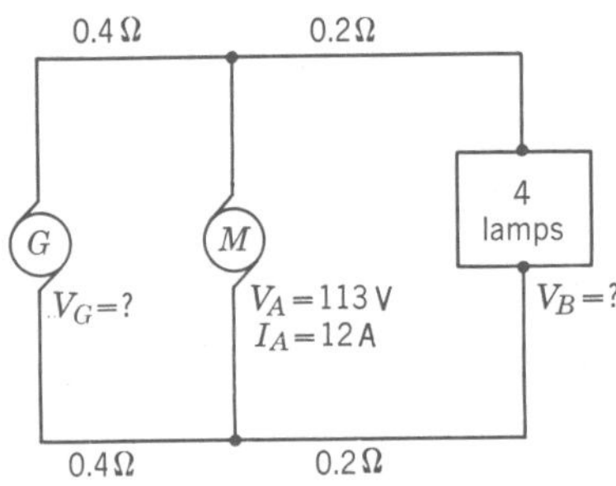


FIGURE 6-48

6. In the circuit shown in Fig. 6-49, each lamp takes 1 A. $V_A = 114$ V, and $V_G = 117$ V. Find $V_B$ and $I_A$.

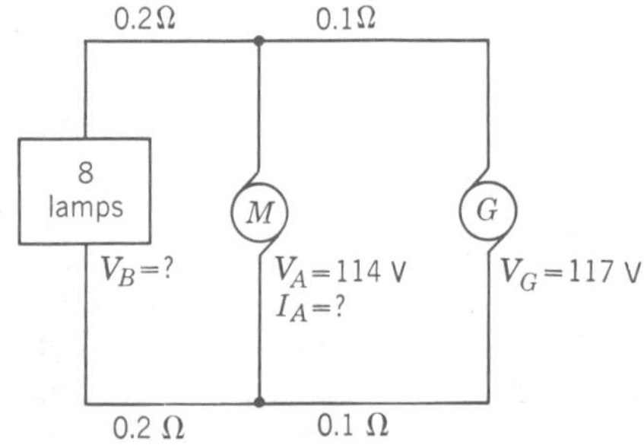


FIGURE 6-49

7. Find the generator voltage in the circuit shown in Fig. 6-50.

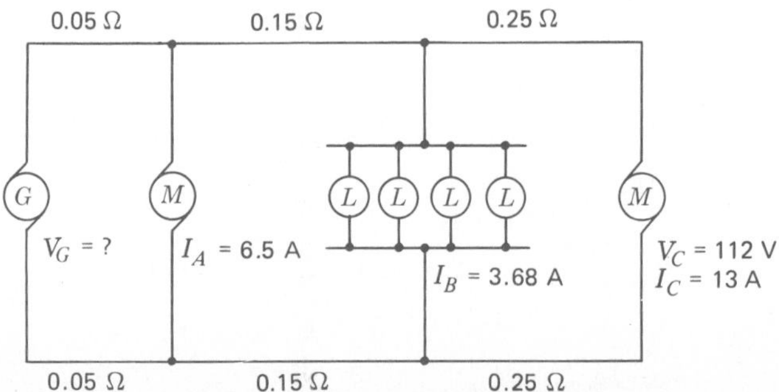


FIGURE 6-50

8. Each line wire of a two-wire distribution system has a resistance of 0.2 Ω. A motor drawing 10 A is connected by two of these wires to a generator delivering 121 V. Two more of these line wires continue from the motor and carry current to a lamp bank drawing 4.5 A. Find (*a*) the voltage at the motor and (*b*) the voltage at the lamp bank.
9. In Fig. 6-51, find (*a*) the voltage across motor 1 and (*b*) the voltage across motor 2. *Hint:* $V_{M_1} = V_G - (V_{l_1} + V_{l_2} + V_{l_4})$, and $V_{M_2} = V_G - (V_{l_1} + V_{l_3} + V_{l_2})$.

(See Instructor's Manual for Test 6-3, Distribution Systems)

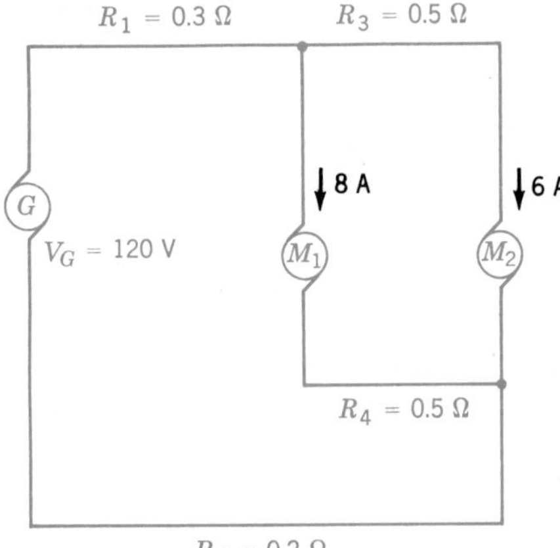


FIGURE 6-51

## JOB 7-1 ELECTRIC POWER IN SIMPLE CIRCUITS

**Meaning of electric power.** Did you know that it is possible to push a candle through a wooden board? You may not be successful when you try it because you may omit an essential element necessary to do this. This missing quantity is *speed.* Power depends on *how fast* a certain amount of work is done. Thus, if you *shoot* the candle from a shotgun, it will have the necessary speed to penetrate the board.

More than speed, however, is necessary to increase the power. In recognition of this fact, the Roman gladiators of long ago wrapped strips of lead around their fists to increase the power of their blows. Our modern boxers are not permitted to increase the weight of their gloves, but they increase their hitting power by punching with short, *fast* blows rather than with looping, slow swings.

Power, then, depends on two quantities—the work done (or the weight moved) and the speed of doing it.

$$\text{Power} = \text{work done} \times \text{speed}$$

In electrical work,

$$\text{Work done} = Q \times V$$

$$\text{Speed} = \frac{1}{t}$$

Therefore,

$$P = Q \times V \times \frac{1}{t}$$

or

$$P = \frac{Q \times V}{t} = \frac{Q}{t} \times V$$

but since $Q/t = I$

$$P = I \times V$$

**RULE** The electric power in any part of a circuit is equal to the current in that part multiplied by the voltage across that part of the circuit.

**FORMULA**

$$P = I \times V \tag{7-1}$$

where $P$ = power, W
$I$ = current, A
$V$ = voltage, V

A *watt* of electric power is the power used when one volt causes one ampere of current to flow in a circuit. A *kilowatt* of power is equal to 1000 watts.

EXAMPLE 7-1. A variable resistor in the volume control of a code practice oscillator passes 0.05 A with a voltage drop of 30 V. Find the power consumed.

SOLUTION. Given: $V = 30$ V Find: $P = ?$
$I = 0.05$ A

$$P = I \times V \tag{7-1}$$

$$P = 0.05 \times 30$$

$$P = 1.5 \text{ W} \quad \textit{Ans.}$$

EXAMPLE 7-2. A 1-kΩ resistor in a power-supply filter circuit carries 0.06 A at 60 V. How many watts of power are developed in the resistor? What must be the wattage rating of the resistor in order to dissipate this power safely as heat?

SOLUTION. Given: $I = 0.06$ A Find: $P = ?$
$V = 60$ V Wattage rating = ?

$$P = I \times V \tag{7-1}$$

$$P = 0.06 \times 60 = 3.6 \text{ W} \quad \textit{Ans.}$$

The wattage rating of a resistor describes its ability to dissipate the heat produced in it by the passage of an electric current without itself overheating. For example, a 2-W resistor could dissipate 2 W of heat energy without overheating. However, if 3 W of power were to be developed in it, it would overheat because of the 1 W of power which it could not dissipate. Owing to the lack of ventilation in the close quarters of most television receivers, the wattage ratings of these resistors is usually at least twice the wattage developed in them.

$$\text{Wattage rating} = 2 \times P$$

$$\text{Wattage rating} = 2 \times 3.6 = 7.2 \text{ W} \quad \textit{Ans.}$$

SELF-TEST 7-3. A 56-Ω resistor is shunted with a 168-Ω resistor in a 0.3-A circuit as shown in Fig. 7-1. Find the wattage rating of the 168-Ω resistor.

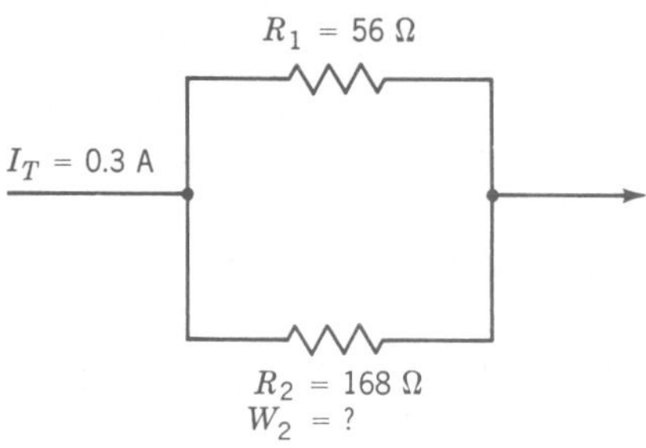


**FIGURE 7-1**
The wattage rating of a resistor is equal to at least twice the wattage developed in it.

SOLUTION

1. In order to find the wattage developed in $R_2$ we must know the value of $V_2$ and ______. | $I_2$

$$I_2 = \frac{?}{R_1 + R_2} \times I_T \qquad (5\text{-}7)$$

$$I_2 = \frac{56}{56 + ?} \times 0.3$$ | 168

$$I_2 = \frac{56}{224} \times 0.3 = \underline{\qquad} \text{ A}$$ | 0.075

2. Find $V_2$.

$$V_2 = I_2 \times \underline{\qquad} \qquad (4\text{-}5)$$ | $R_2$

$$V_2 = \underline{\qquad} \times 168 = \underline{\qquad} \text{ V}$$ | 0.075 12.6

3. Find the wattage developed in $R_2$.

$$P = I \times V \qquad (7\text{-}1)$$

$$P = 0.075 \times \underline{\qquad} = \underline{\qquad} \text{ W}$$ | 12.6 0.945

4. Find the wattage rating.

Wattage rating = ______ × wattage | 2

Wattage rating = 2 × 0.945 = ______ W | 1.89

or Wattage rating = ______ W *Ans.* | 2

**PROBLEMS**

1. An automobile starting motor draws 80 A at 6 V. How much power is drawn from the battery?
2. A 20-hp motor takes 74 A at 230 V when operating at full load. Find the power used in kilowatts.
3. What is the wattage dissipated as heat by a 550-Ω resistor operating at 110 V and 0.2 A?
4. What is the power consumed by an automobile headlight if it takes 2.8 A at 6 V?

5. Find the power used by a 3.4-A soldering iron at 110 V.
6. The high contact resistance of a poorly wired electric toaster plug reduced the current by 1 A on a 110-V line. Find the power wasted in the plug.
7. How much power is consumed by an electric clock using 0.02 A at 110 V?
8. If the voltage drop across a spark-plug air gap is 30 V, find the power consumed in sending 2 mA across the gap.
9. Find the power dissipated by the collector of a transistor that passes 0.25 A at 9.2 V.
10. A power supply delivers 0.16 A at 250 V to a public-address amplifier. Find the watts of power delivered.
11. A window air conditioner is rated at 7.5 A and is operated on a 120-V line. Find the wattage used by the conditioner.
12. An electric oven uses 36.3 A at 117 V. Find the power used in kilowatts.
13. An emitter bias resistor carries 45 mA at 10 V. What must be its wattage rating?
14. An electric enameling kiln takes 9.2 A from a 117-V line. Find the power used.
15. How much power is used by a $^3/_4$-ton air conditioner drawing 11.4 A from a 220-V line?
16. Fifteen lamps in parallel each take 1.67 A when connected to a 120-V line. Find the wattage of each lamp and the total wattage used.

## JOB 7-2 TOTAL POWER IN AN ELECTRIC CIRCUIT

When using Ohm's law, we found that it could be used for total values in a circuit as well as for the individual parts of the circuit. In the same way, the formula for power may be used for total values.

**FORMULA**

$$P_T = I_T \times V_T \qquad \text{(7-2)}$$

where $P_T$ = total power, W
$I_T$ = total current, A
$V_T$ = total voltage, V

EXAMPLE 7-4. Two 0.62-A lamps, each drawing 120 V, are connected in series. Find the total power used.

SOLUTION. Given: $I_1 = I_2 = 0.62$ A Find: $P_T = ?$
$V_1 = V_2 = 120$ V

1. Find the total voltage.

$$V_T = V_1 + V_2 = 120 + 120 = 240 \text{ V} \qquad \text{(4-2)}$$

2. Find the total current.

$$I_T = I_1 = I_2 = 0.62 \text{ A} \qquad (4\text{-}1)$$

3. Find the total power.

$$P_T = I_T \times V_T \qquad (7\text{-}2)$$
$$P_T = 0.62 \times 240 = 148.8 \text{ W} \qquad \textit{Ans.}$$

EXAMPLE 7-5. If the same lamps were connected in parallel across 120 V, find the total power used and compare it with the power used when the lamps were connected in series as in Example 7-4.

SOLUTION. Given: $I_1 = I_2 = 0.62$ A Find: $P_T = ?$
$V_1 = V_2 = 120$ V

1. Find the total voltage. $V_T = 120$ V (given)
2. Find the total current.

$$I_T = I_1 + I_2 = 0.62 + 0.62 = 1.24 \text{ A} \qquad (5\text{-}2)$$

3. Find the total power. $P_T = I_T \times V_T$ (7-2)

$$P_T = 1.24 \times 120 = 148.8 \text{ W} \qquad \textit{Ans.}$$

The total power in parallel is the same as the total power in series.

If the power used by the parts of a circuit is known, the total power may be found by the following

**FORMULA** $$P_T = P_1 + P_2 + P_3 \qquad \boxed{7\text{-}3}$$

where $P_T$ = total power, W
$P_1, P_2, P_3$ = power used by the parts of the circuit, W

SELF-TEST 7-6. Two resistors in series form the base-bias voltage divider for an audio amplifier. The voltage drops across them are 2.4 V and 6.6 V, respectively, in the 1.5-mA circuit. Find the power used by the circuit.

SOLUTION. Given: $V_1 = 2.4$ V
$V_2$ = ______ V — 6.6
$I_T = I_1 = I_2 = 1.5$ mA = ______ A Find: $P_T = ?$ — 0.0015

1. Find $P_1$. $P_1 = I_1 \times V_1$ (7-1)
$P_1 = 0.0015 \times$ ______ = ______ W — 2.4 0.0036

2. Find $P_2$. $P_2 = I_2 \times V_2$
$P_2 = 0.0015 \times 6.6$ = ______ W — 0.0099

3. Find the total power. $P_T = P_1 + P_2$ (7-3)
$P_T = 0.0036 + 0.0099$ = ______ W *Ans.* — 0.0135

The problem can be solved by another method.

1. Find the total voltage used by both resistors. Since they are in series,

$$V_T = V_1 + ?$$ (4-2)

$$V_T = 2.4 + \_\_\_\_\_ = \_\_\_\_\_ \text{ V}$$

$V_2$
6.6  9

2. Find the total power.

$$P_T = I_T \times \_\_\_\_\_$$ (7-2)

$$P_T = \_\_\_\_\_ \times 9 = \_\_\_\_\_ \text{ W}$$

$V_T$
0.0015  0.0135

**PROBLEMS**

1. Seven Christmas-tree lamps are connected in series. Each lamp requires 16 V and 0.1 A. Find the total power used.
2. If the same lamps were connected in parallel across a 16-V source, what would be the power taken?
3. A certain transistor must pass a maximum of 0.3 A. If it is rated at 0.15 W maximum, will it be able to withstand a 0.45-V collector-emitter voltage?
4. What is the total power used by a 4.5-A electric iron, a 0.85-A fan, and a 2.2-A refrigerator motor if they are all connected in parallel across a 115-V line?
5. Find the power drawn from a 6-V battery by a parallel circuit of two headlights (4 A each) and two taillights (0.9 A each).
6. In Fig. 6-48, the generator voltage is 120 V. The motor takes 12.4 A, and each lamp takes 0.92 A. Neglecting the line drop, find the total power used by the devices.
7. An electric percolator drawing 9 A and an electric toaster drawing 10.2 A are connected in parallel to the 117-V house line. Find the total power consumed.
8. In a circuit similar to that shown in Fig. 6-1, $V_T = 200$ V, $R_1 = 10\ \Omega$, $R_2 = 15\ \Omega$, $R_3 = 40\ \Omega$, and $R_4 = 60\ \Omega$. Find (*a*) the total current, (*b*) the power used by each resistor, and (*c*) the total power.
9. In a circuit similar to that shown in Fig. 6-15, $V_T = 13$ V, $R_1 = 2\ \Omega$, $R_2 = 1\ \Omega$, $R_3 = 6\ \Omega$, $R_4 = 3\ \Omega$, $R_5 = 40\ \Omega$, and $R_6 = 3\ \Omega$. Find (*a*) the total current, (*b*) the power used by each resistor, and (*c*) the total power.

## JOB 7-3 SOLVING THE POWER FORMULA FOR CURRENT OR VOLTAGE

In Job 3-9 we learned how to solve Ohm's law for the current or the resistance. The power formula is the same general type of equation, and we can use the methods learned in Job 3-9 to solve it for values of current or voltage.

EXAMPLE 7-7. What is the operating voltage of an electric toaster rated at 600 W if it draws 5 A?

SOLUTION. Given: $P = 600$ W  Find: $V = ?$

$$I = 5\text{ A}$$

$$P = I \times V \qquad (7\text{-}1)$$

$$600 = 5 \times V$$

$$V = \frac{600}{5} = 120\text{ V} \qquad \textit{Ans.}$$

SELF-TEST 7-8. A stabilizing resistor in the emitter circuit of a 2N109 output transistor develops a voltage drop of 1.2 V while consuming 0.06 W of power. Find the current through the resistor.

SOLUTION. Given: $P = 0.06$ W Find: ______ = ? | $I$

V = ______ V | 1.2

$$P = I \times V \qquad (7\text{-}1)$$

______ $= I \times 1.2$ | 0.06

$$I = \frac{0.06}{?}$$ | 1.2

$I =$ ______ A *Ans.* | 0.05

## PROBLEMS

1. A 550-W neon sign operates on a 110-V line. Find the current drawn.
2. What current is drawn by a 480-W soldering iron from a 120-V line?
3. Find the current drawn by a 1200-W aircraft system from a 24-V source.
4. A resistor is capable of dissipating 10 W of power. If the current is 0.3 A, what is the maximum voltage drop permitted across the resistor?
5. The quiescent point of a transistor is at a collector current of 0.06 A and a collector voltage of 20 V. Find the transistor power dissipation with no signal applied.
6. What current is drawn by a 1500-W electric ironing machine from a 120-V line?
7. What current is drawn by a 55-cd 110-V lamp if it uses 1 W per candela?
8. Two resistors, each dissipating 2 W, are connected in series with a 40-V source. What is the total current drawn? What is the current in each resistor?
9. Twenty 60-W lamps are connected in parallel to light a stage. Find the current drawn from a 220-V source.
10. The nameplate on a resistance welder reads 25 kVA at 440 V. What is its rated primary current?
11. An electric broiler rated at 1550 W operates from a 117-V line. The available fuses are rated at 20, 30, 50, and 60 A. Which fuse should be used in the circuit to protect the broiler?

## JOB 7-4 INTERMEDIATE REVIEW OF POWER

In some problems, the power cannot be found because either the voltage or the current is unknown. In these instances, the unknown value is found by Ohm's law.

EXAMPLE 7-9. Find the power taken by a soldering iron of 60 Ω resistance if it draws a current of 2 A.

SOLUTION. Given: $R = 60\ \Omega$ Find: $P = ?$
$I = 2$ A

1. Find the voltage. $V = I \times R$ (2-1)
   $V = 2 \times 60 = 120$ V
2. Find the power. $P = I \times V$ (7-1)
   $P = 2 \times 120 = 240$ W *Ans.*

SELF-TEST 7-10. Find the power used by the 11-Ω resistance element of an electric furnace if the voltage is 110 V.

SOLUTION. Given: $R = 11\ \Omega$ Find: $P = ?$
$V =$ _______ V | 110

1. Find the current. $V = I \times R$ (2-1)
   $110 = I \times$ _______ | 11
   $I = \frac{110}{11} =$ _______ A | 10
2. Find the power. $P =$ _______ $\times V$ (6-1) | $I$
   $P = 10 \times 110 =$ _______ W *Ans.* | 1100

### PROBLEMS

1. A 20-Ω neon sign operates on a 120-V line. Find the power used by the sign.
2. Find the power used by a 55-Ω electric light which draws 2 A.
3. What is the wattage dissipated by a 10,000-Ω voltage divider if the voltage across it is 250 V? What is its wattage rating?
4. What is the maximum power obtainable from a Grenet cell of 2 V which has an internal resistance of 0.02 Ω?
5. A 240-Ω resistor in the emitter circuit of the output stage of a transistor radio carries 0.005 A. Find the wattage developed in the resistor.
6. The resistance of an ammeter is 0.025 Ω. Find the power used by the meter when it reads 4 A.
7. What is the power consumed by a 90-Ω subway-car heater if the operating voltage is 550 V?
8. A 40-Ω pilot light is to be operated in a 0.15-A circuit. How many watts are developed in the lamp?

9. A 20-Ω toaster operates on a 115-V line. Find the power used.
10. A voltmeter has an internal resistance of 220,000 Ω. How much power does it use when the meter reads 110 V?
11. In the circuit shown in Fig. 7-2, find (*a*) the total resistance, (*b*) the total current, and (*c*) the total power.

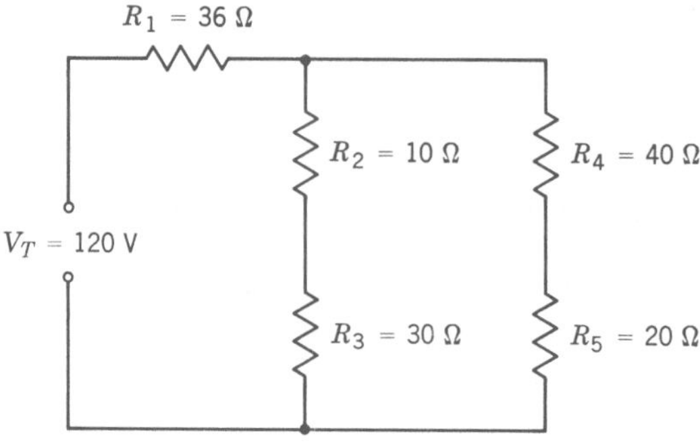


FIGURE 7-2

12. In a circuit similar to that shown in Fig. 6-15, $V_T = 188$ V, $R_1 = 22$ Ω, $R_2 = 18$ Ω, $R_3 = 70$ Ω, $R_4 = 80$ Ω, $R_5 = 56$ Ω, and $R_6 = 30$ Ω. Find (*a*) the total current, (*b*) the power used by each resistor, and (*c*) the total power.
13. In Fig. 7-3, find (*a*) the total resistance, (*b*) the total current, and (*c*) the total power.

(See Instructor's Manual for Test 7-1, Power)

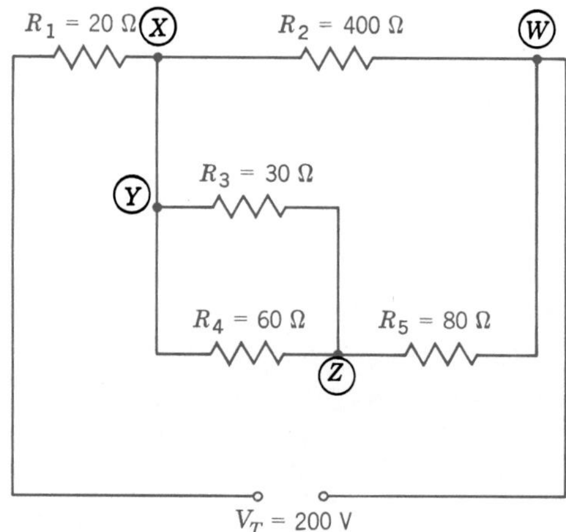


FIGURE 7-3

## JOB 7-5 SQUARING POWERS OF 10

The next job, as well as many others, will require us to find the square of very large or very small numbers. The calculations will be simplified if we express these numbers as powers of 10.

EXAMPLE 7-11. Find the value of $(10^3)^2$.

SOLUTION. An exponent means to multiply its base by itself as many times as the exponent indicates. If the base happens to be $10^3$, then

$$(10^3)^2 \text{ means } 10^3 \times 10^3$$

and, since exponents are added when multiplying (see page 68),

$$(10^3)^2 = 10^3 \times 10^3 = 10^6 \qquad \textit{Ans.}$$

The answer may be obtained more directly by applying the following rule.

| RULE | To find the power of a power, express the base to a power obtained by multiplying the exponents. |
|---|---|

For example:

1. $(10^3)^2 = (10)^{(3\times 2)} = 10^6$
2. $(10^4)^2 = (10)^{(4\times 2)} = 10^8$
3. $(10^{-3})^2 = (10)^{(-3\times 2)} = 10^{-6}$
4. $(10^{-6})^2 = 10^{-12}$

EXAMPLE 7-12. Express the number $(12{,}000)^2$ as a power of 10.

SOLUTION

1. Express 12,000 as a power of 10.

$$12{,}000 = 12 \times 10^3$$

2. Square this number.

$$\begin{aligned}(12{,}000)^2 &= (12 \times 10^3)^2 \\ &= (12)^2 \times (10^3)^2\end{aligned}$$

Notice that both the 12 and the $10^3$ are squared because each is a part of the base $(12 \times 10^3)$ that is being squared.

$$(12{,}000)^2 = 144 \times 10^6 \qquad \textit{Ans.}$$

EXAMPLE 7-13. Express $(0.000\,08)^2$ as a power of 10.

SOLUTION $0.000\,08 = 8 \times 10^{-5}$

Therefore,

$$\begin{aligned}(0.000\,08)^2 &= (8 \times 10^{-5})^2 \\ &= 8^2 \times (10^{-5})^2 \\ &= 64 \times 10^{-10} \qquad \textit{Ans.}\end{aligned}$$

EXAMPLE 7-14. Find the value of $(2.5 \times 10^2)^2$.

SOLUTION

$$\begin{aligned}(2.5 \times 10^2)^2 &= (2.5)^2 \times (10^2)^2 \\ &= 6.25 \times 10^4 = 62{,}500 \qquad \textit{Ans.}\end{aligned}$$

**PROBLEMS**

Perform the indicated operations.

1. $(10^5)^2$
2. $(10^{-4})^2$
3. $(3 \times 10^3)^2$
4. $(5 \times 10^{-3})^2$
5. $(10^2 \times 10^3)^2$
6. $(10^6 \times 10^{-2})^2$
7. $(8000)^2$
8. $(15{,}000)^2$
9. $(1100)^2$
10. $(0.007)^2$
11. $(0.000\,25)^2$
12. $(0.015)^2$
13. $(1.5 \times 10^3)^2$
14. $(5.1 \times 10^{-4})^2$
15. $(1.2 \times 10^{-3})^2$

## JOB 7-6 THE EXPONENTIAL POWER FORMULAS

In Job 7-4, either the voltage or the current was unknown. These values were found by Ohm's law and then used to find the power. The two steps involved in solving problems of this type may be combined into a single formula. The power is given by

$$P = I \times V \qquad (7\text{-}1)$$

But by Ohm's law,

$$V = I \times R \qquad (2\text{-}1)$$

Therefore, we may substitute the quantity $I \times R$ for $V$. This gives

$$P = I \times I \times R$$

**FORMULA**

$$P = I^2R \qquad \boxed{7\text{-}4}$$

Also by Ohm's law, $I = V/R$. Therefore, we may substitute the quantity $V/R$ for $I$. Thus,

$$P = I \times V \qquad (7\text{-}1)$$

Substituting,

$$P = \frac{V}{R} \times V$$

**FORMULA**

$$P = \frac{V^2}{R} \qquad \boxed{7\text{-}5}$$

EXAMPLE 7-15. A 4-kΩ base-bias resistor in a silicon NPN amplifier transistor carries a base current of 5 mA. Find the power consumed and the wattage rating required by the resistor.

SOLUTION. The diagram for the circuit is shown in Fig. 7-4.

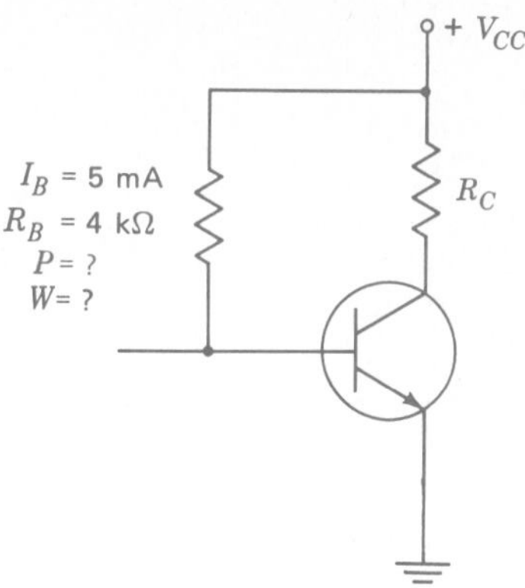


FIGURE 7-4

$$P = I^2R \qquad (7\text{-}4)$$
$$P = (5 \times 10^{-3})^2 \times 4 \times 10^3$$
$$P = 25 \times 10^{-6} \times 4 \times 10^3$$
$$P = 100 \times 10^{-3} = 0.1 \text{ W} \qquad \textit{Ans.}$$
$$\text{Wattage rating} = 2 \times P$$
$$= 2 \times 0.1 = 0.2 \text{ W} \qquad \textit{Ans.}$$

EXAMPLE 7-16. A motor has a total resistance of 20 Ω and operates on a 120-V line. Find the power used.

SOLUTION. Given: $R = 20\ \Omega$ Find: $P = ?$
$V = 120$ V

$$P = \frac{V^2}{R} \qquad (7\text{-}5)$$

$$P = \frac{(120)^2}{20} = \frac{120 \times 120}{20} = 720 \text{ W} \qquad \textit{Ans.}$$

SELF-TEST 7-17. Find the wattage rating of $R_1$ and $R_2$ in the voltage-divider circuit shown in Fig. 7-5.

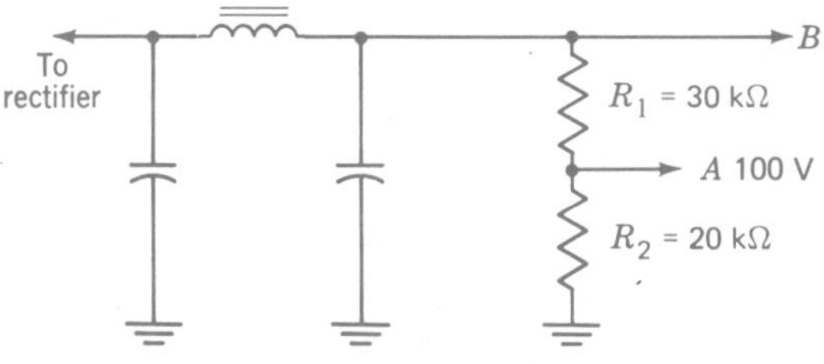


FIGURE 7-5
Simple voltage-divider circuit.

SOLUTION. The voltage drop across $R_1$ is equal to the difference between the voltage $V$ at $B$ (250) and the voltage $V$ at $A$ (100). Therefore, $V_1$ = ________ V. 150

$$P_1 = \frac{V_1^2}{R_1}$$

$$P_1 = \frac{(150)^2}{?}$$ 30,000

$$P_1 = \frac{225 \times 10^2}{?}$$ $3 \times 10^4$

$$P_1 = 75 \times \underline{\qquad} = \underline{\qquad}\ \text{W}$$ $10^{-2}$ 0.75

$$\text{Wattage rating} = \underline{\qquad} \times P_1$$ 2

$$= 2 \times 0.75 = 1.5\ \text{W}$$

However, since the 1.5-W size is not usually stocked, we must use a 2-W 30-kΩ resistor. *Ans.*

For $R_2$, the voltage drop $V_2$ = ______ V 100

$$P_2 = \frac{(100)^2}{2 \times 10^4} = \frac{1 \times 10^4}{2 \times 10^4} = \underline{\qquad}\ \text{W}$$ 0.5

Therefore, Wattage rating = ______ W *Ans.* 1

## PROBLEMS

1. A 90-Ω device draws 5 A. Find the power used in kilowatts.
2. If the voltage drop across a 10,000-Ω voltage divider is 90 V, find the power used.
3. Find the power consumed by a 100-Ω electric iron operating on a 115-V line.
4. A poorly soldered joint has a contact resistance of 100 Ω. What is the power lost in the joint if the current is 0.5 A?
5. Two 2N406 transistors operating in push-pull deliver an average current of 0.05 A through a 2600-Ω load resistance. Find the power developed.
6. Find the power used by a 15-Ω neon sign on a 110-V line.
7. The motor shown in Fig. 7-6 takes 1400 W at 220 V. Find the current drawn. How many watts are consumed in the line wires if each has a resistance of 0.2 Ω?

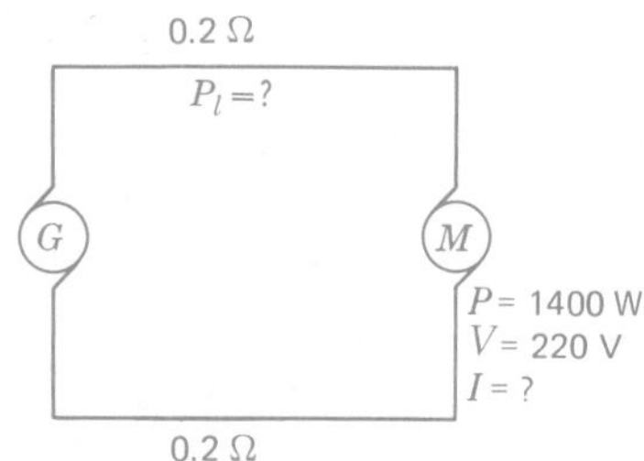


FIGURE 7-6

8. How much power is dissipated in the form of heat in a ballast resistor of 60 Ω if the current is 0.3 A?
9. A 100- and a 260-Ω resistor are connected in series to a 120-V source. Find the power used by each resistor.
10. Find the total voltage across the circuit shown in Fig. 7-7. What is

the voltage across each resistor? What is the power taken by each resistor?

(See Instructor's Manual for Test 7-2)

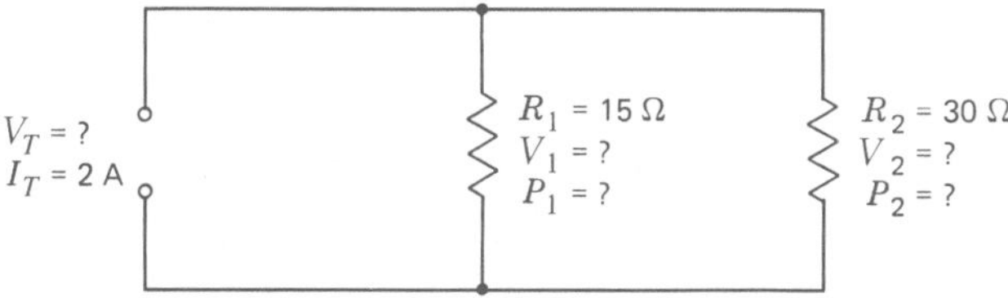


FIGURE 7-7

## JOB 7-7 CHECKUP ON SQUARE ROOT (DIAGNOSTIC TEST)

In the last job we learned two new formulas which use the exponent 2. This exponent is read as the word "square." We can use these formulas to find the current, voltage, or resistance if we know the power. To do this, we must be able to do the opposite of "squaring" a number. This is called "finding the square root." We shall also use "square root" when we get to the study of ac circuits. Can you do the following problems? If you have any difficulty with them, turn to Job 7-8 which follows.

### PROBLEMS

Find the square root of each of the following numbers:

| | | | |
|---|---|---|---|
| 1. 64 | 2. 169 | 3. 17.64 | 4. 3481 |
| 5. 12,544 | 6. 57.76 | 7. 14 | 8. 567.9 |
| 9. 76,432 | 10. 30 | 11. 0.652 | 12. 870,000 |

## JOB 7-8 BRUSHUP ON SQUARE ROOT

The *square root* of a number is that number which must be multiplied by *itself* in order to obtain the original number. The symbol for the square root is $\sqrt{\ }$. When a number appears under this symbol, it means that we are to find a number which can be multiplied by itself to give the number under the square-root symbol. For example, the square root of 9 ($\sqrt{9}$) must be the number 3 because only $3 \times 3 = 9$. Also,

$$\sqrt{16} = 4 \text{ because only } 4 \times 4 \text{ will equal } 16$$
$$\sqrt{25} = 5 \text{ because only } 5 \times 5 \text{ will equal } 25$$
$$\sqrt{36} = 6 \text{ because only } 6 \times 6 \text{ will equal } 36$$

These numbers under the square-root sign are called *perfect squares* because the square root of each is a whole number.

We can continue to find the square roots of numbers in this manner until our knowledge of the multiplication table is insufficient to keep up with the large numbers involved. For example,

$$\sqrt{49} = 7 \qquad \sqrt{121} = 11$$
$$\sqrt{64} = 8 \qquad \sqrt{144} = 12$$
$$\sqrt{81} = 9 \qquad \sqrt{169} = 13$$
$$\sqrt{100} = 10$$

Somewhere along here we begin to forget whether $14 \times 14$ is 196 or not or whether $16 \times 16$ is 256 or not. You can see that we will not get very far if we rely on just our knowledge of the multiplication table. Besides, what about all those smaller numbers between the perfect squares, like

$$\sqrt{7} = ?$$
$$\sqrt{15} = ?$$
$$\sqrt{32} = ?$$

There doesn't seem to be any number that we can multiply by itself to get 7 or 15 or 32. No, there aren't any *whole* numbers which are the square roots of these numbers, but there are decimal numbers which will satisfy.

Obviously, as the numbers get beyond the range of the ordinary multiplication table, we feel the need for some system that will help us to find the square root of *any* number—large or small, whole number or decimal.

In the following system, it will help if we mark off the numbers into groups of two digits to a group. We start marking off the groups *at the decimal point.* If there is no decimal point indicated, it may be assumed to be at the *end* of the number.

EXAMPLE 7-18. Group the digits in the number 3456.

SOLUTION. The decimal point is at the end of the number. Starting at the decimal point, the numbers are grouped two to a group as we move to the left.

$$\underbrace{34}\ \underbrace{56}.$$

EXAMPLE 7-19. Group the digits in the number 54,819.8.

SOLUTION. Starting at the decimal point and proceeding to the left, we find that the digit 5 is left over. In situations like this, the single digit at the extreme left is considered to be a group. Now, return to the decimal point and group the digits to the *right* of the point—two to a group. A zero must be added after the 8 at the right to complete the group.

$$\underbrace{5}\ \underbrace{48}\ \underbrace{19}\ .\ \underbrace{80}$$

EXAMPLE 7-20. Find the square root of 5776.

SOLUTION

1. Locate the decimal point. In a whole number the decimal point is at the end of the number. Place the point in the answer directly above its position in the number.

```
       .
√5 7 7 6 .
```

2. Separate the digits into groups—two digits to a group. Start at the decimal point, and group the digits to the left.

```
      .
√57 76 .
```

3. Start with the first group (57). Find a number which when multiplied by itself will give an answer close to or equal to but not larger than 57. 8 squared is 64, but that is too large. 7 squared is 49, which is just right. Place the 7 over the first group in the answer, and the 49 under the 57. Draw a line and subtract the 49 from the 57, leaving a remainder of 8.

```
   7     .
√57 76 .
– 49
    8
```

4. Bring down the next *group*. *Never* bring down a single number. Make a little box to the left of this new number 876. *Double* the answer at this point (the 7), and place it in this box. This will be the number 14.

```
         7     .
      √57 76 .
      – 49  ↓
14   |  8 76
```

5. Place your finger over the *last digit* in the number 876. The number there will now appear to be 87. Divide the number in the box (14) into this 87. It will go about 6 times. Place this 6 in the answer above the second group, *and also place it next to the* 14 *in the box.*

```
           7   6 .
        √57 76 .
        – 49  ↓
14 6   |  8 76
```

6. Multiply the 6 by the number just formed (the 146). If the product is larger than 876, we shall be forced to change the 6 to a smaller number. However, $6 \times 146 = 876$. Write this 876 under the 876 already there and subtract. Since there is no remainder, 76 is the exact square root of 5776. *Check:* $76 \times 76 = 5776$.

```
           7   6 .
        √57 76 .
        – 49  ↓
14 6   |  8 76
          8 76
             0
```

EXAMPLE 7-21. Find the square root of 930.25.

SOLUTION

1. Locate the decimal point. Place the point in the answer directly above its position in the number.

```
      .
 √9 30.25
```

2. Separate the digits into groups—two digits to a group. Start at the decimal point, and group the digits to the left. Return to the decimal point, and group the digits to the right.

```
      .
 √9 30 .25
  ‿  ‿  ‿
```

3. Start with the first group (9). Find a number which when multiplied by itself will give an answer close to or equal to but not larger than 9. 3 squared is 9, which is exactly right. Place the 3 over the first group in the answer, and the 9 under the 9 in the number. Draw a line and subtract, leaving a remainder of 0.

```
   3    .
 √9 30 . 25
  ‿       ‿
 −9
 ----
```

4. Bring down the next *group* (30). *Never* bring down a single digit. Make a little box to the left of this number. *Double* the answer at this point (the 3), and place it in this box. This will be the number 6.

```
       3    .
     √9 30 . 25
      ‿  ‿    ‿
      9  ↓
 6  |   30
```

5. Place your finger over the *last digit* in the number 30. The number there will now appear to be 3. Divide the number in the box (6) into this 3. It will go 0 times. Place this 0 in the answer above the second group, *and also place it next to the 6 in the box.*

```
        3  0 .
      √9 30 . 25
       ‿  ‿    ‿
      −9  ↓
 6 0  |  30
```

6. Multiply the 0 by the number just formed (the 60). The answer is 00. Write this 00 under the 30 and subtract, leaving a remainder of 30.

```
        3  0 .
      √9 30 . 25
       ‿  ‿    ‿
      −9  ↓
 6 0  |  30
        −00
        ---
         30
```

7. Bring down the next group (25). Make a little box to the left of this new number (3025). *Double* the answer up to this point (the 30), and place the product (60) in this box.

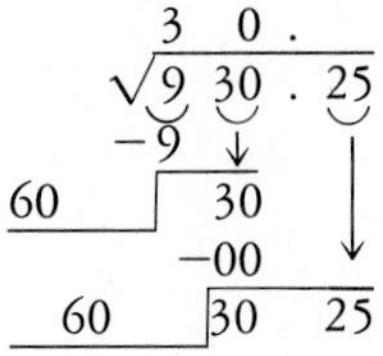

```
            3  0 .
          √9 30 . 25
          -9  ↓
   60    |   30     |
            -00     ↓
     60    |30    25
```

8. Place your finger over the *last digit* in the number 3025. The number there will now appear to be 302. Divide the number in the box (60) into this 302. It will go 5 times. Place this 5 in the answer above the third group, *and also place it next to the 60 in the box.*

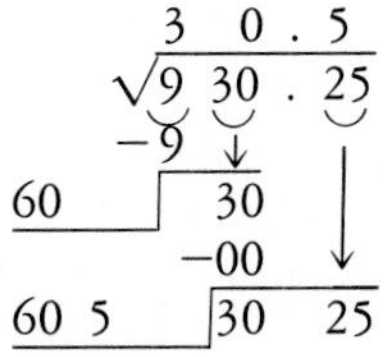

```
            3  0 . 5
          √9 30 . 25
          -9  ↓
   60    |   30     |
            -00     ↓
   60 5    |30    25
```

9. Multiply the 5 by the number just formed (the 605). The answer is 3025. Write this 3025 under the 3025 already there and subtract. Since there is no remainder, 30.5 is the exact square root of 930.25. *Check:* $30.5 \times 30.5 = 930.25$.

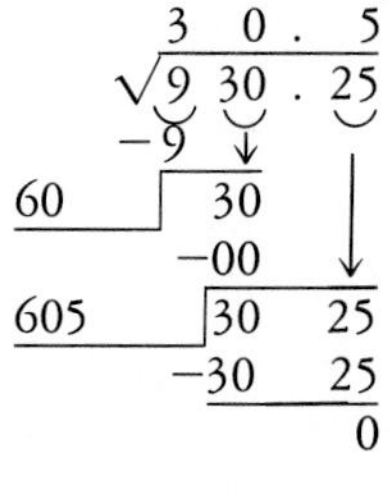

```
            3  0 .  5
          √9 30 . 25
          -9  ↓
   60    |   30     |
            -00     ↓
   605     |30    25
           -30    25
                   0
```

EXAMPLE 7-22. Find the square root of 12.

SOLUTION

1. Locate the decimal point. In a whole number the decimal point is at the end of the number. Place the point in the answer directly above its position in the number.

$$\sqrt{12\,.}$$

2. Separate the digits into groups—two digits to a group. Start at the decimal point, and group the digits to the left.

$$\sqrt{\underbrace{1\,2}\,.}$$

3. Start with the first group (12). Find a number which when multiplied by

itself will give an answer close to or equal to but not larger than 12. 4 squared is 16, but that is too large. 3 squared is 9, which is less than 12 and so is just right. Place the 3 over the first group in the answer, and the 9 under the 12 in the number. Draw a line and subtract, leaving a remainder of 3.

```
  3 .
√1 2 .
 −9
  3
```

4. Since there is a remainder, the square root will be a decimal. Add two *pairs* of zeros after the decimal point. Bring down the next group (00). Make a little box to the left of this number (300). *Double* the answer up to this point (the 3), and place the product (6) in this box.

```
        3 .
     √1 2 . 00 00
      −  9   ↓
  6  |   3   00
```

5. Place your finger over the *last digit* in the number 300. The number there will now appear to be 30. Divide the number in the box (6) into this 30. It will go 5 times. Place this 5 in the answer above the second group, *and also place it next to the 6 in the box.* Multiply the 5 by the number just formed (the 65), and place the product (325) under the 300 already there. *But* 325 *is larger than* 300, *and we have evidently made an error in using the* 5. Since 5 was too large, use 4 instead of 5.

```
        3 .  5
     √1 2 . 00 00
      −  9   ↓
  6 5 |  3   00
         3   25     Too large!
```

6. Be sure to change *both* the 5 in the answer and the 5 in the 65 in the box to the number 4. Now multiply the 4 in the answer by the 64 in the box. Place the product (256) under the 300 already there and subtract, leaving a remainder of 44.

```
        3 .  4
     √1 2   00 00
      −  9   ↓
  6 4 |  3   00
        −2   56
             44
```

7. Bring down the next group (00). Make a little box to the left of this new number (4400). Double the answer up to this point (the 34), and place the product (68) in this box. Always disregard the decimal point in this step.

```
        3 .  4
     √1 2 . 00 00
      −  9   ↓   |
  6 4 |  3   00  |
        −2   56  ↓
   68  |     44 00
```

8. Place your finger over the *last digit* in the number 4400. The number there will now appear to be 440. Divide the number in the box (68) into this 440. It will go 6 times. Place this 6 in the answer over the third group, *and also place it next to the* 68 *in the box.*

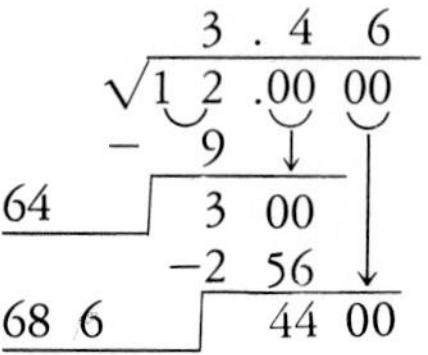

9. Multiply the 6 by the number just formed (the 686), and place the product (4116) under the 4400 already there. Draw a line and subtract, leaving a remainder of 284. This remainder may be disregarded, as we shall rarely need an answer more accurate than two decimal places. The problem may be worked out to any number of decimal places and then "rounded off" to suit.

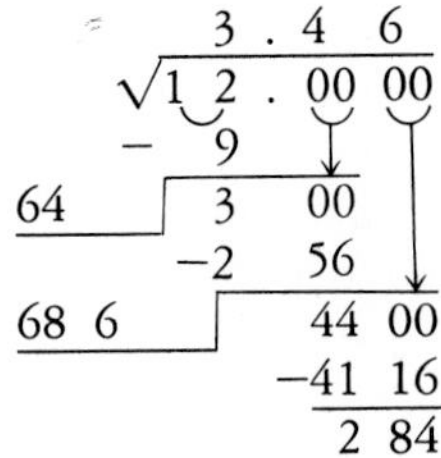

*Check:* $3.46 \times 3.46 = 11.97$, or practically 12

### PROBLEMS

Find the square roots of the following numbers:

| | | | | |
|---|---|---|---|---|
| 1. 3481 | 2. 17.64 | 3. 15.21 | 4. 12,544 | 5. 18,769 |
| 6. 151.29 | 7. 40 | 8. 267 | 9. 65 | 10. 53.87 |

## JOB 7-9 THE SQUARE ROOT OF A POWER OF 10

**RULE** To find the square root of a power, express the base to a power obtained by dividing the exponent by 2.

For example:

1. $\sqrt{10^6} = 10^{(6\div 2)} = 10^3$
2. $\sqrt{10^{10}} = 10^{(10\div 2)} = 10^5$
3. $\sqrt{10^{-12}} = 10^{(-12\div 2)} = 10^{-6}$
4. $\sqrt{10^{-8}} = 10^{(-8\div 2)} = 10^{-4}$

EXAMPLE 7-23. Express $\sqrt{250{,}000}$ as a power of 10.

SOLUTION $250{,}000 = 25 \times 10^4$

Find the square root of this number. Be careful to find the square root of *both* the 25 and the $10^4$.

$$\sqrt{250{,}000} = \sqrt{25 \times 10^4} = \sqrt{25} \times \sqrt{10^4}$$
$$= 5 \times 10^{(4 \div 2)}$$
$$= 5 \times 10^2 \quad \textit{Ans.}$$

EXAMPLE 7-24. Express $\sqrt{0.0004}$ as a power of 10.

SOLUTION
$$\sqrt{0.0004} = \sqrt{4 \times 10^{-4}} = \sqrt{4} \times \sqrt{10^{-4}}$$
$$= 2 \times 10^{(-4 \div 2)}$$
$$= 2 \times 10^{-2} \quad \textit{Ans.}$$

Now let's see what happens if we get an exponent which is *not* exactly divisible by 2. For example, $\sqrt{10^3} = 10^{1.5}$. This decimal exponent, although it has wide use in all the sciences, is very inconvenient, and should be avoided if at all possible. Let's see how this may be done.

EXAMPLE 7-25. Express $\sqrt{500{,}000}$ as a power of 10.

SOLUTION. When 500,000 is expressed as a power of 10, it *must* be written so that the power of 10 is an *even* number that can be divided by 2. Thus, 500,000 written as $5 \times 10^5$ is not good because the exponent 5 is not evenly divisible by 2. And 500,000 written as $500 \times 10^3$ is wrong for the same reason. Therefore, we must write 500,000 as a number $\times$ 10 to an *even* exponent.

$$\sqrt{500{,}000} = \sqrt{50 \times 10^4} = \sqrt{50} \times \sqrt{10^4}$$
$$= 7.07 \times 10^2 \quad \textit{Ans.}$$

EXAMPLE 7-26. Express $\sqrt{0.000\ 64}$ as a power of 10.

SOLUTION
$$\sqrt{0.000\ 64} = \sqrt{6.4 \times 10^{-4}}$$
$$= \sqrt{6.4} \times \sqrt{10^{-4}}$$
$$= 2.53 \times 10^{-2} \quad \textit{Ans.}$$

EXAMPLE 7-27

$$\sqrt{0.9 \times 10^3} = \sqrt{9 \times 10^2}$$
$$= 3 \times 10 = 30 \quad \textit{Ans.}$$

EXAMPLE 7-28

$$\sqrt{0.0016 \times 0.0004} = \sqrt{16 \times 10^{-4} \times 4 \times 10^{-4}}$$

$$= 4 \times 10^{-2} \times 2 \times 10^{-2}$$
$$= 8 \times 10^{-4} \qquad \textit{Ans.}$$

EXAMPLE 7-29

$$\sqrt{8000 \times 400} = \sqrt{8 \times 10^3 \times 4 \times 10^2}$$
$$= \sqrt{32 \times 10^5}$$
$$= \sqrt{320 \times 10^4}$$
$$= 17.9 \times 10^2 \qquad \textit{Ans.}$$

EXAMPLE 7-30. This example will be particularly important when we study the impedance of an ac circuit. Make note of the fact that quantities involving *different* powers of 10 may *not* be added or subtracted. If a problem involves different powers of 10, we must change the power in one quantity to agree with the other before addition or subtraction.

Find the value of $\sqrt{4 \times 10^4 + 41 \times 10^2}$.

SOLUTION. Since the powers of 10 are different, we change $4 \times 10^4$ to $400 \times 10^2$ so that both quantities contain the same power $10^2$.

$$\sqrt{4 \times 10^4 + 41 \times 10^2} = \sqrt{400 \times 10^2 + 41 \times 10^2}$$
$$= \sqrt{(400 + 41) \times 10^2}$$
$$= \sqrt{441 \times 10^2}$$
$$= 21 \times 10 = 210 \qquad \textit{Ans.}$$

**PROBLEMS**

Perform the indicated operation.

1. $\sqrt{10^8}$
2. $\sqrt{10^{-6}}$
3. $\sqrt{16 \times 10^4}$
4. $\sqrt{9 \times 10^{-6}}$
5. $\sqrt{160{,}000}$
6. $\sqrt{0.0064}$
7. $\sqrt{0.000\ 025}$
8. $\sqrt{20 \times 80}$
9. $\sqrt{120 \times 30}$
10. $\sqrt{200{,}000}$
11. $\sqrt{800 \times 500}$
12. $\sqrt{0.0006}$
13. $\sqrt{0.016 \times 10^3}$
14. $\sqrt{0.25 \times 10^8}$
15. $\sqrt{0.4 \times 10^5}$
16. $\sqrt{0.0049 \times 0.0009}$
17. $\sqrt{6 \times 10^3 \times 500}$
18. $\sqrt{8 \times 10^{-3} \times 0.08}$
19. $\sqrt{14.4 \times 10^7}$
20. $\sqrt{0.081 \times 10^9}$
21. $\sqrt{0.225 \times 10^{-3}}$
22. $\sqrt{0.625 \times 10^{-11}}$
23. $\sqrt{6 \times 10^4 + 25 \times 10^2}$
24. $\sqrt{4 \times 10^3 + 900}$

(See Instructor's Manual for Test 7-3)

## JOB 7-10 APPLICATIONS OF THE EXPONENTIAL POWER FORMULA

The formula $P = I^2R$ may be used to find the current $I$ or the resistance $R$. The formula $P = V^2/R$ may be used to find the voltage $V$ or the resistance $R$.

EXAMPLE 7-31. A 2N1479 transistor delivers an output power of 4 W at an average current of 0.2 A. What is the value of the load resistance?

SOLUTION. Given: $P = 4$ W Find: $R = ?$
$I = 0.2$ A

$$P = I^2R \tag{7-4}$$
$$4 = (0.2)^2 \times R$$
$$4 = 0.04 \times R$$
$$R = \frac{4}{0.04} = 100\ \Omega \qquad \textit{Ans.}$$

EXAMPLE 7-32. What is the maximum current-carrying capacity of a resistor marked 1000 Ω and 10 W?

SOLUTION. Given: $R = 1000\ \Omega$ Find: $I = ?$
$P = 10$ W

$$P = I^2R \tag{7-4}$$
$$10 = I^2 \times 1000$$
$$I^2 = \frac{10}{1000} = 0.01 \text{ A}$$

Now $I^2 = 0.01$ means that some number $I$ multiplied by itself will equal 0.01. Another way to say this is "What number multiplied by itself will equal 0.01?" This can be written as

$$I = \sqrt{0.01}$$

Actually, we have transformed the equation $I^2 = 0.01$ into $I = \sqrt{0.01}$ by taking the square root of both sides of the equality sign as shown in the following step. Since $\sqrt{I^2} = I$,

$$\sqrt{I^2} = \sqrt{0.01}$$
$$I = \sqrt{0.01}$$
$$I = 0.1 \text{ A} \qquad \textit{Ans.}$$

EXAMPLE 7-33. The total resistance of the field coils of a 240-W motor is 60 Ω. Find the voltage needed to operate the motor at its rated power.

SOLUTION. Given: $R = 60\ \Omega$ Find: $V = ?$
$P = 240$ W

1. Write the formula.

$$P = \frac{V^2}{R} \qquad (7\text{-}5)$$

2. Substitute numbers.

$$\frac{240}{1} = \frac{V^2}{60}$$

3. Cross-multiply.

$$V^2 = 240 \times 60$$

4. Multiply.

$$V^2 = 14{,}400$$

5. Take the square root of both sides.

$$\sqrt{V^2} = \sqrt{14{,}400}$$

6. Since $\sqrt{V^2} = V$,

$$V = \sqrt{14{,}400}$$

7. The voltage is

$$V = 120 \text{ V} \qquad \textit{Ans.}$$

SELF-TEST 7-34. What is the maximum current-carrying capacity of a resistor marked 10 kΩ and 4 W?

SOLUTION. Given: $P =$ ______ W  Find: $I = ?$
$R =$ ______ Ω

$P = I^2 \times$ ______
$4 = I^2 \times$ ______

$$I^2 = \frac{4}{?}$$

$I^2 =$ ______
$\sqrt{I^2} =$ ______
$I =$ ______ A = ______ mA  *Ans.*

| Answers |
|---|
| 4 |
| 10 k |
| $R$ |
| $10^4$ |
| $10^4$ |
| $4 \times 10^{-4}$ |
| $\sqrt{4 \times 10^{-4}}$ |
| $2 \times 10^{-2}$  20 |

**PROBLEMS**

1. What current flows through a line supplying 1500 W of power to an electric range of 15 Ω resistance?
2. What voltage is necessary to operate an 18-W automobile headlight bulb of 2 Ω resistance?
3. What is the current flowing through a 50-Ω electromagnet drawing 200 W?
4. What is the maximum current-carrying capacity of a resistor marked 500 Ω and 10 W?

5. Find the voltage drop across a corroded connection if its contact resistance is 100 Ω and it uses 4 W of power.
6. Find the internal resistance of a 2-W electric clock which operates on a 110-V line.
7. A 60- and a 40-W lamp are in parallel across 120 V. Find the combined resistance of the lamps.
8. What is the voltage necessary to operate a 600-W neon sign whose resistance is 20 Ω?
9. An ammeter shunt has a resistance of 0.01 Ω and is rated at 15 W. Find the maximum safe current it can carry.
10. The secondary of a filament transformer delivers 1.89 W to a tube filament whose resistance is 21 Ω. Find the voltage drop across the filament.
11. What is the maximum current-carrying capacity of a resistor marked 4 kΩ and 10 W?

## JOB 7-11 REVIEW OF ELECTRIC POWER

The power used by any part of a circuit is equal to the ______ in that part multiplied by the ______ across that part. — current, voltage

The formula for power is

$$P = I \times \_\_\_\_\_\_$$ — $V$

where $P$ = ______, measured in ______ — power, watts
$I$ = ______, measured in ______ — current, amperes
$V$ = ______, measured in ______ — voltage, volts

This formula may be used to find

$P$ if ______ and $V$ are known — $I$
$I$ if ______ and $V$ are known — $P$
$V$ if $P$ and ______ are known — $I$

The wattage rating of a resistor is equal to ______ times the wattage developed in the resistor. — 2

The total power in a circuit may be found by the formula

$$P_T = \_\_\_\_\_\_ \times V_T$$ — $I_T$

or

$$P_T = P_1 + \_\_\_\_\_\_ + \_\_\_\_\_\_$$ — $P_2$ $P_3$

In the expression $10^3$, the exponent is the number ______. — 3
$5^3$ means $5 \times 5 \times$ ______ or ______. — 5 125
$10^3 =$ ______ and $10^6 =$ ______. — 1000 1,000,000

Some other formulas for power are:

$$P = \_\_\_\_\_\_ \times R$$ — $I^2$

and

$$P = \frac{V^2}{?}$$ — $R$

The square root of a number is that number which must be multiplied by ______ to get the original number. — itself

The symbol for square root is ______. | $\sqrt{\ }$

$\sqrt{16}$ = ______ | 4

$\sqrt{V^2}$ = ______ | $V$

$\sqrt{I^2}$ = ______ | $I$

When finding the square root of a number, the digits should be marked off, ______ digits to a group, starting at the ______ point. | 2 decimal

The decimal point in a whole number is at the ______ of the number. | end

Group the digits in the following numbers preparatory to finding the square root.

*a.* 469.4 ______ | 4 69. 40
*b.* 8062 ______ | 80 62.
*c.* 12,345 ______ | 1 23 45.
*d.* 0.012 ______ | 0.01 20
*e.* 0.002 ______ | 0.00 20
*f.* 6.05 ______ | 6. 05
*g.* 0.000 06 ______ | 0.00 00 60

To find the power of a power, the base is expressed to a power obtained by (adding/multiplying) the powers. | multiplying

To find the square root of a power, the base is expressed to a power obtained by (multiplying/dividing) the exponent by 2. | dividing

## PROBLEMS

1. Find the power used by an electric toaster if it draws 6 A from a 110-V line.
2. Find the power used by a 22-Ω motor if the current is 10 A.
3. Find the total power drawn by the four lamps shown in Fig. 7-8.

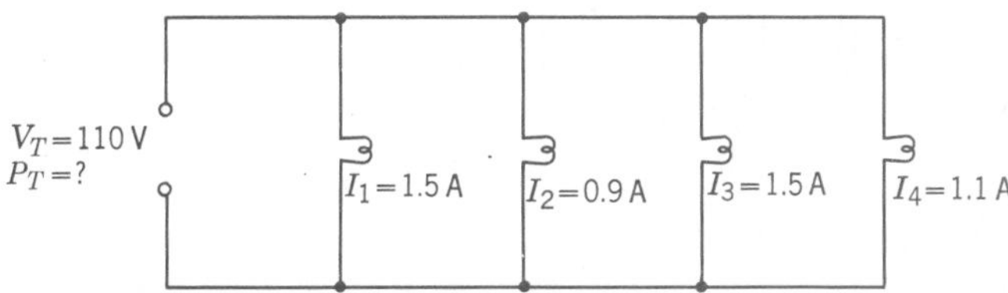


FIGURE 7-8

4. What current is drawn by a 250-W electric vacuum cleaner when operated on 110 V?
5. Three 18-V 0.8-A bells are in series. Find the total power.
6. How many watts of power are dissipated in a 100-kΩ voltage divider if the voltage across it is 300 V?
7. What voltage is required to operate a 25-W automobile headlight bulb properly if the current drawn is 4 A?
8. Find the resistance of a 1000-W electric ironing machine if it uses 5 A.
9. A toy electric-train semaphore is made of a 28-Ω lamp in parallel with a solenoid coil with an effective resistance of 42 Ω. If the total current drawn is 0.6 A, find (*a*) the total resistance and (*b*) the total power used.

10. The combined resistance of a coffee percolator and toaster in parallel is 22 Ω. Find the total power used if the line voltage is 110 V.
11. What power is dissipated in the form of heat in a 130-Ω ballast resistor designed to use up 40 V of excess voltage?
12. What is the voltage needed to operate a 10-W electric-train accessory whose resistance is 15 Ω?
13. A 100- and a 60-W lamp are connected in parallel across 120 V. Find the combined resistance.
14. What is the maximum current-carrying capacity of a resistor marked 5000 Ω and 20 W?
15. A number of incandescent lamps in parallel are supplied by a generator delivering 112 V at its brushes. The resistance of each of the two leads carrying current to the lamps is 0.05 Ω and causes a total voltage drop of 2 V. If each lamp draws 50 W, how many lamps are lit? *Hint:*

 1. Find the line current.

$$V_l = I_l \times R_l \tag{6-1}$$

 2. Find the voltage at the load.

$$V_L = V_G - V_l \tag{6-3}$$

 3. Find the power supplied to the load.

$$P_L = I_l \times V_L \tag{7-1}$$

 4. The number of lamps = $P_L$ ÷ wattage per lamp.

16. A dc generator supplies a 5-A lamp bank and a 1.5-A motor as shown in Fig. 7-9. The resistor $R$ reduces the voltage to that required by the motor. Find (*a*) the terminal voltage of the generator, (*b*) the wattage dissipated by the resistor $R$, (*c*) the wattage lost in all the line wires, and (*d*) the total power supplied to the circuit.

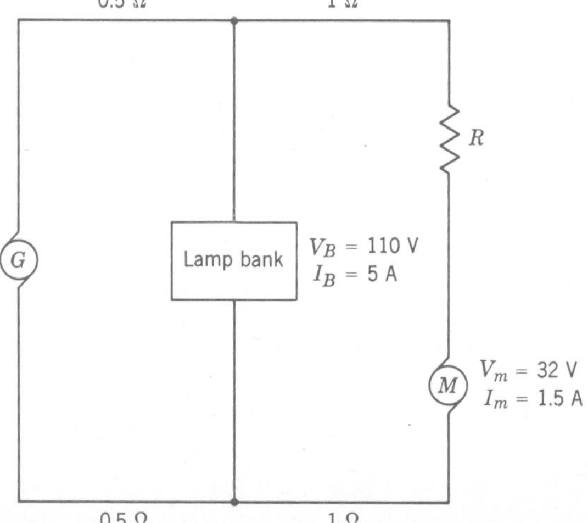


**FIGURE 7-9**

17. In a circuit similar to that shown in Fig. 6-21, $V_T = 100$ V, $R_1 = 50$ Ω, $R_2 = 25$ Ω, $R_3 = 8$ Ω, $R_4 = 2$ Ω, $R_5 = 10$ Ω, $R_6 = 15$ Ω, $R_7 = 105$ Ω, $R_8 = 450$ Ω, and $R_9 = 20$ Ω. Find (*a*) the total resistance of the circuit, (*b*) the total current in the circuit, and (*c*) the total power used by the circuit.

(See Instructor's Manual for Test 7-4, Power)

# ALGEBRA FOR COMPLEX ELECTRIC CIRCUITS

There are many circuits which cannot be solved by the methods used in Chap. 6. These extremely complicated circuits must be solved by the application of Kirchhoff's laws, which will be discussed in the next chapter. However, the solution of these circuits by Kirchhoff's laws requires an extension of our knowledge of algebra.

## JOB 8-1 COMBINING LIKE TERMS

Different quantities of the same item may be added or subtracted. Thus, 2 apples plus 3 apples will equal 5 apples. Similarly, 3 pencils subtracted from 5 pencils will equal 2 pencils. If we use the symbol $a$ for apples and $p$ for pencils, these statements are shortened to read

$$2a + 3a = 5a \qquad \text{and} \qquad 5p - 3p = 2p$$

Quantities involving the *same* letter or letter combinations are called *like terms.* The process of adding or subtracting these like terms is called *combining* terms.

**RULE** To combine like terms, combine the numerical quantities and place the result before the common letter.

EXAMPLE 8-1. Combine the following like terms:

$$3x + 4x = 7x$$
$$9y - 2y = 7y$$
$$4I_1 + 5I_1 = 9I_1$$

*Note:* A letter that stands alone such as $x$, $R$, or $T$ means $1x$, $1R$, or $1T$. Thus, $4x + x$ means $4x + 1x$ or $5x$.

$$6R + 3R - R = 8R \qquad 1.2x + 3.4x = 4.6x$$
$$4.7R + 2R = 6.7R \qquad 7.8x - 4x = 3.8x$$

SELF-TEST 8-2. Combine the following like terms:

| | |
|---|---|
| $2\text{ mA} + 8\text{ mA} =$ ______ mA | 10 |
| $8R + R + 0.2R = 9.2$ ______ | $R$ |
| $5K - 2K + 4K =$ ______ | $7K$ |
| $8I_2 + 4I_2 - 3I_2 =$ ______ | $9I_2$ |

**PROBLEMS**

Combine the following like terms:

1. $3x + 5x$
2. $6y - 4y$
3. $8R - 2R$
4. $2I + 5I$
5. $4x + 7x + x$
6. $8y - y + 4y$
7. $4x + 3x - 2x$
8. $2.5R + 1.2R$
9. $5.6R - 1.2R$
10. $2I + 3.7I$
11. $7.5x - 4x$
12. $\frac{1}{2}y + \frac{1}{4}y$
13. $\frac{1}{2}T + \frac{1}{3}T$
14. $x - 0.2x$
15. $1.2R + 5.4R + 2.4R$
16. $3I + 5I + 0.2I$
17. $5.6x + 3.7x - 1.8x$
18. $x + 5x - 0.6x$
19. $7x - 3.5x + x$
20. $y - 0.4y$
21. $5.2R + 1.8R - 0.4R$

Using Fig. 8-1, express the total length $D$ of the block in terms of $x$, if

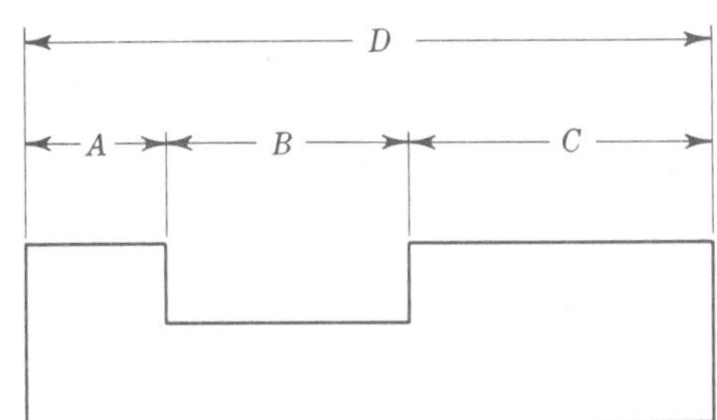


FIGURE 8-1

22. $A = x$, $B = 2x$, and $C = 3x$
23. $A = \frac{3}{4}x$, $B = 1\frac{1}{2}x$, and $C = \frac{5}{8}x$
24. $A = 0.05x$, $B = 1.25x$, and $C = 0.4x$

Using Fig. 8-2, find the measurement $B$, if

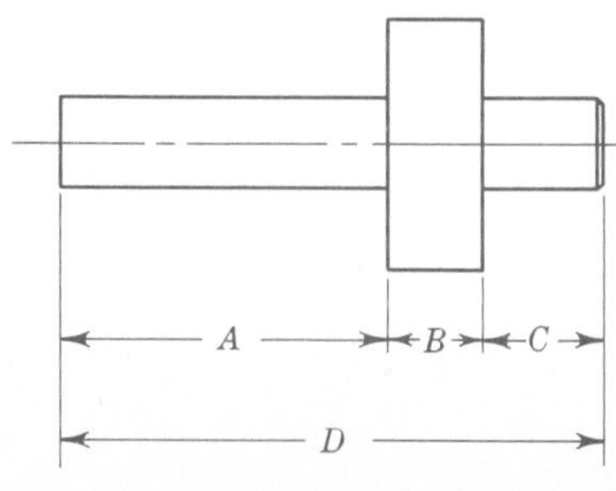


FIGURE 8-2

25. $A = 3x$, $C = x$, and $D = 7x$
26. $A = 2\frac{1}{2}y$, $C = \frac{3}{4}y$, and $D = 5\frac{7}{8}y$
27. $A = 1.6x$, $C = 0.28x$, and $D = 3x$

## JOB 8-2 COMBINING UNLIKE TERMS

*Unlike terms* are those in which the letter portions are *different.* $4R$, $3X$, $3Y$, and $4I$ are all unlike terms. Similarly, $4X$ and the number 6 are unlike terms because the number 6 has no letter.

**RULE** To combine several quantities involving unlike terms, combine each group of like terms separately.

EXAMPLE 8-3. Combine terms.

$$2x + 4y + 5x + 7y$$

SOLUTION

1. Combine the like $x$ terms. $2x + 5x = 7x$
2. Combine the like $y$ terms. $4y + 7y = 11y$
3. Since unlike terms may *not* be combined, state the answer.

$$7x + 11y \quad \textit{Ans.}$$

EXAMPLE 8-4. Combine terms.

$$2R + 4I + 7 + 5R - I - 3$$

SOLUTION

1. Combine the like $R$ terms. $2R + 5R = 7R$
2. Combine the like $I$ terms. $4I - I = 3I$
3. Combine the like "number" terms. $7 - 3 = 4$
4. State the answer. $7R + 3I + 4$ *Ans.*

EXAMPLE 8-5. Combine terms.

$$8I_1 + 5I_2 + 7I_3 + 2I_1 - I_3 - 3I_2$$

SOLUTION

1. Combine the like $I_1$ terms. $8I_1 + 2I_1 = 10I_1$
2. Combine the like $I_2$ terms. $5I_2 - 3I_2 = 2I_2$
3. Combine the like $I_3$ terms. $7I_3 - I_3 = 6I_3$
4. State the answer. $10I_1 + 2I_2 + 6I_3$ *Ans.*

SELF-TEST 8-6. Combine terms.

$3I + 10 + 6I - 2 = 9I +$ ______ | 8

$5R + 3I + 4I - R = \_\_\_\_\_\_ R + \_\_\_\_\_\_$ | 4 $7I$

$12 - 2R - 3 + 7R = \_\_\_\_\_\_ + \_\_\_\_\_\_$ | 9 $5R$

$4I_1 + 2I_2 - I_1 + 3I_2 = 3\_\_\_\_\_\_ + \_\_\_\_\_\_ I_2$ | $I_1$ 5

$3I_1 + 4I_2 - 3I_1 + 5I_2 + 7 = \_\_\_\_\_\_$ | $9I_2 + 7$

## PROBLEMS

Combine the terms in the following algebraic expressions:

1. $2R + 4I + 3I + 8R$
2. $3R + 2I - R + 5I$
3. $3x + 7 - x + 2$
4. $6x + 2y - y + x$
5. $4R + 3R + 6 - R + 2$
6. $1.5I + 7 + 0.5I - 3$
7. $40 + 30 + 3R - 10$
8. $9I + 16 - 4 + 2I - I$
9. $4.2x + 6y + 1.8x - y$
10. $6R + R - 0.5R + 2I$
11. $1.5I + 0.7R - 0.6I + 1.2R$
12. $2I + 60 - 20 + 3I - 5$
13. $3I_1 + 4I_2 + 5I_1 + 6I_2$
14. $5I_2 + I_2 + 3I_1 - 0.5I_1$
15. $5x + 2y + 7 - x + 3y + 3$
16. $2.5I_1 + 40 - 0.4I_1 + 20$
17. $2.5I_2 + 10 - 1.3I_2 - 8$
18. $8.6R + 7.4 + 2.6 - 1.3R$
19. $3.1 + 4I - 0.7 - 0.6I$
20. $8R - 0.3R + 26 - 9.2$

Find the total voltage $V_T$ of the circuit shown in Fig. 8-3 if

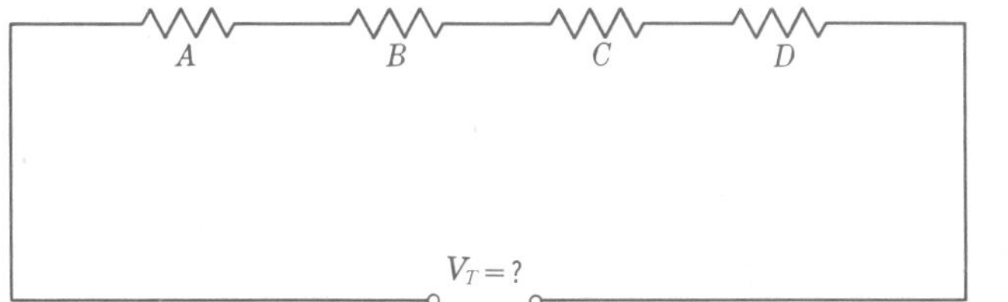


FIGURE 8-3

21. $A = 2x$ V, $B = 40$ V, $C = 4x$ V, and $D = 15$ V
22. $A = 12.6$ V, $B = 2.5y$ V, $C = y$ V, and $D = 35$ V
23. $A = 0.15R$ V, $B = 0.15R$ V, $C = 25$ V, and $D = 0.15R$ V

In Fig. 8-4, find the sum of the voltage drops by tracing around the circuit named in the following problems:

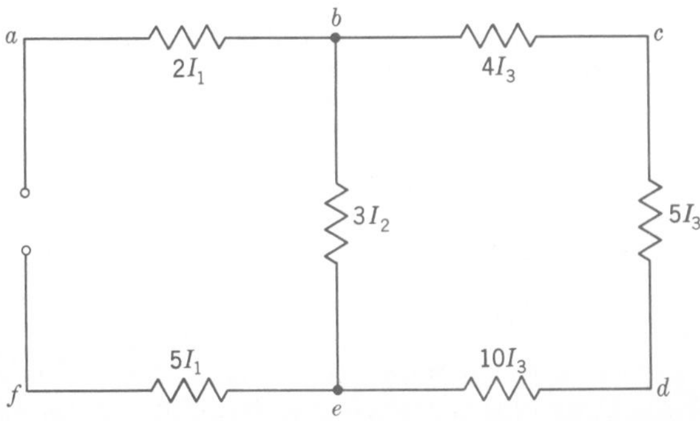


FIGURE 8-4

24. Circuit *abef*
25. Circuit *bcdeb*
26. Circuit *abcdef*

(See Instructor's Manual for Test 8-1)

## JOB 8-3 SOLVING SIMPLE ALGEBRAIC EQUATIONS

In Job 3-9 we learned to solve the standard equation of the type $3x = 12$. In Job 4-6 we learned to solve equations of the type $R + 2 = 10$ and $x - 3 = 7$. In Job 5-7 we learned to solve equations of the type $V/4 = 5$.

These are the three basic types of equations. The more difficult equations are solved by simplifying them step by step into one or the other of these basic types. The concept of combining like terms should be used whenever possible, as this combination will often simplify the equation into one of the basic types which are readily solved.

EXAMPLE 8-7. Solve the equation

$$8x - 2x - 4x = 10$$

SOLUTION

| | | |
|---|---|---|
| 1. Write the equation. | $8x - 2x - 4x = 10$ | |
| 2. Combine like terms. | $2x = 10$ | |
| 3. Solve the basic equation. | $x = 5$ | *Ans.* |

EXAMPLE 8-8. Solve the equation

$$21 = 0.7x - 0.4x$$

SOLUTION

| | |
|---|---|
| 1. Write the equation. | $21 = 0.7x - 0.4x$ |
| 2. Combine like terms. | $21 = 0.3x$ |
| 3. Solve the basic equation. | $\dfrac{21}{0.3} = x$ |
| 4. Write the answer. | $70 = x$ or $x = 70$ *Ans.* |

EXAMPLE 8-9. Solve the equation

$$3x + 5 = 50$$

SOLUTION. In this equation, as in all equations, our main objective is to get the unknown letter *all alone on one side of the equality sign.* The number that remains on the other side will be the answer. The letter $x$ will remain alone on the left side of the equality sign if we can eliminate the $+5$ and the 3 from that side. Since transposing is a very simple operation, your *first* step will *always* be to transpose any single + or − quantity which is not part of a group. Thus, our first step in this problem will be to transpose the +5 from the left side of the equality sign to the right side as −5.

1. Write the equation. $3x + 5 = 50$
2. Transpose the +5. $3x = 50 - 5$
3. Combine like terms. $3x = 45$
4. Solve the basic equation. $x = 15$ *Ans.*

EXAMPLE 8-10. Solve the equation

$$3I - 4 = 20$$

SOLUTION

1. Write the equation. $3I - 4 = 20$
2. Transpose the −4. $3I = 20 + 4$
3. Combine like terms. $3I = 24$
4. Solve the basic equation. $I = 8$ *Ans.*

EXAMPLE 8-11. Solve the equation

$$25 = 4x - 3$$

SOLUTION

1. Write the equation. $25 = 4x - 3$
2. Transpose the −3 to the *left* side of the equality sign.

$$25 + 3 = 4x$$

3. Combine like terms. $28 = 4x$
4. Solve the basic equation. $7 = x$ or $x = 7$ *Ans.*

SELF-TEST 8-12. Solve the equation $3x + 2 + x = 14$.

SOLUTION

1. Write the equation. $3x + 2 + x = 14$
2. Combine like terms. ______ $+ 2 = 14$
3. Transpose the +2 $4x = 14$ ______
4. Combine like terms. $4x =$ ______
5. Solve the basic equation. $x =$ ______ *Ans.*

| $4x$
| $-2$
| 12
| 3

SELF-TEST 8-13. Solve the equation $4x - 6.2 - x = 1.3$.

SOLUTION

1. Write the equation. $4x - 6.2 - x = 1.3$
2. Combine like terms. ______ $- 6.2 = 1.3$
3. Transpose the −6.2. $3x = 1.3$ ______

| $3x$
| $+ 6.2$

4. Combine like terms $3x =$ ______ 7.5
5. Solve for $x$. $x =$ ______ *Ans.* 2.5

SELF-TEST 8-14. An architect's plan calls for five lights laid out as shown in Fig. 8-5. If the total length of electrical conduit used was 58 ft, find the length from $B$ to $C$.

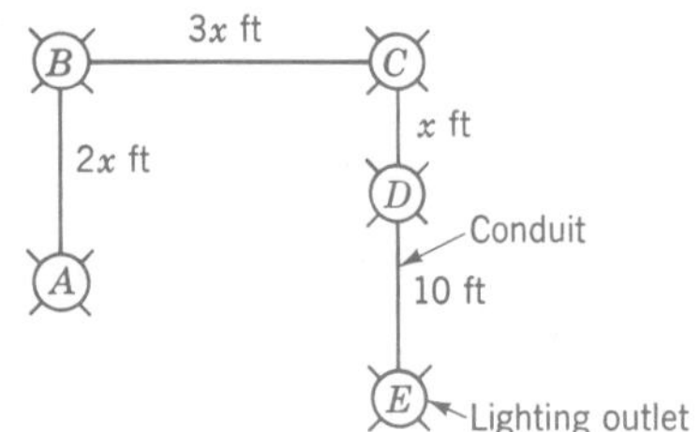

FIGURE 8-5

SOLUTION

1. The total length is equal to the ______ of the individual lengths. sum
2. Write an equation to describe this.

$$2x + 3x + \_\_\_\_\_\_ + 10 = 58$$ $x$

3. Combine like terms. $\_\_\_\_\_\_ + 10 = 58$ $6x$
4. Transpose the +10. $6x = 58 -$ ______ 10
5. Combine like terms. $6x =$ ______ 48
6. Solve the basic equation. $x =$ ______ 8
7. The length from $B$ to $C$ is ______ ft. $3x$
8. By substituting our answer 8 for $x$, we get the length from $B$ to $C = 3 \times$ ______ = ______ ft *Ans.* 8 24

SELF-TEST 8-15. Solve the circuit shown in Fig. 8-6 for the value of $R_1$.

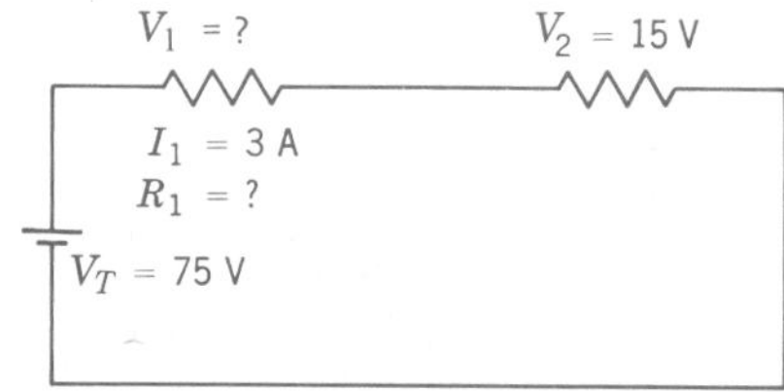

FIGURE 8-6

SOLUTION. We can obtain the value of $V_1$ by using the formula

$$V_1 = I_1 \times \_\_\_\_\_\_ \quad (4\text{-}4)$$ $R_1$

$$V_1 = \_\_\_\_\_\_ \times R_1$$ 3

$$V_1 = 3R_1$$

1. In this series circuit, the total voltage is equal to the ______ of all the voltages. sum

$$V_T = V_1 + \_\_\_\_\_\_$$ $V_2$

| | | | |
|---|---|---|---|
| 2. Substitute numbers. | $75 = \_\_\_\_\_ + 15$ | | $3R_1$ |
| 3. Transpose the +15. | $75 - 15 = \_\_\_\_\_$ | | $3R_1$ |
| 4. Combine like terms. | $\_\_\_\_\_ = 3R_1$ | | 60 |
| 5. Solve the basic equation. | $20 = R_1$ or $R_1 = 20\ \Omega$ | *Ans.* | |

## PROBLEMS

Solve the following equations:

1. $3x = 21$
2. $14 = 2R$
3. $2a + 3a = 25$
4. $3I + I = 24$
5. $40 = 3R + 5R$
6. $\frac{I}{5} = 6$
7. $\frac{2R}{5} = \frac{8}{4}$
8. $x - 4 = 6$
9. $\frac{2x}{3} = 8$
10. $\frac{2x}{3} = 24$
11. $20 = 6x - 2x$
12. $x + 3 = 9$
13. $3x = 16$
14. $10R = 2$
15. $0.2x = 40$
16. $14 = R - 4$
17. $9 = 0.3x$
18. $9R - 5R = 38$
19. $0.3a + 0.2a = 35$
20. $26 = 1.1R + 0.2R$
21. $2.4x - 0.4x = 10$
22. $0.4R + 0.3R = 4.9$
23. $\frac{3y}{4} = 0.6$
24. $R - 5.6 = 14$
25. $5x + 2x + x = 40$
26. $3I + 7I - 2I = 56$
27. $26 = 8x + 6x - 2$
28. $0.12x = 0.06$
29. $\frac{2R}{7} = 7$
30. $5a - 6 = 4$
31. $25 = 3R + 4$
32. $6 = \frac{3x}{5}$
33. $1.9x - 0.4x = 4.5$
34. $3x + 2 = 11$
35. $42 = 6x - x + 2x$
36. $34 = 7R - 8$
37. $29 = 3x + 11$
38. $2R + 3 = 3$
39. $8x + 1 = 3$
40. $2x - 0.3 = 1.1$
41. $4x + 2.6 = 5.4$
42. $4R - R + 2 = 20$
43. $3R - 9 = 11$
44. $59 = 5x - 6$
45. $5y - 17 = 37$
46. $12x - 1 = 8$
47. $5x + 0.2x = 10.4$
48. $x + 1.2x = 4.4$
49. $x + 1.2x + 9 = 97$
50. $2I + 7I + 30 - I = 70$
51. Find the value of each dimension shown in Fig. 8-7.

Using Fig. 8-1, find the value of each dimension if

52. $A = x$, $B = 3x$, $C = 4x$, and $D = 32$

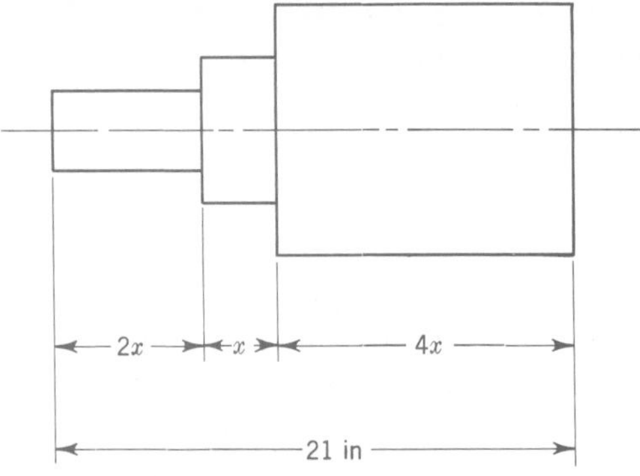


FIGURE 8-7

53. $A = 1.3x$, $B = 0.04x$, $C = 2.16x$, and $D = 35$
54. $A = x + 4$, $B = 5$, $C = 2x$, and $D = 21$
55. $A = 1.2y + 8$, $B = 0.8y$, $C = 0.4y$, and $D = 56$
56. The current in a transistor circuit divides according to the formula $I_E = I_B + \beta I_B$. Find the current gain $\beta$ of the transistor shown in Fig. 8-8.

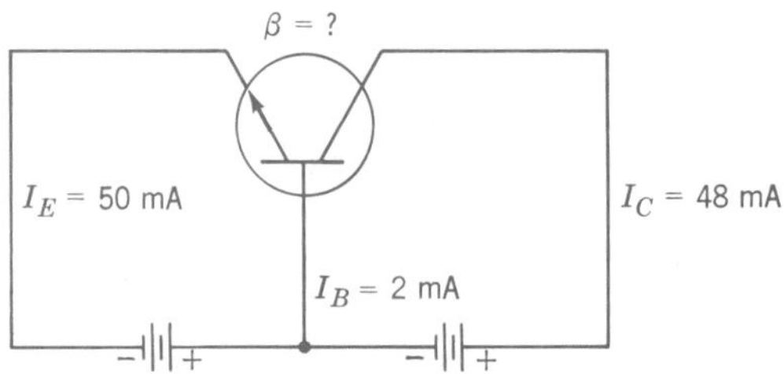


FIGURE 8-8
Division of current in an NPN transistor circuit.

57. Two equal resistors are connected in series with a 150-Ω resistor to make a total resistance of 200 Ω as shown in Fig. 8-9. (*a*) Form an equation which can be used to solve for $R$. (*b*) Solve for $R$.

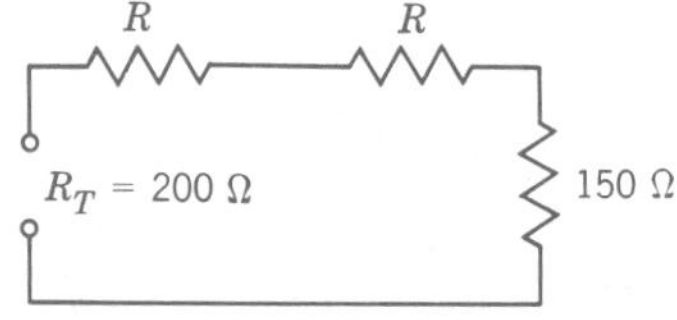


FIGURE 8-9

58. Two holes, 2.25 in apart on center are to be drilled equidistant from the ends of a steel plate 4.75 in long as shown in Fig. 8-10. How far from the ends should the centers of the holes be marked?

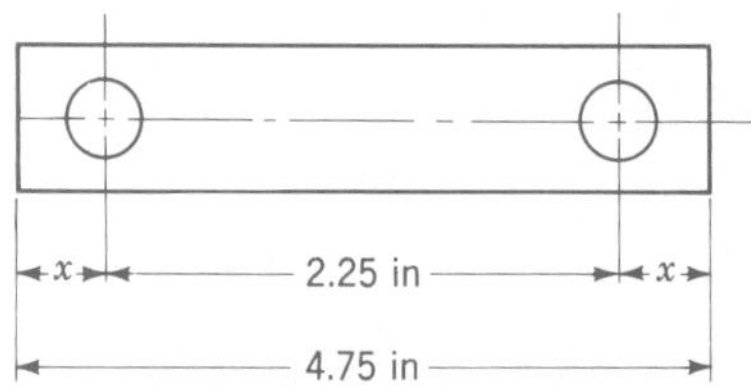


FIGURE 8-10

59. A motor, a lamp, and a soldering iron are in parallel. The motor draws twice as much current as the lamp, and the soldering iron draws $3^1/_2$ times as much current as the lamp. Find the current drawn by each item if the total current drawn is 13 A.
60. A voltage divider is made in three sections connected in series. If $R_2$ is three times $R_1$, and $R_3$ is twice $R_2$, find the resistance of each section when the total resistance is 100,000 Ω.
61. In a series circuit, the voltage across $V_2$ is twice that across $V_1$, and the voltage $V_3$ is one-half of $V_2$. If the total voltage across the circuit is 120 V, find the voltage across each resistor.

(See Instructor's Manual for Test 8-2)

## JOB 8-4 SOLVING EQUATIONS BY TRANSPOSITION AND CROSS MULTIPLICATION

EXAMPLE 8-16. Solve the equation

$$\frac{2x}{5} + 3 = 7$$

SOLUTION

1. Write the equation. $\quad \frac{2x}{5} + 3 = 7$
2. Transpose the + 3. $\quad \frac{2x}{5} = 7 - 3$
3. Combine like terms. $\quad \frac{2x}{5} = 4$
4. Cross-multiply. $\quad 2x = 20$
5. Solve the basic equation. $\quad x = 10$ *Ans.*

EXAMPLE 8-17. Solve the equation

$$23 = \frac{2R}{3} + 7$$

SOLUTION

1. Write the equation. $\quad 23 = \frac{2R}{3} + 7$
2. Transpose the + 7. $\quad 23 - 7 = \frac{2R}{3}$
3. Combine like terms. $\quad 16 = \frac{2R}{3}$
4. Cross-multiply. $\quad 2R = 48$
5. Solve the basic equation. $\quad R = 24$ *Ans.*

SELF-TEST 8-18. Solve the equation

$$1\frac{1}{4}x + 1\frac{1}{2}x + 3 = 14$$

SOLUTION

1. Write the equation. $1\frac{1}{4}x + 1\frac{1}{2}x + 3 = 14$
2. Combine like terms. $\_\_\_\_\_\_ + 3 = 14$ — $2\frac{3}{4}x$
3. Change the mixed number to an improper fraction.

   $\_\_\_\_\_\_ + 3 = 14$ — $\frac{11x}{4}$
4. Transpose the +3. $\frac{11x}{4} = 14\_\_\_\_\_\_$ — $-3$
5. Combine like terms. $\frac{11x}{4} = \_\_\_\_\_\_$ — 11
6. Cross-multiply. $11x = \_\_\_\_\_\_$ — 44
7. Solve the basic equation. $x = \_\_\_\_\_\_$ *Ans.* — 4

**PROBLEMS**

Solve the following equations.

1. $\frac{x}{2} + 4 = 9$
2. $\frac{x}{3} + 7 = 12$
3. $\frac{R}{3} - 2 = 8$
4. $\frac{y}{5} - 6 = 3$
5. $9 = \frac{x}{2} + 1$
6. $5 = \frac{y}{3} - 2$
7. $\frac{2R}{3} - 4 = 6$
8. $\frac{3R}{5} + 4 = 10$
9. $\frac{x}{2} - 1\frac{1}{2} = 3\frac{1}{2}$
10. $5 = \frac{3x}{2} - 7$
11. $\frac{x}{3} - 1.2 = 3.2$
12. $\frac{x}{5} + 0.2 = 1.8$
13. $\frac{5x}{3} + 4 = 9$
14. $\frac{4x}{5} - 18 = 22$
15. $32 = \frac{3x}{2} - 10$
16. $\frac{x}{3} + \frac{1}{4} = 1\frac{1}{4}$
17. $1\frac{1}{4}x - 3 = 12$
18. $2\frac{1}{3}x + 4 = 18$
19. In Fig. 8-5, $AB = 1^1/_2x$, $BC = 3x$, $CD = x$, and $DE = 10$. If the total length of conduit is 87 ft, find the length of $CD$.
20. A lamp, a toaster, and a coffee maker are connected in parallel and draw a total of 12 A. The lamp draws one-sixth as much current as the toaster, and the coffee maker draws 1 A less than the toaster. Find the current drawn by each appliance.

## INTERMEDIATE REVIEW

SELF-TEST 8-19. Find the distance $x$ in Fig. 8-11.

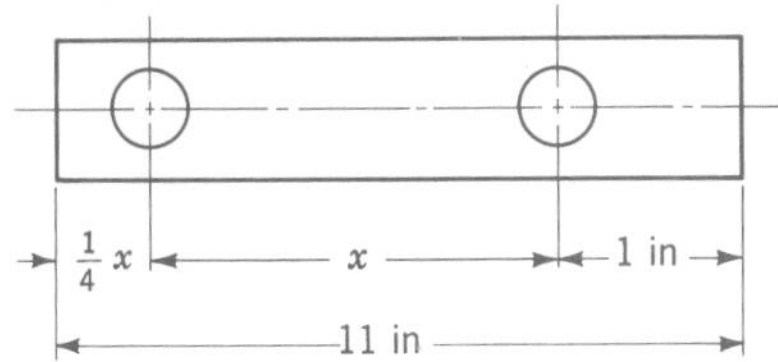


FIGURE 8-11

SOLUTION. The total length is equal to the ______ of the parts. | sum

$$\frac{1}{4}x + x + 1 = \_\_\_\_$$ | 11

$$\underset{?}{\_\_\_\_}x + 1 = 11$$ | $1\frac{1}{4}$

$$\frac{5}{4}x = 11 - \_\_\_\_$$ | 1

$$\frac{5}{4}x = \_\_\_\_$$ | 10

$$5x = \_\_\_\_$$ | 40

$$x = \_\_\_\_ \text{ in} \quad \textit{Ans.}$$ | 8

SELF-TEST 8-20. Three identical resistors are in series with a 100- and a 300-Ω resistor. If the total resistance of the circuit is 1600 Ω, find the resistance of each unknown resistor.

SOLUTION. In a series circuit, the total resistance is equal to the ______ of the individual resistances. | sum

$$R_T = R_1 + R_2 + R_3 + \text{etc.} \qquad (4\text{-}3)$$

$$1600 = R + R + \_\_\_\_ + 100 + 300$$ | $R$

$$1600 = \_\_\_\_ + 400$$ | $3R$

$$1600 - \_\_\_\_ = 3R$$ | 400

$$\_\_\_\_ = 3R$$ | 1200

$$R = \_\_\_\_\ \Omega \quad \textit{Ans.}$$ | 400

(See Instructor's Manual for Test 8-3, Solving Equations)

## JOB 8-5 COMBINING UNLIKE TERMS INVOLVING SIGNED NUMBERS

The procedure is exactly the same as that outlined in Job 8-2 except that the *signs* of the quantities must now be taken into account in the addition.

EXAMPLE 8-21. Combine the following terms:

$$40 - 55 + 2x - 10 - 5x$$

SOLUTION

1. Combine the like "number" terms. $+40 - 55 - 10 = -25$
2. Combine the like $x$ terms. $+2x - 5x = -3x$
3. State the answer. $-25 - 3x$ *Ans.*

EXAMPLE 8-22. Combine the following terms:

$$5x - 2y - 5y - 7x + 14 - y - 6$$

SOLUTION

1. Combine the like $x$ terms. $+5x - 7x = -2x$
2. Combine the like $y$ terms. $-2y - 5y - y = -8y$
3. Combine the like "number" terms. $+14 - 6 = +8$
4. State the answer. $-2x - 8y + 8$ *Ans.*

EXAMPLE 8-23. Combine the following terms:

$$2I_1 - 6I_2 - 4I_1 - 30 - 4I_1 + 2I_2$$

SOLUTION

1. Combine the like $I_1$ terms. $+2I_1 - 4I_1 - 4I_1 = -6I_1$
2. Combine the like $I_2$ terms. $-6I_2 + 2I_2 = -4I_2$
3. State the answer. $-6I_1 - 4I_2 - 30$ *Ans.*

SELF-TEST 8-24. An equation that might result from an application of Kirchhoff's laws (which we shall study in the next chapter) is

$$-10 - 2x + 2y - 3x - 5y = 0$$

Combine the terms on the left side of the equation.

SOLUTION

1. Combine the like $x$ terms. $-2x - 3x =$ ______ | $-5x$
2. Combine the like $y$ terms. $+2y - 5y =$ ______ | $-3y$
3. State the simplified equation. $-5x$ ______ $-10 = 0$ *Ans.* | $-3y$

## PROBLEMS

Combine the following terms:

1. $3x + 4 - 5x + 3$
2. $16 - 4y - 2y - 7$
3. $2I_1 + 3I_2 - 5I_2 - 6I_1$
4. $3 - 8 + 4x - 2 - x$
5. $2x - 3y - 2y - 4x + y$
6. $8 - 2I_2 - 3 - 4I_2 - 3I_2$
7. $4x - 13 - 5x + 6 - 2x$
8. $2x - y - 3x - 4y + 7 - x$
9. $20 - 35 + 4x - 5 - 6x$
10. $-x - y - 4 + 4x + 2 - 3y$

## JOB 8-6 SOLVING EQUATIONS WHICH HAVE UNKNOWNS AND NUMBERS ON BOTH SIDES OF THE EQUALITY SIGN

As noted in Job 8-3, our main objective is to get the unknown letter all alone on one side of the equality sign. If the unknowns appear on *both sides* of the equality sign, we must gather them together as our first step. This will be accomplished by the normal process of transposing. In general then, transpose *all* letters to one side of the equality sign and transpose *all* numbers to the other side. Letters are collected on the left side or on the right side, the side chosen depending on the whim of the solver.

EXAMPLE 8-25. Solve the equation

$$7R - 84 = 4R$$

SOLUTION

1. Write the equation. $7R - 84 = 4R$
2. Transpose the $4R$ to the left side, and the $-84$ to the right side.

$$7R - 4R = 84$$

3. Combine like terms. $3R = 84$
4. Solve the basic equation. $R = 28$ *Ans.*

EXAMPLE 8-26. Solve the equation

$$8x - 7 = 3x + 8$$

SOLUTION

1. Write the equation. $8x - 7 = 3x + 8$
2. Transpose the $3x$ to the left side and the $-7$ to the right.

$$8x - 3x = 8 + 7$$

3. Combine like terms. $5x = 15$
4. Solve the basic equation. $x = 3$ *Ans.*

EXAMPLE 8-27. Solve the equation

$$3x - 3 = 8x - 18$$

SOLUTION

1. Write the equation. $3x - 3 = 8x - 18$

2. Since the greater number of $x$'s appear on the right side of the equality sign, we shall transpose *all* $x$'s to this right side and *all* numbers to the left side.

$$-3 + 18 = 8x - 3x$$

3. Combine like terms. $15 = 5x$
4. Solve the basic equation. $3 = x$ or $x = 3$ *Ans.*

**PROBLEMS**

Solve the following equations:

1. $7R = 2R + 10$
2. $8x = 10 + 6x$
3. $4x = 27 + x$
4. $5R = -2R + 14$
5. $3y + 24 = 9y$
6. $28 - 2I = 5I$
7. $5x - 20 = 3x$
8. $7R - 48 = 3R$
9. $2x = 35 - 5x$
10. $-x - 19 = -2x$
11. $-2y - 21 = -5y$
12. $6T = 4 - 2T$
13. $3 - R = 8R$
14. $2x = 5x - 39$
15. $y = 9y - 40$
16. $x = 3.9 - 2x$
17. $5R + 1 = 3R + 9$
18. $6x - 4 = 2x + 28$
19. $26 + x = 4x - 1$
20. $7x - 25 = 4x + 23$
21. $R + 10 = 45 - 4R$
22. $2x - 8 = 9x - 50$
23. $7x - x + 8 = 2x + 40$
24. $12R - 5 = 6R - R + 23$
25. $8x - 3 + 9 = 2x + x + 31$
26. $12 + 2x = 8x - 60$
27. $8I - 1 = 3 - 4I$
28. $1 - 3R = 7R - 4$
29. $5T = 3T + 4.2$
30. $1.4x - 12 = -0.6x$

(See Instructor's Manual for Test 8-4)

## JOB 8-7 SOLVING EQUATIONS WITH POSSIBLE NEGATIVE ANSWERS

Consider the equation

$$2R = -10$$

Solving,

$$R = \frac{-10}{+2}$$

or

$$R = -5 \qquad \textit{Ans.}$$

The signs in this problem might appear as shown below.

$$-2R = 10 \qquad\qquad -2R = -10$$

Solving,

$$R = \frac{+10}{-2} \qquad\qquad R = \frac{-10}{-2}$$

or

$$R = -5 \quad \textit{Ans.} \qquad\qquad R = +5 \quad \textit{Ans.}$$

The answers above are not incorrect. It is quite possible for an answer to be a negative number. In the next chapter on complex circuits and in our future study of ac electricity, we shall meet these frequently. The basic equation is solved in a normal manner, but care must be used in dividing the signed numbers. Another method for handling these signs is illustrated in the next example.

EXAMPLE 8-28. Solve the equation

$$-2R = -10$$

SOLUTION. Since all parts of an equation may be multiplied by the same number without destroying the equality, we can eliminate the cumbersome negative sign of the unknown by multiplying the entire equation by $-1$. Therefore,

$$-1(-2R = -10)$$

becomes $$2R = 10$$

or $$R = 5 \quad Ans.$$

This operation may be described in the following simple rule.

**RULE** All the signs of the individual parts of an equation may be changed without destroying the equality.

EXAMPLE 8-29. Solve the following equations.

| $-R = -6$ | $-I = 25$ | $7 = -R$ | |
|---|---|---|---|
| $R = 6$ | $I = -25$ | $-7 = R$ | *Ans.* |

EXAMPLE 8-30. Solve the equation

$$4x - 5 = 5x + 1$$

SOLUTION

| | |
|---|---|
| 1. Write the equation. | $4x - 5 = 5x + 1$ |
| 2. Transpose. | $4x - 5x = 1 + 5$ |
| 3. Combine like terms. | $-x = 6$ |
| 4. Change signs throughout. | $x = -6$ *Ans.* |

**PROBLEMS**

Solve the following equations:

1. $-x = +8$
2. $12 = -R$
3. $3x = -15$
4. $-21 = 3T$
5. $8x = -4$
6. $-5 = 10x$
7. $-2x = 16$
8. $40 = -4x$
9. $-2x = -14$
10. $-24 = -3R$
11. $3x + 17 = 5$
12. $6 - 3x = 18$
13. $9 = 8 - R$
14. $y = 5y + 28$
15. $3R + 10 = R$
16. $5x + 9 = 4x + 1$
17. $5R + 7 = 6R + 18$
18. $3I + 22 = 9I - 20$
19. $-4R - 11 = 6R + 19$
20. $0.4x = -20$
21. $-1.2I = 2.4$
22. $-39 = 0.3T$
23. $1.6R + 10 = 0.6R + 3$
24. $-2I_2 = 10 - 8.58$

## JOB 8-8 REMOVING PARENTHESES

In order to solve the equations in the next job, we must be able to simplify an equation by removing any parentheses in it. A parenthesis is used to indicate that the quantities within it represent a single idea. For example, if a transformer costs \$30 and the tax is \$1, then the total actual cost is represented as the quantity $(30 + 1)$. Of course, this would be written as \$31 because we would naturally combine the like terms. However, if the cost were unknown, we would represent it as $x$ dollars. If the tax remains constant, then the only way to represent the total cost would be as the quantity $x + 1$ or $(x + 1)$.

Now, if we bought six transformers, the total cost would be six times the cost of one transformer or $6 \times (x + 1)$. This is usually written as $6(x + 1)$. This means that the 6 is multiplied by the $x$ *and also* by the number 1. Thus,

$$6(x + 1) = 6x + 6$$

We can check the accuracy of this method by using the actual cost.

$$6(30 + 1) = 6(31) = 186$$

or $$6(30 + 1) = 6 \times 30 + 6 \times 1 = 180 + 6 = 186$$

**RULE**

To multiply a parenthesis by a single quantity:
1. Multiply each term of the parenthesis by the multiplier.
2. Combine like terms.

EXAMPLE 8-31. Multiply $+3(2R - 7)$.

SOLUTION

1. Multiply each part of the parenthesis by $+3$.

$$(+3) \times (2R) = +6R$$
$$(+3) \times (-7) = -21$$

2. State the answer.

$$6R - 21 \quad \textit{Ans.}$$

EXAMPLE 8-32. Remove parentheses and collect like terms.

*a.* $8x - 3(2 + x)$
$8x - 6 - 3x$
$5x - 6$ *Ans.*

*b.* $11 + 2(3x - 9)$
$11 + 6x - 18$
$6x - 7$ *Ans.*

*c.* $-3(I + 4) - 5$
$-3I - 12 - 5$
$-3I - 17$ *Ans.*

EXAMPLE 8-33. Remove parentheses and collect like terms.

$$5x + (3 - 8x)$$

SOLUTION. When the parenthesis is preceded by just a plus sign or a minus sign, the number 1 is understood to be present.

1. Write the problem.

$$5x + (3 - 8x)$$

2. Insert the number 1 after the sign.

$$5x + 1(3 - 8x)$$

3. Multiply the parenthesis by +1.

$$5x + 3 - 8x$$

4. Collect like terms.

$$-3x + 3 \quad \text{or} \quad 3 - 3x \quad \textit{Ans.}$$

Certain problems can lead to errors. Be careful to note the difference between the following examples.

EXAMPLE 8-34. Remove parentheses and collect like terms.

| | | |
|---|---|---|
| $-9(4x - 3)$ | | $-9 - (4x - 3)$ |
| $-36x + 27$ | *Ans.* | $-9 - 1(4x - 3)$ |
| | | $-9 - 4x + 3$ |
| | | $-6 - 4x$ *Ans.* |

SELF-TEST 8-35. An equation that results from an application of Kirchhoff's laws to the circuit shown in Fig. 9-22 in the next chapter is

$$-10 - 2(x - y) + 3y = 0$$

Simplify the equation by removing parentheses and collecting like terms.

SOLUTION

$$-10 - 2x \_\_\_\_\_\_ + 3y = 0$$
$$-10 - 2x + \_\_\_\_\_\_ = 0 \quad \textit{Ans.}$$

| $+ 2y$
| $5y$

**PROBLEMS**

Remove parentheses and collect like terms.

1. $2(3 - 4x)$
2. $3(5x - 6)$
3. $-3(x - 4)$
4. $4(2x - 3) - 3x$
5. $7 + 3(x - 4)$
6. $2y - (y - 3)$
7. $9I - 3(I + 8)$
8. $6 + (R - 7)$
9. $+(8 - 2R) - 7$
10. $-(2 + 3x) + 5$
11. $14 - 3(x - 5)$
12. $60 - 3(I_1 - 5)$
13. $20 + 2(I_1 + I_2)$
14. $30 - 4(6 - I_2)$
15. $5R - 2(10 - 4 - R)$
16. $(x - 2) + 2(x - 4)$

17. $-3(-2 + R) + (R - 1)$ 18. $7x - 3x - 2(x + 4)$
19. $-(I_1 - I_2) + 4(I_1 - 6)$ 20. $3x - 7 - 2(x - 3) + 20$

(See Instructor's Manual for Test 8-5, Combining Terms and Solving Equations)

## JOB 8-9 SOLVING EQUATIONS CONTAINING PARENTHESES

EXAMPLE 8-36. Solve the equation

$$6x - 2(x - 4) = 32$$

SOLUTION. In order to solve an equation we collect *all* the unknowns on one side of the equality sign and *all* the numbers on the other side. We can't do this in this equation until we release the $x$ and the $-4$ from the parenthesis. Once this has been done, the solution proceeds normally.

1. Write the equation. $6x - 2(x - 4) = 32$
2. Remove parentheses. $6x - 2x + 8 = 32$
3. Transpose the +8. $6x - 2x = 32 - 8$
4. Combine like terms. $4x = 24$
5. Solve the basic equation. $x = 6$ *Ans.*

SELF-TEST 8-37. Find $R_2$ in the circuit shown in Fig. 8-12.

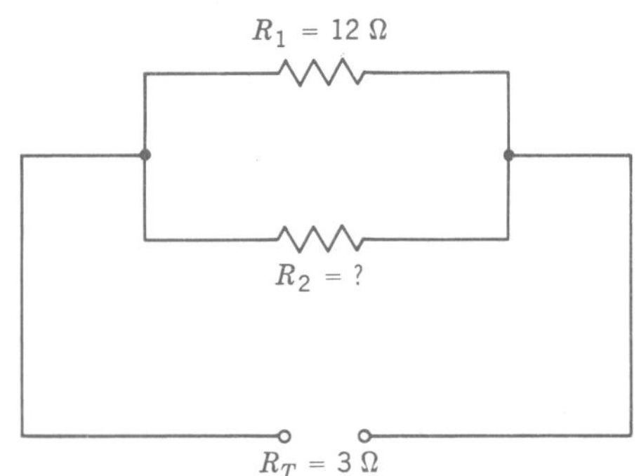


FIGURE 8-12

SOLUTION. This is a ______ circuit. | parallel

1. Write the formula for $R_T$. $R_T = \dfrac{(R_1 \times R_2)}{(R_1 + ?)}$ | $R_2$
2. Substitute numbers. $\dfrac{3}{1} = \dfrac{(12 \times ?)}{(? + R_2)}$ | $R_2$, 12
3. Cross-multiply. $3(12 + R_2) =$ ______ | $12R_2$
4. Remove parentheses. $36 +$ ______ $= 12\,R_2$ | $3R_2$
5. Transpose the $3R_2$. $36 = 12R_2 -$ ______ | $3R_2$
6. Combine like terms. $36 =$ ______ | $9R_2$
7. Solve the basic equation. $R_2 =$ ______ Ω *Ans.* | 4

SELF-TEST 8-38. An electric iron uses 90 W more than twice the power used by an electric toaster. Three times the toaster power equals 460 W less than twice the iron wattage. Find the power used by each.

SOLUTION. In order to describe the problem in the language of algebra, we need algebraic names for the power used by each appliance.

Let $x$ = the power used by the toaster. Then twice this power = ________, and | $2x$

90 W more than this = (________) | $2x + 90$

or $(2x + 90)$ = the power used by the ________. | iron

Now to make the equation.

3 times the toaster power = twice the iron power − 460

$3 \times$ ________ = 2(________) − 460 | $x$ $2x + 90$

Solve this equation.

$$3x = 2(2x + 90) - 460$$

1. Remove parentheses.

$3x$ = ________ + 180 − 460 | $4x$

2. Transpose the $4x$.

$3x$ − ________ = 180 − 460 | $4x$

3. Combine like terms.

$-x$ = ________ | −280

4. Solve for $x$.

$x$ = ________ | 280

5. State the answer. Since $x$ was our algebraic name for the toaster power,

Toaster power = ________ W *Ans.* | 280

Since $(2x + 90)$ was our algebraic name for the iron power, by substituting 280 for $x$, we get

Iron power = $(2 \times 280) + 90$

= ________ + 90 | 560

= ________ W *Ans.* | 650

**PROBLEMS**

Solve the following equations:

1. $2(x + 3) = 8$
2. $3(R - 4) = 3$
3. $4(R + 3) = 4$
4. $2(3 - R) = 4$
5. $5(2x + 3) = 35$
6. $7(3y - 2) = 28$
7. $3(2x - 2) = -24$
8. $3(3x - 1) - 4 = 11$
9. $5(3R - 2) + 8 = 43$
10. $3(a - 4) = 2(a + 1)$

11. $7(x + 1) = 5(x + 1)$
12. $2(I - 4) = 5(I + 2)$
13. $3(3x - 2) + 5 = 26$
14. $2(y + 4) + y = 23$
15. $4(x - 3) - 2x = -20$
16. $6x - (x - 3) = 28$
17. $4x - 5(x - 2) = 3$
18. $8R - (R - 4) = 11$
19. $6x - 2(x + 6) = x$
20. $10x = 38 + (4x - 2)$
21. $2R = 15 + 3(R + 2)$
22. $6y = 20 - (y - 8)$
23. $8y = 1 - 2(y - 2)$
24. $5x - 2(x - 3) = 17$
25. $2y + (y - 3) = 18 - 2(y + 3)$
26. $4(R - 2) - 7 = 9 - (R + 4)$
27. $0.4(x - 2) = 1.4$
28. $0.3(5R - 6) = 2.7$
29. $4(x - 0.2) = 2(x + 0.7)$
30. $2(x - 1.2) = 0.2(x - 3)$
31. In a circuit similar to that shown in Fig. 8-8, $I_E = 52$ mA and $I_B = 2$ mA. Find $\beta$ using the transistor formula $I_E = I_B\ (\beta + 1)$.
32. In a circuit similar to that shown in Fig. 8-12, find $R_1$ if $R_2 = 60\ \Omega$ and $R_T = 15\ \Omega$. Use the formula given in Example 8-37.
33. In a parallel circuit of two resistances, $R_1 = 16\ \Omega$, $I_2 = 2$ A, $I_T = 8$ A. Find $R_2$ using the formula

$$I_2 = \frac{R_1}{(R_1 + R_2)} \times I_T$$

34. Using the formula to change Fahrenheit temperature to Celsius temperature, $C = \dfrac{5(F - 32)}{9}$, find the number of degrees Fahrenheit $F$ which is equivalent to $C = 20°\text{C}$.
35. Each heating element in a cafeteria grill uses 130 W of power less than three times the wattage used by a toaster. When two toasters and three heating elements are in use, they require a total of 2470 W of power. Find the wattage used by one heating element.
36. Using the transistor formula $\beta = \dfrac{\alpha}{1 - \alpha}$, find $\alpha$ if $\beta = 49$.

(See Instructor's Manual for Test 8-6)

## JOB 8-10 SOLVING EQUATIONS CONTAINING FRACTIONS

In Job 5-7 we learned how to solve some simple fractional equations. These equations were limited to those containing only two fractions which were equal to each other. They were solved by cross multiplication, which is merely a short cut to be used only in the situation where two fractions equal each other. If there are *more* than two fractions, we are forced to use the least common denominator method discussed in Examples 5-17 to 5-19 of Job 5-7.

**RULE**

If an equation contains more than two fractions, eliminate the denominators by multiplying all parts of the equation by the least common denominator.

EXAMPLE 8-39. Solve the equation

$$\frac{x}{2} - \frac{x}{3} = \frac{1}{2}$$

SOLUTION

1. Write the equation.

$$\frac{x}{2} - \frac{x}{3} = \frac{1}{2}$$

2. Multiply all terms by the LCD, which is 6.

$$\frac{\overset{3}{(\not{6})}x}{\underset{1}{\not{2}}} - \frac{\overset{2}{(\not{6})}x}{\underset{1}{\not{3}}} = \frac{\overset{3}{(\not{6})}1}{\underset{1}{\not{2}}}$$

3. Multiply.

$$3x - 2x = 3$$

4. Combine like terms.

$$x = 3 \qquad Ans.$$

EXAMPLE 8-40. Solve the equation

$$\frac{x}{4} + \frac{x}{8} = 6$$

SOLUTION

1. Write the equation.

$$\frac{x}{4} + \frac{x}{8} = 6$$

2. Complete all fractions by placing whole numbers or decimals over 1.

$$\frac{x}{4} + \frac{x}{8} = \frac{6}{1}$$

3. Multiply all terms by the LCD, which is 8.

$$\frac{\overset{2}{(\not{8})}x}{\underset{1}{\not{4}}} + \frac{\overset{1}{(\not{8})}x}{\underset{1}{\not{8}}} = \frac{\overset{8}{(\not{8})}6}{\underset{1}{\not{1}}}$$

4. Multiply.

$$2x + x = 48$$
$$3x = 48$$
$$x = 16 \qquad Ans.$$

EXAMPLE 8-41. Solve the equation

$$\frac{2R}{5} + 3 = \frac{R}{4}$$

SOLUTION

1. Write the equation.

$$\frac{2R}{5} + 3 = \frac{R}{4}$$

2. Complete all fractions by placing whole numbers or decimals over 1.

$$\frac{2R}{5} + \frac{3}{1} = \frac{R}{4}$$

3. Multiply all terms by the LCD, which is 20.

$$\overset{4}{(\cancel{20})}\,\frac{2R}{\underset{1}{\cancel{5}}} + \overset{20}{(\cancel{20})}\,\frac{3}{\underset{1}{\cancel{1}}} = \overset{5}{(\cancel{20})}\,\frac{R}{\underset{1}{\cancel{4}}}$$

4. Multiply.

$$8R + 60 = 5R$$

5. Transpose.

$$8R - 5R = -60$$
$$3R = -60$$
$$R = -20 \qquad \textit{Ans.}$$

## PROBLEMS

Solve the following equations.

1. $\frac{x}{2} - \frac{x}{3} = \frac{1}{3}$
2. $\frac{x}{2} - \frac{x}{4} = \frac{1}{2}$
3. $\frac{x}{4} - \frac{x}{5} = 2$
4. $\frac{3R}{4} - \frac{2R}{3} = 1$
5. $\frac{x}{3} - \frac{x}{6} = 5$
6. $\frac{x}{2} + \frac{x}{3} = 5$
7. $\frac{V}{2} + \frac{V}{8} = 5$
8. $\frac{V}{5} + \frac{V}{3} = \frac{8}{15}$
9. $\frac{10}{3} = \frac{x}{3} + \frac{x}{7}$
10. $\frac{R}{4} + \frac{2R}{5} = 13$
11. $\frac{5x}{3} - 7 = \frac{x}{2}$
12. $\frac{V}{30} - 2 = \frac{V}{40}$
13. $\frac{V}{20} - \frac{V}{30} = 5$
14. $\frac{V}{20} + \frac{V}{10} + \frac{V}{60} = 5$

15. $\frac{3V}{9} - \frac{3V}{10} = \frac{1}{3}$

16. $\frac{R}{10} + \frac{R}{30} + \frac{R}{60} = \frac{9}{10}$

17. $\frac{6x}{50} + \frac{3x}{50} = \frac{9}{25}$

18. $\frac{V}{200} - \frac{V}{500} = \frac{3}{2}$

19. $\frac{R}{8} - \frac{R}{15} = 1$

20. $\frac{5x}{6} + \frac{x}{9} - \frac{2x}{3} = \frac{5}{9}$

21. $\frac{V}{6} + \frac{V}{9} + \frac{V}{12} = 39$

22. $\frac{V}{10} + \frac{V}{8} + \frac{V}{6} = 18.8$

(See Instructor's Manual for Test 8-7, Equations with Parentheses and Fractions)

## JOB 8-11 SOLVING SIMULTANEOUS EQUATIONS BY ADDITION

Simultaneous equations are those that arise at the same time (simultaneously) from the same problem. In simple problems, an equation can usually be found to describe the conditions of the problem. The solution of the equation finds the unknown. In the complex circuits to be discussed in the next chapter, there will be *two* unknowns to find. Suppose that the two unknown currents are represented as $x$ and $y$ and an equation connecting these currents is found to be

$$x + y = 12$$

There is an infinite variety of answers which will fit this equation. If $x = 2$, then $y = 10$. If $x = 7$, then $y = 5$. If $y = 3$, then $x = 9$, etc. Obviously, then, an equation which contains two unknown quantities *cannot* be solved. However, if a *second* equation connecting the *same* two quantities can be found, we might be able to find a solution by working the two equations simultaneously. In general, *two* unknowns will require *two* equations. *Three* unknowns will require *three* equations, etc.

EXAMPLE 8-42. Solve the pair of simultaneous equations given below for the value of $x$ and $y$.

SOLUTION

1. The equations are:

$$x + y = 12 \tag{1}$$
$$x - y = 4 \tag{2}$$

2. Since both equations contain two unknowns and therefore cannot be solved separately, let us *add* the equations. Since only like terms may be added, we must be sure to have the equations arranged so that the like terms are one under the other. If they are not so arranged, we must transpose one equation so as to agree with the other. Adding,

$$2x = 16$$

Note that the sum of $+y$ and $-y$ equals zero, and that the letter $y$ disappears. The resulting equation contains *only one unknown,* and is easily solved. Solving,

$$x = 8 \qquad Ans.$$

3. Substitute this value of $x$ in *either* of the original equations to find the value of $y$. Using equation (1),

$$x + y = 12 \qquad (1)$$

Substituting, $8 + y = 12$

Transposing, $y = 12 - 8$

Combining like terms. $y = 4 \qquad Ans.$

We were very lucky in the last problem. The simple act of adding the two equations *eliminated* one letter and provided a *third* equation which contained only *one* unknown which was easily solved. Will this occur all the time, or are certain conditions necessary for the elimination of one unknown by addition?

Consider the following additions: $(+7) + (-7) = 0$; $(+4) + (-4) = 0$; $(-5) + (+5) = 0$; $(+19) + (-19) = 0$. Apparently, the total is zero whenever we add *identical* quantities with *opposite* signs. This is what happened when we added $+y$ and $-y$ in the last problem. The addition totaled zero, which therefore eliminated the unknown $y$.

The number portion of an algebraic expression is called its *numerical coefficient.* For example,

2 is the numerical coefficient of the expression $2x$.
7 is the numerical coefficient of the expression $7xy$.
1 is the numerical coefficient of the expression $ab$.

**RULE** To eliminate an unknown by addition, the numerical coefficients of the unknown must be identical, but with opposite signs.

If necessary, one or *both* equations may be multiplied by the proper number and sign to create identical coefficients with opposite signs. Let's see how this works out in the next example.

EXAMPLE 8-43. Solve the pair of simultaneous equations given below for the value of $x$ and $y$.

SOLUTION

1. The equations are:

$$x + 2y = 10 \qquad (1)$$
$$x - y = 1 \qquad (2)$$

You can see that neither $x$ nor $y$ will be eliminated by adding these two equations. ($x$ added to $x$ gives $2x$ and *not* zero. $+2y$ added to $-y$ gives $+y$ and *not* zero.) However, if the $-y$ in equation (2) were $-2y$, then the total of $+2y$ in equation (1) and $-2y$ in equation (2) *would* total zero. We shall therefore *change* the $-y$ in equation (2) into $-2y$ by multiplying the *entire* equation by the quantity which will change $-y$ into $-2y$. This multiplier must be $+2$.

2. Rewrite the problem as shown.

$$\begin{array}{r} x + 2y = 10 \\ \underline{+2(x - \quad y = 1)} \end{array}$$

3. Multiplication gives

$$\begin{array}{r} x + 2y = 10 \\ \underline{2x - 2y = 2} \end{array}$$

4. Add the equations.

$$3x = 12$$
$$x = 4 \qquad \textit{Ans.}$$

5. Substitute this value for $x$ in *either* original equation.

$$x + 2y = 10 \qquad (1)$$
$$4 + 2y = 10$$
$$2y = 10 - 4$$
$$2y = 6$$
$$y = 3 \qquad \textit{Ans.}$$

EXAMPLE 8-44. Solve the pair of simultaneous equations given below for the value of $a$ and $b$.

SOLUTION

1. The equations are:

$$a + b = 5 \qquad (1)$$
$$2a + b = 7 \qquad (2)$$

In order to eliminate the unknown $b$, we must multiply equation (1) by $-1$.

$$\begin{array}{r} -1(a + b = 5) \\ \underline{2a + b = 7} \end{array}$$

2. Multiplication gives

$$\begin{array}{r} -a - b = -5 \\ \underline{2a + b = 7} \end{array}$$

3. Add the equations.

$$a = 2 \qquad \textit{Ans.}$$

4. Substitute 2 for $a$ in equation (1).

$$a + b = 5 \qquad (1)$$
$$2 + b = 5$$
$$b = 5 - 2$$
$$b = 3 \qquad \textit{Ans.}$$

EXAMPLE 8-45. Solve the pair of simultaneous equations given below for the value of $x$ and $y$.

SOLUTION

1. The equations are:

$$5x + 2y = 4 \qquad (1)$$
$$4x - 3y = 17 \qquad (2)$$

The elimination of *either* unknown will solve the problem. However, since the signs of the unknown $y$ are exactly what they should be, we shall attempt to eliminate $y$. It is very difficult to change a 2 into a 3 or a 3 into a 2 by multiplication. However, it is not necessary that the coefficients of $y$ remain what they were in the original problem.

*All that is necessary is that the coefficients be the same number, but with opposite signs.*

We shall therefore change $+2y$ in equation (1) into $+6y$ by multiplying it by 3, and the $-3y$ in equation (2) into $-6y$ by multiplying it by 2.

$$3(5x + 2y = 4)$$
$$2(4x - 3y = 17)$$

2. Multiplication gives

$$15x + 6y = 12$$
$$8x - 6y = 34$$

3. Add the equations.

$$23x = 46$$

4. Solve for $x$.

$$x = 2 \qquad \textit{Ans.}$$

5. Substitute this value for $x$ in equation (1).

$$5x + 2y = 4 \qquad (1)$$
$$5(2) + 2y = 4$$
$$10 + 2y = 4$$
$$2y = 4 - 10$$
$$2y = -6$$
$$y = -3 \qquad \textit{Ans.}$$

SELF-TEST 8-46. Solve the pair of simultaneous equations given below for the value of $x$ and $y$.

SOLUTION

1. The equations are:

$$\begin{aligned} 2x + 3y &= -8 \quad (1)\\ 7x + 4y &= \ \ 11 \quad (2) \end{aligned}$$

| | |
|---|---|
| 2. We plan to eliminate one letter by ______ the equations. | adding |
| 3. The letter will disappear if the addition of the coefficients of that letter is equal to ______. | zero |
| 4. The addition of these coefficients will equal zero if *a.* The coefficients are ______ in value. *b.* The signs are ______. | equal<br>opposite |
| 5. The coefficients of $x$ (do/do not) satisfy these conditions. The coefficients of $y$ (do/do not) satisfy these conditions. | do not<br>do not |
| 6. Yet, if we are going to solve the equations, the coefficients of one of these letters *must* be ______ and with ______ signs. | equal opposite |
| 7. Let us try to attain this condition for the letter $y$. | |
| 8. The smallest number into which we can change both 3 and 4 by multiplication is ______. | 12 |
| 9. If we change the $3y$ of equation (1) into $12y$, then we must change the $4y$ of equation (2) into ______. | $-12y$ |
| 10. To get $12y$, we must multiply equation (1) by ______ because only $4 \times 3y = 12y$. To get $-12y$, we must multiply equation (2) by ______ because only $-3 \times 4y = -12y$. | 4<br>$-3$ |
| 11. Indicate the plan by placing the entire equation in parentheses with the proper multiplier in front. ______ $(2x + 3y = -8)$ ______ $(7x + 4y = 11)$ | 4<br>3 |
| 12. Multiply each equation by its multiplier to give $8x + 12y =$ ______; ______ $- 12y = -33$ | $-32$<br>$-21x$ |
| 13. Add the equations. $-13x =$ ______ | $-65$ |
| 14. Solve this equation. $x =$ ______ *Ans.* | $+5$ |
| 15. Substitute this value for $x$ in equation (1). $2x + 3y = -8$; $2($______$) + 3y = -8$; ______ $+ 3y = -8$; $3y = -8$ ______; $3y =$ ______; $y =$ ______ *Ans.* | 5<br>10<br>$-10$<br>$-18$<br>$-6$ |

16. *Check:* If we are correct, then the substitution of these values in the original equations should make true statements. Substitute $x = 5$ and $y = -6$ in equation (1).

$$2x + 3y = -8 \qquad (1)$$
$$2(5) + 3(\_\_\_\_) = -8$$
$$\_\_\_\_ - 18 = -8$$
$$\_\_\_\_ = -8 \qquad \textit{Check}$$

−6
10
−8

Substitute $x = 5$ and $y = -6$ in equation (2).

$$7x + 4y = 11 \qquad (2)$$
$$7(\_\_\_\_) + 4(-6) = 11$$
$$35 - \_\_\_\_ = 11$$
$$\_\_\_\_ = 11 \qquad \textit{Check}$$

5
24
11

## PROBLEMS

Solve the following sets of equations:

1. $2x + y = 8$
   $x - y = 1$
2. $x + y = 6$
   $x - y = 4$
3. $2x + y = 12$
   $x + y = 7$
4. $x + y = 1$
   $x - y = 5$
5. $I_1 + I_2 = 8$
   $I_1 - 2I_2 = 2$
6. $x + y = 8$
   $x - y = 0$
7. $V - 4R = -1$
   $V - 2R = 3$
8. $x + 2y = 7$
   $x - 2y = 3$
9. $2I + 3V = 8$
   $4I - 3V = 34$
10. $7x + 4y = 26$
    $2x - 4y = -8$
11. $5a + 4b = -7$
    $a - 2b = 7$
12. $3I_1 - I_2 = 5$
    $2I_1 + 3I_2 = 18$
13. $2x - 3y = 7$
    $3x + 4y = 19$
14. $3I_1 + 7I_2 = 5$
    $2I_1 + 3I_2 = 5$
15. $3a + 4b = 58$
    $5a - 2b = 10$
16. $12I_2 + 3I_1 = 75$
    $4I_2 - I_1 = 7$
17. $4I_2 + I_3 = 25$
    $-3I_2 + 8I_3 = -45$
18. $20x - 7y = -36$
    $-4x - 2y = -20$
19. $6I_2 + 5I_3 = 40$
    $-4I_2 + 7I_3 = -6$
20. $5I_1 + 2I_2 = 36$
    $3I_1 - 4I_2 = -20$

(See Instructor's Manual for Test 8-8)

## JOB 8-12 SOLVING SIMULTANEOUS EQUATIONS BY SUBSTITUTION

There are some pairs of equations which may be solved more efficiently by this method of substitution than by the addition method discussed in Job 8-11.

EXAMPLE 8-47. Solve the pair of simultaneous equations given below for the value of $x$ and $y$.

SOLUTION

1. The equations are:

$$x = 4 \quad (1)$$
$$2x + y = 11 \quad (2)$$

2. Since $x$ is given equal to 4, we may *substitute* this value for $x$ in equation (2).

   *a.* Write the equation. $2x + y = 11$ (2)
   *b.* Substitute 4 for $x$. $2(4) + y = 11$
   *c.* Multiply. $8 + y = 11$
   *d.* Transpose. $y = 11 - 8$
   *e.* Combine like terms. $y = 3$ *Ans.*

EXAMPLE 8-48. Solve the pair of simultaneous equations given below for the value of $x$ and $y$.

SOLUTION

1. The equations are:

$$y = 2x \quad (1)$$
$$x + 2y = 20 \quad (2)$$

2. Since $y$ is given equal to $2x$, we may *substitute* this value for $y$ in equation (2).

   *a.* Write the equation. $x + 2y = 20$ (2)
   *b.* Substitute $2x$ for $y$. $x + 2(2x) = 20$
   *c.* Multiply. $x + 4x = 20$
   *d.* Combine like terms. $5x = 20$
   *e.* Solve the basic equation. $x = 4$ *Ans.*

3. Substitute this value for $x$ in equation (1) to find $y$.

   *a.* Write the equation. $y = 2x$ (1)
   *b.* Substitute 4 for $x$. $y = 2(4)$
   *c.* Multiply. $y = 8$ *Ans.*

EXAMPLE 8-49. Solve the pair of simultaneous equations given below for the value of $x$ and $y$.

SOLUTION

1. The equations are:

$$3x + y = 6 \quad (1)$$
$$y = 2x + 1 \quad (2)$$

2. In equation (2), $y$ is given equal to the *quantity* $2x + 1$, which is best written as $(2x + 1)$. We can substitute *this* value for $y$ in equation (1).
   *a.* Write the equation. $3x + y = 6$ (1)
   *b.* Substitute $(2x + 1)$ for $y$. $3x + (2x + 1) = 6$
   *c.* Remove parentheses. $3x + 2x + 1 = 6$
   *d.* Transpose. $3x + 2x = 6 - 1$
   *e.* Combine like terms. $5x = 5$
   *f.* Solve the basic equation. $x = 1$ *Ans.*
3. Substitute this value for $x$ in equation (2) to find $y$.
   *a.* Write the equation. $y = 2x + 1$ (2)
   *b.* Substitute 1 for $x$. $y = 2(1) + 1$
   *c.* Multiply. $y = 2 + 1$
   *d.* Combine like terms. $y = 3$ *Ans.*

EXAMPLE 8-50. Solve the pair of simultaneous equations given below for the value of $x$ and $y$.

SOLUTION

1. The equations are:

$$2x + y = 6 \quad (1)$$
$$3x + 0.5y = 5 \quad (2)$$

2. Solve equation (1) so as to obtain a value for $y$ expressed in terms of $x$.
   *a.* Write the equation.

$$2x + y = 6 \quad (1)$$

   *b.* Transpose *all* quantities *except* $y$ to the other side.

$$y = 6 - 2x$$

3. Since $y$ is now expressed as the *quantity* $(6 - 2x)$, we may substitute *this* value for $y$ in equation (2).
   *a.* Write the equation. $3x + 0.5y = 5$ (2)
   *b.* Substitute $(6 - 2x)$ for $y$. $3x + 0.5(6 - 2x) = 5$
   *c.* Remove parentheses. $3x + 3 - 1x = 5$
   *d.* Transpose. $3x - x = 5 - 3$
   *e.* Combine like terms. $2x = 2$
   *f.* Solve the basic equation. $x = 1$ *Ans.*
4. Substitute this value for $x$ in the equation found in step 2.
   *a.* Write the equation. $y = 6 - 2x$
   *b.* Substitute 1 for $x$. $y = 6 - 2(1)$
   *c.* Multiply. $y = 6 - 2$
   *d.* Combine like terms. $y = 4$ *Ans.*

## SUMMARY OF METHOD

1. Write the equations.
2. Solve either equation for one of the unknowns in terms of the other.
3. Substitute this value in the other equation.
4. Solve the resulting equation.

5. Substitute the value found in step 4 in the equation found in step 2.
6. Solve the resulting equation for the value of the second unknown.
7. Check.

SELF-TEST 8-51. Solve the pair of simultaneous equations given below for the value of $x$ and $y$.

SOLUTION. The equations are:

$$5x + 3y = 6.5 \quad (1)$$
$$2x - y = 7 \quad (2)$$

1. Solve one of these equations for one letter in terms of the other. The better equation to use for this purpose is equation (______) in which we shall solve the equation for the letter ______ because the coefficient of $y$ is already 1. | 2 | $y$

2. *a.* Write the equation.

$$2x - y = 7 \quad (2)$$

*b.* Transpose to get $+y$ even if it is on the right side.

$$2x - \_\_\_\_\_\_ = y$$ 7

or

$$y = \_\_\_\_\_\_$$ $2x - 7$

3. Since $y$ is now expressed as the quantity ______, we may now ______ this value for $y$ in equation (1). | $(2x - 7)$ | substitute

*a.* Write the equation. $5x + 3y = 6.5 \quad (1)$

*b.* Substitute. $5x + 3(\_\_\_\_\_\_) = 6.5$ | $2x - 7$

*c.* Remove parentheses.

$$5x + \_\_\_\_\_\_ - \_\_\_\_\_\_ = 6.5$$ $6x$ 21

*d.* Transpose. $5x + 6x = 6.5 + \_\_\_\_\_\_$ | 21

*e.* Combine like terms. $11x = \_\_\_\_\_\_$ | 27.5

*f.* Solve the basic equation. $x = \frac{27.5}{?}$ | 11

*g.* Divide. $x = \_\_\_\_\_\_$ *Ans.* | 2.5

4. Substitute this value for $x$ in the equation found in step 2.

*a.* Write the equation. $y = 2x - 7$

*b.* Substitute. $y = 2(\_\_\_\_\_\_) - 7$ | 2.5

*c.* Multiply. $y = \_\_\_\_\_\_ - 7$ | 5

*d.* Combine like terms. $y = \_\_\_\_\_\_$ *Ans.* | $-2$

5. *Check:* Substitute these values for $x$ and $y$ in the original equations (1) and (2).

Substitute $x = \_\_\_\_\_\_$ and $y = \_\_\_\_\_\_$ in equation (1). | 2.5 | $-2$

$$5x + 3y = 6.5 \quad (1)$$

$$5(\_\_\_\_\_\_) + 3(\_\_\_\_\_\_) = 6.5$$ 2.5 | $-2$

$$\_\_\_\_\_\_ - 6 = 6.5$$ 12.5

$$\_\_\_\_\_\_ = 6.5 \quad \textit{Check}$$ 6.5

Substitute $x = 2.5$ and $y = -2$ in equation (2).

$$2x - y = 7 \quad (2)$$

$2(\_\_\_\_\_\_) - (\_\_\_\_\_\_) = 7$ | 2.5 −2

$5 \_\_\_\_\_\_ = 7$ | +2

$\_\_\_\_\_\_ = 7$ *Check* | 7

**PROBLEMS**

Solve the following sets of equations:

1. $y = 3$, $4x + y = 23$
2. $4y - x = 9$, $x = 3$
3. $x = 5$, $4x - y = 18$
4. $2x + y = 10$, $y = 3x$
5. $y = 3x + 1$, $2x + y = 11$
6. $3x + 2y = 17$, $y = x - 4$
7. $2y - x = 1$, $x = y + 2$
8. $2x + 5y = 31$, $x + y = 8$
9. $y - x = 3$, $2x + y = 6$
10. $2x - y = 7$, $x - y = 3$
11. $x + 2y = 9$, $x + 3y = 13$
12. $11x - y = 18$, $5x + 3y = 22$

## JOB 8-13 REVIEW OF ALGEBRA

1. *Like* terms are those which contain the ______ letter portions. | same
2. *Unlike* terms are those in which the letter portions are ______. | different
3. Adding signed numbers.
   *a.* To add two signed numbers of the *same* sign, ______ the numbers and use the common sign. | add
   *b.* To add two signed numbers of *different* sign, ______ the smaller from the larger and use the sign of the ______ number. | subtract | larger
4. To combine a string of signed numbers means to ______ the individual signed numbers using the rules for ______. | add | addition
5. Multiplying and dividing signed numbers.
   *a.* Two signed numbers with the *same* sign yield a (plus/minus) answer. | plus
   *b.* Two signed numbers with *different* signs yield a (plus/minus) answer. | minus
6. A parenthesis is removed by ______ the signed number in front of the parenthesis by each and every quantity within the parenthesis. | multiplying
7. To combine several quantities involving *unlike* terms, combine each group of ______ terms separately. | like
8. Solving equations.
   *a.* Fractional equations containing one fraction.
      (1) Transpose any other number to the other side of the equality sign.

(2) Combine like terms.
(3) Make two fractions by placing any whole numbers or decimals over the number ________. | 1
(4) Cross-multiply.
(5) Solve the basic equation.

*b.* Fractional equations containing two fractions.
(1) Cross-________. | multiply
(2) Solve the basic equation.

*c.* Fractional equations with more than two fractions.
(1) Find the least common ________. | denominator
(2) Multiply each term of the equation by this ________. | LCD
(3) Transpose *all* unknowns to one side of the ________ sign and *all* numbers to the other side. | equality
(4) Combine the ________ terms on each side separately. | like
(5) Solve the basic equation.

*d.* Equations with parentheses.
(1) Remove parentheses.
(2) ________ *all* unknowns to one side of the equality sign, and *all* numbers to the other side. | Transpose
(3) ________ the like terms on each side separately. | Combine
(4) Solve the basic equation.

*e.* Simultaneous equations solved by addition.
(1) In each equation, transpose *all* unknowns to one side of the equality sign, and *all* ________ to the other side. | numbers
(2) Combine the ________ terms on each side separately. | like
(3) Arrange the ________ letters under each other in both equations. | same
(4) Eliminate one letter by ________ one or both equations by the proper number and sign so as to obtain the ________ coefficient with ________ signs. | multiplying | same | different or opposite
(5) ________ the two equations to eliminate the letter. | Add
(6) Solve the resulting equation, which will now contain only ________ letter. | one
(7) Substitute this value in *either* of the two ________ equations. | original
(8) Solve this equation to obtain the value of the second letter.

*f.* Simultaneous equations solved by substitution.
(1) Solve either equation for one of the unknowns in terms of the ________ unknown. | other
(2) Substitute this expression for the letter in the other ________. | equation
(3) Solve the resulting equation.
(4) Substitute the value found in step 3 in the equation found in step ________. | 1
(5) Solve the resulting equation for the value of the ________ unknown. | second

(See Instructor's Manual for Test 8-9, Algebraic Equations)

# 9 KIRCHHOFF'S LAWS

## JOB 9-1 KIRCHHOFF'S FIRST LAW

**RULE** At any point in an electric circuit, the sum of the currents that enter a junction is equal to the sum of the currents that leave it.

**FORMULA**

$$\Sigma I_e = \Sigma I_l \qquad \boxed{9\text{-}1}$$

where $\Sigma I_e$ = summation of all currents *entering* a point
$\Sigma I_l$ = summation of all currents *leaving* a point

In Fig. 9-1, three resistors meet at point $X$. The figure shows three ways in which this condition may be drawn. In each, the current $I_3$ that *enters* point $X$ is equal to the *sum* of the currents $I_1$ and $I_2$ that *leave* the junction. This fact may be expressed as the formula $I_3 = I_1 + I_2$. You will probably recognize this as the formula for the total current in a parallel circuit.

$I_3 = I_1 + I_2$ = $I_3 = I_1 + I_2$ = $I_3 = I_1 + I_2$

**FIGURE 9-1** The current that enters a point is equal to the current that leaves it.

### FINDING THE CURRENT IN BRANCH WIRES

EXAMPLE 9-1. Find the current in $R_2$ of the circuit shown in Fig. 9-2.

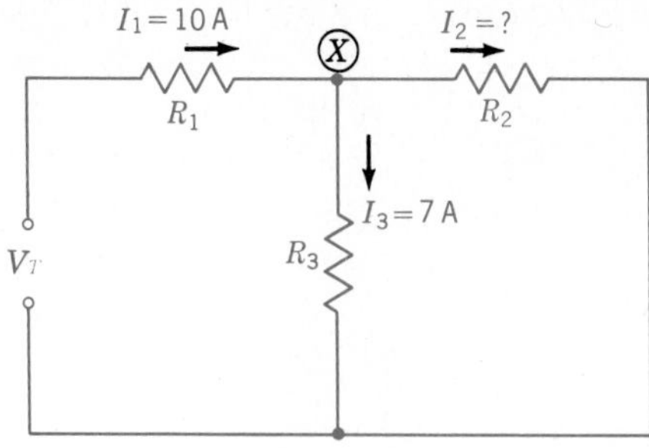


FIGURE 9-2

SOLUTION. At point $X$,

$$\Sigma I_e = \Sigma I_l \qquad (9\text{-}1)$$
$$I_1 = I_2 + I_3$$
$$10 = I_2 + 7$$
$$10 - 7 = I_2$$
$$3 = I_2 \qquad \text{or} \qquad I_2 = 3 \text{ A} \qquad \textit{Ans.}$$

EXAMPLE 9-2. Find the current $I_2$ in the circuit shown in Fig. 9-3.

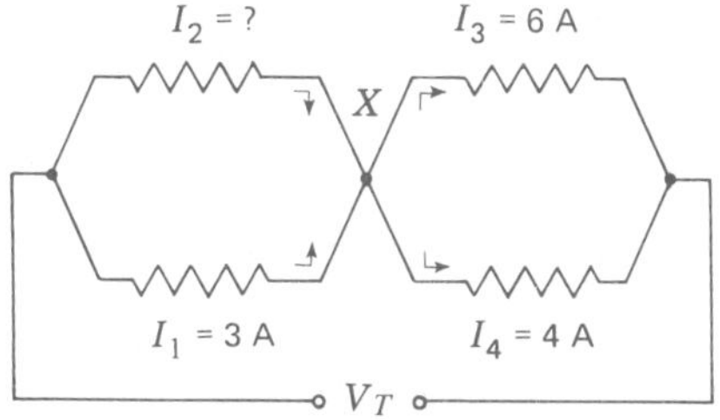


FIGURE 9-3

SOLUTION. The current that enters point $X$ is equal to the current that leaves it. Since $I_1$ and $I_2$ enter point $X$, and $I_3$ and $I_4$ leave point $X$, we can apply Eq. (9-1).

$$\Sigma I_e = \Sigma I_l \qquad (9\text{-}1)$$
$$I_1 + I_2 = I_3 + I_4$$
$$3 + I_2 = 6 + 4$$
$$I_2 = 10 - 3 = 7 \text{ A} \qquad \textit{Ans.}$$

## JOB 9-2 KIRCHHOFF'S SECOND LAW

**RULE** The voltage supplied to a circuit is always equal to the sum of the voltage drops across the individual parts of the circuit.

This fact was used when we studied series circuits and was expressed as the formula $V_T = V_1 + V_2 + V_3$. Kirchhoff's second law explains that the sum of the voltage *gains* supplied to a circuit from a generator, battery, etc., is always equal to the sum of the voltage *drops* ($V_1$, $V_2$, $V_3$, etc.) in the circuit. Since the gains are equal to the losses, the total sum of the gains plus the losses must equal zero. The quantities involved must be added *algebraically* in order to include the *direction* of the various voltages. Kirchhoff's second law may now be expressed as follows.

**RULE** The *algebraic* sum of all the voltages in any complete electric circuit is equal to zero.

**FORMULA** $$V_1 + V_2 + V_3 + \text{etc.} = 0 \qquad (9\text{-}2)$$

**Direction of voltages.** If two equal voltages oppose each other, they may be expressed as $+V_1$ and $-V_2$, where the minus sign in $-V_2$ represents the opposition to $+V_1$. Their algebraic sum would be indicated as

$$+V_1 + (-V_2)$$

or

$$V_1 - V_2$$

Therefore, as we go around a circuit, adding the voltages, we must be very careful to indicate voltage *gains* as *plus* values and voltage *losses* as *minus* values. The following procedure will determine the sign of the voltage.

1. Note the direction of the electron flow around the circuit. Electrons *leave* the negative terminal of the voltage source and *enter* the positive terminal. This is illustrated in Fig. 9-4*a*.
2. Mark the circuit to indicate the electron flow through *each* resistance. This is shown in Fig. 9-4*b*. In more complicated circuits, the direction may be unknown. In this event, we shall *assume* a direction. If our assumption is incorrect, the error will be indicated by a minus sign when the resulting equation is solved. A minus current therefore tells us that the true direction is *opposite* to the direction originally assumed. The value of the current will *not* be affected.
3. We intend to move around a circuit to combine all the voltages encountered. We shall start at some convenient point and move around the *complete* circuit until we return to the original starting point. The direction of this movement is usually in the direction of the electron flow.

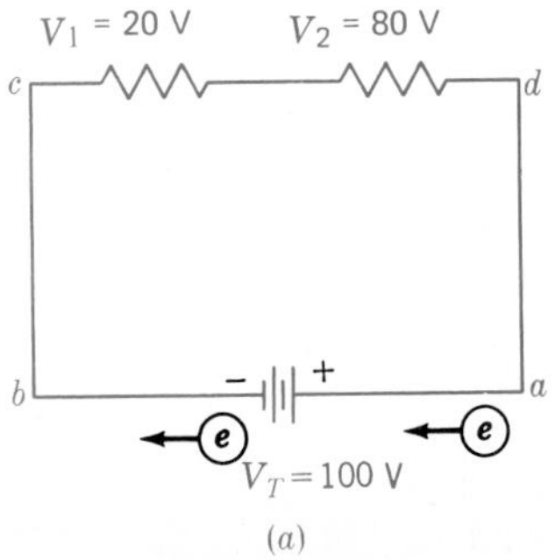


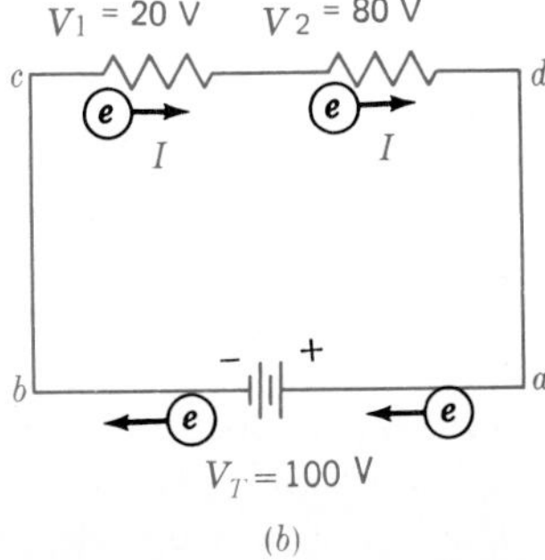


**FIGURE 9-4** Electrons *leave* the negative terminal of a voltage source and *enter* the positive terminal.

4. *Source voltages* are *positive* (+) if we move through them from the positive (+) to the negative (−) terminal. *Source voltages* are *negative* (−) if we move through them from the negative (−) to the positive (+) terminal. Therefore, in Fig. 9-4*b*, if we start at point *a* and move around the circuit in the direction *abcd*, we shall go through $V_T$ from + to − and $V_T$ will equal +100 V. However, if we start at point *a* and move around the circuit in the direction *adcb*, we shall go through $V_T$ from − to + and $V_T$ will equal −100 V.

5. The voltage across any resistance will be *negative* if we go through it *in the direction of the assumed electron flow*. Thus, in Fig. 9-4*b*, if we go through the circuit in the direction *abcd*, $V_1 = -20$ V and $V_2 = -80$ V.

6. The voltage in any resistance will be *positive* if we go through it in a direction *opposite* to the assumed electron flow. Thus, in Fig. 9-4*b*, if we go through the circuit in the direction *adcb*, $V_2 = +80$ V and $V_1 = +20$ V.

EXAMPLE 9-3. Find the signs of the voltages encountered when tracing the circuits shown in Fig. 9-5*a*.

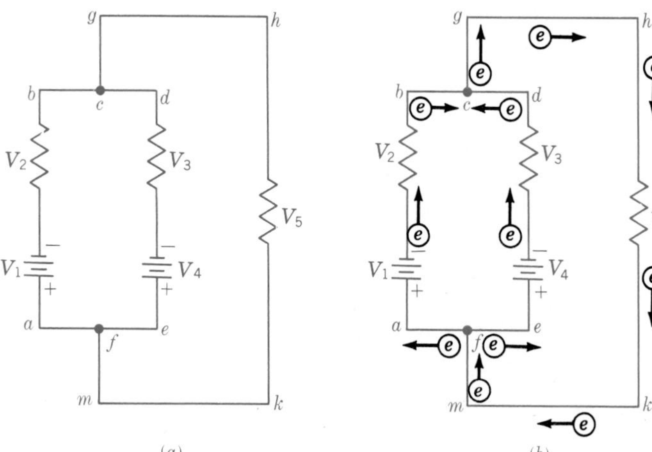


FIGURE 9-5
Electrons leave the negative terminal of a voltage source.

SOLUTION

1. Determine the direction of the electron flow around the circuit. Since electrons leave the negative terminal, the direction of flow will be *up* to point *c* from both $V_1$ and $V_4$. At point *c*, since the electrons that enter point *c* must also leave it, the electrons must flow *out* of point *c* *up* to point g and continue around to point *f*. At *f*, the current divides, part going to point *a* and the rest to point *e*. This is indicated in Fig. 9-5*b*.

2. Determine the direction of the voltages around the circuit *abcdef*.
   *a*. $V_1$ is +, since we go through it from + to −.
   *b*. $V_2$ is −, since we go through it in the direction of the electron flow.
   *c*. $V_3$ is +, since we go through it in a direction *opposite* to the electron flow.
   *d*. $V_4$ is −, since we go through it from − to +.

3. Determine the direction of the voltages around the circuit *abcghkmf*.
   *a*. $V_1$ is +, since we go through it from + to −.
   *b*. $V_2$ is −, since we go through it in the direction of the electron flow.
   *c*. $V_5$ is −, since we go through it in the direction of the electron flow.

4. Determine the direction of the voltages around the circuit *fedcghkm.*
   *a.* $V_4$ is +, since we go through it from + to −.
   *b.* $V_3$ is −, since we go through it in the direction of the electron flow.
   *c.* $V_5$ is −, since we go through it in the direction of the electron flow.

SELF-TEST 9-4. Find the signs of the voltages encountered when tracing the circuit *abcdefa* shown in Fig. 9-6.

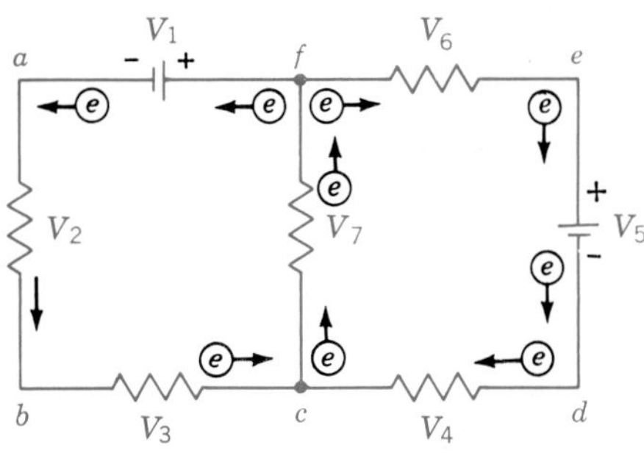


**FIGURE 9-6**
The direction of the electron flow is assumed.

SOLUTION

1. Assume that the direction of the electron flow is that shown on the diagram.
2. Determine the signs of the voltages. The direction of our trace is in a counterclockwise direction starting from point *a.*
   *a.* $V_2$ and $V_3$ are both ______, since we are going through them both in the ___(same/opposite)___ direction as the electron flow. — same
   *b.* $V_4$ and $V_6$ are both ______, since we are going through them both in the ___(same/opposite)___ direction as the electron flow. + opposite
   *c.* $V_5$ is ______, since we go through it from ______ to ______. — — +
   *d.* $V_1$ is ______, since we go through it from ______ to ______. + + —

## JOB 9-3 USING KIRCHHOFF'S LAWS IN SERIES CIRCUITS

EXAMPLE 9-5. Find the current in the series circuit shown in Fig. 9-7*a*.

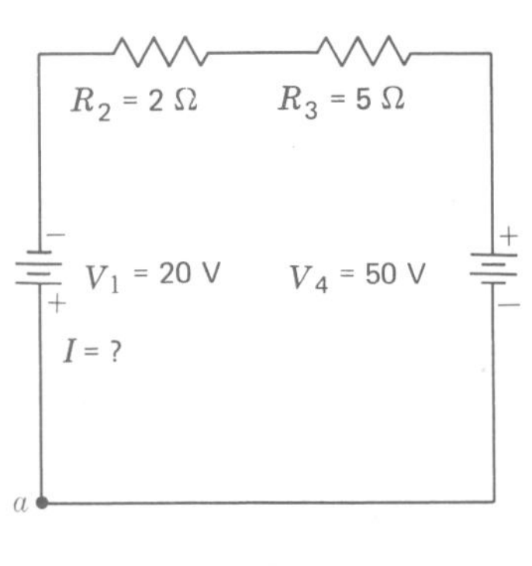


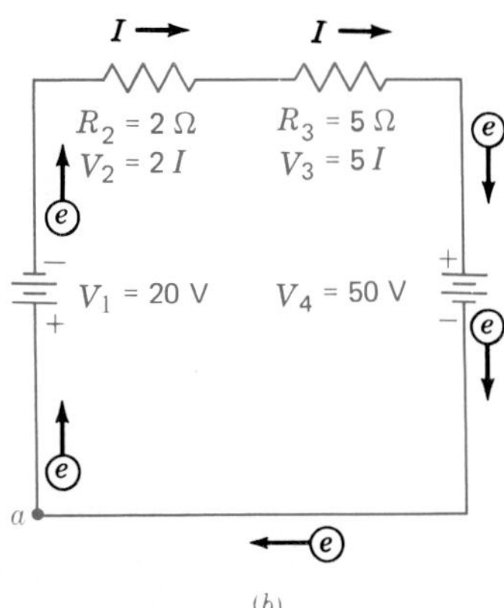


**FIGURE 9-7**
The sources are connected so as to aid each other.

SOLUTION

1. Determine the direction of the electron flow. $V_1$ and $V_4$ are connected so that the electron flow comes out of one source and continues through the other in the same direction. This connection of sources aiding each other causes an electron flow moving *clockwise* around the circuit.

2. Mark this direction on the circuit as shown in Fig. 9-7*b*.

3. By Ohm's law, the voltages in $R_2$ and $R_3$ will equal $2I$ and $5I$, respectively.

4. Determine the direction of the voltages. Trace the circuit in a clockwise direction starting at point *a*.

   *a*. $V_1$ and $V_4$ are both +, since we go through them from + to −.

   *b*. $V_2$ and $V_3$ are both −, since we go through them in the direction of the electron flow.

5. Apply Kirchhoff's second law. The *algebraic* sum of all the voltages around a circuit will equal zero. Add the voltages starting at point *a*.

$$V_1 + V_2 + V_3 + V_4 = 0$$
$$+20 - 2I - 5I + 50 = 0$$
$$-2I - 5I = -20 - 50$$
$$-7I = -70$$
$$7I = 70$$
$$I = 10 \text{ A} \qquad \textit{Ans.}$$

6. *Check:* If we trace the circuit in a counterclockwise direction starting from point *a*, the voltages through the sources are now negative and the voltages through the resistors are positive, since we are now going through them in a direction *opposite* to the electron flow.

$$V_4 + V_3 + V_2 + V_1 = 0$$
$$-50 + 5I + 2I - 20 = 0$$
$$5I + 2I = 50 + 20$$
$$7I = 70$$
$$I = 10 \text{ A} \qquad \textit{Check}$$

EXAMPLE 9-6. Find the current in the series circuit shown in Fig. 9-8*a*.

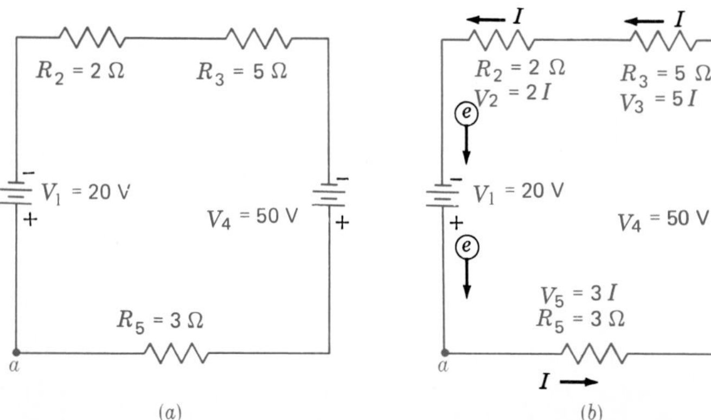


**FIGURE 9-8**
The sources are connected so as to oppose each other.

SOLUTION

1. Determine the direction of the electron flow. $V_1$ and $V_4$ are connected so that they *oppose* each other. $V_4$ (50 V) is stronger than $V_1$ (20 V), and the electron flow is therefore governed by $V_4$. This will cause the electron flow to be in a *counterclockwise* direction.

2. Mark this direction on the circuit as shown in Fig. 9-8*b*.

3. By Ohm's law, the voltages in $R_2$, $R_3$, and $R_5$ will equal $2I$, $5I$, and $3I$, respectively.

4. Determine the direction of the voltages. Trace the circuit in a counterclockwise direction starting at point *a*. $V_5$, $V_3$, and $V_2$ are all $-$, since we go through them in the direction of the electron flow. $V_4$ is $+$, since we go through it from $+$ to $-$. $V_1$ is $-$, since we go through it from $-$ to $+$.

5. Apply Kirchhoff's second law.

$$V_5 + V_4 + V_3 + V_2 + V_1 = 0$$
$$-3I + 50 - 5I - 2I - 20 = 0$$
$$-3I - 5I - 2I = 20 - 50$$
$$-10I = -30$$
$$10I = 30$$
$$I = 3 \text{ A} \qquad \textit{Ans.}$$

EXAMPLE 9-7. Use the circuit shown in Fig. 9-9*a*. Find (*a*) the total current and (*b*) the voltage drop across each resistance.

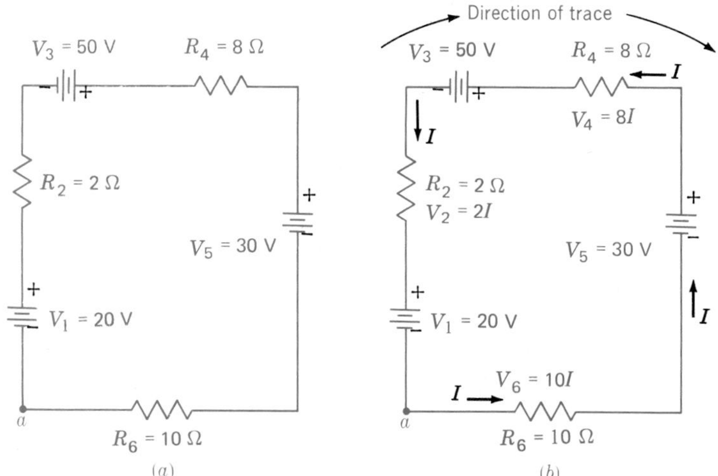


FIGURE 9-9

SOLUTION

*a.* Find the total current.

1. Determine the direction of the electron flow. $V_1$ and $V_3$ aid each other, and since their total (70 V) is greater than the opposing $V_5$ (30 V), the flow will be in a counterclockwise direction.

2. Mark this direction on the circuit as shown in Fig. 9-9*b*.

3. By Ohm's law, the voltages in $R_2$, $R_4$, and $R_6$ will equal $2I$, $8I$, and $10I$, respectively.

4. Determine the direction of the voltages. Trace the circuit in a *clockwise* direction starting at point *a*.

$V_1$ and $V_3$ are $-$, since we go through them from $-$ to $+$.

$V_5$ is $+$, since we go through it from $+$ to $-$.

$V_2$, $V_4$, and $V_6$ are all $+$, since we go through them in a direction *opposite* to the electron flow.

5. Apply Kirchhoff's second law.

$$V_1 + V_2 + V_3 + V_4 + V_5 + V_6 = 0$$
$$-20 + 2I - 50 + 8I + 30 + 10I = 0$$
$$2I + 8I + 10I = 20 + 50 - 30$$
$$20I = 40$$
$$I = 2 \text{ A} \qquad \textit{Ans.}$$

*b.* Find the voltage drop across each resistance.

$$V_2 = IR_2 \qquad V_4 = IR_4 \qquad V_6 = IR_6$$
$$V_2 = 2 \times 2 \qquad V_4 = 2 \times 8 \qquad V_6 = 2 \times 10$$
$$V_2 = 4 \text{ V} \qquad V_4 = 16 \text{ V} \qquad V_6 = 20 \text{ V} \qquad \textit{Ans.}$$

*Check:* The sum of the voltages around the complete circuit should equal zero. Add the voltages in a counterclockwise direction starting from point *a*.

$$V_6 + V_5 + V_4 + V_3 + V_2 + V_1 = 0$$
$$-20 - 30 - 16 + 50 - 4 + 20 = 0$$
$$0 = 0 \qquad \textit{Check}$$

SELF-TEST 9-8. In Fig. 9-10, $R_2$ represents the internal resistance of the generator. Find (*a*) the total current in the circuit, (*b*) the voltage drop across each resistance, and (*c*) the emf of the generator.

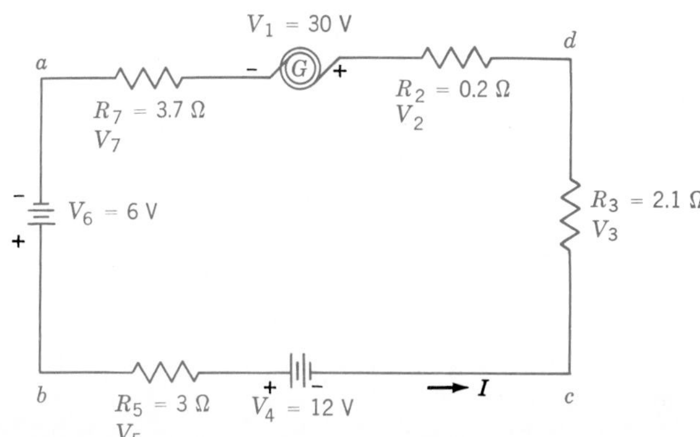


**FIGURE 9-10**
$R_2$ represents the internal resistance of the generator.

SOLUTION

*a.* Find the total current.

1. Determine the direction of the electron flow. $V_1$ and $V_4$ ___(aid/oppose)___ each other, and since their total ($30 + 12 = 42$) is greater than $V_6$ (6 V), which ___(aids/opposes)___ them, the flow will be in a ___(clockwise/counterclockwise)___ direction.

aid
opposes
counterclockwise

2. Mark this direction on the circuit diagram.

3. By Ohm's law, the voltages in the resistances are

$$V_2 = 0.2I \qquad V_3 = \_\_\_\_\_\_$$
$$V_5 = \_\_\_\_\_\_ \qquad V_7 = \_\_\_\_\_\_$$

2.1*I*
3*I* 3.7*I*

4. Determine the direction of the voltages. Trace the circuit in the same direction as the electron flow, which will be ______. Start at point *a*.
$V_6$ is ______, since we go through it from ______ to ______.
$V_4$ and $V_1$ are both ______, since we go through them both from ______ to ______.
$V_5$, $V_3$, $V_2$, and $V_7$ are all ______, since we go through them all in the ___(same/opposite)___ direction as the electron flow.

counterclockwise
− − +
+
+ −
−
same

5. Apply Kirchhoff's second law.

$$V_6 + V_5 + V_4 + V_3 + V_2 + V_1 + V_7 = \_\_\_\_\_\_$$
$$-6 - 3I + 12 - 2.1I - 0.2I + 30 - \_\_\_\_\_\_ = 0$$
$$36 - \_\_\_\_\_\_ = 0$$
$$36 = \_\_\_\_\_\_$$
$$I = \_\_\_\_\_\_ \text{ A} \qquad \textit{Ans.}$$

0
3.7*I*
9*I*
9*I*
4

*b.* Find the voltage drop across each resistance.

$$V_2 = IR_2 \qquad V_3 = IR_3$$
$$V_2 = 4 \times 0.2 = 0.8 \text{ V} \qquad V_3 = 4 \times 2.1 = \_\_\_\_\_\_ \text{ V}$$
$$V_5 = IR_5 \qquad V_7 = IR_7$$
$$V_5 = 4 \times 3 = 12 \text{ V} \qquad V_7 = 4 \times \_\_\_\_\_\_ = \_\_\_\_\_\_ \text{ V}$$

8.4
3.7 14.8

*Check:* The sum of the voltages around the circuit should equal ______. Add the voltages in a clockwise direction starting from point *a*. In this direction, $R_7$, $R_2$, $R_3$, and $R_5$ will all be ______ because we go through them in a direction ______ to the flow of the electrons.
$V_1$ and $V_4$ will both be ______ because we go through them from ______ to ______.
$V_6$ will be ______ because we go through it from ______ to ______.

$$V_7 + V_1 + V_2 + V_3 + V_4 + V_5 + V_6 = 0$$
$$14.8 - 30 \_\_\_\_\_\_ + 8.4 - 12 + 12 + 6 = 0$$
$$0 = 0 \qquad \textit{Check}$$

zero
+
opposite
− −
+
+ + −
+ 0.8

*c.* Find the emf of the generator.

$$\text{emf} = V_1 + V_2 = 30 + 0.8 = \_\_\_\_\_\_ \text{ V} \qquad \textit{Ans.}$$

30.8

## PROBLEMS

In each of the circuits shown, find (*a*) the total current and (*b*) the voltage drop across each resistor.

1.

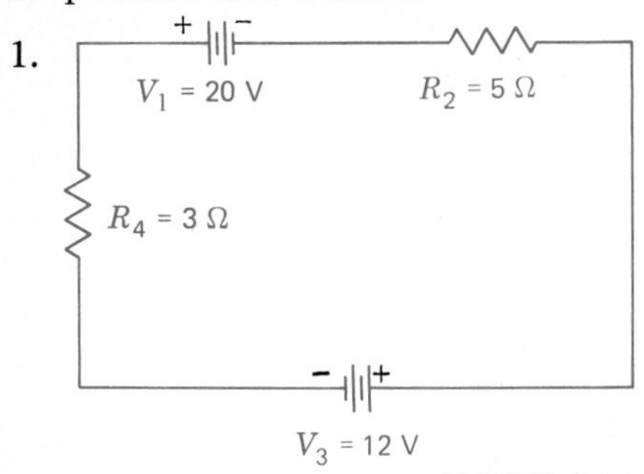


FIGURE 9-11

2.

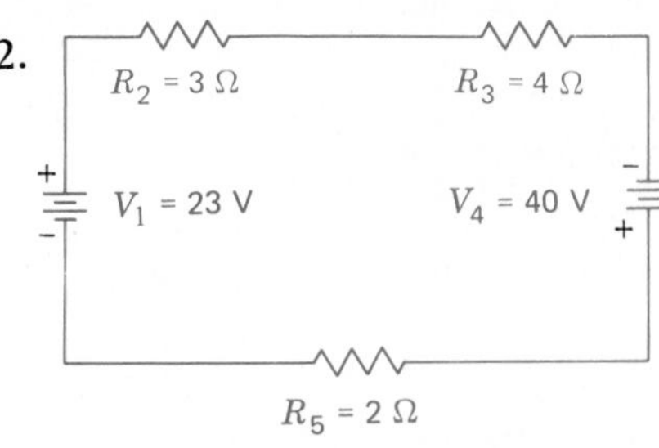


FIGURE 9-12

3. Use a circuit similar to that shown in Fig. 9-12. $V_1 = 20$ V, $R_2 = 6\ \Omega$, $R_3 = 4\ \Omega$, $V_4 = 16$ V, and $R_5 = 5\ \Omega$.
4. Use a circuit similar to that shown in Fig. 9-12. $V_1 = 45$ V, $R_2 = 2\ \Omega$, $R_3 = 4\ \Omega$, $R_5 = 6\ \Omega$, and $V_4 = 9$ V with the polarity reversed.

5. 6.

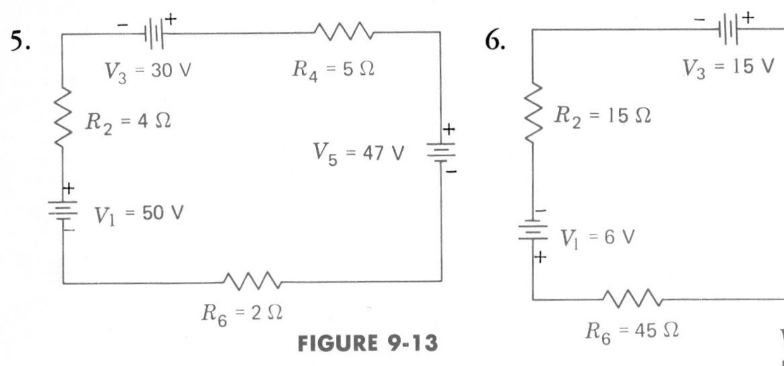


FIGURE 9-13

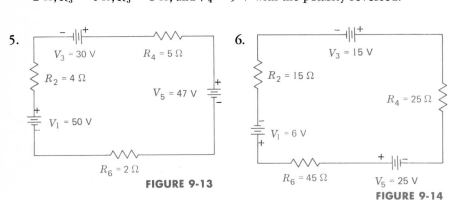

FIGURE 9-14

7. Find the total current flowing in the circuit shown in Fig. 9-15. Be sure to include the internal resistance of the batteries and generator.

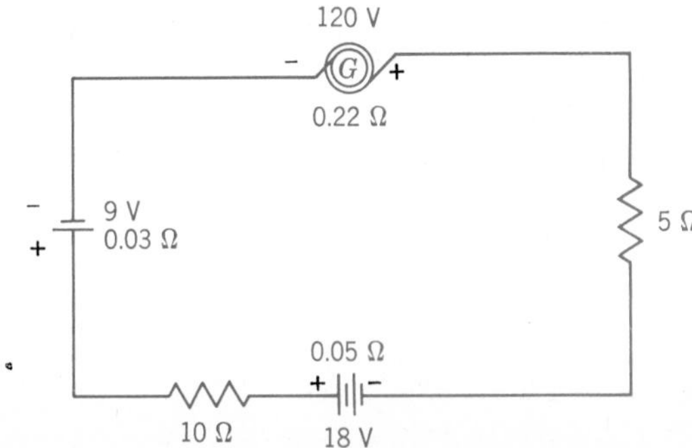


**FIGURE 9-15**
The circuit includes the internal resistances of the voltage sources.

8. A current of 4 A flows in the circuit shown in Fig. 9-16. Find the value of $R$.

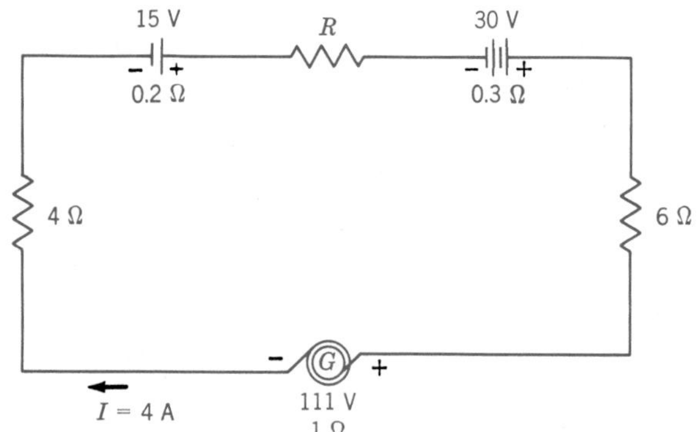


FIGURE 9-16

## JOB 9-4 USING KIRCHHOFF'S LAWS IN PARALLEL CIRCUITS

EXAMPLE 9-9. Find all missing values of voltage, current, and resistance in the circuit shown in Fig. 9-17.

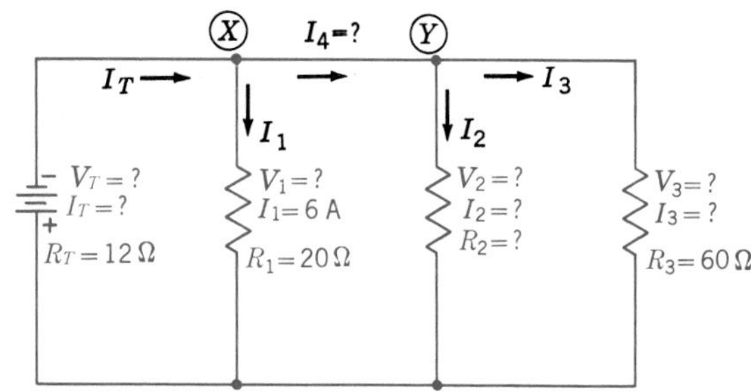


FIGURE 9-17
Using Kirchhoff's first law in parallel circuits.

SOLUTION

1. Start at the point where the greatest amount of information is available. This will be branch $R_1$. Find $V_1$.

$$V_1 = I_1 \times R_1$$
$$V_1 = 6 \times 20 = 120 \text{ V} \quad \textit{Ans.}$$

2. Find the other voltages. Since this is a parallel circuit,

$$V_T = V_1 = V_2 = V_3 = 120 \text{ V} \quad \textit{Ans.}$$

3. Find $I_T$ and $I_3$ by Ohm's law.

$$V_T = I_T \times R_T \qquad V_3 = I_3 \times R_3$$
$$120 = I_T \times 12 \qquad 120 = I_3 \times 60$$
$$I_T = 10 \text{ A} \qquad I_3 = 2 \text{ A}$$

4. Find the current $I_2$ by applying Kirchhoff's first law.
   *a.* Assume the direction of the currents at point $X$ to be as shown on the

diagram. $I_T = 10$ A *entering* point X, $I_1 = 6$ A *leaving* point X, and $I_4 = ?$ *leaving* point X.

$$\Sigma I_e = \Sigma I_l \qquad (9\text{-}1)$$
$$I_T = I_4 + I_1$$
$$10 = I_4 + 6$$
$$10 - 6 = I_4$$
$$I_4 = 4 \text{ A } \textit{leaving} \text{ point } X$$

*b.* At point Y, $I_4$ enters the point, and $I_2$ and $I_3$ leave the point.

$$\Sigma I_e = \Sigma I_l \qquad (9\text{-}1)$$
$$I_4 = I_2 + I_3$$
$$4 = I_2 + 2$$
$$I_2 = 2 \text{ A } \textit{leaving} \text{ point } Y \qquad \textit{Ans.}$$

5. Find $R_2$ by Ohm's law.

$$V_2 = I_2 \times R_2$$
$$120 = 2 \times R_2$$
$$R_2 = 60\ \Omega \qquad \textit{Ans.}$$

EXAMPLE 9-10. Find the total voltage of the circuit shown in Fig. 9-18.

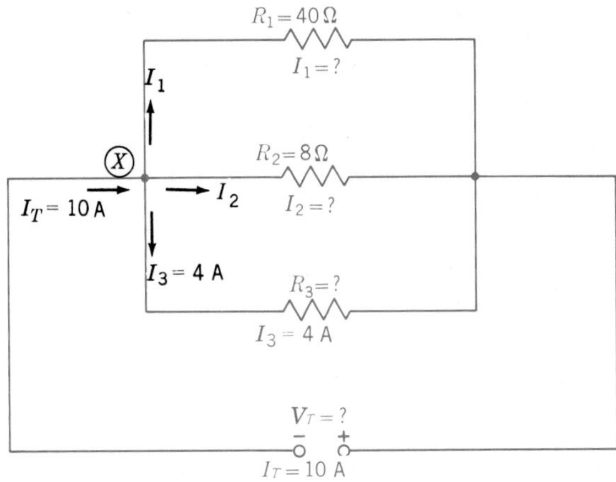


FIGURE 9-18

SOLUTION. The total voltage $V_T$ in this parallel circuit will be equal to any of the branch voltages. $V_1$ or $V_2$ may be found if we can determine the value of either $I_1$ or $I_2$.

1. Find $I_2$. Assume the direction of the currents to be as shown on the diagram. At point X, by Kirchhoff's first law,

$$\Sigma I_l = \Sigma I_e \qquad (9\text{-}1)$$
$$4 + I_1 + I_2 = 10$$
$$I_2 = 10 - 4 - I_1$$

or

$$I_2 = 6 - I_1 \qquad (1)$$

2. Apply Kirchhoff's second law. Trace around the circuit through $R_1$ in a clockwise direction starting at $V_T$.

$$+V_T - 40I_1 = 0$$

or

$$V_T = 40I_1 \tag{2}$$

Trace the circuit again, but this time through $R_2$.

$$+V_T - 8I_2 = 0$$

or

$$V_T = 8I_2 \tag{3}$$

3. In equation (1), we found that $I_2 = 6 - I_1$. Substitute this value for $I_2$ in equation (3).

$$V_T = 8I_2 \qquad (3)$$
$$V_T = 8(6 - I_1)$$
$$V_T = 48 - 8I_1 \tag{4}$$

4. Set the values for $V_T$ found in equations (2) and (4) equal to each other.

$$40I_1 = 48 - 8I_1$$
$$40I_1 + 8I_1 = 48$$
$$48I_1 = 48$$
$$I_1 = 1 \text{ A} \qquad \textit{Ans.}$$

5. Find $V_1$ by Ohm's law.

$$V_1 = I_1 \times R_1 = 1 \times 40 = 40 \text{ V}$$

6. Find $V_T$. Since this is a parallel circuit,

$$V_T = V_1 = V_2 = V_3 = 40 \text{ V} \qquad \textit{Ans.}$$

SELF-TEST 9-11. A voltage source supplies 12 A to a parallel combination of 60 Ω and 20 Ω, as shown in Fig. 9-19. Find $I_1$, $I_2$, and $V_T$.

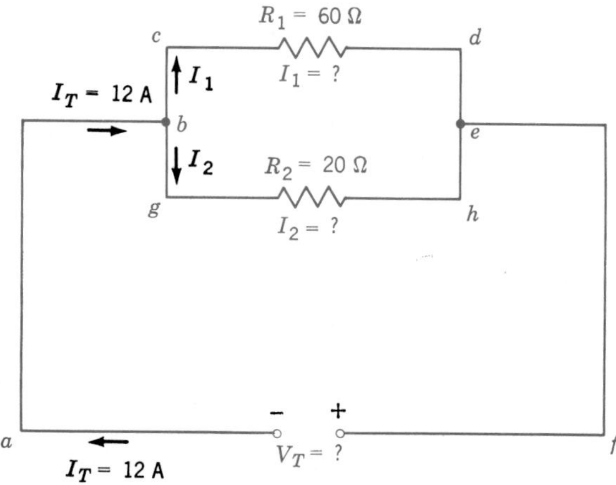


FIGURE 9-19

SOLUTION

1. Apply Kirchhoff's second law. Trace around the circuit in the direction *fabcdef*.

$$V_T - 60I_1 = 0$$

$V_T =$ ______ (1) | $60I_1$

Trace around the circuit in the direction *fabghef.*

______ $- 20I_2 = 0$ | $V_T$

$V_T =$ ______ (2) | $20I_2$

These two expressions for $V_T$ may be set ______ to each other, which gives | equal

$20I_2 =$ ______ | $60I_1$

$$I_2 = \frac{60I_1}{?}$$ | 20

$I_2 =$ ______ $I_1$ (3) | 3

2. Now apply Kirchhoff's first law.

$$\Sigma I_i = \Sigma I_e \qquad (9\text{-}1)$$

At point *b*,

$I_1 + I_2 =$ ______ (4) | 12

Now, according to equation (3), we can substitute $3I_1$ for $I_2$ in equation (4). This will give

$I_1 +$ ______ $= 12$ | $3I_1$

______ $= 12$ | $4I_1$

$I_1 =$ ______ A *Ans.* | 3

Now we can go back to equation (3) to find $I_2$.

$I_2 = 3I_1$

$I_2 = 3 \times$ ______ $=$ ______ A *Ans.* | 3 9

3. Find $V_2$. By Ohm's law,

$V_2 = I_2 \times R_2$

$V_2 = 9 \times$ ______ $=$ ______ V | 20 180

But in a parallel circuit,

$$V_2 = V_T \qquad (5\text{-}1)$$

$V_T =$ ______ V *Ans.* | 180

*Note:* You have already learned a much simpler method for solving this problem. Can you recall it? If not, see Job 5-9.

## PROBLEMS

1. In a circuit similar to that shown in Fig. 9-17, $R_T = 6\ \Omega$, $I_1 = 9$ A, $R_1 = 10\ \Omega$, and $R_3 = 60\ \Omega$. Find all missing values of voltage, current, and resistance.
2. In a circuit similar to that shown in Fig. 9-17, $R_T = 8\ \Omega$, $I_1 = 5$ A, $I_2 = 1$ A, and $R_2 = 80\ \Omega$. Find all missing values of voltage, current, and resistance.

3. In a circuit similar to that shown in Fig. 9-17, $R_T = 15\ \Omega$, $I_1 = 10$ A, $R_1 = 30\ \Omega$, and $R_3 = 40\ \Omega$. Find all missing values of voltage, current, and resistance.
4. In a circuit similar to that shown in Fig. 9-18, $I_T = 40$ A, $R_1 = 4\ \Omega$, $R_2 = 12\ \Omega$, and $I_3 = 20$ A. Find the total voltage $V_T$ and $R_3$.
5. In a circuit similar to that shown in Fig. 9-18, $I_T = 39$ A, $R_1 = 12\ \Omega$, $R_2 = 9\ \Omega$, and $I_3 = 18$ A. Find the total voltage $V_T$ and $R_3$.
6. In a circuit similar to that shown in Fig. 9-18, $I_T = 15$ A, $I_1 = 9$ A, $R_2 = 10\ \Omega$, and $R_3 = 30\ \Omega$. Find the total voltage $V_T$ and $R_1$.
7. In a circuit similar to that shown in Fig. 9-19, $R_1 = 4\ \Omega$, $R_2 = 12\ \Omega$, and $I_T = 2.6$ A. Find $I_1$, $I_2$, and $V_T$.
8. In a circuit similar to that shown in Fig. 9-19, $R_1 = 100\ \Omega$, $R_2 = 45\ \Omega$, and $I_T = 7.975$ A. Find $I_1$, $I_2$, and $V_T$.
9. Check Prob. 7 using the method outlined in Job 5-9.
10. Check Prob. 8 using the method outlined in Job 5-9.
11. A generator supplies 12 A to three resistors of 16 Ω, 48 Ω, and 24 Ω connected in parallel. Find (*a*) the current in each resistor and (*b*) the generator voltage.
12. In Fig. 9-20, find (*a*) $I_1$, (*b*) $I_2$, and (*c*) $V_T$. Check this problem using the method outlined in Job 6-3.

(See Instructor's Manual for Test 9-1)

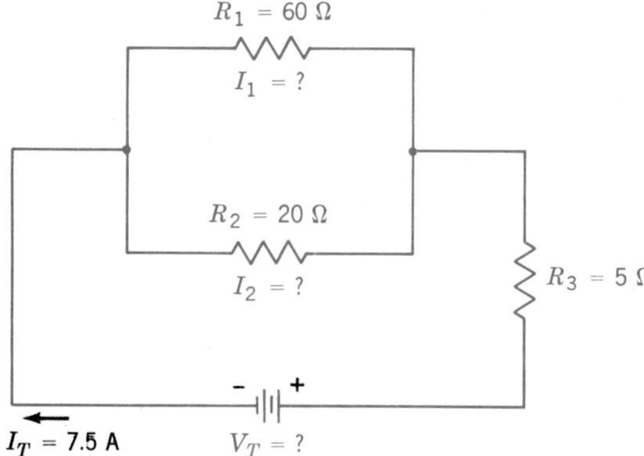


**FIGURE 9-20**

## JOB 9-5 USING KIRCHHOFF'S LAWS IN COMPLEX CIRCUITS

As the circuits under consideration become more complicated, it is generally easier to designate the currents in the various branches as *x*, *y*, and *z*, rather than $I_1$, $I_2$, or $I_3$.

EXAMPLE 9-12. Solve the circuit shown in Fig. 9-21 for the values of the currents *x*, *y*, and *z*.

SOLUTION

1. Assume the direction of the electron flow to be that indicated on the diagram.

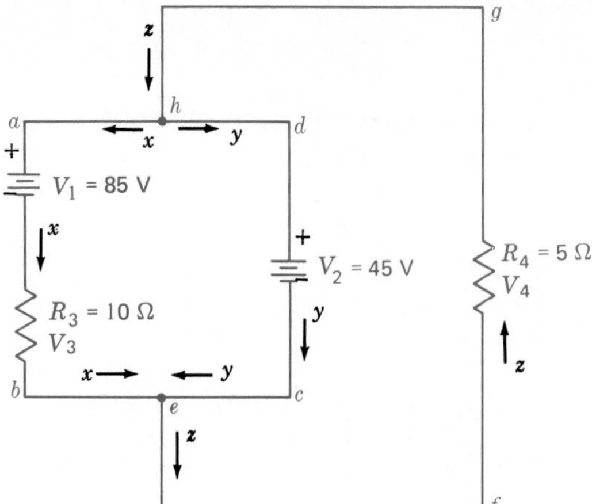


**FIGURE 9-21**
In complex circuits, the branch currents are designated as x, y, and z A.

2. Apply Kirchhoff's second law. Trace around the circuit in the direction *abcd.*

$$V_1 + V_3 + V_2 = 0$$
$$85 - 10x - 45 = 0$$
$$-10x = 45 - 85$$
$$-10x = -40$$
$$x = 4\text{ A} \qquad \textit{Ans.}$$

Trace around the circuit in the direction *abefgh.*

$$V_1 + V_3 + V_4 = 0$$
$$85 - 10x - 5z = 0$$

Substitute 4 for $x$.

$$85 - 10(4) - 5z = 0$$
$$85 - 40 - 5z = 0$$
$$-5z = 40 - 85$$
$$-5z = -45$$
$$z = 9\text{ A} \qquad \textit{Ans.}$$

3. Find the current $y$ by applying Kirchhoff's first law. At point $e$,

$$\Sigma I_e = \Sigma I_l \tag{9-1}$$
$$x + y = z$$
$$4 + y = 9$$
$$y = 9 - 4 = 5\text{ A} \qquad \textit{Ans.}$$

4. Check by tracing the circuit *dcefgh,* using Kirchhoff's second law.

$$V_2 + V_4 = 0 \tag{9-2}$$
$$45 - 5z = 0$$
$$45 - 5(9) = 0$$
$$45 - 45 = 0$$
$$0 = 0 \qquad \textit{Check}$$

EXAMPLE 9-13. Solve the circuit shown in Fig. 9-22 for the values of the currents $x$, $y$, and $z$.

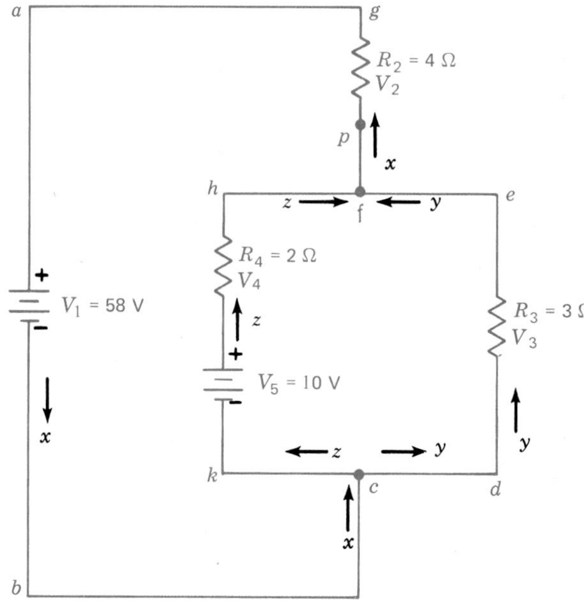


**FIGURE 9-22**
Since $V_1$ is stronger than $V_5$, the electron flow is *in* to point c from the source $V_1$.

SOLUTION

1. Since $V_1$ is greater than $V_5$, the electron flow $x$ is assumed to be *in* to point $c$ from $V_1$. At point $c$, the flow is assumed to be as shown on the diagram. If we are incorrect in these assumptions, a minus sign for a calculated current will advise us to change the direction.
2. Express the current $z$ in terms of currents $x$ and $y$. At point $c$,

$$\Sigma I_l = \Sigma I_e \tag{9-1}$$

$$y + z = x$$

$$z = x - y \tag{a}$$

3. Apply Kirchhoff's second law. Trace around the circuit in the direction *abcdefg*.

$$58 - 3y - 4x = 0$$

$$-3y - 4x = -58$$

or

$$3y + 4x = 58 \tag{1}$$

Trace around the circuit in the direction *khfedck*.

$$-10 - 2z + 3y = 0$$

Substitute $(x - y)$ for $z$ from equation $(a)$.

$$-10 - 2(x - y) + 3y = 0$$

$$-10 - 2x + 2y + 3y = 0$$

$$5y - 2x - 10 = 0$$

or

$$5y - 2x = 10 \tag{2}$$

4. Solve equations (1) and (2) simultaneously.

$$3y + 4x = 58 \qquad (1)$$
$$5y - 2x = 10 \qquad (2)$$

Multiply equation (2) by 2.

$$3y + 4x = 58$$
$$10y - 4x = 20$$

Add:

$$13y = 78$$
$$y = 6 \text{ A} \qquad \textit{Ans.}$$

5. Substitute $y = 6$ in equation (2) to find $x$.

$$5y - 2x = 10 \qquad (2)$$
$$5(6) - 2x = 10$$
$$30 - 2x = 10$$
$$-2x = 10 - 30$$
$$-2x = -20$$
$$x = 10 \text{ A} \qquad \textit{Ans.}$$

6. Substitute the values for $x$ and $y$ in equation ($a$) to find the current $z$.

$$z = x - y \qquad (a)$$
$$z = 10 - 6 = 4 \text{ A} \qquad \textit{Ans.}$$

7. Find the voltage drop in each resistor.

$$V_2 = 4x \qquad V_3 = 3y \qquad V_4 = 2z$$
$$V_2 = 4(10) \qquad V_3 = 3(6) \qquad V_4 = 2(4)$$
$$V_2 = 40 \text{ V} \qquad V_3 = 18 \text{ V} \qquad V_4 = 8 \text{ V}$$

8. Check by tracing the circuit *abckhfg,* using Kirchhoff's second law.

$$V_1 + V_5 + V_4 + V_2 = 0$$
$$58 - 10 - 8 - 40 = 0$$
$$58 - 58 = 0$$
$$0 = 0 \qquad \textit{Check}$$

In order to illustrate the technique of handling problems in which the assumed direction of electron flow proves to be incorrect, let us investigate the following example.

EXAMPLE 9-14. Solve the circuit shown in Fig. 9-23 for the values of the currents $x$, $y$, and $z$.

SOLUTION

1. Assume the direction of the electron flow to be that shown on the diagram.
2. Express the current $z$ in terms of currents $x$ and $y$. At point $c$,

$$\Sigma I_l = \Sigma I_e \qquad (9\text{-}1)$$
$$y + z = x$$
$$z = x - y \qquad (a)$$

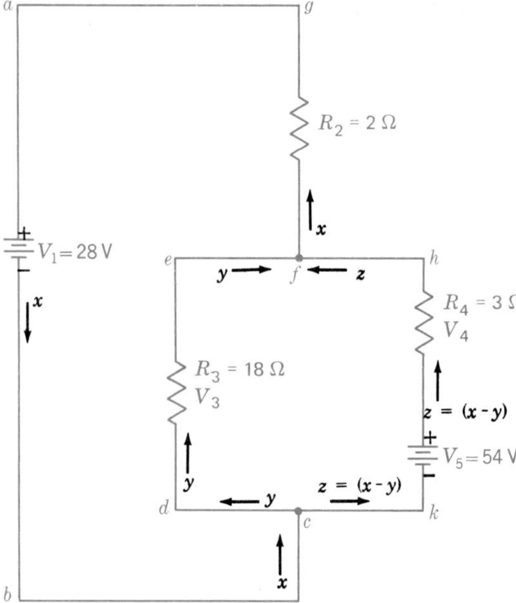


FIGURE 9-23

3. Apply Kirchhoff's second law. Trace around the circuit in the direction *abcdefg*.

$$28 - 18y - 2x = 0$$
$$-2x - 18y = -28$$

Divide the equation by $-2$ in order to simplify. This gives

$$x + 9y = 14 \qquad (1)$$

Trace around the circuit in the direction *khfedc*.

$$-54 - 3z + 18y = 0$$

Substitute $(x - y)$ for $z$ from equation $(a)$.

$$-54 - 3(x - y) + 18y = 0$$
$$-54 - 3x + 3y + 18y = 0$$
$$-3x + 21y = 54$$

Divide the equation by 3 in order to simplify. This gives

$$-x + 7y = 18 \qquad (2)$$

4. Solve equations (1) and (2) simultaneously.

$$x + 9y = 14 \qquad (1)$$
$$-x + 7y = 18 \qquad (2)$$

Add:

$$16y = 32$$
$$y = 2\text{ A} \qquad \textit{Ans.}$$

5. Substitute $y = 2$ in equation (1) to find $x$.

$$x + 9y = 14 \qquad (1)$$

$$x + 9(2) = 14$$
$$x + 18 = 14$$
$$x = 14 - 18 = -4 \text{ A} \qquad \textit{Ans.}$$

The fact that the current $x$ is *minus* 4 A means that current $x$ does *not* flow *into* point $c$ but *does* flow *out* of this point. Similarly, the current $x$ does *not* flow *out* of point $f$ but *does* flow *into* this point.

6. Substitute the values for $x$ and $y$ in equation ($a$) to find the current $z$.

$$z = x - y \qquad (a)$$
$$z = -4 - 2 = -6 \text{ A} \qquad \textit{Ans.}$$

Since $z$ is *minus* 6 A, the actual direction of flow of current $z$ is *into* point $c$ and *out* of point $f$.

7. Find the voltage drop in each resistor.

| | | |
|---|---|---|
| $V_2 = 2x$ | $V_3 = 18y$ | $V_4 = 3z$ |
| $V_2 = 2(-4)$ | $V_3 = 18(2)$ | $V_4 = 3(-6)$ |
| $V_2 = -8 \text{ V}$ | $V_3 = 36 \text{ V}$ | $V_4 = -18 \text{ V}$ |

8. Check by tracing the circuit *abckhfg,* using Kirchhoff's second law.

$$V_1 + V_5 + V_4 + V_2 = 0$$

*Note:*

$V_5 = -54$ V because we go through it from − to +.

$V_4 = -(-18)$ V because we go through $R_4$ in the direction of the assumed current flow.

$V_2 = -(-8)$ V because we go through $R_2$ in the direction of the assumed current flow.

$$28 - 54 - (-18) - (-8) = 0$$
$$28 - 54 + 18 + 8 = 0$$
$$54 - 54 = 0$$
$$0 = 0 \qquad \textit{Check}$$

EXAMPLE 9-15. Solve the circuit shown in Fig. 9-24 for the values of the currents $x$, $y$, and $z$.

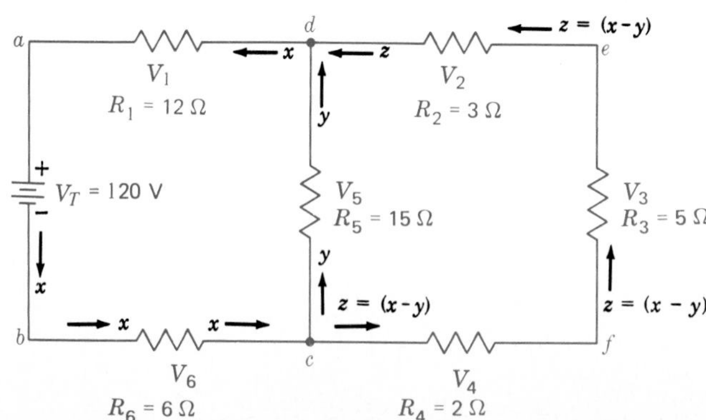


FIGURE 9-24

SOLUTION

1. Assume the direction of the electron flow to be that shown on the diagram. Please remember that these directions could be assumed to point in *other* directions without affecting the validity of the calculations. As we have seen, the truth of our assumptions will be determined by the signs of the calculated currents.

2. Express current $z$ in terms of currents $x$ and $y$. By Kirchhoff's first law, operating at point $c$,

$$y + z = x$$

or

$$z = x - y \qquad (a)$$

3. Apply Kirchhoff's second law. Trace around the circuit in the direction *cdef*. Note that in tracing from point $d$ around to point $c$, we shall be tracing a *series* circuit made of the resistors $R_2$, $R_3$, and $R_4$. We can therefore simplify our calculations by considering these three resistors to be one resistor of $3 + 5 + 2$, or $10\ \Omega$.

$$-15y + 10z = 0$$

Substitute $(x - y)$ for $z$ from equation $(a)$.

$$-15y + 10(x - y) = 0$$
$$-15y + 10x - 10y = 0$$
$$-25y + 10x = 0$$
$$-25y = -10x$$
$$y = \frac{-10}{-25}x = \frac{2}{5}x \qquad (1)$$

4. Trace around the circuit in the direction *abcd*.

$$120 - 6x - 15y - 12x = 0$$
$$-18x - 15y = -120$$

Divide the equation by $-3$ in order to simplify. This gives

$$6x + 5y = 40 \qquad (2)$$

5. Substitute the value of $y$ from equation (1) in equation (2).

$$6x + 5y = 40 \qquad (2)$$
$$6x + 5\left(\frac{2}{5}x\right) = 40$$
$$6x + 2x = 40$$
$$8x = 40$$
$$x = 5\ \text{A} \qquad \textit{Ans.}$$

6. Substitute $x = 5$ in equation (1).

$$y = \frac{2}{5}x \qquad (1)$$
$$y = \frac{2}{5}(5) = 2\ \text{A} \qquad \textit{Ans.}$$

7. Substitute the values for $x$ and $y$ in equation $(a)$ to find the current $z$.

$$z = x - y \qquad (a)$$

$$z = 5 - 2 = 3 \text{ A} \qquad \textit{Ans.}$$

8. Find the voltage drop in each resistor.

$$V_1 = 12x \qquad V_2 = 3z \qquad V_3 = 5z$$
$$V_1 = 12(5) \qquad V_2 = 3(3) \qquad V_3 = 5(3)$$
$$V_1 = 60 \text{ V} \qquad V_2 = 9 \text{ V} \qquad V_3 = 15 \text{ V}$$

$$V_4 = 2z \qquad V_5 = 15y \qquad V_6 = 6x$$
$$V_4 = 2(3) \qquad V_5 = 15(2) \qquad V_6 = 6(5)$$
$$V_4 = 6 \text{ V} \qquad V_5 = 30 \text{ V} \qquad V_6 = 30 \text{ V}$$

9. Check by tracing the circuit *abcfeda,* using Kirchhoff's second law.

$$V_T + V_6 + V_4 + V_3 + V_2 + V_1 = 0$$
$$120 - 30 - 6 - 15 - 9 - 60 = 0$$
$$120 - 120 = 0$$
$$0 = 0 \qquad \textit{Check}$$

EXAMPLE 9-16. Solve the circuit shown in Fig. 9-25 for the values of the currents $w$, $x$, $y$, and $z$.

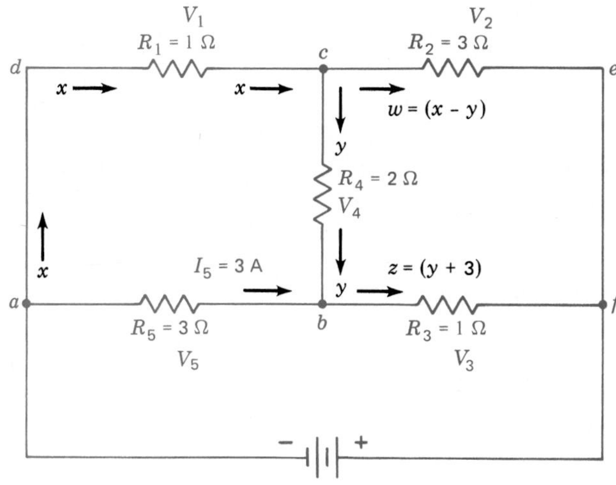


FIGURE 9-25

SOLUTION

1. Assume the direction of the electron flow to be that shown on the diagram.

2. Express current $w$ in terms of currents $x$ and $y$. By Kirchhoff's first law, operating at point $c$,

$$w + y = x$$

or

$$w = x - y \qquad (a)$$

At point $b$,

$$z = y + 3 \qquad (b)$$

3. Apply Kirchhoff's second law. Trace around the circuit in the direction *abcd.*

$$-3(3) + 2y + 1(x) = 0$$
$$-9 + 2y + x = 0$$

or
$$x + 2y = +9 \quad (1)$$

4. Trace around the circuit in the direction *bcef.*

$$+2y - 3w + 1z = 0$$

Substitute $(x - y)$ for $w$ from equation $(a)$ and $(y + 3)$ for $z$ from equation $(b)$.

$$+2y - 3(x - y) + 1(y + 3) = 0$$
$$+2y - 3x + 3y + y + 3 = 0$$
$$-3x + 6y = -3$$

Divide the equation by 3 in order to simplify. This gives

$$-x + 2y = -1 \quad (2)$$

5. Solve equations (1) and (2) simultaneously.

$$x + 2y = 9 \quad (1)$$
$$-x + 2y = -1 \quad (2)$$

Add:
$$4y = 8$$
$$y = 2 \text{ A} \quad \textit{Ans.}$$

6. Substitute this value for $y$ in equation (1) to find $x$.

$$x + 2y = 9 \quad (1)$$
$$x + 2(2) = 9$$
$$x + 4 = 9$$
$$x = 9 - 4 = 5 \text{ A} \quad \textit{Ans.}$$

7. Substitute these values for $x$ and $y$ in equation $(a)$ to find $w$.

$$w = x - y \quad (a)$$
$$w = 5 - 2 = 3 \text{ A} \quad \textit{Ans.}$$

8. Substitute $y = 2$ in equation $(b)$ to find $z$.

$$z = y + 3 \quad (b)$$
$$z = 2 + 3 = 5 \text{ A} \quad \textit{Ans.}$$

9. Find the voltage drop in each resistor.

$$V_1 = 1x \qquad V_2 = 3w \qquad V_3 = 1z$$
$$V_1 = 1(5) \qquad V_2 = 3(3) \qquad V_3 = 1(5)$$
$$V_1 = 5 \text{ V} \qquad V_2 = 9 \text{ V} \qquad V_3 = 5 \text{ V}$$

$$V_4 = 2y \qquad V_5 = 3(3)$$
$$V_4 = 2(2) = 4 \text{ V} \qquad V_5 = 9 \text{ V}$$

10. Check by tracing the circuit *abfecd* using Kirchhoff's second law.

$$V_5 + V_3 + V_2 + V_1 = 0$$
$$-9 - 5 + 9 + 5 = 0$$
$$0 = 0 \quad \textit{Check}$$

EXAMPLE 9-17. Solve the unbalanced bridge circuit shown in Fig. 9-26 for the values of the currents $x$, $y$, $v$, and $w$ and the total voltage.

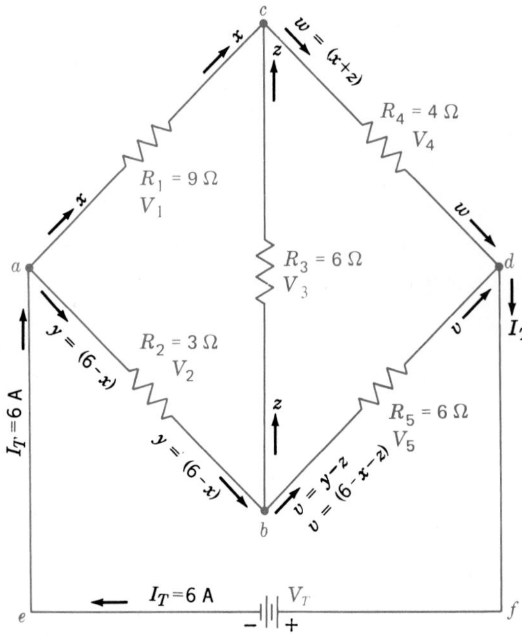


FIGURE 9-26
Unbalanced bridge circuit.

SOLUTION

1. Assume the direction of the electron flow to be that shown on the diagram.

2. Express current $y$ in terms of $x$ and the total current. By Kirchhoff's first law, operating at point $a$,

$$y + x = 6$$

or

$$y = 6 - x \qquad (a)$$

At point $c$:

$$w = x + z \qquad (b)$$

At point $b$:

$$v + z = y$$

or

$$v = y - z$$

Substituting $(6 - x)$ for $y$ from equation $(a)$,

$$v = 6 - x - z \qquad (c)$$

3. Apply Kirchhoff's second law. Trace around the circuit in the direction $abca$.

$$-3y - 6z + 9x = 0$$

Substitute $(6 - x)$ for $y$ from equation $(a)$.

$$-3(6 - x) - 6z + 9x = 0$$

$$-18 + 3x - 6z + 9x = 0$$

$$12x - 6z = 18$$

Divide the equation by 6 in order to simplify. This gives

$$2x - z = 3 \tag{1}$$

4. Trace around the circuit in the direction *bcdb.*

$$-6z - 4w + 6v = 0$$

Substitute $(x + z)$ for $w$ from equation $(b)$ and $(6 - x - z)$ for $v$ from equation $(c)$.

$$-6z - 4(x + z) + 6(6 - x - z) = 0$$
$$-6z - 4x - 4z + 36 - 6x - 6z = 0$$
$$-10x - 16z = -36$$

Divide the equation by $-2$ in order to simplify. This gives

$$5x + 8z = 18 \tag{2}$$

5. Solve equations (1) and (2) simultaneously.

$$2x - z = 3 \tag{1}$$
$$5x + 8z = 18 \tag{2}$$

Multiply equation (1) by 8.

$$16x - 8z = 24$$
$$5x + 8z = 18$$

Add: $21x = 42$

$$x = 2 \text{ A} \qquad \textit{Ans.}$$

6. Substitute this value for $x$ in equation (1) to find $z$.

$$2x - z = 3 \tag{1}$$
$$2(2) - z = 3$$
$$4 - z = 3$$
$$-z = 3 - 4$$
$$-z = -1$$
$$z = 1 \text{ A} \qquad \textit{Ans.}$$

7. Substitute $x = 2$ in equation $(a)$ to find $y$.

$$y = 6 - x$$
$$y = 6 - 2 = 4 \text{ A} \qquad \textit{Ans.}$$

8. Substitute $x = 2$ and $z = 1$ in equation $(b)$ to find $w$.

$$w = x + z \tag{b}$$
$$w = 2 + 1 = 3 \text{ A} \qquad \textit{Ans.}$$

9. Substitute $x = 2$ and $z = 1$ in equation $(c)$ to find $v$.

$$v = 6 - x - z \tag{c}$$
$$v = 6 - 2 - 1 = 3 \text{ A} \qquad \textit{Ans.}$$

10. Find the voltage drop in each resistor.

$V_1 = 9x$ $\qquad V_2 = 3y$ $\qquad V_3 = 6z$

$V_1 = 9(2) = 18$ V $\qquad V_2 = 3(4) = 12$ V $\qquad V_3 = 6(1) = 6$ V

$$V_4 = 4w \qquad V_5 = 6v$$
$$V_4 = 4(3) = 12 \text{ V} \qquad V_5 = 6(3) = 18 \text{ V}$$

11. Check by tracing the circuit in the direction *abdca*, using Kirchhoff's second law.

$$V_2 + V_5 + V_4 + V_1 = 0$$
$$-12 - 18 + 12 + 18 = 0$$
$$0 = 0 \qquad \textit{Check}$$

12. Find the total voltage. Trace the circuit in the direction *feabd*.

$$V_T + V_2 + V_5 = 0$$
$$V_T - 12 - 18 = 0$$
$$V_T = 12 + 18 = 30 \text{ V} \qquad \textit{Ans.}$$

SELF-TEST 9-18. In Fig. 9-27, find (*a*) the currents $x$, $y$, and $z$ and (*b*) the voltage drops across all the resistors. (*c*) Check the problem.

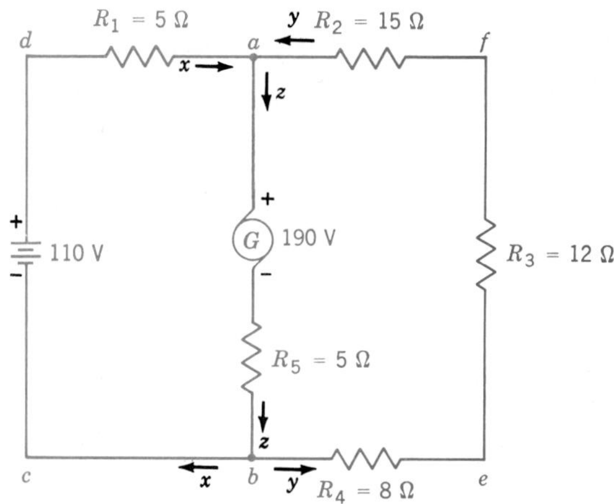


**FIGURE 9-27**
Opposing voltage sources.

SOLUTION

1. Since the generator voltage (190 V) is greater than the battery voltage (110 V), it is assumed that the current $z$ will flow ____(up/down)____ from point $a$ to point $b$. Assume the remaining currents to be directed as shown. down

2. Apply Kirchhoff's first law to point $a$.

$$x + \_\_\_\_ = \_\_\_\_$$ $y \quad z$

or $$y = z - \_\_\_\_ \qquad (a)$$ $x$

3. Apply Kirchhoff's second law. Trace the circuit *abcda*.

$$190 - 5z - 110 - 5\_\_\_\_ = 0$$ $x$

$$-5z - 5x = \_\_\_\_$$ $-80$

Divide the equation by $-5$ in order to simplify. This gives

$$z + x = \_\_\_\_ \qquad (1)$$ 16

4. Trace the circuit *abefa*. In this path, $R_4$, $R_3$, and $R_2$ are in ____ and may be considered to be one resistance of ____ Ω. series 35

$$190 - 5z - \_\_\_\_ = 0$$ 35y

$$-5z - 35y = \_\_\_\_$$ −190

Divide the equation by ____ in order to simplify. This gives −5

$$z + 7y = \_\_\_\_ \qquad (2)$$ 38

5. At this point we would solve equations (1) and (2) simultaneously. However, these two equations contain three letters at this point—$z$, $x$, and $y$. One of them must be expressed in terms of the other two letters. We shall express $y$ in terms of $z$ and $x$ by using equation ($a$), and substituting $z - x$ for $y$ in equation (2),

$$z + 7y = 38 \qquad (2)$$

$$z + 7(\_\_\_\_) = 38$$ z − x

$$z + 7z - \_\_\_\_ = 38$$ 7x

$$\_\_\_\_ - 7x = 38 \qquad (3)$$ 8z

6. Now, since they contain the same letters, $z$ and $x$, we can solve equations (____) and (____) simultaneously. 1 3

$$z + x = 16 \qquad (1)$$

$$8z - 7x = 38 \qquad (3)$$

Multiplying equation (1) by ____, 7

$$7z + 7x = \_\_\_\_$$ 112

$$8z - 7x = 38$$

Adding, $15z = \_\_\_\_$ 150

$$z = \_\_\_\_ \text{ A} \qquad \textit{Ans.}$$ 10

7. Substitute this value for $z$ in equation (1) to find $x$.

$$z + x = 16 \qquad (1)$$

$$\_\_\_\_ + x = 16$$ 10

$$x = 16 - \_\_\_\_$$ 10

$$x = \_\_\_\_ \text{ A} \qquad \textit{Ans.}$$ 6

8. Substitute these values for $x$ and $z$ in equation ($a$) to find $y$.

$$y = z - x \qquad (a)$$

$$y = 10 - 6 = \_\_\_\_ \text{ A} \qquad \textit{Ans.}$$ 4

9. Find the voltage drops across the resistors.

$V_1 = x \times R_1$ $\qquad V_5 = z \times R_5$

$V_1 = \_\_\_\_ \times 5$ $\qquad V_5 = \_\_\_\_ \times 5$ 6 10

$V_1 = 30$ V *Ans.* $\qquad V_5 = \_\_\_\_$ V *Ans.* 50

| | | | |
|---|---|---|---|
| $V_2 = y \times R_2$ | $V_3 = y \times R_3$ | $V_4 = y \times R_4$ | |
| $V_2 =$ ______ $\times 15$ | $V_3 =$ ______ $\times 12$ | $V_4 =$ ______ $\times 8$ | 4 4 4 |
| $V_2 = 60$ V *Ans.* | $V_3 = 48$ V *Ans.* | $V_4 =$ ______ V *Ans.* | 32 |

10. *Check:* Trace the circuit *abcda.*

$190 - 50 - 110 -$ ______ $= 0$ — 30

______ $= 0$ *Check* — 0

Trace the circuit *abefa.*

$190 - 50 - 32 -$ ______ $- 60 = 0$ — 48

______ $= 0$ *Check* — 0

## PROBLEMS

### Set No. 1

1. In a circuit similar to that shown in Fig. 9-21, $V_1 = 22$ V, $V_2 = 20$ V, $R_3 = 1\ \Omega$, and $R_4 = 4\ \Omega$. Find the currents *x, y,* and *z,* and check your answer.
2. In a circuit similar to that shown in Fig. 9-22, $V_1 = 75$ V, $R_2 = 3\ \Omega$, $R_3 = 12\ \Omega$, $R_4 = 4\ \Omega$, and $V_5 = 28$ V. Find the currents *x, y,* and *z,* and check your answer.
3. In a circuit similar to that shown in Fig. 9-24, $V_T = 25$ V, $R_1 = 1\ \Omega$, $R_2 = 3\ \Omega$, $R_3 = 6\ \Omega$, $R_4 = 1\ \Omega$, $R_5 = 5\ \Omega$, and $R_6 = 4\ \Omega$. Find the currents *x, y,* and *z,* and check your answer.
4. In a circuit similar to that shown in Fig. 9-25, $R_1 = 2\ \Omega$, $R_2 = 7\ \Omega$, $R_3 = 9\ \Omega$, $R_4 = 4\ \Omega$, $R_5 = 5\ \Omega$, and $I_5 = 4$ A. Find the currents *w, x, y,* and *z,* and check your answer.
5. In a circuit similar to that shown in Fig. 9-26, $R_1 = 10\ \Omega$, $R_2 = 8\ \Omega$, $R_3 = 15\ \Omega$, $R_4 = 18\ \Omega$, $R_5 = 2\ \Omega$, and $I_T = 15$ A. Find the currents *v, w, x, y,* and *z* and the total voltage.
6. Solve the circuit shown in Fig. 9-28 for the values of the currents *x, y,* and *z.*

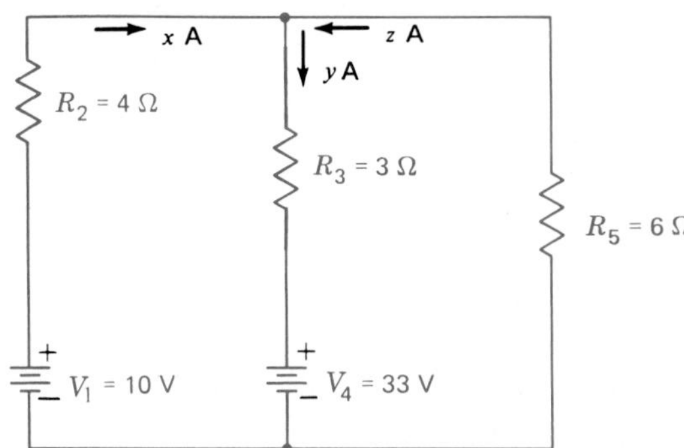


FIGURE 9-28

Solve the following circuits for the values of the currents *x, y,* and *z.*

7.

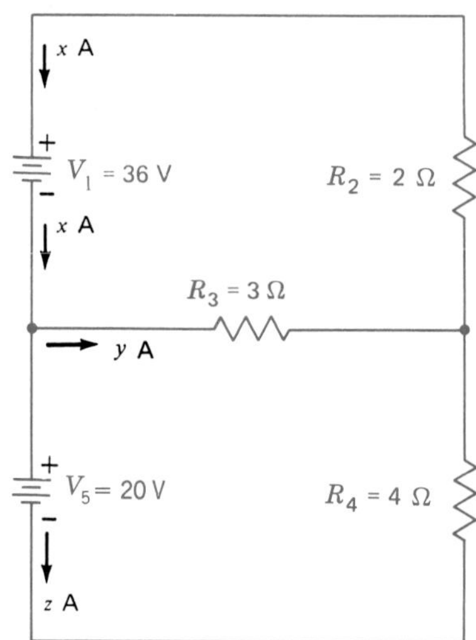


FIGURE 9-29

8.

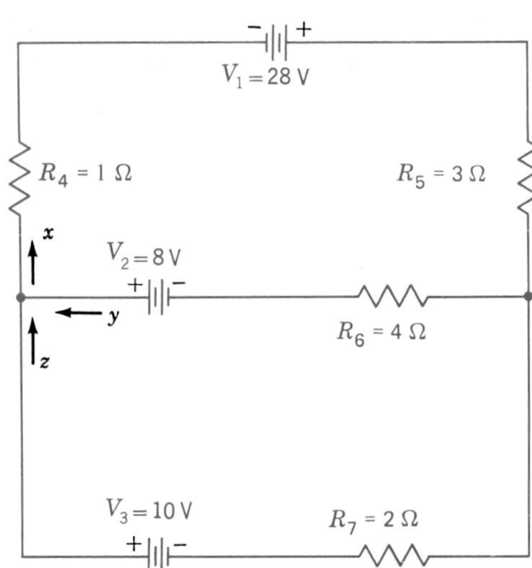


FIGURE 9-30

Set No. 2

1. In a circuit similar to that shown in Fig. 9-21, $V_1 = 32$ V, $V_2 = 12$ V, $R_3 = 5\ \Omega$, and $R_4 = 2\ \Omega$. Find the currents *x, y,* and *z,* and check your answer.
2. In a circuit similar to that shown in Fig. 9-22, $V_1 = 28$ V, $R_2 = 1\ \Omega$, $R_3 = 15\ \Omega$, $R_4 = 5\ \Omega$, and $V_5 = 12$ V. Find the currents *x, y,* and *z,* and check your answer.
3. In a circuit similar to that shown in Fig. 9-24, $V_T = 11$ V, $R_1 = 2\ \Omega$, $R_2 = 3\ \Omega$, $R_3 = 1\ \Omega$, $R_4 = 6\ \Omega$, $R_5 = 10\ \Omega$, and $R_6 = 4\ \Omega$. Find the currents *x, y,* and *z,* and check your answer.
4. In a circuit similar to that shown in Fig. 9-25, $R_1 = 10\ \Omega$, $R_2 = 18\ \Omega$, $R_3 = 2\ \Omega$, $R_4 = 15\ \Omega$, $R_5 = 8\ \Omega$, and $I_5 = 10$ A. Find the currents *w, x, y,* and *z,* and check your answer.
5. In a circuit similar to that shown in Fig. 9-26, $R_1 = 30\ \Omega$, $R_2 = 10\ \Omega$, $R_3 = 5\ \Omega$, $R_4 = 16\ \Omega$, $R_5 = 15\ \Omega$, and $I_T = 11$ A. Find the currents *v, w, x, y,* and *z* and the total voltage.
6. In a circuit similar to that shown in Fig. 9-28, $V_1 = 22$ V, $R_2 = 2\ \Omega$, $R_3 = 4\ \Omega$, $V_4 = 60$ V, and $R_5 = 16\ \Omega$. Find the currents *x, y,* and *z.*
7. In a circuit similar to that shown in Fig. 9-29, $V_1 = 76$ V, $R_2 = 10\ \Omega$, $R_3 = 2\ \Omega$, $R_4 = 15\ \Omega$, and $V_5 = 54$ V. Find the currents *x, y,* and *z.*
8. In a circuit similar to that shown in Fig. 9-30, $V_1 = 23$ V, $V_2 = 11$ V, $V_3 = 8$ V, $R_4 = 5\ \Omega$, $R_5 = 2\ \Omega$, $R_6 = 2\ \Omega$, and $R_7 = 3\ \Omega$. Find the currents *x, y,* and *z.*

## Set No. 3

1. In a circuit similar to that shown in Fig. 9-21, $V_1 = 10$ V, $V_2 = 5$ V, $R_3 = 10\ \Omega$, and $R_4 = 2.5\ \Omega$. Find the currents $x$, $y$, and $z$, and check your answer.
2. In a circuit similar to that shown in Fig. 9-22, $V_1 = 12$ V, $R_2 = 2.5\ \Omega$, $R_3 = 3.4\ \Omega$, $R_4 = 3\ \Omega$, and $V_5 = 38$ V. Find the currents $x$, $y$, and $z$, and check your answer.
3. In a circuit similar to that shown in Fig. 9-24, $V_T = 10$ V, $R_1 = 5\ \Omega$, $R_2 = 6\ \Omega$, $R_3 = 8\ \Omega$, $R_4 = 4\ \Omega$, $R_5 = 6\ \Omega$, and $R_6 = 3\ \Omega$. Find the currents $x$, $y$, and $z$, and check your answer.
4. In a circuit similar to that shown in Fig. 9-25, $R_1 = 5\ \Omega$, $R_2 = 40\ \Omega$, $R_3 = 10\ \Omega$, $R_4 = 10\ \Omega$, $R_5 = 3\ \Omega$, and $I_5 = 2$ A. Find the currents $w$, $x$, $y$, and $z$, and check your answer.
5. In a circuit similar to that shown in Fig. 9-26, $R_1 = 32\ \Omega$, $R_2 = 2\ \Omega$, $R_3 = 5\ \Omega$, $R_4 = 4\ \Omega$, $R_5 = 20\ \Omega$, and $I_T = 3.5$ A. Find the currents $v$, $w$, $x$, $y$, and $z$ and the total voltage.
6. In a circuit similar to that shown in Fig. 9-28, $V_1 = 12$ V, $R_2 = 10\ \Omega$, $R_3 = 30\ \Omega$, $V_4 = 41$ V, and $R_5 = 20\ \Omega$. Find the currents $x$, $y$, and $z$.
7. In a circuit similar to that shown in Fig. 9-29, $V_1 = 31$ V, $R_2 = 2\ \Omega$, $R_3 = 10\ \Omega$, $R_4 = 4\ \Omega$, and $V_5 = 6$ V *with polarity opposite to that shown in Fig. 9-29.* Find the currents $x$, $y$, and $z$.
8. In a circuit similar to that shown in Fig. 9-29, $V_1 = 17$ V, $R_2 = 20\ \Omega$, $R_3 = 10\ \Omega$, $R_4 = 40\ \Omega$, and $V_5 = 13$ V. Find the currents $x$, $y$, and $z$.

## JOB 9-6 REVIEW OF KIRCHHOFF'S LAWS

1. Kirchhoff's first law.

**RULE** At any point in an electric circuit, the sum of the currents that ______ a junction is equal to the sum of the currents that ______ it.

enter
leave

**FORMULA**

$$\Sigma I_e = \Sigma I_l \tag{9-1}$$

2. Direction of electron flow.

   *a.* If the total known electron flow *in* to a point is *greater* than the total known flow *out* of the point, then the direction of any unknown flow must be ___(into/out of)___ the point.

   out of

   *b.* If the total known electron flow *into* a point is *less* than the total known flow *out of* the point, then the direction of any unknown flow must be ___(into/out of)___ the point.

   into

3. Kirchhoff's second law.

**RULE** The ______ sum of all the voltages in any complete electric circuit is equal to ______.

algebraic
zero

**FORMULA** $V_1 + V_2 + V_3 + \text{etc.} = 0$ 9-2

4. Direction of voltages.
   *a.* *Source* voltages are positive (+) if we move through them from the ________ (________) terminal to the negative (−) terminal. — positive +
   *b.* *Source* voltages are ________ (________) if we move through them from the negative (−) terminal to the positive (+) terminal. — negative −
   *c.* A voltage across a resistance is *negative* (−) if we go through it in the (same/opposite) direction as that of the assumed electron flow. — same
   *d.* A voltage across a resistance is ________ (________) if we go through it in a direction *opposite* to the assumed electron flow. — positive +
5. Using Kirchhoff's laws.
   *a.* Determine the ________ of the electron flow. If insufficient data are given, *assume* directions of flow in each branch of the circuit. A minus sign for the solution indicates that the assumed direction was (correct/incorrect). The value of the current found (is/is not) affected by an incorrect choice of direction. — direction; incorrect; is not
   *b.* Mark the diagram to indicate all ________ of the electron flow and all polarities of sources of voltage. — directions
   *c.* If possible, express the currents in the various branches in terms of the ________ in other branches. — currents
   *d.* Trace around a ________ circuit, totaling the ________ sum of *all* voltages encountered. Set the sum equal to ________. Be careful to note the correct sign for each voltage. — complete algebraic; zero
   *e.* Trace around a *second complete* circuit, obtaining a second equation.
   *f.* Solve the set of ________ obtained in steps *d* and *e* simultaneously. — equations
   *g.* Find all voltage drops in all resistances by ________ law. — Ohm's
   *h.* Check the solution by tracing around a *complete third* circuit, setting the algebraic sum of the voltages equal to ________. — zero

SELF-TEST 9-19. In Fig. 9-31, find the generator voltage $V$ which will reduce the current through $R_3$ to zero.

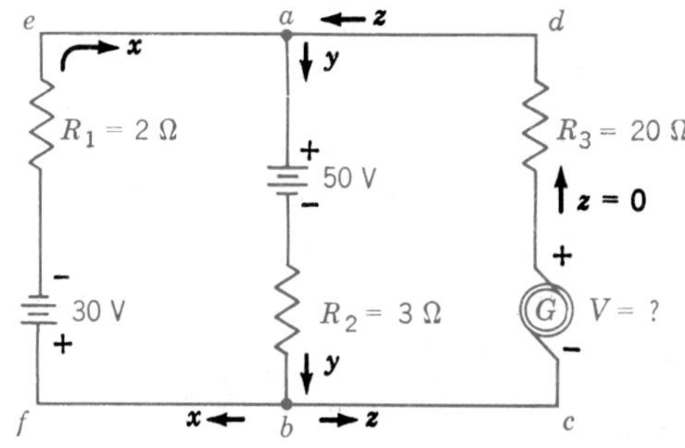


**FIGURE 9-31**
Find the generator voltage that will reduce the current through $R_3$ to zero.

SOLUTION

1. Since the two batteries (aid/oppose) each other, it is assumed that the current $y$ will flow (up/down) from point $a$ to point $b$. Assume the currents $x$ and $z$ to be directed as shown. — aid; down

2. Apply Kirchhoff's first law to point *b*.

$y = x +$ ______     $z$

$y = x +$ ______     0

or $y = x$     ($a$)

3. Apply Kirchhoff's second law to the circuit *efbae*.

$+2x - 30$ ______ (+/−) $3y - 50 = 0$     +

$2x + 3y =$ ______     80

Now, since $y =$ ______ from equation ($a$),     $x$

$2x + 3$______ $= 80$     $x$

______ $= 80$     $5x$

$x =$ ______ A     16

And, since $y = x$, $y =$ ______ A.     16

4. Trace the circuit *abcda*.

$50 - 3y -$ ______ $- 20z = 0$     $V$

Substitute 0 for $z$ as given.

$$50 - 3y - V - 20(0) = 0$$
$$50 - 3y - V = 0$$

Substitute 16 for $y$ in this equation.

$$50 - 3(16) - V = 0$$

$50 - 48 =$ ______     $V$

$2 = V$    or    $V = 2$ V    *Ans.*

5. *Check:* Trace the circuit *efbcdae*.

$$2x - 30 - V - 20z = 0$$

$2(16) - 30 -$ ______ $- 20($______$) = 0$     2   0

$32 - 30 - 2 -$ ______ $= 0$     0

______ $= 0$    *Check*     0

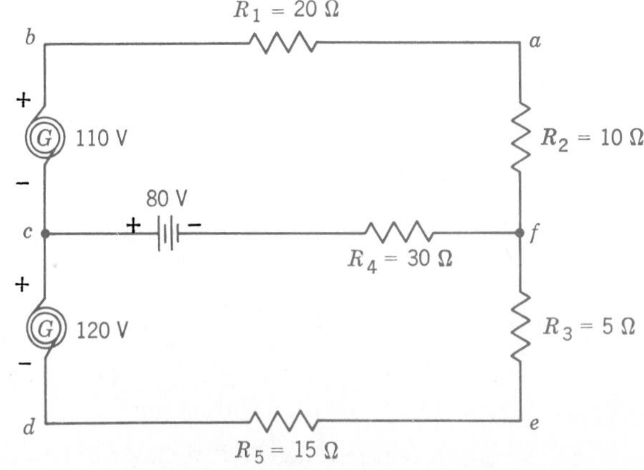


FIGURE 9-32

Now try just one more problem.

9. In Fig. 9-32, find the power consumed in $R_5$.

(See Instructor's Manual for Test 9-2, Kirchhoff's Laws)

## JOB 9-7 EQUIVALENT DELTA AND STAR CIRCUITS

Example 9-17 of the last job can be greatly simplified by the application of the theory of delta and star circuits. We shall solve this problem again at the conclusion of this job.

The triangle formed by the three resistors $R_2$, $R_1$, and $R_3$ of Fig. 9-26 has been turned through one-fourth turn and is shown in Fig. 9-33*a*. The resistors are said to be connected in *delta* because the connection resembles the Greek letter delta ($\Delta$). The resistors $R_a$, $R_b$, and $R_c$ in Fig. 9-33*b* are connected in *star,* or Y, formation. Most delta circuits will be much easier to solve if we can change them into *equivalent* star circuits.

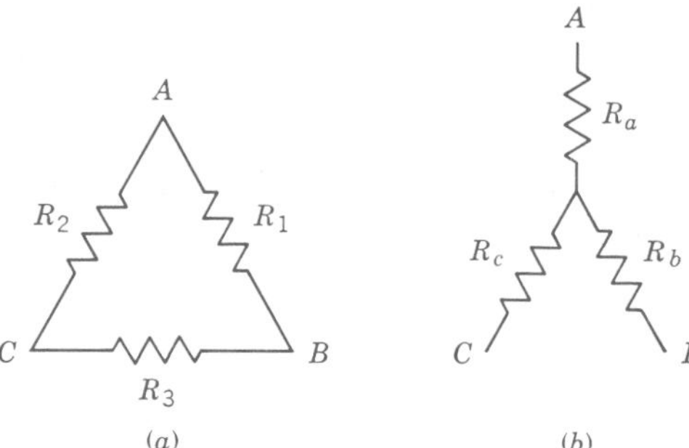


**FIGURE 9-33** (*a*) $R_1$, $R_2$, and $R_3$ are connected in *delta* formation. (*b*) $R_a$, $R_b$, and $R_c$ are connected in *star* or Y formation.

*Note:* Although the same nomenclature is used, these circuits are not to be confused with those of three-phase systems.

Now the word "equivalent" means that the circuits must do the same job as they did before. That is, the resistance between *A* and *B*, *B* and *C*, and *C* and *A* must be the same in each circuit.

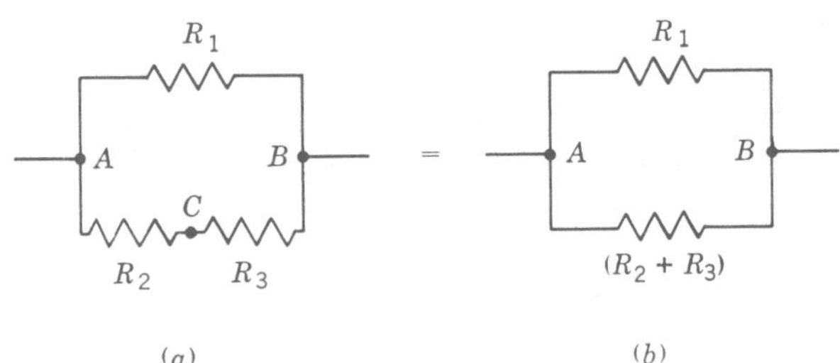


**FIGURE 9-34** (*a*) Figure 9-33*a* redrawn in standard form. (*b*) $R_2$ and $R_3$ are added since they are in series.

The resistance from *A* to *B* in Fig. 9-33*a* is found by redrawing the circuit as shown in Fig. 9-34*a*. This reduces to a circuit in which $R_1$ is in parallel with the series combination of $R_2 + R_3$, as in Fig. 9-34*b*. Thus,

$$R_{AB} = \frac{R_1(R_2 + R_3)}{R_1 + R_2 + R_3} \quad (1)$$

In Fig. 9-33*b*, the resistance from *A* to *B* is simply the series combination of $R_a + R_b$, or

$$R_{AB} = R_a + R_b \tag{2}$$

Since the two resistances must be equal, we can set equation (2) equal to equation (1) and get

$$R_a + R_b = \frac{R_1R_2 + R_1R_3}{R_1 + R_2 + R_3}$$

If we use the symbol $\Sigma R_\Delta$ (read as "the summation of resistances in delta") to mean $R_1 + R_2 + R_3$, we get

$$R_a + R_b = \frac{R_1R_2 + R_1R_3}{\Sigma R_\Delta} \tag{3}$$

In a similar manner,

$$R_b + R_c = \frac{R_1R_3 + R_2R_3}{\Sigma R_\Delta} \tag{4}$$

$$R_a + R_c = \frac{R_1R_2 + R_2R_3}{\Sigma R_\Delta} \tag{5}$$

When equations (3) to (5) are solved simultaneously, we get the following

**FORMULAS**

$$R_a = \frac{R_1R_2}{\Sigma R_\Delta} \tag{9-3}$$

$$R_b = \frac{R_1R_3}{\Sigma R_\Delta} \tag{9-4}$$

$$R_c = \frac{R_2R_3}{\Sigma R_\Delta} \tag{9-5}$$

Do not attempt to memorize these formulas. They depend on the positions of the letters *A*, *B*, and *C* and the positions of $R_1$, $R_2$, and $R_3$.

An easy way to develop these formulas is shown in Fig. 9-35*a*. Each equivalent *Y* resistance is obtained by multiplying the two adjacent delta resistances and then dividing by the sum of the delta resistances ($\Sigma R_\Delta$). For example, the resistors $R_2$ and $R_3$ are adjacent (next) to $R_c$. Therefore,

$$R_c = \frac{R_2 \times R_3}{\Sigma R_\Delta}$$

But, in Fig. 9-35*b*, in which the positions of $R_1$ and $R_2$ have been interchanged, $R_1$ and $R_3$ are adjacent to $R_c$. Therefore,

$$R_c = \frac{R_1R_3}{\Sigma R_\Delta}$$

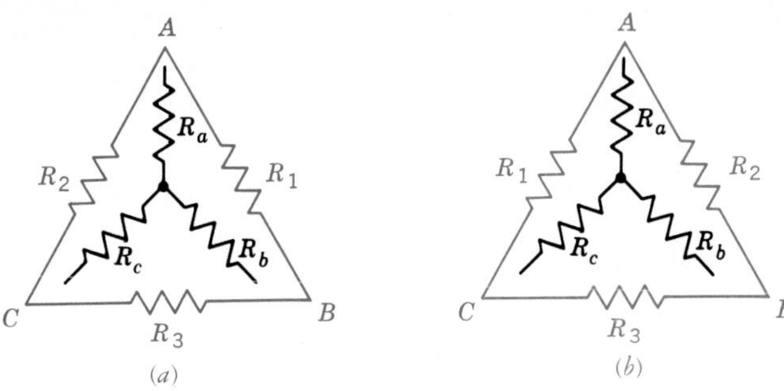


**FIGURE 9-35** (a) $R_1$ and $R_2$ are adjacent to $R_a$; $R_1$ and $R_3$ are adjacent to $R_b$; $R_2$ and $R_3$ are adjacent to $R_c$. (b) $R_1$ and $R_2$ are adjacent to $R_a$; $R_2$ and $R_3$ are adjacent to $R_b$; $R_1$ and $R_3$ are adjacent to $R_c$.

EXAMPLE 9-20. In Fig. 9-33*a*, $R_1 = 10\ \Omega$, $R_2 = 8\ \Omega$, and $R_3 = 2\ \Omega$. Find the resistances of the equivalent Y circuit of Fig. 9-33*b*.

SOLUTION

1.

$$\Sigma R_\Delta = R_1 + R_2 + R_3$$
$$\Sigma R_\Delta = 10 + 8 + 2 = 20\ \Omega$$

2.

$$R_a = \frac{R_1 R_2}{\Sigma R_\Delta} \quad (9\text{-}3) \qquad R_b = \frac{R_1 R_3}{\Sigma R_\Delta} \quad (9\text{-}4) \qquad R_c = \frac{R_2 R_3}{\Sigma R_\Delta} \quad (9\text{-}5)$$

$$R_a = \frac{10 \times 8}{20} \qquad R_b = \frac{10 \times 2}{20} \qquad R_c = \frac{8 \times 2}{20}$$

$$R_a = 4\ \Omega \qquad R_b = 1\ \Omega \qquad R_c = 0.8\ \Omega \qquad \textit{Ans.}$$

SELF-TEST 9-21. Find the total resistance between points *A* and *D* in the circuit shown in Fig. 9-36*a*.

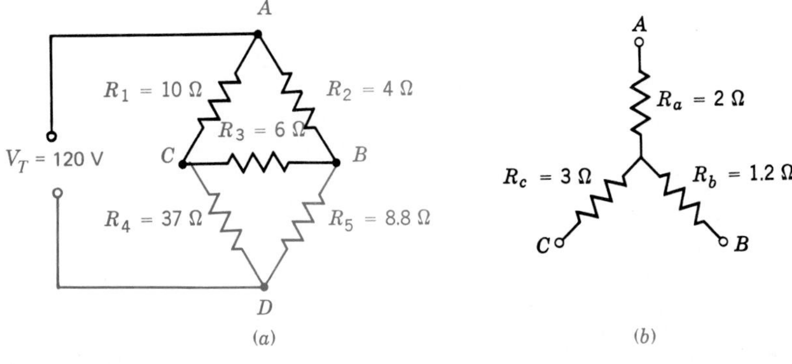


**FIGURE 9-36** (a) An unbalanced bridge circuit. (b) The Y equivalent of the delta circuit *ABC*.

SOLUTION

1. Change the delta circuit *ABC* into its equivalent Y circuit as shown in Fig. 9-36*b*. Note that $R_1$ and $R_2$ are *not* in the same position as they were in Fig. 9-33*a*. Develop the correct formulas for $R_a$, $R_b$, and $R_c$ as shown above.

$$\Sigma R_\Delta = R_1 + R_2 + R_3$$

$$\Sigma R_\Delta = 10 + 4 + 6 = \_\_\_\_\_\_ \ \Omega$$ 20

$$R_a = \frac{R_1R_2}{\Sigma R_\Delta} \qquad R_b = \frac{R_2R_3}{\Sigma R_\Delta} \qquad R_c = \frac{R_1R_3}{\Sigma R_\Delta}$$

$$R_a = \frac{10 \times ?}{20} \qquad R_b = \frac{4 \times ?}{20} \qquad R_c = \frac{10 \times ?}{20}$$ 4 6 6

$$R_a = \_\_\_\_\_\_ \ \Omega \qquad R_b = \_\_\_\_\_\_ \ \Omega \qquad R_c = \_\_\_\_\_\_ \ \Omega$$ 2 1.2 3

2. Redraw the delta circuit as a Y circuit and connect it to the remainder of the original circuit as shown in Fig. 9-37*a*.

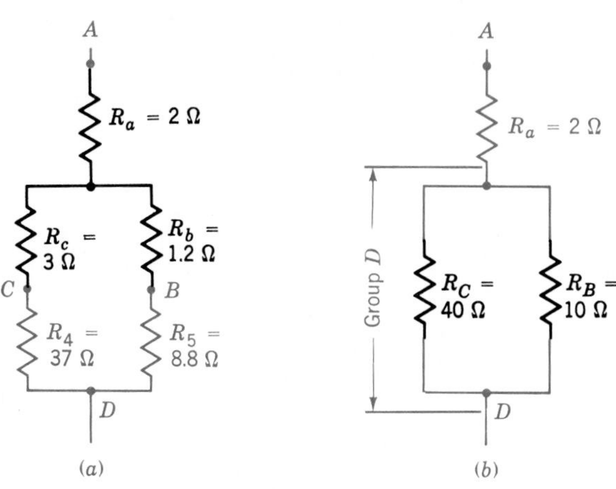


**FIGURE 9-37**
(*a*) The Y equivalent of delta *ABC* is attached to the remainder of the bridge circuit *BDC*. (*b*) $R_C$ is the series equivalent of $R_c + R_4$; $R_B$ is the series equivalent of $R_b + R_5$.

3. In Fig. 9-37*b*, $R_C$ represents the series resistance of $R_c$ plus $R_4$, which equals $3 + 37 =$ ______ Ω. Also, $R_B$ represents the series resistance of ______ plus $R_5$, which equals $1.2 + 8.8 =$ ______ Ω. 40 $R_b$ 10

4. In Fig. 9-37*b*, $R_C$ and $R_B$ are connected in ______. The total resistance for this group is parallel

$$R_D = \frac{R_C \times R_B}{R_C + R_B} \qquad (5\text{-}5)$$

$$R_D = \frac{40 \times 10}{40 + 10} = \_\_\_\_\_\_ \ \Omega$$ 8

5. In Fig. 9-37*b*, the total resistance for the circuit from *A* to *D* is made of the ___(series/parallel)___ combination of $R_a$ and $R_D$. series

$$R_T = R_a + \_\_\_\_\_\_$$ $R_D$

$$R_T = 2 + 8 = \_\_\_\_\_\_ \ \Omega \quad \textit{Ans.}$$ 10

6. Since the voltage across points *A* and *D* equals 120 V, we can find the total current by ______ law. Ohm's

$$E_T = I_T \times R_T \qquad (4\text{-}7)$$

$$120 = I_T \times \_\_\_\_\_\_$$ 10

$$I_T = \_\_\_\_\_\_ \ \text{A} \quad \textit{Ans.}$$ 12

SELF-TEST 9-22. Solve the unbalanced bridge circuit given in Example 9-17 for the values of the currents $x$, $y$, $v$, $z$, and $w$ and the total voltage.

SOLUTION

The circuit is redrawn for your convenience as Fig. 9-38.

1. Find the total resistance between points $a$ and $d$.

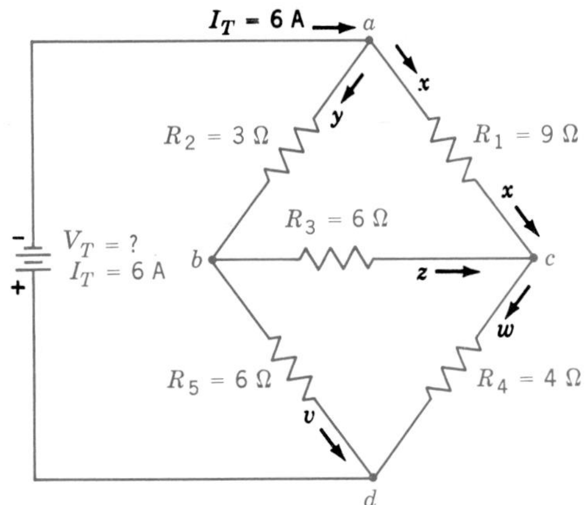


**FIGURE 9-38**
Figure 9-26 of Example 9-17 is redrawn.

*a.* Change the delta circuit *abc* into its equivalent Y circuit as shown in Fig. 9-39*a*.

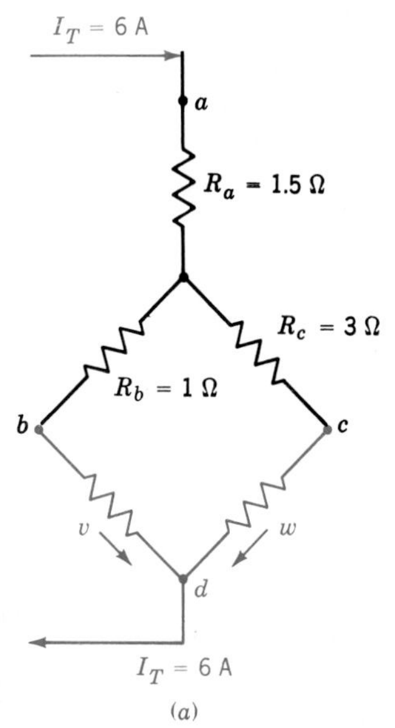


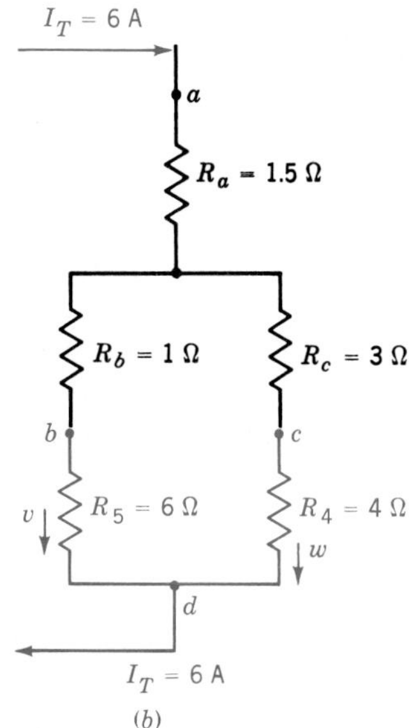


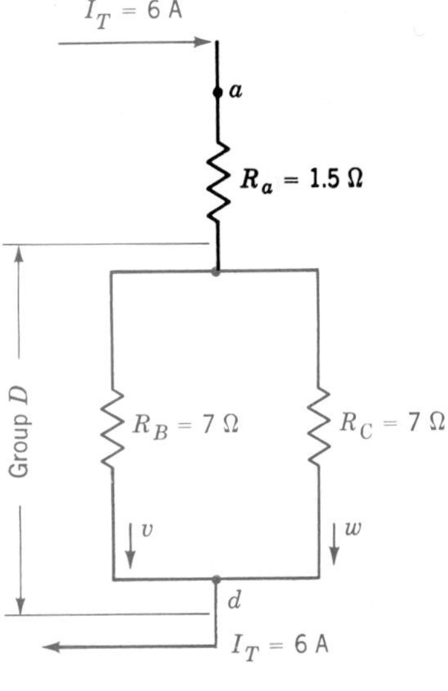


**FIGURE 9-39**
(*a*) Delta *abc* converted to Y formation. (*b*) Y formation of *abc* connected to *bdc* in standard form. (*c*) $R_B$ is the series equivalent of $R_b + R_5$; $R_C$ is the series equivalent of $R_c + R_4$.

$$\Sigma R_\Delta = R_1 + R_2 + R_3$$
$$\Sigma R_\Delta = 9 + \_\_\_\_\_\_ + 6 = \_\_\_\_\_\_ \ \Omega$$ 3 18

$$R_a = \frac{3 \times ?}{?} \qquad R_b = \frac{? \times 6}{18} \qquad R_c = \frac{6 \times ?}{18}$$ 9 3 9 18

$$R_a = \_\_\_\_\_\_ \ \Omega \qquad R_b = \_\_\_\_\_\_ \ \Omega \qquad R_c = \_\_\_\_\_\_ \ \Omega$$ 1.5 1 3

*b.* Connect the equivalent Y circuit to the remainder of the circuit as shown in Fig. 9-39*b*. Label all the parts with their respective values.

*c.* In Fig. 9-39*c*, $R_C$ represents the ___(series/parallel)___ resistance of $R_c$ and $R_4$, which equals ______ Ω. Also, $R_B$ represents the series resistance of ______ and $R_5$, which equals ______ Ω. series 7 $R_b$ 7

*d.* In Fig. 9-39*c*, $R_C$ and $R_B$ are connected in ______. parallel

$$R_D = \_\_\_\_\_\_ \ \Omega$$ 3.5

*e.* The total resistance for the circuit from *a* to *d* equals $R_a + R_D = 1.5 + 3.5 = \_\_\_\_\_\_$ Ω. 5

2. Find the total voltage $V_T$.

$$V_T = I_T \times R_T = 6 \times 5 = \_\_\_\_\_\_ \text{ V} \quad \textit{Ans.}$$ 30

3. Find the currents *v* and *w*.

In Fig. 9-39*c*, since $R_C = R_B = 7$ Ω, the total current $I_T = 6$ A will divide so that

$$v = w = \_\_\_\_\_\_ \text{ A} \quad \textit{Ans.}$$ 3

4. Find $V_5$ and $V_4$. In Fig. 9-39*b*,

$$V_5 = v \times R_5 \qquad V_4 = w \times R_4$$
$$V_5 = 3 \times 6 = \_\_\_\_\_\_ \text{ V} \qquad V_4 = 3 \times 4 = \_\_\_\_\_\_ \text{ V}$$ 18 12

5. Trace the circuit *cbd* in Fig. 9-38.

$$6z - V_5 + V_4 = 0$$
$$6z - \_\_\_\_\_\_ + 12 = 0$$ 18
$$6z = \_\_\_\_\_\_$$ 6
$$z = \_\_\_\_\_\_ \text{ A} \quad \textit{Ans.}$$ 1

6. Find the currents *x* and *y* in Fig. 9-38.

Apply Kirchhoff's first law to point *c*.

$$x + z = \_\_\_\_\_\_$$ *w*
$$x + 1 = 3 \text{ (from step 3)}$$
$$x = \_\_\_\_\_\_ \text{ A} \quad \textit{Ans.}$$ 2

Apply Kirchhoff's first law to point *a*.

$$y + x = \_\_\_\_\_\_$$ 6
$$y + \_\_\_\_\_\_ = 6$$ 2
$$y = \_\_\_\_\_\_ \text{ A} \quad \textit{Ans.}$$ 4

7. *Check:*

$$V_1 = 9x = 9(\_\_\_\_\_\_) = 18 \text{ V}$$ 2
$$V_2 = 3\_\_\_\_\_\_ = 3(4) = 12 \text{ V}$$ *y*

$V_3 = \_\_\_\_\_\_ z = 6(1) = 6$ V | 6

$V_4 = (\_\_\_\_\_\_)(\_\_\_\_\_\_) = 4(3) = 12$ V | 4 $w$

$V_5 = 6(\_\_\_\_\_\_) = 6(3) = 18$ V | $v$

Trace the circuit *abdca.*

$$-V_2 - V_5 + V_4 + V_1 = 0$$
$$-12 - 18 + 12 + 18 = 0$$
$$\_\_\_\_\_\_ = 0 \qquad \textit{Check}$$

| 0

## PROBLEMS

1. In the circuit shown in Fig. 9-33*a*, $R_1 = R_2 = R_3 = 60\ \Omega$. Find the resistances of the equivalent Y circuit.
2. In the circuit shown in Fig. 9-33*a*, $R_1 = 50\ \Omega$, $R_4 = 40\ \Omega$, and $R_3 = 10\ \Omega$. Find the resistances of the equivalent Y circuit.
3. In the circuit shown in Fig. 9-33*a*, $R_1 = 20\ \Omega$, $R_2 = 12\ \Omega$, and $R_3 = 8\ \Omega$. Find the resistances of the equivalent Y circuit.
4. In the circuit shown in Fig. 9-33*a*, $R_1 = 20\ \Omega$, $R_2 = 10\ \Omega$, and $R_3 = 15\ \Omega$. Find the resistances of the equivalent Y circuit.
5. In a circuit similar to that shown in Fig. 9-36*a*, $R_1 = 2\ \Omega$, $R_2 = 4\ \Omega$, $R_3 = 6\ \Omega$, $R_4 = 5\ \Omega$, and $R_5 = 4\ \Omega$. Find the total resistance from $A$ to $D$.
6. In a circuit similar to that shown in Fig. 9-36*a*, $R_1 = 12\ \Omega$, $R_2 = 18\ \Omega$, $R_3 = 10\ \Omega$, $R_4 = 1\ \Omega$, and $R_5 = 1.5\ \Omega$. Find the total resistance from $A$ to $D$.
7. In a circuit similar to that shown in Fig. 9-36*a*, $R_1 = 30\ \Omega$, $R_2 = 12\ \Omega$, $R_3 = 8\ \Omega$, $R_4 = 10.2\ \Omega$, and $R_5 = 3.08\ \Omega$. Find the total resistance from $A$ to $D$.
8. In a circuit similar to that shown in Fig. 9-36*a*, $R_1 = 10\ \Omega$, $R_2 = 30\ \Omega$, $R_3 = 5\ \Omega$, $R_4 = 15\ \Omega$, and $R_5 = 16\ \Omega$. Find the total resistance from $A$ to $D$.
9. In a circuit similar to that shown in Fig. 9-38, $R_1 = 10\ \Omega$, $R_2 = 8\ \Omega$, $R_3 = 2\ \Omega$, $R_4 = 1\ \Omega$, and $R_5 = 1.2\ \Omega$. If $I_T = 10$ A, find (*a*) the total resistance of the circuit, (*b*) the currents $v$, $w$, $x$, $y$, and $z$, and (*c*) the total voltage.
10. In a circuit similar to that shown in Fig. 9-38, $R_1 = 10\ \Omega$, $R_2 = 10\ \Omega$, $R_3 = 20\ \Omega$, $R_4 = 19\ \Omega$, and $R_5 = 3\ \Omega$. If $I_T = 16$ A, find (*a*) the total resistance of the circuit, (*b*) the currents $v$, $w$, $x$, $y$, and $z$, and (*c*) the total voltage.
11. In a circuit similar to that shown in Fig. 9-38, $R_1 = 10\ \Omega$, $R_2 = 50\ \Omega$, $R_3 = 15\ \Omega$, $R_4 = 8\ \Omega$, and $R_5 = 5\ \Omega$. If $I_T = 10$ A, find (*a*) the total resistance of the circuit, (*b*) the currents $v$, $w$, $x$, $y$, and $z$, and (*c*) the total voltage.
12. Figure 9-40 shows the original circuit and the transformations necessary to find the total resistance. Find (*a*) the total resistance, (*b*) the total current, (*c*) the current in $R_5$, and (*d*) the current in $R_6$.

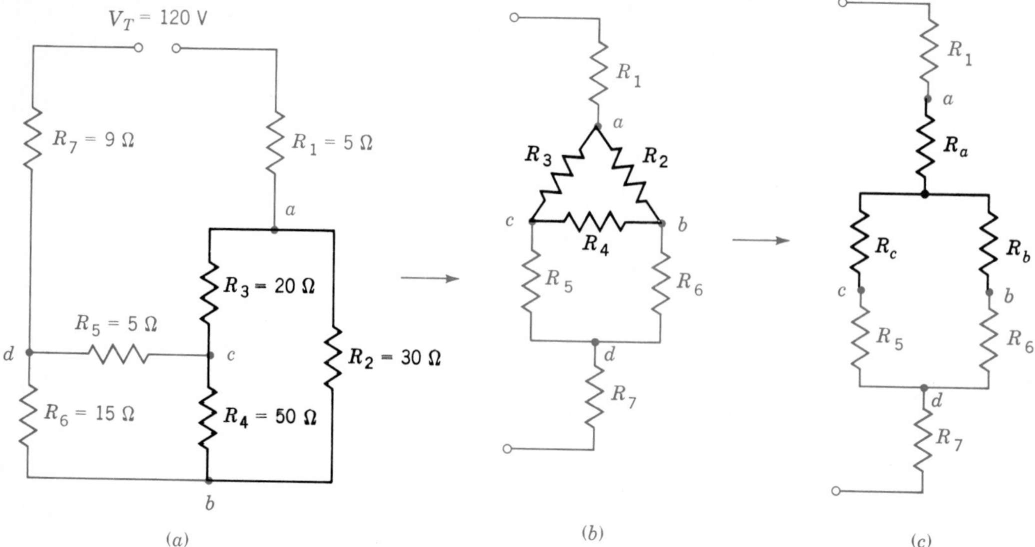


**FIGURE 9-40**
(*a*) Original circuit. (*b*) $R_2$, $R_3$, and $R_4$ in delta formation. (*c*) Delta *abc* in Y formation and connected to the remainder of the circuit.

## JOB 9-8 THEVENIN'S THEOREM

The main objective in the solution of a circuit is usually the calculation of the values for the load voltage and the load current. As we have seen, this usually involves many intermediate steps. Thevenin's theorem will eliminate many of these intermediate steps and permit us to calculate the load values directly.

Figure 9-41*a* is the same as Fig. 9-24 of Example 9-15 except that the resistances $R_2$, $R_3$, and $R_4$ have been combined into the load resistance $R_L = 10\ \Omega$. In the solution of Example 9-15, we found

$$z = I_L = 3\text{ A}$$

$$V_2 + V_3 + V_4 = V_L$$

$$9 + 15 + 6 = V_L = 30\text{ V}$$

Now if this same load $R_L = 10\ \Omega$ were connected by a switching arrangement to become the load of circuit Fig. 9-41*b*, then $R_T = 10 + 15 = 25\ \Omega$, and

$$I_L = \frac{V_T}{R_T} = \frac{75}{25} = 3\text{ A}$$

and

$$V_L = I_L \times R_L$$

$$V_L = 3 \times 10 = 30\text{ V}$$

Thus, the *same* values for $V_L$ and $I_L$ were obtained when $R_L$ was connected as part of circuit (*a*) or when $R_L$ was connected as part of circuit (*b*).

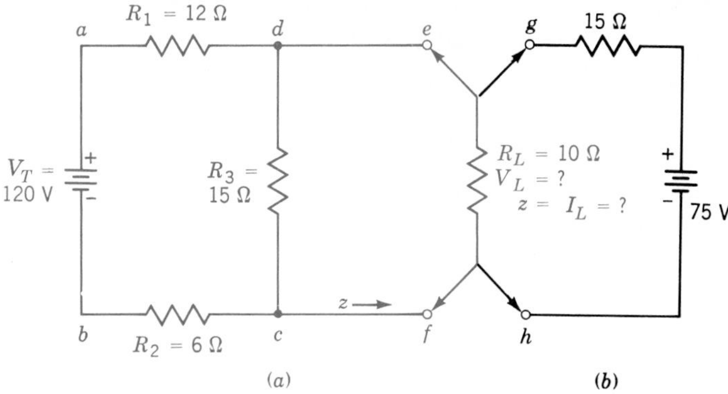


**FIGURE 9-41** As far as the load $R_L$ is concerned, circuit (*a*) is equivalent to circuit (*b*).

Therefore, as far as $R_L$ is concerned, circuit (*b*) could be substituted for circuit (*a*) and the same results obtained. Since the two circuits produce the same results, they may be said to be equivalent. Obviously, if we had a choice, we would rather work with circuit (*b*) than with circuit (*a*).

Thevenin's theorem is a method for changing a complex circuit into a simple *equivalent* circuit. Then the simple circuit may be solved with a minimum of effort.

**Thevenin's theorem.** Any linear network of voltage sources and resistances, if viewed from any two points in the network, can be replaced by an equivalent resistance $R_{TH}$ in series with an equivalent source $V_{TH}$. Actually, then, the problem consists of three steps.

1. Find the equivalent voltage $V_{TH}$.
2. Find the equivalent resistance $R_{TH}$.
3. Place $R_{TH}$ in series with the load, and solve the simple series circuit.

## METHOD

1. Disconnect the part of the circuit which is considered as the load.

2. Find the voltage that would appear across the load terminals when the load is disconnected. In Fig. 9-41*a* this would be the voltage across terminals *e* and *f*. This open-circuit voltage is called the *Thevenin voltage* ($V_{TH}$). See Fig. 9-42*a*.

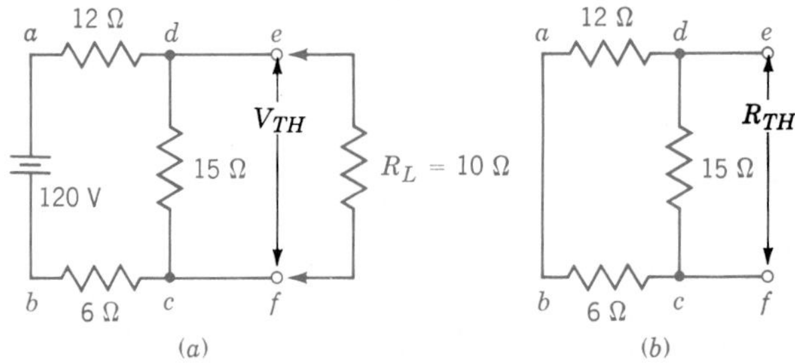


**FIGURE 9-42** (*a*) When the load is removed, the equivalent voltage $V_{TH}$ equals the voltage across the terminals *ef* or *dc*. (*b*) The equivalent resistance $R_{TH}$ equals the total resistance with all sources of emf shorted and the load removed.

3. Replace each voltage source with a short, reducing the voltage to zero, and remove the load as shown in Fig. 9-42*b*.

4. Find the total resistance that would appear across the load terminals. This will be the *Thevenin resistance* ($R_{TH}$), as shown in Fig. 9-42*b*.

5. Draw the equivalent circuit consisting of $R_{TH}$ in series with $R_L$ and connected across the equivalent voltage $V_{TH}$, as shown in Fig. 9-43.

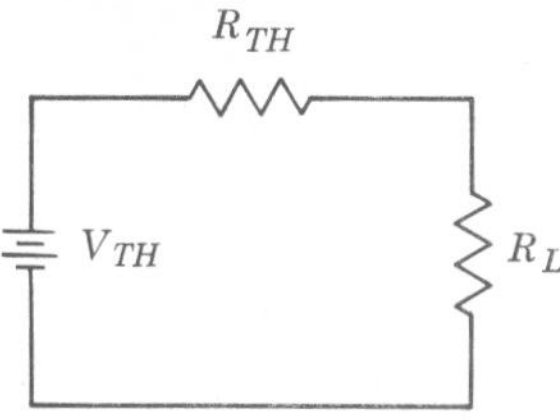


**FIGURE 9-43**
Thevenin equivalent circuit supplying a load $R_L$.

6. Solve for the load current and the load voltage.

EXAMPLE 9-23. Solve the circuit shown in Fig. 9-41*a* for the values of $I_L$ and $V_L$.

SOLUTION

1. Disconnect the load $R_L$ as shown in Fig. 9-42*a*.
2. Find the voltage across the terminals *ef*. This is equal to the voltage across terminals *dc*. In the circuit *abcd*,

$$R_t = 6 + 15 + 12 = 33\ \Omega \qquad (4\text{-}3)$$

$$I_t = \frac{120}{33} = 3.63\ \text{A}$$

Therefore,

$$V_{dc} = 3.63 \times 15 = 54.45\ \text{V}$$

or

$$V_{TH} = 54.45\ \text{V}$$

3. Replace each voltage source with a short, and remove the load as shown in Fig. 9-42*b*.
4. Find the value of $R_{TH}$. In Fig. 9-42*b*, the 12 Ω is in series with the 6 Ω for a total of 18 Ω. This 18 Ω is in parallel with the 15 Ω.

$$R_{TH} = \frac{18 \times 15}{18 + 15} = 8.18\ \Omega \qquad (5\text{-}5)$$

5. Draw the equivalent circuit as shown in Fig. 9-44.

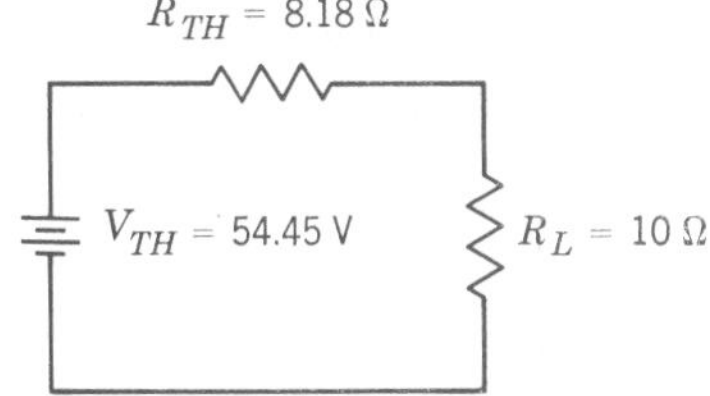


**FIGURE 9-44**
$V_{TH}$ is in series with $R_{TH}$ and $R_L$.

6. Solve for $I_L$.

$$R_t = 8.18 + 10 = 18.18\ \Omega$$

$$I_L = \frac{V_{TH}}{R_t} = \frac{54.45}{18.18} = 3\ \text{A} \qquad \textit{Ans.}$$

7. Solve for $V_L$.

$$V_L = I_L \times R_L$$

$$V_L = 3 \times 10 = 30\ \text{V} \qquad \textit{Ans.}$$

The true value of Thevenin's theorem is apparent when we compare the small effort involved here with the lengthy calculations needed to solve the same problem of Example 9-15.

EXAMPLE 9-24. Solve the circuit of Example 9-13 by Thevenin's theorem.

SOLUTION. Figure 9-22 is repeated here for your convenience.

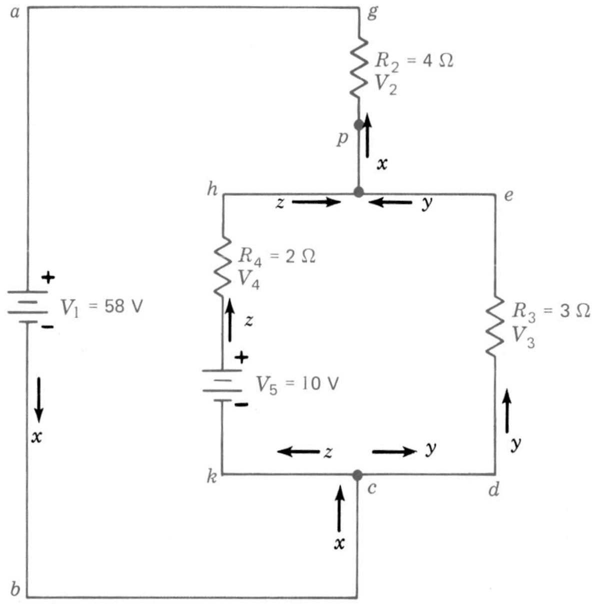


1. We must redraw the figure so that the load may be separated from the rest of the circuit at only two terminals. See Fig. 9-45. Consider the load to be $R_3$.

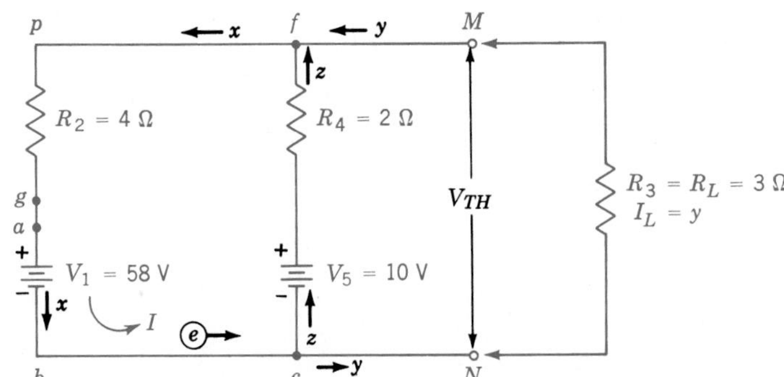


**FIGURE 9-45**
The circuit of Fig. 9-22 is redrawn to present only two terminals to the load.

2. Find $V_{TH}$ with the load removed. This will be the voltage across the terminals $M$ and $N$ which is equal to the voltage across the terminals $f$ and $c$.

*a.* In the series circuit *pbcf*, the circulating current $I$ is

$$I = \frac{V}{R} = \frac{58 - 10}{4 + 2} = \frac{48}{6} = 8 \text{ A}$$

*b.* Find the voltage from $f$ to $c$.

$$V_{fc} = IR_4 + V_5$$
$$V_{fc} = 8(2) + 10 = 26 \text{ V}$$

Therefore, $$V_{TH} = 26 \text{ V}$$

3. Replace each voltage source with a short, and remove the load as shown in Fig. 9-46.

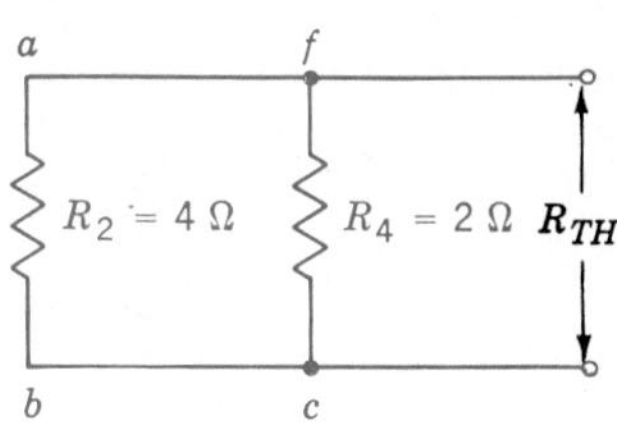


**FIGURE 9-46**
The equivalent resistance $R_{TH}$ consists of $R_2$ and $R_4$ in parallel.

4. Find $R_{TH}$.

$$R_{TH} = \frac{4 \times 2}{4 + 2} = 1.33\ \Omega$$

5. Draw the equivalent circuit as shown in Fig. 9-47.

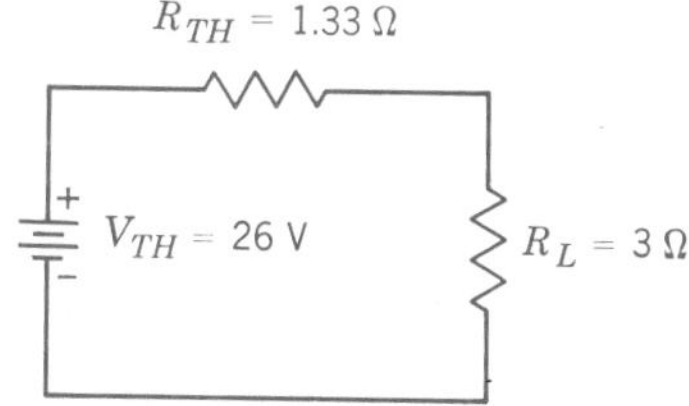


**FIGURE 9-47**
Thevenin equivalent circuit. $V_{TH}$ is in series with $R_{TH}$ and $R_L$.

6. Solve for $I_L$.

$$R_t = 1.33 + 3 = 4.33\ \Omega$$
$$I_L = \frac{V_{TH}}{R_t} = \frac{26}{4.33} = 6 \text{ A} \qquad \textit{Ans.}$$

Solve for $V_L$.

$$V_L = I_L \times R_L$$
$$V_L = 6 \times 3 = 18 \text{ V} \qquad \textit{Ans.}$$

7. Find the currents $x$ and $z$. In Fig. 9-45, since $V_{pb} = V_{fc} = V_L = 18$ V (all in parallel),

$$V_{pb} = -4x + 58 = 18$$
$$-4x = -40$$
$$x = 10 \text{ A} \qquad \textit{Ans.}$$

Also, since $V_{fc} = 18$ V,

$$2z + 10 = 18$$
$$2z = 8$$
$$z = 4 \text{ A} \qquad \textit{Ans.}$$

EXAMPLE 9-25. Find the load current and the load voltage in the circuit shown in Fig. 9-48*a*.

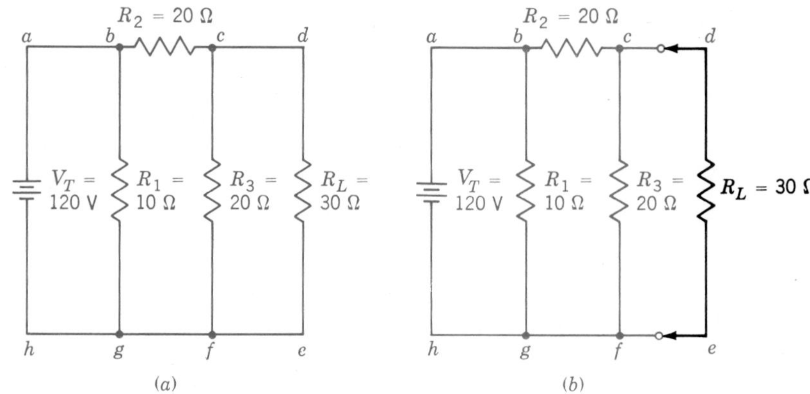


**FIGURE 9-48**
(*a*) Original circuit. (*b*) The circuit with the load removed.

SOLUTION

1. Remove the load from the circuit as shown in Fig. 9-48*b*.
2. Find $V_{TH}$ with the load removed. This will be the voltage across $R_3$.
   *a.* The resistors $R_2$ and $R_3$ between points *b* and *f* are in series. Therefore, $R_{bf} = 20 + 20 = 40\ \Omega$.
   *b.* The path from *b* to *f* is in parallel with the 120-V source. Therefore,

$$I_{bf} = \frac{120}{40} = 3\ \text{A}$$

$$V_3 = V_{TH} = I_3 \times R_3$$

$$V_{TH} = 3 \times 20 = 60\ \text{V}$$

3. Replace each voltage source with a short and remove the load as shown in Fig. 9-49*a*. Be careful to note that *when the battery is replaced with a short, it also shorts out the resistor* $R_1$, which results in the circuit shown in Fig. 9-49*b*.

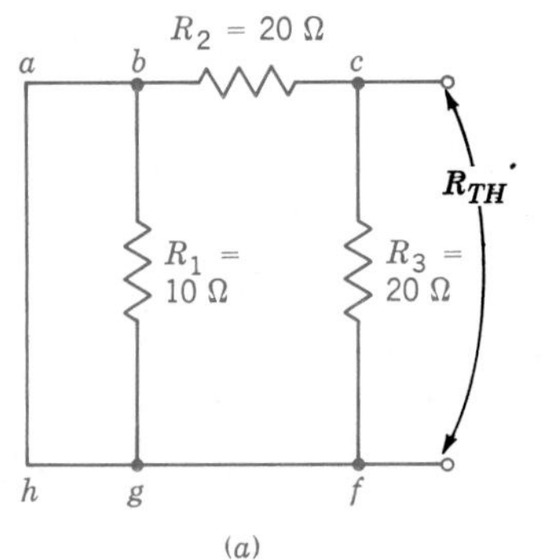


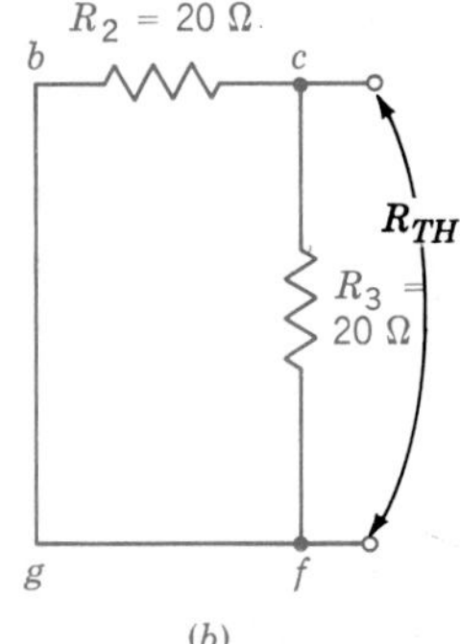


**FIGURE 9-49**
(*a*) The equivalent $R_{TH}$ equals the total resistance with the source shorted and the load removed. (*b*) The short across *ab* of Fig. 9-49*a* also shorts out $R_1$; $R_{TH}$ is found from this circuit.

4. Find $R_{TH}$. Since $R_2$ and $R_3$ are now in parallel,

$$R_{TH} = \frac{R_2 \times R_3}{R_2 + R_3} \qquad (5\text{-}5)$$

$$R_{TH} = \frac{20 \times 20}{20 + 20} = 10\ \Omega$$

5. Draw the equivalent circuit as shown in Fig. 9-50.

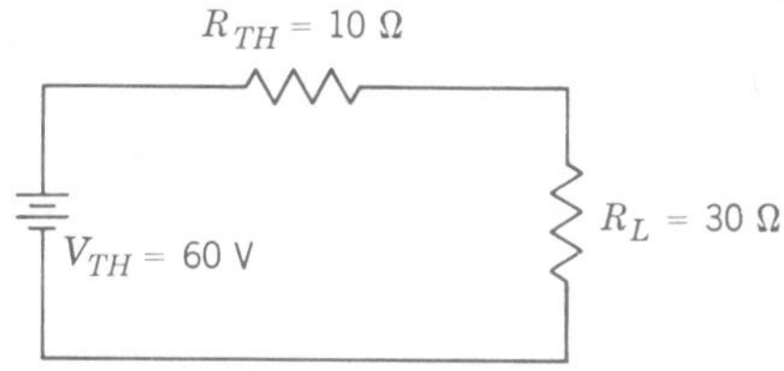


**FIGURE 9-50**
Thevenin equivalent circuit. $V_{TH}$ is in series with $R_{TH}$ and $R_L$.

6. Solve for $I_L$.

$$R_t = 10 + 30 = 40\ \Omega$$

$$I_L = \frac{V_{TH}}{R_t} = \frac{60}{40} = 1.5\text{ A} \qquad \textit{Ans.}$$

Solve for $V_L$.

$$V_L = I_L \times R_L$$
$$V_L = 1.5 \times 30 = 45\text{ V} \qquad \textit{Ans.}$$

EXAMPLE 9-26. Solve the circuit shown in Fig. 9-29 for the load current $y$ and the load voltage $V_3$ by Thevenin's theorem.

SOLUTION. Figure 9-29 is repeated here for your convenience.

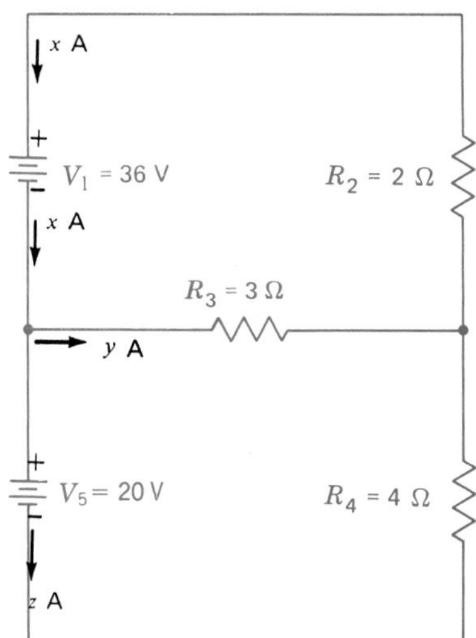


1. Remove the load $R_3$ from the circuit as shown in Fig. 9-51*a*.
2. Find $V_{TH}$ with the load removed. This will be the voltage between points *e* and *b*.

   *a.* The circuit *abcdef* is a simple series circuit in which

$$V_t = 36 + 20 = 56\text{ V}$$
$$R_t = 4 + 2 = 6\ \Omega$$

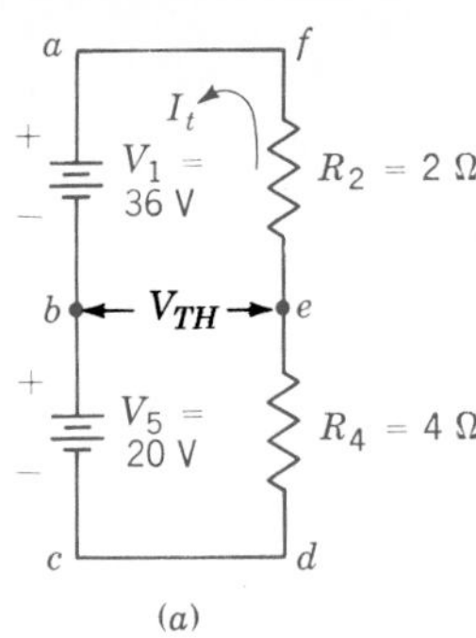


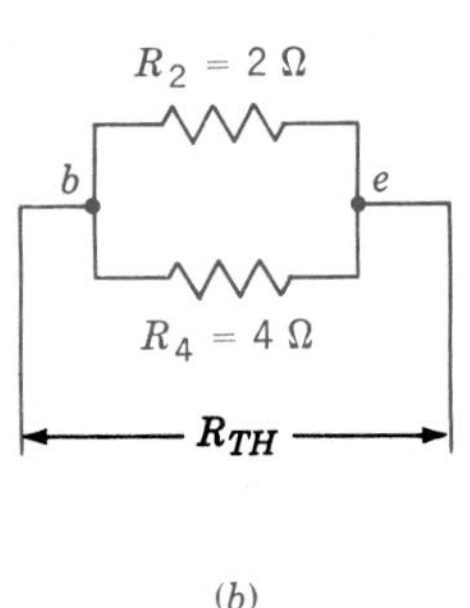


**FIGURE 9-51**
(a) $V_{TH}$ is the voltage across the terminals from which the load has been removed. (b) The equivalent $R_{TH}$ is the *total* resistance with all sources of emf shorted and the load removed.

$$I_t = \frac{56}{6} = 9.33 \text{ A}$$

*b.* The voltage from *e* to *b* around the path *efab* $= V_{TH}$.

$$V_{TH} = -I_tR_2 + V_1$$
$$V_{TH} = -(9.33 \times 2) + 36$$
$$V_{TH} = -18.66 + 36 = 17.34 \text{ V}$$

3. Replace each voltage source with a short and remove the load as shown in Fig. 9-51*b*. In this parallel circuit,

$$R_{TH} = \frac{2 \times 4}{2 + 4} = 1.33 \ \Omega$$

4. Draw the equivalent circuit as shown in Fig. 9-52.

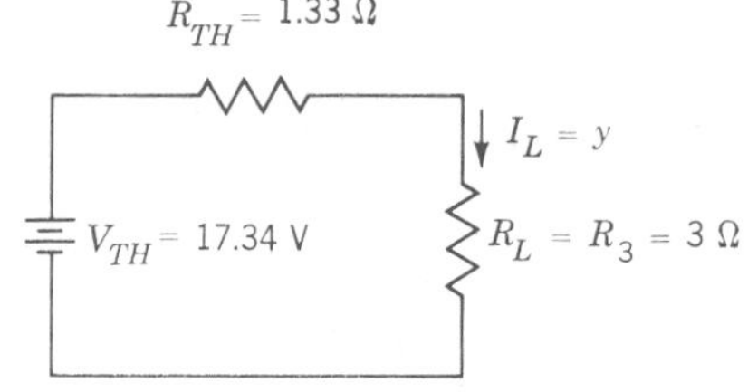


**FIGURE 9-52**
Thevenin equivalent circuit. $V_{TH}$ is in series with $R_{TH}$ and $R_3$.

5. Solve for the load current *y*.

$$R_t = 1.33 + 3 = 4.33 \ \Omega$$

$$y = I_L = \frac{V_{TH}}{R_t} = \frac{17.34}{4.33} = 4 \text{ A} \qquad \textit{Ans.}$$

Solve for the load voltage $V_3$.

$$V_3 = y \times R_3 = 4 \times 3 = 12 \text{ V} \qquad \textit{Ans.}$$

SELF-TEST 9-27. Find the load current and the load voltage in the circuit shown in Fig. 9-53.

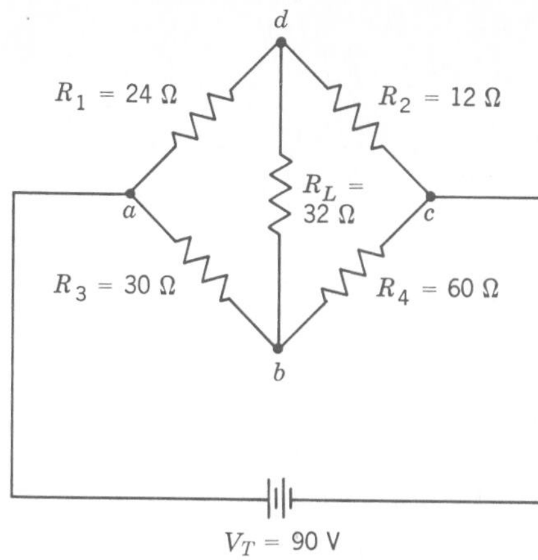


FIGURE 9-53

SOLUTION

1. Remove the load from the circuit as shown in Fig. 9-54. Move $V_T = 90$ V to the inside of the figure for greater clarity.

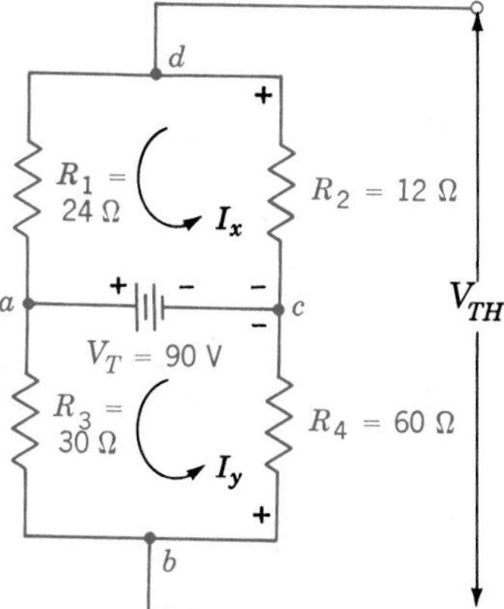


**FIGURE 9-54**
$V_{TH}$ is the voltage across the terminals from which the load has been removed.

2. Find $V_{TH}$ with the load removed. This will be the voltage across terminals ______ and ______. — *b d*

*a.* In the circuit *acd*, $R_1$ and $R_2$ are in ______. — series

$$R_x = 24 + \_\_\_\_\_\_ = \_\_\_\_\_\_\ \Omega$$ — 12 36

$$I_x = \frac{V_T}{R_x} = \frac{90}{?} = 2.5 \text{ A}$$ — 36

*b.* In circuit *abc*, $R_3$ and $R_4$ are in ______. — series

$$R_y = \_\_\_\_\_\_ + 60 = \_\_\_\_\_\_\ \Omega$$ — 30 90

$$I_y = \frac{?}{90} = 1 \text{ A}$$ — 90

*c.* Find $V_2$ and $V_4$.

$V_2 = \_\_\_\_\_\_ \times R_2$ $V_4 = \_\_\_\_\_\_ \times R_4$ — $I_x$ $I_y$

$V_2 = \_\_\_\_\_\_ \times 12$ $V_4 = \_\_\_\_\_\_ \times 60$ — 2.5 1

$V_2 = 30$ V $V_4 = 60$ V

*d.* $V_{TH}$ is equal to the *algebraic* sum of the voltages around path *bcd*. Be careful to note the polarities as indicated on the figure.

$$V_{bcd} = V_{TH} = V_4 \underline{\quad (+/-) \quad} V_2$$ —

$$V_{TH} = 60 - 30 = \underline{\qquad} \text{ V}$$ 30

3. Replace each voltage source with a ______ and remove the ______ as shown in Fig. 9-55*a*. The circuit should now be redrawn in standard form as shown in Fig. 9-55*b*.

short load

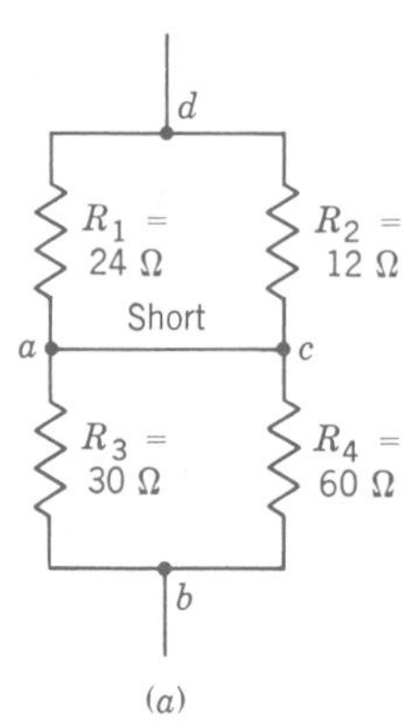


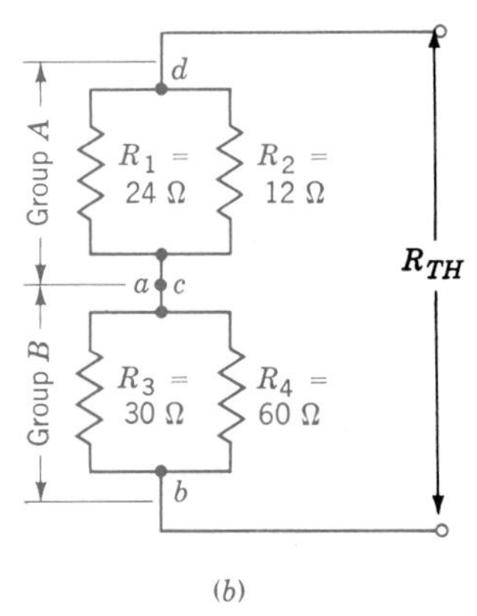


**FIGURE 9-55**
(*a*) Figure 9-53 with the load removed and the battery shorted. (*b*) Figure 9-55*a* redrawn in standard form.

4. Find $R_{TH}$. In Fig. 9-55*b*,
$R_1$ and $R_2$ form a __(series/parallel)__ group *A*.
$R_3$ and $R_4$ form a __(series/parallel)__ group *B*.
Group *A* is in __(series/parallel)__ with group *B*.

parallel
parallel
series

$$R_{TH} = \frac{R_1 \times R_2}{R_1 + R_2} + \underline{\qquad}$$

$$\frac{R_3 \times R_4}{R_3 + R_4}$$

$$R_{TH} = \frac{24 \times 12}{24 + 12} + \frac{30 \times 60}{30 + 60}$$

$$R_{TH} = \underline{\qquad} + \underline{\qquad}$$ 8 20

$$R_{TH} = \underline{\qquad} \ \Omega$$ 28

5. Complete the drawing of the equivalent circuit of Fig. 9-56.

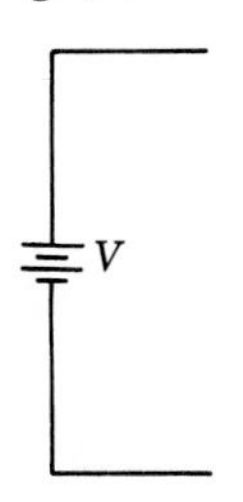


**FIGURE 9-56**
Incomplete Thevenin circuit.

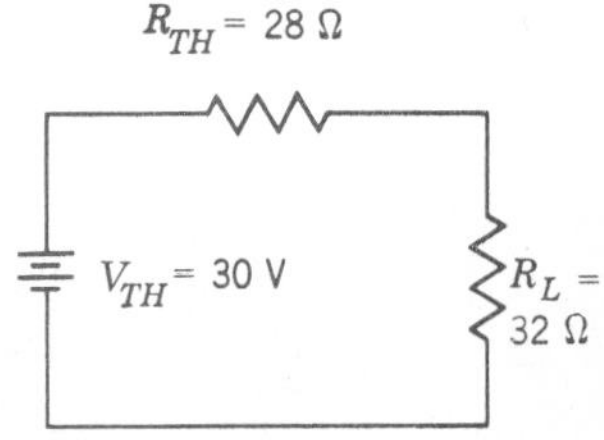


**FIGURE 9-56*a***
Thevenin equivalent circuit.

6. Solve for $I_L$.

$$R_t = \underline{\qquad} + \underline{\qquad} = 60 \ \Omega$$ 28 32

$$I_L = \frac{?}{R_t}$$ $V_{TH}$

$$I_L = \frac{?}{60} = 0.5 \text{ A} \qquad \textit{Ans.}$$ 30

Solve for $V_L$.

$V_L =$ ______ $\times$ ______

$V_L = 0.5 \times$ ______ $=$ ______ V *Ans.*

| $I_L$ | $R_L$ |
|---|---|
| 32 | 16 |

## PROBLEMS

Solve the following problems by applying Thevenin's theorem to the circuit.

1. In a circuit similar to that shown in Fig. 9-41*a*, $V_T = 25$ V, $R_1 = 1$ Ω, $R_2 = 4$ Ω, $R_3 = 5$ Ω, and $R_L = 10$ Ω. Find $I_L$ and $V_L$.
2. In a circuit similar to that shown in Fig. 9-22, $V_1 = 75$ V, $R_2 = 3$ Ω, $R_4 = 4$ Ω, and $V_5 = 28$ V. If the load $R_3 = 12$ Ω, find the currents *x*, *y*, and *z* and the load voltage $V_3$.
3. In a circuit similar to that shown in Fig. 9-48*a*, $V_T = 120$ V, $R_1 = 2$ Ω, $R_2 = 4$ Ω, $R_3 = 6$ Ω, and $R_L = 3.6$ Ω. Find $I_L$ and $V_L$.
4. In a circuit similar to that shown in Fig. 9-29, $V_1 = 76$ V, $R_2 = 10$ Ω, $R_3 = 2$ Ω, $R_4 = 15$ Ω, and $V_5 = 54$ V. Find the current *y* and the voltage across $R_3$.
5. In a circuit similar to that shown in Fig. 9-53, $R_1 = 30$ Ω, $R_2 = 70$ Ω, $R_3 = 10$ Ω, $R_4 = 15$ Ω, and $V_T = 100$ V. If $R_L = 23$ Ω, find $I_L$ and $V_L$.
6. Solve Example 9-14 by Thevenin's theorem.
7. Using Fig. 9-28, find $I_5$ and $V_5$ by Thevenin's theorem.
8. Using Fig. 9-25, find the current *w* and $V_2$ by Thevenin's theorem.
9. Solve the circuit shown in Fig. 9-57 for the values of $I_L$ and $V_L$.

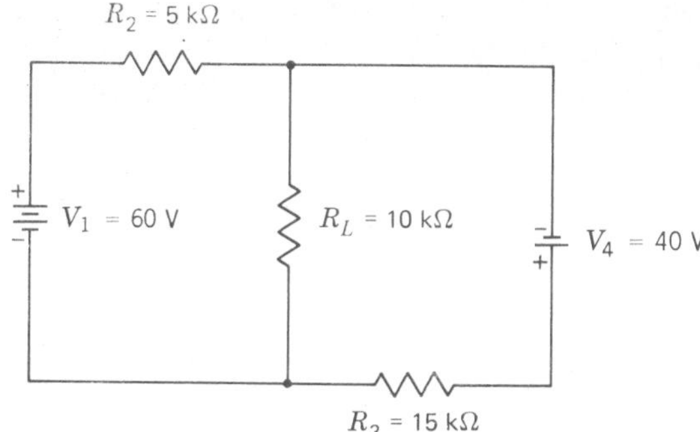


FIGURE 9-57

10. Solve the circuit shown in Fig. 9-58 for the values of $I_L$ and $V_L$.

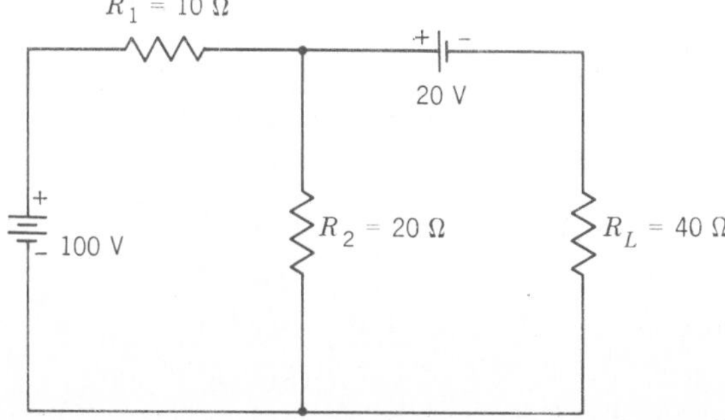


FIGURE 9-58

## JOB 9-9. NORTON'S THEOREM

Norton's theorem is another theorem like Thevenin's which is used to simplify circuit analysis. The two theorems give identical results, and the two equivalent circuits are very easily related. The difference is that Norton's theorem uses a *current* source instead of the voltage source used by Thevenin. This is more convenient for many of the newer semiconductor devices which are current-operated rather than voltage-operated.

The form of the Norton equivalent is shown in Fig. 9-59.

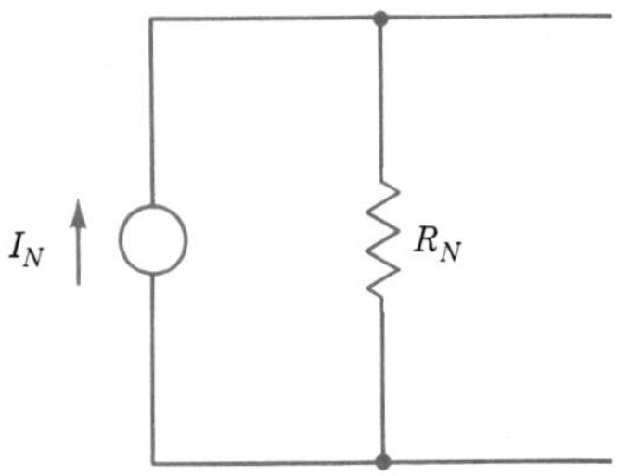


**FIGURE 9-59**
Norton equivalent circuit.

The $R_N$ of the Norton equivalent is the same as the $R_{TH}$ of the Thevenin equivalent. When making a Thevenin equivalent for a given pair of points in a network, the voltage at the two points is the $V_{TH}$. For a Norton equivalent, the two points are shorted together and the current through them is the Norton equivalent current $I_N$. The relationships between the parameters of the two equivalents is shown in Fig. 9-60.

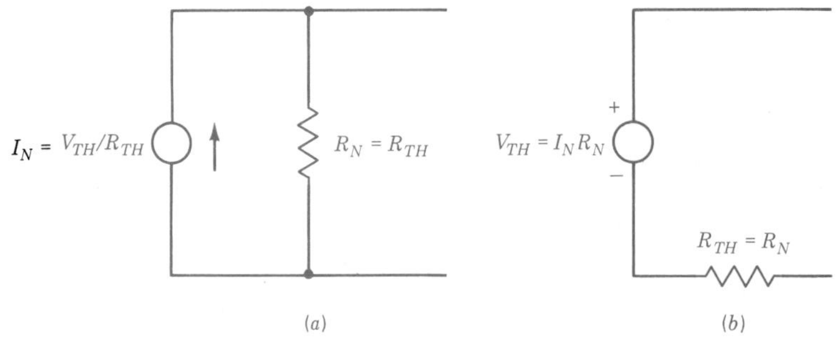


**FIGURE 9-60**
Relationship between the (a) Norton and (b) Thevenin equivalent circuits.

In order to emphasize the close relationship between the two concepts, Norton's theorem will first be stated in the manner in which Thevenin's theorem was stated in Job 9-8. The equivalence of the two forms will then be shown.

### NORTON'S THEOREM

Any linear network of voltage sources and resistances, if viewed from any two points in the network, can be replaced by an equivalent resistance $R_N$ in parallel with an equivalent current source $I_N$.

## PROCEDURE

1. Find the equivalent current source $I_N$.
2. Find the equivalent resistance $R_N$.
3. Place the $R_N$ in parallel with the load, and solve the simple parallel circuit.

EXAMPLE 9-28. Convert the Thevenin equivalent circuit shown in Fig. 9-43 to a Norton equivalent.

SOLUTION.

1. Disconnect the load $R_L$ from the circuit.
2. Short the open ends of the circuit together and determine the current flowing through the terminals.

$$I = \frac{V_{TH}}{R_{TH}} = I_N$$

3. Short the voltage sources and determine the resistance between the two open-circuit points.

$$R = R_{TH} = R_N$$

These equations are exactly the same as those shown in Fig. 9-60. The final circuit is shown on the left of that figure.

EXAMPLE 9-29. Solve the circuit in Example 9-24 by Norton's theorem.

SOLUTION. The original problem was redrawn as in Fig. 9-45, which is repeated here for your convenience.

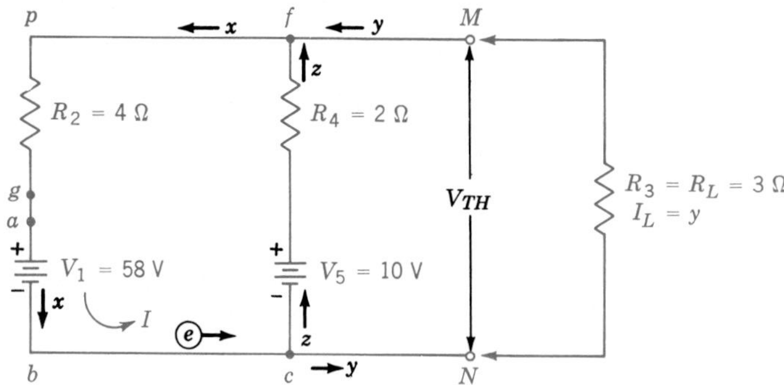


1. Find the current through terminals M and N when they are shorted. By Kirchoff's law, the Norton current $I_N$ will be the sum of the currents in $R_2$ and $R_4$.

$$\begin{aligned} I_N &= \frac{V_5}{R_4} + \frac{V_1}{R_2} \\ &= \frac{10}{2} + \frac{58}{4} \end{aligned}$$

$$= 5 + 14.5 = 19.5 \text{ A}$$

2. Find $R_N$. This value is determined exactly as is the $R_{TH}$ found in step 4 on page 288. $R_N = 1.33$
3. Draw the equivalent circuit as shown in Fig. 9-61.

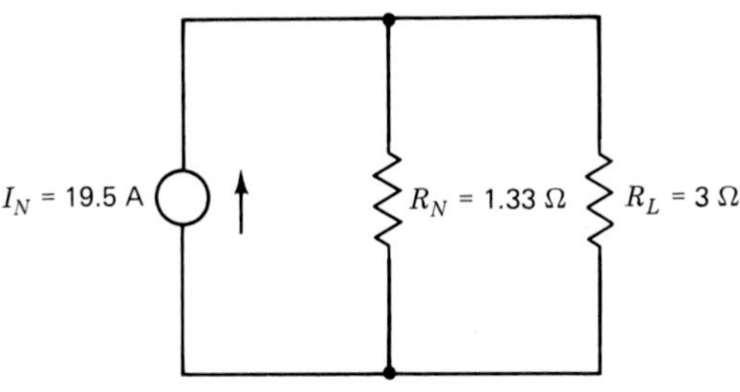


**FIGURE 9-61**
Norton equivalent circuit for Example 9-29.

4. Solve for $I_L$. Use the current divider Eqs. (5-6) and (5-7)

$$I_1 = \frac{R_2}{R_1 + R_2} \times I_T \tag{5-6}$$

$$I_L = \frac{R_N}{R_N + R_L} \times I_N$$

$$I_L = \frac{1.33}{1.33 + 3} \times 19.5$$

$$I_L = 6 \text{ A} \qquad \textit{Ans.}$$

This is exactly the same answer as that obtained with Thevenin's theorem, and it was obtained much more easily. The reason is that once the equivalent circuit shown in Fig. 9-45 was obtained, it was very easy to see what the Norton equivalent current would be. This may not be true for all circuits, which requires that we understand both theorems.

SELF-TEST 9-30. Solve the circuit of Example 9-25, using Norton's theorem. Use exactly the same configuration as shown in Fig. 9-48, which is repeated here for your convenience.

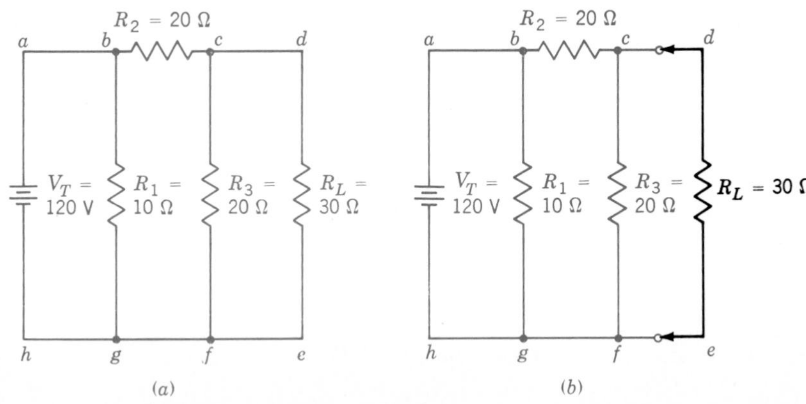

SOLUTION

1. Find the Norton equivalent current.
   *a.* Disconnect the load $R_L$ from the circuit.
   *b.* Short the open ends *c* and *f* together and determine the current flowing through the terminals. This will be the current through ______. $R_2$

$$I_N = \frac{V_T}{R_2} = \frac{120}{20} = 6 \text{ A}$$

2. Find $R_N$. This is exactly the same as $R_{TH}$, as shown in step 4 on page 289. $R_{TH} = 10\ \Omega$, and $R_N =$ ______ Ω. 10
3. Draw the equivalent circuit as shown in Fig. 9-62.

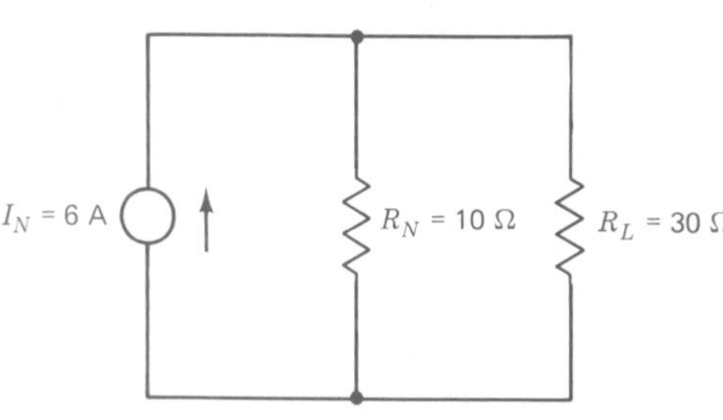


**FIGURE 9-62**
Norton equivalent circuit for Example 9-30.

4. Solve for $I_L$. Use the current divider equations.

$$I_L = \frac{?}{R_N + R_L} \times I_N$$ $R_N$

$$= \frac{10}{10 + 30} \times 6 = \underline{\qquad} \text{ A} \quad \textit{Ans.}$$ 1.5

**PROBLEMS**

Repeat the problems given in Job 9-8, using Norton's theorem. All the answers are the same. Compare the two methods and determine when one is better than the other.

## JOB 9-10 REVIEW

In Fig. 9-38, circuit *abc* is in (delta/Y) formation. delta

$$\Sigma R_\Delta = \underline{\qquad}$$ $R_1 + R_2 + R_3$

$$R_a = \frac{?}{\Sigma R_\Delta}$$ $R_1 \times R_2$

$$R_b = \frac{?}{\Sigma R_\Delta}$$ $R_2 \times R_3$

$$R_c = \frac{?}{\Sigma R_\Delta}$$ $R_1 \times R_3$

THEVENIN'S THEOREM

Any two-terminal network can be replaced by an ______ simple series circuit of one resistance and one voltage. equivalent

1. The equivalent voltage is the terminal voltage of the complex circuit when the ______ is removed. Its symbol is ______. — load $V_{TH}$

2. The equivalent resistance is the resistance between the terminals of the complex circuit when the load is ______ and all sources of emf are ______. Its symbol is ______. — removed shorted $R_{TH}$

## PROCEDURE

1. Determine the load whose current $I_L$ and whose voltage (______) are to be found. To simplify the problem, a voltage source (may/may not) be included in the load. — $V_L$ may

2. Disconnect the ______ from the circuit at two points, and label these points as Thevenin terminals. — load

3. Find the open-circuit output voltage that would appear across these terminals. The symbol for this voltage is ______. — $V_{TH}$

4. Replace each voltage source with a ______ and ______ the load. — short remove

5. Redraw the remaining circuit in standard form.

6. Find the total resistance that would appear across the Thevenin terminals. This is the symbol ______. — $R_{TH}$

7. Draw the equivalent circuit consisting of $V_{TH}$ in ______ with $R_{TH}$ and the ______. — series load

8. If the "load" has included any voltage source, it must be (included/excluded) in the equivalent Thevenin circuit with its original polarity. — included

9. Solve the circuit for the load ______ and the load ______. The total resistance of this circuit is the sum of $R_{TH}$ and the ______ resistance. — current voltage load

## NORTON'S THEOREM

1. The Norton equivalent current is the current through the circuit terminals when the output is ______. — shorted

2. The symbol for the Norton equivalent current is ______. — $I_N$

3. The symbol for the Norton equivalent resistance is ______. — $R_N$

4. The Norton equivalent resistance is calculated in exactly the same manner as the ______. — $R_{TH}$

5. The Norton equivalent is made by putting the $R_N$ in ______ with ______. — parallel $R_L$

6. The $V_{TH}$ and the $I_N$ are related by the equation ______. — $V_{TH} = I_N \times R_N$

(See Instructor's Manual for Test 9-3, Delta Circuits and Thevenin and Norton Theorems)

# APPLICATIONS OF SERIES AND PARALLEL CIRCUITS

## JOB 10-1 EXTENDING THE RANGE OF AN AMMETER

An ammeter is a device which is used to measure the current in an electric circuit. To do this, it must always be inserted directly into the line whose current is to be measured. It is essentially a coil of very fine wire which turns in proportion to the current flowing through it. A pointer attached to the coil moves over a dial and indicates the value of the current. The weight of the coil must be very small in order that it be able to react to the current in it. This weight factor necessitates the use of very fine wire which can carry only about 0.05 A at best. Yet we can use this ammeter to measure much larger currents by taking advantage of the fact that current in a parallel circuit will divide—the large current flowing through the small resistance and the small current flowing through the large resistance.

**Range of an ammeter.** The *range* of an ammeter indicates the value of current required to cause the pointer to swing over the entire scale. This is called "full-scale deflection." Ranges are indicated as "0–1 mA," "0–100 mA," etc. Thus:

0–1 mA requires 1 mA for full-scale deflection
0–100 mA requires 100 mA for full-scale deflection
0–10 A requires 10 A for full-scale deflection

**Extending the range of an ammeter.** Suppose we had a milliammeter whose range was 0–1 mA. It is desired to extend its range to read up to 50 mA. This can be accomplished by connecting a low-resistance resistor called a *shunt* in parallel with the meter as shown in Fig. 10-1. The object is to use a shunt of such a value that the current $I$ in the line will divide at point $A$ so as to keep the current flowing through the coil unchanged. Thus, using a 0–1-mA milliammeter, if the range is to be extended to read 0–50 mA, the shunt should carry 49 mA and the coil should carry its original 1 mA. If the range is to be extended to read 100 mA, then the shunt should carry 99 mA and the coil of the meter should

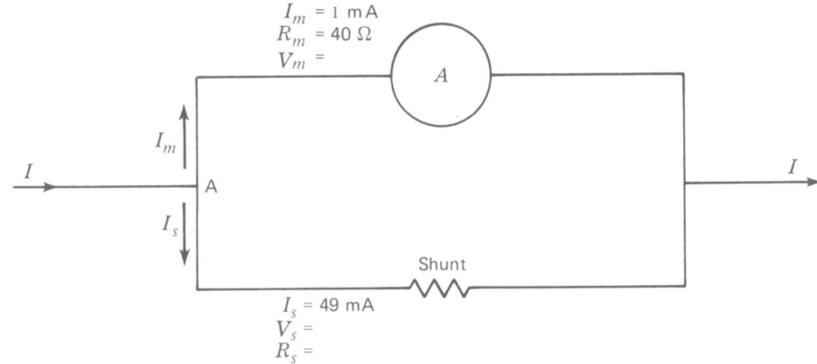


**FIGURE 10-1**
A shunt is placed in parallel with an ammeter. $I$ = line current to be measured; $V_m$ = voltage drop across meter; $I_m$ = current through meter; $R_m$ = resistance of meter; $V_s$ = voltage across shunt; $I_s$ = current through shunt; $R_s$ = resistance of shunt.

carry its original 1 mA. In general, with *any* shunt, the current through the coil for full-scale deflection is the current necessary for full-scale deflection *before* using the shunt.

In Fig. 10-1, to extend the meter to 50 mA, we require a shunt $R_s$ of such a value that the original 1 mA will flow through the meter and the remaining 49 mA will flow through the shunt. In effect, the pointer will cover the full scale (owing to the 1 mA), but it will be indicating the full 50 mA.

$$V_m = I_m \times R_m = 0.001 \times 40 = 0.04 \text{ V}$$

but since the shunt is in parallel with the meter,

$$V_s = V_m = 0.04 \text{ V}$$

$$R_s = \frac{V_s}{I_s} = \frac{0.04}{0.049} = 0.816\ \Omega \qquad \textit{Ans.}$$

These calculations may be combined to produce the

**FORMULA** $$R_s = \frac{I_m \times R_m}{I_s} \tag{10-1}$$

And since $$I = I_m + I_s \tag{5-2}$$

by transposing, $$I_s = I - I_m \tag{10-2}$$

By substituting this value of $I_s$ in formula (10-1), we obtain the

**FORMULA** $$R_s = \frac{I_m \times R_m}{I - I_m} \tag{10-3}$$

In this formula, the resistance of the meter must be known, and it can usually be obtained from the manufacturer.

EXAMPLE 10-1. A Weston model 430 0–1-mA milliammeter has a resistance of 92 Ω. What shunt is necessary to extend its range up to 50 mA?

SOLUTION. Given: $I_m = 1\text{ mA} = 0.001\text{ A}$ Find: $R_s = ?$
$R_m = 92\ \Omega$
$I = 50\text{ mA} = 0.05\text{ A}$

$$R_s = \frac{I_m \times R_m}{I - I_m} = \frac{0.001 \times 92}{0.05 - 0.001} = \frac{0.092}{0.049} = 1.88\ \Omega \qquad \textit{Ans.}$$

A simpler formula to extend the range of ammeters may be found by a series of algebraic manipulations on formula (10-3).

$$R_s = \frac{I_m \times R_m}{I - I_m} \tag{10-3}$$

or

$$R_s = R_m \frac{I_m}{I - I_m}$$

Dividing by 1 expressed as $I_m/I_m$,

$$R_s = R_m \frac{I_m \div I_m}{(I - I_m) \div I_m}$$

$$R_s = R_m \frac{1}{\left(\dfrac{I}{I_m} - 1\right)} \tag{a}$$

The fraction $I/I_m$ is the ratio of the maximum *new* range as compared with the maximum *old* range. As such, it represents the amount by which the range of the meter has been multiplied. If $N$ is used to represent this multiplying factor,

$$N = \frac{I}{I_m}$$

**FORMULA** $$N = \frac{\text{max new range}}{\text{max old range}} \tag{10-4}$$

By substituting $N$ for $I/I_m$ in equation ($a$) above, we obtain

**FORMULA** $$R_s = \frac{R_m}{N - 1} \tag{10-5}$$

where $R_s$ = shunt resistor
$R_m$ = resistance of the meter
$N$ = multiplying factor

SELF-TEST 10-2. A Weston model 433 0–100-mA milliammeter has a resistance of 49 Ω. Find the shunt resistor necessary to extend the range to give maximum deflection at 750 mA.

SOLUTION. Given: Old range = ________ mA Find: $R_s = ?$ | 100

New range = ______ mA | 750

$R_m$ = ______ Ω | 49

$$\text{Multiplying factor } N = \frac{\text{max ? range}}{\text{max ? ránge}}$$ | new / old

$$N = \frac{750}{?} = 7.5 \qquad (10\text{-}4)$$ | 100

$$R_s = \frac{R_m}{N-1} \qquad (10\text{-}5)$$

$$R_s = \frac{?}{7.5 - 1}$$ | 49

$R_s$ = ______ Ω *Ans.* | 7.5

## PROBLEMS

1. A D'Arsonval meter movement has a resistance of 50 Ω and requires 1 mA of current for full-scale deflection. Find the shunt resistor needed to convert this movement to a milliammeter with a range of 1–10 mA.
2. A 0–1-mA milliammeter has a resistance of 100 Ω. If the meter is to be converted to read 100 mA, which of the following shunts should be used: (*a*) 0.1 Ω, (*b*) 10 Ω, (*c*) 1.01 Ω, (*d*) 1010 Ω?
3. Find the shunt necessary to extend the range of a 0–1-mA milliammeter whose resistance is 27 Ω to read 0–10 mA.
4. A 0–1-mA milliammeter has a resistance of 5 Ω. What shunt resistor is necessary to give a full-scale reading of 51 mA?
5. Find the shunt needed for the milliammeter of Prob. 4 which will permit full-scale deflection at 0.1 A.
6. A Weston model 45, 0–100-mA milliammeter has a resistance of 0.5 Ω. Find the shunt needed to extend its range to read 0–300 mA.
7. A Weston model 1, 0–15-mA milliammeter has a resistance of 2.24 Ω. Find the shunt needed to extend its range to read 0–60 mA.
8. Full-scale deflection results when 25 mA flows through an ammeter of 25 Ω resistance. Find the shunt resistance needed to extend its range to read (*a*) 250 mA, (*b*) 500 mA, and (*c*) 1 A.
9. A Weston model 45 0–3-A ammeter has a resistance of 5 Ω. Find the shunt needed to extend its range to read 0–75 A.
10. When the coil current in a meter is 10 mA, the pointer deflects past a full scale of 100 divisions. If the coil resistance is 15 Ω, find the shunts needed to extend its range to read (*a*) 0–1 A and (*b*) 0–5 A.

## JOB 10-2 READING AMMETERS OF EXTENDED RANGE

EXAMPLE 10-3. The range of a 0–100-mA milliammeter was increased to read 0 to 1 A. Find the true current flowing when the meter reads 60 mA.

SOLUTION. Given: Old range = 100 mA  Find: $I$ = ?

New range = 1 A = 1000 mA

$I_m$ = 60 mA

The increase in range was from 100 to 1000 mA.

$$N = \frac{\text{max new range}}{\text{max old range}} = \frac{1000}{100} = 10 \qquad (10\text{-}4)$$

Therefore all readings will be increased 10 times, or a reading of 60 mA really means

$$60 \times 10 = 600 \text{ mA} \quad \text{or} \quad 0.6 \text{ A} \quad \textit{Ans.}$$

EXAMPLE 10-4. A model 221-T Triplett 0–1-mA milliammeter has a resistance of 55 Ω. When it is shunted with a 2-Ω resistor, the meter reads 0.5 mA. Find the true current.

SOLUTION. Given: $I_m$ = 0.5 mA  Find: $I$ = ?

$R_m$ = 55 Ω

$R_s$ = 2 Ω

In this problem we do not know the increase in range and cannot use the method shown in the last example. However, we can find the shunt current by formula (10-1). The line current $I$ may then be found by formula (5-2).

$$R_s = \frac{I_m \times R_m}{I_s} \qquad (10\text{-}1)$$

$$\frac{2}{1} = \frac{0.5 \times 55}{I_s}$$

$$2 \times I_s = 0.5 \times 55$$

$$I_s = \frac{27.5}{2} = 13.8 \text{ mA}$$

The line current may now be found.

$$I = I_m + I_s = 0.5 + 13.8 = 14.3 \text{ mA} \quad \textit{Ans.} \qquad (5\text{-}2)$$

*Note:* Since the shunt usually carries most of the line current, the shunt current is very often used as the line current.

**PROBLEMS**

1. The range of a 0–1-mA milliammeter was increased to read 0–50 mA. Find the true current when the meter reads 0.3 mA.

2. The range of a 0–0.1-A ammeter was increased to read 0–1 A. Find the true current when the meter reads 0.07 A.
3. A 0–1-mA milliammeter of 25 Ω resistance is used with a 0.25-Ω shunt. What is the true current when the meter reads 0.4 mA?
4. A Weston model 45 0–3-A ammeter of 5 Ω resistance is used with a 0.1-Ω shunt. What is the true current when the meter reads 0.6 A?
5. A 0–1-A ammeter of 10 Ω resistance is used with a 0.1-Ω shunt. What is the true current when the meter reads 0.7 A?
6. A 0–10-mA milliammeter of 10 Ω resistance is used with a 0.2-Ω shunt. What is the true current when the meter reads 2 mA?

## JOB 10-3 EXTENDING THE RANGE OF A VOLTMETER

A voltmeter is a device which is used to measure the difference in electrical pressure across two points in a circuit. To do this, the instrument must be placed in parallel with the portion of the circuit being tested as shown in Fig. 10-2. If the resistance of the voltmeter is low, the current

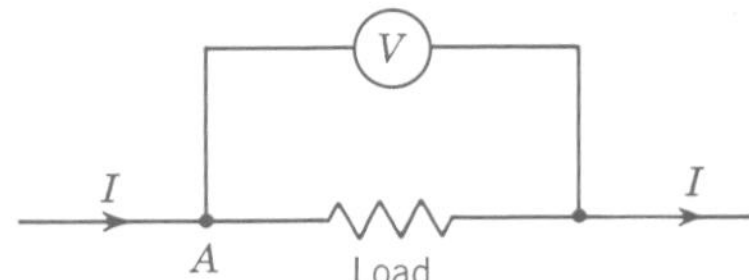


**FIGURE 10-2**
A voltmeter is always connected in parallel across the terminals of the part being tested.

in the line will divide at $A$ and the large current will flow through the moving coil of the meter and burn it out. To avoid this, the resistance of the moving coil is increased by adding a *high* resistance in *series* with the coil. These resistors are called *multipliers.* Essentially, a voltmeter is really an ammeter with a series resistor. This extra resistance holds down the current to the value required for full-scale deflection when the full-scale voltage is applied to it. It is possible to use one meter to read many different voltage ranges if the correct resistances can be placed in the instrument with an arrangement for disconnecting one resistance and connecting another in series. Commercial meters are available which do this by a variety of switching arrangements.

**Ohms per volt.** A perfect meter should measure the current but should not use any of the current in the circuit for itself. This is actually impossible, since some current, however little, is needed to operate the meter. The best, or most sensitive, meters are those which use as little current as possible. The more resistance that can be placed in a meter and still have it read 1 V, the less the current that is drawn. This means that the meter will use very little current for itself and will have a high sensitivity. Sensitivity is usually expressed as the number of ohms needed to read 1 V, or the ohms per volt $R/v$.

**FORMULA**

$$R/v = \frac{R_T}{V_T} \qquad \boxed{10\text{-}6}$$

where $R/v$ = sensitivity, Ω/V
$R_T$ = total meter resistance, Ω
$V_T$ = maximum scale voltage, V

EXAMPLE 10-5. A 50,000-Ω voltmeter reads 50 V at full scale. What is its sensitivity in ohms per volt?

SOLUTION. Given: $R_T = 50{,}000\ \Omega$ Find: $R/v = ?$
$V_T = 50\ \text{V}$

$$R/v = \frac{R_T}{V_T} = \frac{50{,}000}{50} = 1000\ \Omega/\text{V} \quad \textit{Ans.} \qquad (10\text{-}6)$$

EXAMPLE 10-6. A 1000-Ω/V voltmeter has a range of 0–10 V. What is the total resistance of the meter?

SOLUTION. Given: $R/v = 1000\ \Omega/\text{V}$ Find: $R_T = ?$
$V_T = 10\ \text{V}$

$$R/v = \frac{R_T}{V_T} \qquad (10\text{-}6)$$

$$1000 = \frac{R_T}{10}$$

$$R_T = 1000 \times 10$$

$$R_T = 10{,}000\ \Omega \quad \textit{Ans.}$$

**Extending the range of a voltmeter.** A 0–150-V voltmeter has a 150,000-Ω resistor in series with its moving coil. How can it be altered so as to read up to 750 V?

SOLUTION. Since 750 V is five times as much as 150 V, we shall require five times as much resistance as is in the meter for the 150-V range. The resistance already in the meter may then be subtracted from the total required resistance to find the needed *additional* resistance to increase the range to 750 V. The increase in voltage (the five times) represents the multiplying power we desire. Therefore, the series resistor $R_s$ which must be added to extend the range of an existing voltmeter is given by

$$R_s = (\text{multiplying power} \times R_{\text{meter}}) - R_{\text{meter}}$$

Another way to write this is the

**FORMULA**

$$R_s = (MP - 1) \times R_{\text{meter}} \qquad \boxed{10\text{-}7}$$

where $R_s$ = series resistor to be added, Ω
$R_{meter}$ = resistance of meter, Ω

and $$MP = \frac{\text{max new voltage}}{\text{max old voltage}} \quad \boxed{10\text{-}8}$$

Let us use this formula to solve the problem above.

$$MP = \frac{\text{max new voltage}}{\text{max old voltage}} = \frac{750}{150} = 5 \qquad (10\text{-}8)$$

$$R_s = (MP - 1) \times R_m \qquad (10\text{-}7)$$

$$= (5 - 1) \times 150{,}000 = 4 \times 150{,}000 = 600{,}000\ \Omega \qquad \textit{Ans.}$$

*Note:* The resistance of the meter must be known in order to use formula (10-7). If it is not given, it may be found by connecting the meter in series with a milliammeter across some voltage within its range as shown in Fig. 10-3. The resistance of the meter may then be found by the

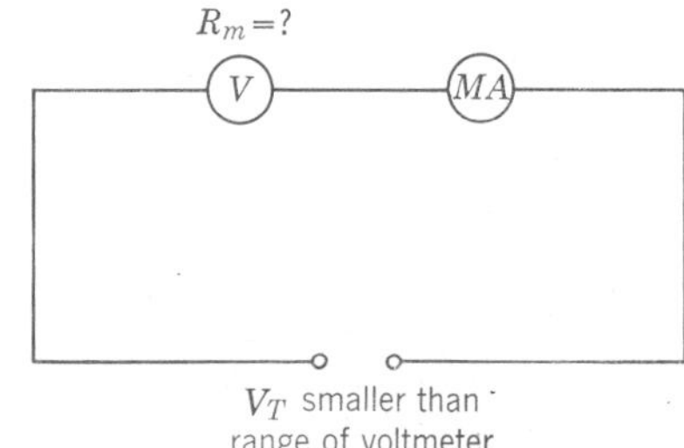


**FIGURE 10-3**
Circuit for finding the resistance of a voltmeter.

**FORMULA** $$R_m = \frac{\text{voltmeter reading (V)}}{\text{ammeter reading (A)}} \quad \boxed{10\text{-}9}$$

EXAMPLE 10-7. A 0–10-V 10,000-Ω voltmeter is to be extended to read 100 V. What additional series resistance is needed?

SOLUTION. Given: Max new voltage = 100 V Find: $R_s$ = ?
Max old voltage = 10 V
$R_{meter}$ = 10,000 Ω

$$MP = \frac{\text{max new voltage}}{\text{max old voltage}} = \frac{100}{10} = 10 \qquad (10\text{-}8)$$

$$R_s = (MP - 1) \times R_m \qquad (10\text{-}7)$$

$$= (10 - 1) \times 10{,}000 = 9 \times 10{,}000 = 90{,}000\ \Omega \qquad \textit{Ans.}$$

EXAMPLE 10-8. A 1000-Ω/V 0–10-V voltmeter is to be extended to read 0–100 V. What is the needed series resistance?

SOLUTION. Given: $R/v = 1000\ \Omega$ Find: $R_s = ?$
Max new voltage = 100 V
Max old voltage = 10 V

Before we can find the series resistance, we must know the resistance of the meter and the multiplying power.

$$R/v = \frac{R_T}{V_T} \qquad (10\text{-}6)$$

$$1000 = \frac{R_T}{10}$$

$$R_T = 10 \times 1000 = 10{,}000\ \Omega$$

$$R_m = 10{,}000\ \Omega \qquad \textit{Ans.}$$

$$MP = \frac{\text{max new voltage}}{\text{max old voltage}} = \frac{100}{10} = 10 \qquad \textit{Ans.} \qquad (10\text{-}8)$$

$$R_s = (MP - 1) \times R_m \qquad (10\text{-}7)$$

$$= (10 - 1) \times 10{,}000 = 9 \times 10{,}000 = 90{,}000\ \Omega \qquad \textit{Ans.}$$

SELF-TEST 10-9. A 0–10-V voltmeter of 1000 Ω resistance is extended to read 100 V. Find the true voltage when the meter reads 5.5 V.

SOLUTION. Given: $R_m$ = ________ Ω Find: True voltage = ? | 1000
Max new voltage = ________ V | 100
Max old voltage = ________ V | 10
Meter reading = 5.5 V

$$MP = \frac{\text{max new voltage}}{\text{max old voltage}} = \frac{?}{10} = 10$$ | 100

Therefore, all readings will be increased 10 times, so a reading of 5.5 V means

5.5 × ________ = 55 V *Ans.* | 10

## PROBLEMS

1. A 1000-Ω/V 0–10-V voltmeter is to be extended to read 300 V. Find (*a*) the series resistor needed and (*b*) the true voltage when the meter reads 5.6 V.
2. A 0–7.5-mV millivoltmeter whose resistance is 1 Ω is to be extended to read 15 V. Find (*a*) the series resistor needed and (*b*) the true voltage when the meter reads 6 mV.
3. A 0–50-mV millivoltmeter of 5 Ω resistance is to be extended to read up to 30 V. Find (*a*) the series resistor needed and (*b*) the true voltage when the meter reads 36 mV.
4. A 0–100-V voltmeter of 12,000 Ω resistance is to read up to 300 V. Find the multiplying resistor needed.
5. A voltmeter with a full-scale deflection of 150 V has a resistance of

17,000 Ω. Find the external series resistance necessary to extend it to read (*a*) 300 V and (*b*) 600 V.

6. A 1000-Ω/V 0–1-V voltmeter is to be extended to read up to 100 V. Find (*a*) the series resistor needed and (*b*) the true voltage when the meter reads 0.7 V.
7. If the current taken by a 5000-Ω meter is 1 mA, find the series resistor needed to extend its range to read up to (*a*) 50 V, (*b*) 100 V, and (*c*) 150 V. *Hint:* Find $V$ of the meter by $V = I \times R$.
8. Find the series resistor to be used with a 15,000-Ω voltmeter in order to have its readings multiplied by 5.

## JOB 10-4 REVIEW OF METERS

### EXTENDING THE RANGE OF AN AMMETER

Ammeters may be used to measure currents larger than their full-scale deflections by connecting appropriate shunts in parallel with the meter. The value of this shunt resistor may be found by the following formulas:

$$R_s = \frac{I_m \times R_m}{I_s} \qquad \boxed{10\text{-}1}$$

$$R_s = \frac{I_m \times R_m}{I - I_m} \qquad \boxed{10\text{-}3}$$

$$R_s = \frac{R_m}{N - 1} \qquad \boxed{10\text{-}5}$$

where $R_s$ = resistance of the ________, Ω — shunt
$I_s$ = current in the ________, A — shunt
$R_m$ = resistance of the ________, Ω — meter
$I_m$ = current in the ________, A — meter
$I$ = line current to be measured, A
$N$ = multiplying factor which indicates the number of times that the range of the meter is increased

$$N = \frac{\text{max \_\_\_\_\_\_\_\_ range}}{\text{max \_\_\_\_\_\_\_\_ range}} \qquad \boxed{10\text{-}4}$$

new
old

### Reading ammeters of extended range

1. If the multiplying factor $N$ is known, the true current is equal to the actual reading multiplied by ________. — $N$

**True reading = actual reading × $N$**

2. If $N$ is unknown,
   *a.* Find the shunt current by

$$R_s = \frac{I_m \times R_m}{?} \qquad \boxed{10\text{-}1}$$

$I_s$

   *b.* Find the true current by

$$I = I_m + ? \qquad \boxed{5\text{-}2}$$

$I_s$

## EXTENDING THE RANGE OF A VOLTMETER

The range of a voltmeter may be extended by placing it in series with the appropriate series multiplier resistor.

$$R_s = (MP - 1) \times \_\_\_\_\_\_ \qquad \boxed{10\text{-}7}$$

$R_{meter}$

where $R_s$ = series resistor to be added, Ω
$R_{meter}$ = resistance of meter, Ω

$$MP = \frac{\text{max } \_\_\_\_\_\_ \text{ voltage}}{\text{max } \_\_\_\_\_\_ \text{ voltage}} \qquad \boxed{10\text{-}8}$$

new
old

**Meter sensitivity.** The sensitivity of a meter is expressed as its resistance in ohms for every volt measured on its scale. For the same reading of 1 V, the meter with the larger resistance will draw __(more/less)__ current for itself and therefore give more accurate measurements.

less

$$R/v = \frac{R_T}{V_T} \qquad \boxed{10\text{-}6}$$

where $R/v$ = ______, Ω/V
$R_T$ = total meter resistance, Ω
$V_T$ = maximum scale voltage, V

sensitivity

### Finding the total resistance of the meter

1. Use formula (10-6) to find $R_T$ of the meter, or
2. Connect the voltmeter in series with an ammeter across some voltage within its range. The meter resistance is given by

$$R_m = \frac{\text{voltmeter reading (V)}}{\text{ammeter reading (A)}} \qquad \boxed{10\text{-}9}$$

## PROBLEMS

1. A 0–2-mA milliammeter has a resistance of 18 Ω. Find the parallel shunt resistor which will permit measurement of currents up to 200 mA.
2. The range of a 12-Ω 0–5-mA milliammeter was increased to read 0–50 mA. Find the true current when the meter reads 4.5 mA.
3. A 0–1-mA milliammeter of 20 Ω resistance is used with a 0.2-Ω shunt in parallel. Find the true current when the meter reads 0.6 mA.
4. A 10,000-Ω voltmeter reads 100 V at full-scale deflection. Find its sensitivity in ohms per volt.
5. A D'Arsonval meter movement with a resistance of 250 Ω requires 3 mA for full-scale deflection. Find its sensitivity in ohms per volt.
6. The range of a 15-V 1000-Ω/V voltmeter is to be extended to read 150 V. What value of resistance should be added?
7. Which of the following represents the full-scale sensitivity of a 20,000-Ω/V meter: (*a*) 20 μA, (*b*) 50 μA, (*c*) 2 mA, (*d*) 5 mA?

8. A 500-Ω meter movement has a full-scale current rating of $400 \times 10^{-6}$ A. If this movement is used in a 200-V voltmeter with a multiplier, find the ohms per volt rating of the voltmeter.
9. A Weston model 5 dc voltmeter has a sensitivity of 100 Ω/V and a range of 0–15 V. Find the total resistance of the meter.
10. A 0–30-V voltmeter is to be extended to read 150 V. If the resistance of the meter is 3000 Ω, find (*a*) the sensitivity, (*b*) the series resistor needed, and (*c*) the true voltage when the meter reads 12 V.
11. A 100-Ω/V 0–7.5-V voltmeter is to be extended to read 0–30 V. Find (*a*) the series multiplier needed and (*b*) the true voltage when the meter reads 3.2 V.
12. If the current taken by a 1000-Ω meter is 15 mA, find (*a*) the voltage indicated by the full-scale deflection, (*b*) the sensitivity in ohms per volt, and (*c*) the series multiplier resistor needed to extend the range to read (1) 120 V and (2) 180 V.
13. The resistance of a 0–15-mA milliammeter is 3.2 Ω. Find the parallel shunt required to extend the range to 0–100 mA.
14. The range of an 18-Ω 0–3-mA milliammeter was increased to read up to 50 mA. Find the true current when the meter reads 1.7 mA.

(See Instructor's Manual for Test 10-1, Meters)

## JOB 10-5 VOLTAGE DIVIDERS

The filter circuit of a power supply is designed to smooth out the dc voltage delivered from the rectifier. This voltage may be 30 V. However, although some circuits in the receiver may require these 30 V, other circuits may need only 18 or only 9 V. These different voltages may be obtained by placing a large resistor across the filter output as shown in Fig. 10-4. This resistor, known as a *bleeder* resistor, serves as a

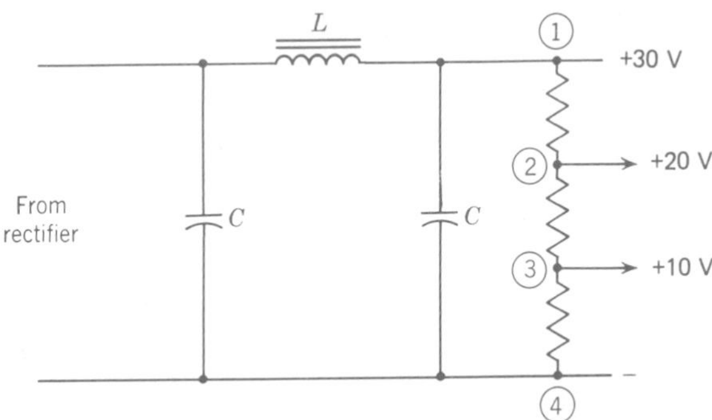


**FIGURE 10-4**
Power-supply voltage divider.

fixed load on the power supply. Also, any charge remaining on the filter capacitors when the power supply is turned off bleeds through this resistor and eliminates the danger of shock. Some current (about 11 percent of the total current) will always flow through the bleeder as long as the rectifier is operating. The different voltages required may be obtained by tapping the bleeder at various points, which changes the bleeder into a *voltage divider.*

The total voltage of 30 V appears across points 1 and 4. If the resistor is divided into three equal parts by taps at points 2 and 3, at point 2, which is one-third of the resistor, one-third of the total voltage ($^1/_3 \times 30 = 10$) has been used up. Therefore only

$$30 - 10 = 20 \text{ V}$$

will be available at point 2. At point 3, which is two-thirds of the resistor, two-thirds of the 30 V ($^2/_3 \times 30 = 20$) has been used up, and at point 3 only $30 - 20 = 10$ V will be available. This will all be true only if the current in all parts of the voltage divider is the same current. However, the different voltages must be available at different currents, and so the calculations for a working voltage divider are a bit more complicated.

EXAMPLE 10-10. Calculate the resistances for the sections of a voltage divider which is to deliver 40 mA at 300 V, 30 mA at 180 V, and 20 mA at 90 V.

SOLUTION. The diagram for the circuit is shown in Fig. 10-5.

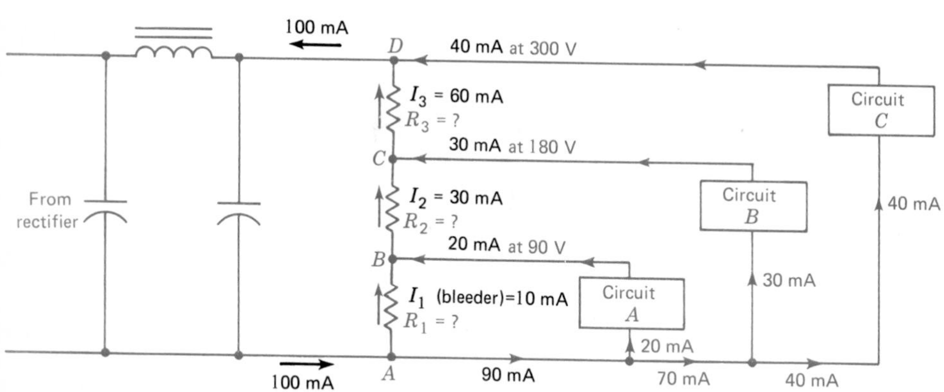


FIGURE 10-5

1. Determine the current distribution. The total current required equals $20 + 30 + 40 = 90$ mA. The bleeder current = 11 percent of 90 mA = $0.11 \times 90 = 10$ mA (approximately).

At *A*:
$$I_1 = I \text{ from rectifier} - \text{current to loads}$$
$$I_1 = 100 - 90 = 10 \text{ mA}$$

At *B*:
$$I_2 = I_1 + \text{current from load}$$
$$I_2 = 10 + 20 = 30 \text{ mA}$$

At *C*:
$$I_3 = I_2 + \text{current from load}$$
$$I_3 = 30 + 30 = 60 \text{ mA}$$

At *D*:
$$I_T = I_3 + \text{current from load}$$
$$I_T = 60 + 40 = 100 \text{ mA}$$

2. Find the sectional resistances.

For $R_3$: The voltage drop from $D$ to $C$ = 300 − 180 = 120 V.

$$V_3 = I_3 \times R_3 \qquad (4\text{-}6)$$

$$120 = 0.06 \times R_3$$

$$R_3 = \frac{120}{0.06} = 2000\ \Omega \qquad \textit{Ans.}$$

For $R_2$: The voltage drop from $C$ to $B$ = 180 − 90 = 90 V.

$$V_2 = I_2 \times R_2 \qquad (4\text{-}5)$$

$$90 = 0.03 \times R_2$$

$$R_2 = \frac{90}{0.03} = 3000\ \Omega \qquad \textit{Ans.}$$

For $R_1$: The voltage drop from $B$ to A = 90 − 0 = 90 V.

$$V_1 = I_1 \times R_1 \qquad (4\text{-}4)$$

$$90 = 0.01 \times R_1$$

$$R_1 = \frac{90}{0.01} = 9000\ \Omega \qquad \textit{Ans.}$$

SELF-TEST 10-11. Find the resistances of the sections $R_1$, $R_2$, $R_3$, and $R_4$ of the voltage divider shown in Fig. 10-6. The −50-V bias terminal draws no current, and the bleeder current is 10 percent of the total load current.

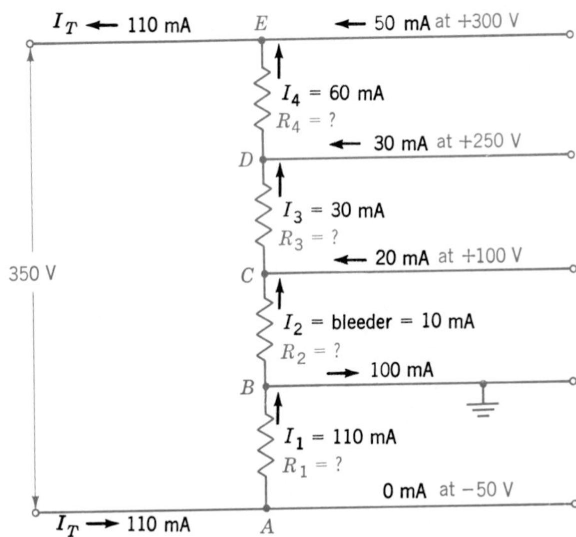


**FIGURE 10-6**
Voltage-divider circuit for Self-test 10-11.

SOLUTION

1. Determine the current distribution.

Total load current = 50 + 30 + ______ mA 20

| | | |
|---|---|---|
| | $I_L =$ ______ mA | 100 |
| | Bleeder current = 10 percent × 100 | |
| | = ______ × 100 | 0.1 |
| | $I_b =$ ______ mA | 10 |
| | Total current = bleeder current + load current | |
| At $B$: | $I_T = I_1 = I_b$ + load current | |
| | $I_T = 10 +$ ______ | 100 |
| | $I_T = 110$ mA = ______ A | 0.11 |
| At $C$: | $I_3 = I_2$ + current from load | |
| | = 10 + ______ | 20 |
| | $I_3 =$ ______ mA | 30 |
| At $D$: | $I_4 =$ ______ + current from load | $I_3$ |
| | = 30 + ______ | 30 |
| | $I_4 =$ ______ mA | 60 |
| At $E$: | $I_T =$ ______ + current from load | $I_4$ |
| | = 60 + ______ | 50 |
| | $I_T =$ ______ mA = ______ A | 110 0.11 |

2. Find the sectional resistances.

For $R_1$: The voltage drop between $A$ and $B$ = 50 − 0.

$$V_1 = 50 \text{ V}$$

$$R_1 = \frac{50}{?}$$ 0.11

$R_1 =$ ______ Ω *Ans.* 455

For $R_2$: The voltage drop between $B$ and $C$ = ______ V. 100

$$R_2 = \frac{100}{?}$$ 0.01

$R_2 =$ ______ Ω *Ans.* 10,000

For $R_3$: The voltage drop between ______ and ______ $C$ $D$

= 250 − ______ 100

= ______ V 150

$$R_3 = \frac{?}{0.03}$$ 150

$R_3 =$ ______ Ω *Ans.* 5000

For $R_4$: The voltage drop between $D$ and $E$ = ______ V. 50

$$R_4 = \frac{?}{0.06}$$ 50

$R_4 =$ ______ Ω *Ans.* 833

**PROBLEMS**

1. Find the resistances of sections $R_1$, $R_2$, and $R_3$ of the voltage divider shown in Fig. 10-7*a*.
2. Find the resistances of sections $R_1$, $R_2$, $R_3$, and $R_4$ of the voltage

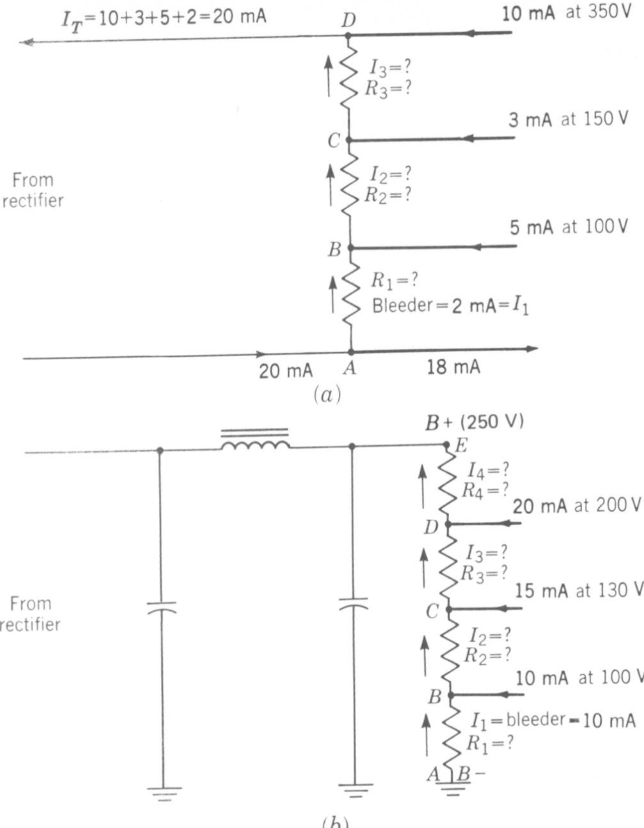


FIGURE 10-7

divider shown in Fig. 10-7*b*. *Hint:* Find the current distribution starting from point *A* and working up to point *E*.

3. A power supply is to deliver 45 mA at 150 V, 30 mA at 100 V, and 10 mA at 50 V. Assuming a bleeder current of 5 mA, calculate the resistances of the sections of the voltage divider needed.
4. Design a voltage divider to deliver 70 mA at 350 V, 40 mA at 250 V, and 20 mA at 100 V if the power-supply transformer is rated at 155 mA and operates at 90 percent of its rated value.

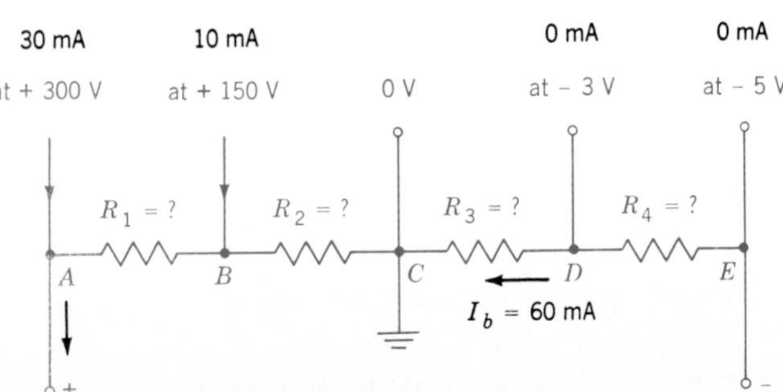


FIGURE 10-8

5. A power supply is to deliver 60 mA at 250 V, 5 mA at 100 V, and 2 mA at 50 V. The transformer is rated at 80 mA and operates at 90 percent of its rated value. Design a voltage divider to deliver the required currents and voltages.
6. In the voltage divider shown in Fig. 10-8, the bias bleeder current is 60 mA. Find the values of $R_1$, $R_2$, $R_3$, and $R_4$.

## JOB 10-6 RESISTANCE MEASUREMENT BY THE VOLTAGE-COMPARISON METHOD. THE POTENTIOMETER RULE

In Fig. 10-9, the value of the unknown resistor $R_x$ may be found by connecting it in series with a known standard resistance $R_k$ and measuring the voltage drops across each.

$$V_x = I_x \times R_x \tag{4-4}$$
$$V_k = I_k \times R_k \tag{4-5}$$
$$V_T = I_T \times R_T \tag{4-7}$$
$$I_T = I_x = I_k \tag{4-1}$$

Dividing Eq. (4-4) by Eq. (4-5) and canceling out the equal currents $I_x = I_k$,

$$\frac{V_x = \overset{1}{\cancel{I_x}} \times R_x}{V_k = \underset{1}{\cancel{I_k}} \times R_k}$$

we obtain the

**FORMULA**

$$\frac{V_x}{V_k} = \frac{R_x}{R_k} \tag{10-10}$$

Similarly, by dividing Eq. (4-4) by Eq. (4-7), we obtain the

**FORMULA**

$$\frac{V_x}{V_T} = \frac{R_x}{R_T} \tag{10-11}$$

These formulas mean that, in a series circuit, the voltage across any resistance depends on the value of the resistance—a large voltage appearing across a large resistance and a small voltage appearing across a small resistance. This is discussed in greater detail in Job 12-2.

EXAMPLE 10-12. Using Fig. 10-9, find the value of the unknown resistance $K$ if the known resistance $R_k = 1000\ \Omega$, $V_x = 100$ V, and $V_k = 25$ V.

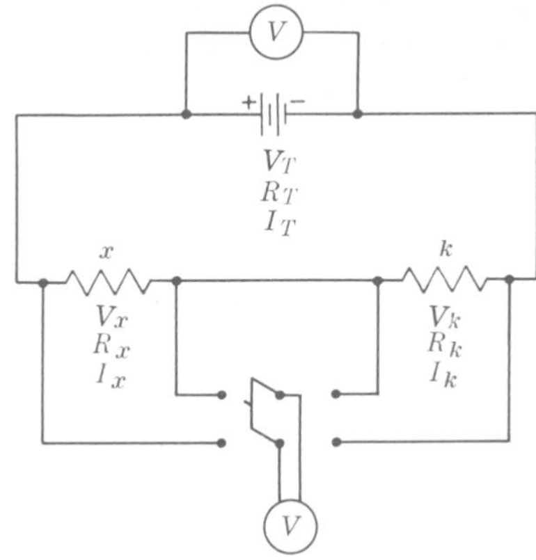


**FIGURE 10-9**
Voltage-comparison method of measuring resistance. $X$ = unknown resistance; $K$ = known standard resistance; $R_T$ = total resistance.

SOLUTION. Given: $R_k = 1000\ \Omega$ Find: $R_x = ?$

$V_x = 100$ V

$V_k = 25$ V

$$\frac{V_x}{V_k} = \frac{R_x}{R_k} \tag{10-10}$$

$$\frac{100}{25} = \frac{R_x}{1000}$$

$$25 \times R_x = 100 \times 1000$$

$$R_x = \frac{100{,}000}{25} = 4000\ \Omega \qquad Ans.$$

EXAMPLE 10-13. A horizontal-hold control circuit of a television receiver is essentially that shown in Fig. 10-10. Find (*a*) the voltage drop across the 27-kΩ resistor, (*b*) the voltage drop across the 75-kΩ resistor, and (*c*) the voltage drop across the 68-kΩ resistor.

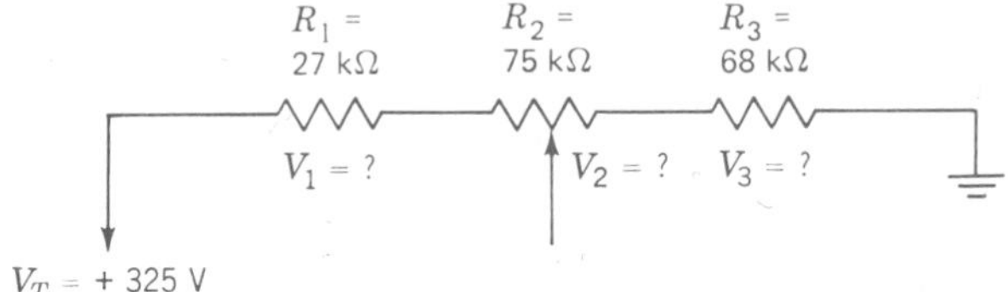


**FIGURE 10-10**
Horizontal-hold control circuit in a television receiver.

SOLUTION

*a.* Find the total resistance of the circuit.

$$R_T = R_1 + R_2 + R_3 \tag{4-3}$$

$$R_T = 27\ \text{k}\Omega + 75\ \text{k}\Omega + 68\ \text{k}\Omega = 170\ \text{k}\Omega \qquad Ans.$$

Find $V_1$.

$$\frac{V_1}{V_T} = \frac{R_1}{R_T} \tag{10-11}$$

$$\frac{V_1}{325} = \frac{27\ \not{k\Omega}}{170\ \not{k\Omega}}$$

$$170 \times V_1 = 325 \times 27$$

$$V_1 = \frac{8775}{170} = 51.6\ \text{V} \qquad \textit{Ans.}$$

*b.* Find $V_2$.

$$\frac{V_1}{V_2} = \frac{R_1}{R_2} \tag{10-10}$$

$$\frac{51.6}{V_2} = \frac{27\ \not{k\Omega}}{75\ \not{k\Omega}}$$

$$27 \times V_2 = 51.6 \times 75$$

$$V_2 = \frac{3870}{27} = 143.3\ \text{V} \qquad \textit{Ans.}$$

*c.* Find $V_3$.

$$\frac{V_1}{V_3} = \frac{R_1}{R_3} \tag{10-10}$$

$$\frac{51.6}{V_3} = \frac{27\ \not{k\Omega}}{68\ \not{k\Omega}}$$

$$27 \times V_3 = 51.6 \times 68$$

$$V_3 = \frac{3508.8}{27} = 129.9\ \text{V} \qquad \textit{Ans.}$$

## THE POTENTIOMETER RULE

When a number of resistances are connected in series, the voltage drop across any one of them may be found by using another form of formula (10-11).

$$\frac{V_x}{V_T} = \frac{R_x}{R_T} \tag{10-11}$$

By cross-multiplying, we get

$$V_x \times R_T = R_x \times V_T$$

Dividing both sides by $R_T$ will give the

**FORMULA**

$$V_x = \frac{R_x}{R_T} \times V_T \tag{10-12}$$

In effect, this formula says that the voltage across any part of a series circuit is some fractional part of the total voltage. The numerator of the fraction is the resistance of the part of the circuit and the denominator is the total resistance of the circuit.

SELF-TEST 10-14. A voltage divider in the base circuit of a transistor is shown in Fig. 10-11. Find the voltage drop across $R_{B'}$.

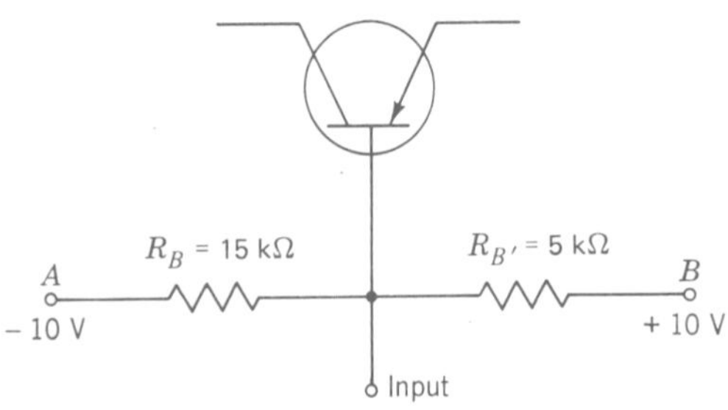


**FIGURE 10-11**
Voltage divider in the base circuit of a transistor.

SOLUTION. Since $R_B$ and $R_{B'}$ are in ______, the potentiometer rule may be used. The total voltage between the points $B$ and $A$ is the difference in potential between these points. Thus, — series

$$V_T = +10 - (-10) = \_\_\_\_\_\_ \text{ V}$$ — +20

Formula (10-12) may now be rewritten using the symbols given in our circuit. Thus,

$$V_x = \frac{R_x}{R_T} \times V_T \qquad (10\text{-}12)$$

will become

$$V_{RB'} = \frac{R_{B'}}{R_T} \times \_\_\_\_\_\_$$ — $V_T$

Substituting values will give

$$V_{RB'} = \frac{5}{5 + ?} \times 20$$ — 15

$$V_{RB'} = \frac{5 \text{ k}\Omega}{20 \text{ k}\Omega} \times 20$$

$$V_{RB'} = \_\_\_\_\_\_ \text{ V} \qquad Ans.$$ — 5

## PROBLEMS

In Probs. 1 to 6, $R_k$ and $R_x$ are connected in series. Find the missing values in each problem.

| PROBLEM | $R_k$, Ω | $V_k$, V | $V_x$, V | $R_x$, Ω | $V_T$, V |
|---|---|---|---|---|---|
| 1 | 500 | 10 | 25 | ? | |
| 2 | 50 | 15.5 | 70.2 | ? | |
| 3 | 10 | 36.4 | 40.3 | ? | |
| 4 | 100 | 15 | ? | 250 | |
| 5 | 1000 | ? | ? | 2000 | 100 |
| 6 | 15,000 | ? | ? | 20,000 | 120 |

7. In a circuit similar to that shown in Fig. 10-10, $R_1 = 22{,}000\ \Omega$, $R_2 = 50{,}000\ \Omega$, and $R_3 = 100{,}000\ \Omega$. Find (*a*) $V_1$, (*b*) $V_2$, and (*c*) $V_3$ if $V_T = 200$ V.

8. In the picture-control circuit shown in Fig. 10-12, find the voltage drop across $R_1$ and $R_2$.

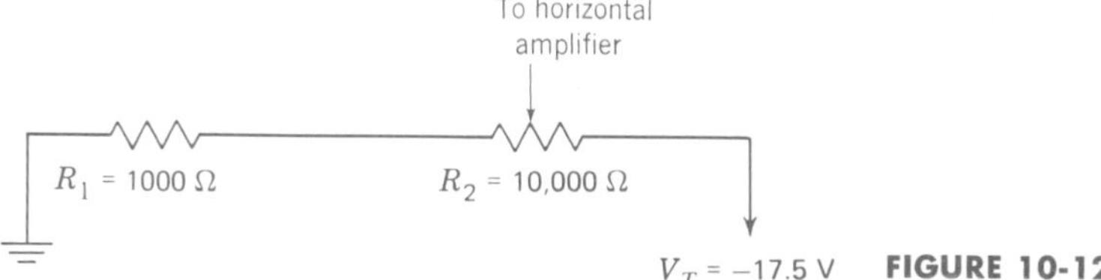


**FIGURE 10-12**

9. A brightness-control circuit in the Admiral TV chassis model NA1-1A is essentially that shown in Fig. 10-13. Find the voltage between ground and point $A$.

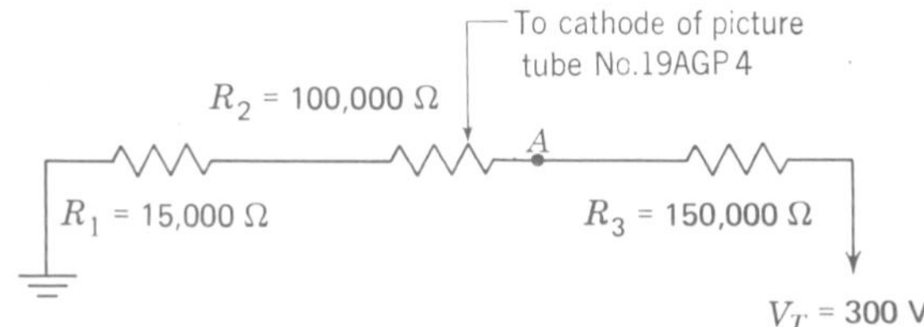


**FIGURE 10-13**
Brightness-control circuit in a television receiver.

10. In a circuit similar to that shown in Fig. 10-11, $R_B = 35$ kΩ, $R_{B'} = 5$ kΩ, and the total voltage from $A$ to $B$ is 30 V. Find the voltage drop across $R_{B'}$.
11. The volume control of a transistorized phono amplifier is essentially a 470-kΩ and a 5-kΩ resistor in series across 9.5 V. Find the voltage drop across the 5-kΩ resistor.
12. In a brightness-control circuit, the total resistance of the potentiometer equals 5 MΩ, and $V_T$ equals 145 V. Find the cathode voltage when the potentiometer arm has covered (*a*) 1 MΩ and (*b*) 3.5 MΩ.

## JOB 10-7 RESISTANCE MEASUREMENT USING THE WHEATSTONE BRIDGE

The Wheatstone bridge is a device for obtaining accurate measurements of resistance. A diagrammatic sketch of the circuit is shown in Fig. 10-14. The resistors $R_1$, $R_2$, $R_3$, and $R_x$ are arranged in two parallel branches. $R_1$, $R_2$, and $R_3$ are variable *known* resistors, and $R_x$ is the *unknown* resistance. A galvanometer $G$ is connected across the two branches between points $B$ and $D$. A battery supplies the total current $I_T$.

### OPERATION OF THE BRIDGE

1. When the circuit is closed, a current $I_T$ will flow through the circuit which

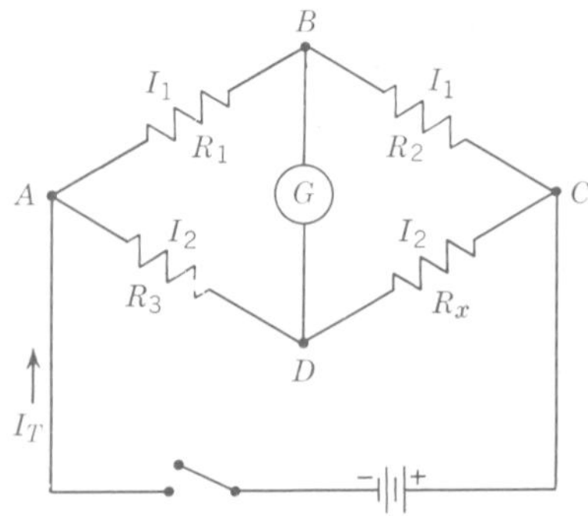


**FIGURE 10-14**
Wheatstone-bridge circuit.

divides at point $A$, current $I_1$ flowing through the branch $ABC$ ($R_1$ and $R_2$), and current $I_2$ flowing through the branch $ADC$ ($R_3$ and $R_x$).

2. The variable resistors are adjusted until the galvanometer indicates no deflection. $R_1$ and $R_2$ are usually fixed at some convenient ratio, and then $R_3$ is varied until the galvanometer reads zero.

3. When there is no deflection of the galvanometer, it indicates that there is no current flowing between points $B$ and $D$ and therefore $B$ and $D$ are at the same voltage level. Under these conditions, we find that

Voltage across $AB$ = voltage across $AD$ or $I_1 \times R_1 = I_2 \times R_3$
Voltage across $BC$ = voltage across $DC$ or $I_1 \times R_2 = I_2 \times R_x$

Dividing one equation by the other gives us the fundamental equation for the Wheatstone bridge.

$$\frac{I_1 \times R_1}{I_1 \times R_2} = \frac{I_2 \times R_3}{I_2 \times R_x}$$

Canceling out the currents $I_1$ and $I_2$, we obtain

**FORMULA**

$$\frac{R_1}{R_2} = \frac{R_3}{R_x} \qquad \boxed{10\text{-}13}$$

EXAMPLE 10-15. When a Wheatstone bridge was used, the following readings were obtained: $R_1 = 1000\ \Omega$, $R_2 = 10{,}000\ \Omega$, and $R_3 = 67.4\ \Omega$. Find the unknown resistance.

SOLUTION. Given: $R_1 = 1\ \text{k}\Omega$ Find: $R_x = ?$
$R_2 = 10\ \text{k}\Omega$
$R_3 = 67.4\ \Omega$

$$\frac{R_1}{R_2} = \frac{R_3}{R_x} \qquad (10\text{-}13)$$

$$\frac{1\ \cancel{\text{k}\Omega}}{10\ \cancel{\text{k}\Omega}} = \frac{67.4}{R_x}$$

$$R_x = 67.4 \times 10 = 674\ \Omega \qquad \textit{Ans.}$$

## PROBLEMS

Find the value of $R_x$ for each problem.

| PROBLEM | $R_1$, Ω | $R_2$, Ω | $R_3$, Ω | $R_x$, Ω |
|---|---|---|---|---|
| 1 | 1000 | 10,000 | 84.3 | ? |
| 2 | 10,000 | 1000 | 952.7 | ? |
| 3 | 100 | 10 | 60.4 | ? |
| 4 | 10 | 1 | 175.3 | ? |
| 5 | 1 | 10 | 128.9 | ? |
| 6 | 5000 | 10,000 | 823.7 | ? |
| 7 | 100 | 1000 | 46.5 | ? |
| 8 | 1000 | 100 | 9.16 | ? |
| 9 | 10 | 1 | 1.6 | ? |
| 10 | 1 | 10 | 35.79 | ? |

(See Instructor's Manual for Test 10-2, Voltage Dividers and Resistance Measurement)

## JOB 10-8 EQUIVALENT CIRCUITS

The Thevenin and the Norton equivalent circuits are very useful in reducing the complexity of circuits for analysis. Each of these equivalents has its own advantages, depending on the particular circuit that is being analyzed. Since both equivalents have the same resistance, the factor determining which to use is whether the terminals analyzed will be a short or an open circuit.

The Thevenin equivalent finds the voltage with the terminals being analyzed open. The Norton equivalent finds the current with the terminals shorted.

Our first example (Example 10-16) has three objectives: (1) to develop skills with the Thevenin equivalent, (2) to increase our knowledge about the output characteristics of circuits, and (3) to apply the Thevenin equivalent to a voltage divider.

The problem as phrased is for a voltage divider power supply. The general concepts and procedures would be the same whether it were an amplifier or any other type of linear circuit rather than a power supply. The Thevenin equivalent resistance that is obtained is sometimes called the *output resistance* of the power supply. This equivalent resistance is important because it defines how loads interact with the power supply.

EXAMPLE 10-16. Figure 10-15 shows the circuit of a power supply that delivers two different voltages; 30 and 9 V. Each voltage has its own load which is represented by the load resistors $R_L9$ and $R_L30$.

Find the Thevenin equivalent circuit at output terminals $A$ and $B$ of the 9-V supply.

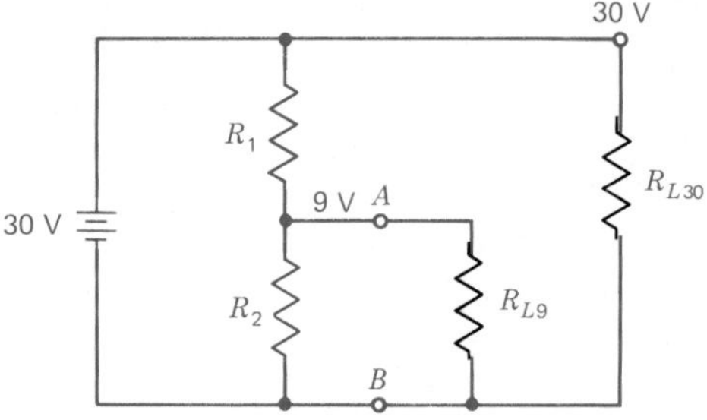


**FIGURE 10-15**
Dual voltage supply with voltage divider output.

SOLUTION

1. First redraw the circuit to that shown in Fig. 10-16. This shows the circuit in the form in which it will be analyzed and the final form that is sought.

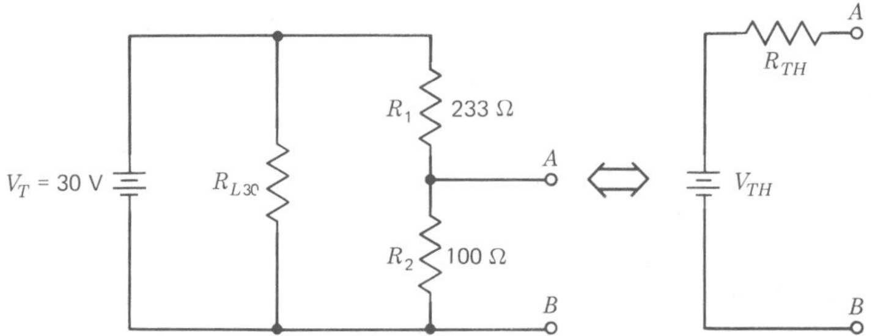


**FIGURE 10-16**
Power supply circuit with Thevenin equivalent for terminals *A* and *B*.

2. Find the Thevenin equivalent resistance. Short the 30-V battery. Notice that this removes the second load, $R_L30$. The resistance between terminals $A$ and $B$ will be the result of $R_1$ in parallel with $R_2$.

$$R_{TH} = \frac{R_1 \times R_2}{R_1 + R_2} = \frac{100 \times 233}{100 + 233} = 70\ \Omega$$

3. Find the Thevenin equivalent voltage. This will be the voltage across $R_2$.

$$V_{TH} = V_2 = \frac{R_2}{R_1 + R_2} \times V_T \qquad (10\text{-}12)$$

$$V_{TH} = \frac{100}{233 + 100} \times 30$$

$$= \frac{100}{333} \times 30 = 9\ \text{V}$$

The Thevenin equivalent circuit shown in Fig. 10-16 is now completely specified. The output voltage is the 9 V that is desired. It will be easy to determine the effects of loading on the output voltage. The Thevenin equivalent is used as in Fig. 10-17. The load is placed between the two terminals $A$ and $B$. This forms a very simple one-loop circuit which can be analyzed for any value of load.

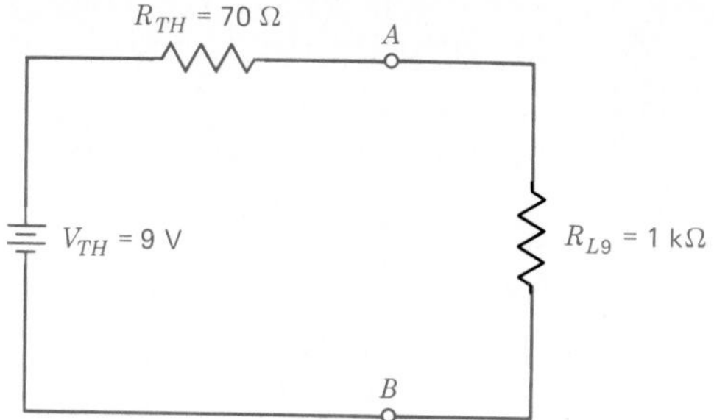


**FIGURE 10-17**
Power supply equivalent circuit with load attached.

4. Find the output voltage and the current through the load if the load value is 1 kΩ.

$$I_L = \frac{V_{TH}}{R_{TH} + R_L}$$

$$I_L = \frac{9}{70 + 1\ \text{k}\Omega} = 8.4\ \text{mA}$$

$$E_{RL}9 = I_L \times R_L = 8.4\ \text{mA} \times 1\ \text{k}\Omega = 8.4\ \text{V}$$

This example shows how to use the Thevenin theorem to greatly reduce the complexity of a very practical circuit. It also shows that if you ignore 70 Ω relative to 1 kΩ, the result may be misleading. The change in output voltage from 9 V unloaded to 8.4 V loaded could be significant. It also shows that the output resistance of a power supply must be considered.

Example 10-16 only analyzed the effects of the voltage divider on the output of the power supply. The internal impedance of the power supply was ignored. Figure 10-18 shows the change in the circuit if a source resistance $R_S$ for the power supply is included. Example 10-17 shows the effect of this impedance and the use of both theorems in the same problem.

EXAMPLE 10-17. Find the Norton equivalent for the output of the power supply in Fig. 10-18.

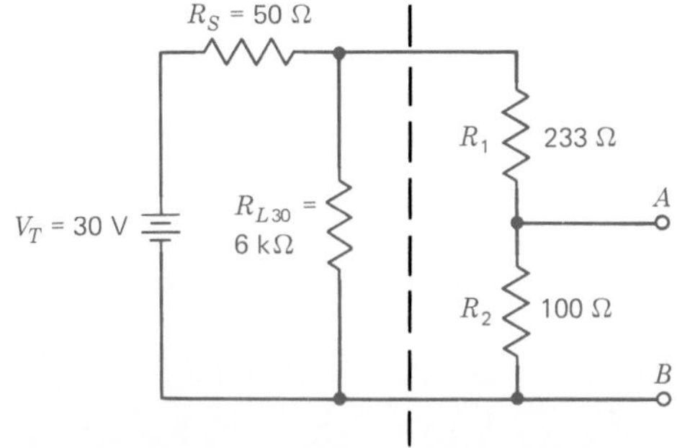


**FIGURE 10-18**
Voltage divider power supply with internal resistance $R_S$.

SOLUTION. This is a complicated series-parallel network. Consider the ways in which the circuit could be simplified. The dotted line divides the circuit into two parts. If the part on the left were converted into a Thevenin equivalent with Thevenin voltage $V_{TH}$ and Thevenin resistance $R_{TH}$, the entire circuit would be only one loop. This would greatly simplify the analysis and would improve the understanding of the circuit operation (see Fig. 10-19).

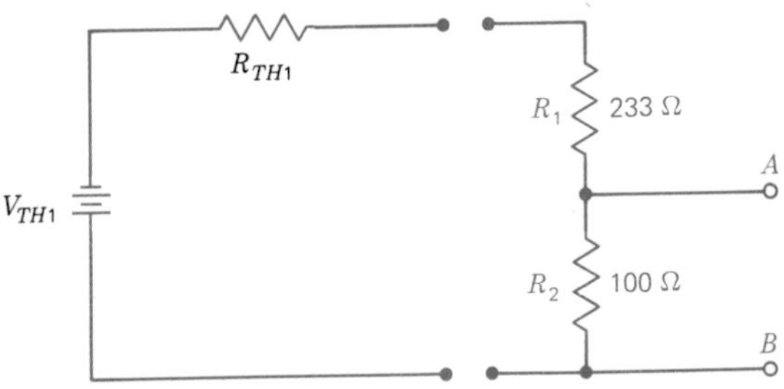


**FIGURE 10-19**
Power supply with Thevenin equivalent for part of circuit.

1. Make a Thevenin equivalent circuit for the left part of the circuit shown in Fig. 10-18. Assume that the load on the 30-V line is 6 kΩ.

$$V_{TH} = \frac{R_L}{R_L + R_S} \times V_T = \frac{6\text{ k}\Omega}{6\text{ k}\Omega + 50} \times 30 = 29.75\text{ V}$$

$$R_{TH} = \frac{R_S \times R_L 30}{R_S + R_L 30} = \frac{50 \times 6\text{ k}\Omega}{50 + 6\text{ k}\Omega} = 49.59$$

Notice that in the above calculations the load on the 30-V line affects the values of the Thevenin equivalent circuit. This means that if the 30-V load changes, some of this change will show up in the 9-V power supply. In Example 10-16 the two supplies were independent. Now we find that the inclusion of an internal resistance in the power supply causes a connection between the two loads.

2. Connect the Thevenin equivalent to the remainder of the circuit and determine the Norton equivalent for the power supply. The resulting circuit is shown in Fig. 10-19. When the Norton current is found by shorting terminals $A$ and $B$, the resulting circuit is only the series connection of $V_{TH}$, $R_{TH}$, and $R_1$. This is easy to analyze.

$$I_N = \frac{V_{TH}}{R_{TH} + R_1} = \frac{29.75}{49.59 + 233} = 105\text{ mA}$$

$$R_N = \frac{R_2 \times (R_1 + R_{TH})}{R_2 + R_1 + R_{TH}} = \frac{100 \times (233 + 49.59)}{100 + 233 + 49.59} = 73.86\ \Omega$$

The final Norton equivalent circuit is shown in Fig. 10-20.

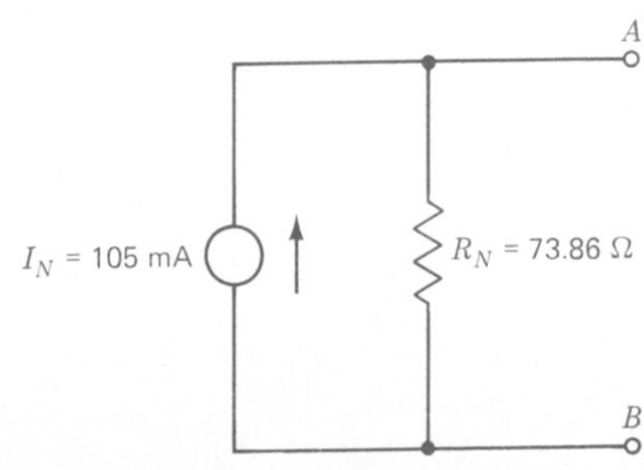


**FIGURE 10-20**
Norton equivalent for 9-V power supply.

## PROBLEMS

1. Solve for current $z$ in Fig. 9-21. *Hint:* Make an equivalent circuit for circuit points $g$ and $f$.
2. Solve Example 9-13 for current $x$ by making a Thevenin equivalent of the parallel network between points $p$ and $b$. This makes the circuit a simple one-loop problem.
3. In Fig. 9-28, solve the circuit for current $z$ through $R_5$. *Hint:* Use a Norton equivalent.
4. Make a Norton equivalent of Fig. 6-9 to find the current $I_3$ in $R_3$.
5. In Fig. 6-16, find the current in $R_5$ using a Norton's equivalent.

## JOB 10-9 DC EQUIVALENT CIRCUITS FOR SELF-BIASED TRANSISTOR CIRCUITS

**Solid-state diodes.** As we learned in Job 1-1, many metals have free electrons in the outer shells of their atoms. The ability of these free electrons to move about makes most metals fine conductors of electricity. On the other hand, insulators such as glass and rubber do not have many free electrons. A semiconductor material such as germanium has some free electrons whose number can be considerably increased by the addition of small amounts of arsenic atoms as an impurity. These many loosely held electrons in the outer shells of the germanium atoms cause it to be referred to as N germanium. This does not mean that it is negatively charged, but merely that there are free electrons present.

If gallium is added as an impurity instead of arsenic, the outer shells of the germanium atoms will be filled except for one electron. This missing electron constitutes a "hole" which the atom wants to fill to reach a stable state. This material is known as P germanium. Because N germanium wants to "give" electrons and P germanium wants to "accept" them, both materials act as fairly good electron conductors.

Now suppose that we press these two materials together, as shown in Fig. 10-21. The free electrons at the junction will meet with the positive holes and cancel each other's charge. After a very short time, the field produced by this zero-charged area repels both the electrons and the holes and acts as a barrier to both. Of course, no current will flow. In Fig. 10-22*a*, we have connected the − terminal of a battery to the N material, and the + terminal to the P material. The forward push of the many electrons from the battery helps the free electrons to overcome the

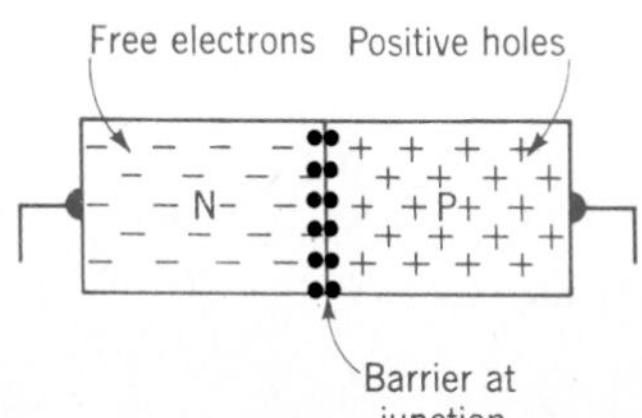


**FIGURE 10-21**
A barrier is formed at the junction of N- and P-type materials.

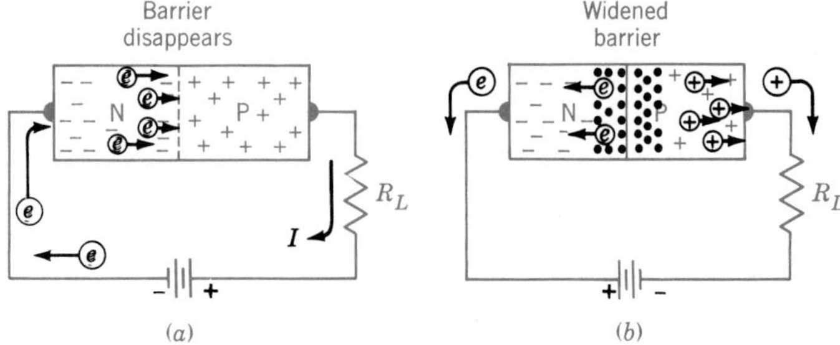


**FIGURE 10-22**
(*a*) The barrier disappears under forward biasing, and the junction conducts. (*b*) The barrier is widened under reverse biasing, and no current flows.

barrier, and the combination acts as a good conductor. Connected in this way, the combination is said to be *forward-biased,* and current flows through the load resistor.

In Fig. 10-22*b,* we have reversed the polarity, connecting the + of the battery to the N material and the − of the battery to the P material. The + battery pulls the free electrons away from the junction, and the − battery pulls the positive holes away from the junction. This has the effect of widening the barrier at the junction, and no current will flow. This is called *reverse bias.*

This combination is called a *diode,* the symbol for which is shown in Fig. 10-23. When connected as shown, current will flow only from left to right.

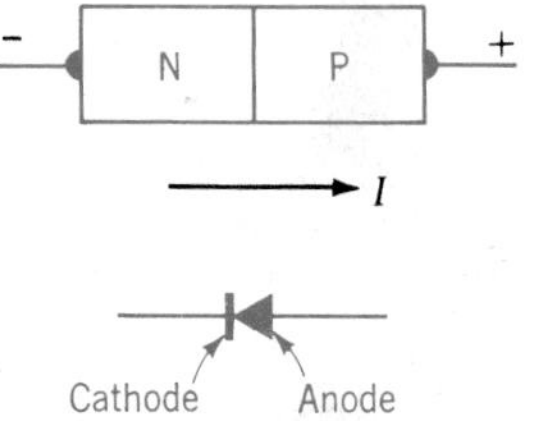


**FIGURE 10-23**
Diode symbol.

**Transistors.** A transistor is very similar to a semiconductor diode except that the transistor has two junctions instead of one. In Fig. 10-24, the *emitter* is an N-type material which forms a junction with a P-type material called the *base.* This base in turn forms another junction with the *collector,* an N-type material. Such a combination is called an *NPN-type* transistor.

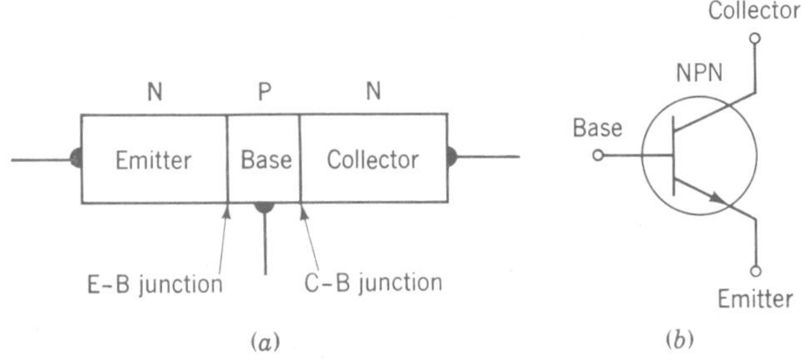


**FIGURE 10-24**
(*a*) NPN transistor. (*b*) Symbol for NPN transistor.

If the emitter and collector are made of P-type material and the base is N-type as shown in Fig. 10-25, the transistor is called a *PNP type.* The only difference between the two types is the direction in which the current flows in the emitter. In the symbols, the electron current is *against* the arrowheads.

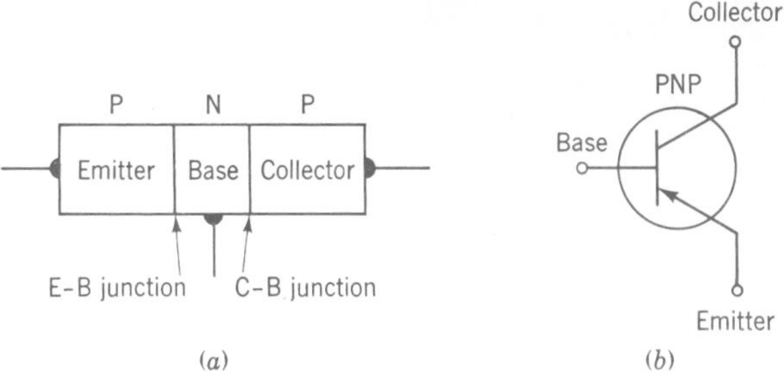


**FIGURE 10-25**
(*a*) PNP transistor. (*b*) Symbol for PNP transistor.

**Transistor operation.** In Fig. 10-26, when the voltage $-V_{EE}$ is applied, the electrons move out into the emitter lead and enter the emitter material. These electrons, together with the free electrons of the emitter material, are pushed toward the base region. The greater the forward bias, the greater the number of electrons that enter the base region. The number of holes in the base region is very small because it is so very thin and because it is so slightly doped. A few electrons will find holes and combine to form a small base current. The rest of the electrons (about 98 percent) will not find a hole and will be pushed across the junction by the electrons newly arriving from the emitter. Once they get past the collector-base junction, they are attracted by the large collector voltage ($V_{CC}$) and become part of the collector current. They return to the battery, join up with $I_B$, enter $V_{EE}$, and start around again. As you can see, the emitter current is the sum of the base current and the collector current. An increase in the forward bias will cause an increase in the emitter current and a corresponding increase in the collector current and the base current. Conversely, a decrease in the forward bias results in a decrease of both base and collector current.

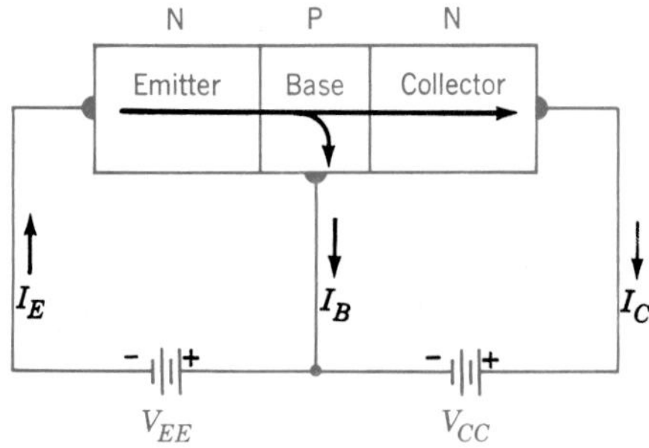


**FIGURE 10-26**
Electron flow in an operating NPN transistor.

**Transistor current gain.** As an amplifying device, the transistor is now replacing many of the functions of the vacuum tube. In Fig. 10-27*a*, a small change in the input signal to the grid controls a large change in the current that flows through the load. Similarly, in Fig. 10-27*b*, a small change in the input signal to the base of the transistor controls a

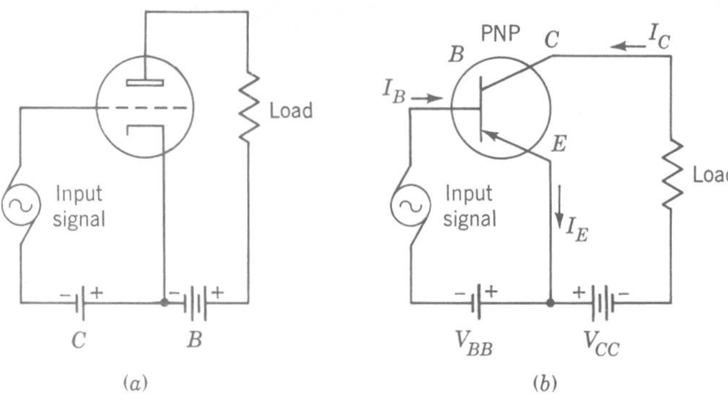


**FIGURE 10-27** Comparison of (a) triode circuit and (b) transistor circuit. Small changes in the input signal control large currents through the load.

large change in the current that flows through the collector and the load. This base-to-collector current gain is defined as $\beta$ (beta).

$$\beta = \frac{\Delta I_C}{\Delta I_B} \tag{10-14}$$

where $\beta$ = current gain
$\Delta I_C$ = change in dc collector current
$\Delta I_B$ = change in dc base current

In order to solve a transistor circuit by the equivalent-resistance circuit-analysis method, we must first find the collector-to-base resistance $R_{CB}$ for the particular transistor. For the self-bias circuit shown in Fig. 10-28*a*, the formula for $R_{CB}$ is

$$R_{CB} = \frac{R_B}{\beta} \tag{10-15}$$

where $R_{CB}$ = collector-to-base resistance
$R_B$ = base-bias resistor
$\beta$ = current gain

## SELF-BIAS CIRCUITS

EXAMPLE 10-18. Solve the circuit shown in Fig. 10-28*a* for (*a*) the load current $I_{RL}$, (*b*) the voltage across $R_L$, and (*c*) the collector voltage $V_C$.

SOLUTION

1. Find $R_{CB}$.

$$R_{CB} = \frac{R_B}{\beta} \qquad (10\text{-}15)$$

$$R_{CB} = \frac{200\ \text{k}\Omega}{100} = 2\ \text{k}\Omega$$

2. Find the load current $I_{RL}$. This is the current that will flow through the

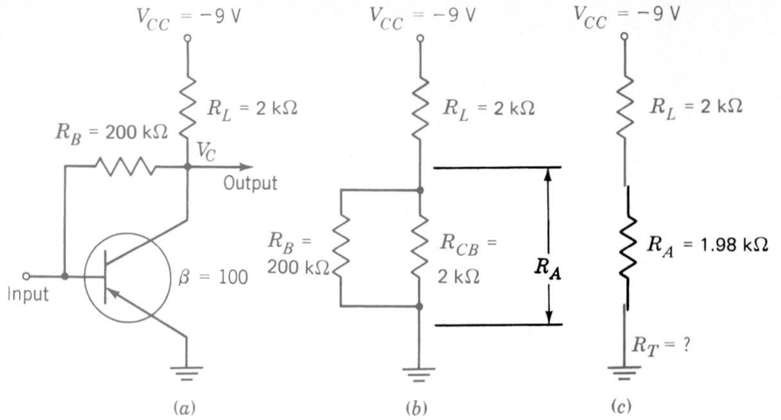


**FIGURE 10-28**
(a) Self-bias circuit. (b) Dc equivalent circuit. (c) $R_A$ replaces the parallel combination of $R_B$ and $R_{CB}$.

series-parallel circuit shown in Fig. 10-28*b*. This circuit is further simplified into that shown in Fig. 10-28*c*, in which $R_A$ replaces the parallel combination of $R_B$ and $R_{CB}$ of Fig. 10-28*b*.

*a.* Find $R_A$ in Fig. 10-28*b*. Since $R_B$ and $R_{CB}$ are in parallel,

$$R_A = \frac{R_B \times R_{CB}}{R_B + R_{CB}} \qquad (5\text{-}5)$$

$$= \frac{200 \text{ k}\Omega \times 2 \text{ k}\Omega}{200 \text{ k}\Omega + 2 \text{ k}\Omega}$$

$$= \frac{400 \times 10^6}{202 \times 10^3} = 1.98 \times 10^3$$

$$R_A = 1.98 \text{ k}\Omega$$

*b.* Find $R_T$ in Fig. 10-28*c*. Since $R_L$ and $R_A$ are in series,

$$R_T = R_L + R_A \qquad (4\text{-}3)$$

$$= 2 + 1.98 = 3.98 \text{ k}\Omega$$

*c.* Find the total current in the circuit of Fig. 10-28*c*.

$$I_T = \frac{V_{CC}}{R_T} = \frac{9}{3.98 \times 10^3} = 2.26 \times 10^{-3}$$

$$I_T = 2.26 \text{ mA}$$

*d.* Find $I_{RL}$.

$$I_{RL} = I_T = 2.26 \text{ mA} \qquad \textit{Ans.} \qquad (4\text{-}1)$$

3. Find the voltage across $R_L$.

$$V_{RL} = I_{RL} \times R_L \qquad (2\text{-}1)$$

$$= 2.26 \times 10^{-3} \times 2 \times 10^3 = 4.52 \text{ V} \qquad \textit{Ans.}$$

4. Find the collector voltage $V_C$. As indicated in Fig. 10-28*a*, the collector voltage $V_C$, referred to ground, is equal to the difference between $V_{CC}$ and $V_{RL}$.

$$V_C = -V_{CC} + V_{RL} = -9.0 + 4.52 = -4.48 \text{ V} \qquad \textit{Ans.}$$

## SELF-BIAS CIRCUITS WITH EMITTER RESISTANCE

EXAMPLE 10-19. Solve the circuit shown in Fig. 10-29*a* for (*a*) the collector current $I_C$, (*b*) the emitter current $I_E$, (*c*) the voltage across $R_L$, (*d*) the collector voltage $V_C$, (*e*) the voltage across $R_E$, and (*f*) the collector-to-emitter voltage $V_{CE}$.

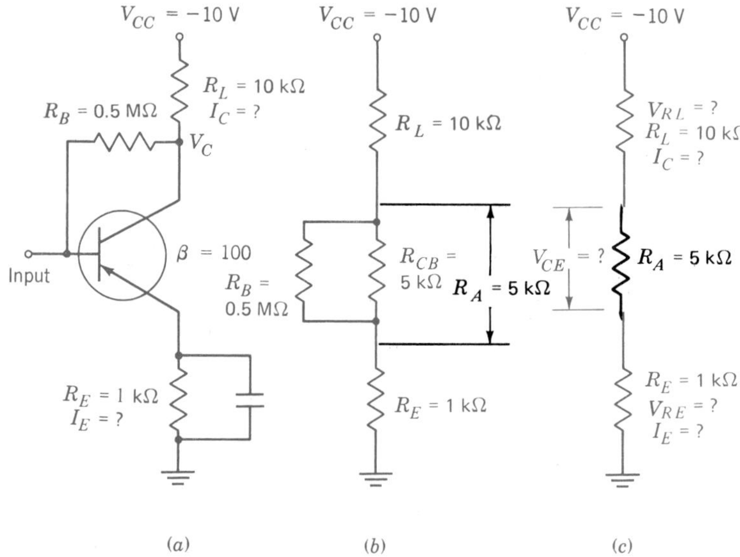


**FIGURE 10-29**
(*a*) Self-bias circuit with $R_E$. (*b*) Dc equivalent circuit. (*c*) $R_A$ equals $R_{CB}$ since $R_B$ is so large in comparison with $R_{CB}$.

SOLUTION

1. Find $R_{CB}$.

$$R_{CB} = \frac{R_B}{\beta} \tag{10-15}$$

$$= \frac{0.5 \times 10^6}{10^2}$$

$$= 0.5 \times 10^4$$

$$= 5 \times 10^3 = 5 \text{ k}\Omega$$

2. Find the collector current $I_C$. This is the current that will flow through the series-parallel circuit shown in Fig. 10-29*b*. This circuit is further simplified into that shown in Fig. 10-29*c*, in which $R_A$ replaces the parallel combination of $R_B$ and $R_{CB}$ of Fig. 10-29*b*.

   *a.* Find $R_A$ in Fig. 10-29*b*. Since $R_B$ (0.5 MΩ = 500,000 Ω) is so much larger than $R_{CB}$ (5 kΩ = 5000 Ω), the effect of $R_B$ on the value of the parallel combination is negligible. We may therefore consider $R_A = R_{CB}$ = 5 kΩ. The student should work out the actual value of $R_A$ to become satisfied on this point.

   *b.* Find $R_T$ in Fig. 10-29*c*. Since $R_L$, $R_A$, and $R_E$ are in series,

$$R_T = R_L + R_A + R_E \qquad (4\text{-}3)$$
$$= 10 + 5 + 1 = 16 \text{ k}\Omega$$

*c.* Find the total current in the circuit of Fig. 10-29*c*.

$$I_T = \frac{V_{CC}}{R_T} = \frac{10}{16 \times 10^3}$$
$$= 0.625 \times 10^{-3} = 0.625 \text{ mA}$$

*d.* In the series circuit of Fig. 10-29*c*,

$$I_T = I_C = I_E = 0.625 \text{ mA} \qquad (4\text{-}1)$$
$$I_C = 0.625 \text{ mA} \qquad \textit{Ans.}$$
$$I_E = 0.625 \text{ mA} \qquad \textit{Ans.}$$

3. Find the voltage across $R_L$.

$$V_{RL} = I_C \times R_L \qquad (2\text{-}1)$$
$$= 0.625 \times 10^{-3} \times 10 \times 10^3 = 6.25 \text{ V} \qquad \textit{Ans.}$$

4. Find the collector voltage $V_C$. As indicated in Fig. 10-29*a*, the collector voltage $V_C$, referred to ground, is equal to the difference between $V_{CC}$ and $V_{RL}$.

$$V_C = -V_{CC} + V_{RL}$$
$$V_C = -10 + 6.25 = -3.75 \text{ V} \qquad \textit{Ans.}$$

5. Find the voltage across the emitter resistor $V_{RE}$.

$$V_{RE} = I_E \times R_E \qquad (2\text{-}1)$$
$$= 0.625 \times 10^{-3} \times 1 \times 10^3 = 0.625 \text{ V} \qquad \textit{Ans.}$$

6. Find the collector-to-emitter voltage drop $V_{CE}$. As indicated in Fig. 10-29*c*, $V_{CE}$ is the drop across $R_{CB}$ and is equal to the difference between $V_{CC}$ and the sum of $V_{RL}$ and $V_{RE}$.

$$V_{CE} = V_{CC} - (V_{RL} + V_{RE})$$
$$= 10 - (6.25 + 0.625)$$
$$V_{CE} = 10 - 6.88 = 3.12 \text{ V} \qquad \textit{Ans.}$$

## PROBLEMS

1. In a circuit similar to that shown in Fig. 10-28*a*, $V_{CC} = -10$ V, $R_L = 5$ kΩ, $R_B = 100$ kΩ, and $\beta = 50$. Find (*a*) the load current $I_{RL}$, (*b*) the voltage across $R_L$, and (*c*) the collector voltage $V_C$.
2. Repeat Prob. 1 with the following values: $V_{CC} = -10$ V, $R_B = 80$ kΩ, $R_L = 2$ kΩ, and $\beta = 20$.
3. In a circuit similar to that shown in Fig. 10-29*a*, $V_{CC} = -13$ V, $R_B = 1$ MΩ, $R_L = 40$ kΩ, $R_E = 2$ kΩ, and $\beta = 100$. Find (*a*) the collector current $I_C$, (*b*) the emitter current $I_E$, (*c*) the voltage across $R_L$, (*d*) the collector voltage $V_C$, (*e*) the voltage across $R_E$, and (*f*) the collector-to-emitter voltage $V_{CE}$.
4. Repeat Prob. 3 with the following values: $V_{CC} = -14$ V, $R_B = 0.5$ MΩ, $R_L = 20$ kΩ, $R_E = 5$ kΩ, and $\beta = 50$.

## JOB 10-10 DC EQUIVALENT CIRCUITS FOR FIXED-BIAS TRANSISTOR CIRCUITS

In the fixed-bias circuit shown in Fig. 10-30*a*, the load resistor $R_L$ is included in the beta-dependent loop, and it must therefore be included in the formula for $R_{CB}$. For the fixed-bias circuit shown in Fig. 10-30*a*, the formula for $R_{CB}$ is

$$R_{CB} = \frac{R_B}{\beta} - R_L \tag{10-16}$$

where $R_{CB}$ = collector-to-base resistance
$R_B$ = base-bias resistor
$R_L$ = collector or load resistance
$\beta$ = current gain

### FIXED-BIAS CIRCUITS

EXAMPLE 10-20. Solve the circuit shown in Fig. 10-30*a* for (*a*) the collector current $I_C$, (*b*) the voltage across $R_L$, and (*c*) the collector voltage $V_C$.

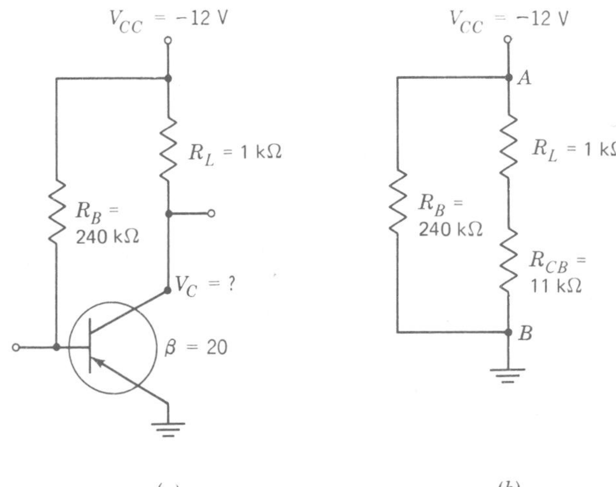


**FIGURE 10-30** (*a*) Fixed-bias circuit. (*b*) Dc equivalent circuit.

SOLUTION

1. Find $R_{CB}$.

$$R_{CB} = \frac{R_B}{\beta} - R_L \qquad (10\text{-}16)$$

$$= \frac{240}{20} - 1$$

$$R_{CB} = 12 - 1 = 11 \text{ k}\Omega$$

2. Find $I_C$. The collector current is the current through the series group of

$R_L$ and $R_{CB}$ as shown in Fig. 10-30*b*. Since $R_B$ is in parallel with this series group, the voltage across this group is equal to the total voltage $V_{CC} = -12$ V.

$$I_C = \frac{V}{R_T}$$

$$I_C = \frac{12}{11 + 1} = \frac{12}{12 \times 10^3}$$

$$= 1 \times 10^{-3} = 1 \text{ mA} \qquad Ans.$$

3. Find $V_{RL}$.

$$V_{RL} = I_C \times R_L$$

$$V_{RL} = 0.001 \times 1000 = 1 \text{ V} \qquad Ans.$$

4. Find $V_C$. As shown in Fig. 10-30*a*, the collector voltage $V_C$, referred to ground, is equal to the difference between $V_{CC}$ and $V_{RL}$.

$$V_C = -12 + 1 = -11 \text{ V} \qquad Ans.$$

## FIXED-BIAS CIRCUITS WITH EMITTER RESISTANCE

EXAMPLE 10-21. Solve the circuit shown in Fig. 10-31*a* for (*a*) the emitter current $I_E$, (*b*) the emitter voltage $V_E$, (*c*) the collector current $I_C$, (*d*) the voltage across $R_L$, and (*e*) the collector voltage $V_C$.

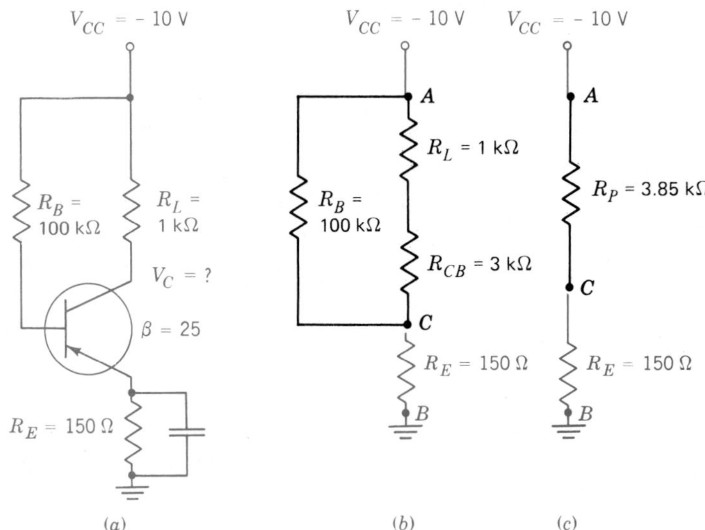


**FIGURE 10-31** (*a*) Fixed-bias circuit with $R_E$. (*b*) Dc equivalent circuit. (*c*) $R_P$ replaces the series-parallel combination of Fig. 10-31*b*.

SOLUTION

1. Find $R_{CB}$.

$$R_{CB} = \frac{R_B}{\beta} - R_L \tag{10-16}$$

$$= \frac{100}{25} - 1$$

$$R_{CB} = 4 - 1 = 3\ \text{k}\Omega$$

2. Find $I_E$, the emitter current. This is the current that flows through the series-parallel circuit shown in Fig. 10-31*b* and which must be simplified into the circuit shown in Fig. 10-31*c*. In Fig. 10-31*b*, the series resistance of $R_{CB}$ and $R_L$ is equal to 3 kΩ + 1 kΩ = 4 kΩ. This 4 kΩ is in parallel with the 100 kΩ of $R_B$. The total resistance of this parallel combination is

$$R_P = \frac{100 \times 4}{100 + 4} \qquad (5\text{-}5)$$

$$R_P = \frac{400}{104} = 3.85\ \text{k}\Omega$$

Now we can find the total resistance between points $A$ and $B$.

$$R_T = R_P + R_E = 3.85 + 0.15 = 4\ \text{k}\Omega$$

The emitter current $I_E$ is equal to the total current that flows between points $A$ and $B$.

$$I_E = I_T = \frac{V_{CC}}{R_T}$$

$$I_E = \frac{10}{4 \times 10^3} = 2.5\ \text{mA} \qquad \textit{Ans.}$$

3. Find the emitter voltage $V_E$. This is the voltage, referred to ground, which is equal to the drop across $R_E$.

$$V_E = I_E \times R_E$$
$$V_E = 0.0025 \times 150 = 0.375\ \text{V} \qquad \textit{Ans.}$$

4. In Fig. 10-31*b*, the collector current is the current flowing between points $A$ and $C$, and is equal to the voltage between $A$ and $C$ divided by the resistance of the series group made of $R_L$ and $R_{CB}$. The voltage between $A$ and $C$ ($V_{AC}$) is equal to the difference between $V_{CC}$ and $V_E$.

$$V_{AC} = V_{CC} - V_E$$
$$V_{AC} = 10 - 0.375 = 9.625\ \text{V}$$

The resistance of the series group between $A$ and $C$ was found to be 4 kΩ. Therefore,

$$I_C = \frac{V_{AC}}{R_{AC}} = \frac{9.625}{4000} = 2.4\ \text{mA} \qquad \textit{Ans.}$$

5. Find $V_{RL}$. In Fig. 10-31*a*,

$$V_{RL} = I_C \times R_L$$
$$V_{RL} = 0.0024 \times 1000 = 2.4\ \text{V} \qquad \textit{Ans.}$$

6. Find $V_C$.

$$V_C = -V_{CC} + V_{RL}$$
$$V_C = -10 + 2.4 = -7.6\ \text{V} \qquad \textit{Ans.}$$

## PROBLEMS

1. In a circuit similar to that shown in Fig. 10-30*a*, $V_{CC} = -10$ V, $R_L = 1$ kΩ, $R_B = 200$ kΩ, and $\beta = 20$. Find (*a*) the collector current $I_C$, (*b*) the voltage across $R_L$, and (*c*) the collector voltage $V_C$.
2. Repeat Prob. 1 with the following values: $V_{CC} = -8$ V, $R_L = 2$ kΩ, $R_B = 120$ kΩ, and $\beta = 30$.
3. In a circuit similar to that shown in Fig. 10-31*a*, $V_{CC} = -8$ V, $R_L = 1$ kΩ, $R_B = 200$ kΩ, $R_E = 200$ Ω, and $\beta = 100$. Find (*a*) the emitter current $I_E$, (*b*) the emitter voltage $V_E$, (*c*) the collector current $I_C$, (*d*) the voltage across $R_L$, and (*e*) the collector voltage $V_C$.
4. Repeat Prob. 3 with the following values: $V_{CC} = -10$ V, $R_L = 2$ kΩ, $R_B = 100$ kΩ, $R_E = 140$ Ω, and $\beta = 20$.

## JOB 10-11 TRANSISTOR EQUIVALENT CIRCUITS

The electrical properties of semiconductors are very complicated. The real properties are represented in models so that reasonable analysis can be made. The most popular model is called the *H-parameter model* (Fig. 10-32). This model is a mathematical model that relates the voltages and currents at the three terminals of the transistor.

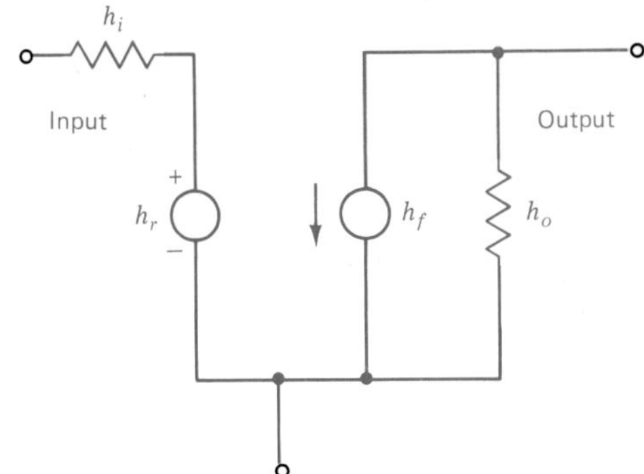


**FIGURE 10-32**
General H-parameter model for a transistor.

In the model, one of the terminals is common between the input and the output. If you look at the input terminal pair, you will notice that the model is a Thevenin equivalent circuit. The model on the output pair is a Norton equivalent circuit.

The terms for each of the components of the model are:

$h_i$—input impedance
$h_r$—reverse transfer voltage ratio
$h_f$—forward current transfer ratio
$h_o$—output conductance

Any of the three terminals of the transistor (the emitter, the base, or the collector) may be taken as the common terminal in the model. These three possible configurations are shown in Fig. 10-33. Beside

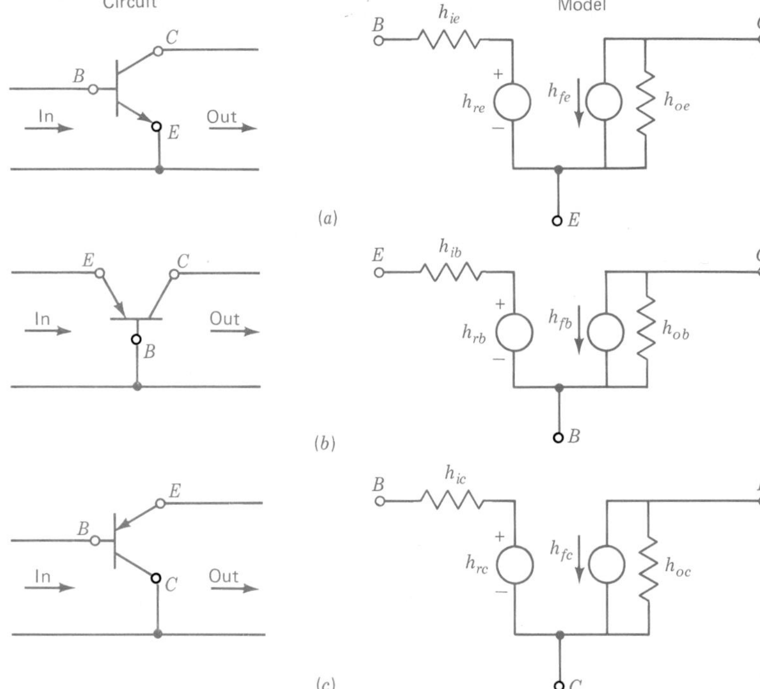


**FIGURE 10-33**
Three general transistor configurations and their associated H-parameter models: (a) common emitter; (b) common base; (c) common collector.

each model is shown the $H$-parameter equivalent circuit. Notice that the general form of the model does not change. The only difference is that the second subscript is added to each of the parameter names to indicate which terminal is common. For example, the input impedance is labeled $h_{ie}$ for the common emitter, $h_{ib}$ for the common base, and $h_{ic}$ for the common collector. The common emitter configuration is the most commonly used. To give a better understanding of the model, this configuration is further analyzed.

In the common-emitter configuration the input is between the base and the emitter. The input impedance, $h_{ie}$, is the impedance that would be measured between the base and the emitter terminals at the operating point of the transistor. This is expressed as:

$$h_{ie} = \frac{v_{be}}{i_b}$$

This is simply Ohm's law: $R = V/I$.

The reverse voltage transfer ratio $h_{re}$ is the Thevenin voltage that is measured between the base and the emitter due to the voltage that is between the collector and the emitter. It is called "reverse" because the voltage comes from the output and is measured at the input. It is expressed as

$$h_{re} = \frac{v_{be}}{v_{ce}}$$

The forward current transfer ratio is the Norton output current due to the current that comes into the input (base). It is the ratio

$$h_{fe} = \frac{i_c}{i_b}$$

For the common emitter, the forward current transfer ratio $h_{fe}$, is the beta of the transistor.

The last parameter of the model is the output conductance. This is also just like the conductance in Ohm's law:

$$h_{oe} = \frac{i_c}{v_{ce}}$$

This would be the reciprocal of the Norton equivalent resistance at the output. This is the only subtle difference between the Norton model and the $H$-paramenter model—$h_o$ is expressed as a conductance in the $H$-parameter model and as a resistance in the Norton model. There is no essential difference between the two.

Both upper- and lowercase subscripts are used for transistor $H$-parameter values. The uppercase subscripts represent the dc values of the parameter, and the lowercase represents the ac or signal parameter values.

Typical values of $H$ parameters for a modern transistor can be seen from those for the 2N3726 silicon NPN transistor. There are two classifications for the parameters, dc and signal (ac). The dc beta for the transistor varies with collector current as shown in the table below.

| $I_c$ mA | $H_{FE}$ min | $H_{FE}$ max |
|---|---|---|
| 0.01 | 80 | — |
| 0.1 | 120 | — |
| 1.0 | 135 | 350 |
| 50 | 115 | — |

Notice that there is a value of collector current for which the $H_{FE}$ peaks. Not all of the $H$ parameters are specified for dc; $H_{FE}$ is usually the only one of interest.

The $H$ parameters for the signal frequencies are:

$h_{ie}$—11.5 kΩ max
$h_{re}$—1500 × $E_6$
$h_{fe}$—135 min, 420 max at 1 mA $I_c$
$h_{oe}$—80$E_6$ mhos

The values of $h_{re}$ and $h_{oe}$ are very small. This leads to a further approximation that is often made to the $H$-parameter model for the common-emitter configuration. The reverse transfer voltage $h_{re}$ is so small for modern transistors that it is ignored. The output conductance $h_{oe}$ is also very small (output resistance is very high), and hence it also is neglected. This leaves the model shown in Fig. 10-34.

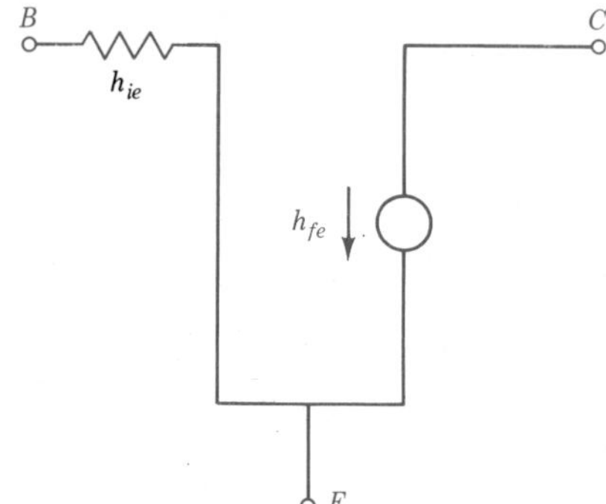


**FIGURE 10-34**
Simplified $H$-parameter model for the common-emitter circuit.

Use of the simplified model of Fig. 10-34 is very easy, except that it has in it a symbol that is rather unusual, $h_{ie}$. In the previous discussion of transistors, the concept of beta was well covered; however, when it is written as a circuit symbol as above, it is called a "dependent source". This means that the value of the current cannot be set independently but is determined by some other value in the circuit, in this case the base current. Some caution must be taken when calculating with it. Example 10-23 will show how the $H$-parameter model is used in a circuit and for calculation.

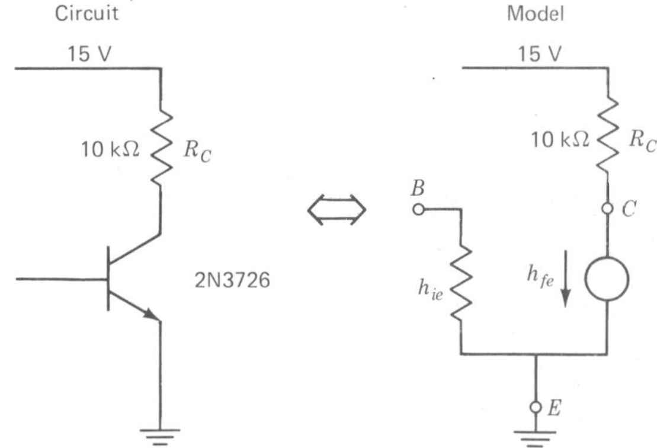


**FIGURE 10-35**
A common-emitter amplifier circuit and its associated $H$-parameter model.

EXAMPLE 10-23. Figure 10-35 shows a very elementary common-emitter amplifier. It is designed so that there is 1 mA of dc current flowing in the collector. Find the dc values of $V_{ce}$, $I_b$, and $I_e$.
Use a 2N3726 transistor and assume that the transistor has the minimum beta.

SOLUTION

1. If 1 mA of current is the collector, that current must flow through $R_c$. The voltage drop across $R_c$ is:

$$V_{rc} = I_c \times R_c = 1 \text{ mA} \times 10 \text{ k}\Omega$$
$$= 10 \text{ V}$$

the supply voltage is 15 V, and $V_{rc}$ is 10 V; therefore:

$$V_{ce} = 15 - 10 = 5 \text{ V}$$

2. The current flowing in the collector is $H_{fe}$ times the base current; therefore, to find the base current, perform the following calculation:

$$I_b = \frac{I_c}{H_{fe}} = \frac{1 \text{ mA}}{135} = 7.4\ \mu\text{A}$$

3. The emitter current is the sum of the collector current and the base current.

$$I_e = 1 \text{ mA} + 7\ \mu\text{A} = 1.007 \text{ mA}$$

For a modern high-gain transistor of this type, the example shows that the base current can usually be ignored with respect to the collector current. This gives the approximation that the emitter current is equal to the collector current. As shown above, this is a very good approximation.

EXAMPLE 10-24. Using the same circuit as that in Example 10-23, find the $V_{be}$ and $V_{rc}$ for a signal of 1 $\mu$A into the base of the transistor. Notice that in this example we are interested in the ac, or signal values. The parameters used are thus the signal parameters that are expressed with lowercase symbols.

SOLUTION

1. First find the collector current. Assume the minimum value of $h_{fe}$:

$$i_c = h_{fe} \times i_b = 135 \times 1E_6 \text{ A} = 0.135 \text{ mA}$$

2. Now find $v_{rc}$.

$$v_{rc} = i_c \times R_c = 0.135 \text{ mA} \times 10 \text{ k}\Omega = 1.35 \text{ V}$$

3. The $v_{be}$ is caused by the base current flowing through $h_{ie}$. Therefore:

$$v_{be} = i_b \times h_{ie} = 1E_6 \text{ A} \times 11.5 \text{ k}\Omega = 11.5 \text{ mV}$$

For this amplifier an 11.5-mV input signal is amplified to 1.35 V across the collector resistor.

## SELF-TEST 10-20

1. The most popular model of a transistor is the ______ model. | *H*-parameter

2. In this model the input is a ______ equivalent and the output is a ______ equivalent. | Thevenin  Norton

3. On the input the value of $h_i$ is the input ______. It is the ratio ______. | impedance  $V_{in}/I_{in}$

4. The output parameter $h_o$ is the output ______ obtained from the ratio ______. | conductance  $I_{out}/V_{out}$

5. For the common-emmiter configuration, the forward current gain $h_{ie}$ is also known as ______. | beta

6. The Thevenin voltage on the input is called ______ and is the ratio ______. | $h_r$  $V_{in}/V_{out}$

7. When the parameter values are given, the uppercase values represent ______ values. | dc

8. Parameter values given in lowercase are ______ or ______ values. | ac signal

9. The most popular configuration of a transistor is the ______. | common emmiter

## PROBLEMS

1. Assume that the $H_{fe}$ of the transistor is 250. Use the circuit in Example 10-23 to find the base current.
2. If the base current is 5 $\mu$A and the collector-to-emitter voltage in Example 10-23 is 10 V, what is the beta of the transistor?
3. Using the circuit shown in Fig. 10-33, find the ac base to emitter voltage $V_{be}$ if the beta of the transistor is 250 and the base signal is 1 $\mu$A.
4. Using the conditions stated in Prob. 3, find the signal voltage across $R_c$. (Notice that for the elementary common-emitter amplifier, the beta of the transistor does not affect the base voltage but does affect the collector circuit values.)

(See Instructor's Manual for Test 10-3, Transistor Equivalent Circuits)

## JOB 10-12 ATTENUATORS

An *attenuator pad* is a combination of resistors whose purpose is (1) to reduce the source voltage to the lower voltage required by the load but (2) to keep the original source current unchanged.

The first of these requirements may be met very easily by inserting a resistance in series with the source as explained in Example 4-24 in Job 4-7. In this example, a lamp rated at 12 V and 40 Ω is to operate from a 24-V source. Therefore, the original 24-V source voltage must be reduced to the 12 V required by the lamp. This reduction is accomplished by the insertion of a 40-Ω resistor in series with the lamp. If the lamp had been connected directly to the 24-V source, the current would have been

$$I_L = \frac{V_L}{R_L} = \frac{24}{40} = 0.6 \text{ A}$$

But, with the 40-Ω series resistor added,

$$R_T = R_1 + R_L \tag{4-3}$$

$$R_T = 40 + 40 = 80\ \Omega$$

and
$$I_L = \frac{V_L}{R_T} = \frac{24}{80} = 0.3\ \text{A}$$

Notice that although the voltage was reduced from 24 to 12 V, the current drawn from the source was *also reduced* from 0.6 to 0.3 A. An attenuator circuit would also reduce the voltage from 24 to 12 V but *not* change the original current. Actually, this means that the resistance that the load presents to the source would remain unchanged. Thus, if the lamp resistance were 40 Ω, an attenuator circuit would reduce the voltage but would *not* change the total resistance of the circuit, and would therefore *not* change the original current. Now how can this be done?

Consider the circuit shown in Fig. 10-36. $V_T = 24$ V, and $I_T = V_T \div R_T$ or $^{24}/_{40} = 0.6$ A. We wish to have $V_L = 12$ V, but *keep* $I_T = 0.6$ A.

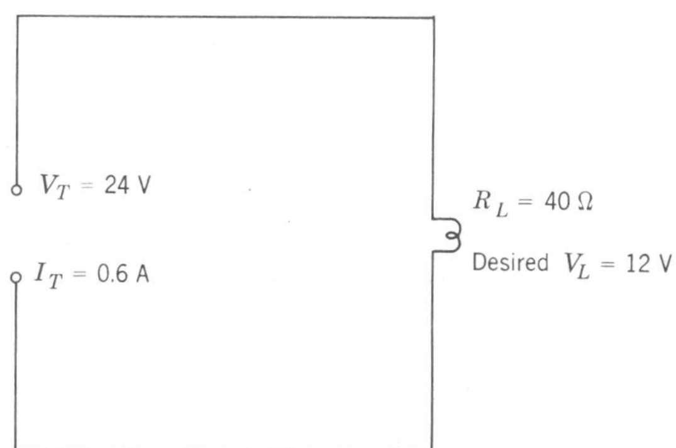


**FIGURE 10-36**
A 24-V source is to produce a load voltage of 12 V.

In order to do this, we shall insert resistors $R_1$ and $R_2$ as shown in Fig. 10-37. The current drawn by the lamp when $V_L = 12$ V is

$$I_L = \frac{V_L}{R_L} = \frac{12}{40} = 0.3\ \text{A}$$

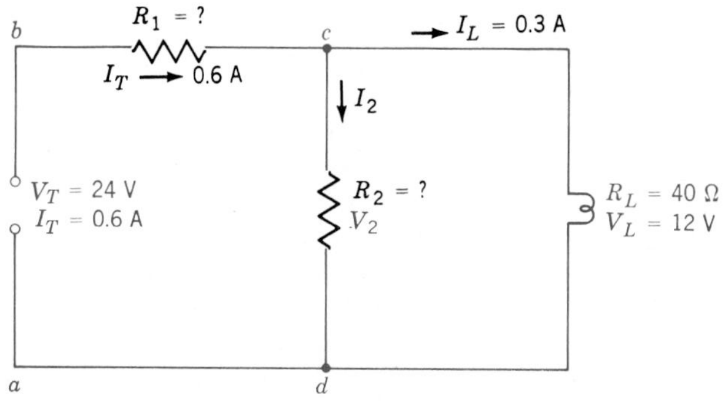

$R_1$ = ?
$I_T$ → 0.6 A
$I_L$ = 0.3 A
$I_2$
$V_T$ = 24 V
$I_T$ = 0.6 A
$R_2$ = ?
$V_2$
$R_L$ = 40 Ω
$V_L$ = 12 V
a b c d

**FIGURE 10-37**
An L pad is inserted in the circuit of Fig. 10-36.

By Kirchhoff's first law,

$$I_2 = I_T - I_L$$
$$I_2 = 0.6 - 0.3 = 0.3\ \text{A}$$

Since $V_2$ and $V_L$ are in parallel, $V_2 = V_L = 12$ V. Therefore,

$$R_2 = \frac{V_2}{I_2} = \frac{12}{0.3} = 40\ \Omega \qquad Ans.$$

In order to find $R_1$, apply Kirchhoff's second law, and trace around the circuit in the direction *abcd.*

$$V_T - V_1 - V_2 = 0$$
$$24 - V_1 - 12 = 0$$
$$24 - 12 = V_1$$
$$V_1 = 12\text{ V}$$

$R_1$ may now be found.

$$R_1 = \frac{V_1}{I_1} = \frac{12}{0.6} = 20\ \Omega \qquad Ans.$$

*Check:* If we are correct, the total resistance of the series-parallel combination of $R_1$, $R_2$, and $R_L$ will exactly equal the original $R_L$. Since $R_2$ and $R_L$ are in parallel, the effective resistance of the group equals

$$\frac{R_2 \times R_L}{R_2 + R_L} = \frac{40 \times 40}{40 + 40} = \frac{1600}{80} = 20\ \Omega$$

This 20-Ω effective resistance is in series with $R_1 = 20\ \Omega$. $R_T$ therefore equals $20 + 20 = 40\ \Omega$, which is exactly what the total resistance was *before* the insertion of the attenuator resistors. If the total resistance of the entire combination is 40 Ω, then $I_T = V_T \div R_T$, or $I_T = {}^{24}/_{40} = 0.6$ A. Notice that $I_T$ is *still* 0.6 A but that $V_L$ is now reduced to 12 V. This combination of $R_1$ and $R_2$ when inserted as shown in Fig. 10-37 is called an L *pad* owing to its resemblance to an inverted letter L.

EXAMPLE 10-25. Insert an L pad in the circuit shown in Fig. 10-38 which will reduce the load voltage to 4 V while maintaining a constant resistance to the source.

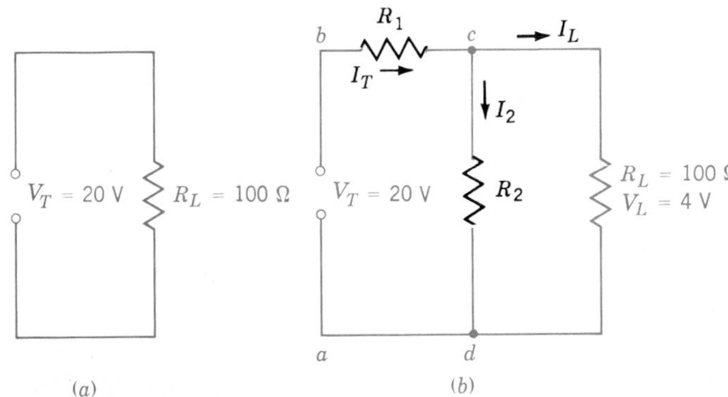


**FIGURE 10-38** (a) Original circuit. (b) Circuit with L pad inserted.

SOLUTION. Insert $R_1$ and $R_2$ as shown in Fig. 10-38*b*. The total current $I_T$ *before* the insertion of the pad was

$$I_T = \frac{V_T}{R_T} = \frac{20}{100} = 0.2 \text{ A}$$

If the voltage at the load is to be 4 V,

$$I_L = \frac{V_L}{R_L} = \frac{4}{100} = 0.04 \text{ A}$$

Find $I_2$ by Kirchhoff's first law.

$$I_2 = I_T - I_L = 0.2 - 0.04 = 0.16 \text{ A}$$

Find $R_2$. Since $R_2$ and $R_L$ are in parallel, $V_2 = V_L = 4$ V.

$$R_2 = \frac{V_2}{I_2} = \frac{4}{0.16} = 25\ \Omega \qquad \textit{Ans.}$$

Find $R_1$. Apply Kirchhoff's second law to find $V_1$ by tracing around the circuit in the direction *abcd*.

$$V_T - V_1 - V_2 \doteq 0$$
$$20 - V_1 - 4 = 0$$
$$V_1 = 16 \text{ V}$$

Therefore,

$$R_1 = \frac{V_1}{I_1} = \frac{16}{0.2} = 80\ \Omega \qquad \textit{Ans.}$$

*Check:* The total resistance of $R_1$, $R_2$, and $R_L$ connected as shown must equal $R_L$ or 100 Ω.

$$R_T = R_1 + \frac{R_2 \times R_L}{R_2 + R_L}$$
$$= 80 + \frac{25 \times 100}{25 + 100}$$
$$= 80 + \frac{2500}{125}$$
$$R_T = 80 + 20 = 100\ \Omega \qquad \textit{Check}$$

SELF-TEST 10-26. The resistance of a 40-V source is 50 Ω. In order to reduce the volume to a 50-Ω speaker, an L pad is inserted to reduce the voltage to 4 V. Find $R_1$ and $R_2$.

SOLUTION. The diagram for the circuit is shown in Fig. 10-39. In Fig. 10-39*a*, the total current *before* the pad is inserted is

$$I_T = \frac{V_T}{R_T} = \frac{V_T}{R_S + ?}$$

$$I_T = \frac{40}{?}$$

$I_T =$ ______ A

$R_L$

100

0.4

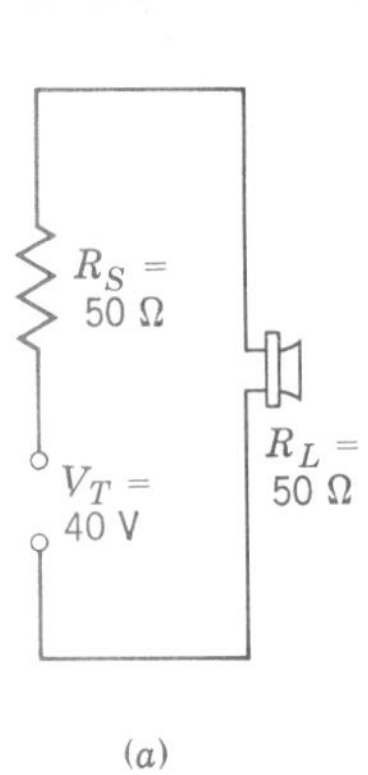


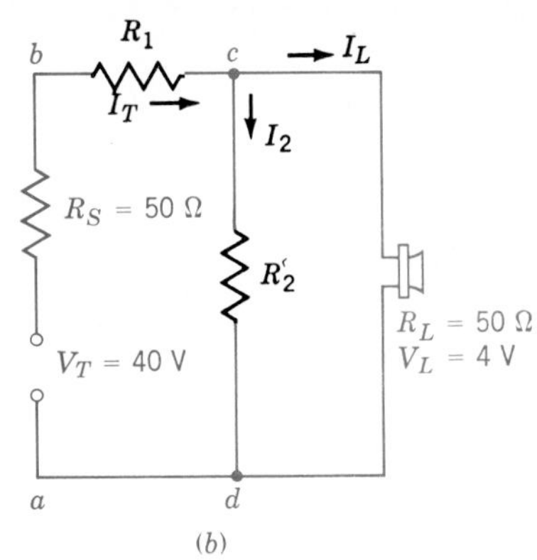


**FIGURE 10-39**
(a) Original circuit. (b) Circuit with L pad inserted.

In Fig. 10-39*b*, after $R_1$ and $R_2$ are in the circuit,

$$I_L = \frac{V_L}{R_L} = \frac{4}{50} = \_\_\_\_\_\_ \text{ A}$$ 0.08

Find $I_2$ by Kirchhoff's first law. At point *c*,

$$I_2 = I_T - I_L = 0.4 - \_\_\_\_\_\_ = 0.32 \text{ A}$$ 0.08

Find $R_2$. Since $V_2 = V_L =$ ______ V, 4

$$R_2 = \frac{V_2}{I_2} = \frac{4}{0.32} = \_\_\_\_\_\_ \ \Omega$$ 12.5

Find $R_1$. Apply Kirchhoff's second law to find $V_1$ by tracing around the circuit in the direction *abcd*.

$$V_T - V_S - V_1 - \_\_\_\_\_\_ = 0$$ $V_2$

$$V_T - I_T R_S - V_1 - V_2 = 0$$

$$40 - (\_\_\_\_\_\_)(50) - V_1 - 4 = 0$$ 0.4

$$40 - \_\_\_\_\_\_ - V_1 - 4 = 0$$ 20

$$\_\_\_\_\_\_ - V_1 = 0$$ 16

$$V_1 = \_\_\_\_\_\_ \text{ V}$$ 16

Therefore,

$$R_1 = \frac{V_1}{I_1} = \frac{V_1}{I_T} = \frac{16}{?}$$ 0.4

$$R_1 = \_\_\_\_\_\_ \ \Omega \qquad \textit{Ans.}$$ 40

*Check:* The total resistance of $R_1$, $R_2$, and $R_L$ connected as shown must equal $R_L$ or 50 Ω.

$$R_T = R_1 + \frac{R_2 \times ?}{R_2 + ?}$$ $R_L$ $R_L$

$$= \_\_\_\_\_\_ + \frac{12.5 \times 50}{12.5 + 50}$$ 40

$$= 40 + \_\_\_\_\_\_$$ 10

$$R_T = 50 \ \Omega \qquad \textit{Check}$$

## PROBLEMS

Insert an L pad into each of the following circuits. Find the values of $R_1$ and $R_2$ which will provide the required reduction in voltage.

| PROBLEM | CIRCUIT FIG. NO. | $V_T$, V | $R_S$, Ω | $R_L$, Ω | $V_L$, V |
|---|---|---|---|---|---|
| 1 | 10-38*a* | 60 | .... | 15 | 12 |
| 2 | 10-39*a* | 60 | 60 | 60 | 15 |
| 3 | 10-38*a* | 120 | .... | 10 | 60 |
| 4 | 10-39*a* | 120 | 100 | 100 | 30 |
| 5 | 10-39*a* | 120 | 60 | 60 | 24 |
| 6 | 10-39*a* | 100 | 20 | 20 | 5 |
| 7 | 10-39*a* | 250 | 100 | 100 | 100 |
| 8 | 10-38*a* | 6 | .... | 10 | 1.5 |
| 9 | 10-39*a* | 40 | 100 | 100 | 5 |
| 10 | 10-39*a* | 6.3 | 20 | 20 | 2.1 |

## T-TYPE ATTENUATORS

As we have seen, the L-type attenuator maintains a constant resistance to the source even though the voltage across the load is changed. However, the position of the source and the load may not be interchanged, because the resistance presented on one side of the pad is different from the resistance on the other side. If it is desired that the pad present a constant resistance on *each* side of it, a *T type* of pad is needed. This is a symmetrical pad and offers the *same* resistance on both sides. When this type of pad is used, the source and the load *may* be interchanged, because the resistance presented by each side of the pad is constant.

In the T type, $R_1 = R_3$ and $R_S = R_L$.

EXAMPLE 10-27. In Fig. 10-40*a*, it is desired to reduce the voltage across $R_L$ to 20 V. Design a T-type attenuator pad to achieve this result.

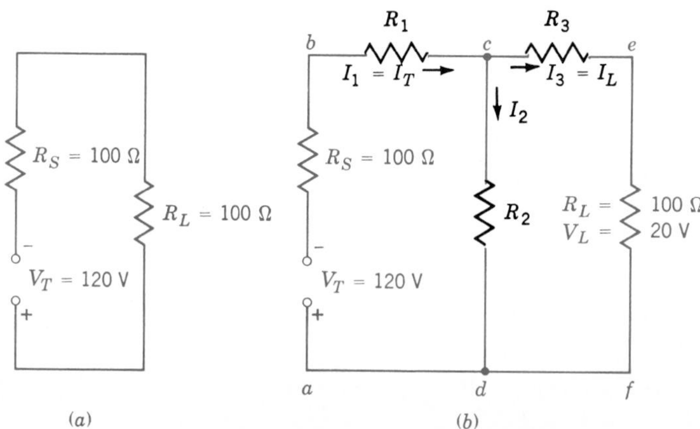


**FIGURE 10-40**
(*a*) Original circuit. (*b*) Circuit with T pad inserted; $R_1 = R_3$.

SOLUTION. In Fig. 10-40*a*, the total current *before* the pad is inserted is

$$I_T = \frac{V_T}{R_T} = \frac{V_T}{R_S + R_L} = \frac{120}{100 + 100} = \frac{120}{200} = 0.6 \text{ A}$$

In Fig. 10-40*b*, after $R_1$, $R_2$, and $R_3$ have been inserted,

$$I_L = \frac{V_L}{R_L} = \frac{20}{100} = 0.2 \text{ A}$$

Find $I_2$ by Kirchhoff's first law operating at point *c*.

$$I_2 = I_1 - I_L = 0.6 - 0.2 = 0.4 \text{ A}$$

$R_1$, $R_2$, and $R_3$ may be found by applying Kirchhoff's second law to different loops of the circuit of Fig. 10-40*b*. Trace the circuit *abcd*, assuming the currents to be in the directions indicated on the diagram.

$$\begin{aligned} 120 - 0.6(100) - 0.6R_1 - 0.4R_2 &= 0 \\ 120 - 60 - 0.6R_1 - 0.4R_2 &= 0 \\ -0.6R_1 - 0.4R_2 &= -60 \qquad (1) \end{aligned}$$

Now trace the circuit *cefdc*.

$$\begin{aligned} -0.2R_3 - 20 + 0.4R_2 &= 0 \\ -0.2R_3 + 0.4R_2 &= 20 \end{aligned}$$

but since $R_1 = R_3$, we can substitute $R_1$ for $R_3$ in this equation.

$$-0.2R_1 + 0.4R_2 = 20 \qquad (2)$$

Solve equations (1) and (2) simultaneously.

$$\begin{aligned} -0.6R_1 - 0.4R_2 &= -60 \qquad (1) \\ -0.2R_1 + 0.4R_2 &= 20 \qquad (2) \\ \text{Add:} \quad -0.8R_1 \qquad &= -40 \end{aligned}$$

$$R_1 = \frac{-40}{-0.8} = 50 \ \Omega \qquad \textit{Ans.}$$

and

$$R_3 = 50 \ \Omega \qquad \textit{Ans.}$$

Substitute this value for $R_1$ in equation (2).

$$\begin{aligned} -0.2R_1 + 0.4R_2 &= 20 \\ -0.2(50) + 0.4R_2 &= 20 \\ -10 + 0.4R_2 &= 20 \\ 0.4R_2 &= 20 + 10 \\ 0.4R_2 &= 30 \end{aligned}$$

$$R_2 = \frac{30}{0.4} = 75 \ \Omega \qquad \textit{Ans.}$$

*Check:* The total resistance of $R_1$, $R_2$, and $R_L$ connected as shown must equal $R_L$, or 100 Ω.

$$\text{The resistance from } c \text{ to } f = R_3 + R_L$$
$$= 50 + 100 = 150\ \Omega$$

This 150 Ω is in parallel with $R_2$. The resistance of this group is equal to

$$\frac{75 \times 150}{75 + 150} = \frac{11{,}250}{225} = 50\ \Omega$$

This 50 Ω is in series with $R_1$. The total resistance is

$$R_T = 50 + R_1$$
$$R_T = 50 + 50 = 100\ \Omega$$

which is equal to $R_L$. *Check*

## PROBLEMS

Insert a T pad into the following circuits similar to that shown in Fig. 10-40*a*. Find the values of $R_1$, $R_2$, and $R_3$ which will provide the required reduction in voltage.

| PROBLEM | $V_T$, V | $R_S$, Ω | $R_L$, Ω | $V_L$, V |
|---|---|---|---|---|
| 1 | 120 | 60 | 60 | 30 |
| 2 | 12 | 200 | 200 | 4 |
| 3 | 60 | 60 | 60 | 15 |
| 4 | 100 | 400 | 400 | 30 |
| 5 | 60 | 10 | 10 | 5 |
| 6 | 200 | 400 | 400 | 20 |
| 7 | 24 | 100 | 100 | 1.5 |

(See Instructor's Manual for Test 10-4, Attenuator Circuits)

## JOB 10-13 POWER AND RESISTOR SIZE

In all the previous work there has been only small concern for the power that is dissipated in a resistor and the size of that resistor. The greater the power used by a device, the greater the heat produced. If this heat cannot be dissipated to the surrounding atmosphere, the resistor becomes hotter and hotter until it burns up. To overcome this difficulty, resistors are manufactured and rated according to their ability to dissipate the heat generated within them. High-power resistors are much larger than low-power resistors.

Resistors are manufactured in standard power sizes. These standards are developed by groups such as the military and the Electronics Industry Associations (EIA). These standards are necessary so that the design and manufacture of electronic hardware will be simplified.

Resistor standards are grouped in many ways. One group is according to the level of power that they can handle. This general grouping has three major categories. The first is called "general-purpose" resistors.

These are resistors that can handle up to 2 W. The second group is called "medium power" resistors. This group can handle 2 to 6 W. The third group is called "high-power" resistors. This group will handle 6 to 210 W.

The general-purpose group is probably the most standard of the three groups. The most popular type of resistor in this group is the carbon composition type. The standard sizes are $^1/_8$, $^1/_4$, $^1/_2$, 1, and 2 W. Notice that each size is twice as high in heat dissipation ability as the next-lower size.

For the next two groups, one of the most popular types of resistor is the wirewound resistor. Standards for this group give the following sizes: 3, 5, 7, 10, 15, 26, 55, 113, 159, and 210 W.

It is usually possible to find a manufacturer who either produces another standard or manufactures nonstandard devices. This may be an advantage in some situations, but replacement and cost may be a problem.

Once you have determined the power in a resistor by use of Eqs. (7-1), (7-4), or (7-5), you must select the size of resistor that will handle that, or more power. This is similar to the selection of wire sizes.

EXAMPLE 10-28. Find the size of resistor needed for the resistors shown in Fig. 6-20.

SOLUTION

1. For resistor $R_1$, the power is 30 V × 3 A = 90 W. From the list of standard sizes, the next-*larger* size is 113 W. Thus, although the resistor generates only 90 W, we must choose a 10-Ω 113-W resistor.
2. For resistor $R_2$, the power is 3.75 V × 0.75 A = 2.81 W. Choose the next-larger size of standard resistor—a 5-Ω 3-W resistor.

Continuing as above, we can develop the following tabulation of resistor powers and sizes:

| *R*, No. | *R*, Ω | POWER, W | SIZE, W |
|---|---|---|---|
| 1 | 10 | 90 | 113 |
| 2 | 5 | 2.81 | 3 |
| 3 | 10 | 5.63 | 7 |
| 4 | 24 | 5.06 | 7 |
| 5 | 8 | 40.5 | 55 |
| 6 | 20 | 180 | 210 |

## PROBLEMS

1. Find the proper resistor power size for the resistors shown in Fig. 10-37.
2. Find the proper resistor power size for the resistors shown in Fig. 10-39.

3. Find the proper resistor power size for the resistors shown in Fig. 4-8.
4. Find the proper resistor power size for the resistors shown in Fig. 4-9.
5. Find the proper resistor power size for the resistors shown in Fig. 5-21.
6. Find the proper resistor power size for the resistors shown in Fig. 6-12.

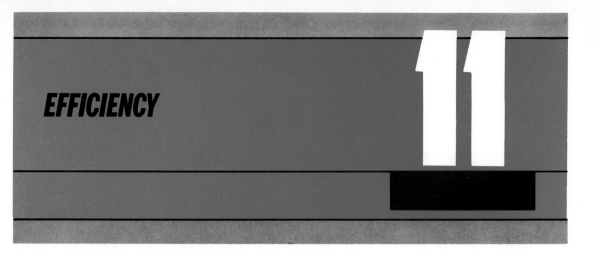

## JOB 11-1 CHECKUP ON PERCENT (DIAGNOSTIC TEST)

The efficiency of electric machinery is expressed as a percent. The concept of percent is used throughout the fields of electricity and electronics to describe and compare electrical effects. Can you solve the following problems which use percent? If you have difficulty with any of these problems, turn to Job 11-2, which follows.

### PROBLEMS

1. What is the possible error in a resistor marked 1000 Ω if the indicated error (tolerance) is 10 percent?
2. A solenoid exerts a pull of 0.9 lb at its rated voltage. What pull is exerted at 85 percent of its rated voltage?
3. A 15-A fuse carried a 15 percent overload for 3 s. What current flowed through the fuse during this time?
4. The voltage to a relay might vary up to 10 percent of the rated value. Is a voltage of 200 V within the normal variation if the rated voltage is 230 V?
5. The peak voltage of an ac wave is 141 percent of the ac meter reading. Find the peak voltage if the meter reads 50 V.
6. What is the efficiency of a motor if it uses 20 units of energy and delivers 18 units of energy?
7. What is the efficiency of a transmission line if the power supplied is 4500 W and the power delivered is 4200 W?
8. A transformer has an efficiency of 96 percent. If the transformer uses 60 W, what power will it deliver?
9. The National Electrical Code specifies a maximum of 2 percent voltage drop in a house line. What is the minimum voltage at a load if the supply voltage is 120 V?
10. A television mechanic bought $125 worth of capacitors at a 35 percent discount. What did he pay for the tubes?

11. If he was also allowed 2 percent discount for payment within 10 days, what was his actual cost?
12. A television mechanic charged the list price of \$2.25 for a capacitor. If the item cost her \$1.80, what was her percent of profit?
13. The television mechanic estimated that her overhead expenses equaled 20 percent of her cost. If she charges the list price of \$4 for an item, what is the most that she should pay for it?
14. Some storage batteries are shipped dry and filled with electrolyte by the seller. In making up a batch of electrolyte to contain 40 percent acid, how much acid should be used to make 32 oz of electrolyte?

## JOB 11-2 BRUSHUP ON PERCENT

The word *percent* is used to indicate some portion of 100. A mark of 92 percent on an examination indicates that the student got 92 points out of a possible 100 points. This may also be written as a decimal fraction (0.92) or as a common fraction ($^{92}/_{100}$), both of which are read as 92 *hundredths.* A percent indicated by a percent sign (%) cannot be used in a calculation until it has been changed to an equivalent decimal fraction.

**Changing percents to decimals.** Since 92 percent means $^{92}/_{100}$, the change to a decimal is made by dividing 92 by 100. This is easily accomplished by moving the decimal point two places to the left as shown in Job 2-6.

**RULE** To change a percent to a decimal, move the decimal point two places to the left and drop the percent sign.

EXAMPLE 11-1. Change the following percents to decimals.

SOLUTION

$$32\% = 32.\% = 0.32 \quad \textit{Ans.}$$
$$8\% = 8.\% = 0.08 \quad \textit{Ans.}$$
$$16^1/_2\% = 16.5\% = 0.165 \quad \textit{Ans.}$$

**PROBLEMS**

Change the following percents to decimals:

| | | | | |
|---|---|---|---|---|
| 1. 38% | 2. 60% | 3. 6% | 4. 19% | 5. 4% |
| 6. 26.4% | 7. 3.6% | 8. 100% | 9. 125% | 10. 0.5% |
| 11. $16^2/_3$% | 12. $^3/_4$% | 13. 62.5% | 14. 1% | 15. 12.5% |
| 16. 0.9% | 17. 2.25% | 18. $5^1/_2$% | 19. $4^1/_4$% | 20. $^1/_2$% |

**Changing decimals to percents.** To change a decimal to a percent is to reverse the process of changing a percent to a decimal.

**RULE** To change a decimal to a percent, move the decimal point two places to the right and add a percent sign.

EXAMPLE 11-2. Change the following decimals to percents:

SOLUTION

$0.20 = 20.\% = 20\%$ *Ans.*

$0.06 = 6.\% = 6\%$ *Ans.*

$0.125 = 12.5\%$ *Ans.*

$1.1 = 110.\% = 110\%$ *Ans.*

$2 = 2.00 = 200.\% = 200\%$ *Ans.*

**PROBLEMS**

Express the following decimals as percents:

| | | | | |
|---|---|---|---|---|
| 1. 0.50 | 2. 0.04 | 3. 0.2 | 4. 0.075 | 5. 1.45 |
| 6. 0.008 | 7. 1.00 | 8. 0.87 | 9. 0.625 | 10. 0.092 |
| 11. 0.222 | 12. 0.15 | 13. 0.7 | 14. 3 | 15. 0.055 |

**Changing common fractions to percents**

**RULE** To change a common fraction to a percent, express the fraction as a decimal; then move the decimal point two places to the right and add a percent sign.

The decimal equivalents for many of the common fractions may be found in Table 2-3.

EXAMPLE 11-3. Express the following fractions as percents.

SOLUTION

$^3/_4 = 0.75 = 75\%$ *Ans.*

$^1/_2 = 0.50 = 50\%$ *Ans.*

$^5/_8 = 0.625 = 62.5\%$ *Ans.*

If the fraction is not on the decimal equivalent chart, change the fraction to a decimal by dividing the numerator by the denominator as shown in Job 2-5.

EXAMPLE 11-4. Express $^8/_{25}$ as a percent.

SOLUTION

$$\frac{8}{25} = 25\overline{)8.00} = 0.32 = 32\% \quad Ans.$$

$$\begin{array}{r} 0.32 \\ 25\overline{)\,8.00} \\ -7\,5 \\ \hline 50 \\ -50 \\ \hline 0 \end{array}$$

**PROBLEMS**

Express the following common fractions as percents:

| | | | | |
|---|---|---|---|---|
| 1. $^1/_4$ | 2. $^3/_8$ | 3. $^2/_5$ | 4. $^4/_5$ | 5. $^5/_8$ |
| 6. $^3/_{16}$ | 7. $^3/_{10}$ | 8. $^3/_5$ | 9. $^3/_6$ | 10. $^9/_{10}$ |
| 11. $^{13}/_{20}$ | 12. $^3/_{32}$ | 13. $^2/_3$ | 14. $^5/_6$ | 15. $^5/_{11}$ |
| 16. $^3/_7$ | 17. $^{16}/_{48}$ | 18. $^{12}/_{30}$ | 19. $^2/_9$ | 20. $^{13}/_{15}$ |

**Using percent in problems.** There are three parts to every problem involving percent.

1. The *base* $B$ is the entire amount.
2. The *rate* $R$ is the percent of the base.
3. The *part* $P$ is the portion of the base.

These three parts are combined in the following

**FORMULA** $B \times R = P$ **11-1**

## Finding the part

EXAMPLE 11-5. An electrician earning \$340 per week got a $6^1/_2$ percent increase in pay. Find the amount of the increase.

SOLUTION. Given: $B = \$340$ Find: $P = ?$
$R = 6^1/_2\% = 0.065$

The rate of $6^1/_2$ percent must be changed to a working decimal, as shown above. In the formula, $R = 0.065$.

$$P = B \times R$$
$$P = 340 \times 0.065 = \$22.10 \quad Ans.$$

EXAMPLE 11-6. The resistance of a resistor depends on the accuracy with which it is manufactured. This is shown on the resistor as a fourth band of color called the *tolerance value.* What are the upper and lower values of resistance that might be expected on a resistor marked 3000 Ω and 5 percent tolerance?
*Note:* The tolerance colors are: gold = 5%; silver = 10%; black (or no color) = 20%.

SOLUTION

Given: $B = 3000\ \Omega$ Find: Tolerance $= P = ?$

$R =$ tolerance $= 5\% = 0.05$ Upper value = ?

Lower value = ?

1. Find the permitted tolerance.

$$P = B \times R$$

$$P = 3000 \times 0.05 = 150\ \Omega \quad \textit{Ans.}$$

2. Find the upper and lower values.

$$\text{Upper value} = 3000 + 150 = 3150\ \Omega \quad \textit{Ans.}$$

$$\text{Lower value} = 3000 - 150 = 2850\ \Omega \quad \textit{Ans.}$$

SELF-TEST 11-7. The voltage lost in a line supplying a motor is 5 percent of the generator voltage of 220 V. Find (*a*) the voltage lost and (*b*) the voltage supplied to the motor load.

SOLUTION. Given: $V_G =$ base $B =$ ______ Find: $V_l =$ part $P = ?$ | 220

$R = 5\% =$ __(decimal)__ $V_L = ?$ | 0.05

*a.* $B \times R = P$ (11-1)

$220 \times$ ______ $= P$ | 0.05

$P =$ ______ V | 11

$V_l$ ______ V *Ans.* | 11

*b.* $V_L = V_G -$ ______ (6-3) | $V_l$

$V_L$ ______ $- 11$ | 220

$V_L$ ______ V *Ans.* | 209

**PROBLEMS**

1. How much is 22 percent of 150?
2. How much is 6 percent of $500?
3. How much is 2.5 percent of $300?
4. What is the amount of sales tax on an order of $7.50 if the tax rate is 3 percent?
5. A television benchman earned $375 per week. He received an increase of 15 percent. Find the amount of the increase.
6. A TV mechanic used 30 percent of a 1000-ft roll of push-back wire. How many feet of wire did she use?
7. A 15-A fuse carried a 15 percent overload for 2 s. What current flowed through the fuse during this time?
8. What is the possible error in a 500-Ω resistor if it is marked with a silver band indicating only 10 percent accuracy?
9. The effective value of an ac wave is 70.7 percent of the peak value.

What is the effective voltage of a wave which reaches a peak of 165 V?

10. In a brightness-control circuit, 6 percent of the 525 horizontal lines were blanked out. How many lines were blanked out?
11. Out of a lot of 200 transistors, $2^1/_2$ percent were rejected as being defective. How many were rejected?
12. The voltage drop in a line supplied by a 220-V generator is 2 percent of the generator voltage. Find the voltage drop and the voltage supplied to the load.
13. How many watts of power are lost in a transformer rated at 250 W if 3 percent of the energy is lost as heat?
14. What is the interest for 1 year at $14^1/_2$ percent on \$125?
15. If you are allowed a discount of 15 percent on a bill amounting to \$127.85, how much is the discount?
16. A zener diode rated at 10 W is used as a regulator in a power supply as shown in Fig. 11-1. The power supply delivers 20 V to the load. Find the largest current which the zener should be allowed to pass if only 70 percent of the rated current is permitted.

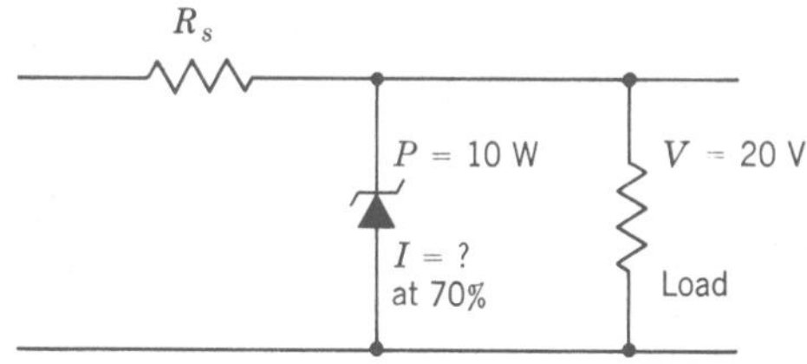


**FIGURE 11-1**
A zener diode used as a regulator in a power supply.

### Finding the rate

EXAMPLE 11-8. 20 is what percent of 80?

SOLUTION Given: $B = 80$ Find: $R = ?$
$P = 20$

$$B \times R = P \tag{11-1}$$

$$80 \times R = 20$$

$$R = \frac{20}{80} = \frac{1}{4} = 0.25 = 25\% \quad \textit{Ans.}$$

EXAMPLE 11-9. In transmitting 50 hp by a belt system, 1.2 hp was lost due to slippage. What percent of the power was lost?

SOLUTION. Given: $B = 50$ hp Find: $R = ?$
$P = 1.2$ hp

$$B \times R = P$$

$$50 \times R = 1.2$$

$$R = \frac{1.2}{50} = 0.024 = 2.4\% \qquad Ans.$$

**PROBLEMS**

1. 30 is what percent of 120?
2. 30 is what percent of 80?
3. What percent of 60 is 24?
4. What percent of 85 is 12?
5. What percent of 25 is 0.5?
6. What percent of 30 is 40?
7. 150 is what percent of 100?
8. 24.5 is what percent of 70?
9. What percent of $25^1/_2$ is $8^1/_2$?
10. What percent of 18.5 is 3.5?
11. A discount of \$2 was given on a bill of \$16. Find the rate of discount.
12. If three tubes out of a lot of 90 tubes are defective, what percent are defective?
13. If 100 m of a 250-m roll of wire has been used, what percent has been used?
14. A signal generator costing \$65 was sold at a loss of \$25. Find the rate of loss.
15. In transmitting 32 hp by a belt system, 0.72 hp was lost due to slippage. What percent of the power was lost?
16. A team played 18 games and won 15 of them. What percent of the games played did they win?
17. An electrician used 40 ft of BX cable from a 200-ft-long coil. What percent did he use?
18. A woman earning \$380 per week received an increase of \$45 per week. Find the percent of increase.
19. The voltage loss in a supply line is 5.5 V. If the generator voltage is 110 V, find the rate of loss.
20. A 15-A fuse carried 18 A for 2 s. What was the percent of overload?

## Finding the base

EXAMPLE 11-10. 5 is 25 percent of what number?

SOLUTION. Given: $P = 5$ Find: $B = ?$

$R = 25\% = 0.25$

$$B \times R = P \qquad (11\text{-}1)$$

$$B \times 0.25 = 5$$

$$B = \frac{5}{0.25} = 20 \qquad Ans.$$

EXAMPLE 11-11. A mechanic is able to save \$18.50 each week. This amount is equal to $12^1/_2$ percent of his weekly wages. How much does he earn each week?

SOLUTION

Given: $P$ = amount saved = \$18.50 Find: $B$ = total wages = ?

$R$ = percent saved = $12\frac{1}{2}\%$ = 0.125

$$B \times R = P \tag{11-1}$$

$$B \times 0.125 = 18.50$$

$$B = \frac{18.50}{0.125} = \$148 \qquad \textit{Ans.}$$

**PROBLEMS**

1. 10 is 50 percent of what number?
2. 20 is 4 percent of what number?
3. 16 is 40 percent of what number?
4. 25 is 2.5 percent of what number?
5. 70 is $3\frac{1}{2}$ percent of what number?
6. The voltage loss in a line is 3 V. If this is 2 percent of the generator voltage, what is the generator voltage?
7. A mechanic received a 15 percent increase in her wages amounting to \$35.75. What were her wages before the increase?
8. A motor has an output of 5 hp. This amount is 85 percent of the power put into the motor. Calculate the power input.
9. A fuse carried 16.5 A. This was 110 percent of the fuse rating. What is the fuse rating?
10. What must be the generator voltage if only 98 percent of the generator voltage is delivered to a 117-V line?

## REVIEW OF PERCENT PROBLEMS

**RULE** To change a percent to a decimal, move the decimal point ______ places to the ______ and drop the percent sign. | two left

**RULE** To change a decimal to a percent, move the decimal point ______ places to the ______ and add a percent sign. | two right

**RULE** To change a common fraction to a percent, express the fraction as a decimal; then move the decimal point ______ places to the right and add a ______ sign. | two percent

All percent problems contain three parts:

1. The base $B$ is the ______ amount. | total

2. The rate $R$ is a percent of the ________. | base
3. The part $P$ is the portion of the ________. | base

The relationship among these parts is given by the formula

$$\_\_\_\_\_ \times \_\_\_\_\_ = \_\_\_\_\_ \qquad \boxed{11\text{-}1}$$

$B$ $R$ $P$

**PROBLEMS**

1. Change the following percents to decimals: (*a*) 62%, (*b*) 3%, (*c*) 5.6%, (*d*) 0.8%, (*e*) 116%, (*f*) $4^1/_2$%, and (*g*) $6^1/_4$%.
2. Change the following decimals to percents: (*a*) 0.4, (*b*) 0.08, (*c*) 0.6, (*d*) 2.00, (*e*) 0.625, (*f*) 0.045, and (*g*) 5.
3. Change the following fractions to percents: (*a*) $^3/_4$, (*b*) $^3/_5$, (*c*) $^3/_7$, (*d*) $^3/_{10}$, (*e*) $^3/_{13}$, (*f*) $^1/_6$, and (*g*) $^{24}/_{52}$.
4. What is 18 percent of 96?
5. How much is 4.5 percent of $2000?
6. 20 is what percent of 120?
7. What percent of 30 is 12?
8. 8 is 40 percent of what number?
9. How much is $3^1/_4$ percent of $1500?
10. 10 is $2^1/_2$ percent of what number?
11. What percent of 38.4 is 8?
12. How much is 0.6 percent of 75?
13. In the transmission of 6 hp by a belt system, 0.09 hp was lost owing to slippage. What percent was lost?
14. Of the 60 workers in a shop, 80 percent got production bonuses. How many got a bonus?
15. According to the National Electrical Code, the maximum allowable voltage drop to a power load is 3 percent. If the supply voltage is 120 V, find the minimum voltage at the input terminals of a motor requiring 20 A and located 200 ft from the supply.
16. A woman received 150 percent of her hourly wage for every hour that she worked above 40 h. If she worked 50 h in a certain week. at a base rate of $6.50, what was her overtime pay? What was her total salary? How much was withheld from her salary for social security at 5.85 percent tax rate?
17. A man earning $285/wk received an increase of 8 percent. Find the amount of the increase and his new salary.
18. Emitter bias resistors usually have a wattage rating about 80 percent higher than the calculated wattage. What should be the wattage rating of an emitter bias resistor developing 0.5 W?
19. The cost of rewinding an armature is $74. If the electrician wishes to make a profit of 20 percent, at what price must the customer by billed?

(See Instructor's Manual for Test 11-1, Percent)

## JOB 11-3 CONVERSION FACTORS FOR ELECTRIC AND MECHANICAL POWER

A *motor* is a device which uses electric power and converts this power into the mechanical power of a rotating shaft. The electric power supplied to a motor is measured in watts or kilowatts; the mechanical power delivered by a motor is measured in horsepower. One horsepower is equivalent to 746 W of electric power. For most calculations, it is sufficiently accurate to consider 1 hp equal to 750 W.

A *generator* is a device which uses mechanical power and converts this power into electric power. The mechanical power supplied to a generator is measured in horsepower; the electric power delivered by a generator is measured in watts or kilowatts.

**Tables of conversion factors.** Tables 11-1 and 11-2 are based on the following relationships.

$$1 \text{ kW} = 1000 \text{ W}$$
$$1 \text{ hp} = 750 \text{ W}$$

TABLE 11-1

| | TO CHANGE | TO | MULTIPLY BY |
|---|---|---|---|
| 1 | Kilowatts (kW) | Watts (W) | 1000 |
| 2 | Horsepower (hp) | Watts (W) | 750 |
| 3 | Kilowatts (kW) | Horsepower (hp) | $^4/_3$ |
| 4 | Horsepower (hp) | Kilowatts (kW) | $^3/_4$ |

TABLE 11-2

| | TO CHANGE | TO | DIVIDE BY |
|---|---|---|---|
| 5 | Watts (W) | Kilowatts (kW) | 1000 |
| 6 | Watts (W) | Horsepower (hp) | 750 |

Items 3 and 4 are derived as follows:

Since hp = W/750 (item 6) and W = kW × 1000 (item 1), we can substitute (kW × 1000) for watts in item 6. This gives

$$\text{hp} = \frac{\text{kW} \times 1000}{750}$$

or

$$\text{hp} = \text{kW} \times \frac{4}{3}$$

Similarly, since kW = W/1000 (item 5) and W = hp × 750 (item 2), we can substitute (hp × 750) for watts in item 5. This gives

$$\text{kW} = \frac{\text{hp} \times 750}{1000}$$

or

$$\text{kW} = \text{hp} \times \frac{3}{4}$$

EXAMPLE 11-12

*a.* 1.5 kW = 1.5 × 1000 = 1500 W *Ans.*
*b.* $^1/_2$ hp = $^1/_2$ × 750 = 375 W *Ans.*
*c.* $1^1/_2$ kW = $1^1/_2 \times {}^4/_3 = {}^3/_2 \times {}^4/_3$ = 2 hp *Ans.*
*d.* $1^1/_2$ hp = $1^1/_2 \times {}^3/_4 = {}^3/_2 \times {}^3/_4 = {}^9/_8$ = 1.125 kW *Ans.*
*e.* 1800 W = 1800 ÷ 1000 = 1.8 kW *Ans.*
*f.* 1125 W = 1125 ÷ 750 = 1.5 hp *Ans.*

**PROBLEMS**

Change the following units of measurement.

1. 6.5 kW to watts
2. 2 hp to watts
3. 2300 W to kilowatts
4. 2625 W to horsepower
5. 0.05 kW to watts
6. 1 kW to horsepower
7. 1 hp to kilowatts
8. $1^3/_4$ kW to horsepower
9. $2^1/_4$ hp to kilowatts
10. 7.5 kW to horsepower
11. $^1/_8$ hp to kilowatts
12. $^3/_4$ hp to watts
13. 4000 W to horsepower
14. $4^1/_2$ hp to kilowatts
15. 10 kW to horsepower
16. 8.75 kW to horsepower

## JOB 11-4 EFFICIENCY OF ELECTRIC APPARATUS

Most machines are designed to do a specific job. A motor takes in electric energy and delivers mechanical energy in the form of a rotating shaft. If it could change *all* the electric energy into mechanical energy, it would be said to be 100 percent efficient. Unfortunately, not all the electric energy put into the motor appears as mechanical energy at the shaft. Some of the energy is used in overcoming friction, and some appears as heat energy. *No energy is lost*—it merely does not appear at the shaft as *useful* energy. The amount of useful energy (the power output) is therefore always *less* than the energy received by the machine (the power input). This means that the power output is always some fractional part of the power input. This fraction, obtained by dividing the output by the input, is called the *efficiency* of the machine. It is always expressed as a percent by multiplying the fraction by 100. In general, efficiency means the ability of a device to pass on the energy it receives without loss. The efficiency of electric devices is generally large, ranging from 75 to 98 percent.

**FORMULA**

$$\text{Efficiency} = \frac{\text{output}}{\text{input}} \qquad \boxed{11\text{-}2}$$

In this formula, the output and the input must *both* be expressed in the *same units of measurement.* This will ordinarily necessitate a change in the units of measurement, since for a motor,

Output is measured in horsepower.
Input is measured in watts or kilowatts.

For a generator,

Output is measured in watts or kilowatts.
Input is measured in horsepower.

EXAMPLE 11-13. Find the efficiency of a motor which receives 4 kW and delivers 4 hp.

SOLUTION. Given: Output = 4 hp    Find: Percent eff = ?
Input = 4 kW

1. Express all measurements in the same kind of units.

Output = 4 hp = 4 × 750 = 3000 W
Input = 4 kW = 4 × 1000 = 4000 W

2. Find the efficiency.

$$\text{Eff} = \frac{\text{output}}{\text{input}} = \frac{3000}{4000} = 0.75 = 75\% \qquad \textit{Ans.}$$

SELF-TEST 11-14. A generator receives 7 hp and delivers 20 A at 230 V. Find its efficiency.

SOLUTION. Given: Output $\begin{cases} V = \_\_\_\_\_ \text{ V} \\ I = \_\_\_\_\_ \text{ A} \end{cases}$    Find: Percent eff = ?    230 20
Input = _____ hp    7

1. Express all measurements in the same kind of unit.

Output: $P = I \times V = 20 \times 230 = 4600$ _____    W
Input: 7 hp = 7 × 750 = 5250 _____    W

2. Find the efficiency.

$$\text{Eff} = \frac{?}{?}$$    output input

$$\text{Eff} = \frac{4600}{5250} = 0.876 = \_\_\_\_\_ \% \qquad \textit{Ans.}$$    87.6

SELF-TEST 11-15. A transformer requires a current of 46 mA at 120 V. If its power output is 5 W, find the efficiency of the transformer.

SOLUTION. Given: Output = _____ W    Find: Plate eff = ?    5
Input = $\begin{cases} V = \_\_\_\_\_ \text{ V} \\ I = \_\_\_\_\_ \text{ mA} \end{cases}$    120 46

Express all measurements in the same kind of unit.

Input: $P = I \times V =$ ______ $\times 125$ | 0.046

Input $P = 5.52$ ______ | W

Output: $P = 5$ ______ | W

$$\text{Plate eff} = \frac{\text{output}}{?}$$ | input

$$\text{Plate eff} = \frac{5}{5.52} = 0.40 = \text{\_\_\_\_\_\_}\ \% \qquad \textit{Ans.}$$ | 90.6

**PROBLEMS**

1. A motor rated at 2 hp (delivers 2 hp) receives 1.8 kW of energy. Find its efficiency.
2. A motor delivers 3 hp and receives 2.4 kW. Find its efficiency.
3. A "power" transformer draws 0.5 kW from a line and delivers 480 W. Find its efficiency.
4. A filament transformer draws 60 W and supplies 50 W to the tube filaments. Find its efficiency.
5. A generator rated at 10 kW (delivers 10 kW) receives 15 hp. Find its efficiency.
6. A generator delivers $1^1/_4$ kW and receives 2 hp. What is its efficiency?
7. A transmission line receives 230 kW and delivers 210 kW. What is the efficiency of transmission?
8. A shunt motor takes 24 A at 220 V and delivers 5 hp. Find the watt output of the motor and its efficiency.
9. A 2-hp motor requires 17 A at 110 V. What percent of the input is delivered at the shaft? What percent is wasted?
10. A "power" transformer draws 1.4 A from a 117-V line. What is the efficiency of the transformer if it delivers 235 V at 0.65 A?

## JOB 11-5 FINDING THE OUTPUT AND INPUT OF AN ELECTRIC DEVICE

The formula for the percent efficiency may be used to find the values of both the output and the input of electric devices. The percent efficiency must be expressed as a decimal by moving the decimal point two places to the left.

### Finding the output

EXAMPLE 11-16. Find the kilowatt output of a generator if it receives 6 hp and operates at an efficiency of 90 percent.

SOLUTION. Given: Input = 6 hp Find: Output = ?

Eff = 90% = 0.90

1. Write the formula. $\text{Eff} = \frac{\text{output}}{\text{input}}$ (11-2)

2. Substitute. $\frac{0.90}{1} = \frac{\text{output}}{6}$

3. Cross-multiply. $\text{Output} = 6 \times 0.90 = 5.4 \text{ hp}$

But the output of a generator must be measured in kilowatts. Therefore,

$$5.4 \text{ hp} = 5.4 \times \frac{3}{4} = \frac{16.2}{4} = 4.05 \text{ kW} \qquad \textit{Ans.}$$

EXAMPLE 11-17. A motor has an efficiency of 87 percent. If it draws 20 A from a 220-V line, what is its hp output?

SOLUTION. Given: Input $I = 20$ A Find: Output = ?
Input $V = 220$ V
Eff = 87% = 0.87

1. Find the wattage input.

$$P = I \times V = 20 \times 220 = 4400 \text{ W input}$$

2. Find the wattage output.

$$\text{Eff} = \frac{\text{output}}{\text{input}} \qquad (11\text{-}2)$$

$$\frac{0.87}{1} = \frac{\text{output}}{4400}$$

$$\text{Output} = 4400 \times 0.87 = 3828 \text{ W output}$$

But the output of a motor must be measured in horsepower. Therefore,

$$3828 \text{ W} = 3828 \div 750 = 5.1 \text{ hp} \qquad \textit{Ans.}$$

## Finding the input

EXAMPLE 11-18. How much power is needed to operate a 5-kW generator if its efficiency is 92 percent?

SOLUTION. Given: Output = 5 kW Find: Input = ?
Eff = 92% = 0.92

$$\text{Eff} = \frac{\text{output}}{\text{input}} \qquad (11\text{-}2)$$

$$\frac{0.92}{1} = \frac{5}{\text{input}}$$

$$0.92 \times \text{input} = 5$$

$$\text{Input} = \frac{5}{0.92} = 5.43 \text{ kW}$$

But the input to a generator is measured in horsepower. Therefore,

$$5.43 \text{ kW} = 5.43 \times \frac{4}{3} = \frac{21.72}{3} = 7.24 \text{ hp} \qquad \textit{Ans.}$$

SELF-TEST 11-19. How much current is drawn by a $2^1/_2$-hp motor if it has an efficiency of 90 percent and operates on 110 V?

SOLUTION. Given: Output = $2^1/_2$ ______ Find: Input $I$ = ? | hp

Eff = 90% = ___(decimal)___ | 0.90

Input $V$ = 110 V

1. Find the power input.

$$\text{Eff} = \frac{\text{output}}{\text{input}} \qquad (11\text{-}2)$$

$$\frac{0.90}{1} = \frac{2.5}{?}$$ | input

$0.90 \times$ ______ $= 2.5$ | input

$$\text{Input} = \frac{2.5}{?}$$ | 0.90

Input = 2.78 ______ | hp

But the input to a motor is measured in watts or ______. Therefore, | kilowatts

2.78 hp = 2.78 × ______ W | 750

or 2.78 hp = ______ W *Ans.* | 2085

2. Find the current input.

$$P = I \times V \qquad (7\text{-}1)$$

______ $= I \times 110$ | 2085

$I$ = ______ A *Ans.* | 19

## PROBLEMS

1. Find the horsepower output of a motor drawing 3 kW and operating at an efficiency of 90 percent.
2. An 80 percent efficient "power" transformer delivers 50 W. Find its power input.
3. Find the horsepower output of a motor drawing 5 A at 110 V and operating at an efficiency of 80 percent.
4. A motor operates at an efficiency of 92 percent and draws 5.5 A from a 110-V line. Find the horsepower output.
5. A transmission line operating at an efficiency of 98 percent receives 20 A at 230 V. Find the power delivered.

6. How much power is delivered by a transformer operating at an efficiency of 88 percent if it draws 0.08 kW?
7. Find the input to a 3-hp motor if its efficiency is 85 percent.
8. Find the horsepower input to a generator delivering 5 kW at an efficiency of 90 percent.
9. A generator delivers 20 A at a voltage of 110 V. Find its output in watts. If it has an efficiency of 90 percent, find the horsepower input.
10. How much current is drawn by a 2-hp motor if it has an efficiency of 87 percent and operates at 110 V?
11. What voltage is necessary to operate a 3-hp motor if it draws 11.37 A and its efficiency is 90 percent?
12. Find the kilowatt output of a generator which uses 9 hp and has an operating efficiency of 92 percent.

## JOB 11-6. LOAD MATCHING FOR POWER TRANSFER

This section combines the concepts of Thevenin equivalents, power, and efficiency to give a better understanding of how power is transferred from one circuit to another. Assume that you have an amplifier, or battery, or generator, or any other power source. Make a Thevenin equivalent at the output terminals of the device. This would result in the circuit shown in Fig. 11-2. The Thevenin resistance is usually called the *output impedance* or the *source resistance.* This represents all the resistors inside the power source.

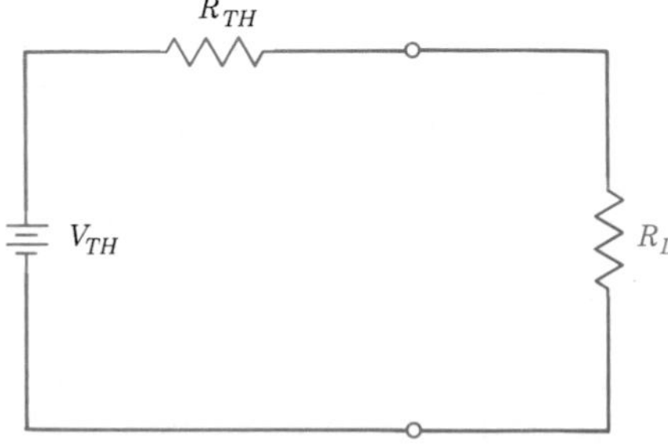


**FIGURE 11-2**
Power source Thevenin equivalent with load attached.

A load $R_L$ is attached to the power output terminals. Since we have made a Thevenin equivalent of the power source, the resulting circuit is a simple one-loop circuit. The Thevenin voltage drives the output and puts current through $R_{TH}$ and the load $R_L$. This current causes power losses in $R_{TH}$ and $R_L$. The power loss in $R_{TH}$ is lost as heat inside the power source. The power loss in the load is the desired output power. The efficiency is the ratio of the desired output power $P_L$ to the total power delivered by the voltage source ($P_{TH} + P_L$).

Assume that $V_{TH}$ is 12 V and $R_{TH}$ is 10 Ω. Table 11-3 shows how the power delivered to the load and the efficiency change as the load resistance changes.

**TABLE 11-3**

| $R_L$, Ω | $R_{TH}$, Ω | $P_{R_{TH}}$, W | $P_{R_L}$, W | Eff, % |
|---|---|---|---|---|
| 6 | 10 | 5.63 | 3.38 | 38 |
| 8 | 10 | 4.44 | 3.56 | 45 |
| 10 | 10 | 3.60 | 3.60 | 50 |
| 12 | 10 | 2.98 | 3.57 | 55 |
| 14 | 10 | 2.58 | 3.50 | 58 |

Table 11-3 indicates that as the load increases, the efficiency of the circuit increases. This occurs because as the load resistance increases, the voltage across the load also increases. Since it is a series circuit, the current through $R_L$ and $R_{TH}$ is equal. Using the power equation $P = V \times I$, we find that the power output increases relative to the power lost in the source. Careful now! This does not mean that the power keeps increasing!

As $R_L$ increases, the current in the loop will decrease, thus producing the decreasing power in $R_{TH}$ as seen in Table 11-3. Notice that the power in the load reaches a maximum (3.60 W) when $R_{TH}$ is equal to $R_L$. (See Example 15-4.)

If $R_L$ is less than $R_{TH}$, increasing $R_L$ increases the voltage across $R_L$ and the power in $R_L$ increases. When $R_L = R_{TH}$, the power in the load is equal to the power lost in the source. When $R_L$ becomes larger than $R_{TH}$, the current decreases so quickly that the power in the load decreases. This experiment helps us to formulate the following rule:

**RULE**

For maximum transfer of power from a source to a load, the source resistance must equal the load resistance.

SELF-TEST 11-20

1. The output impedance of the source is the same as the ________ impedance.
2. As the load impedance increases the power ________ increases.
3. The maximum power is delivered to the load when the load resistance ________ or ________ the output resistance.

Thevenin
efficiency

equals matches

**PROBLEMS**

1. Calculate the loop current, the power in $R_L$ and $R_{TH}$ and the efficiency for the circuit shown in Fig. 11-2, using the values of $R_L$ given in the table.
2. A power supply has an open circuit voltage of 9 V and an output

impedance of 2 Ω. What is the maximum power that it can deliver to a load?

3. A motor runs on 5 V. At maximum speed it requires $^1/_4$ W. What is the largest source impedance that the supply could have and still run the motor at maximum speed?

## JOB 11-7 REVIEW OF CONVERSION FACTORS AND EFFICIENCY

1 kW = ______ W | 1000

1 hp = 750 ______ | W

W = kW × ______ | 1000

W = hp × ______ | 750

hp = kW × ______ | $^4/_3$

kW = hp × ______ | $^3/_4$

kW = W ÷ ______ | 1000

hp = W ÷ ______ | 750

The *efficiency* of a machine is the ratio of the power ______ to the power ______ and is usually expressed as a percent. | output input

$$\text{Eff} = \frac{\text{output}}{\text{input}} \qquad \boxed{11\text{-}2}$$

**Units of measurement.** For a generator:

Output is in kilowatts or ______ or voltamperes | watts
Input is in ______ | horsepower

For a motor:

Output is in ______ | horsepower
Input is in kilowatts or watts or ______ | voltamperes

### PROBLEMS

1. Change (*a*) 500 W to kilowatts; (*b*) 500 W to horsepower; (*c*) 1.5 kW to watts; (*d*) 1.5 hp to watts; (*e*) 10 kW to horsepower; and (*f*) 10 hp to kilowatts.
2. Find the efficiency of a $1^1/_2$-hp induction motor using 1.3 kW of power.
3. Find the efficiency of a generator if it delivers 20 A at 120 V and uses 3.5 hp.
4. What is the efficiency of a $^3/_4$-hp motor if it draws 5.5 A from a 110-V line?
5. Find the horsepower output of a motor drawing 5 A at 110 V if it operates at an efficiency of 85 percent.

6. Find the kilowatt output of a generator which uses 12 hp and operates at an efficiency of 92 percent.
7. Find the power needed to operate a $^3/_4$-hp motor if its efficiency is 90 percent.
8. How much current is drawn from a 120-V line by a 3-hp motor if it operates at an efficiency of 87 percent?

(See Instructor's Manual for Test 11-2, Conversion Factors, Power, and Efficiency)

# 12

# RESISTANCE OF WIRE

## JOB 12-1 CHECKUP ON RATIO AND PROPORTION (DIAGNOSTIC TEST)

Many complicated ideas may be expressed quite simply when written in formula form. For example, the rule "In a series circuit, the voltage is directly proportional to the resistance" may be written as $V_1/V_2 = R_1/R_2$. Or "The resistance of a wire is inversely proportional to its area" may be written as $R_1/R_2 = A_2/A_1$. The mathematical concepts of *ratio and proportion* enable us to compare voltages, currents, etc., and to show how they depend on each other. We shall be using these ideas in the next and succeeding jobs. Let us check up on what we know about ratio and proportion. If you have any difficulty with any of the following problems, turn to Job 12-2 which follows.

### PROBLEMS

1. What is the ratio of 12 to 30 cm?
2. Two pulleys have diameters of 20 and 5 in, respectively. What is the ratio of their diameters?
3. If the number of turns of wire on the primary of a bell-ringing transformer is 360 turns and the secondary has 40 turns, what is the turns ratio of primary to secondary?
4. The power factor (PF) of an ac circuit is described as the ratio of its resistance $R$ to its impedance $Z$. Find the PF if $R = 500\ \Omega$ and $Z = 550\ \Omega$.
5. What is the ratio of 1 hp to 1 kW?
6. A wire 30 ft long has a resistance of 0.02 Ω. If the resistance is directly proportional to the length, find the resistance of 10 ft of this wire.
7. If 60 ft of conduit costs $9.50, how much does 150 ft cost?
8. The ratio of acid to water in the electrolyte of a storage battery is 2:3. If 30 oz of acid is used to prepare a batch of electrolyte, how much water should be mixed with it?

9. In a simple transformer, the voltage is directly proportional to the number of turns. In a simple step-down transformer, the primary has 240 turns and the secondary has 30 turns. Find the voltage of the secondary if the primary voltage is 120 V.
10. The resistance of a wire is inversely proportional to its cross-sectional area. If a wire whose area is 5200 units has a resistance of 2 Ω, find the resistance of a wire of the same material whose area is 100 units.
11. In a parallel circuit, the current is inversely proportional to the resistance. Write a formula to state this fact.
12. In a series circuit, the voltage drops are directly proportional to the resistances. What is the voltage across a 100-Ω resistance if the voltage across a 500-Ω resistance is 10 V?

## JOB 12-2 BRUSHUP ON RATIO AND PROPORTION

**Ratio.** When two quantities are compared by division, the quotient is called the *ratio* of the quantities. The two quantities must be the same *kind* of thing. That is, we can compare two weights, two lengths, or two voltages, but we cannot compare a current with a voltage, or a length with a weight.

**Units of a ratio.** The comparison, or ratio, of 2 yd to 1 ft is not 2 to 1. This would mean that 2 yd is only twice as large as 1 ft, when actually it is six times as large. Therefore, in order to be comparable, two quantities must be measured in the *same units* of measurement. Since 2 yd and 1 ft are both lengths, we may compare them by division.

$$\frac{2 \text{ yd}}{1 \text{ ft}} = \frac{6 \cancel{\text{ft}}}{1 \cancel{\text{ft}}} = \frac{6}{1}$$

Since the units of measurement are identical, they cancel out. Thus, a ratio itself has no units of measurement. Another way to write this ratio is 6:1, in which the ratio symbol (:) replaces the fraction bar. However it is written, the ratio is read as "6 to 1."

**RULE**

To find the ratio of two similar quantities:

1. Express them in the same units of measurement.
2. Form a fraction using the first quantity as the numerator and the second quantity as the denominator.
3. Express in lowest terms.

**Order of a ratio.** The order in which a ratio is expressed is as important as the numbers themselves. The ratio *must* be expressed in the same order in which the comparison is made.

EXAMPLE 12-1. The pinion in Fig. 12-1 has 15 teeth and is in mesh with a gear having 60 teeth. Find (*a*) the ratio of gear teeth to pinion teeth, and (*b*) the ratio of pinion teeth to gear teeth.

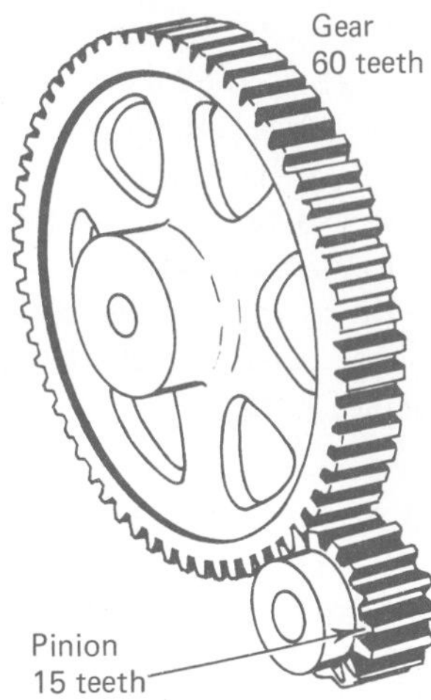


**FIGURE 12-1**
Two spur gears in mesh.

SOLUTION

*a.* $$\frac{\text{Gear}}{\text{Pinion}} = \frac{60}{15} = \frac{4}{1} = 4{:}1 \qquad \textit{Ans.}$$

This means that for every 4 teeth on the gear, there is 1 tooth on the pinion.

*b.* $$\frac{\text{Pinion}}{\text{Gear}} = \frac{15}{60} = \frac{1}{4} = 1{:}4 \qquad \textit{Ans.}$$

This means that for every 1 tooth on the pinion, there are 4 teeth on the gear.

EXAMPLE 12-2. Find the ratio of 1 mV to 1 V.

SOLUTION. Given: The quantities 1 mV and 1 V Find: Ratio = ?

$$\text{Ratio} = \frac{1\text{ mV}}{1\text{ V}} = \frac{1\text{ mV}}{1000\text{ mV}} = \frac{1}{1000} \text{ or 1:1000 (read as “1 to 1000”)}$$

EXAMPLE 12-3. Find the ratio of 1 hp to 1 kW.

SOLUTION. Given: The quantities 1 hp and 1 kW Find: Ratio = ?

750 To 1000

If it is difficult to change one unit to the other or vice versa, then change both units to a third unit. Since 1 hp = 750 W and 1 kW = 1000 W,

$$\text{Ratio} = \frac{1\text{ hp}}{1\text{ kW}} = \frac{750\text{ W}}{1000\text{ W}} = \frac{3}{4} \text{ or 3:4 (read as “3 to 4”)} \qquad \textit{Ans.}$$

Not all ratios are expressed as fractions. It is sometimes more convenient to express a ratio as a decimal or as a percent.

EXAMPLE 12-4. The power factor (PF) of an ac circuit is the ratio of its resistance $R$ to its impedance $Z$. Find the PF of a circuit if $R = 2.46\ \Omega$ and $Z = 30\ \Omega$.

SOLUTION. Given: $R = 2.46\ \Omega$ Find: PF = ?
$Z = 30\ \Omega$

$$\text{PF} = \frac{R}{Z} = \frac{2.46}{30} = 0.082$$

Since PF is often expressed as a percent,

$$\text{PF} = 0.082 \text{ or } 8.2\% \qquad \textit{Ans.}$$

EXAMPLE 12-5. The efficiency of a motor is the ratio of the output to the input and is described as a percent. Find the efficiency of a motor whose output is 5 hp and whose input is 4 kW.

SOLUTION. Given: Output = 5 hp Find: Eff = ?
Input = 4 kW

$$\text{Eff} = \frac{\text{output}}{\text{input}} = \frac{5\text{ hp}}{4\text{ kW}}$$

$$= \frac{5 \times 750}{4 \times 1000} = \frac{3750\text{ W}}{4000\text{ W}} = 0.9375$$

$$\text{Eff} = 0.9375 = 93.75\% \qquad \textit{Ans.}$$

**PROBLEMS**

Find the ratio of the quantities in each problem.

1. 3 in to 12 in
2. 6 ft to 2 ft
3. 18 V to 27 V
4. 105 turns to 20 turns
5. 3 ft to 18 in
6. 2 kHz to 500 Hz
7. 0.4 MΩ to 100,000 Ω
8. 2 hp to 3 kW
9. In Fig. 12-2, find (*a*) the ratio of $D_2$ to $D_1$ and (*b*) the ratio of $D_1$ to $D_2$.

10. Find the teeth ratio of two gears if the first has 35 teeth and the second has 30 teeth.

11. In Fig. 12-3, the turns ratio of the transformer is the ratio of the number of turns on the primary coil $N_p$ to the number of turns on the secondary coil $N_s$. Find the turns ratio.

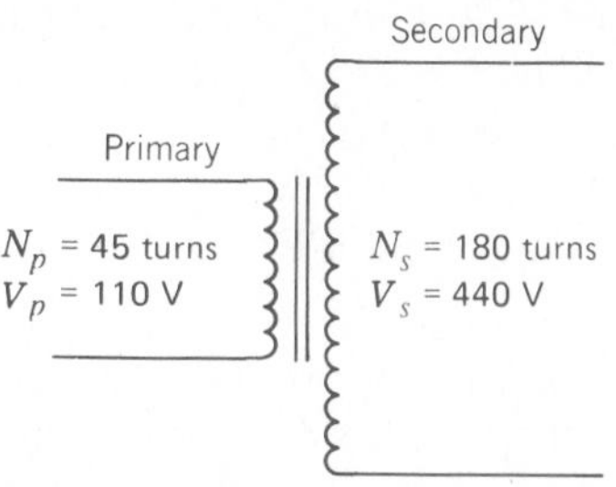


**FIGURE 12-3**
Step-up transformer.

12. In Fig. 12-3, find the ratio of the secondary turns $N_s$ to the primary turns $N_p$.
13. In Fig. 12-3, find the ratio of the primary voltage $V_p$ to the secondary voltage $V_s$.
14. If the primary voltage $V_p$ of a transformer is 125 V and the secondary voltage $V_s$ is 20 V, find the ratio of $V_p$ to $V_s$.
15. Find the efficiency of a motor if the output is 2 hp and the input is 1.8 kW.
16. The quality $Q$ of a coil is the ratio of its reactance $X$ to its resistance $R$. Find the $Q$ of an RF coil if its reactance is 2500 Ω and its resistance is 25 Ω.
17. Find the $Q$ of a tuned circuit if $X = 12{,}000\ \Omega$ and $R = 70\ \Omega$.
18. If there are 10 mm in each centimeter (cm) of metric length, find the ratio of 14 mm to 49 cm.
19. $\beta$, the current gain of a transistor, is defined as the ratio of the collector current $I_C$ to the base current $I_B$. Find the $\beta$ of a transistor if $I_B = 5\ \mu\text{A}$ and $I_C = 200\ \mu\text{A}$.
20. Find the PF of an ac circuit if $R = 56.8\ \Omega$ and $Z = 65\ \Omega$.
21. The multiplying power of a meter is the ratio of the new voltage to the old voltage. Find the multiplying power of a voltmeter reading 10 V if it is extended to read 100 V.
22. Find the multiplying power of a 150-V 150,000-Ω voltmeter which has been extended to read 750 V.
23. What is the ratio of the distance on a drawing to the actual distance if the scale of the drawing is ¼ in equals 1 ft?
24. If 34 oz of acid is mixed with 51 oz of water to make the electrolyte for a storage battery, find (*a*) the ratio of acid to water and (*b*) the ratio of water to acid.
25. A generator rated at 117 V supplies 115 V to a motor some distance away. Find (*a*) the line loss in volts, (*b*) the ratio of the line loss to the rated voltage in percent, and (*c*) the ratio of the load voltage to the rated voltage in percent.

**Proportion.** Consider the following statement. If 3 pencils cost 8 cents, then 6 pencils will cost 16 cents. At the given rate, the cost depends only on the number of pencils bought. The more pencils bought, the larger the cost. The fewer pencils bought, the smaller the cost. When two quantities depend on each other, the relationship between them may be stated as a *proportion.*

A proportion is a mathematical statement that two ratios are equal. In our problem, the two quantities are the number of pencils $N$ and the cost $C$.

| | | |
|---|---|---|
| First purchase: | $N_1 = 3$ pencils | $C_1 = 8$ cents |
| Second purchase: | $N_2 = 6$ pencils | $C_2 = 16$ cents |

The ratio of the number of pencils is

$$\frac{N_1}{N_2} = \frac{3}{6} = \frac{1}{2}$$

The ratio of the costs is

$$\frac{C_1}{C_2} = \frac{8}{16} = \frac{1}{2}$$

Since the two ratios are equal, the proportion may be written mathematically as

$$\frac{N_1}{N_2} = \frac{C_1}{C_2}$$

and is an example of a *direct proportion.*

**Direct proportion.** When two quantities depend on each other so that one increases as the other increases or one decreases as the other decreases, they are said to be *directly proportional* to each other. Notice that the two ratios are compared in the *same order* so that the items in the first situation ($N_1$ and $C_1$) are in the numerator and the items in the second situation ($N_2$ and $C_2$) are in the denominator. The proportion may be written as

$$\frac{C_1}{C_2} = \frac{N_1}{N_2} \quad \text{or} \quad \frac{C_2}{C_1} = \frac{N_2}{N_1} \quad \text{or} \quad \frac{N_2}{N_1} = \frac{C_2}{C_1} \quad \text{or} \quad \frac{N_1}{N_2} = \frac{C_1}{C_2}$$

It is not important which ratio is written first. However, once the first ratio is written, the second ratio must be written *in the same order.*

**RULE**

To set up a direct proportion between two variables:

1. Make a ratio of one of the variables.
2. Make a ratio of the second variable in the same order.
3. Set the two ratios equal to each other.

EXAMPLE 12-6. Write as a proportion: The weight of a pipe is directly proportional to its length.

SOLUTION

| First pipe | Second pipe |
|---|---|
| Length $= L_1$ | Length $= L_2$ |
| Weight $= W_1$ | Weight $= W_2$ |

Since the ratio of the lengths ($L_1/L_2$) and the ratio of the weights ($W_1/W_2$) are equal, the proportion may be written as

$$\frac{L_1}{L_2} = \frac{W_1}{W_2} \quad \text{or} \quad \frac{W_1}{W_2} = \frac{L_1}{L_2}$$

or, since the ratio of the lengths ($L_2/L_1$) and the ratio of the weights ($W_2/W_1$) are equal, the proportion may also be written as

$$\frac{L_2}{L_1} = \frac{W_2}{W_1} \quad \text{or} \quad \frac{W_2}{W_1} = \frac{L_2}{L_1}$$

**RULE**

To solve problems involving direct proportion:

1. Set up the proportion.
2. Substitute the values.
3. Solve by cross multiplication.

EXAMPLE 12-7. In the series circuit shown in Fig. 12-4, the voltage across any resistor is directly proportional to the resistance of the resistor. Find the value of $V_2$.

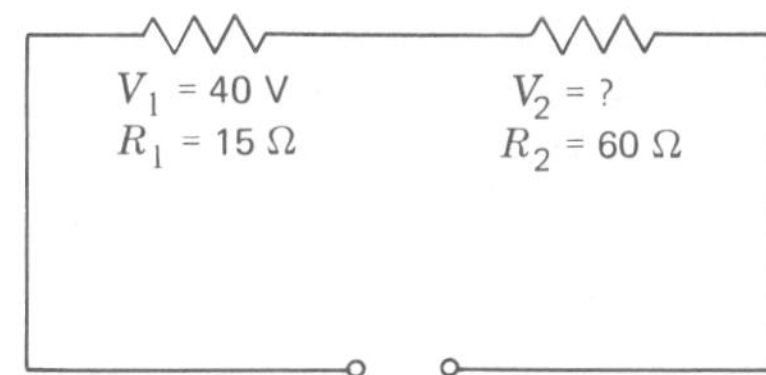


**FIGURE 12-4**

SOLUTION

1. Set up the direct proportion.

$$\frac{V_1}{V_2} = \frac{R_1}{R_2}$$

2. Substitute numbers.

$$\frac{40}{V_2} = \frac{15}{60}$$

3. Cross-multiply.

$$15 \times V_2 = 40 \times 60$$

4. Solve for $V_2$.

$$V_2 = \frac{2400}{15} = 160 \text{ V} \qquad Ans.$$

SELF-TEST 12-8. An automobile can travel 280 mi in 7 h. At the same rate of speed, how long will it take to travel 200 mi?

SOLUTION. This is an example of a direct proportion, since the distance increases as the time ______ and the distance traveled depends on the ______ spent traveling. increases time

Given: $D_1 =$ ______ mi Find: $T_2 = ?$ 280

$T_1 =$ ______ h 7

______ $= 200$ mi $D_2$

1. Set up the direct proportion.

$$\frac{D_1}{D_2} = \frac{?}{?}$$ $T_1$ $T_2$

2. Substitute numbers.

$$\frac{280}{200} = \frac{7}{?}$$ $T_2$

3. Cross-multiply.

$$280 \times T_2 = 200 \times \_\_\_\_$$ 7

4. Solve for $T_2$.

$$T_2 = \frac{1400}{?}$$ 280

$$T_2 = \_\_\_\_ \text{ h} \qquad Ans.$$ 5

**PROBLEMS**

1. An 8-ft-long steel beam weighs 2200 lb. What would be the weight of a similar beam 10 ft long?
2. If 1 gross (144) of resistors costs \$3.90, how much do 36 resistors cost?
3. An airplane can travel 2040 mi in 6 h. At the same rate, how long would it take the plane to travel 1530 mi?
4. The shadows cast by vertical poles are directly proportional to the heights of the poles. If a pole 9 ft high casts a shadow 12 ft long, how high is a tree that casts a shadow 36 ft long?
5. The volume of a gas is directly proportional to the temperature. If a gas occupies 100 ft$^3$ at 70°F, what volume will it occupy at 50°F, assuming the pressure to be constant?
6. If 14 lb of cement is used to make 70 lb of concrete, how many pounds of concrete can be made with 30 lb of cement?

7. Using a diagram similar to Fig. 12-4, find $R_1$ if $V_1 = 20$ V, $V_2 = 36$ V, and $R_2 = 1800\ \Omega$.
8. If a 500-ft length of wire has a resistance of 30 Ω, find the resistance of an 850-ft length of the same wire. Use the fact that the resistance of a wire is directly proportional to its length.
9. In a transformer, the voltage is directly proportional to the number of turns. In a diagram similar to that shown in Fig. 12-3, find the secondary voltage $V_s$ if $N_p = 160$ turns, $V_p = 18$ V, and $N_s = 400$ turns.
10. The Elco stereo amplifier model 3080 uses a series-circuit base-bias voltage divider similar to that shown in Fig. 12-5. Find the value of the bias voltage $V_2$.

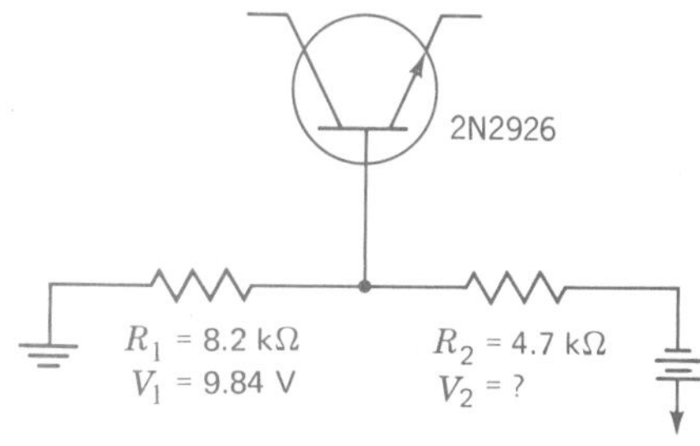


**FIGURE 12-5**

Series-circuit base-bias voltage divider.

**Inverse proportion.** When two gears are meshed as shown in Fig. 12-6, the smaller the gear, the faster it turns. When gear 1 turns once, its 40 teeth must engage 40 teeth on gear 2. Therefore, gear 2 must turn *twice* in order for 40 teeth (2 × 20) to mesh with the 40 teeth on gear 1.

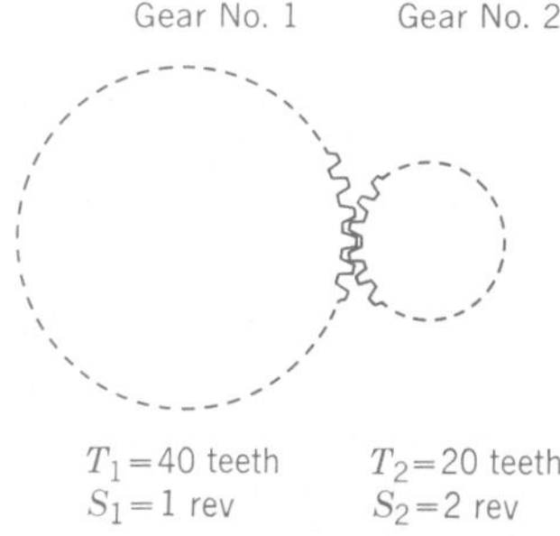


**FIGURE 12-6**

The speed is inversely proportional to the number of teeth.

When two quantities depend on each other so that one *increases* as the other *decreases* or one *decreases* as the other *increases,* they are said to be *inversely proportional* to each other. This may be written mathematically as

$$\frac{T_1}{T_2} = \frac{S_2}{S_1}$$

Notice that the two ratios are compared in the *opposite,* or *inverse,* order. The proportion will be correct regardless of which ratio is written first, *provided* the second ratio is written in the *opposite order.* Thus, the proportion may be written as

$$\frac{T_1}{T_2} = \frac{S_2}{S_1} \quad \text{or} \quad \frac{T_2}{T_1} = \frac{S_1}{S_2} \quad \text{or} \quad \frac{S_1}{S_2} = \frac{T_2}{T_1} \quad \text{or} \quad \frac{S_2}{S_1} = \frac{T_1}{T_2}$$

**RULE**

To set up an inverse proportion between two variables:

1. Make a ratio of one variable.
2. Make a ratio of the second variable in the opposite order.
3. Set the two ratios equal to each other.

EXAMPLE 12-9. Write as a proportion: "The speeds of pulleys are inversely proportional to the diameters."

SOLUTION. The diagram for the problem is shown in Fig. 12-7.

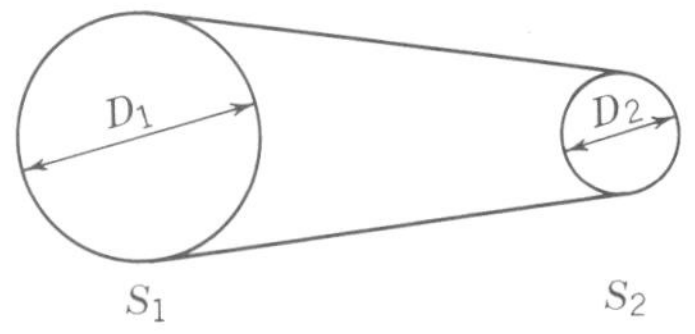


FIGURE 12-7

$$\frac{D_1}{D_2} = \frac{S_2}{S_1} \quad \text{or} \quad \frac{S_1}{S_2} = \frac{D_2}{D_1} \quad \text{or} \quad \frac{D_2}{D_1} = \frac{S_1}{S_2} \quad \text{or} \quad \frac{S_2}{S_1} = \frac{D_1}{D_2}$$

**RULE**

To solve problems involving inverse proportion:

1. Set up the proportion.
2. Substitute the values.
3. Solve by cross multiplication.

EXAMPLE 12-10. The resistance of a length of wire is inversely proportional to its cross-sectional area. A certain copper wire has a resistance of 32 Ω and a cross-sectional area of 20 units. Find the resistance of another wire of the same length whose area is 100 units.

SOLUTION. Given: $R_1 = 32\ \Omega$ Find: $R_2 = ?$

$A_1 = 20$

$A_2 = 100$

$$\frac{R_1}{R_2} = \frac{A_2}{A_1}$$

$$\frac{32}{R_2} = \frac{100}{20}$$

$$100R_2 = 32 \times 20$$

$$R_2 = \frac{640}{100} = 6.4\ \Omega \qquad Ans.$$

SELF-TEST 12-11. In a parallel circuit, the currents are inversely proportional to the resistances. In the circuit shown in Fig. 12-8, find $I_2$.

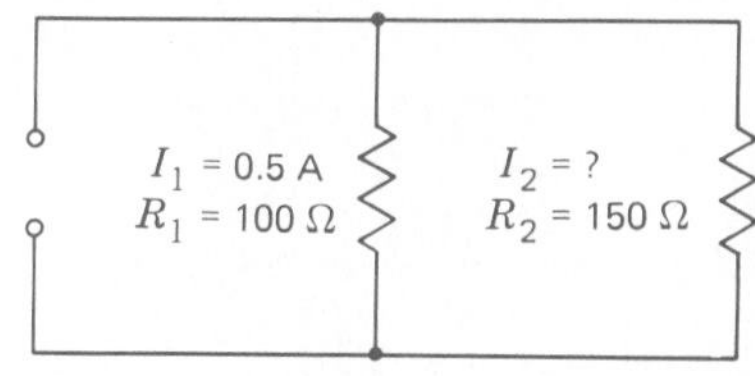


FIGURE 12-8

SOLUTION

1. The variables in this problem are the current and the ______. — resistance
2. Set up a ratio comparing the current $I_1$ to $I_2$.

$$\frac{?}{?}$$

$I_1$ / $I_2$

3. Since the currents are *inversely* proportional to the resistances, set up an *inverse* ratio of resistances.

$$\frac{?}{?}$$

$R_2$ / $R_1$

4. Since the two variables are proportional, the ratios may be set ______ to each other. — equal

$$\frac{I_1}{I_2} = \frac{?}{?}$$

$R_2$ / $R_1$

5. Substitute numbers.

$$\frac{0.5}{I_2} = \frac{150}{?}$$

100

6. Solve for $I_2$.

$$150I_2 = 0.5 \times 100$$

$$I_2 = \frac{50}{?}$$

$$I_2 = \underline{\quad\quad} \text{ A} \qquad Ans.$$

150

0.33

**PROBLEMS**

The speeds of pulleys are inversely proportional to their diameters. Using Fig. 12-7,

1. Find $D_1$ if $S_1 = 40$ rpm, $S_2 = 100$ rpm, and $D_2 = 2$ in.
2. Find $S_1$ if $D_1 = 5$ in, $D_2 = 6$ in, and $S_2 = 120$ rpm.

The resistance of a wire is inversely proportional to its cross-sectional area.

3. Find $R_2$ if $A_1 = 200$, $R_1 = 50\ \Omega$, and $A_2 = 150$.
4. Find $A_2$ if $A_1 = 200$, $R_1 = 50\ \Omega$, and $R_2 = 10\ \Omega$.
5. Find $A_1$ if $R_1 = 20\ \Omega$, $A_2 = 10{,}000$, and $R_2 = 1\ \Omega$.
6. Find $R_1$ if $A_1 = 30$, $A_2 = 600$, and $R_2 = 16\ \Omega$.

In a parallel circuit, the currents are inversely proportional to the resistances. Using a diagram similar to Fig. 12-8,

7. Find $I_1$ if $I_2 = 0.06$ A, $R_1 = 30\ \Omega$, and $R_2 = 40\ \Omega$.
8. Find $R_2$ if $I_1 = 2$ A, $I_2 = 4.5$ A, and $R_1 = 90\ \Omega$.
9. Find $R_1$ if $I_1 = 2.25$ A, $I_2 = 3.5$ A, and $R_2 = 0.6\ \Omega$.

In a transformer, the currents in the primary and secondary coils are inversely proportional to the voltages. Using a diagram similar to that shown in Fig. 12-3,

10. Find $V_s$ if $V_p = 120$ V, $I_p = 5$ A, and $I_s = 2$ A.
11. Find $I_p$ if $V_p = 120$ V, $V_s = 15$ V, and $I_s = 2$ A.
12. Find $I_s$ if $V_p = 120$ V, $I_p = 4$ A, and $V_s = 6$ V.

## JOB 12-3 REVIEW OF RATIO AND PROPORTION

A *ratio* is the ________ of two quantities of the same kind.

comparison

A ratio is indicated by the ________ of the quantities.

division

A ratio may be written as (1) 8 to 2, (2) 8:2, or (3) 8/2 and is always read as ________.

"8 to 2"

The two quantities of a ratio must always be measured in the ________ units of measurement.

same

A *proportion* is a mathematical statement that two ratios are ________.

equal

In a *direct proportion,* the two quantities increase or decrease together. This is indicated by comparing the two ratios in the ____(same/opposite)____ order. Thus, if $A$ is directly proportional to $B$,

same

$$\frac{A_1}{A_2} = \frac{B_1}{B_2} \quad \text{or} \quad \frac{B_1}{B_2} = \frac{?}{?}$$

$A_1$ / $A_2$

or

$$\frac{A_2}{A_1} = \frac{?}{?}$$

$B_2$ / $B_1$

or

$$\frac{B_2}{B_1} = \frac{?}{?}$$

$A_2$ / $A_1$

In an *inverse proportion,* as one quantity increases, the other ________, and vice versa. This is indicated by comparing the two ratios in the ____(same/opposite)____ order. Thus, if $C$ is inversely proportional to $D$,

decreases

opposite

$$\frac{C_1}{C_2} = \frac{D_2}{D_1} \quad \text{or} \quad \frac{D_2}{D_1} = \frac{?}{?}$$

$C_1$ / $C_2$

or

$$\frac{C_2}{C_1} = \frac{?}{?}$$

$D_1$ / $D_2$

or

$$\frac{D_1}{D_2} = \frac{?}{?}$$

$C_2$ / $C_1$

**PROBLEMS**

1. Find the ratio of (*a*) 6 to 18 in, (*b*) 1 ft to 2 yd, (*c*) 100 mA to 1 A, (*d*) 500 Hz to 0.8 kHz, and (*e*) 4 hp to 4 kW.
2. In a series circuit, the voltage across a resistance is directly proportional to the resistance. Using a circuit similar to that shown in Fig. 12-4, find $R_1$ if $V_1 = 90$ V, $V_2 = 48$ V, and $R_2 = 80\ \Omega$.

In Probs. 3 to 6, $A$ is directly proportional to $B$.

3. Find $B_2$ if $A_1 = 15$, $A_2 = 6$, and $B_1 = 90$.
4. Find $B_1$ if $A_1 = 35$, $A_2 = 45$, and $B_2 = 63$.
5. Find $A_2$ if $A_1 = 27$, $B_1 = 3$, and $B_2 = 8$.
6. Find $A_1$ if $A_2 = 56$, $B_1 = 4$, and $B_2 = 7$.

In a parallel circuit similar to that shown in Fig. 12-8, the current in each branch is inversely proportional to the resistance.

7. Find $I_1$ if $I_2 = 0.4$ A, $R_1 = 100\ \Omega$, and $R_2 = 150\ \Omega$.
8. Find $I_2$ if $I_1 = 1.8$ A, $R_1 = 225\ \Omega$, and $R_2 = 75\ \Omega$.

In a transformer, the currents in the primary and secondary coils are inversely proportional to the voltages. Using a diagram similar to that shown in Fig. 12-3,

9. Find $V_p$ if $V_s = 2200$ V, $I_p = 50$ A, and $I_s = 2.5$ A.
10. Find $I_p$ if $V_p = 24$ V, $V_s = 120$ V, and $I_s = 0.2$ A.

(See Instructor's Manual for Test 12-1, Ratio and Proportion)

## JOB 12-4 THE AMERICAN WIRE GAGE TABLE

Wires are manufactured in standard sizes which are listed in Table 12-1 and are known as the American Wire Gage (AWG). It has been found that calculations involving these wires may be simplified by expressing their diameters and cross-sectional areas in new units of measurement.

**Mils.** The diameters of wires are expressed in terms of *mils* instead of inches.

$$1000 \text{ mil} = 1 \text{ in} \qquad \text{or} \qquad 1 \text{ mil} = \frac{1}{1000} \text{ in}$$

CHANGING UNITS

*To change inches to mils:* Multiply the number of inches by 1000.
*To change mils to inches:* Divide the number of mils by 1000.

EXAMPLE 12-12. A wire has a diameter of 0.162 in. Change the diameter to mils and find the AWG number.

TABLE 12-1

AMERICAN WIRE GAGE TABLE

Resistance of bare annealed copper wire at 20°C (68°F)

| AWG NO. | DIAMETER $d$, MIL | AREA $d^2$, CMIL | RESISTANCE, Ω/1000 FT |
|---|---|---|---|
| 0000 | 460.0 | 211,600 | 0.0490 |
| 000 | 409.6 | 167,800 | 0.0618 |
| 00 | 364.8 | 133,100 | 0.0779 |
| 0 | 324.9 | 105,500 | 0.0983 |
| 1 | 289.3 | 83,690 | 0.1239 |
| 2 | 257.6 | 66,360 | 0.1563 |
| 3 | 229.4 | 52,630 | 0.1970 |
| 4 | 204.3 | 41,740 | 0.2485 |
| 5 | 181.9 | 33,100 | 0.3133 |
| 6 | 162.0 | 26,250 | 0.3951 |
| 7 | 144.3 | 20,820 | 0.4982 |
| 8 | 128.5 | 16,510 | 0.6282 |
| 9 | 114.4 | 13,090 | 0.7921 |
| 10 | 101.9 | 10,380 | 0.9989 |
| 11 | 90.74 | 8234 | 1.260 |
| 12 | 80.81 | 6530 | 1.588 |
| 13 | 71.96 | 5178 | 2.003 |
| 14 | 64.08 | 4107 | 2.525 |
| 15 | 57.07 | 3257 | 3.184 |
| 16 | 50.82 | 2583 | 4.016 |
| 17 | 45.26 | 2048 | 5.064 |
| 18 | 40.30 | 1624 | 6.385 |
| 19 | 35.89 | 1288 | 8.051 |
| 20 | 31.96 | 1022 | 10.15 |
| 21 | 28.46 | 810.1 | 12.80 |
| 22 | 25.35 | 642.4 | 16.14 |
| 23 | 22.57 | 509.5 | 20.36 |
| 24 | 20.10 | 404.0 | 25.67 |
| 25 | 17.90 | 320.4 | 32.37 |
| 26 | 15.94 | 254.1 | 40.81 |
| 27 | 14.20 | 201.5 | 51.47 |
| 28 | 12.64 | 159.8 | 64.90 |
| 29 | 11.26 | 126.7 | 81.83 |
| 30 | 10.03 | 100.5 | 103.2 |
| 31 | 8.928 | 79.70 | 130.1 |
| 32 | 7.950 | 63.21 | 164.1 |
| 33 | 7.080 | 50.13 | 206.9 |
| 34 | 6.305 | 39.75 | 260.9 |
| 35 | 5.615 | 31.52 | 329.0 |
| 36 | 5.000 | 25.00 | 414.8 |
| 37 | 4.453 | 19.83 | 523.1 |
| 38 | 3.965 | 15.72 | 659.6 |
| 39 | 3.531 | 12.47 | 831.8 |
| 40 | 3.145 | 9.888 | 1049 |

SOLUTION

$$0.162 \text{ in} = 0.162 \times 1000 = 162 \text{ mil} \qquad \textit{Ans.}$$

In the second column of the AWG table marked "Diameter, mil," read down until you find 162 mil. Read to the left to the column marked "AWG No." to find No. 6 wire. *Ans.*

EXAMPLE 12-13. Express the diameter of No. 18 wire in mils and inches.

SOLUTION. Read down in the first column until you find No. 18. Read to the right to the second column to find 40.30 mil.

$$40.30 \text{ mil} = 40.30 \div 1000 = 0.0403 \text{ in} \qquad \textit{Ans.}$$

**PROBLEMS**

Using Table 12-1, find the missing values in the following problems.

| PROBLEM | GAGE NO. | DIAMETER, MIL | DIAMETER, IN |
|---|---|---|---|
| 1 | ? | 28.46 | ? |
| 2 | 10 | ? | ? |
| 3 | ? | ? | $^1/_4$ |
| 4 | ? | ? | 0.050 82 |
| 5 | ? | ? | $^{162}/_{1000}$ |
| 6 | ? | 204.3 | ? |
| 7 | 14 | ? | ? |
| 8 | 12 | ? | ? |
| 9 | ? | 40.30 | ? |
| 10 | ? | ? | $^1/_{16}$ |

**Circular mils.** The circular mil (cmil) is the unit of area which is used to measure the cross-sectional area of wires. The circular-mil area is used rather than the ordinary units of area because of the ease of obtaining the circular-mil area when the diameter is given in mils.

**RULE** To obtain the circular-mil area of a wire, express the diameter in mils and then square the resulting number.

**FORMULA**

$$A = D_m^2 \qquad (12\text{-}1)$$

where $A$ = area, cmil
$D_m$ = diameter, mil

In order to reduce the number of digits needed to write the circular-mil areas of conductors over 4/0, we use the term MCM, which means *thousands* of circular mils. Thus, 360 MCM means 360,000 cmil.

EXAMPLE 12-14. Find the circular-mil area $A$ of No. 12 wire.

SOLUTION. Read across to the right from No. 12 in the first column to find the circular-mil area in column 3.

$$\text{Area of No. 12 wire} = 6530 \text{ cmil} \qquad \textit{Ans.}$$

EXAMPLE 12-15. Find the circular-mil area $A$ of a wire whose diameter is 0.5 in.

SOLUTION

$$0.5 \text{ in} = 0.5 \times 1000 = 500 \text{ mil} = D_m$$

$$A = D_m^2 \qquad (12\text{-}1)$$

$$A = (500)^2 = 25 \times 10^4$$

$$= 250 \times 10^3 = 250 \text{ MCM} \qquad \textit{Ans.}$$

EXAMPLE 12-16. A cable is to be replaced by a rectangular busbar 1 in wide and 1/4 in thick. Find the circular-mil area.

SOLUTION

1. Change inches to mils.

$$1 \text{ in} = 1 \times 1000 = 1000 \text{ mil}$$

$$1/4 \text{ in} = 0.25 \times 1000 = 250 \text{ mil}$$

2. Find the area.

$$A = L \times W = 1000 \times 250 = 250{,}000 \text{ mil}^2$$

3. Convert square mils to circular mils. The area of any circle is given by the formula

$$0.7854d^2 = A$$

In mils,

$$0.7854d_m^2 = A_{\text{mil}^2}$$

Substituting $A_{\text{cmil}}$ for $d_m^2$ will give

$$0.7854A_{\text{cmil}} = A_{\text{mil}^2}$$

or

$$A_{\text{cmil}} = \frac{A_{\text{mil}^2}}{0.7854} \qquad \boxed{12\text{-}2}$$

$$A_{\text{cmil}} = \frac{250{,}000}{0.7854} = 318{,}000 \text{ cmil} = 318 \text{ MCM} \qquad \textit{Ans.}$$

**RULE** To obtain the diameter of a wire in mils, find the square root of the circular-mil area.

**FORMULA** $D_m = \sqrt{\text{cmil}}$ **12-3**

EXAMPLE 12-17. Find the diameter of a wire whose cross-sectional area is 25 MCM.

SOLUTION

$$D_m = \sqrt{\text{cmil}} \qquad (12\text{-}3)$$

$$D_m = \sqrt{25{,}000} = 158 \text{ mil} \qquad \textit{Ans.}$$

(Review Job 7-8 on square root.)

If the diameter is wanted in inches,

$$158 \text{ mil} = \frac{158}{1000} = 0.158 \text{ in} \qquad \textit{Ans.}$$

### PROBLEMS

Find the missing values in the following problems:

| PROBLEM | DIAMETER, IN | DIAMETER, MIL | AREA, CMIL |
|---|---|---|---|
| 1 | ? | 10 | ? |
| 2 | 0.05 | ? | ? |
| 3 | ? | ? | 4096 |
| 4 | ? | 60 | ? |
| 5 | 0.032 | ? | ? |
| 6 | ? | ? | 9500 |
| 7 | ? | 87.2 | ? |
| 8 | $^1/_4$ | ? | ? |
| 9 | ? | ? | 30,000 |
| 10 | 0.102 | ? | ? |

11. A rectangular busbar measures $1^1/_4$ in by $^1/_4$ in. Find its circular-mil area.
12. A rectangular busbar measures $^3/_4$ in by $^3/_8$ in. Find the diameter (in mils) of an equivalent round wire.

### JOB 12-5 RESISTANCES OF WIRES OF DIFFERENT MATERIALS

All materials differ in their atomic structure and therefore in their ability to resist the flow of an electric current. The measure of the ability of a specific material to resist the flow of electricity is called its *specific resistance.*

The *specific resistance* $K$ of a material is the resistance offered by a wire of this material which is 1 ft long with a diameter of 1 mil. The specific resistances of different materials are given in Table 12-2.

TABLE 12-2
SPECIFIC RESISTANCES OF MATERIALS IN OHMS PER MIL-FOOT AT 20°C

| MATERIAL | $K$ | MATERIAL | $K$ |
|---|---|---|---|
| Silver | 9.7 | Tantalum | 93.3 |
| Copper | 10.4 | German silver (18%) | 200 |
| Gold | 14.7 | Monel metal | 253 |
| Aluminum | 17.0 | Manganin | 265 |
| Tungsten | 34.0 | Magnesium | 276 |
| Brass | 43.0 | Constantan | 295 |
| Iron (pure) | 60.0 | Nichrome | 600 |
| Tin | 69.0 | Nickel | 947 |

**RULE** The resistance of a wire is directly proportional to the specific resistance of the material.

**FORMULA**

$$\frac{R_1}{R_2} = \frac{K_1}{K_2} \qquad \boxed{12\text{-}4}$$

where $R_1$ and $K_1$ = resistance and specific resistance of a wire of one material, respectively

$R_2$ and $K_2$ = resistance and specific resistance of a wire of another material, respectively

EXAMPLE 12-18. A copper wire has a resistance of 5 Ω. What is the resistance of a Nichrome wire of the same length and cross-sectional area?

SOLUTION

Given: Copper wire, $R_1 = 5\ \Omega$ Find: $R_2$ of Nichrome = ?

$K_1 = 10.4$

Nichrome wire, $K_2 = 600$

$$\frac{R_1}{R_2} = \frac{K_1}{K_2} \qquad (12\text{-}4)$$

$$\frac{5}{R_2} = \frac{10.4}{600}$$

$$10.4R_2 = 5 \times 600$$

$$10.4R_2 = 3000$$

$$R_2 = \frac{3000}{10.4} = 288\ \Omega \qquad \textit{Ans.}$$

## PROBLEMS

1. A copper wire has a resistance of 12 Ω. What is the resistance of an aluminum wire of the same length and diameter?
2. A tungsten wire has a resistance of 40 Ω. What is the resistance of a Nichrome wire of the same length and diameter?
3. A Nichrome wire has a resistance of 125 Ω. What is the resistance of an iron wire of the same length and diameter?
4. A copper wire has a resistance of 2.5 Ω. What is the resistance of a manganin wire of the same length and diameter?.

## JOB 12-6 RESISTANCES OF WIRES OF DIFFERENT LENGTHS

Consider two wires of the same material and diameter but of different lengths. A length of wire may be considered to be made of a large number of small lengths all connected in series. The total resistance of the length of wire is then equal to the sum of the resistances of all the small lengths. Therefore, the longer the wire, the greater its resistance.

**RULE** The resistance of a wire is directly proportional to its length.

**FORMULA**

$$\frac{R_1}{R_2} = \frac{L_1}{L_2} \tag{12-5}$$

where $R_1$ and $L_1$ = resistance and length of first wire, respectively
$R_2$ and $L_2$ = resistance and length of second wire, respectively

EXAMPLE 12-19. If 1000 ft of No. 16 copper wire has a resistance of 4 Ω, find the resistance of 1800 ft of the same wire.

SOLUTION. Given: $R_1 = 4\ \Omega$ Find: $R_2 = ?$
$L_1 = 1000$ ft
$L_2 = 1800$ ft

$$\frac{R_1}{R_2} = \frac{L_1}{L_2} \qquad (12\text{-}5)$$

$$\frac{4}{R_2} = \frac{10\not{0}\not{0}}{18\not{0}\not{0}}$$

$$10R_2 = 4 \times 18$$

$$R_2 = \frac{72}{10} = 7.2\ \Omega \qquad \textit{Ans.}$$

SELF-TEST 12-20. Find the resistance of 700 ft of No. 20 copper wire. What current flows through the wire when there is a voltage drop of 14.2 V across the ends of the wire?

SOLUTION. Given: $L =$ ______ ft of No. 20 copper wire Find: $R = ?$ 700
$V =$ ______ V $I = ?$ 14.2

1. Find the resistance of No. 20 copper wire. From Table 12-1, No. 20 wire has a resistance of ______ Ω/1000 ft. 10.15
2. Find the resistance of 1 ft of this wire.

$$\frac{R}{\text{ft}} = \frac{10.15}{1000} = \underline{\qquad\qquad} \ \Omega/\text{ft}$$ 0.010 15

3. Find the resistance of ______ ft of this wire. 700

$$R = \frac{R}{\text{ft}} \times \text{ft} = 0.010\ 15 \times 700 = \underline{\qquad\qquad} \ \Omega \qquad \textit{Ans.}$$ 7.1

4. Find the current.

$$V = IR \qquad (2\text{-}1)$$

$$14.2 = I \times \underline{\qquad\qquad}$$ 7.1

$$I = \underline{\qquad\qquad} \ \text{A} \qquad \textit{Ans.}$$ 2

## PROBLEMS

1. If 1000 ft of No. 12 copper wire has a resistance of 1.6 Ω, find the resistance of 2300 ft of this wire.
2. Find the resistance of 800 ft of No. 16 copper wire.
3. Find the resistance of 400 ft of No. 20 copper wire.
4. What is the resistance of the primary coil of a transformer if it is wound with 600 ft of No. 20 copper wire?
5. The coil of an electromagnet is wound with 300 ft of No. 16 copper wire. What is the resistance of the coil? What current will be drawn from a 6-V battery?
6. The coil of an outboard motor is wound with 100 ft of No. 18 copper wire. What current will it draw from a 6-V battery?
7. The four field windings of a motor are connected in series. If each winding uses 50 ft of No. 32 copper wire, find the total resistance of the field windings. What current is drawn when the motor is used on a 110-V line?
8. Find the resistance of a two-wire service line 100 ft long if No. 8 copper wire is used.
9. What is the resistance of a telephone line 1 mi long if No. 18 copper wire is used? What is the voltage drop across this line if the current is 200 mA?
10. Find the resistance of 10 ft of No. 22 copper wire. What is the resistance of the same length of No. 22 Nichrome wire? If this Nichrome wire is used as the resistance element in an electric furnace, what current is drawn from a 110-V line?

## JOB 12-7 RESISTANCES OF WIRES OF DIFFERENT AREAS

The flow of electricity in a wire is very similar to the flow of water in a pipe. A large-diameter pipe can carry more water than a small-diameter pipe. Similarly, a large-diameter wire can carry more current than a small-diameter wire. But the ability of a wire to carry a large current means that its resistance is small. Thus, a large-diameter wire has a small resistance while a small-diameter wire has a large resistance. As we have learned, when two quantities depend on each other so that one increases as the other decreases, and vice versa, the relationship is called an *inverse ratio*.

**RULE** The resistance of a wire is inversely proportional to its cross-sectional area.

**FORMULA**

$$\frac{R_1}{R_2} = \frac{A_2}{A_1} \qquad \boxed{12\text{-}6}$$

where $R_1$ and $A_1$ = resistance and area of first wire, respectively
$R_2$ and $A_2$ = resistance and area of second wire, respectively

EXAMPLE 12-21. A wire has a resistance of 20 Ω and a cross-sectional area of 320 cmil. Find the resistance of a wire of the same length but with an area of 800 cmil.

SOLUTION. Given: $R_1 = 20\ \Omega$ Find: $R_2 = ?$
$A_1 = 320$ cmil
$A_2 = 800$ cmil

$$\frac{R_1}{R_2} = \frac{A_2}{A_1} \qquad (12\text{-}6)$$

$$\frac{20}{R_2} = \frac{800}{320}$$

$$800R_2 = 320 \times 20$$

$$R_2 = \frac{6400}{800} = 8\ \Omega \qquad \textit{Ans.}$$

EXAMPLE 12-22. A certain length of No. 36 copper wire has a resistance of 200 Ω. Find the resistance of the same length of No. 30 wire.

SOLUTION. Given: First wire: No. 36 = 200 Ω = $R_1$ Find $R_2 = ?$
Second wire: No. 30

1. From Table 12-1, find the circular-mil area of each wire.

$A_1$ of No. 36 = 25 cmil
$A_2$ of No. 30 = 100 cmil (approx)

2. Set up the inverse ratio.

$$\frac{R_1}{R_2} = \frac{A_2}{A_1} \qquad (12\text{-}6)$$

$$\frac{200}{R_2} = \frac{100}{25}$$

$$100R_2 = 200 \times 25$$

$$R_2 = \frac{5000}{100} = 50\ \Omega \qquad \textit{Ans.}$$

**PROBLEMS**

1. If $R_1 = 21\ \Omega$, $A_1 = 10$ MCM, and $A_2 = 4$ MCM, find $R_2$.
2. A wire has a resistance of 30 Ω and a cross-sectional area of 100 cmil. Find the resistance of the same length of a wire whose area is 320 cmil.
3. A certain length of No. 30 wire has a resistance of 50 Ω. Find the resistance of the same length of No. 40 wire.
4. A certain length of No. 36 wire has a resistance of 1.5 Ω. Find the resistance of the same length of No. 34 wire.
5. A certain length of No. 12 wire has a resistance of 3.2 Ω. Find the resistance of the same length of No. 18 wire.

(See Instructor's Manual for Test 12-2)

## JOB 12-8 RESISTANCES OF WIRES OF ANY DIAMETER, ANY LENGTH, OR ANY MATERIAL

In the last three jobs we learned that

1. The resistance is directly proportional to the specific resistance $K$ of the material.
2. The resistance is directly proportional to the length $L$.
3. The resistance is inversely proportional to the area $A$.

These three concepts may be combined into one rule.

**RULE** The resistance $R$ of a wire is equal to the specific resistance $K$ multiplied by the length in feet $L$, and divided by the circular-mil area of the wire $A$.

**FORMULA**

$$R = \frac{K \times L}{A} \qquad \boxed{12\text{-}7}$$

where $R$ = resistance of wire, Ω
$K$ = specific resistance for a particular material
$L$ = length of wire, ft
$A$ = area of wire, cmil

EXAMPLE 12-23. Find the resistance of a tungsten wire 10 ft long whose diameter is 0.005 in.

SOLUTION. Given: Tungsten $K = 34$ Find: $R = ?$

$$L = 10 \text{ ft}$$
$$D = 0.005 \text{ in}$$

1. Change the diameter to mils.

$$0.005 \text{ in} = 0.005 \times 1000 = 5 \text{ mil}$$

2. Find the circular-mil area.

$$A = D_m^2 = 5^2 = 25 \text{ cmil} \qquad (12\text{-}1)$$

3. Find the resistance of the wire.

$$R = \frac{K \times L}{A} \qquad (12\text{-}7)$$

$$R = \frac{34 \times 10}{25} = \frac{340}{25} = 13.6\ \Omega \qquad \textit{Ans.}$$

EXAMPLE 12-24. An electric heater is made with a heating element of 4 ft of No. 30 Nichrome wire. (*a*) What is the resistance of the heating coil? (*b*) What current will it draw from a 120-V line? (*c*) How much power will it use?

SOLUTION. The diagram for the problem is shown in Fig. 12-9.

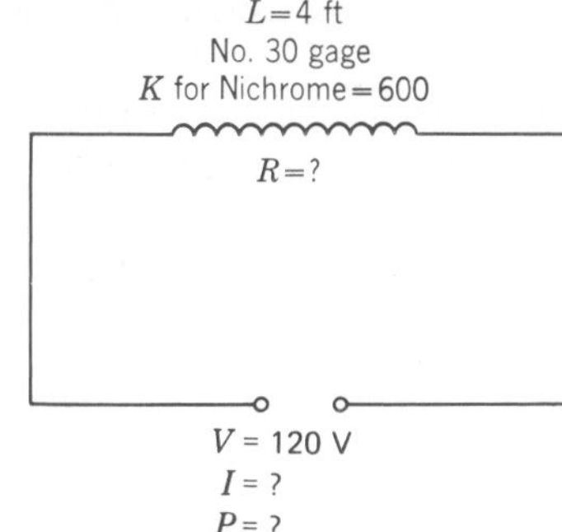


FIGURE 12-9

*a*. Find the resistance of the heating coil.

1. Find the circular-mil area. From Table 12-1,

$$\text{No. } 30 = 100.5 \text{ cmil} = 100 \text{ cmil} \qquad \text{(approx)}$$

2. Find the resistance of the coil.

$$R = \frac{K \times L}{A} \qquad (12\text{-}7)$$

$$R = \frac{600 \times 4}{100} = \frac{2400}{100} = 24\ \Omega \qquad \textit{Ans.}$$

*b*. Find the current drawn.

$$V = IR \qquad (2\text{-}1)$$
$$120 = I \times 24$$

$$I = \frac{120}{24} = 5 \text{ A} \qquad Ans.$$

*c.* Find the power used.

$$P = I \times V \tag{7-1}$$

$$P = 5 \times 120 = 600 \text{ W} \qquad Ans.$$

## PROBLEMS

1. Find the resistance of a 0.1-in-diameter copper wire which is 100 ft long.
2. Find the resistance of 150 ft of No. 32 copper wire.
3. Find the resistance of 10 ft of $^1/_8$-in-diameter silver wire.
4. What is the resistance of 20 ft of No. 20 gage manganin wire? What current will flow through the wire if the voltage is 6 V?
5. Find the resistance of 30 ft of No. 24 iron wire. What current will flow through the wire if it is used as a resistance element across 110 V? What is the power used?
6. What is the resistance of $^1/_2$ ft of No. 40 Nichrome wire?
7. A winding made of 400 ft of No. 30 copper wire is placed across 6 V. Find (*a*) the current drawn and (*b*) the power used.
8. Find the resistance of 200 ft of copper annunciator wire whose diameter is 0.04 in.
9. The $^1/_2$-in-diameter copper leads from a welding control are each 1 ft long. What is the voltage drop along these leads when the welding current is 2500 A?
10. What is the resistance of an aluminum conductor 162 mil in diameter and 500 ft long?

## JOB 12-9 FINDING THE LENGTH OF WIRE NEEDED TO MAKE A CERTAIN RESISTANCE

EXAMPLE 12-25. How many feet of No. 20 gage Nichrome wire are required to make a heater coil of 10 Ω resistance?

SOLUTION. Given: $K$ for Nichrome = 600 Find: $L$ = ?
Gage = No. 20
$R$ = 10 Ω

1. Find the circular-mil area. From Table 12-1, for No. 20 gage,

$$A = 1022 \text{ cmil}$$

2. Find the length of the wire.

$$R = \frac{K \times L}{A} \tag{12-7}$$

$$\frac{10}{1} = \frac{600 \times L}{1022}$$

$$600L = 1022 \times 10$$

$$L = \frac{10{,}220}{600} = 17.03 \text{ ft} \qquad \textit{Ans.}$$

EXAMPLE 12-26. How much resistance must be placed in series with a 50-V 50-Ω lamp in order to operate it from a 110-V line? How many feet of No. 18 German silver wire are required to make this limiting resistor?

SOLUTION. The diagram for the circuit is shown in Fig. 12-10.

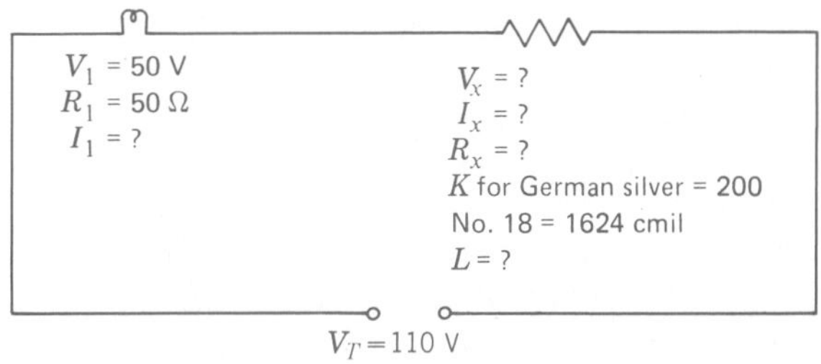


FIGURE 12-10

1. Find the series current.

$$V_1 = I_1 \times R_1 \qquad (4\text{-}4)$$

$$50 = I_1 \times 50$$

$$I_1 = \frac{50}{50} = 1 \text{ A}$$

Therefore,

$$I_x = I_1 = 1 \text{ A} \qquad (4\text{-}1)$$

2. Find $V_x$.

$$V_T = V_1 + V_x \qquad (4\text{-}2)$$

$$110 = 50 + V_x$$

$$V_x = 110 - 50 = 60 \text{ V}$$

3. Find $R_x$.

$$V_x = I_x \times R_x \qquad (4\text{-}5)$$

$$60 = 1 \times R_x$$

$$R_x = 60\ \Omega \qquad \textit{Ans.}$$

4. Find the length of wire needed to make a 60-Ω resistor.

$$R = \frac{K \times L}{A} \qquad (12\text{-}7)$$

$$\frac{60}{1} = \frac{200 \times L}{1624}$$

$$200L = 1624 \times 60$$

$$L = \frac{97{,}440}{200} = 487.2 \text{ ft} \qquad \textit{Ans.}$$

## PROBLEMS

1. What is the resistance of an aluminum wire 200 ft long if its circular-mil area is 2000?
2. What is the resistance of a 100-ft-long constantan wire if its diameter is 80 mil?
3. How many feet of 0.01-in-diameter copper wire are required to make a resistance of 10 Ω?
4. A 0.02-in-diameter Nichrome wire is used for the coils of an electric heater. If the resistance of the coils is to be 18 Ω, how many feet of wire are needed? What are the current and power drawn when the heater is used on a 110-V line?
5. How many feet of No. 28 copper wire are needed to make an ammeter shunt of 0.2 Ω?
6. What would be the length of the wire in Prob. 5 if manganin wire were used instead of copper?
7. A subway-car heater is made of No. 24 iron wire. It is to operate on 550 V and 10 A. How many feet of wire are needed to make the heater?
8. What resistance must be inserted in series with an MZ4616 zener diode in order to operate it from two dry cells connected in series? The diode takes 2.3 V at 0.3 A. What length of No. 20 iron wire is needed to make the resistance?
9. Which of the following has the greater resistance: (*a*) 50 ft of No. 20 iron wire or (*b*) 10 ft of No. 18 Nichrome wire?
10. How many feet of No. 14 copper wire are required to make 0.8 Ω of resistance?

## JOB 12-10 FINDING THE DIAMETER OF WIRE NEEDED TO MAKE A CERTAIN RESISTANCE

EXAMPLE 12-27. What must be the diameter and gage number of a Nichrome heating element 40 ft long if the resistance is to be 22 Ω?

SOLUTION. Given: $K$ for Nichrome $= 600$  Find: $D = ?$
$R = 22\ \Omega$  Gage No. $= ?$
$L = 40$ ft

1. Find the required circular-mil area.

$$R = \frac{K \times L}{A} \qquad (12\text{-}7)$$

$$\frac{22}{1} = \frac{600 \times 40}{A}$$

$$22A = 24{,}000$$
$$A = \frac{24{,}000}{22} = 1091 \text{ cmil}$$

2. Find the diameter in mils.

$$D_m = \sqrt{\text{cmil}} \qquad (12\text{-}2)$$
$$D_m = \sqrt{1091} = 33.03 \text{ mil} \qquad \textit{Ans.}$$

3. Find the gage number. This may be found by looking up the table for either the diameter of 33.03 mil or the area of 1091 cmil.

| | | | |
|---|---|---|---|
| No. 19 | = 35.89 mil | No. 19 | = 1288 cmil |
| ? | = 33.03 mil | ? | = 1091 cmil |
| No. 20 | = 31.96 mil | No. 20 | = 1022 cmil |

Using either method, the closest gage number is No. 20. However, always choose the *smaller* gage number so as to select a wire of *larger* area, and therefore smaller resistance and larger current-carrying ability. Therefore, use No. 19 wire. *Ans.*

EXAMPLE 12-28. Find the size of copper wire to be used in a two-wire system 100 ft long if the total resistance of the system is to be 1.3 Ω.

FIGURE 12-11

SOLUTION. The diagram for the problem is shown in Fig. 12-11. When the wire is made of copper, we can use Table 12-1 to simplify our calculations. If we can find the resistance per 1000 ft, we can look in the table to find the wire which has this resistance. The total length $L_1$ of the two-wire system equals 2 × 100 = 200 ft.

$$\Omega/\text{ft} = \frac{1.3}{200} = 0.0065 \ \Omega/\text{ft}$$
$$\Omega/1000 \text{ ft} = 0.0065 \times 10^3 = 6.5 \ \Omega/1000 \text{ ft}$$

In column 4 of Table 12-1, the next *smaller* resistance per 1000 ft is 6.385 Ω. Reading to the left to column 1, we find the gage number to be No. 18. *Ans.*

## PROBLEMS

1. What diameter of constantan wire should be used to make a coil of 10 Ω resistance if only 3 ft of wire is used?
2. What must be the diameter of an aluminum wire 2000 ft long if its resistance is to be 3 Ω?

3. What must be the diameter of an iron wire 40 ft long if the resistance is to be 50 Ω?
4. What size of copper wire should be used for a single line 200 ft long if the allowable resistance is 1 Ω?
5. What diameter of copper wire should be used for a two-wire system 500 ft long if the total resistance is not to exceed 10 Ω?
6. The heating coil of an electric heater using 4 ft of Nichrome wire is to have a resistance of 24 Ω. What gage wire should be used?
7. A rheostat is to be wound with 500 ft of German silver wire to produce a resistance of 60 Ω. What gage wire should be used?

## JOB 12-11 REVIEW OF RESISTANCE OF WIRES

A *mil* is the unit used to measure the ________ of round wires. A *mil* is equal to one-________ of an inch. A *circular mil* (cmil) is the unit used to measure the cross-sectional ________ of wires. diameters thousandth areas

**FORMULAS**

$$A = D_m^2 \qquad \boxed{12\text{-}1}$$

$$D_m = \sqrt{\text{cmil}} \qquad \boxed{12\text{-}3}$$

where $A$ = area of the wire, ________ cmil
$D_m$ = diameter of the wire, ________ mil

To convert square mils to circular mils, we use the formula

$$A_{\text{cmil}} = \frac{A_{\text{mil}^2}}{?} \qquad \boxed{12\text{-}2}$$ 0.7584

The specific resistance $K$ of a wire is the resistance of a wire 1 ft long with a diameter of 1 ________. The specific resistances of different materials are given in Table 12-2. mil

The resistance of a wire is directly proportional to its specific resistance.

$$\frac{R_1}{R_2} = \frac{K_?}{K_?} \qquad \boxed{12\text{-}4}$$ 1 2

The resistance of a wire is directly proportional to its length.

$$\frac{R_1}{R_2} = \frac{L_?}{L_?} \qquad \boxed{12\text{-}5}$$ 1 2

The resistance of a wire is inversely proportional to its area.

$$\frac{R_1}{R_2} = \frac{A_?}{A_?} \qquad \boxed{12\text{-}6}$$ 2 1

The general formula for the resistance of any wire is

$$R = \frac{K \times L}{?} \qquad \boxed{12\text{-}7}$$ A

where $R$ = resistance of the wire, Ω
$K$ = specific resistance of the material
$L$ = length of the wire, ______ | ft
$A$ = area of the wire, ______ | cmil

## PROBLEMS

1. Find the diameter in mils and gage number of the wires whose diameters are (*a*) 0.025 in, (*b*) 0.102 in, (*c*) 1/8 in, and (*d*) 128/1000 in.
2. Find the area in circular mils of wires whose diameters are (*a*) 15 mil, (*b*) 40 mil, (*c*) 0.01 in, and (*d*) 0.204 in.
3. A constantan wire has a resistance of 50 Ω. What is the resistance of a manganin wire of the same length and area?
4. Find the resistance of a two-wire line 500 ft long using No. 20 copper wire. Find the voltage drop across this line if the current is 150 mA.
5. A length of No. 12 copper wire has a resistance of 3.6 Ω. What is the resistance of the same length of No. 8 wire?
6. Find the resistance of 6 in of No. 20 Nichrome wire.
7. A resistance element designed to operate at 110 V and 6 A is to be made of No. 30 Nichrome wire. Find the length of wire needed.
8. What diameter of iron wire 12 ft long is needed to make a resistance of 80 Ω?

(See Instructor's Manual for Test 12-3, Resistance of Wires)

# SIZE OF WIRING

## JOB 13-1 MAXIMUM CURRENT-CARRYING CAPACITIES OF WIRES

Heat is produced whenever a current flows in a wire. The larger the current, the greater the amount of heat that is produced. This heat must be given up to the surrounding atmosphere, or the wire may get hot enough to burst into flame.

The heat produced in a wire depends not only on the current but also on the resistance to that flow of current. Thick wires, because of their low resistance, can carry more current than thin wires before they dangerously overheat. But exactly how many amperes can be safely carried by a particular wire? Can a No. 20 wire carry 30 A, or is a No. 12 or larger wire required? Many tests have been made by the National Board of Fire Underwriters to determine the largest current which may be safely carried by wires of different sizes and different insulation. The National Electrical Code has set forth these values as shown in Table 13-1. This table gives the *maximum* current permitted in any size wire insulated as indicated.

### USING THE TABLE OF ALLOWABLE CURRENT CAPACITIES

EXAMPLE 13-1. What is the largest current that may be safely carried by No. 12 rubber-covered wire?

SOLUTION

1. Locate No. 12 wire in column 1 of Table 13-1.
2. Read across to the right to column 5 for rubber-covered wire.
3. The answer (25 A) means that No. 12 rubber-covered wire should never be permitted to carry more than 25 A. *Ans.*

EXAMPLE 13-2. What is the allowable current-carrying capacity of No. 14 varnished-cambric-insulated wire?

SOLUTION

1. Locate No. 14 wire in column 1.
2. Read across to the right to column 6 for varnished-cambric-insulated wire.
3. The answer is 25 A.

EXAMPLE 13-3. What is the smallest rubber-covered wire that should be used to carry 16 A?

SOLUTION

1. Read down column 5 for rubber-covered wire until you find a number *larger* than 16 A. This will be 20 A.
2. Read to the left until you get to column 1. No. 14 gage is the smallest wire that can be used. *Ans.*

EXAMPLE 13-4. What gage wire should be used if the circular-mil area required is 3000 cmil?

SOLUTION

1. Read down column 3 until you get to the number which is *just larger* than 3000 cmil. This number is 4107 cmil.
2. Read to the left until you get to column 1. No. 14 gage is the smallest wire that may be used. *Ans.*

TABLE 13-1

ALLOWABLE CURRENT-CARRYING CAPACITIES OF INSULATED COPPER CONDUCTORS (AMPACITIES)

National Electrical Code 1981

| | | | | ALLOWABLE CARRYING CAPACITY, A | | |
|---|---|---|---|---|---|---|
| (1) AWG NO. | (2) DIAMETER, MIL | (3) CROSS SECTION, CMIL | (4) RESISTANCE, Ω/1000 FT | (5) RUBBER-COVERED, TYPE RH | (6) VARNISHED CAMBRIC, TYPE V | (7) ASBESTOS, TYPE A |
| 14 | 64.1 | 4107 | 2.53 | 20 | 25 | 30 |
| 12 | 80.8 | 6530 | 1.59 | 25 | 30 | 40 |
| 10 | 101.9 | 10,380 | 0.999 | 35 | 40 | 55 |
| 8 | 128.5 | 16,510 | 0.628 | 50 | 55 | 70 |
| 6 | 162.0 | 26,250 | 0.395 | 65 | 70 | 95 |
| 4 | 204.3 | 41,740 | 0.249 | 85 | 95 | 120 |
| 3 | 229.4 | 52,630 | 0.197 | 100 | 110 | 145 |
| 2 | 257.6 | 66,370 | 0.156 | 115 | 125 | 165 |
| 1 | 289.3 | 83,690 | 0.124 | 130 | 145 | 190 |
| 0 | 325 | 105,500 | 0.0983 | 150 | 165 | 225 |
| 00 | 364.8 | 133,100 | 0.0779 | 175 | 190 | 250 |
| 000 | 409.6 | 167,800 | 0.0618 | 200 | 215 | 285 |
| 0000 | 460 | 211,600 | 0.0490 | 230 | 250 | 340 |

EXAMPLE 13-5. What gage wire should be used if the resistance required is 0.5 Ω/1000 ft?

SOLUTION

1. Read down column 4 until you get to a number which is *just smaller* than 0.5 Ω. This number is 0.395 Ω.
2. Read to the left until you get to column 1. No. 6 gage is the smallest wire that may be used. *Ans.*

SELF-TEST 13-6. Find the voltage drop along 1000 ft of No. 10 rubber-covered wire if it is carrying its maximum current.

SOLUTION

1. Find the maximum current for No. 10 rubber-covered wire. This is found in column 5 to be ______ A. | 35
2. Find the resistance per 1000 ft of No. 10 rubber-covered wire. Read across to the right from No. 10 in column 1 to find ______ Ω in column ______. | 0.999 4
3. Find the voltage drop.

$$V = IR \qquad (2\text{-}1)$$

$V =$ ______ $\times 0.999 =$ ______ V *Ans.* | 35 34.97

SUMMARY. Always choose the wire

1. Which can carry (more/less) current than is required | more
2. Which has the (smaller/larger) circular-mil area | larger
3. Which has the (smaller/larger) resistance | smaller

**PROBLEMS**

1. What is the maximum current that may be safely carried by No. 8 varnished-cambric-insulated wire?
2. What is the smallest rubber-covered wire that should be used to carry 50 A?
3. What gage wire should be used if the circular-mil area required is 18,000 cmil?
4. What gage wire should be used if the required resistance is not to exceed 0.91 Ω/1000 ft?
5. Find the voltage drop along 1000 ft of No. 8 varnished-cambric-insulated wire when it is carrying its maximum current.
6. What is the smallest varnished-cambric-insulated wire that should be used to carry 62 A safely?
7. What is the resistance per foot of the smallest rubber-covered wire that should be used to carry 66 A safely?

8. What size of asbestos-covered wire should be used to carry 42 A safely?
9. What size of wire should be used if the circular-mil area required is 42,000 cmil?
10. What is the diameter in mils and inches of the smallest rubber-covered wire that can be used to carry 80 A safely?

## JOB 13-2 FINDING THE MINIMUM SIZE OF WIRE TO SUPPLY A GIVEN LOAD

The minimum size of wire that may be used in any installation depends on two factors: (1) the total current load and (2) the voltage drop in the wire. In this job we shall determine the size of wire needed to carry the given current. In the next job, we shall take the voltage drop into consideration.

### PROCEDURE

1. Determine the total current load.
2. Consult Table 13-1 to find the smallest wire to carry this load.

EXAMPLE 13-7. What is the smallest size of rubber-covered wire that may be used to supply two 10-Ω heating elements operated in parallel from a 220-V line?

SOLUTION

Given: $R_1 = 10\ \Omega$ Find: Smallest size of rubber-covered wire = ?

$R_2 = 10\ \Omega$

$V_T = 220$ V

1. Find the total resistance.

$$R_T = \frac{R}{N} = \frac{10}{2} = 5\ \Omega \qquad (5\text{-}4)$$

2. Find the total current.

$$I_T = \frac{V_T}{R_T} = \frac{220}{5} = 44\ \text{A} \qquad (4\text{-}7)$$

3. Find the gage number to handle 44 A.
   *a.* Read down column 5 until you get to the first number *larger* than 44 A. This number is 45 A.
   *b.* Read to the left from 45 A until you get to column 1, where we read the gage number as No. 8. *Ans.*

EXAMPLE 13-8. What is the smallest size of slow-burning weatherproof cable that may be used to carry current to a 10-hp motor from a 100-V source?

SOLUTION

Given: $P = 10$ hp Find: Smallest size of slow-burning cable = ?
$V = 100$ V

1. Find the watts of power to be delivered by the cable. Assume an input to the motor equal to 110 percent of the rated horsepower.

$$\text{Input hp} = 1.1 \times 10 = 11 \text{ hp}$$
$$\text{Watts} = \text{hp} \times 750 = 11 \times 750 = 8250 \text{ W}$$

2. Find the current drawn.

$$P = I \times V \qquad (7\text{-}1)$$
$$8250 = I \times 100$$
$$I = \frac{8250}{100} = 82.5 \text{ A}$$

3. Find the gage number to handle 82.5 A. *Note:* For any wire other than rubber-covered or varnished-cambric-insulated wire, use the data for asbestos-covered wire in column 7.

*a.* Read down column 7 until you get to the first number *larger* than 82.5 A. This number is 95 A.

*b.* Read to the left from 95 A until you get to column 1, where we read the gage number as No. 6. *Ans.*

## PROBLEMS

1. What is the minimum size of rubber-covered wire that may be used to supply a load of 60 lamps each drawing 0.91 A?
2. What is the minimum size of asbestos-covered wire that may be used to supply a load of three 5-Ω electric ovens operated in parallel from a 110-V line?
3. What is the minimum size of varnished-cambric-covered wire that may be used to supply a load of thirty 200-W lamps and twenty 100-W lamps if they are all operated in parallel from the same 110-V line?
4. A 30-hp motor is to be used on a 220-V line. If the efficiency of the motor is 90 percent, what is the smallest size of varnished-cambric-covered wire that may be used?
5. A 220-V generator supplies motors drawing 5, 10, and 12 kW of power. What is the minimum size of rubber-covered wire that may be used?

## JOB 13-3 FINDING THE SIZE OF WIRE NEEDED TO PREVENT EXCESSIVE VOLTAGE DROPS

A wire not only must be able to carry a current without overheating but must also be able to carry this current without too many volts being used up in the wire itself. If the voltage drop in the line is too large, the voltage available at the load will be too small for proper operation of the load. For example, a 117-V generator is used to supply current to a bank of lamps some distance away. If the supply line should use up 17 V in supplying this current, then only 100 V would be available at the lamps and they would be dim. Not only that, but the loss in the line represents wasted power and a very inefficient system. To cover this, the National Electrical Code stipulates that the voltage drop in a line must not be more than 2 percent of the rated voltage for lighting purposes and not more than 5 percent of the rated voltage for power installations.

EXAMPLE 13-9. A bank of lamps draws 32 A from a 117-V source 300 ft away. Find the smallest gage of rubber-covered wire that may be safely used.

SOLUTION. The diagram for the circuit is shown in Fig. 13-1.

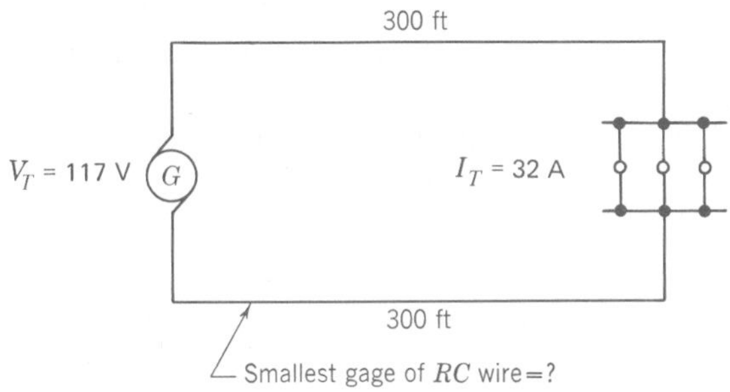


FIGURE 13-1

1. Find the allowable voltage drop. For lighting purposes, the Code permits only a 2 percent drop. Therefore,

$$\text{Allowable drop} = 2\% \text{ of } 117 = 0.02 \times 117 = 2.34 \text{ V}$$

2. Find the line resistance for this drop.

$$V_l = I_l \times R_l \qquad (6\text{-}1)$$

$$2.34 = 32 \times R_l$$

$$R_l = \frac{2.34}{32} = 0.0731\ \Omega$$

3. Find the circular-mil area of the wire that has this resistance. Since there are two wires, $L = 2 \times 300 = 600$ ft. Also, the $K$ for copper is 10.4.

$$R = \frac{K \times L}{A} \qquad (12\text{-}7)$$

$$\frac{0.0731}{1} = \frac{10.4 \times 600}{A}$$

$$0.0731 \times A = 10.4 \times 600$$

$$0.0731A = 6240$$

$$A = \frac{6240}{0.0731} = 85{,}360 \text{ cmil}$$

4. Find the gage number of the wire whose circular-mil area is *larger* than 85,360 cmil.

*a.* Read down column 3 until you get to a number *just larger* than 85,360. This number is 105,500.

*b.* Read across to the left to find gage No. 0 in column 1. *Ans.*

The steps involved in the last problem may be combined as follows: The voltage drop in the line $V_d$ is given as

$$V_d = I \times R \qquad (2\text{-}1)$$

Also,

$$R = \frac{K \times L}{A} \qquad (12\text{-}7)$$

Substituting,

$$V_d = I \times \frac{KL}{A}$$

For two wires, and substituting $CM$ for $A$,

$$V_d = \frac{2KIL}{CM} \qquad \boxed{13\text{-}1}$$

or, solving for $CM$,

$$CM = \frac{2KIL}{V_d} \qquad \boxed{13\text{-}2}$$

where $V_d$ = allowable voltage drop, V
$K$ = specific resistance
$L$ = length of the line, ft
$I$ = line current, A
$CM$ = area of the wire, cmil

EXAMPLE 13-10. A 117-V circuit is run 300 ft to serve a load of 10 A. Using No. 12 wire, find (*a*) the voltage drop, and (*b*) the voltage at the load.

SOLUTION. From Table 12-1, for No. 12 wire, $CM = 6530$.

*a.*

$$V_d = \frac{2KIL}{CM} \qquad (13\text{-}1)$$

$$= \frac{2(10.4)(10)(300)}{6530}$$

$$V_d = 9.6 \text{ V} \qquad \textit{Ans.}$$

*b.*

$$V_L = V_G - V_1 \qquad (6\text{-}3)$$

$$= 117 - 9.6 = 107.4 \text{ V} \qquad \textit{Ans.}$$

EXAMPLE 13-11. A bank of lathes is operated by individual motors in a machine shop. The motors draw a total of 60 A at 110 V from the distributing panel box. What size of rubber-covered wire is required for the two-wire line between the panel box and the switchboard located 100 ft away if the switchboard voltage is 115 V?

SOLUTION. The diagram for the circuit is shown in Fig. 13-2.

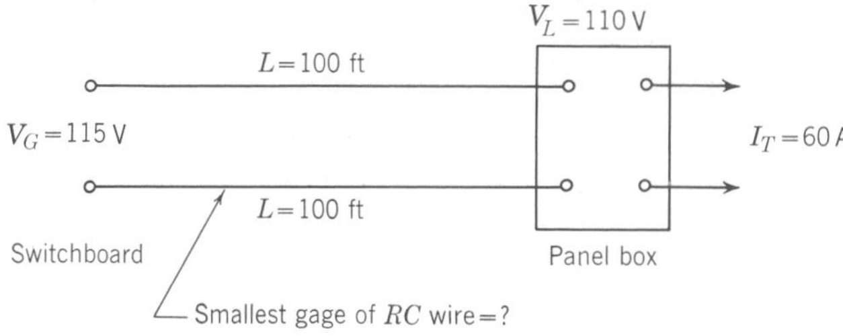


FIGURE 13-2

1. Find the line drop between the switchboard and the panel box.

$$V_G = V_l + V_L \tag{6-2}$$
$$115 = V_l + 110$$
$$V_l = 115 - 110 = 5 \text{ V}$$

2. Find the circular-mil area of the wire.

$$CM = \frac{2KIL}{V_d} \tag{13-2}$$
$$= \frac{2(10.4)(60)(100)}{5}$$
$$= 25{,}000 \text{ cmil}$$

3. Find the gage number of the wire whose circular-mil area is *larger* than 25,000 cmil.

*a.* Read down column 3 until you get to a number *just larger* than 25,000 cmil. This number is 26,250.

*b.* Read across to the left to find gage No. 6 in column 1. This No. 6 wire can safely carry the 60 A without overheating, since its maximum current-carrying capacity is 65 A as found in column 5 for rubber-covered wire. *Ans.*

SELF-TEST 13-12. A 10-hp motor is operated at an efficiency of 90 percent from a 232-V source 100 ft away. What is the minimum size of slow-burning insulated wire that may be used for the line supplying the motor?

SOLUTION. The diagram for the circuit is shown in Fig. 13-3.

1. Find the allowable voltage drop. For power installations, the Code permits a 5 percent drop. Therefore,

$V_l$ = allowable drop = 5 percent of ______ | 232

$V_l$ = ______ × 232 = ______ V | 0.05 11.6

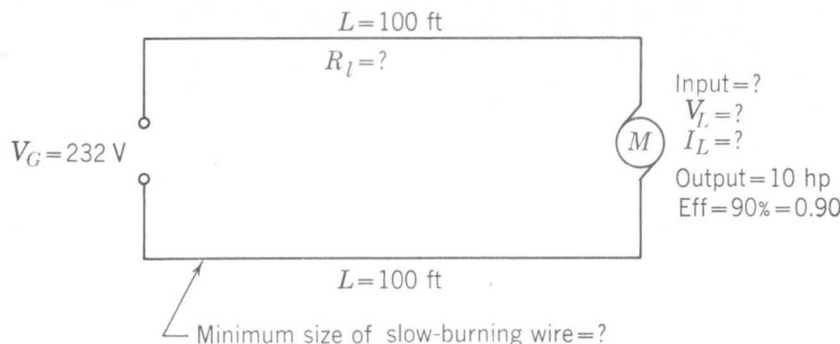


FIGURE 13-3

2. Find the voltage at the motor load.

$$V_L = V_G - \_\_\_\_ \quad (6\text{-}3)$$
$$V_L = 232 - \_\_\_\_ = 220.4 \text{ V}$$

| $V_l$ |
| 11.6 |

3. Find the input to the motor.

$$\text{Eff} = \frac{\text{output}}{\text{input}} \quad (11\text{-}2)$$
$$\_\_\_\_ = \frac{10 \times 750}{\text{input}}$$
$$0.90 \text{ input} = \_\_\_\_$$
$$\text{Input} = \frac{7500}{0.90} = \_\_\_\_ \text{ W}$$

0.90
7500
8333

4. Find the current drawn.

$$P_L = I_L \times V_L \quad (7\text{-}1)$$
$$8333 = I_L \times \_\_\_\_$$
$$I_L = \frac{8333}{220.4} = \_\_\_\_ \text{ A}$$

220.4
37.8

5. Find the circular-mil area of the wire.

$$CM = \frac{?}{V_d}$$
$$= \frac{2(10.4)(37.8)(100)}{?}$$
$$= \_\_\_\_ \text{ cmil}$$

$2KIL$
11.6
6778

6. Find the gage number of the wire whose circular-mil area is ____(smaller/larger)____ than 6778 cmil. — larger

*a.* Read down column 3 until you get to a number just ____(smaller/larger)____ than 6778 cmil. This number is ____. — larger, 10,380

*b.* Read across to the left to find gage No. ____ in column ____. — 10, 1

This No. 10 wire can safely carry the required 37.8 A without overheating, since its maximum current-carrying capacity is ____ A as found in column ____ for slow-burning wire. *Ans.* — 55, 7

**PROBLEMS**

1. What is the minimum size of rubber-covered wire that may be used

to carry 40 A if the allowable voltage drop is 10 V/1000 ft of wire?

2. A 50-ft two-wire extension line is to carry 30 A with a voltage drop of only 5 V. What is the smallest size of weatherproof cable that may be used?
3. A bank of lamps draws 90 A from a 110-V line. What size of rubber-covered wire should be used if the voltage drop along 150 ft of the two-wire system must not exceed 3 V?
4. A bank of lamps draws 50 A from a 117-V source 40 ft away. Find the smallest gage rubber-covered wire that may be used for this two-wire system if the voltage drop must not exceed 2 percent of the voltage at the source.
5. A 20-kW motor load is 100 ft from a 230-V source. If the allowable voltage drop is 5 percent, what is the smallest size of varnished-cambric-covered wire that may be used?
6. A load 400 ft from a generator requires 80 A. The generator voltage is 115.6 V, and the load requires 110 V. What is the smallest size of rubber-covered wire that may be used?
7. A 10-hp motor is located 20 ft from a 225-V generator. What is the smallest size of slow-burning wire that may be used?
8. A 10-hp motor operates at an efficiency of 85 percent at a distance of 100 ft from a 122-V generator. What is the smallest size of varnished-cambric-covered wire that may be used?
9. A generator supplies 5.5 kW at 125 V to a motor 500 ft away. The motor delivers 6.3 hp at an efficiency of 90 percent. Find (*a*) the watt input to the motor, (*b*) the watts lost in the line, (*c*) the current delivered to the line, (*d*) the voltage drop in the line, and (*e*) the smallest size of rubber-covered wire that may be used.
10. Calculate the smallest size of weatherproof wire required to conduct current to one hundred 220-Ω lamps in parallel which are located 125 ft from a generator delivering 112 V. The lamps must receive 110 V.
11. A portable tree-cutting saw requires 4 A at 120 V. If the line drop in the No. 10 cable used must not exceed 3 percent of the motor voltage, what is the maximum length of cable that may be used? *Hint:* Use Table 13-1 to get the resistance per foot.
12. A two-wire No. 10 AWG branch circuit carries 10 A from the main distribution panel to a load 100 ft away. Find (*a*) the total resistance of the line wires, (*b*) the voltage at the load when the voltage at the panel is 120 V, and (*c*) the maximum distance that this branch could carry a peak load of 15 A if the line drop must not exceed 3 percent.

## JOB 13-4 REVIEW OF SIZE OF WIRING

The size of a wire depends on (1) the total current load and (2) the voltage drop in the wires.

To find the smallest wire that may be used in a given job:

1. Find the total current drawn by the load.
2. Find the voltage drop in the supply wires.
3. Find the circular-mil area of the required wire.
4. Look up Table 13-1 to find the wire which has the next larger circular-mil area.
5. Look up the table to find the wire which can carry *more* current than that required by the load.
6. Choose the larger of the two wires found in steps 4 and 5.

## PROBLEMS

1. What is the maximum current that may be safely carried by No. 10 varnished-cambric-covered wire?
2. What is the smallest asbestos-covered (type A) wire that should be used to carry 56 A?
3. What gage number wire should be used if the circular-mil area required is 52,000 cmil?
4. What is the diameter in mils of the smallest rubber-covered wire that should be used to carry 22 A?
5. What size of varnished-cambric-insulated wire is needed to supply a 5-hp motor operating at an efficiency of 90 percent from a 220-V line?
6. What size of rubber-covered wire is needed to supply a bank of ten 200-W lamps and two $^{1}/_{2}$-hp motors connected in parallel across a 110-V line?
7. Find the size of rubber-covered wire needed to supply two 8-Ω heaters in parallel across 122 V.
8. What is the smallest gage of rubber-covered wire that may be used to carry a current of 32 A if the allowable voltage drop is 6 V/1000 ft of wire?
9. A 50-A 112-V load is located 200 ft from a switchboard delivering 115 V. Find the smallest varnished-cambric-covered wire that may be safely used in this two-wire system.
10. A lamp bank draws 60 A from a 230-V source 500 ft away. Assuming a maximum voltage drop of 2 percent, what is the smallest rubber-covered wire that may be used?
11. A 5-hp motor operates at an efficiency of 87 percent at a distance of 50 ft from a 232-V source. Find the smallest size of rubber-covered wire that may be safely used.
12. The $1^{1}/_{4}$-hp motor of an industrial floor-scraping machine operates at an efficiency of 87 percent from a 220-V line. If the line drop in the No. 14 gage wire used must not exceed 5 percent of the line voltage, find the maximum radius of operation of the scraper.

(See Instructor's Manual for Test 13-1, Size of Wiring)

# TRIGONOMETRY FOR ALTERNATING-CURRENT ELECTRICITY

## JOB 14-1 TRIGONOMETRIC FUNCTIONS OF A RIGHT TRIANGLE

Trigonometry is the study of the relationships that exist among the sides and angles of a triangle. These relationships will form a basic mathematical tool used in the solution of ac electrical problems.

**Angles.** An angle is a figure formed when two lines meet at a point. In Fig. 14-1, the lines $OA$ and $OC$ are the sides of the angle. They meet at the point $O$, which is the vertex of the angle.

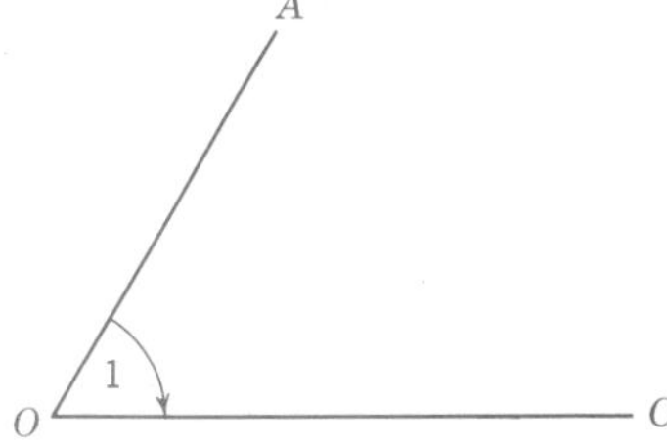


**FIGURE 14-1**
Naming an angle. Angle $AOC$ or angle O or angle 1.

**Naming angles.** The angle shown in Fig. 14-1 may be named in three ways. (1) Use three capital letters, setting them down in order from one end of the angle to the other. Thus the angle may be named $\angle AOC$ or $\angle COA$. (2) Use only the capital letter at the vertex of the angle, as $\angle O$. (3) Use a small letter or number inside the angle, as $\angle 1$.

**Measuring angles.** When a straight line *turns* about a point from one position to another, an angle is formed. When the minute hand of a clock turns from the 1 to the 3, the hand has turned through an angle approximately equal to $\angle AOC$. The size of an angle is measured only by the *amount of rotation of a line about a fixed point.* Suppose we place a small watch on the face of a large clock, as shown in Fig. 14-2. When the hand of the large clock turns from 1 to 3 it forms the angle $AOC$. But in this *same time,* the hands of the small watch have formed the angle $BOD$,

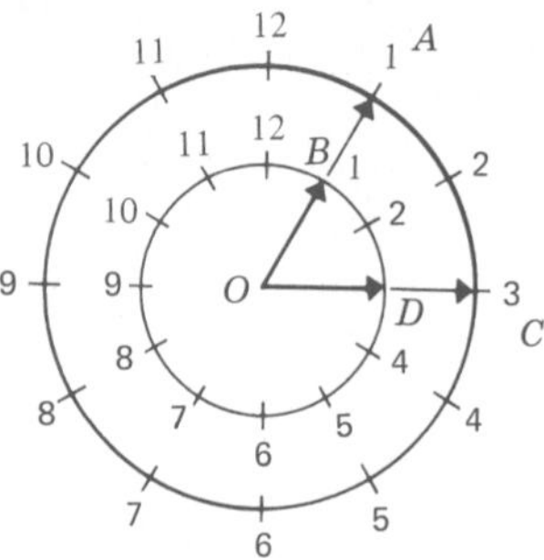


**FIGURE 14-2**
The angle formed by the rotation of the minute hand from 1 to 3 o'clock is the same on any clock face. $\angle BOD = \angle AOC$.

which is exactly equal to $\angle AOC$. Thus we see that the lengths of the sides have no effect on the size of the angle. The size of an angle depends *only* on the amount of rotation between the sides of the angle. The end of a line that makes one complete turn will describe a circle. In Fig. 14-3, a circle is divided into 360 parts. The angle formed by a rotation through $^1/_{360}$ part of the circle is called 1 *degree* (1°).

A *right angle* is an angle formed by a rotation through one-fourth of a circle, as $\angle ABC$ in Fig. 14-4. Since a complete circle contains 360°, a right angle contains $^1/_4 \times 360$, or 90°.

An *acute angle* is an angle containing *less* than 90°, as $\angle AOC$ in Fig. 14-1.

**Triangles.** A triangle is a closed plane figure of three sides. Triangles are named by naming the three vertices of the triangle using capital letters. The letters are then given in order around the triangle. Thus in Fig. 14-4, the triangle is named $\triangle ACB$ or $\triangle CBA$ or $\triangle BAC$ or $\triangle ABC$ or $\triangle BCA$ or $\triangle CAB$.

A *right triangle* is a triangle which contains a right angle, as $\triangle ABC$ in Fig. 14-4. The little square at $\angle B$ is used to indicate a right angle.

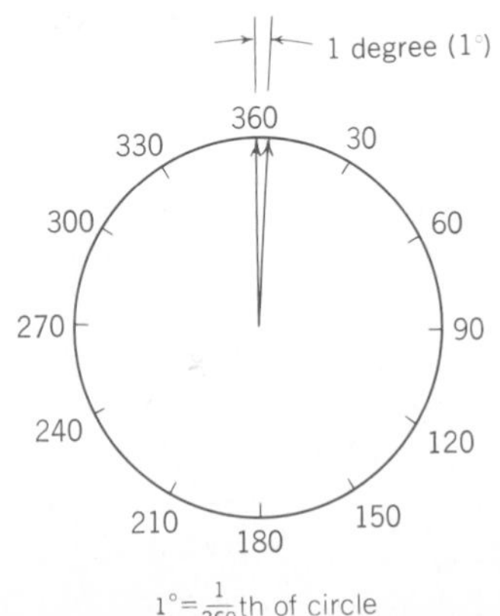


**FIGURE 14-3**
An angle is measured by the amount of rotation of a line about a fixed point.

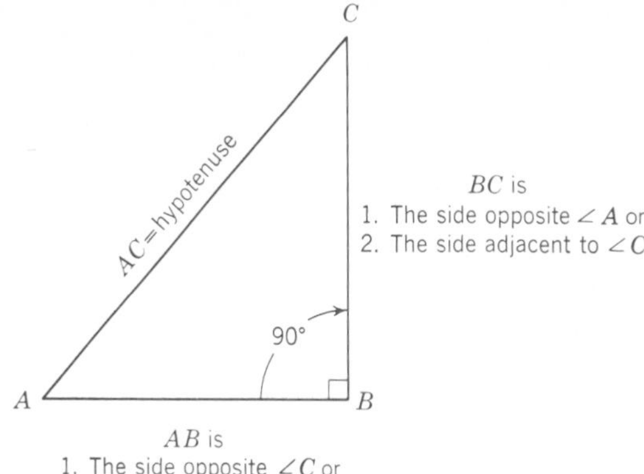


**FIGURE 14-4**
Naming the sides of a right triangle.

**Naming the sides of a right triangle.** In trigonometry, the sides of a right triangle are named depending on which of the *acute* angles are used. In Fig. 14-4,

For angle $A$:

1. The side opposite the right angle is called the *hypotenuse* ($AC$).
2. The side opposite angle $A$ is called the *opposite side* ($BC$).
3. The side of angle $A$ which is *not* the hypotenuse is called the *adjacent side* ($AB$)

For angle $C$:

1. The side opposite the right angle is called the *hypotenuse* ($AC$).
2. The side opposite angle $C$ is called the *opposite side* ($AB$).
3. The side of angle $C$ which is *not* the hypotenuse is called the *adjacent side* ($BC$).

## THE TRIGONOMETRIC FORMULAS

**The tangent of an angle.** In addition to degrees, the size of an angle may also be described in terms of the lengths of the sides of a right triangle formed from the angle. Refer to Fig. 14-5, in which each space represents one unit of length. From various points on one side of angle $A$, lines have been drawn making right angles with the other side. These lines have formed the right triangles $ABE$, $ACF$, and $ADG$. Notice that all these triangles contain the same angle $A$.

| | In $\triangle ABE$ | In $\triangle ACF$ | In $\triangle ADG$ |
|---|---|---|---|
| Hypotenuse | $= AB = 5$ | $AC = 10$ | $AD = 12.5$ |
| Side opposite $\angle A$ | $= BE = 3$ | $CF = 6$ | $DG = 7.5$ |
| Side adjacent $\angle A$ | $= AE = 4$ | $AF = 8$ | $AG = 10$ |

In each triangle, let us divide the side *opposite* $\angle A$ by the side *adjacent* to $\angle A$.

In $\triangle ABE$:
$$\frac{\text{Side opposite } \angle A}{\text{Side adjacent } \angle A} = \frac{BE}{AE} = \frac{3}{4} = 0.75$$

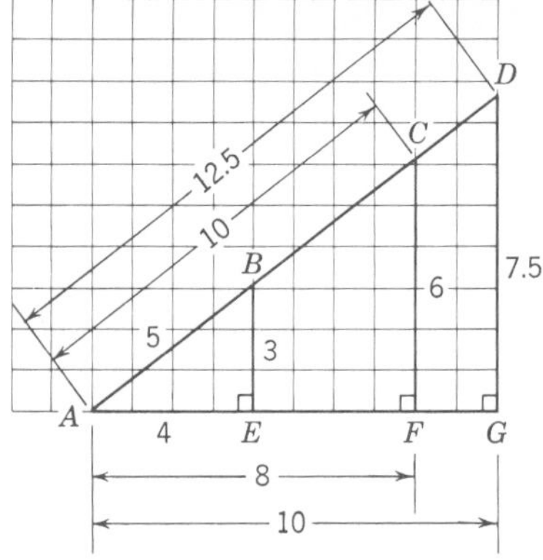


**FIGURE 14-5**
Diagram used to obtain the trigonometric formulas.

In $\triangle ACF$: $\dfrac{\text{Side opposite } \angle A}{\text{Side adjacent } \angle A} = \dfrac{CF}{AF} = \dfrac{6}{8} = 0.75$

In $\triangle ADG$: $\dfrac{\text{Side opposite } \angle A}{\text{Side adjacent } \angle A} = \dfrac{DG}{AG} = \dfrac{7.5}{10} = 0.75$

Notice that this particular division of the opposite side by the adjacent side always results in the *same* answer *regardless* of the size of the triangle. This is true because *all* the triangles contain the same angle $A$. The ratio of these sides remains constant because they all describe the same angle $A$. This constant number describes the size of angle $A$ and is called the *tangent of angle A.* Therefore, if the tangent of an *unknown* angle were calculated to be 0.75, then the angle would be equal to angle $A$, or about 37°. If the angle changes, then the number for the tangent of the angle will also change. However, the tangent of every angle is a specific number which never changes. These numbers are found in Table 14-1.

**RULE** The tangent of an angle $= \dfrac{\text{side opposite the angle}}{\text{side adjacent to the angle}}$

**FORMULA**

$$\tan \angle = \frac{o}{a} \qquad \boxed{14\text{-}1}$$

**The sine of an angle.** In each triangle shown in Fig. 14-5, let us divide the side *opposite* angle $A$ by the side called the *hypotenuse.*

In $\triangle ABE$: $\dfrac{\text{Side opposite } \angle A}{\text{Hypotenuse}} = \dfrac{BE}{AB} = \dfrac{3}{5} = 0.6$

In $\triangle ACF$: $\dfrac{\text{Side opposite } \angle A}{\text{Hypotenuse}} = \dfrac{CF}{AC} = \dfrac{6}{10} = 0.6$

In $\triangle ADG$: $\dfrac{\text{Side opposite } \angle A}{\text{Hypotenuse}} = \dfrac{DG}{AD} = \dfrac{7.5}{12.5} = 0.6$

Once again, the resulting quotients are fixed regardless of the size of the triangles. This number is called the *sine of angle* A and is *another* way to describe the size of the angle. If the angle changes, then the number for the sine of the angle will also change. However, the sine of every angle is a specific number which never changes. These numbers are also found in Table 14-1.

**RULE** The sine of an angle $= \dfrac{\text{side opposite the angle}}{\text{hypotenuse}}$

**FORMULA** $$\sin \angle = \frac{o}{h}$$ 14-2

**The cosine of an angle.** In each triangle shown in Fig. 14-5, let us divide the side *adjacent* to angle *A* by the *hypotenuse*.

In $\triangle ABE$: $\dfrac{\text{Side adjacent } \angle A}{\text{Hypotenuse}} = \dfrac{AE}{AB} = \dfrac{4}{5} = 0.8$

In $\triangle ACF$: $\dfrac{\text{Side adjacent } \angle A}{\text{Hypotenuse}} = \dfrac{AF}{AC} = \dfrac{8}{10} = 0.8$

In $\triangle ADG$: $\dfrac{\text{Side adjacent } \angle A}{\text{Hypotenuse}} = \dfrac{AG}{AD} = \dfrac{10}{12.5} = 0.8$

Here, too, the resulting quotients are fixed regardless of the size of the triangles. This number is called the *cosine of angle* A and is a third way to describe the size of the angle. If the angle changes, then the number for the cosine of the angle will also change. However, the cosine of every angle is a specific number which never changes. These numbers are found in Table 14-1.

**RULE** The cosine of an angle $= \dfrac{\text{side adjacent to the angle}}{\text{hypotenuse}}$

**FORMULA** $$\cos \angle = \frac{a}{h}$$ 14-3

EXAMPLE 14-1. Find the sine, cosine, and tangent of $\angle A$ in the triangle shown in Fig. 14-6.

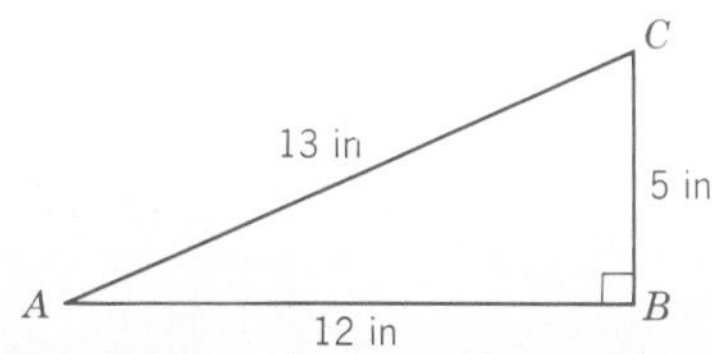


**FIGURE 14-6**

SOLUTION

1. Name the sides using $\angle A$ as the reference angle.

$$\text{Hypotenuse} = AC = 13 \text{ in}$$
$$\text{Opposite side} = BC = 5 \text{ in}$$
$$\text{Adjacent side} = AB = 12 \text{ in}$$

2. Find the values of the three functions.

$$\sin \angle A = \frac{o}{h} \qquad \cos \angle A = \frac{a}{h} \qquad \tan \angle A = \frac{o}{a}$$
$$\sin \angle A = \frac{5}{13} \qquad \cos \angle A = \frac{12}{13} \qquad \tan \angle A = \frac{5}{12}$$
$$\sin \angle A = 0.384 \qquad \cos \angle A = 0.923 \qquad \tan \angle A = 0.417 \qquad \textit{Ans.}$$

EXAMPLE 14-2. Find the sine, cosine, and tangent of $\angle B$ in the triangle shown in Fig. 14-7.

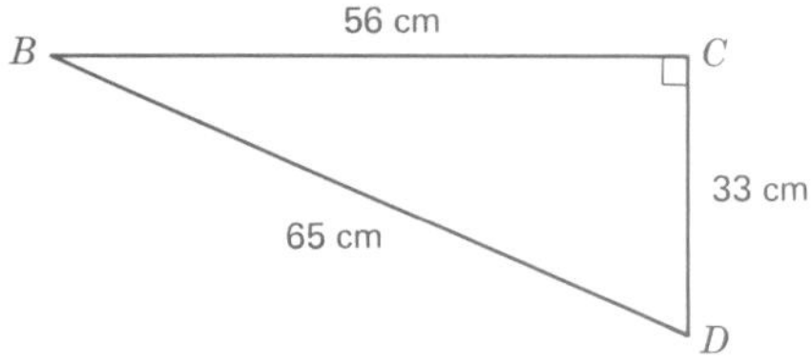


FIGURE 14-7

SOLUTION

1. Name the sides, using $\angle B$ as the reference angle.

$$\text{Hypotenuse} = BD = 65 \text{ cm}$$
$$\text{Opposite side} = CD = 33 \text{ cm}$$
$$\text{Adjacent side} = BC = 56 \text{ cm}$$

2. Find the values of the three functions.

$$\sin \angle B = \frac{o}{h} \qquad \cos \angle B = \frac{a}{h} \qquad \tan \angle B = \frac{o}{a}$$
$$\sin \angle B = \frac{33}{65} \qquad \cos \angle B = \frac{56}{65} \qquad \tan \angle B = \frac{33}{56}$$
$$\sin \angle B = 0.5077 \qquad \cos \angle B = 0.8615 \qquad \tan \angle B = 0.5893 \qquad \textit{Ans.}$$

**PROBLEMS**

Calculate the sine, cosine, and tangent of the angles named in each triangle in Fig. 14-8.

## JOB 14-2 USING THE TABLE OF TRIGONOMETRIC FUNCTIONS

Since the values of the trigonometric functions of an angle never change, it is possible to set these values down in a table such as Table 14-1. (Functions of angles for decimal degrees may be found in the appendix.)

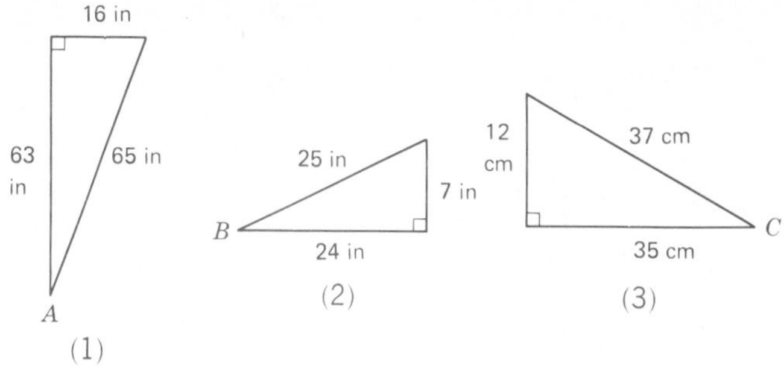


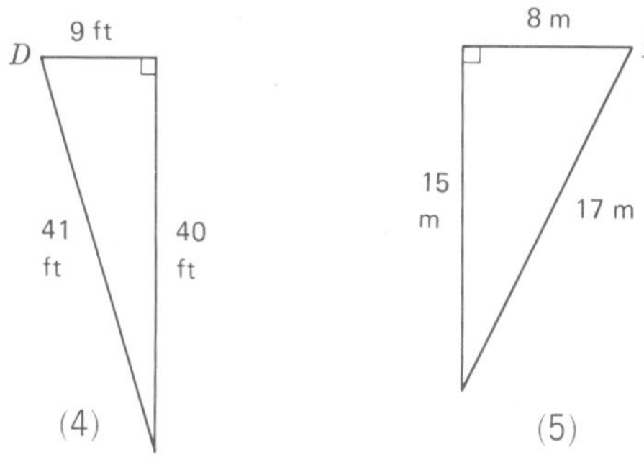


FIGURE 14-8

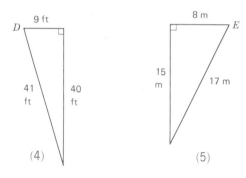

EXAMPLE 14-3. Find the value of sin 50°.

SOLUTION. Follow down the column marked "Angle" in Table 14-1 until you reach 50°. Read across to the right to find the number 0.7660 in the sine column. Therefore,

$$\sin 50° = 0.7660 \qquad \textit{Ans.}$$

EXAMPLE 14-4. Find the value of tan 30°.

SOLUTION. Follow down the column marked "Angle" until you get to 30°. Read across to the right to find the number 0.5774 in the tangent column. Therefore,

$$\tan 30° = 0.5774 \qquad \textit{Ans.}$$

## FINDING THE ANGLE WHEN THE FUNCTION IS GIVEN

EXAMPLE 14-5. Find $\angle A$ if $\cos A = 0.7314$.

SOLUTION. Follow down the column marked "cos" until you find the number 0.7314. Read across to the left to find 43° in the column marked "Angle." Therefore,

$$\angle A = 43° \qquad \textit{Ans.}$$

EXAMPLE 14-6. Find $\angle B$ if $\tan B = 1.8807$.

SOLUTION. Follow down the column marked "tan" until you find the number 1.8807.

TABLE 14-1

VALUES OF THE TRIGONOMETRIC FUNCTIONS

| ANGLE | SIN | COS | TAN | ANGLE | SIN | COS | TAN |
|---|---|---|---|---|---|---|---|
| 0° | 0.0000 | 1.0000 | 0.0000 | 46° | 0.7193 | 0.6947 | 1.0355 |
| 1° | 0.0175 | 0.9998 | 0.0175 | 47° | 0.7314 | 0.6820 | 1.0724 |
| 2° | 0.0349 | 0.9994 | 0.0349 | 48° | 0.7431 | 0.6691 | 1.1106 |
| 3° | 0.0523 | 0.9986 | 0.0524 | 49° | 0.7547 | 0.6561 | 1.1504 |
| 4° | 0.0698 | 0.9976 | 0.0699 | 50° | 0.7660 | 0.6428 | 1.1918 |
| 5° | 0.0872 | 0.9962 | 0.0875 | 51° | 0.7771 | 0.6293 | 1.2349 |
| 6° | 0.1045 | 0.9945 | 0.1051 | 52° | 0.7880 | 0.6157 | 1.2799 |
| 7° | 0.1219 | 0.9925 | 0.1228 | 53° | 0.7986 | 0.6018 | 1.3270 |
| 8° | 0.1392 | 0.9903 | 0.1405 | 54° | 0.8090 | 0.5878 | 1.3764 |
| 9° | 0.1564 | 0.9877 | 0.1584 | 55° | 0.8192 | 0.5736 | 1.4281 |
| 10° | 0.1736 | 0.9848 | 0.1763 | 56° | 0.8290 | 0.5592 | 1.4826 |
| 11° | 0.1908 | 0.9816 | 0.1944 | 57° | 0.8387 | 0.5446 | 1.5399 |
| 12° | 0.2079 | 0.9781 | 0.2126 | 58° | 0.8480 | 0.5299 | 1.6003 |
| 13° | 0.2250 | 0.9744 | 0.2309 | 59° | 0.8572 | 0.5150 | 1.6643 |
| 14° | 0.2419 | 0.9703 | 0.2493 | 60° | 0.8660 | 0.5000 | 1.7321 |
| 15° | 0.2588 | 0.9659 | 0.2679 | 61° | 0.8746 | 0.4848 | 1.8040 |
| 16° | 0.2756 | 0.9613 | 0.2867 | 62° | 0.8829 | 0.4695 | 1.8807 |
| 17° | 0.2924 | 0.9563 | 0.3057 | 63° | 0.8910 | 0.4540 | 1.9626 |
| 18° | 0.3090 | 0.9511 | 0.3249 | 64° | 0.8988 | 0.4384 | 2.0503 |
| 19° | 0.3256 | 0.9455 | 0.3443 | 65° | 0.9063 | 0.4226 | 2.1445 |
| 20° | 0.3420 | 0.9397 | 0.3640 | 66° | 0.9135 | 0.4067 | 2.2460 |
| 21° | 0.3584 | 0.9336 | 0.3839 | 67° | 0.9205 | 0.3907 | 2.3559 |
| 22° | 0.3746 | 0.9272 | 0.4040 | 68° | 0.9272 | 0.3746 | 2.4751 |
| 23° | 0.3907 | 0.9205 | 0.4245 | 69° | 0.9336 | 0.3584 | 2.6051 |
| 24° | 0.4067 | 0.9135 | 0.4452 | 70° | 0.9397 | 0.3420 | 2.7475 |
| 25° | 0.4226 | 0.9063 | 0.4663 | 71° | 0.9455 | 0.3256 | 2.9042 |
| 26° | 0.4384 | 0.8988 | 0.4877 | 72° | 0.9511 | 0.3090 | 3.0777 |
| 27° | 0.4540 | 0.8910 | 0.5095 | 73° | 0.9563 | 0.2924 | 3.2709 |
| 28° | 0.4695 | 0.8829 | 0.5317 | 74° | 0.9613 | 0.2756 | 3.4874 |
| 29° | 0.4848 | 0.8746 | 0.5543 | 75° | 0.9659 | 0.2588 | 3.7321 |
| 30° | 0.5000 | 0.8660 | 0.5774 | 76° | 0.9703 | 0.2419 | 4.0108 |
| 31° | 0.5150 | 0.8572 | 0.6009 | 77° | 0.9744 | 0.2250 | 4.3315 |
| 32° | 0.5299 | 0.8480 | 0.6249 | 78° | 0.9781 | 0.2079 | 4.7046 |
| 33° | 0.5446 | 0.8387 | 0.6494 | 79° | 0.9816 | 0.1908 | 5.1446 |
| 34° | 0.5592 | 0.8290 | 0.6745 | 80° | 0.9848 | 0.1736 | 5.6713 |
| 35° | 0.5736 | 0.8192 | 0.7002 | 81° | 0.9877 | 0.1564 | 6.3138 |
| 36° | 0.5878 | 0.8090 | 0.7265 | 82° | 0.9903 | 0.1392 | 7.1154 |
| 37° | 0.6018 | 0.7986 | 0.7536 | 83° | 0.9925 | 0.1219 | 8.1443 |
| 38° | 0.6157 | 0.7880 | 0.7813 | 84° | 0.9945 | 0.1045 | 9.5144 |
| 39° | 0.6293 | 0.7771 | 0.8098 | 85° | 0.9962 | 0.0872 | 11.4300 |
| 40° | 0.6428 | 0.7660 | 0.8391 | 86° | 0.9976 | 0.0698 | 14.3010 |
| 41° | 0.6561 | 0.7547 | 0.8693 | 87° | 0.9986 | 0.0523 | 19.0810 |
| 42° | 0.6691 | 0.7431 | 0.9004 | 88° | 0.9994 | 0.0349 | 28.6360 |
| 43° | 0.6820 | 0.7314 | 0.9325 | 89° | 0.9998 | 0.0175 | 57.2900 |
| 44° | 0.6947 | 0.7193 | 0.9657 | 90° | 1.0000 | 0.0000 | |
| 45° | 0.7071 | 0.7071 | 1.0000 | | | | |

Read across to the left to find 62° in the column marked "Angle." Therefore,

$$\angle B = 62° \qquad \textit{Ans.}$$

**PROBLEMS**

Find the number of degrees in each angle.

| | | |
|---|---|---|
| 1. $\tan A = 0.3249$ | 2. $\sin B = 0.6428$ | 3. $\cos C = 0.1736$ |
| 4. $\cos D = 0.9877$ | 5. $\sin E = 0.5000$ | 6. $\tan F = 1.0724$ |
| 7. $\tan G = 0.5774$ | 8. $\cos H = 0.7071$ | 9. $\sin J = 0.8660$ |

## FINDING THE ANGLE WHEN THE FUNCTION IS NOT IN THE TABLE

EXAMPLE 14-7. Find $\angle A$ if $\tan A = 0.5120$.

SOLUTION. The number 0.5120 is not in the table under the column "tan" but lies between 0.5095 (27°) and 0.5317 (28°). Choose the number closest to 0.5120.

$$\begin{aligned} \tan 27 &= 0.5095 \\ \tan A &= 0.5120 \\ \tan 28 &= 0.5317 \end{aligned} \quad \begin{aligned} &\text{difference} = 0.0025 \\ &\text{difference} = 0.0197 \end{aligned}$$

The smaller difference indicates the closer number. Therefore,

$$\angle A = 27° \qquad \textit{Ans.}$$

**PROBLEMS**

Find the number of degrees in each angle correct to the nearest degree.

| | | |
|---|---|---|
| 1. $\tan A = 0.2700$ | 2. $\cos B = 0.7500$ | 3. $\sin C = 0.8500$ |
| 4. $\sin D = 0.2350$ | 5. $\cos E = 0.4172$ | 6. $\tan F = 1.9120$ |
| 7. $\tan G = 0.2783$ | 8. $\cos H = 0.1645$ | 9. $\sin J = 0.7250$ |

## JOB 14-3 FINDING THE ACUTE ANGLES OF A RIGHT TRIANGLE

In order to find the value of an angle in any problem, it is only necessary to find the number for the sine *or* the cosine *or* the tangent of the angle. If we know *any one* of these values, we can determine the angle by finding the number in the appropriate column of the table.

EXAMPLE 14-8. Find $\angle A$ and $\angle B$ in the triangle of Fig. 14-9.

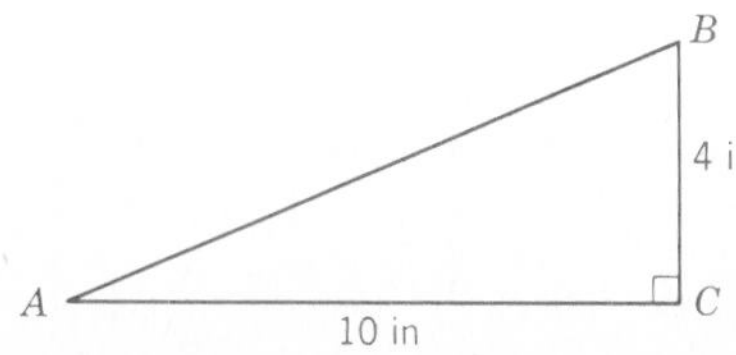


FIGURE 14-9

SOLUTION

1. Name the sides which have values, using $\angle A$ as the reference angle.

$$4 \text{ in} = \text{the side } \textit{opposite } \angle A$$
$$10 \text{ in} = \text{the side } \textit{adjacent } \angle A$$

2. We can find $\angle A$ if we can find *any* of the three functions of the angle. However, the only function that can *actually* be found is the one for which we have *known* values. In this problem, the only known values are those for the opposite and adjacent sides. The only formula that uses these particular sides is the *tangent* formula.

$$\tan A = \frac{o}{a} = \frac{4}{10} = 0.4000 \qquad (14\text{-}1)$$

3. Find the number in the tangent table closest to 0.4000.

$\angle A = 22°$ (nearest angle) *Ans.*

4. Since the two acute angles of any right triangle always total 90°,

$$\angle B = 90° - \angle A \qquad \boxed{14\text{-}4}$$
$$\angle B = 90° - 22° = 68° \quad \textit{Ans.}$$

EXAMPLE 14-9. The phase angle in an ac circuit may be represented by $\angle\theta$ (angle theta), as shown in Fig. 14-10. Find $\angle\theta$ and $\angle B$.

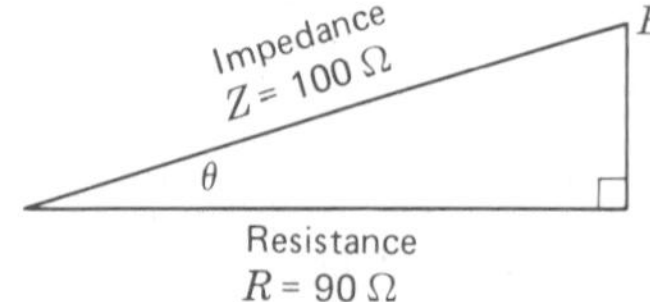


**FIGURE 14-10**
The angle theta ($\theta$) is the phase angle in an inductive ac circuit.

SOLUTION

1. Name the sides which have values, using $\angle\theta$ as the reference angle.

$$100\ \Omega = \text{hypotenuse}$$
$$90\ \Omega = \text{side adjacent } \angle\theta$$

2. Choose the trigonometric formula which uses the adjacent side and the hypotenuse.

$$\cos\theta = \frac{a}{h} = \frac{90}{100} = 0.9000 \qquad (14\text{-}3)$$

3. Find the number in the cosine table closest to 0.9000.

$\angle\theta = 25°$ (nearest angle) *Ans.*

4. Since the two acute angles of a right triangle always total 90°,

$$\angle B = 90° - \angle\theta \qquad (14\text{-}4)$$
$$\angle B = 90° - 25° = 65° \quad . \textit{Ans.}$$

SELF-TEST 14-10. An electrician must bend a pipe to make a $3^1/_2$-ft rise in a 5-ft horizontal distance. What is the angle at each bend?

SOLUTION. The diagram describing the conditions is shown in Fig. 14-11.

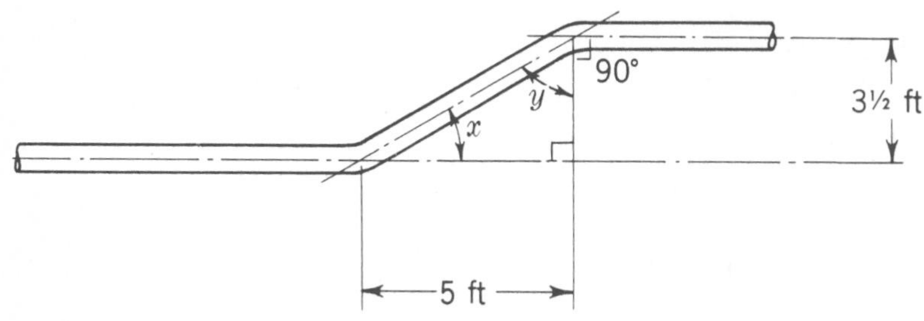


FIGURE 14-11

1. Name the sides which have values, using $\angle x$ as the reference angle.

Side opposite $\angle x$ = ______ ft — $3^1/_2$

Side adjacent $\angle x$ = ______ ft — 5

2. Choose the trigonometric formula which uses the opposite side and the adjacent side. The only formula that uses these two sides is the ______. — tangent

$$\tan x = \frac{?}{?} \qquad (14\text{-}1)$$ — $\frac{o}{a}$

$$\tan x = \frac{3.5}{5} = \_\_\_\_\_\_$$ — 0.7000

3. Find the number in the ______ table that is closest to 0.7000. — tangent

$\angle x$ = ______° *Ans.* — 35

4. Since the two acute angles of a right triangle always total ______°, — 90

$$\angle y = 90° - \_\_\_\_\_\_ \qquad (14\text{-}4)$$ — $\angle x$

$$\angle y = 90° - 35° = \_\_\_\_\_\_$$ — 55°

5. The angle at the top bend = $\angle y$ + ______, or — 90°

Angle at top bend = 55° + 90°

= ______° *Ans.* — 145

## PROBLEMS

1. Find $\angle A$ and $\angle B$ in the right triangle shown in Fig. 14-12.

Using Fig. 14-12, find $\angle A$ and $\angle B$ if

2. $AC$ = 100 ft and $BC$ = 70 ft
3. $BC$ = 4 in and $AB$ = 8 in
4. $BC$ = 40 Ω and $AC$ = 25 Ω
5. $AC$ = 200 Ω and $AB$ = 350 Ω
6. $BC$ = 300 W and $AB$ = 1000 W
7. $AC$ = 600 Ω and $AB$ = 960 Ω

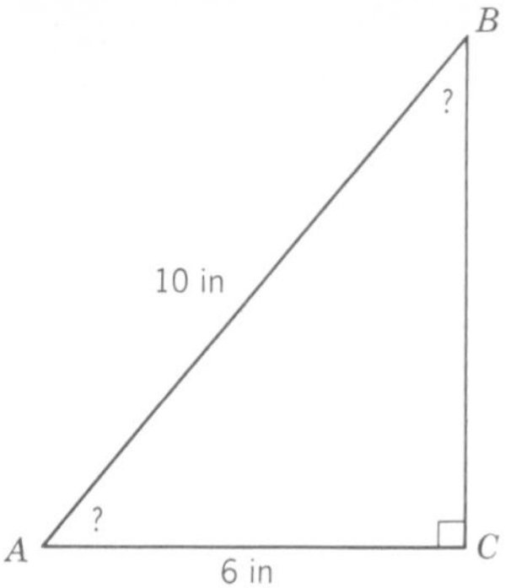


FIGURE 14-12

8. $BC = 7.5$ V and $AC = 12.5$ V
9. $BC = 12.4$ A and $AB = 67.8$ A
10. $AC = 62.5\ \Omega$ and $BC = 100\ \Omega$

11. At what angles must a pipe be bent in order to make a 7-ft rise in a 3-ft horizontal distance?
12. At what angles must a pipe be bent in order to make a 4-ft 6-in rise in a 2-ft 3-in horizontal distance?
13. A car rises 50 ft while traveling along a road 1000 ft long. At what angle is the road inclined to the horizontal?
14. Find angle $x$ in the taper shown in Fig. 14-13.

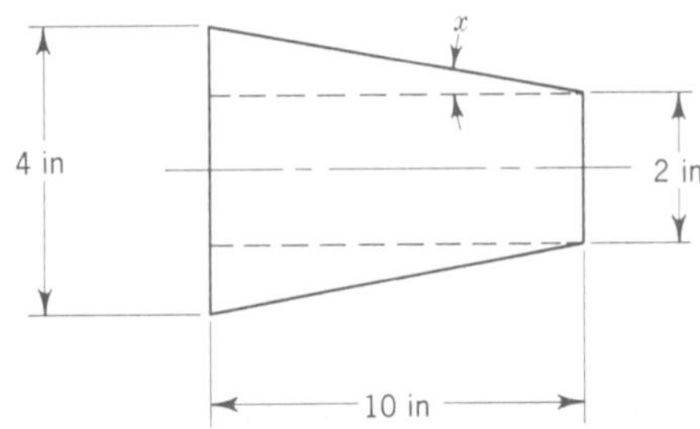


FIGURE 14-13
Angle x is equal to one-half the taper angle.

15. A guy wire 120 ft long reaches from the top of a pole to a point 64 ft from the foot of the pole. What angle does the wire make with the ground?

(See Instructor's Manual for Test 14-1)

## JOB 14-4 FINDING THE SIDES OF A RIGHT TRIANGLE

EXAMPLE 14-11. Find (*a*) side $BC$ and (*b*) $\angle B$ of the right triangle shown in Fig. 14-14.

SOLUTION

*a*. Find the side $BC$.

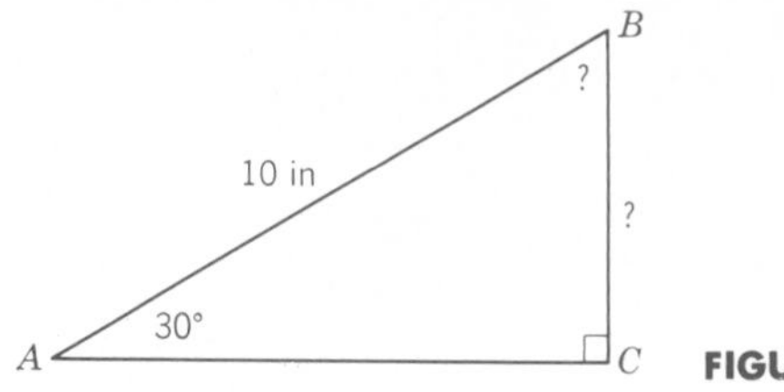

FIGURE 14-14

1. Select the angle to be used ($\angle A = 30°$).
2. Select the side to be found ($BC$).
3. Select one other side whose value is known ($AB = 10$ in).
4. Name these two sides.

$$BC = \text{side } \textit{opposite } \angle A$$
$$10 \text{ in} = AB = \textit{hypotenuse}$$

5. Select the correct trigonometric formula. It will be the formula that uses these two sides—the *opposite* and the *hypotenuse*. Only the *sine* formula uses the *opposite* side and the *hypotenuse*.

$$\sin A = \frac{o}{h} \tag{14-2}$$

$$\sin 30° = \frac{BC}{10}$$

From Table 14-1, sin 30° = 0.5000. Therefore,

$$\frac{0.5000}{1} = \frac{BC}{10}$$
$$BC = 0.5000 \times 10 = 5 \text{ in} \qquad \textit{Ans.}$$

*b.* Find $\angle B$.

$$\angle B = 90° - \angle A \tag{14-4}$$
$$\angle B = 90° - 30° = 60° \qquad \textit{Ans.}$$

EXAMPLE 14-12. Find side $BC$ in the right triangle shown in Fig. 14-15.

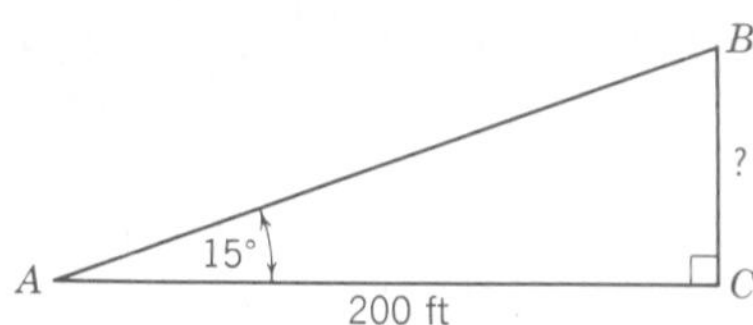

FIGURE 14-15

SOLUTION. Find the side $BC$.

1. Select the angle to be used ($\angle A = 15°$).
2. Select the side to be found ($BC$).
3. Select one other side whose value is known ($AC = 200$ ft).
4. Name these two sides.

$$BC = \text{the side } \textit{opposite } \angle A$$
$$200 \text{ ft} = AC = \text{the side } \textit{adjacent } \angle A$$

5. Select the correct trigonometric formula. It will be the formula that uses these two sides—the *opposite* and the *adjacent.* Only the tangent formula uses the *opposite* side and the *adjacent* side.

$$\tan A = \frac{o}{a} \tag{14-1}$$

$$\tan 15° = \frac{BC}{200}$$

From Table 14-1, tan 15° = 0.2679. Therefore,

$$\frac{0.2679}{1} = \frac{BC}{200}$$
$$BC = 0.2679 \times 200 = 53.58 \text{ ft} \qquad \textit{Ans.}$$

EXAMPLE 14-13. The relationship among the impedance $Z$, the resistance $R$, and the capacitive reactance $X_C$ of an ac circuit is shown in Fig. 14-16. Find the impedance.

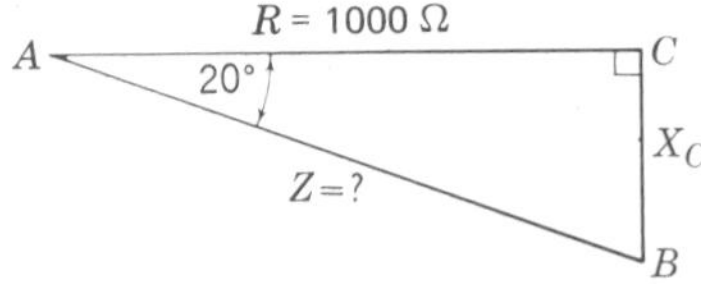


FIGURE 14-16

SOLUTION. Find the impedance $Z$ represented by side $AB$.

1. Select the angle to be used ($\angle A = 20°$).
2. Select the side to be found ($AB$).
3. Select one other side whose value is known ($AC = 1000\ \Omega$).
4. Name these two sides.

$$AB = \textit{hypotenuse}$$
$$1000\ \Omega = AC = \text{side } \textit{adjacent } \angle A$$

5. Select the correct trigonometric formula. It will be the formula that uses these two sides—the *adjacent* and the *hypotenuse.* Only the cosine formula uses the *adjacent* side and the *hypotenuse.*

$$\cos A = \frac{a}{h} \tag{14-3}$$

$$\cos 20° = \frac{1000}{AB}$$

From Table 14-1, cos 20° = 0.9397. Therefore,

$$\frac{0.9397}{1} = \frac{1000}{AB}$$
$$AB \times 0.9397 = 1000$$

$$AB = \frac{1000}{0.9397} = 1064\ \Omega$$

$$\text{Impedance } Z = 1064\ \Omega \qquad \textit{Ans.}$$

SELF-TEST 14-14. As shall be discussed in Job 18-1, the current and voltage of an ac circuit do not always appear at the same time. They are said to be "out of phase." In Fig. 14-17*a*, the current "leads" the voltage by a phase angle of 37°. In order to cope with this situation, the current "phasor" $I = 20$ A is resolved into its "in-phase" component $I_x$ and its "reactive" component $I_y$, as shown in Fig. 14-17*b*. Find $I_x$ and $I_y$. A further discussion of the resolution of phasors may be found in Job 19-5.

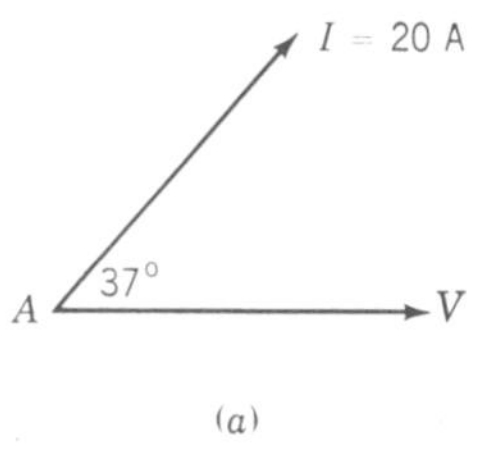


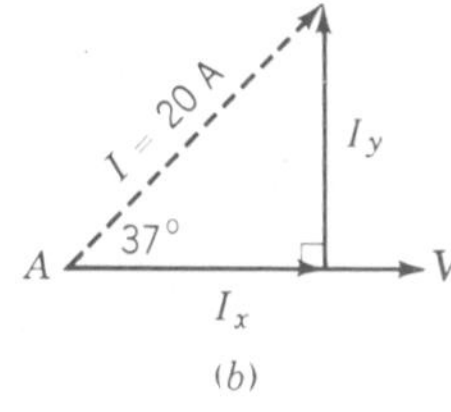


**FIGURE 14-17** (*a*) The current $I$ "leads" the voltage $V$ by 37°. (*b*) $I$ is resolved into its "in-phase" component $I_x$ and its "reactive" component $I_y$.

SOLUTION. Find $I_x$. In Fig. 14-17*b*,

1. The angle to be used = ________. | 37°
2. The side to be found = ________. | $I_x$
3. Another side whose value is known is $I$ = ________. | 20
4. Name these two sides using $\angle A = 37°$ as the reference angle.

$I_x$ = the side ________ $\angle A$ | adjacent

$I = 20$ = the ________ | hypotenuse

5. Select the correct trigonometric formula that uses these two particular sides. The adjacent side and the hypotenuse are connected by the ________ formula. | cosine

$$\cos \angle A = \frac{a}{h} \qquad (14\text{-}2)$$

$$\cos 37° = \frac{?}{20}$$ | $I_x$

From Table 14-1, cos 37° = ________. Therefore, | 0.7986

$$\frac{0.7986}{1} = \frac{I_x}{20}$$

and $I_x = 0.7986 \times 20$ = ________ A *Ans.* | 16

Find $I_y$.

1. Using $\angle A = 37°$, the sides to be used are $I_y$ and ________. | 20
2. Using $\angle A$ as the reference angle, these sides are named as follows:

$I_y$ = the side ________ $\angle A$ | opposite

$I = 20$ = the ________ | hypotenuse

3. The only formula that uses these two sides is the ______ formula. | sine

$$\sin A = \frac{?}{h}$$ | o

$$\sin 37° = \frac{I_y}{?}$$ | 20

From Table 14-1, sin 37° = ______. Therefore, | 0.6018

$$\frac{0.6018}{1} = \frac{I_y}{20}$$

and $I_y = 0.6018 \times 20 =$ ______ A *Ans.* | 12

## PROBLEMS

Use the triangle shown in Fig. 14-14 for the following problems.

1. Find $AC$ and $\angle B$ if $AB = 20$ in and $\angle A = 60°$.
2. Find $BC$ and $\angle B$ if $AB = 26$ ft and $\angle A = 40°$.
3. Find $AC$ and $\angle A$ if $BC = 75$ ft and $\angle B = 50°$.
4. Find $AC$ and $\angle B$ if $AB = 400$ W and $\angle A = 28°$.
5. Find $AB$ and $\angle B$ if $BC = 75.5\ \Omega$ and $\angle A = 30°$.
6. Find $AB$ and $\angle B$ if $AC = 30\ \Omega$ and $\angle A = 25°$.
7. Find $BC$ and $\angle A$ if $AC = 500$ ft and $\angle B = 28°$.
8. Find $AC$ and $\angle A$ if $AB = 36.8$ in and $\angle B = 35°$.
9. Find $BC$ and $\angle B$ if $AB = 475\ \Omega$ and $\angle A = 15°$.
10. Find $AC$ and $\angle A$ if $AB = 92.8$ V and $\angle B = 15°$.
11. A guy wire reaches from the top of an antenna pole to a point 12 m from the foot of the pole and makes an angle of 70° with the ground. How long is the wire? How tall is the pole?
12. Use Fig. 14-13. If the small diameter equals 3 in, the length equals 12 in, and angle $x$ equals 4°, find the large diameter.
13. Four holes are drilled evenly spaced around a 12-cm-diameter circle. Find the distance between the centers of the holes.
14. How long is each side of the largest square bar that can be made from a piece of 6-cm-diameter round stock?
15. A ladder 10 m long leans against the side of a building and makes an angle of 65° with the ground. How high up the building does it reach?
16. In an ac circuit, the current $I = 10$ A leads the voltage by 53°. Resolve the current into its in-phase current $I_x$ and its reactive current $I_y$, and calculate their values.
17. In an ac circuit, the current $I = 14$ A leads the voltage by 25°. Resolve the current into its in-phase current $I_x$ and its reactive current $I_y$, and calculate their values.
18. In an impedance triangle similar to that shown in Fig. 14-16, find the reactance $X_C$ if $R = 2000\ \Omega$ and angle $A = 20°$.
19. Find the impedance $Z$ in the triangle used for Prob. 18.
20. In Fig. 14-18, find dimensions *a, b, c, d,* and *e.*

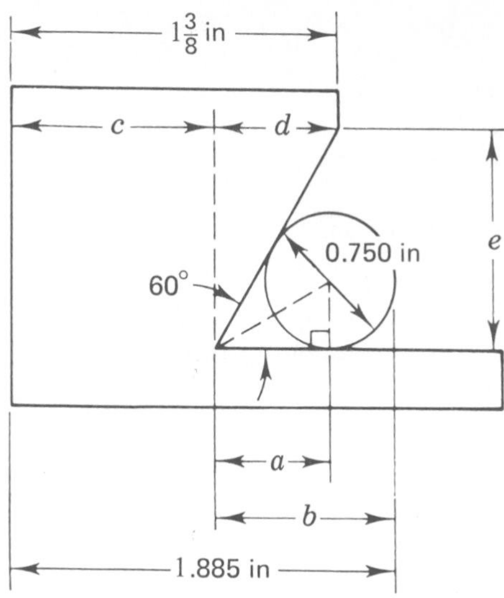


FIGURE 14-18

## JOB 14-5 REVIEW OF TRIGONOMETRY

### FORMULAS

1. Tangent of an angle $= \dfrac{?\quad\text{side}}{?\quad\text{side}}$. | opposite adjacent

$$\tan \angle = \frac{o}{a} \qquad \boxed{14\text{-}1}$$

2. Sine of an angle $= \dfrac{?\quad\text{side}}{?}$. | opposite hypotenuse

$$\sin \angle = \frac{o}{h} \qquad \boxed{14\text{-}2}$$

3. Cosine of an angle $= \dfrac{?\quad\text{side}}{?}$. | adjacent hypotenuse

$$\cos \angle = \frac{a}{h} \qquad \boxed{14\text{-}3}$$

4. The sum of the two acute angles of a right triangle equals ______. | 90°

$$\angle A + \angle B = 90^\circ \qquad \boxed{14\text{-}4}$$

### PROCEDURE FOR FINDING ANGLES IN A RIGHT TRIANGLE

1. Name the sides which have values, using the angle to be found as the reference angle.
2. Choose the trigonometric formula which uses these sides.
3. Substitute values, and divide.
4. Find the number in the table closest to this quotient. Be certain to look in the column indicated by step 2.

5. Find the angle in the angle column corresponding to this number.
6. Use formula (14-4) to find the other acute angle.

## PROCEDURE FOR FINDING SIDES IN A RIGHT TRIANGLE

1. Select the angle to be used.
2. Select the side to be found.
3. Select one other side whose value is known.
4. Name these two sides.
5. Select the correct trigonometric formula which uses these sides.
6. Substitute values using Table 14-1.
7. Solve the equation.

## PROBLEMS

Find the values of the following functions.

1. sin 36° 2. cos 78° 3. tan 69° 4. sin 52° 5. cos 25°

Find angle $A$, correct to the nearest degree.

6. $\sin A = 0.5878$ 7. $\cos A = 0.4226$ 8. $\tan A = 5.7500$
9. $\cos A = 0.9800$ 10. $\sin A = 0.7200$ 11. $\tan A = 0.3113$
12. $\tan A = 0.1340$ 13. $\cos A = 0.2868$ 14. $\sin A = 0.7240$

Use the triangle shown in Fig. 14-14 for the following problems:

15. Find $\angle A$ if $BC = 30$ and $AC = 40$.
16. Find $\angle B$ if $AC = 50$ and $AB = 100$.
17. Find $\angle B$ if $BC = 25$ and $AB = 75$.
18. Find $\angle A$ if $BC = 16$ and $AB = 65$.
19. Find $\angle B$ if $AC = 22.5$ and $BC = 14$.
20. Find $\angle A$ if $AC = 4.5$ and $AB = 20.5$
21. Find $BC$ if $AB = 2200$ W and $\angle A = 25°$.
22. Find $AC$ if $AB = 600$ W and $\angle A = 8°$.
23. Find $AB$ if $BC = 28.6$ and $\angle B = 42°$.
24. Find $AB$ if $AC = 9.3\ \Omega$ and $\angle A = 34°$.
25. Find $BC$ if $AC = 750\ \Omega$ and $\angle A = 48°$.
26. Find $AC$ if $BC = 17.6$ ft and $\angle B = 60°$.
27. Find $AC$ if $AB = 4000$ W and $\angle B = 45°$.
28. Find $BC$ if $AB = 2500$ W and $\angle A = 45°$.
29. Find $AB$ if $BC = 142$ V and $\angle A = 37°$.
30. Find $AC$ if $AB = 85.8\ \Omega$ and $\angle A = 83°$.
31. A pipe must be bent to provide a 6-ft rise in a 1-ft horizontal distance. Find the angle at each bend.
32. The foot of a ladder 30 ft long rests on the ground 12 ft from the side of a building. What angle does the ladder make with the ground?
33. In a taper similar to that shown in Fig. 14-13, the large diameter equals 0.75 in and the small diameter equals 0.47 in. Find angle $x$ if the length equals 2.8 in.

34. How long is a ladder that reaches 8 m up the side of a building if the angle between the ladder and the ground is 65°?
35. Use Fig. 14-13. If the small diameter equals 1.25 in, the length equals 2.25 in, and angle $x$ equals 6°, find the large diameter.
36. How many inches of wire are needed to wind a single-layer coil of 20 turns around a core whose cross section is a square with a diagonal equal to 1.414 in?
37. A rectangle measures 30 by 70 cm. Find the angle that the diagonal makes with the longer side.
38. Find the length of the diagonal in Prob. 37.
39. In an impedance triangle similar to that shown in Fig. 14-16, find the impedance $Z$ if $R = 150\ \Omega$ and angle $A = 50°$.
40. In an inductive ac circuit, the current $I = 15$ A leads the voltage by 30°. Resolve the current into its in-phase current $I_x$ and its reactive current $I_y$, and calculate their values.

(See Instructor's Manual for Test 14-2, Trigonometry)

# INTRODUCTION TO AC ELECTRICITY

# 15

## JOB 15-1 GRAPHS

In Fig. 15-1, the current that flows through the resistor depends on the voltage that is applied across the ends of the resistor. If the voltage changes, then the current will also change. Let us prepare a table of values of different voltages and the resulting currents using Ohm's law and show them in Table 15-1.

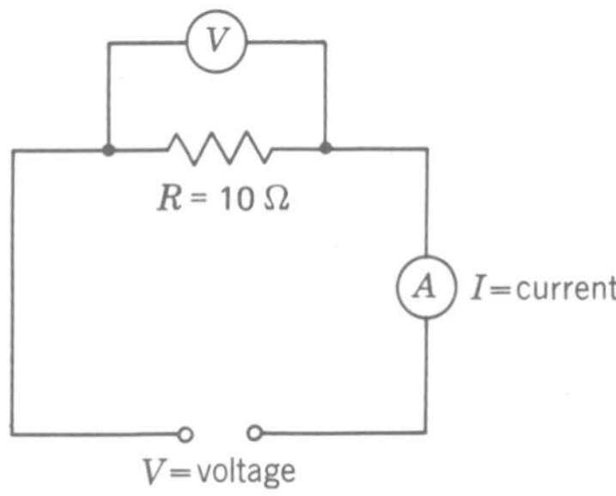


**FIGURE 15-1**
Changes in the independent variable (voltage) cause corresponding changes in the dependent variable (current).

Quantities like voltages and currents whose values may change are called *variables*. An *independent variable,* like voltage, is one whose value is changed in order to observe the effect on another *dependent variable,* like the current. Tables of information like Table 15-1 are fairly easy to

TABLE 15-1

| VOLTAGE $V$ | RESISTANCE $R$ | CURRENT $I$ |
|---|---|---|
| 0 | 10 | 0 |
| 20 | 10 | 2 |
| 40 | 10 | 4 |
| 60 | 10 | 6 |
| 80 | 10 | 8 |
| 100 | 10 | 10 |

understand and interpret. A glance at the data tells us that the current will double if the voltage is doubled. However, as the information given in a table becomes more complicated, it becomes more difficult to understand and interpret. The relationships between the two variables may be made more evident by presenting the data in the form of a picture, or *graph*.

A graph is a picture that shows the effect of one variable on another. Graphs are used throughout the electronics industry to present information in simple form, to describe the operation of circuits, and to illustrate relationships that cannot be shown by data presented in tabular form.

**The scale of a graph.** Graphs are drawn on paper ruled with uniformly spaced horizontal and vertical lines. Every fifth line or every tenth line may be heavier than the rest in order to make it easier to read the graph. Figure 15-2, which shows the graph of the data in Table 15-1, is drawn on paper ruled 10 boxes to the inch. The two heavy lines at right angles to each other are the *base lines,* or *reference lines.* The horizontal line, or *abscissa,* is generally used to describe the independent variable, while the vertical base line, or *ordinate,* is used for the dependent variable. The value of each box along each base line must be indicated on the graph. This is called the *scale* for that variable. The highest value of the variable will determine the scale to be used.

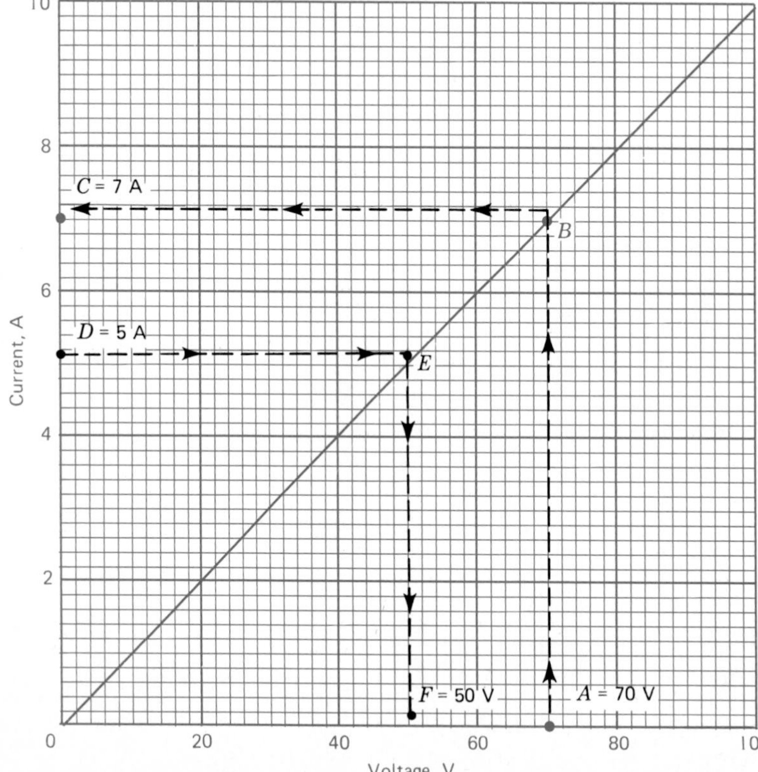


**FIGURE 15-2**
The current through a 10-Ω resistor depends on the voltage across it.

For the voltage or abscissa: If we had 100 boxes, we could make each box equal to 1 V and indicate the scale up to 100 V. However, only 50 boxes are available. Therefore, in order to show 100 V, each box must have a value of 100 ÷ 50, or 2 V for every box. Only every tenth box is numbered in order to keep the graph neat and uncluttered.

For the current or ordinate: The vertical current scale need not be the same as the horizontal scale. Actually, since the highest value of current that must be indicated is only 10 A, we would have a graph only 5 boxes high if we used the same scale. This would cramp the graph and make it difficult to read. Since 50 boxes are available, in order to show 10 A, each box must have a value of 10 ÷ 50, or 0.2 A for each box. Only every tenth box is numbered.

**Reading graphs.** As one variable changes, the corresponding value of the other variable may be read from the graph.

EXAMPLE 15-1. Using Fig. 15-2, find (*a*) the value of the current when the voltage is 70 V, and (*b*) the value of the voltage when the current is 5 A.

SOLUTION. Notice that this information is not given in the table. Without a graph, the information could be obtained only by substitution in the formula for Ohm's law. To get these data from the graph,

*a*. 1. Locate 70 V on the horizontal voltage scale at *A*.
2. Read straight up along this vertical line until the graph is reached at *B*.
3. Read horizontally to the left along this line to reach the vertical current scale at *C*.
4. Read the value of the current at point *C* as 7 A. *Ans.*

*b*. 1. Locate 5 A on the vertical scale at point *D*.
2. Read horizontally to the right until the graph is reached at point *E*.
3. Follow straight down along this vertical line to reach the horizontal voltage scale at point *F*.
4. Read the value of the voltage at point *F* as 50 V. *Ans.*

EXAMPLE 15-2 If a constant voltage is applied across the ends of different resistors, the resulting current may be calculated by Ohm's law. This information may then be shown as a graph such as Fig. 15-3, in which a constant voltage of 10 V was applied across different resistances. Using this graph, find the current when the resistance is (*a*) 5 Ω and (*b*) 7 Ω.

SOLUTION

*a*. 1. Locate 5 Ω at *A* on the horizontal resistance scale.
2. Read straight up to the graph at *B*.
3. Read horizontally to the left to reach the vertical current scale at *C*.
4. Read the value of the current at point *C* as 2 A. *Ans.*

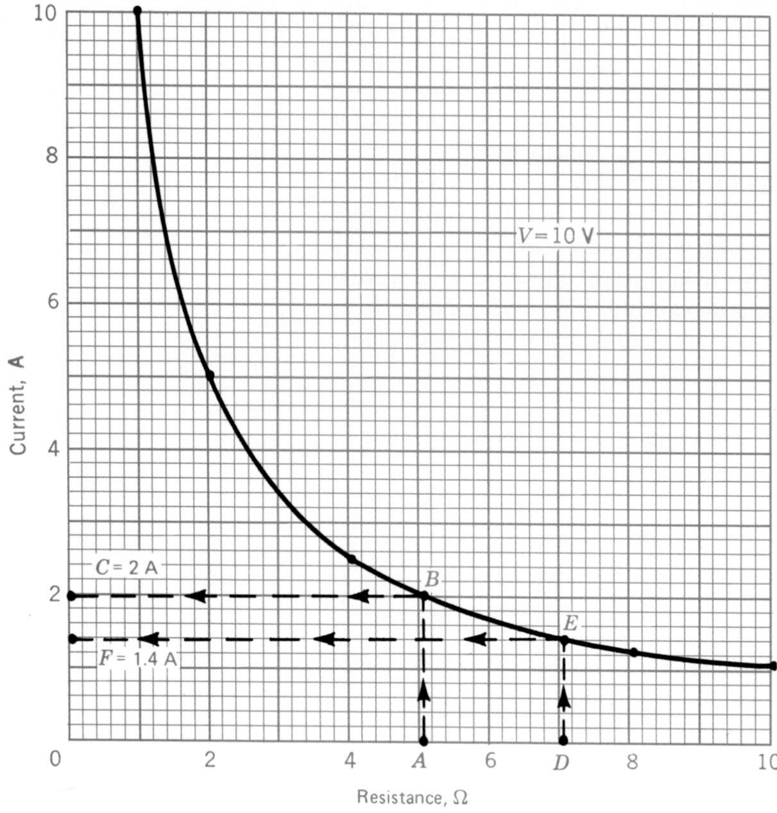


**FIGURE 15-3**
Relationship between current and resistance at a constant voltage.

*b.* 1. Locate 7 Ω at *D* on the horizontal resistance scale.
2. Read straight up to the graph at *E*.
3. Read horizontally to the left to reach the vertical current scale at *F*.
4. Read the value of the current at point *F* as slightly more than 1.4 A. *Ans.*

## PLOTTING GRAPHS

EXAMPLE 15-3. An experiment was performed relating the average forward power and the average forward current for a 1N4719 diode. The data obtained are shown in the following table. Plot the graph.

| Point number | 1 | 2 | 3 | 4 | 5 | 6 |
|---|---|---|---|---|---|---|
| Average Forward power, W | 0 | 1 | 2 | 3 | 4 | 5 |
| Average Forward current, A | 0 | 0.6 | 1.45 | 2.25 | 3.1 | 4 |

SOLUTION. The finished curve is shown in Fig. 15-4.

1. Draw two baselines at right angles to each other.

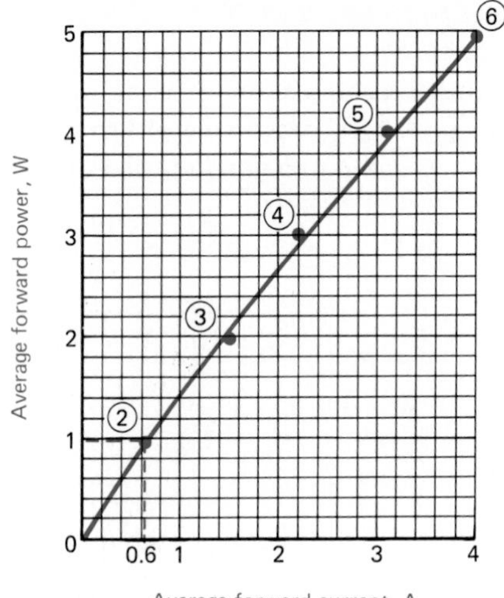


**FIGURE 15-4**
Power dissipation versus current curve for a 1N4719 diode.

2. Label the horizontal line as the average forward current in amperes and the vertical line as the average forward power in watts.

3. Select appropriate scales for each. Since the maximum current is 5 A, we can use 10 boxes to represent 1 A. This will require 5 × 10 = 50 boxes. Number every tenth box. On the power scale we can use 10 boxes to represent 1 W. This will require 40 boxes. Number every tenth box.

4. Plot the individual points. To locate point 1, put a point at the origin where the zero is located. To locate point 2, find 1 W on the vertical scale. Draw a line straight across. Now find the corresponding current (0.6 A) on the horizontal scale. Draw a line straight up. These two lines will intersect at point 1. In the same manner, plot points 3 to 6. Notice that our choice of values for the current scale makes each box represent 0.1 A, and on the vertical scale each box represents 0.1 W.

5. Draw a smooth curve through the points. If not all the points fall on this smooth curve, draw the curve so that the points are as close to the curve as possible and that there are as many points over the line as under it.

**Graphs with positive and negative values.** Many graphs have negative as well as positive values to be considered, and these values must be located on the graph. To do this, the two base lines are extended as shown in Fig. 15-5 to form the horizontal axis $XX'$ and the vertical axis $YY'$. The two axes meet at the *origin* $O$. Values along the $X$ axis measured to the *right* of $YY'$ are *positive;* values measured to the *left* of $YY'$ are *negative.* Values along the $Y$ axis measured *upward* from $XX'$ are *positive;* values *downward* from $XX'$ are *negative.* For example, in Fig. 15-5, the points are located depending on the values of $X$ and $Y$. These values are called the *coordinates* of the point. All points are located starting from the origin $O$. To locate point $A$ ($X = -2, Y = +4$), move two units to the *left* along the $X$ axis, since $X = -2$. Then move four units *up* along this line, since $Y = +4$. To locate point $B$ ($X = -5$, $Y = -3$), move five units to the *left* along the $X$ axis, since $X = -5$.

Then move three units *down* along this line, since $Y = -3$. To locate point $C$ ($X = +3$, $Y = -1$), move three units to the *right* along the $X$ axis, since $X = +3$. Then move one unit *down* along this line, since $Y = -1$. To locate point $D$ ($X = +6$, $Y = +5$), move six units to the *right* along the $X$ axis, since $X = +6$. Then move five units *up* along this line, since $Y = +5$.

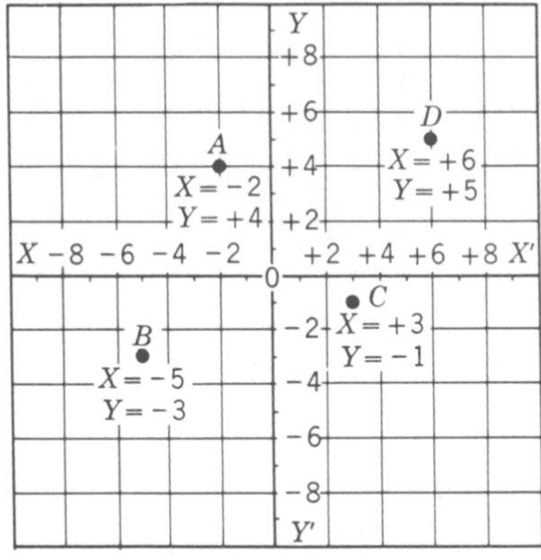


**FIGURE 15-5**
Locating points with positive and negative coordinates.

EXAMPLE 15-4. A storage battery (6.6 V) with an internal resistance of 0.75 Ω is used to supply a variable load. Plot the curve for the power delivered to the load against the load resistance, using the following data.

| Load resistance, Ω | 0.005 | 0.01 | 0.02 | 0.03 | 0.04 |
|---|---|---|---|---|---|
| Power delivered, W | 50 | 67 | 96 | 117 | 130 |
| Point number | 1 | 2 | 3 | 4 | 5 |

| Load resistance, Ω | 0.05 | 0.06 | 0.07 | 0.08 | 0.09 |
|---|---|---|---|---|---|
| Power delivered, W | 140 | 145 | 147 | 146 | 145 |
| Point number | 6 | 7 | 8 | 9 | 10 |

| Load resistance, Ω | 0.10 | 0.11 | 0.12 | 0.13 | 0.14 |
|---|---|---|---|---|---|
| Power delivered, W | 143 | 140 | 137 | 135 | 132 |
| Point number | 11 | 12 | 13 | 14 | 15 |

What conclusion can you reach about the power transferred when the load resistance equals the source resistance?

SOLUTION. The curve is shown in Fig. 15-6. Each fifth box along the load resistance or $X$ axis is worth 0.01 Ω. Each box along the watts or $Y$

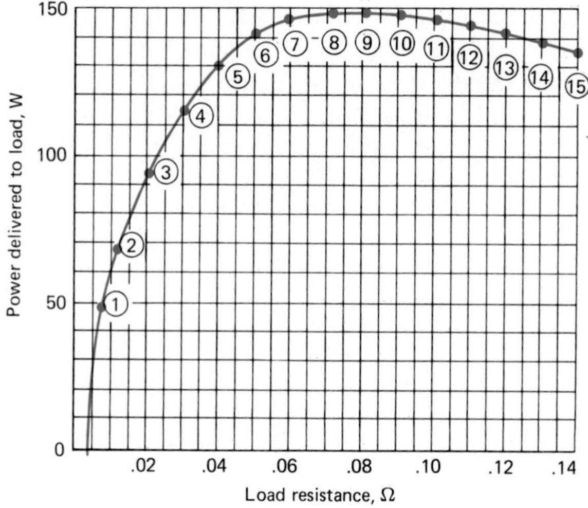


**FIGURE 15-6**
Power delivered to a load plotted against load resistance.

axis is worth 2.5 W. Notice that the power delivered is a maximum when the load resistance is equal to the source resistance.

## PROBLEMS

1. Figure 15-7 is a graph showing the current taken by an incandescent lamp at various voltages. Find the current taken at the following voltages: (*a*) 30 V, (*b*) 45 V, (*c*) 50 V, (*d*) 65 V, (*e*) 80 V, (*f*) 100 V, and (*g*) 120 V.

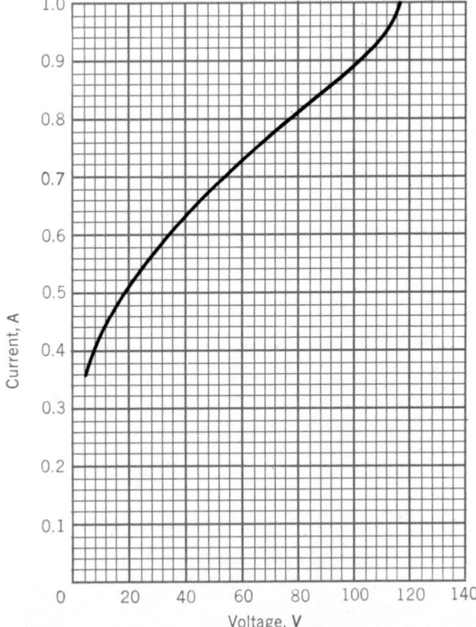


**FIGURE 15-7**
Current taken by an incandescent lamp at various voltages.

2. Figure 15-8 is a graph showing the resistance in ohms per 1000 ft of copper wires of various diameters. Referring to Table 12-1 for the diameter in mils of the gage numbers, find the resistance per 1000 ft of (*a*) No. 0, (*b*) No. 4, (*c*) No. 6, (*d*) No. 8, (*e*) No. 10, and (*f*) No. 12. Find the resistance per 1000 ft of wires whose diameters are (*g*) 150 mil, (*h*) 220 mil, (*i*) 250 mil, and (*j*) 0.3 in.

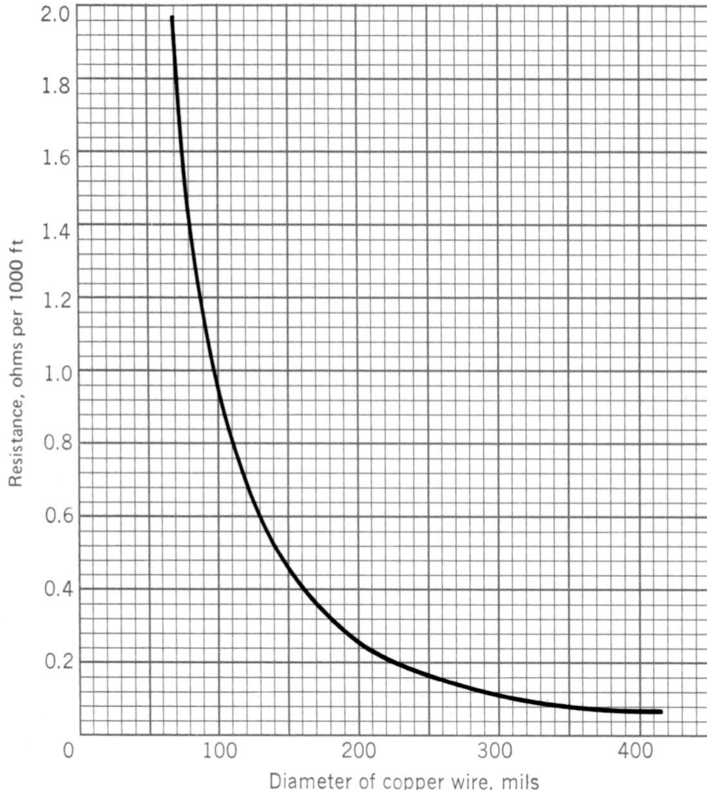


**FIGURE 15-8**
Graph of a portion of the AWG table, Table 12-1.

Plot a graph for each of the following problems:

3. The volume of a cube when the side is changed:

| Side, in | 0 | 1 | 2 | 3 | 4 | 5 |
|---|---|---|---|---|---|---|
| Volume, in$^3$ | 0 | 1 | 8 | 27 | 64 | 125 |

4. The distance covered by a falling body for increasing units of time:

| Time, s | 1 | 2 | 3 | 4 | 5 | 6 | 7 | 8 | 9 | 10 |
|---|---|---|---|---|---|---|---|---|---|---|
| Distance, ft | 16 | 64 | 144 | 256 | 400 | 576 | 784 | 1024 | 1296 | 1600 |

5. The efficiency of an engine against its horsepower output. What is

the maximum efficiency of the engine? At what horsepower output is this efficiency obtained?

| Horsepower output, hp | 0 | 10 | 20 | 30 | 40 | 50 | 60 | 70 |
|---|---|---|---|---|---|---|---|---|
| Efficiency, % | 0 | 30 | 50 | 64 | 78 | 80 | 78 | 64 |

6. The power used by a 100-Ω resistor carrying different currents:

| Current, A | 0.1 | 0.2 | 0.3 | 0.4 | 0.5 | 0.6 | 0.7 | 0.8 | 0.9 | 1.0 |
|---|---|---|---|---|---|---|---|---|---|---|
| Power, W | 1 | 4 | 9 | 16 | 25 | 36 | 49 | 64 | 81 | 100 |

7. In an inductive circuit, the current does not rise to its Ohm's-law value instantaneously. The growth of the current in a certain coil is shown by the following data. Plot the curve.

| Time, s | 0 | 0.1 | 0.2 | 0.3 | 0.4 | 0.5 | 0.6 | 0.7 | 0.8 | 0.9 | 1.0 |
|---|---|---|---|---|---|---|---|---|---|---|---|
| Current, A | 0 | 4.2 | 6.6 | 8.0 | 8.9 | 9.3 | 9.6 | 9.8 | 9.9 | 10 | 10 |

8. The time constant $RC$ of a capacitive circuit describes the time required to charge a capacitor. Plot the charge curve from the following data:

| TIME, s | $0.5RC$ | $1RC$ | $2RC$ | $3RC$ | $4RC$ | $5RC$ |
|---|---|---|---|---|---|---|
| Percentage of applied voltage | 39 | 63 | 86 | 95 | 98 | 99 |

9. The inductive reactance of a 0.01-H coil at different frequencies:

| Frequency, Hz | 60 | 100 | 250 | 500 | 1000 |
|---|---|---|---|---|---|
| Inductive reactance, Ω | 3.7 | 6.3 | 15.7 | 31.4 | 62.8 |

10. The capacitive reactance for a 4-μF capacitor at different frequencies:

| Frequency, Hz | 25 | 50 | 60 | 120 | 240 |
|---|---|---|---|---|---|
| Capacitive reactance, Ω | 1590 | 795 | 663 | 331 | 166 |

11. The collector voltage $V_C$ against the collector current $I_C$ for a 2N265 transistor:

| $V_C$, V | 5 | 10 | 15 | 20 | 25 | 30 |
|---|---|---|---|---|---|---|
| $I_C$, mA | 14.1 | 7.2 | 4.9 | 4.1 | 3.7 | 3.2 |

12. The sine of an angle against the number of degrees in the angle:

| Angle, degrees | 0 | 30 | 45 | 60 | 90 | 120 | 135 | 150 | 180 |
|---|---|---|---|---|---|---|---|---|---|
| Sine of the angle | 0 | 0.5 | 0.7 | 0.86 | 1.0 | 0.86 | 0.7 | 0.5 | 0 |

| Angle, degrees | 210 | 225 | 240 | 270 | 300 | 315 | 330 | 360 |
|---|---|---|---|---|---|---|---|---|
| Sine of the angle | −0.5 | −0.7 | −0.86 | −1.0 | −0.86 | −0.7 | −0.5 | 0 |

## JOB 15-2 THE GENERATION OF AN AC VOLTAGE

**Magnetism.** The phenomenon of magnetism was discovered about 100 B.C. when it was observed that a peculiar stone had the property of attracting bits of iron to it. This natural magnet was called a *lodestone.* The lodestone was used to create other magnets artificially by stroking pieces of iron with it.

**Magnetic poles.** Every magnet has two points opposite each other which attract pieces of iron best. These points are called the *poles* of the magnet: the north pole and the south pole. Just as similar electric charges repel each other and opposite charges attract each other, similar magnetic poles repel each other and unlike poles attract each other.

A magnet evidently attracts a bit of iron because of some force that exists around the magnet. This force is called the *magnetic field.* Although it is invisible to the naked eye, it can be shown to exist by its effect on bits of iron. Place a sheet of glass or some other nonmagnetic material over a bar magnet as shown in Fig. 15-9. Sprinkle some iron filings over the glass, and tap it gently. The filings will fall back into a definite pattern which describes the field of force around the magnet. The field evidently seems to be made up of *lines* of force which appear to leave the magnet at the north pole, travel through the air around the magnet to the south pole, and continue through the magnet to the north pole to form a *closed loop* of force. The stronger the magnet, the greater the number of lines of force and the larger the area covered by the field.

**Electromagnetism.** In 1819, Hans Christian Oersted, a Danish physicist, discovered that a field of magnetic force exists around a wire carry-

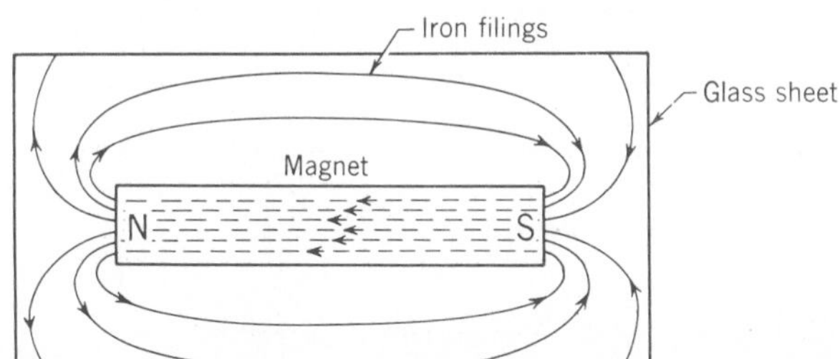


**FIGURE 15-9**
The iron filings indicate the pattern of the lines of force around a bar magnet.

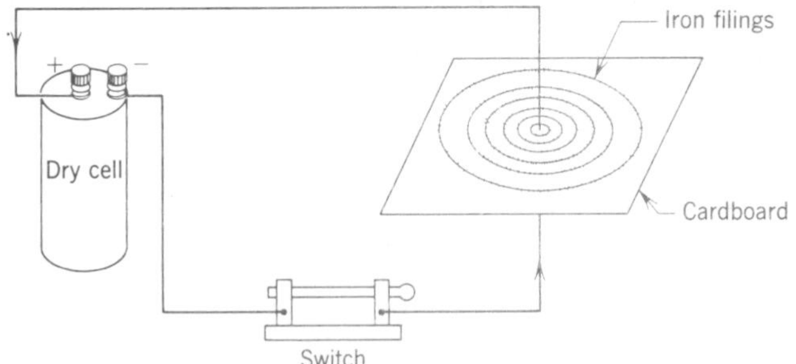

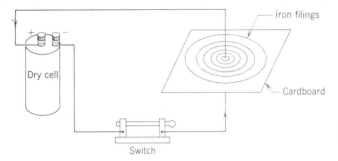

**FIGURE 15-10**
A circular pattern of magnetic force exists around a wire carrying an electric current.

ing an electric current. In Fig. 15-10, a wire is passed through a piece of cardboard and connected through a switch to a dry cell. With the switch open (no current flowing), sprinkle iron filings on the cardboard and tap it gently. The filings will fall back haphazardly. Now close the switch, which will permit a current to flow in the wire. Tap the cardboard again. This time, the magnetic effect of the current in the wire will cause the filings to fall back into a definite pattern of concentric circles with the wire as the center of the circles. Every section of the wire has this field of force around it in a plane perpendicular to the wire, as shown in Fig. 15-11.

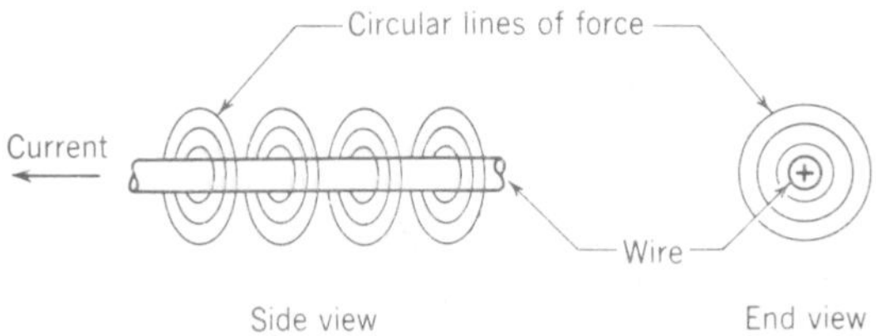


**FIGURE 15-11**
The circular fields of force around a wire carrying a current are in planes which are perpendicular to the wire.

**The strength of the magnetic field.** The ability of the magnetic field to attract bits of iron depends on the number of lines of force present. The strength of the magnetic field around a wire carrying a current depends on the current, since it is the current that produces the field. The greater the current, the greater the strength of the field. A large current will produce many lines of force extending far from the wire, while a small current will produce only a few lines close to the wire as shown in Fig. 15-12.

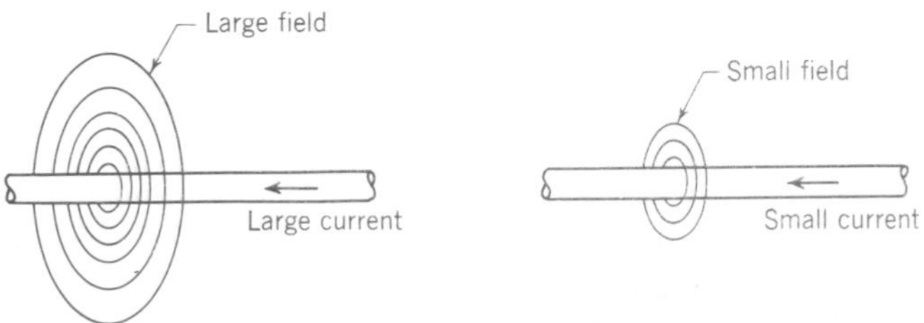


**FIGURE 15-12**
The strength of the magnetic field around a wire carrying a current depends on the amount of current.

**Electromagnetic induction.** Michael Faraday is credited with the discovery, in 1831, of the basic principle underlying operation of ac ma-

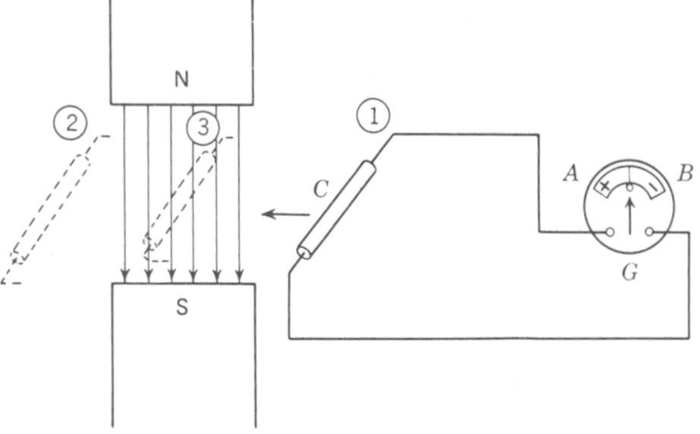

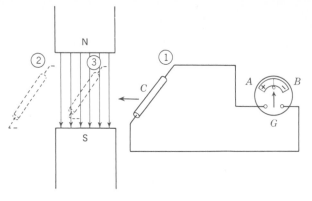

**FIGURE 15-13**
When a conductor cuts lines of force, an emf is induced in the conductor. The direction in which the conductor cuts the lines determines the direction of the induced emf.

chinery. Simply stated, he discovered that if a conductor "cut across" lines of magnetic force, or if lines of force "cut across" a conductor, an electromotive force or voltage would be induced across the ends of the conductor. Figure 15-13 represents a magnet with its lines of force streaming from the north to the south pole. A conductor *C,* which can be moved between the poles, is connected to a galvanometer, which is a very sensitive meter used to indicate the presence of an electromotive force. When the conductor is stationary, the galvanometer indicates zero emf. If the wire is *moved about outside* the magnetic field at position 1, the galvanometer will still indicate zero. However, if the conductor is moved to the *left* to position 2, so that it *cuts across the lines of magnetic force,* the galvanometer pointer will deflect to *A.* This indicates that an emf was induced in the conductor because lines of force were "cut." Upon reaching position 2, the galvanometer pointer will swing back to zero because no lines of force are being cut. Now suppose the conductor is moved to the *right* through the lines of force, back to position 1. During this *movement,* the pointer will deflect to *B,* indicating that an emf has again been induced in the wire, but in the *opposite direction.* If the wire is held *stationary* in the middle of the field of force at position 3, the galvanometer reads zero. If the conductor is moved up or down *parallel* to the lines of force so that *none are cut,* no emf will be induced. From experiments similar to these, Faraday deduced the following:

1. When lines of force are cut by a conductor or lines of force cut a conductor, an emf is induced in the conductor.
2. There must be a relative *motion* between the conductor and the lines of force in order to induce an emf.
3. Changing the direction of the cutting will change the direction of the induced emf.

**Generating an alternating emf.** Since a voltage is induced in a conductor when lines of force are cut, the amount of the induced emf depends on the number of lines cut in a unit time. In order to induce an emf of 1 V, a conductor must cut 100,000,000 lines of force per second. To obtain this great number of "cuttings," the conductor is formed into a loop and rotated on an axis at great speed as shown in Fig. 15-14. The two sides of the loop become individual conductors in series, each side of the loop cutting lines of force and inducing twice the voltage that a single conductor would induce. In commerical generators, the number of "cuttings" and the resulting emf are increased by (1) increasing the number of lines of force by using more magnets or stronger electromagnets, (2) using more conductors or loops, and (3) rotating the loops faster.

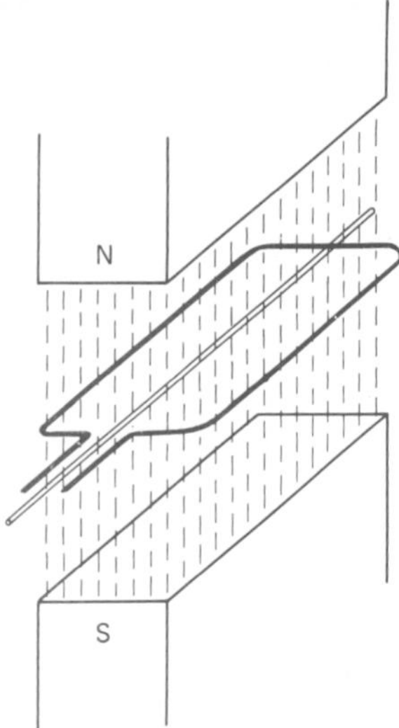


**FIGURE 15-14**
The rotating loop cuts the lines of force to produce an alternating emf.

Let us follow a single conductor as it rotates at a uniform speed through a uniformly distributed field of force. In Fig. 15-15, the lines *AD, BE,* and *CF* are all equal to *OC* and represent the direction in which the conductor is moving at points *A, B,* and *C,* respectively. At *A,* the conductor is moving parallel to the lines of force. No lines of force are cut, and zero volts are induced. At *C,* the conductor has rotated through 90° and is moving in the direction *CF.* This motion is directly across the lines of force and produces the maximum number of cuttings and the maximum voltage $V_{max}$. This maximum voltage may be represented by the distance *CF*. Since *OC* is equal to *CF*, the maximum voltage may be represented by the radius of the circle *OC.* At *B,* the conductor has rotated through some angle $\theta$ (theta) and is moving in the direction shown by *BE.* This motion from *B* to *E* may be considered to be made up of a motion from *B* to *G* and then from *G* to *E.* However, only one of these motions is useful in cutting lines of force. The motion *GE* is parallel to the lines of force and produces no cuttings. Therefore, the distance *BG* represents the total cuttings produced by the motion *BE.* *BG* therefore represents the voltage produced at the instant that the conductor is passing through point *B.* It can be shown by geometry that $BG = BM$. Thus, the vertical line drawn from the conductor at any instant perpendicular to the base represents the voltage induced in the conductor at that instant. The voltage at any instant of time evidently

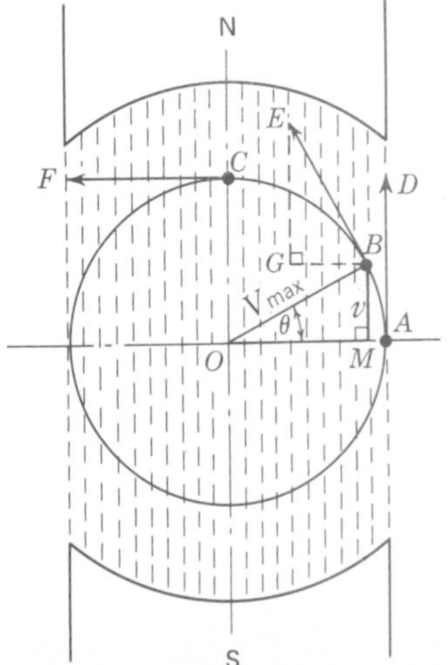


**FIGURE 15-15**
A conductor rotating at a constant speed through a uniform magnetic field.

depends on the position of the conductor at that instant and is known as the *instantaneous* voltage $v$.

1. Only that part of the motion of a conductor directly across the field cuts lines and produces voltage.
2. The number of lines cut and the voltage produced are zero at 0° and increase to a maximum at 90°.
3. The vertical distance drawn from the conductor at any point to the horizontal base line represents the voltage produced at the instant that the conductor is passing through that point.

The complete picture of the voltage produced by a rotating conductor may now be drawn. Draw a circle, and divide the circumference into 12 parts, each 30° apart, as shown in Fig. 15-16. A horizontal baseline is drawn as shown and labeled in degrees to correspond to the positions of the conductor. The vertical distances represent the voltages produced. As the conductor rotates, the vertical distance at each point is drawn at the corresponding point on the graph. Notice that the motion from 0 to 180° was to the *left*. The motion from 180 to 360° was to the *right* and produced a voltage in the *opposite* direction. This is indicated by *negative* voltages drawn *below* the base line. This graph gives a complete picture of all the changes in voltage during one complete rotation.

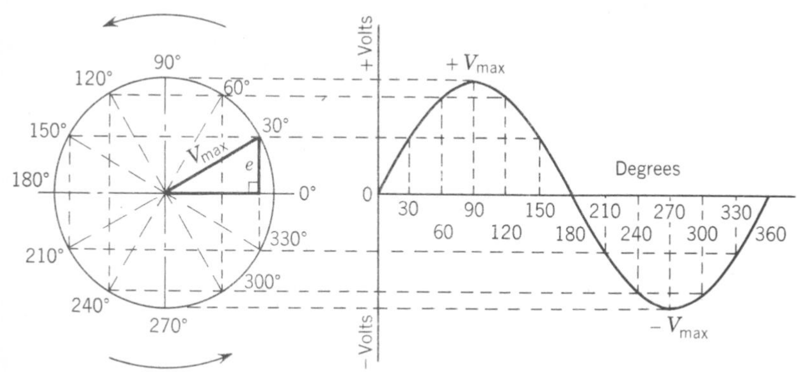


**FIGURE 15-16** The voltage wave produced by a conductor rotating at a constant speed through a uniform magnetic field.

1. At 0°, the emf is 0 V.
2. At 90°, the emf is the greatest and is known as $+V_{max}$.
3. At 180°, the emf is again 0 V. The direction of the rotation and voltage changes at this point.
4. At 270°, the emf is again a maximum, but in the *opposite* direction, and is known as $-V_{max}$.
5. At 360°, the emf is 0 V.
6. The voltage passes twice through both the zero value and the maximum value during the cycle.
7. The maximum voltages are equal in value but are of opposite sign.
8. The instantaneous voltage means the voltage at any instant during an ac cycle. The symbol is $v$.

**The sine wave.** Consider the triangle $OBM$ in Fig. 15-15. $OB$ is a radius of the circle and is equal to the maximum voltage $V_{max}$. $BM$ is the instantaneous voltage at any single $\theta$, given as $v$. By trigonometry,

$$\sin \theta = \frac{v}{V_{max}} \qquad \boxed{15\text{-}1}$$

or

$$v = V_{max} \times \sin \theta \qquad \boxed{15\text{-}2}$$

This formula says that the value of the voltage at any instant depends on the maximum voltage and the *sine of the angle at that instant.* If the maximum voltage is 1 V,

$$v = 1 \times \sin \theta$$

or

$$v = \sin \theta \qquad \boxed{15\text{-}3}$$

We can now plot a graph of the changes in the voltage as a conductor rotates in a circle through a uniform field at a constant speed. Figure 15-17 shows the graph obtained by plotting the angles against the values of the sine of the angles. Compare this curve with the curve of Fig. 15-16. They are identical. Most commercial generators are designed to produce alternating waves of this type. They are called *sine waves* because the voltage at any instant is proportional to the sine of the angle at that instant.

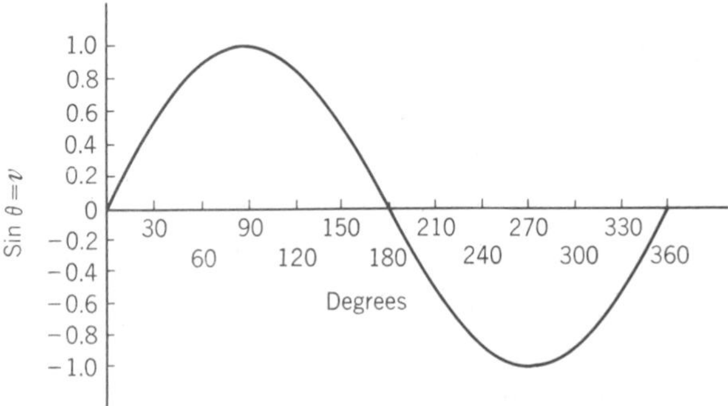


**FIGURE 15-17**
The voltage is proportional to the sine of the angle in a sine wave.

**Cycle, frequency.** An alternating wave may be an ac voltage or an alternating current. During one *cycle,* the wave will pass through a complete series of positive and negative values. In one cycle, as shown in Fig. 15-18, an ac voltage starts at 0°, rises to a positive maximum at 90°, and falls to zero at 180°; then it *reverses its polarity,* rises to a negative maximum at 270°, and falls to zero at 360°. An alternating current starts at 0°, rises to a positive maximum at 90°, and falls to zero at 180°; then it *reverses its direction,* rises to a negative maximum at 270°, and falls to zero at 360°.

Each cycle of an ac wave may be repeated over and over. The frequency $f$ of an ac wave means the number of complete cycles that occur *in one second.* In the United States the standard frequency for light and

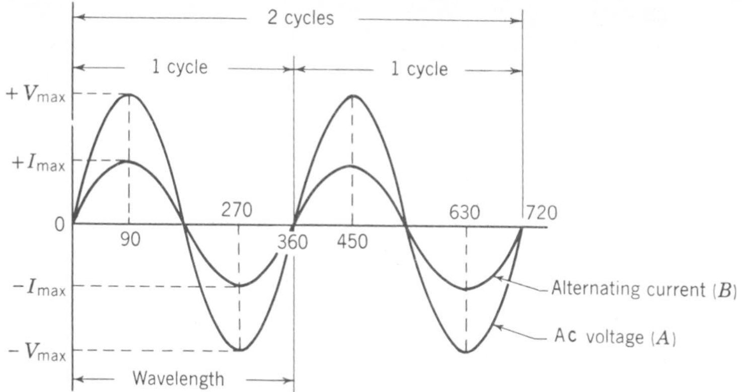


**FIGURE 15-18**
A cycle describes all the changes of voltage or current during a rotation of 360°.

power is 60 cycles/s (60 Hz). Audio frequencies range between 30 and 15,000 Hz. Radio and television frequencies range between 15,000 and 890,000,000 Hz.

**Wavelength.** The distance traveled by an ac wave during one complete cycle is called its *wavelength.* Its symbol is the Greek letter lambda (λ). This distance may be expressed in any convenient unit of length. However, since most scientific work uses the metric system, the wavelength is usually measured in terms of *meters.* One meter is equal to 39.37 in.

The speed of radio waves is the same as the speed of light, 186,000 mi/s. This speed may be expressed in the metric system in terms of meters per second. Since 1 mi = 5820 ft, and 1 m = 39.37 in,

$$\begin{aligned}
186{,}000 \text{ mi} &= 186{,}000 \times 5280 \text{ ft} \\
&= 186{,}000 \times 5280 \times 12 \text{ in} \\
&= \frac{186{,}000 \times 5280 \times 12}{39.37} \text{ m} \\
&= 300{,}000{,}000 \text{ m}
\end{aligned}$$

Therefore, the speed of a radio wave may be expressed as 300,000,000 m/s.

If we divide the speed in meters per second by the frequency in cycles per second, the units of time will cancel out as shown below.

$$\frac{\text{Meters}}{\text{Second}} \div \frac{\text{cycles}}{\text{second}}$$

Inverting and multiplying,

$$\frac{\text{Meters}}{\cancel{\text{Second}}_{1}} \times \frac{\cancel{\text{second}}^{1}}{\text{cycles}} = \frac{\text{meters}}{\text{cycle}}$$

The number of meters per cycle is the wavelength.

**FORMULA**

$$\lambda = \frac{300{,}000{,}000}{f} \qquad \boxed{15\text{-}4}$$

where $\lambda$ = wavelength, m
$f$ = frequency, Hz

If the frequency is expressed in kilohertz,

$$\lambda = \frac{300{,}000}{f} \qquad \boxed{15\text{-}5}$$

where $\lambda$ = wavelength, m
$f$ = frequency, kHz

EXAMPLE 15-5. What is the wavelength of radio station WCBS, which broadcasts at a frequency of 880 kHz?

SOLUTION. Given: Station WCBS Find: $\lambda = ?$
$f = 880$ kHz

$$\lambda = \frac{300{,}000}{f} \qquad (15\text{-}5)$$

$$\lambda = \frac{300{,}000}{880} = 340.9 \text{ m} \qquad \textit{Ans.}$$

EXAMPLE 15-6. What is the frequency of a radio wave if its wavelength is equal to 297 m? What New York City station broadcasts at this frequency?

SOLUTION. Given: $\lambda = 297$ m Find: $f = ?$
Name of New York station = ?

$$\lambda = \frac{300{,}000}{f} \qquad (15\text{-}5)$$

$$297 = \frac{300{,}000}{f}$$

$$297 \times f = 300{,}000$$

$$f = \frac{300{,}000}{297} = 1010 \text{ kHz} \qquad \textit{Ans.}$$

The station is WINS. *Ans.*

## JOB 15-3 INSTANTANEOUS VALUES, MAXIMUM VALUES, AND PHASE ANGLES OF AN AC WAVE

Since alternating voltage and current waves change in amount and direction at regular intervals, the voltage or current at any instant of time must be calculated. These values are called the *instantaneous values.* For example, in Fig. 15-15, when the conductor has rotated through some angle $\theta$ to reach point $B$, the instantaneous value is indicated by the vertical line $BM$. The radius of the circle $OB$ indicates the maximum value of the wave. By trigonometry, we obtain

**FORMULAS**

$$\sin \theta = \frac{v}{V_{max}} \qquad (15\text{-}1)$$

or

$$v = V_{max} \times \sin \theta \qquad (15\text{-}2)$$

If this changing voltage wave is impressed across a resistance, each instantaneous voltage $v$ will produce its own value of instantaneous current $i$. Thus, as shown in Fig. 15-18, the ac voltage wave $A$ produced the ac wave $B$. This current wave will have the same frequency as the voltage wave which produced it. Using this current wave, we obtain

**FORMULAS**

$$\sin \theta = \frac{i}{I_{max}} \qquad \boxed{15\text{-}6}$$

or

$$i = I_{max} \times \sin \theta \qquad \boxed{15\text{-}7}$$

EXAMPLE 15-7. An ac wave has a maximum value of 100 mA. Find the instantaneous current at 70°.

SOLUTION. Given: $I_{max} = 100$ mA Find: $i = ?$
$\theta = 70°$

$$i = I_{max} \times \sin \theta \qquad (15\text{-}7)$$

$$i = 100 \times \sin 70° = 100 \times 0.9397 = 93.97 \text{ mA} \qquad \textit{Ans.}$$

**Maximum values.** The maximum value of an ac voltage ($V_{max}$) or an alternating current ($I_{max}$) is reached twice during each cycle. The positive maximums occur at 90°, and the negative maximums occur at 270°. By solving Eqs. (15-2) and (15-7) for $V_{max}$ and $I_{max}$ we obtain the following formulas:

**FORMULAS**

$$V_{max} = \frac{v}{\sin \theta} \qquad \boxed{15\text{-}8}$$

$$I_{max} = \frac{i}{\sin \theta} \qquad \boxed{15\text{-}9}$$

EXAMPLE 15-8. An ac voltage wave has an instantaneous value of 70 V at 30°. (*a*) Find the maximum value of the wave. (*b*) Can this wave be impressed across a capacitor whose breakdown voltage is 125 V?

SOLUTION. Given: $v = 70$ V Find: $V_{max} = ?$
$\theta = 30°$

*a.* $$V_{max} = \frac{v}{\sin \theta} = \frac{70}{\sin 30°} = \frac{70}{0.5000} = 140 \text{ V} \quad \textit{Ans.}$$

*b.* No, since 140 V is larger than the breakdown voltage of 125 V.

**Phase angles.** The phase angle $\theta$ or phase of an ac cycle refers to the value of the electrical angle at any point during the cycle. These angles may be found by solving formula (15-1) or (15-6).

EXAMPLE 15-9. An ac voltage wave has a maximum value of 155.5 V. Find the phase angle at which the instantaneous voltage is 110 V.

SOLUTION. Given: $V_{max} = 155.5$ V  Find: $\theta = ?$
$v = 110$ V

$$\sin \theta = \frac{v}{V_{max}} = \frac{110}{155.5} = 0.7074 \qquad (15\text{-}1)$$

$$\theta = 45° \quad \textit{Ans.}$$

**PROBLEMS**

1. What is the wavelength of an ac wave whose frequency is 30,000 Hz?
2. Find the frequency of a carrier wave whose wavelength is 600 m.
3. Find the instantaneous voltage at 50° in a wave whose maximum value is 165 V.
4. Find the maximum value of an ac wave if the instantaneous voltage is 50 V at 35°.
5. Find the phase angle at which an instantaneous voltage of 72 V appears in a wave whose maximum value is 250 V.
6. Find the instantaneous current at 85° in a wave whose maximum value is 26 A.
7. Find the maximum value of an ac wave if the instantaneous current is 9 A at 12°.
8. Find the phase angle at which an instantaneous current of 3.5 A appears in a wave whose maximum value is 20 A.
9. The maximum current of the current wave of a transmitter is 10 A. At what instant (angle) will the instantaneous current be (*a*) 5 A, (*b*) 6 A, and (*c*) 8.66 A?
10. The maximum voltage of an ac wave is 100 V. Find the instantaneous voltage at (*a*) 0°, (*b*) 15°, (*c*) 30°, (*d*) 45°, (*e*) 60°, (*f*) 75°, and (*g*) 90°. (*h*) What is the average of these voltages?

## JOB 15-4 EFFECTIVE VALUE OF AN AC WAVE

An alternating current is a current that is continually changing in amount and direction. In Fig. 15-19, the current starts at 0 A at 0 time. After

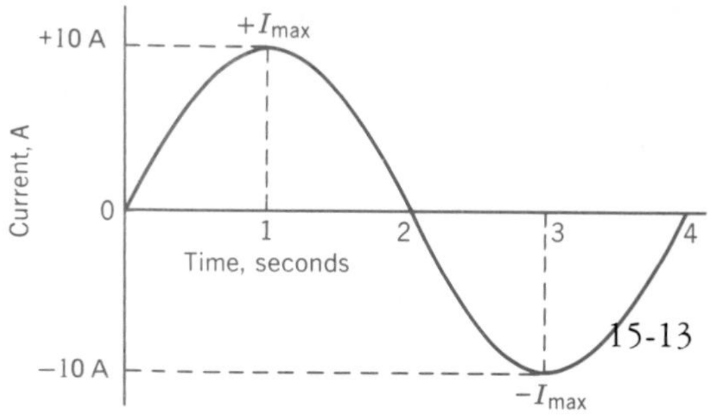


**FIGURE 15-19**
Ac wave.

1 s, it reaches 10 A but drops down to 0 after 2 s. At this point, the current reverses its direction and increases negatively to −10 A at 3 s, after which it drops down to 0 after 4 s. The *average* current during these 4 s equals 0 + 10 + 0 − 10 + 0 divided by 5, or 0 A. Obviously, even though the average of all the currents is zero, this ac wave can be effective in running a motor or lighting a lamp. But how many amperes are actually flowing? If an ac ammeter is used to measure this current, it will read 7.07 A. This meter reading is called the *effective value* of all the changes that occur during one cycle of the wave.

One of the effects of the passage of any electric current through a resistance, whether it be alternating or direct, is the production of heat. To determine the *worth,* or *effectiveness,* of an ac wave, we must compare its effect with the effect of a direct current.

**RULE**

If an ac wave produces as much heat as 1 A of direct current, we say that the ac wave is as effective as 1 A of direct current.

Thus, as in Fig. 15-20, an ac wave may start at zero and pass through many values of current up to a maximum of 10 A, continuing to fall to zero, change direction, rise to a negative maximum of −10 A, and fall to zero again. If the *effect* of all these changes is to produce only as much heat as 7.07 A of direct current would produce, then the wave is said to have an *effective* value of only 7.07 A. The effective value of any current wave may be calculated by averaging the *heating* effects of the many

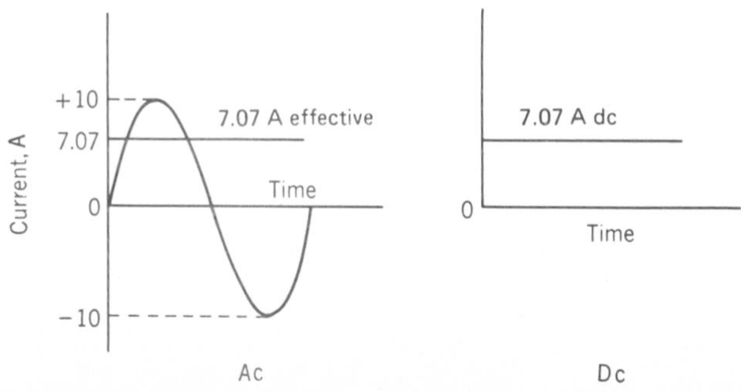


**FIGURE 15-20**
Effective values of alternating and direct currents.

individual instantaneous currents. Since $P = I^2R$, the heating effect on any resistance depends on the *square* of the current. Thus,

$$I_{dc}^2 = \text{average of } i^2$$

By taking the square root of both sides, we obtain

$$I_{dc} = \sqrt{i_{av}^2}$$

Since the effective alternating current is to have the same heating effect as the direct current,

$$I_{ac} = I_{dc} = \sqrt{i_{av}^2}$$

This equation says that the effective value of an ac wave is equal to the square root of the average (mean) of the squares of the instantaneous currents. For this reason it is sometimes called the *rms* (root mean square) value. This gives us a method for calculating the effective value of any ac wave. One-quarter of a cycle is divided into a number of equal parts, and the instantaneous current is calculated at each point. These currents are squared, and the average found. The square root of this average is equal to the effective value of the wave. The results of many calculations for waves of different maximums indicate that the effective value of a sine-wave current is always 0.707 times the maximum value of the wave.

**FORMULA** $$I = 0.707 \times I_{max} \qquad \boxed{15\text{-}10}$$

where $I$ = effective value of an ac wave
$I_{max}$ = maximum value of an ac wave

Since alternating currents are the result of ac voltages, the effective value of a voltage wave has an identical relation to the maximum voltage.

**FORMULA** $$V = 0.707 \times V_{max} \qquad \boxed{15\text{-}11}$$

where $V$ = effective value of an ac voltage wave
$V_{max}$ = maximum value of an ac voltage wave

Notice that the effective values are written as $I$ and $V$; with no subscripts. Alternating-current meters are designed and calibrated to indicate these effective values. Unless otherwise specified, a value of an alternating current or ac voltage always means the effective value.

Formulas (15-10) and (15-11) may be transformed to obtain formulas for $I_{max}$ and $V_{max}$.

$$I = 0.707 \times I_{max}$$

or

$$I_{max} = \frac{I}{0.707}$$

**FORMULA** $$I_{max} = 1.414 \times I \qquad \boxed{15\text{-}12}$$

Similarly, $$V = 0.707 \times V_{max}$$

or $$V_{max} = \frac{V}{0.707}$$

**FORMULA** $$V_{max} = 1.414 \times V \qquad \boxed{15\text{-}13}$$

Thus, an effective voltage of 120 V is the effective voltage of a wave whose maximum equals 1.414 × 120 = 170 V. This means that if an electric heater were used on a 120-V ac line, it would produce only as much heat as would be produced on a 120-V dc line, even though the ac line reaches the maximum of 170 V twice in each cycle.

EXAMPLE 15-10. An alternating current has a maximum value of 50 A. Find (*a*) the effective current and (*b*) the instantaneous current at 10°.

SOLUTION. Given: $I_{max} = 50$ A, $\theta = 10°$ Find: $I = ?$, $i = ?$

*a.*
$$I = 0.707 \times I_{max} \qquad (15\text{-}10)$$
$$I = 0.707 \times 50 = 35.35 \text{ A} \qquad \textit{Ans.}$$

*b.*
$$i = I_{max} \times \sin\theta \qquad (15\text{-}7)$$
$$i = 50 \times \sin 10°$$
$$i = 50 \times 0.1736 = 8.68 \text{ A} \qquad \textit{Ans.}$$

EXAMPLE 15-11. An ac voltage wave has an effective value of 110 V. Find (*a*) the maximum value and (*b*) the instantaneous value at 40°.

SOLUTION. Given: $V = 110$ V, $\theta = 40°$ Find: $V_{max} = ?$, $v = ?$

*a.* $$V_{max} = 1.414 \times V = 1.414 \times 110 = 155 \text{ V} \qquad \textit{Ans.}$$

*b.*
$$v = V_{max} \times \sin\theta \qquad (15\text{-}2)$$
$$v = 155 \times \sin 40°$$
$$v = 155 \times 0.6428 = 99.6 \text{ V} \qquad \textit{Ans.}$$

SELF-TEST 15-12. An ac wave has an instantaneous value of 12.95 A at 15°. Find (*a*) the maximum value and (*b*) the effective value.

SOLUTION. Given: $i = 12.95$ A, $\theta =$ ______ Find: $I_{max} = ?$, $I = ?$

15°

*a.*
$$I_{max} = \frac{i}{?} \qquad (15\text{-}9)$$
$$= \frac{12.95}{\sin 15°} = \frac{12.95}{?}$$
$$I_{max} = \_\_\_\_\_\_ \text{ A} \qquad Ans.$$

*b.*
$$I = 0.707 \times \_\_\_\_\_\_ \qquad (15\text{-}10)$$
$$I = 0.707 \times \_\_\_\_\_\_ = 35.35 \text{ A} \qquad Ans.$$

sin $\theta$

0.2588

50

$I_{max}$

50

**PROBLEMS**

Find the values indicated in each problem.

| PROBLEM | MAXIMUM VALUE | EFFECTIVE VALUE | PHASE ANGLE | INSTANTANEOUS VALUE |
|---|---|---|---|---|
| 1 | 35 A | ? | 30° | ? |
| 2 | 456 V | ? | 40° | ? |
| 3 | ? | 440 V | 50° | ? |
| 4 | ? | 25 A | 60° | ? |
| 5 | 155 V | ? | 30° | ? |
| 6 | 10 A | ? | 45° | ? |
| 7 | ? | 110 V | 65° | ? |
| 8 | ? | 20 A | 75° | ? |
| 9 | ? | ? | 50° | 26.81 A |
| 10 | ? | ? | 15° | 120.3 V |
| 11 | 100 V | ? | ? | 43.84 V |
| 12 | 20 A | ? | ? | 5.5 A |

13. A sine wave has an instantaneous value of 10 V at 30°. What is its value at 60°?
14. What must be the minimum breakdown-voltage rating of a capacitor if it is to be used on a 110-V ac line?
15. An electric stove draws 7.5 A from a 120-V dc source. (*a*) What is the maximum value of an alternating current which will produce heat at the same rate? (*b*) Find the power drawn from the ac line.
16. An underground cable is designed to operate safely at an effective voltage of 2200 V. What dc voltage will the line carry safely?
17. A 50-Ω resistor in an ac circuit is subjected to a peak voltage of 169 V. Find the effective power consumed.

(See Instructor's Manual for Test 15-1, AC Waves)

## JOB 15-5. MEASURING AC WITH AN OSCILLOSCOPE

One of the most widely used electronic instruments for the measurement of ac signals is the oscilloscope. The oscilloscope makes graphs plotting voltage against time. A timing system inside the instrument forms the horizontal time axis. The signal voltage applied to the oscilloscope forms the vertical axis of the graph. The picture on page 452 is similar to that in Fig. 15-19, which is repeated for your convenience.

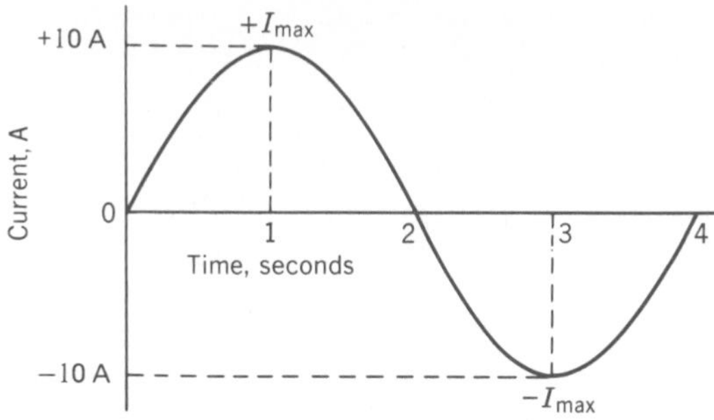


The picture makes it possible to measure the amplitude and the frequency of the applied signal. The amplitude can be measured directly from the graph. The frequency is found by measuring the period of the ac signal. The period is the time it takes for the ac signal to repeat itself. In other words, the period is the time that it takes for the signal to make one complete cycle.

In Fig. 15-19, the period is 4 s. After 4 s the ac signal looks exactly like the part that is shown. If the frequency were twice as high as that in Fig. 15-19, the signal would repeat itself in half the time. This would be represented on the graph by dividing all of the times by 2. The signal would be zero at 0, 1, and 2 s. The maximum positive value would be at $^1/_2$ s and the maximum negative value at $1^1/_2$ s. Evidently, as the frequency increases the period decreases. This leads to the following rule:

**RULE** The frequency equals 1 divided by the period

**FORMULA**

$$f = \frac{1}{\text{period}} \qquad 15\text{-}14.$$

EXAMPLE 15-13. Find the frequency of the signal shown in Fig. 15-19.

SOLUTION

1. Measure the period of the signal from the graph. It measures 4 s.
2. Calculate the frequency.

$$f = \frac{1}{\text{period}} = \tfrac{1}{4} \text{ cycles per second or Hertz} \qquad \textit{Ans.}$$

In power applications, frequency is usually referred to as cycles per second or just cycles. We usually say that the house is 60-cycle power. In electronic applications, the term *Hertz* is used. The two terms are identical in meaning.

When using an oscilloscope, the picture should always be made as large as possible in order to increase the accuracy of the readings. This is accomplished by adjusting the scales on the graph electronically. The horizontal time base is calibrated in time per division. This is adjusted until one cycle takes as much of the screen as possible. The vertical scale is adjusted with the gain control. This is calibrated in volts per division. It should be adjusted until the picture is as large as possible. The adjusted picture should always appear similar to that in Fig. 15-21. The scales are not plotted on the graph as in Fig. 15-19. Instead, the scales are determined from the knob settings of the oscilloscope.

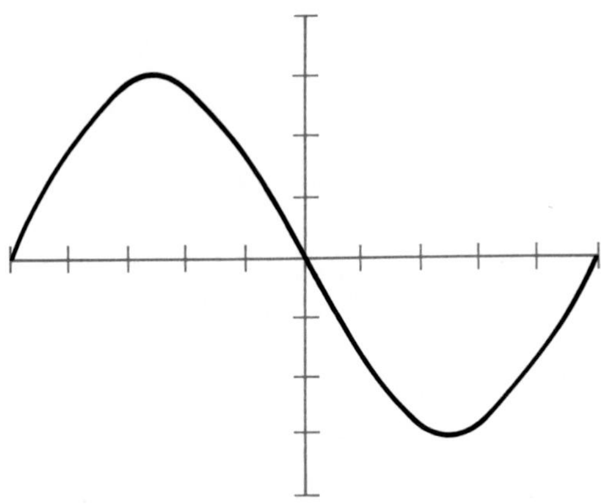

**FIGURE 15-21**
Sine wave displayed on an oscilloscope face.

EXAMPLE 15-14. The knobs of an oscilloscope are adjusted to obtain the picture shown in Fig. 15-21. If the gain is set at 5 V per division, what is the peak voltage?

SOLUTION

1. In order to measure the peak voltage, draw a horizontal line from the peak of the picture to the scale that is in the center of the graph. Count the number of divisions to the point where the line hits the scale. In this example there are three divisions.
2. The knob setting tells us that the gain is 5 V per division. Multiply to obtain the voltage:

$$3 \text{ divisions} \times 5 \text{ V per division} = 15 \text{ V peak} \qquad \textit{Ans.}$$

EXAMPLE 15-15. The time base is set to 1 ms per division. What is the frequency?

SOLUTION

1. Count the number of divisions for one cycle. In this example there are 10 divisions.
2. Get the time scale from the knob setting. It is 1 ms per division.
3. Find the period by multiplying the number of divisions per cycle by the horizontal scale:

$$10 \text{ divisions} \times 1 \text{ ms per division} = 10 \text{ ms} \qquad \textit{Ans.}$$

4. Find the frequency.

$$f = \frac{1}{\text{period}} \qquad (15\text{-}14)$$

$$f = \frac{1}{10\ \text{ms}} = \frac{1}{0.01\ \text{s}}$$

$f = 100$ Hz or cycles per second *Ans.*

SELF-TEST 15-16

1. An ______ draws the graph of an ac signal electronically. — oscilloscope
2. The horizontal axis is the ______ axis of the graph. — time
3. The scale of the horizontal axis is determined by the ______ ______ generator that is inside the oscilloscope. — time-base
4. The vertical axis shows the ______ of the signal as a function of time. — amplitude
5. The scale for the vertical axis is determined by the ______ setting of the oscilloscope vertical amplifier. — gain
6. The frequency of an ac signal is found on an oscilloscope by measuring the ______ of the waveform. — period
7. The frequency is calculated by taking the reciprocal of ______. — period

## PROBLEMS

For the following problems use the graph in Fig. 15-21 to calculate your answers.

1. If the gain setting is 30 V per division, what is the peak voltage of the signal?
2. The time base is set at 200 $\mu$s per division. What is the frequency of the signal?
3. The gain is 50 mV per division and the time base is 500 ms per division. Find the amplitude and the frequency of the signal.
4. If the gain is 30 V per division and the time base is 1 ms per division, find the amplitude and the frequency of the signal.

(See Instructor's Manual for Test 15-2, Oscilloscope Measurements)

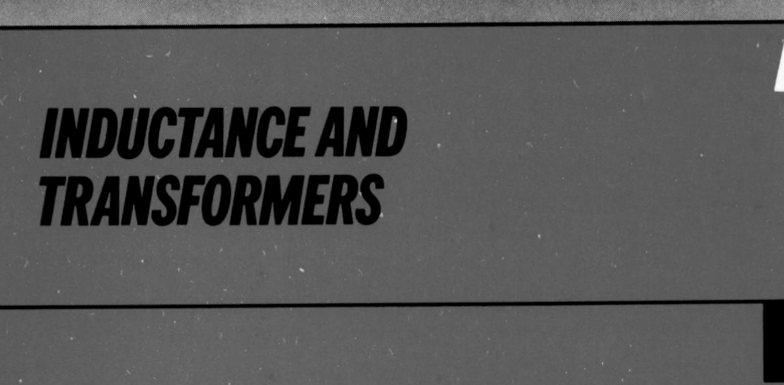

# INDUCTANCE AND TRANSFORMERS

# 16

## JOB 16-1 INDUCTANCE OF A COIL

**Basic ideas.** In Chap. 15 we learned the following:

1. A field of force exists around a wire carrying a current.
2. This field has the form of concentric circles around the wire, in planes perpendicular to the wire, and with the wire at the center of the circles.
3. The strength of the field depends on the current. Large currents produce large fields; small currents produce small fields.
4. When lines of force cut across a conductor, a voltage is induced in the conductor.

### EFFECT OF INDUCTANCE

*Experiment 1.* Connect a 20-V source of direct current across a *straight* wire whose resistance is 10 Ω as shown in Fig. 16-1*a*. When the switch is closed, the direct current

$$I = \frac{V}{R} = \frac{20}{10} = 2 \text{ A}$$

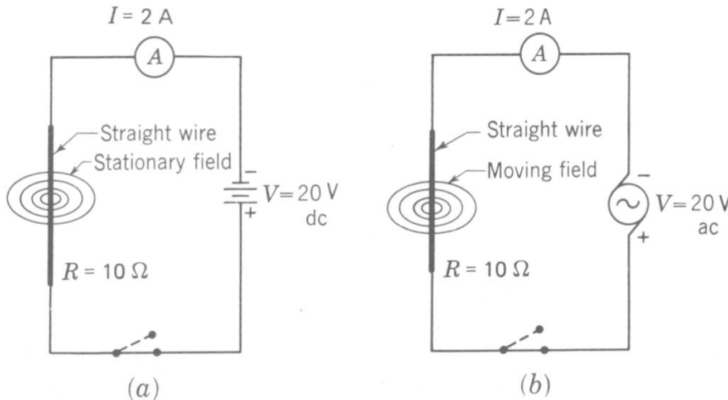


**FIGURE 16-1**
(*a*) Direct current through a straight wire produces 2 A. (*b*) Alternating current through a straight wire produces the same 2 A.

will flow. This current will produce *stationary* lines of force perpendicular to the wire. No cuttings can occur, and therefore no voltage is induced in the wire.

*Experiment 2.* Remove the 20-V dc source from the circuit of Fig. 16-1*a*, and replace it with a 20-V ac source as in Fig. 16-1*b*. When the switch is closed, the alternating current

$$I = \frac{V}{R} = \frac{20}{10} = 2 \text{ A}$$

will flow. This current will produce lines of force that move *out* from the wire as the alternating current increases and collapse *back* toward the wire as the current decreases. The field expands and contracts with the variations in the current. However, since the lines of force are perpendicular to the wire, no lines of force can cut the wire and no voltage is induced in the wire.

*Experiment 3.* Remove the 10-Ω wire, and *twist it into a coil.* The resistance of the wire will still be 10 Ω. Replace the 20-V ac source with a 20-V dc source as shown in Fig. 16-2*a*. When the switch is closed, the direct current

$$I = \frac{V}{R} = \frac{20}{10} = 2 \text{ A}$$

will flow. This current will produce *stationary* lines of force perpendicular to the wire. Since the field is stationary, there can be no *relative motion* between the lines and the wire; therefore no cuttings can occur and no voltage is induced in the wire.

*Experiment 4.* Remove the 20-V dc source from the circuit of Fig. 16-2*a*, and replace it with the 20-V ac source as shown in Fig. 16-2*b*. When the switch is closed, the alternating current

$$I = \frac{V}{R} = \frac{20}{10} = 2 \text{ A}$$

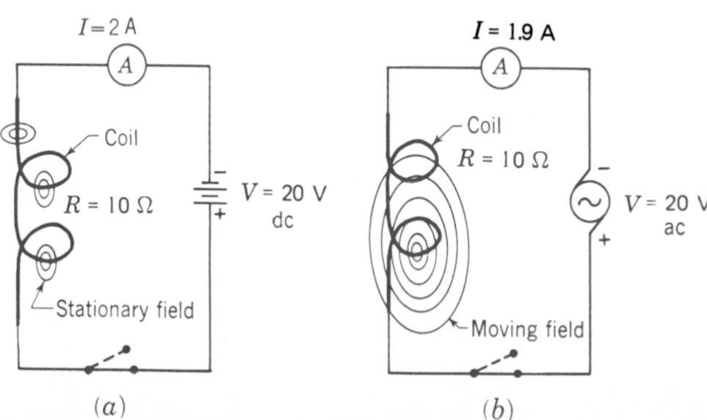


**FIGURE 16-2**
(*a*) Direct current through a coil produces 2 A. (*b*) Alternating current through a coil *reduces* the current to 1.9 A.

*will not flow!* Actually it might be only 1.9 A. This is what happened. The alternating current produced an expanding and collapsing field of force as in Experiment 2, but *because the wire was twisted into a coil, the moving lines of force could cut the turns of the coil.* These cuttings induced a voltage in the coil *and in a direction so as to oppose the original voltage.* For this reason it is called a *back emf.* This back emf is often described as a *counter* emf, and is abbreviated as cemf. If the back emf is 1 V, then the voltage available to force the current through the 10-Ω resistance is only 20 − 1 = 19 V. The current will therefore be

$$I = \frac{V}{R} = \frac{19}{10} = 1.9 \text{ A}$$

Notice that this back emf can occur *only* when a *coil* is carrying a *changing* current.

**Inductance.** The inductance of a coil is a measure of its ability to produce a back emf when an ac voltage is impressed across it. If the voltage and resulting current are *increasing,* the induced emf will be in a direction so as to *reduce* the increase. If the voltage and resulting current are *decreasing,* the induced emf will be in a direction so as to *increase* the current. The net effect of inductance is to slow down the speed at which any change occurs. The symbol for inductance is $L$.

**The unit of inductance.** The back emf induced in a coil depends on the number of "cuttings" of lines of force by conductors. A rapidly changing current will induce a larger back emf than a slowly changing current because the former will produce more cuttings in the same time. In order to compare two coils, we must pass the same kind of current through both and observe the back emf generated in each. If this same current induces 1 V of back emf in coil $A$ and 2 V of back emf in coil $B$, then coil $B$ would have twice the inductance of coil $A$. This enables us to define the unit of inductance.

***Definition.*** If a current changing at the uniform rate of 1 A/s induces a back emf of 1 V, then the inductance is 1 H.

## JOB 16-2 REACTANCE OF A COIL

When an ac voltage is impressed across a coil,

1. The ac voltage will produce an alternating current.
2. When a current flows in a wire, lines of force are produced around the wire.
3. Large currents produce many lines of force; small currents produce only a few lines of force.
4. As the current changes, the number of lines of force will change. The field of force will seem to expand and contract as the current increases and decreases as shown in Fig. 16-3.

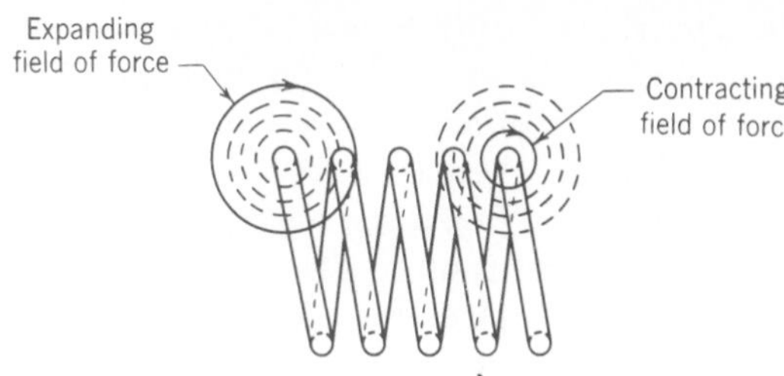


**FIGURE 16-3**
An alternating current produces a *moving* field of force which cuts the wires forming the turns of the coil.

5. As the field expands and contracts, the lines of force must cut across the wires which form the turns of the coil.
6. These "cuttings" induce an emf in the coil.
7. This emf acts in a direction so as to oppose the original voltage and is called a *back emf.*
8. The effect of this back emf is to reduce the original voltage impressed on the coil. The net effect will be to reduce the current below that which would flow if there were no cuttings or back emf.
9. In this sense, the back emf is acting as a resistance in reducing the current.
10. Actually, it is extremely convenient to consider the current-reducing effect of a back emf as a number of ohms of effective resistance. However, since a back emf is not actually a resistance but merely *acts* as a resistance, we use the term *reactance* to describe this effect. The *reactance* of a coil is the number of ohms of resistance which the coil *seems* to offer as a result of a back emf induced in it. Its symbol is $X$ to differentiate it from the dc resistance $R$.

**What factors affect the reactance?** (1) If the frequency of the ac voltage changes, then the number of cuttings and the resulting reactance will change. (2) If the natural ability of the coil to produce a back emf—its inductance—changes, then the reactance will also change. The value of the reactance of a coil is therefore proportional to its inductance and the frequency of the ac circuit in which it is used. Its actual value may be found by the formula $X_L = 2\pi fL$. Since $2\pi = 2 \times 3.14 = 6.28$, the formula for the reactance of a coil becomes

**FORMULA** $$X_L = 6.28fL \qquad \boxed{16\text{-}1}$$

where $X_L$ = inductive reactance, Ω
$f$ = frequency, Hz
$L$ = inductance, H
6.28 = constant of proportionality

EXAMPLE 16-1. A 20-mH coil is in a tank circuit operating at a frequency of 1500 kHz. Find its inductive reactance.

SOLUTION. Given: $L = 20$ mH Find: $X_L = ?$
$f = 1500$ kHz

1. Change units of measurement.

$$20\text{ mH} = 20 \times 10^{-3}\text{ H}$$
$$1500\text{ kHz} = 1500 \times 10^{3}\text{ Hz}$$

2. Find the inductive reactance.

$$X_L = 6.28fL \qquad (16\text{-}1)$$
$$X_L = 6.28 \times 1500 \times 10^{3} \times 20 \times 10^{-3}$$
$$X_L = 6.28 \times 1500 \times 20$$
$$X_L = 188{,}400\ \Omega \qquad \textit{Ans.}$$

EXAMPLE 16-2. The primary of a power transformer has an inductance of 150 mH. (*a*) Find its inductive reactance at a frequency of 60 Hz. (*b*) What current will it draw from a 117-V line?

SOLUTION. Given: $L = 150$ mH Find: $X_L = ?$
$f = 60$ Hz $I_L = ?$
$V = 117$ volts

*a.*

$$X_L = 6.28fL \qquad (16\text{-}1)$$
$$X_L = 6.28 \times 60 \times 150 \times 10^{-3}$$
$$X_L = 56.5\ \Omega \qquad \textit{Ans.}$$

*b.* Since the only resistance in the circuit is the inductive reactance, the formula for Ohm's law, $V = IR$, may be rewritten as the

**FORMULA** $$V_L = I_L \times X_L \qquad \boxed{16\text{-}2}$$

$$117 = I_L \times 56.5$$
$$I_L = \frac{117}{56.5} = 2.07\text{ A} \qquad \textit{Ans.}$$

EXAMPLE 16-3. What must be the inductance of a coil in order that it have a reactance of 942 Ω at a frequency of 60 kHz?

SOLUTION. Given: $X_L = 942\ \Omega$ Find: $L = ?$
$f = 60\text{ kHz} = 60 \times 10^{3}$ Hz

$$X_L = 6.28fL \qquad (16\text{-}1)$$
$$942 = 6.28 \times 60 \times 10^{3} \times L$$
$$942 = 376.8 \times 10^{3} \times L$$
$$L = \frac{942}{376.8 \times 10^{3}}$$
$$L = 2.5 \times 10^{-3}\text{ H}$$
$$L = 2.5\text{ mH} \qquad \textit{Ans.}$$

SELF-TEST 16-4. A tuning coil in a radio transmitter has an inductance of 300 μH. At what frequency will it offer a reactance of 3768 Ω?

SOLUTION. Given: $L = 300\ \mu H$ Find: $f = ?$

$X_L =$ ______ Ω | 3768

$$X_L = 6.28fL \qquad (16\text{-}1)$$

$3768 = 6.28 \times f \times$ ______ | $300 \times 10^{-6}$

$3768 =$ ______ $\times 10^{-6} \times f$ | 1884

$$f = \frac{3768}{1884 \times 10^{-6}}$$

$f = 2 \times$ ______ Hz | $10^6$

$f =$ ______ MHz *Ans.* | 2

**PROBLEMS**

1. A 0.7-H coil is in series with a 5-kΩ resistor in a transistorized bass-boost circuit. Find the inductive reactance of the coil at (*a*) 100 Hz and (*b*) 2 kHz.
2. A loudspeaker coil of 2 H inductance is operating at a frequency of 1 kHz. Find (*a*) its inductive reactance and (*b*) the current flowing if the voltage across the coil is 40 V.
3. A 20-H Stancor C1515 choke in the filter circuit of a power supply operates at a frequency of 60 Hz. Find (*a*) its inductive reactance and (*b*) the current flowing if the voltage across the coil is 150 V.
4. An RF choke coil whose inductance is 5.5 mH operates at a frequency of 1200 kHz. Find (*a*) its inductive reactance and (*b*) the current flowing if the voltage across the coil is 41.5 V.
5. A Miller 7825 line filter choke used in a noise-control circuit of a flasher sign has an inductance of 0.6 mH. Find its inductive reactance at a frequency of 10 kHz.
6. The primary coil of an antenna transformer has an inductance of 50 μH. At what frequency will the reactance equal 314 Ω?
7. What must be the inductance of a coil in order that it have a reactance of 1884 Ω at 60 Hz?
8. An antenna circuit has an inductance of 150 μH. What is its reactance to a 1000-kHz signal?
9. An RF coil in an FM receiver has an inductance of 100 μH. What is its reactance at 100 MHz?
10. What is the reactance of the 10-mH coil in the high-pass filter shown in Fig. 16-4 to (*a*) a 3000-Hz AF current and (*b*) a 600-kHz RF current?

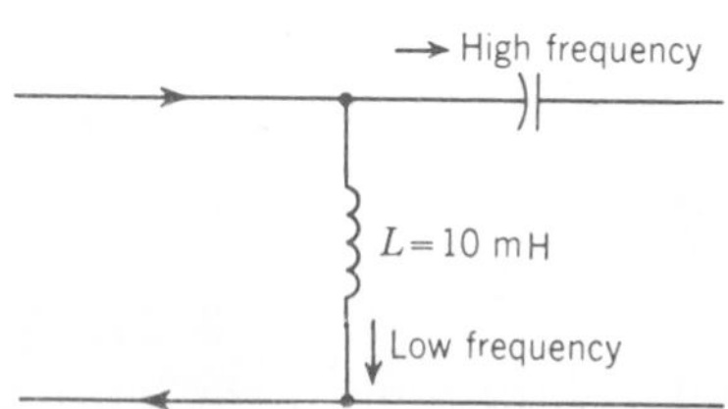


**FIGURE 16-4**
Simple high-pass filter.

11. A 1-mH coil in the primary of an IF transformer is resonant to 456 kHz. Find its inductive reactance at this frequency.
12. A 0.5-mH coil in the oscillator circuit of a continuous-wave transmitter operates at 30 MHz. Find its reactance at this frequency.
13. What must be the inductance of a coil in order that it have a reactance of 10,000 Ω at 600 kHz?
14. A transmitter tuning coil must have a reactance of 95.6 Ω at 3.8 MHz. What must be the inductance of the coil?
15. A choke coil of negligible resistance is to limit the current through it to 25 mA when 40 V are impressed across it at 1000 kHz. Find its inductance.

(See Instructor's Manual for Test 16-1)

## JOB 16-3 IMPEDANCE OF A COIL

In a "pure" coil, the opposition to the flow of an alternating current is its reactance. Actually, of course, every coil is made of wire which has some resistance. If this resistance is small in comparison with the reactance, it may be neglected, and the total opposition to the flow of current through the coil is equal to its inductive reactance. If the ohmic resistance of the coil is large, ir must be added on to the reactance of the coil to obtain the total opposing effect. This total opposition is called the *impedance* $Z$ of the coil. The addition of the ohms of resistance and the ohms of inductive reactance is *not* accomplished by simple addition but by *phasor (vector) addition,* for reasons which will be explained in Job 18-3.

**RULE** The impedance of a coil is the phasor sum of the resistance and the reactance.

**FORMULA**

$$Z^2 = R^2 + X_L^2 \qquad \text{16-3}$$

or

$$Z = \sqrt{R^2 + X_L^2} \qquad \text{16-4}$$

where $Z$ = impedance, Ω
$R$ = dc resistance, Ω
$X_L$ = inductive reactance, Ω

The comparative values of the inductive reactance and the resistance are described by a value called the $Q$, or *quality,* of the coil. This figure of merit expresses the ability of the coil to act as a reactor.

**RULE** The $Q$ of a coil is the ratio of its inductive reactance to its effective resistance.

**FORMULA**

$$Q = \frac{X_L}{R} \qquad \text{16-5}$$

If the $Q$ of a coil is greater than 5, then the resistance may be neglected in the calculation of the impedance. If the $Q$ is smaller than 5, then the resistance must be added to the reactance by formula (16-4) to obtain the impedance. Also, if the ratio $R/X_L$ is larger than 5, the reactance may be neglected and the impedance is equal to the resistance.

EXAMPLE 16-5. A coil has a resistance of 5 Ω and an inductive reactance of 12 Ω at a certain frequency. Find (*a*) the $Q$ of the coil and (*b*) the impedance of the coil.

SOLUTION. Given: $R = 5\ \Omega$, $X_L = 12\ \Omega$ Find: $Q = ?$, $Z = ?$

*a.*
$$Q = \frac{X_L}{R} = \frac{12}{5} = 2.4 \quad \textit{Ans.} \qquad (16\text{-}5)$$

*b.* Since $Q$ is less than 5, the resistance *must* be included in the calculation of the impedance.

$$Z = \sqrt{R^2 + X_L^2} \qquad (16\text{-}4)$$
$$= \sqrt{5^2 + 12^2}$$
$$Z = \sqrt{25 + 144} = \sqrt{169} = 13\ \Omega \quad \textit{Ans.}$$

EXAMPLE 16-6. A coil has a resistance of 10 Ω and an inductive reactance of 70 Ω at a certain frequency. Find (*a*) the $Q$ of the coil and (*b*) the impedance of the coil. (*c*) If the resistance is neglected, what is the percent of error?

SOLUTION. Given: $R = 10\ \Omega$, $X_L = 70\ \Omega$ Find: $Q = ?$, $Z = ?$, Percent of error $= ?$

*a.*
$$Q = \frac{X_L}{R} = \frac{70}{10} = 7 \quad \textit{Ans.} \qquad (16\text{-}5)$$

*b.* Since $Q$ is larger than 5, the resistance may be neglected and the impedance is equal to the inductive reactance.

$$Z = 70\ \Omega \quad \textit{Ans.}$$

*c.* If we include the resistance in the calculation of the impedance,

$$Z = \sqrt{R^2 + X_L^2} \qquad (16\text{-}4)$$
$$Z = \sqrt{10^2 + 70^2} = \sqrt{100 + 4900} = \sqrt{5000}$$
$$Z = 70.7\ \Omega$$

The error introduced by *not* including $R$ is equal to $70.7 - 70 = 0.7\ \Omega$. The percent of error is

$$\frac{0.7}{70.7} \times 100 \text{ equals } 0.99 \text{ or } 1 \text{ percent}$$

This error is well within the normal human error incurred in merely taking measurements and is therefore unimportant.

EXAMPLE 16-7. The field coils of a loudspeaker have a resistance of 6000 Ω and an inductance of 1.592 H. Find (*a*) the inductive reactance at 800 Hz, (*b*) the $Q$ of the coils, (*c*) the impedance of the coils, and (*d*) the current flowing if the voltage across the coils is 40 V.

SOLUTION. Given: $R = 6000\ \Omega$ Find: $X_L = ?$
$L = 1.592$ H $Q = ?$
$f = 800$ Hz $Z = ?$
$V = 40$ V $I = ?$

a. $X_L = 6.28fL = 6.28 \times 800 \times 1.592 = 8000\ \Omega$ *Ans.*

*b.* $$Q = \frac{X_L}{R} = \frac{8000}{6000} = 1.33 \quad \textit{Ans.} \qquad (16\text{-}5)$$

*c.* Since $Q$ is less than 5, the resistance *must* be included in the calculation of the impedance.

$$Z = \sqrt{R^2 + X_L^2} \qquad (16\text{-}4)$$
$$Z = \sqrt{6000^2 + 8000^2}$$
$$Z = \sqrt{(6 \times 10^3)^2 + (8 \times 10^3)^2}$$
$$Z = \sqrt{36 \times 10^6 + 64 \times 10^6}$$
$$Z = \sqrt{(36 + 64) \times 10^6}$$
$$Z = \sqrt{100 \times 10^6}$$
$$Z = 10 \times 10^3 = 10^4 = 10{,}000\ \Omega \quad \textit{Ans.}$$

*d.* Since the total opposition is the impedance, the formula for Ohm's law may be rewritten as the

**FORMULA** $V = IZ$ 16-6

$$40 = I \times 10{,}000$$
$$I = \frac{40}{10{,}000} = 0.004 \text{ A} \quad \textit{Ans.}$$

**PROBLEMS**

1. Find the $Q$ of a coil if $R = 30\ \Omega$ and $X_L = 120\ \Omega$.
2. Find the $Q$ of a coil if $X_L = 6000\ \Omega$ and $R = 1000\ \Omega$.
3. Find the $Q$ of a coil at 100 Hz if $R = 200\ \Omega$ and $L = 10$ H.

4. Find the impedance of a coil if its resistance is 12 Ω and its reactance is 35 Ω.
5. Find (*a*) the $Q$ and (*b*) the impedance of a coil if its resistance is 100 Ω and its reactance is 1000 Ω.
6. An antenna circuit has an inductance of 30 μH and a resistance of 20 Ω. Find the impedance to a 500-kHz signal.
7. A 40-V emf at a frequency of 1 kHz is impressed across a loudspeaker of 5000 Ω resistance and 1.5 H inductance. Find (*a*) the inductive reactance, (*b*) the impedance, and (*c*) the current.
8. A 120-V 60-Hz line is connected across a 10-H choke coil whose resistance is 400 Ω. Find (*a*) the inductive reactance, (*b*) the $Q$ of the coil, (*c*) the impedance, and (*d*) the current.
9. The primary of an AF transformer has a resistance of 100 Ω and an inductance of 50 mH. Find (*a*) the inductive reactance at 1 kHz and (*b*) the impedance.
10. A 3000-Ω resistor has an inductance of 10 mH. Find its impedance at (*a*) 500 Hz, (*b*) 5 kHz, (*c*) 500 kHz, and (*d*) 1500 kHz.
11. The primary of an IF coupling transformer has an inductance of 5 mH and a resistance of 100 Ω. If the voltage across the primary is 10 V at 456 kHz, what is the current flowing in the primary?
12. An AF amplifier circuit uses an audio choke of 100 mH inductance and 3000-Ω resistance. Find the impedance to (*a*) 500 Hz, (*b*) 1000 Hz, and (*c*) 5000 Hz.

(See Instructor's Manual for Test 16-2)

## JOB 16-4 MEASURING THE INDUCTANCE OF A COIL

The inductance of a coil may be calculated by the use of several formulas involving specific dimensions of the coil. However, these dimensions are not always easily obtained, and other methods are substituted. One method uses a standard inductance and a circuit similar to the Wheatstone bridge. Another method obtains the resonant frequency of the combination of the coil with a known capacity, and the inductance is calculated from the formula for the resonant frequency given in Job 18-6. In a third method, called the *impedance* method, an ac voltage is impressed across the coil, and voltage, frequency, and current are measured. The resistance of the coil is obtained by use of an ohmmeter.

EXAMPLE 16-8. What is the inductance of a coil that draws 30 mA from a 120-V 60-Hz ac source? The resistance of the coil is 400 Ω.

SOLUTION. Given: $I = 30 \text{ mA} = 0.03 \text{ A}$ Find: $L = ?$

$$V = 120 \text{ V}$$
$$f = 60 \text{ Hz}$$
$$R = 400\ \Omega$$

1. Find the impedance.

$$V = IZ \qquad (16\text{-}6)$$
$$120 = 0.03 \times Z$$
$$Z = \frac{120}{0.03} = 4000\ \Omega$$

2. Compare the $R$ and the $Z$. When the resistance is small in comparison with the impedance, it may be neglected completely, making $X_L = Z$. The resistance is small when $Z/R$ is more than 5.

$$\frac{Z}{R} = \frac{4000}{400} = 10$$

Therefore, since $Z/R$ is larger than 5, the resistance is small when compared with the impedance and $X_L = Z$. If $Z/R$ had been less than 5, the reactance $X_L$ would have been found by applying the formula $Z^2 = R^2 + X_L^2$ of Eq. (16-3).

3. Find the inductance.

$$X_L = 6.28 \times 60 \times L \qquad (16\text{-}1)$$
$$4000 = 376.8 \times L$$
$$L = \frac{4000}{376.8} = 10.6\ \text{H} \qquad \textit{Ans.}$$

SELF-TEST 16-9. What is the inductance of a coil that draws 0.4 A from a 120-V 60-Hz ac source? The resistance of the coil is 100 Ω.

SOLUTION. Given: $I = 0.4$ A, $V = 120$ V, $f = 60$ Hz, $R = 100\ \Omega$. Find: $L = ?$

1. Find the impedance.

$$V = IZ \qquad (16\text{-}6)$$
$$120 = 0.4 \times Z$$
$$Z = \frac{120}{0.4} = 300\ \Omega$$

2. Compare $Z$ and $R$.

$$\frac{Z}{R} = \frac{300}{100} = 3$$

$R$ ___(must/need not)___ be included in the calculation to find $X_L$. | must

3. Find $X_L$.

$$R^2 + X_L^2 = Z^2 \qquad (16\text{-}3)$$
$$X_L^2 = Z^2 - \_\_\_\_ \qquad R^2$$
$$X_L = \sqrt{Z^2 - R^2}$$
$$= \sqrt{(300)^2 - (100)^2}$$
$$= \sqrt{\_\_\_\_ \times 10^4 - \_\_\_\_ \times 10^4} \qquad 9 \quad 1$$
$$= \sqrt{\_\_\_\_ \times 10^4} \qquad 8$$
$$= \_\_\_\_ \times 10^2 = \_\_\_\_\ \Omega \qquad 2.83 \quad 283$$

4. Find the inductance.

$$X_L = 6.28 \times 60 \times L \qquad (16\text{-}1)$$

$$\_\_\_\_\_ = 376.8 \times L$$

$$L = \frac{283}{376.8} = \_\_\_\_\_ \text{ H} \qquad \textit{Ans.}$$

283
0.75

## PROBLEMS

1. What is the inductance of a coil whose resistance is 200 Ω if it draws 0.1 A from a 120-V 60-Hz line?
2. What is the inductance of a coil whose resistance is 500 Ω if it draws 20 mA from a 110-V 60-Hz line?
3. What is the inductance of a coil whose resistance is 300 Ω if it draws 10 mA from a 50-V 1000-Hz ac source?
4. What is the inductance of a coil whose resistance is 200 Ω if it draws 100 mA from a 50-V 1-kHz source?
5. What is the inductance of a coil whose resistance is 50 Ω if it draws 0.55 A from a 110-V 60-Hz line?
6. What is the inductance of a coil whose resistance is 100 Ω if it draws 0.4 A from a 120-V 100-Hz source?

## JOB 16-5 REVIEW OF COILS AND INDUCTANCE

1. When an ac voltage is impressed across a coil,
   *a.* The resulting current is an _____ current. — alternating
   *b.* This changing current produces changing fields of force which _____ the wires of the coil. — cut
   *c.* These cuttings induce a _____ emf in the coil. — back

2. The *inductance* of a coil is a measure of its ability to produce a _____ emf when the current through it is changing. The symbol for inductance is $L$. — back

3. Unit of inductance. A coil has an inductance of 1 _____ if a current changing at the rate of 1 A/s can induce a back emf of 1 V in the coil. — H

4. The reactance of a coil is the opposition of the coil to the passage of a _____ current. — changing

$$X_L = 6.28fL \qquad \boxed{16\text{-}1}$$

5. The $Q$ of a coil is the comparsion of its inductive reactance with its effective resistance.

$$Q = \frac{X_L}{R} \qquad \boxed{16\text{-}5}$$

6. The impedance $A$ of a coil is the _____ sum of its resistance and its reactance. If $Q$ is _____ than 5, the resistance may be neglected and $Z = X_L$. If $Q$ is _____ than 5, the resistance must be added to the reactance to find $Z$. — phasor / larger / smaller

$$Z^2 = R^2 + X_L^2 \qquad \boxed{16\text{-}3}$$

or $$Z = \sqrt{R^2 + X_L^2} \qquad \boxed{16\text{-}4}$$

7. To measure the inductance of a coil, the ac voltage is impressed across the coil, and the voltage, frequency, and current are measured. The resistance of the coil is measured with an ________. The inductance is calculated by the following steps. ohmmeter

*a.* Find the impedance.

$$V = IZ \qquad \boxed{16\text{-}6}$$

*b.* Find the reactance.

$$R^2 + X_L^2 = Z^2 \qquad \boxed{16\text{-}3}$$

*c.* Find the inductance.

$$X_L = 6.28fL \qquad \boxed{16\text{-}1}$$

*d.* If the ratio of $Z$ to $R$ is ________ than 5, the resistance may be neglected and $X_L = Z$. larger

## PROBLEMS

1. Find the inductive reactance of a 0.2-H choke coil at (*a*) 100 Hz, (*b*) 1000 Hz, (*c*) 10 kHz, and (*d*) 100 kHz.
2. A 5-V 100-kHz ac voltage is impressed across an RF choke whose inductance is 10 mH. Find (*a*) the reactance and (*b*) the current that flows.
3. Find the inductive reactance of a 50-$\mu$H coil at a frequency of 10 MHz.
4. What must be the inductance of a coil in order that it have a reactance of 8000 Ω at 800 Hz?
5. Find the $Q$ of a coil if $R = 80$ Ω and $X_L = 4800$ Ω.
6. Find the impedance of a coil if its resistance is 39 Ω and its reactance is 80 Ω.
7. Find the impedance of a coil to a frequency of 60 Hz if its resistance is 30 Ω and its inductance is 0.2 mH.
8. A coil has an inductance of 50 $\mu$H and a resistance of 5 Ω. Find (*a*) the reactance to a 500-kHz frequency, (*b*) the impedance, and (*c*) the current flowing if the voltage is 3 V.
9. A coil of 100 Ω resistance draws 100 mA from a 25-V 60-Hz source. Find its inductance.
10. A coil having a $Q$ of 50 draws 10 mA when connected to a 15-V 1-kHz power supply. Find its inductance.
11. Use the formula $V_{av} = L\frac{I}{t}$ where $V_{av}$ is the average voltage induced in a circuit of $L$ henries by a *change* of current of $I$ amperes in $t$ seconds. If a current changing at the rate of 200 mA/s in a coil induces a counter emf of 50 mV, what is the inductance of the coil?
12. When the current changes from 0.6 A to 0.9 A in 1.5 s, the average voltage induced in a coil is 0.3 V. Find the inductance of the coil.

(See Instructor's Manual for Test 16-3, Coils)

## JOB 16-6 INTRODUCTION TO TRANSFORMERS

It is cheaper to transmit electric energy at high voltages than at low voltages because of the smaller loss of power in the line at high voltages. For this reason, the 220 V that is delivered by an ac generator may be stepped up to 2200 or even 220,000 V for transmission over long distances. At its destination, the voltage is stepped down to 240 V for industrial users and to 120 V for ordinary home and power users. The changes in the voltage continue. Elsewhere, the 120-V supply is reduced to 20 V to operate a toy train or to 12 V to operate a doorbell. All these changes in voltage are made by an extremely efficient electric device called a *transformer.*

**Basic construction.** As shown in Fig. 16-5, a transformer consists of (1) the *primary coil* which *receives* energy from an ac source, (2) the *secondary coil* which *delivers* energy to an ac load, and (3) a *core* on which the two coils are wound. The core is generally of some highly magnetic material, although cardboard, ceramics, and other nonmagnetic materials are used for the cores of some radio and television transformers.

**Principle of operation.** An alternating current will flow when an ac voltage is applied to the primary coil of a transformer. This current produces a field of force which changes as the current changes. The changing magnetic field is carried by the magnetic core to the secondary coil, where it cuts across the turns of that coil. These "cuttings" induce a voltage in the secondary coil. In this way, an ac voltage in one coil is transferred to another coil, even though there is no electrical connection between them. The number of lines of force available in the primary is determined by the primary voltage and the number of turns on the primary—each turn producing a given number of lines. Now, if there are *many turns on the secondary,* each line of force will cut *many turns* of wire and *induce a high voltage.* If the *secondary contains only a few turns,* there will be few cuttings and a *low induced voltage.* The secondary voltage, then, depends on the number of secondary turns as compared with the number of primary turns. If the secondary has twice as many turns as the primary, the secondary voltage will be twice as large as the

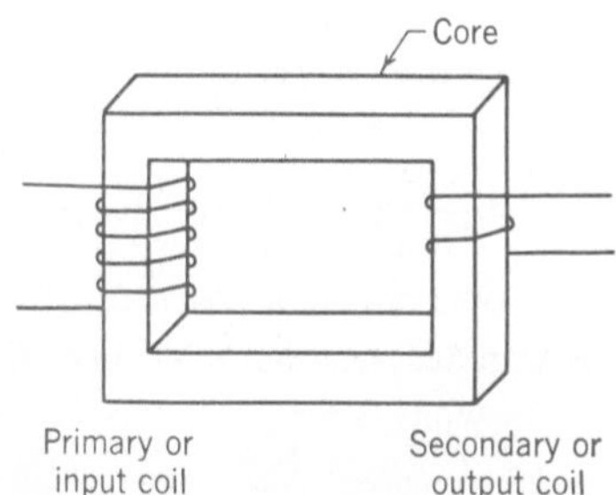


**FIGURE 16-5**
Basic transformer construction.

primary voltage. If the secondary has half as many turns as the primary, the secondary voltage will be one-half as large as the primary voltage. This is stated as the following rule:

**RULE** The voltage on the coils of a transformer is directly proportional to the number of turns on the coils.

**FORMULA**

$$\frac{V_p}{V_s} = \frac{N_p}{N_s} \tag{16-7}$$

where $V_p$ = voltage on primary coil
$V_s$ = voltage on secondary coil
$N_p$ = number of turns on primary coil
$N_s$ = number of turns on secondary coil

The ratio $V_p/V_s$ is called the *voltage ratio* (VR). The ratio $N_p/N_s$ is called the *turns ratio* (TR). By substituting these terms in formula (16-7), we obtain an equivalent statement.

$$\text{VR} = \text{TR} \tag{16-8}$$

**Nomenclature.** A voltage ratio of 1:3 (read as "1 to 3") means that for each volt on the primary, there are 3 V on the secondary. This is called a *step-up* transformer. A step-up transformer *receives a low voltage* on the primary and *delivers a high voltage* from the secondary. A voltage ratio of 3:1 (read as 3 to 1) means that for 3 V on the primary, there is only 1 V on the secondary. This is called a *step-down* transformer. A step-down transformer *receives a high voltage* on the primary and *delivers a low voltage* from the secondary.

**The autotransformer.** Instead of the two-winding, four-terminal standard transformer, the autotransformer (Fig. 16-6) consists of just a

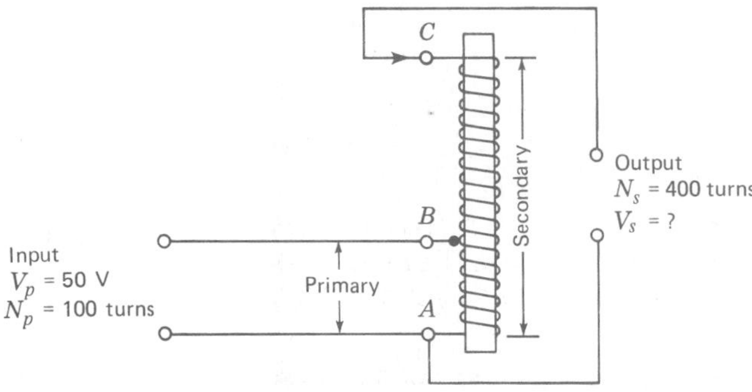


**FIGURE 16-6**
An autotransformer.

*single* winding. This winding may be tapped at any point along its length to provide a set of three terminals (*A, B, C*). The winding *AB* is the primary, and the entire winding *AC* is the secondary. Even though the windings have an electrical connection at *B*, the principle of operation remains the same. When an ac voltage is applied to the primary winding *AB*, the lines of force around *AB* link the turns between *A* and *C*, inducing a higher or lower voltage according to the formula $VR = TR$.

EXAMPLE 16-10. A bell transformer reduces the primary voltage of 120 V to the 18 V delivered by the secondary. If there are 180 turns on the primary and 27 turns on the secondary, find (*a*) the voltage ratio and (*b*) the turns ratio.

SOLUTION. The diagram for the problem is shown in Fig. 16-7.

*a.* $\mathrm{VR} = \dfrac{V_p}{V_s} = \dfrac{120}{18} = \dfrac{20}{3}$ (read as "20 to 3") *Ans.*

*b.* $\mathrm{TR} = \dfrac{N_p}{N_s} = \dfrac{180}{27} = \dfrac{20}{3}$ (read as "20 to 3") *Ans.*

*Note:* The ratios are always expressed in fractional form, even if the fraction can be reduced to a whole number.

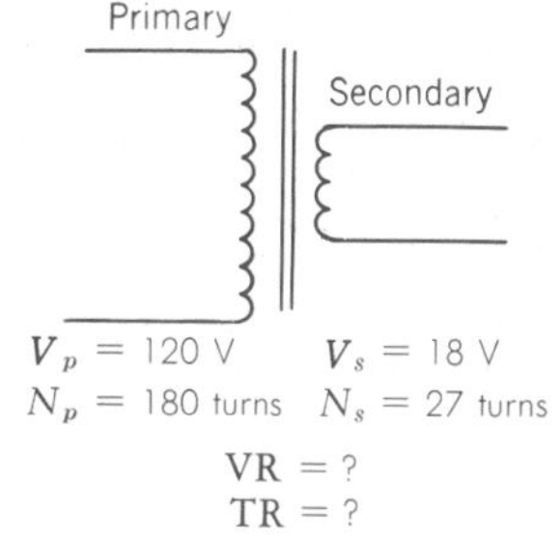


**FIGURE 16-7**

EXAMPLE 16-11. Find the voltage delivered by the secondary of the autotransformer shown in Fig. 16-6.

SOLUTION. Given: $V_p = 50$ V Find: $V_s = ?$

$N_p = 100$ turns

$N_s = 400$ turns

$$\frac{V_p}{V_s} = \frac{N_p}{N_s} \qquad (16\text{-}7)$$

$$\frac{50}{V_s} = \frac{1\not{0}\not{0}}{4\not{0}\not{0}}$$

$V_s = 200$ V *Ans.*

EXAMPLE 16-12. A "power" transformer has 99 turns on the primary and 315 turns on the secondary. What voltage will it deliver if the primary voltage is 110 V?

SOLUTION. Given: $N_p = 99$ turns Find: $V_s = ?$

$N_s = 315$ turns

$V_p = 110$ V

$$\frac{V_p}{V_s} = \frac{N_p}{N_s} \qquad (16\text{-}7)$$

$$\frac{110}{V_s} = \frac{99}{315}$$

$$99V_s = 315 \times 110$$

$$V_s = \frac{315 \times 110}{99} = 350 \text{ V} \qquad \textit{Ans.}$$

SELF-TEST 16-13. A Stancor P-6011 transformer used in a TV set has a voltage ratio of 11:35. If the primary has 242 turns, how many turns must be wound on the secondary?

SOLUTION. Given: VR = 11:35 Find: $N_s$ = ?
$N_p = 242$

$$\text{VR} = \frac{V_p}{?}$$

and since

$$\frac{V_p}{V_s} = \frac{N_p}{N_s} \qquad (16\text{-}7)$$

$$\_\_\_\_\_ = \frac{N_p}{N_s}$$

$$\frac{11}{35} = \frac{242}{N_s}$$

$$N_s = \frac{242 \times 35}{11} = \_\_\_\_\_ \text{ turns} \qquad \textit{Ans.}$$

$V_s$

VR

770

## PROBLEMS

1. A "power" transformer has 85 turns on the primary and 255 turns on the secondary. What voltage will it deliver if the primary is connected to a 120-V source?
2. A filament transformer reduces the 110 V on the primary to 10 V on the secondary. Find (*a*) the voltage ratio and (*b*) the turns ratio.
3. The Stancor P-6293 universal-type power transformer steps down the voltage from 120 to 2.5 V. Find (*a*) the voltage ratio and (*b*) the turns ratio.
4. A Stancor A-4773 transformer whose turns ratio is 1:3 is used as a plate-to-grid coupling transformer in an amplifier circuit. What is the secondary voltage if the primary voltage is 15 V?
5. A 24:1 welding transformer has 25 turns on the secondary. How many turns are there on the primary?
6. A UTC LS-185 plate transformer steps up the voltage from 100 to 2500 V. If there are 50 turns on the primary, how many turns are on the secondary?
7. Find the voltage at the spark plugs if a 6-V alternator is connected to a coil with a primary winding of 50 turns and a secondary winding of 50,000 turns.
8. A coil with a primary winding of 100 turns must deliver 4800 V. If

the primary is connected to a 6-V source, find the number of turns on the secondary.

9. The output from a 2N1097 transistor is to be matched to a 13.9-Ω voice coil by a 12:1 matching transformer. If the primary voltage is 18 V, find the voltage across the voice coil.
10. A transformer whose primary is connected to a 120-V source delivers 10 V. If the number of turns on the secondary is 20 turns, find the number of turns on the primary. How many extra turns must be added to the secondary if it must deliver 35 V?
11. A toy-train transformer is connected to a 120-V 60-Hz source. The secondary has 60 turns and delivers 24 V. How many turns are on the primary?
12. The secondary coil of a transformer has 100 turns, and the secondary voltage is 5 V. If the turns ratio is 22:1, find (*a*) the voltage ratio, (*b*) the primary voltage, and (*c*) the primary turns.
13. A step-down transformer is wound with 3750 turns on the primary and 60 turns on the secondary. What is the delivered voltage if the high-voltage side is 15,000 V?
14. The 117-V primary of a transformer has 250 turns. Two secondaries are to be provided to deliver (*a*) 12.6 V and (*b*) 35 V. How many turns are needed on each secondary?
15. A power transformer with 100 turns on the primary is to be connected to a 120-V source of supply. Separate secondary windings are to deliver (*a*) 2.5 V, (*b*) 6.3 V, and (*c*) 600 V. Find the number of turns on each secondary.
16. A transformer bank is used to transform 2000 kVA (kilovoltamperes) from 14,000 to 4000 V. Find (*a*) the turns ratio of the transformer, and (*b*) the primary current.

## JOB 16-7 CURRENT IN A TRANSFORMER

In the modern transformer, the power delivered to the primary is transferred to the secondary with practically no loss. For all practical purposes, the power input to the primary is equal to the power output of the secondary, and the transformer is assumed to operate at an efficiency of 100 percent. Thus,

$$\text{Power input} = \text{power output} \qquad \boxed{16\text{-}9}$$

Since

$$\text{Power input} = V_p \times I_p \qquad \boxed{16\text{-}10}$$

and

$$\text{Power output} = V_s \times I_s \qquad \boxed{16\text{-}11}$$

$$V_p \times I_p = V_s \times I_s \qquad \boxed{16\text{-}12}$$

By dividing both sides of the equation by $V_s \times I_p$ and canceling out identical terms,

$$\frac{V_p \times \overset{1}{\cancel{I_p}}}{V_s \times \underset{1}{\cancel{I_p}}} = \frac{\overset{1}{\cancel{V_s}} \times I_s}{\underset{1}{\cancel{V_s}} \times I_p}$$

we obtain the

**FORMULA**

$$\frac{V_p}{V_s} = \frac{I_s}{I_p} \qquad \boxed{16\text{-}13}$$

This formula indicates that the current ratio in a transformer is *inversely proportional* to the voltage ratio. If the *voltage* ratio *increases,* the *current* ratio will *decrease.* If the *voltage* ratio *decreases,* the *current* ratio will *increase.*

In addition, since

$$\frac{V_p}{V_s} = \frac{N_p}{N_s}$$

we may substitute $N_p/N_s$ for $V_p/V_s$ in Eq. (16-13). This gives the

**FORMULA**

$$\frac{N_p}{N_s} = \frac{I_s}{I_p} \qquad \boxed{16\text{-}14}$$

EXAMPLE 16-14. The Stancor model P-6293 universal-type power transformer delivers 36 W of power to a rectifier circuit. If the primary voltage is 120 V, how much current is drawn by the transformer?

SOLUTION. Given: Power output = 36 W $\qquad$ Find: $I_p = ?$
$V_p = 120$ V

$$\text{Power input} = \text{power output} \qquad (16\text{-}9)$$

$$V_p \times I_p = \text{power output}$$

$$120 \times I_p = 36$$

$$I_p = \frac{36}{120} = 0.3 \text{ A} \qquad \textit{Ans.}$$

EXAMPLE 16-15. The primary of a transformer is connected to a 110-V 60-Hz line. The secondary delivers 250 V at 0.1 A. Find (*a*) the current in the primary and (*b*) the power input to the primary.

SOLUTION. The diagram for the problem is shown in Fig. 16-8.

*a.*

$$\frac{V_p}{V_s} = \frac{I_s}{I_p} \qquad (16\text{-}13)$$

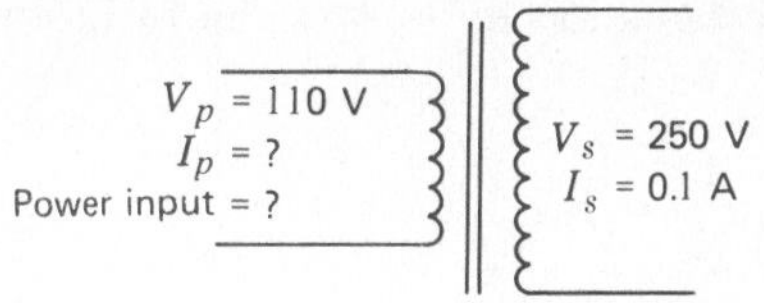

FIGURE 16-8

$$\frac{110}{250} = \frac{0.1}{I_p}$$

$$110 \times I_p = 250 \times 0.1$$

$$I_p = \frac{25}{110} = 0.227 \text{ A} \qquad \textit{Ans.}$$

*b.* Power input $= V_p \times I_p = 110 \times 0.227 = 24.97$ W *Ans.*

EXAMPLE 16-16. A bell transformer with 300 turns on the primary and 45 turns on the secondary draws 0.3 A from the 110 V line. Find (*a*) the current delivered by the secondary, (*b*) the voltage delivered by the secondary, and (*c*) the power delivered by the secondary.

SOLUTION. Given: $N_p = 300$ turns, $N_s = 45$ turns, $I_p = 0.3$ A, $V_p = 110$ V. Find: $I_s = ?$, $V_s = ?$, Power output = ?

*a.*

$$\frac{N_p}{N_s} = \frac{I_s}{I_p} \qquad (16\text{-}14)$$

$$\frac{300}{45} = \frac{I_s}{0.3}$$

$$45 \times I_s = 300 \times 0.3$$

$$I_s = \frac{90}{45} = 2 \text{ A} \qquad \textit{Ans.}$$

*b.*

$$\frac{V_p}{V_s} = \frac{N_p}{N_s} \qquad (16\text{-}7)$$

$$\frac{110}{V_s} = \frac{300}{45}$$

$$300 \times V_s = 110 \times 45$$

$$V_s = \frac{4950}{300} = 16.5 \text{ V} \qquad \textit{Ans.}$$

*c.* Power output $= V_s \times I_s = 16.5 \times 2 = 33$ W *Ans.*

SELF-TEST 16-17. A 120:24-VR transformer draws 1.5 A. Find the secondary current.

SOLUTION. Given: VR = 120:24, $I_p = 1.5$ A. Find: $I_s = ?$

$$\frac{V_p}{V_s} = \frac{I_?}{I_?} \qquad (16\text{-}13)$$

But $$\frac{V_p}{V_s} = \text{VR} = \frac{?}{?}$$

Therefore, we can substitute this ratio in formula (16-13).

$$\frac{120}{24} = \frac{I_s}{?}$$

$$24 \times I_s = 120 \times \underline{\qquad}$$

$$I_s = \frac{180}{24} = \underline{\qquad} \text{ A} \qquad \textit{Ans.}$$

s
p
120
24
1.5
1.5
7.5

**PROBLEMS**

1. A bell transformer draws 20 W from a line. What is the secondary current if the secondary voltage is 10 V?
2. A Thermador model 5A6086 power transformer delivers 22.5 W of power. If the primary voltage is 112.5 V, how much current is drawn by the primary?
3. A 120-V 60-Hz line supplies power to a Stancor model P-6297 universal-type transformer. If the secondary delivers 480 V at 0.04 A, find (*a*) the primary current and (*b*) the power input.
4. A bell transformer with 240 turns on the primary and 30 turns on the secondary draws 0.25 A from a 120-V line. Find (*a*) the secondary current, (*b*) the secondary voltage, and (*c*) the secondary power.
5. A transformer is wound with 2200 turns on the primary and 150 turns on the secondary. (*a*) If it delivers 2 A, what is the primary current? (*b*) If the primary voltage is 110 V, what is the secondary voltage?
6. A 230:110-VR step-down transformer in a stage-lighting circuit draws 10 A from the line. Find the current delivered.
7. A filament transformer delivers 1.5 A at 6.3 V. If $V_p$ is 110 V, find (*a*) $I_p$ and (*b*) the power input.
8. A transformer with 2400 turns on the primary and 480 turns on the secondary draws 8.5 A from a 230-V line. Find (*a*) $I_s$, (*b*) $V_s$, and (*c*) the power output.
9. A transformer has 120 turns on the primary and 1500 turns on the secondary. (*a*) If it delivers 0.4 A, what is the primary current? (*b*) If the primary voltage is 120 V, what is the secondary voltage?
10. A 9:2 step-down transformer draws 1.8 A. Find the $I_s$.
11. A substation transformer reduces the voltage from the transmission-line voltage of 150,000 to 4400 V. If the transmission line carries 15 A, what current will the transformer deliver?
12. A step-down transformer with a turns ratio of 50,000:250 has its primary connected to a 27,000-V transmission line. If the secondary is connected to a 7.5-Ω load, find (*a*) the secondary voltage, (*b*)

the secondary current, (*c*) the primary current, and (*d*) the power output.

## JOB 16-8 EFFICIENCY OF A TRANSFORMER

In Job 11-4 we learned that the efficiency of an electric machine is equal to the ratio of the power output to the power input. This ratio is expressed as a percent by multiplying it by 100. In general, the efficiency of any device is the ratio of its output to its input and describes the effectiveness of the device in utilizing the energy supplied to it. Thus, a transformer which delivers *all* the power put into it would have an efficiency of 100 percent. In the last job, we assumed that transformers have an efficiency of 100 percent and deliver all the energy that they receive. Actually, because of copper and core losses, the efficiency of even the best transformer is less than 100 percent.

**FORMULA**

$$\text{Eff} = \frac{\text{power output}}{\text{power input}} \qquad (11\text{-}2)$$

EXAMPLE 16-18. A plate transformer draws 30 W from a 117-V line and delivers 300 V at 90 mA. Find its percent efficiency.

SOLUTION

Given: Power input = 30 W  Find: Percent eff = ?

$V_s = 300$ V

$I_s = 90$ mA $= 0.09$ A

$$\text{Eff} = \frac{\text{power output}}{\text{power input}} \qquad (10\text{-}2)$$

$$\text{Eff} = \frac{300 \times 0.09}{30} = 0.9 = 90\% \qquad \textit{Ans.}$$

EXAMPLE 16-19. A transformer whose efficiency is 80 percent draws its power from a 120-V line. If it delivers 192 W, find (*a*) the power input and (*b*) the primary current.

SOLUTION. Given: Eff = 80% = 0.80  Find: Power input = ?

$V_p = 120$ V  $I_p = ?$

Power output = 192 W

*a.*

$$\text{Eff} = \frac{\text{power output}}{\text{power input}} \qquad (10\text{-}2)$$

$$0.80 = \frac{192}{\text{power input}}$$

$$\text{Power input} = \frac{192}{0.80} = 240 \text{ W} \qquad \textit{Ans.}$$

*b.*

$$\text{Power input} = V_p \times I_p \qquad (16\text{-}10)$$

$$240 = 120 \times I_p$$

$$I_p = \frac{240}{120} = 2 \text{ A} \qquad \textit{Ans.}$$

## PROBLEMS

1. What is the efficiency of a transformer if it draws 800 W and delivers 700 W?
2. A toy transformer draws 150 W from a 110-V line and delivers 24 V at 5 A. Find its efficiency.
3. A transformer draws 1.5 A at 120 V and delivers 7 A at 24 V. Find (*a*) the power input, (*b*) the power output, and (*c*) the efficiency.
4. A power transformer draws 96 W and delivers 420 V at 200 mA. Find (*a*) the efficiency and (*b*) the primary current if the primary voltage is 120 V.
5. In Fig. 16-9, an impedance-matching transformer couples an output transistor delivering 2.1 W to a voice coil which receives 1.68 W. Find its efficiency.

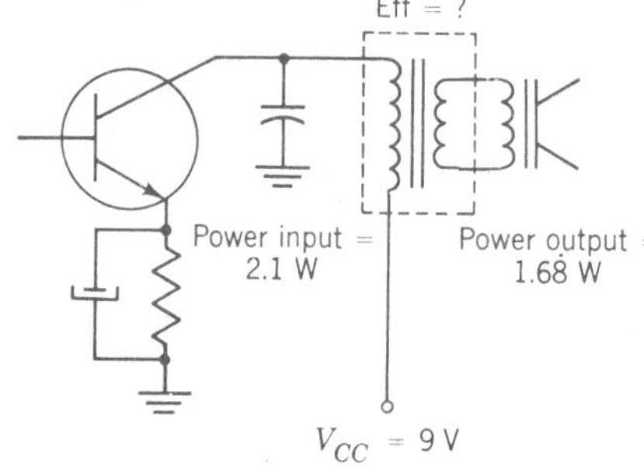


**FIGURE 16-9**
Finding the efficiency of an impedance-matching transformer.

6. A transformer that draws 1000 W from a 230-V line operates at an efficiency of 92 percent and delivers 50 V. Find (*a*) the watts delivered and (*b*) the secondary current.
7. A transformer that delivers 10,000 W at an efficiency of 96 percent draws its power from a 2000-V line. Find (*a*) the power input and (*b*) the primary current.
8. A transformer delivers 750 V at 120 mA at an efficiency of 90 percent. If the primary current is 0.8 A, find (*a*) the power input and (*b*) the primary voltage.
9. A transformer delivers 660 V at 98 mA at an efficiency of 84 percent. If the primary current is 875 mA, find (*a*) the power input and (*b*) the primary voltage.
10. The three secondary coils of a power-supply transformer deliver 100 mA at 350 V, 2 A at 2.5 V, and 1.2 A at 12.6 V. What is the efficiency of the transformer if it draws 60 W from the 117-V line?

## JOB 16-9 IMPEDANCE-MATCHING TRANSFORMERS

The maximum transfer of energy from one circuit to another will occur when the impedances of the two circuits are equal or *matched.* If the two circuits have unequal impedances, a coupling transformer may be used as an intermediate impedance-changing device between the two circuits. In Fig. 16-10, the output of a transistor is used to operate the voice coil of a loudspeaker. The output of the transistor cannot be connected directly to the voice coil because the impedance of the transistor circuit is 4000 Ω and the impedance of the voice coil is only 10 Ω. The circuits may be *matched* by a transformer with an appropriate turns ratio whose value depends on the values of the impedances involved. In Fig. 16-10,

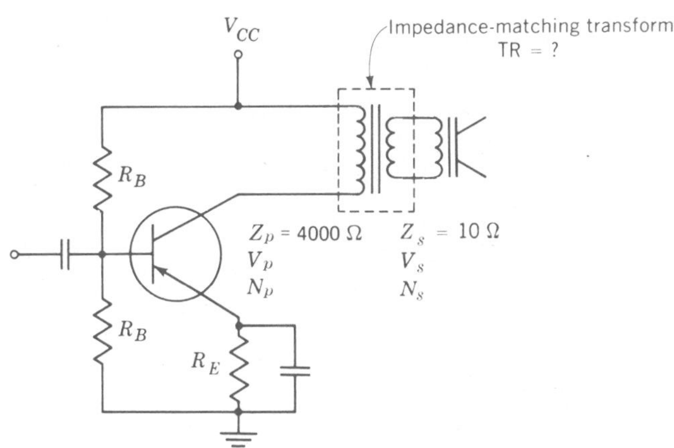


**FIGURE 16-10** An impedance-matching transformer is used to couple the output transistor to the voice coil of the speaker.

$$\frac{N_p}{N_s} = \frac{V_p}{V_s} \quad \text{and} \quad \frac{N_p}{N_s} = \frac{I_s}{I_p}$$

By multiplying the two equations, we obtain

$$\frac{N_p}{N_s} \times \frac{N_p}{N_s} = \frac{V_p}{V_s} \times \frac{I_s}{I_p}$$

or

$$\left(\frac{N_p}{N_s}\right)^2 = \frac{V_p}{I_p} \times \frac{I_s}{V_s}$$

But $V_p/I_p = Z_p$ and $I_s/V_s = 1/Z_s$; therefore,

$$\left(\frac{N_p}{N_s}\right)^2 = \frac{Z_p}{Z_s} \tag{16-15}$$

By taking the square root of both sides we obtain the

**FORMULA**

$$\frac{N_p}{N_s} = \sqrt{\frac{Z_p}{Z_s}} \tag{16-16}$$

where $N_p$ = number of turns on primary
$N_s$ = number of turns on secondary
$Z_p$ = impedance of primary, Ω
$Z_p$ = impedance of secondary, Ω

EXAMPLE 16-20. Find the turns ratio of the transformer shown in Fig. 16-10.

SOLUTION

$$\frac{N_p}{N_s} = \sqrt{\frac{Z_p}{Z_s}} = \sqrt{\frac{4000}{10}} = \sqrt{400} \qquad (16\text{-}16)$$

$$\frac{N_p}{N_s} = \frac{20}{1} \qquad \textit{Ans.}$$

EXAMPLE 16-21. A 2N255A transistor in a telephone amplifier supplies a 490-Ω load. If the load is to be matched to a 10-Ω speaker, find the required turns ratio.

SOLUTION. Given: $Z_p = 490\ \Omega$ Find: $\frac{N_p}{N_s} = ?$
$Z_s = 10\ \Omega$

$$\frac{N_p}{N_s} = \sqrt{\frac{Z_p}{Z_s}} = \sqrt{\frac{490}{10}} = \sqrt{49} \qquad (16\text{-}16)$$

$$\frac{N_p}{N_s} = \frac{7}{1} \qquad \textit{Ans.}$$

EXAMPLE 16-22. A carbon microphone whose impedance is 20 Ω is to be coupled to a grid circuit whose impedance is 72,000 Ω. Find the turns ratio of the coupling transformer.

SOLUTION. Given: $Z_p = 20\ \Omega$ Find: $\frac{N_p}{N_s} = ?$
$Z_s = 72{,}000\ \Omega$

When coupling from a low impedance to a high impedance, it is best to find the turns ratio by comparing $N_s$ with $N_p$. This can be done by inverting both sides of formula (16-16). This gives the

**FORMULA**

$$\frac{N_s}{N_p} = \sqrt{\frac{Z_s}{Z_p}} \qquad \boxed{16\text{-}17}$$

$$\frac{N_s}{N_p} = \sqrt{\frac{72{,}000}{20}} = \sqrt{3600} = 60$$

$$\frac{N_s}{N_p} = \frac{60}{1} \quad \text{or} \quad \frac{N_p}{N_s} = \frac{1}{60} \qquad \textit{Ans.}$$

Thus, the microphone is connected to the primary ($N_p = 1$), and the grid is connected to the secondary ($N_s = 60$).

EXAMPLE 16-23. A 1:20 step-up transformer is used to match a microphone with a grid-circuit impedance of 40,000 Ω. Find the impedance of the microphone.

SOLUTION. Given: $N_p/N_s = 1/20$ Find: $Z_p = ?$

$Z_s = 40{,}000\ \Omega$

$$\left(\frac{N_p}{N_s}\right)^2 = \frac{Z_p}{Z_s} \qquad (16\text{-}15)$$

$$\left(\frac{1}{20}\right)^2 = \frac{Z_p}{40{,}000}$$

$$\frac{1}{400} = \frac{Z_p}{40{,}000}$$

$$Z_p = \frac{40{,}000}{400} = 100\ \Omega \qquad \textit{Ans.}$$

**PROBLEMS**

1. Find the turns ratio of a transformer used to match a 1600-Ω load to a 4-Ω load.
2. Find the turns ratio of a transformer used to match a 60-Ω load to a 540-Ω line.
3. The impedance of the output circuit of a power stage is 8000 Ω. What is the turns ratio of a transformer used to transfer the power to a 500-Ω line supplying a public-address system?
4. A 2N265 transistor works into a load impedance of 9000 Ω. Find the turns ratio of the transformer needed to feed into a 10-Ω voice coil.
5. Find the turns ratio of a microphone transformer required to couple a 20-Ω microphone to a 500-Ω line.
6. Find the turns ratio of a microphone transformer required to couple a 20-Ω microphone to a grid circuit of 50,000 Ω impedance.
7. A 55:1 output transformer is used to match an output tube to a 4-Ω coil. Find the impedance of the output circuit.
8. Two 2N406 output transistors work in push-pull into a load impedance of 14,000 Ω. Find the required turns ratio of a transformer to match the output to an 8-Ω voice coil.
9. Find the turns ratio of a Stancor model A-8101 transformer which is used to match a 500-Ω line to an 8-Ω voice coil.
10. Find the turns ratio of the transformer needed to match a load of 4500 Ω to two 9-Ω speakers in parallel.
11. What would be the turns ratio in Prob. 10 if there were three 9-Ω speakers in parallel?

12. Find the turns ratio of the transformer used to match a 4200-Ω impedance to a 500-Ω line supplying a distant auditorium loudspeaker.
13. Find the turns ratio of the transformer needed to match a 50-Ω Amperite model PGL dynamic microphone to a 500-Ω line.
14. The secondary load of a step-down transformer with a turns ratio of 6 to 1 is 800 Ω. Find the impedance of the primary.

## JOB 16-10 REVIEW OF TRANSFORMERS

1. A transformer transmits energy from one circuit to another by means of electromagnetic induction between two coils.

2. The ______ coil is connected to the source of supply. The *secondary* coil is connected to the ______. — primary; load

3. A step-up transformer ______ the voltage and decreases the current. A step-down transformer decreases the voltage and ______ the current. — increases; increases

4. The turns ratio of a transformer is the comparison (by division) of the number of turns on the ______ with the number of turns on the ______. — primary; secondary

$$TR = \frac{N_p}{N_s}$$

The voltage ratio of a transformer is the comparison (by division) of the voltage on the primary with the voltage on the secondary.

$$VR = \frac{V_p}{V_s}$$

5. In a 100 percent efficient transformer

*a.* Power input = power output **[16-9]**

*b.* Power input $= V_p \times I_p$ **[16-10]**

*c.* Power output $= V_s \times I_s$ **[16-11]**

*d.* The voltage is directly proportional to the number of turns.

$$\frac{V_p}{V_s} = \frac{?}{?} \qquad [16\text{-}7]$$

— $N_p$ / $N_s$

*e.* The voltage is inversely proportional to the current.

$$\frac{V_p}{V_s} = \frac{?}{?} \qquad [16\text{-}13]$$

— $I_s$ / $I_p$

*f.* The number of turns is inversely proportional to the current.

$$\frac{N_p}{N_s} = \frac{?}{?} \qquad [16\text{-}14]$$

— $I_s$ / $I_p$

6. The efficiency of a transformer is the ratio of the power output to the power input and is expressed as a percent.

$$\text{Eff} = \frac{\text{power output}}{\text{power input}}$$ 10-2

7. Impedance-matching transformers permit the transfer of power between loads of different impedance. The turns ratio of such transformers is given by the formulas below:

$$\left(\frac{N_p}{N_s}\right)^2 = \frac{?}{?}$$ 16-15

$Z_p$
$Z_s$

or

$$\frac{N_p}{N_s} = \sqrt{\frac{Z_p}{Z_s}}$$ 16-16

## PROBLEMS

1. A bell transformer reduces the voltage from 120 to 15 V. If there are 22 turns on the secondary, find (*a*) the number of turns on the primary and (*b*) the turns ratio.
2. Find the voltage at the spark plugs if a 12-V alternator is connected to a coil with 80 turns on the primary and 40,000 turns on the secondary.
3. A UTC LS-185 plate transformer has a turns ratio of 1:25. If there are 70 turns on the primary, how many turns are on the secondary?
4. If the turns ratio of a transformer is 20:3, find the primary voltage if the secondary voltage is 24 V.
5. The 110-V primary of a transformer has 500 turns. Two secondaries are to be provided to deliver (*a*) 22 V and (*b*) 5 V. How many turns are needed for each secondary?
6. A power transformer delivers 50 W of power. If the primary voltage is 110 V, find the primary current.
7. A transformer connected to a 120-V 60-Hz line delivers 750 V at 200 mA. Find (*a*) the primary current and (*b*) the power drawn by the primary.
8. A 5:1 transformer draws 0.5 A from a 120-V line. Find (*a*) the secondary current, (*b*) the secondary voltage, and (*c*) the power output.
9. A filament transformer delivers 1.2 A at 6.3 V. If the primary voltage is 120 V, find (*a*) $I_p$ and (*b*) the power input.
10. A transformer with 1500 turns on the primary and 375 turns on the secondary draws 3.5 A from a 115-V line. Find (*a*) the secondary current, (*b*) the secondary voltage, and (*c*) the power output.
11. A transformer draws 250 mA from a 120-V line and delivers 80 mA at 350 V. Find its efficiency.
12. A transformer drawing 150 W from the line operates at an efficiency of 90 percent and delivers 50 V. Find (*a*) the power delivered and (*b*) the secondary current.
13. A 2N321 amplifier feeds into the 500-Ω primary of an audible automobile signal minder. Find the turns ratio of the transformer needed to match it with a 3.2-Ω speaker.

14. A 1:30 step-up transformer is used to match a 50-Ω microphone to a grid circuit. Find the impedance of the grid circuit.
15. A step-down transformer with a turns ratio of 45,000:150 has its primary connected to a 72,000-V transmission line. If the secondary is connected to a 20-Ω load, find (*a*) the secondary voltage, (*b*) the secondary current, (*c*) the primary current, and (*d*) the power output.

(See Instructor's Manual for Test 16-4, Transformers)

## JOB 17-1 INTRODUCTION TO CAPACITANCE

**What is a capacitor?** A capacitor, or "condenser," is formed whenever two pieces of metal are separated by a thin layer of insulating material. To form a capacitor of any appreciable value, however, the area of the metal pieces must be quite large and the thickness of the insulating material, or *dielectric,* must be quite small.

**What does a capacitor do?** A capacitor is an electrical storehouse. When we wish to store up electricity for a little while, we "charge" the capacitor. When we "discharge" a capacitor, we draw the electrons from it to operate some device. The two plates of the capacitor shown in Fig. 17-1 are electrically neutral, since there are as many protons as electrons on each plate. The capacitor has no "charge." Now let us connect a battery across the plates as shown in Fig. 17-2*a*. When the switch is closed (Fig. 17-2*b*), the positive side of the battery pulls electrons from plate *A* of the capacitor and deposits them on plate *B*. Electrons will continue to flow from *A* to *B* until the number of electrons on plate *B* exert a force equal to the electromotive force of the battery. The capacitor is now charged. It will remain in this condition even if the

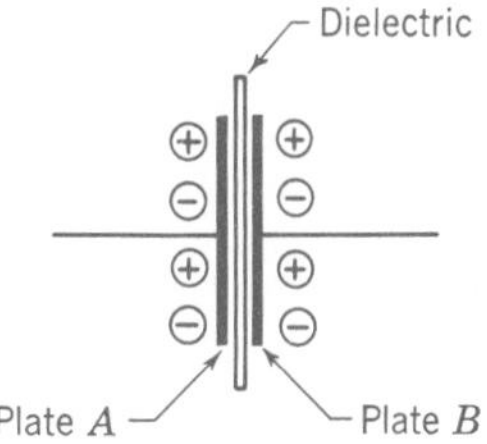


**FIGURE 17-1**
Electrically neutral capacitor.

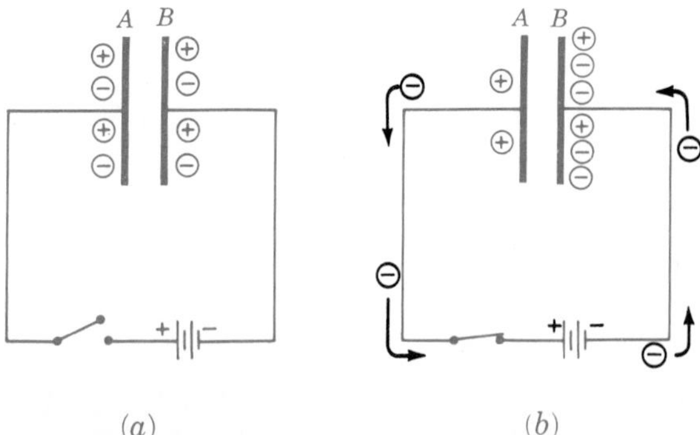


**FIGURE 17-2**
A charged capacitor has an excess of electrons on one plate. (*a*) neutral capacitor; (*b*) charged capacitor.

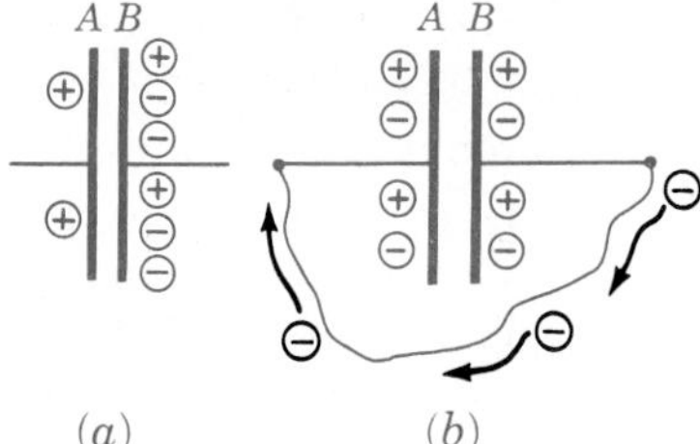


**FIGURE 17-3**
(*a*) A charged capacitor has more electrons on one plate than on the other. (*b*) Discharging a capacitor.

battery is removed as shown in Fig. 17-3*a*. This is a condition of extreme unbalance, and the electrons on plate *B* will attempt to return to plate *A* if they can. The resistance of the dielectric and the surrounding air prevents this from happening, although some electrons do manage to "leak" off plate *B* and return to plate *A*. However, if a conductor is placed across the plates as in Fig. 17-3*b*, then the electrons find an easy path back to plate *A* and they will return in a rush, thus "discharging" the capacitor. There are many uses for capacitors such as in tuning circuits, filter circuits, coupling circuits, bypasses for alternating currents of high frequency, and blocking devices in audio circuits. In each application, the capacitor operates by storing up electrons and discharging them at the proper time. Capacitors may not be used in dc circuits, since the dielectric of the capacitor acts to present an open circuit. Current will flow in an ac circuit containing a capacitor as shown in Fig. 17-4. During the positive half of the ac cycle, the electrons travel through the lamp and pile up on plate *A* of the capacitor. During this time, electrons are drawn off plate *B* of the capacitor by the ac source. During the negative half of the cycle, the direction of the electron flow is reversed. The capacitor discharges through the lamp; the source pulls the electrons from plate *A* and piles them up on plate *B*. This action continues with each reversal of the alternating current. There seems to be a continuous flow of electrons through the capacitor which lights the lamp, but it is actually a flow of electrons *around* the capacitor which operates the lamp.

**Meaning of capacitance.** The measure of the ability of a capacitor to hold electrons is called its *capacitance.* Since an electron is so small, and since there are so many of them, the *coulomb* is used as the measure of

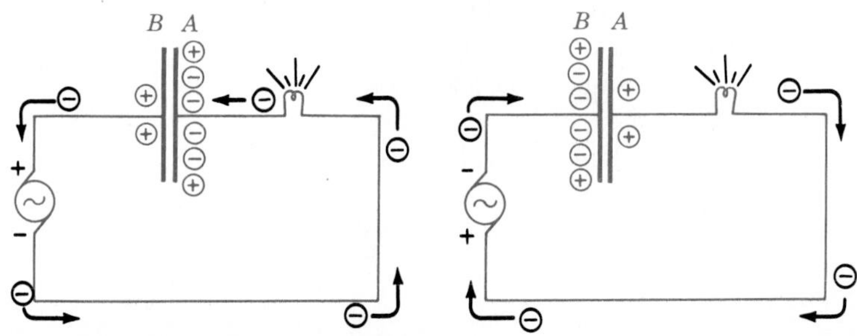


**FIGURE 17-4**
Electron flow around a capacitor in an ac circuit.

electrical quantity. In Job 1-1 we learned that a coulomb is equal to approximately 6 billion billion electrons. The capacitance of a capacitor has been defined as the number of coulombs of electricity that may be held on its plates by a pressure of 1 V. If 1 coulomb is held on the plates by a pressure of 1 V, then the capacitance is called 1 farad (F). If 4 coulombs may be held by 4 V, then 1 coulomb may be held by 1 V and the capacitance will still be 1 F. But if 4 coulombs may be held by only 2 V, then 2 coulombs will be held by 1 V and the capacitance will be 2 F. From this, we can get a formula to find the capacitance of a capacitor.

**FORMULA** $$C = \frac{Q}{V} \qquad \boxed{17\text{-}1}$$

where $C$ = capacitance, F
$Q$ = number of electrons on the plates, coulombs
$V$ = voltage across the plates, V

It is very inconvenient to discuss capacitances in terms of farads, since a farad is such a tremendous unit of measurement. Always change units of capacitance into microfarads ($\mu$F) or picofarads (pF). See Job 3-7 for methods of changing units of measurement.

## JOB 17-2 CAPACITORS IN PARALLEL

In the last job we learned that the capacitance depends on the number of coulombs that can be held on the plates by a pressure of 1 V. A capacitor that can hold 3 coulombs will have three times the capacitance of another that can hold only 1 coulomb if the same voltage is applied to both. What is it about a capacitor that enables one to hold more electrons than another? Obviously, the electrons must be held somewhere, and they are usually distributed on the surface of the capacitor plates. If the area of the plates is large, then many electrons can be placed on the large area, but if the plates are small in area, then only a few electrons can be held there and the capacitance will be small. The capacitance of a capacitor depends on this plate area and also on the thickness of the dielectric. The larger the plate area, the larger the capacitance. The *thinner* the dielectric, the *larger* the capacitance.

**What is the effect of placing capacitors in parallel?** In Fig. 17-5, $C_1$ and $C_2$ are connected in parallel. Since plate $A$ and plate $B$ are connected together, the effect is the same as if we had one big plate whose area is equal to the sum of plates $A$ and $B$. Similarly, plates $C$ and $D$ on the other side are connected together to form one large plate equal to the sum of $C$ and $D$. Since the capacitance increases as we increase the plate area, we get the capacitance of the combination by adding the capacitances of the individual capacitors.

**FORMULA** $$C_T = C_1 + C_2 + C_3 + \text{etc.} \qquad \boxed{17\text{-}2}$$

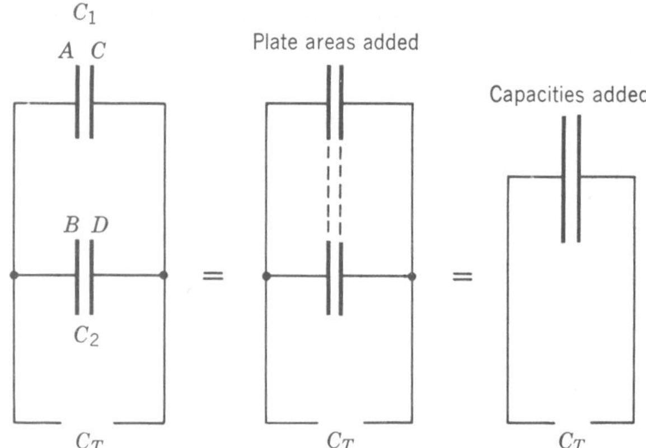


**FIGURE 17-5**
When capacitors are in parallel, the plate areas are added and the total capacitance $C_T$ equals the *sum* of the individual capacitances.

where $C_T$ = total capacitance
$C_1$, $C_2$, $C_3$, etc. = capacitances of the individual capacitors

All capacitances must be measured in the same units.

**Working voltage.** There is a limit to the voltage that may be applied across any capacitor. If too large a voltage is applied, it will overcome the resistance of the dielectric and a current will be forced through it from one plate to the other, sometimes burning a hole in the dielectric. In this event, a short circuit exists and the capacitor must be discarded. This applies only to mica or waxed-paper capacitors. If the dielectric is air, the "short" disappears as soon as the voltage is removed. The maximum voltage that may be applied to a capacitor is known as the *working voltage* and must never be exceeded.

EXAMPLE 17-1. A 0.000 35-μF tuning capacitor is in parallel with a trimmer capacitor of 0.000 075 μF. What is the total capacity?

SOLUTION. The diagram for the circuit is shown in Fig. 17-6.

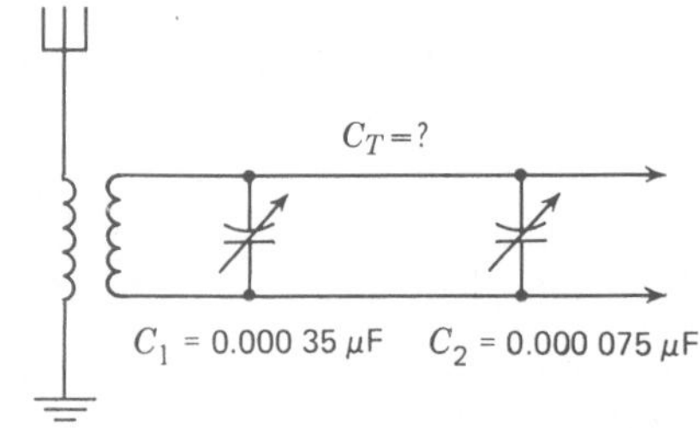


**FIGURE 17-6**

$$C_T = C_1 + C_2 \tag{17-2}$$

$$C_T = 0.000\ 35 + 0.000\ 075$$

$$C_T = 0.000\ 425\ \mu\text{F} \qquad \textit{Ans.}$$

*Note:* Remember to keep the decimal points in line when adding.

EXAMPLE 17-2. What are the total capacitance and working voltage of a 0.0005-$\mu$F 50-V capacitor, a 0.015-$\mu$F 100-V capacitor, and a 0.000 25-$\mu$F 100-V capacitor when they are connected in parallel?

SOLUTION. The total capacitance is the sum of the capacities:

$$\begin{array}{r} 0.000\ 5 \\ 0.015 \\ \underline{0.000\ 25} \\ C_T = 0.015\ 75\ \mu\text{F} \end{array} \qquad \textit{Ans.}$$

Just as a chain is only as strong as its weakest link, the working voltage of a group of parallel capacitors is only as great as the *smallest* working voltage. Therefore, the working voltage of the combination is only 50 V.

## PROBLEMS

1. A capacitor in a tuning circuit has a capacitance of 0.000 32 $\mu$F. When the stage is aligned, the trimmer capacitor in parallel with it is adjusted to a capacitance of 0.000 053 $\mu$F. What is the total capacitance of the combination?
2. A mechanic has the following capacitors available: 0.0003-$\mu$F 75-V; 0.000 25-$\mu$F 50-V; 0.0002-$\mu$F 50-V; 0.000 15-$\mu$F 75-V; and 0.000 05-$\mu$F 75-V. Which of these should be arranged in parallel to form a combination with a capacitance of 0.0005 $\mu$F and 75 V working voltage?
3. What is the total capacitance in parallel of the following capacitors: 35 pF, 0.005 $\mu$F, and 0.000 03 $\mu$F?
4. A capacitor of 0.0003-$\mu$F capacitance is connected in parallel with a 0.000 005-F capacitor. What is the total capacitance?
5. What is the total capacitance of a 25-pF, 0.000 06-$\mu$F, and a 0.000 000 03-F capacitor?
6. A capacitor of 0.003 $\mu$F is in parallel with another which holds a charge of 0.000 004 C at 200 V. What is the total capacitance?
7. What amount of capacitance must be added in parallel with a 0.000 55-$\mu$F capacitor in order to get a total capacitance of 0.0007 $\mu$F?
8. What is the total capacitance in parallel of two capacitors if they are charged at 120 V with 0.000 06 and 0.000 172 8 C, respectively?

## JOB 17-3 CAPACITORS IN SERIES

It has been found that the thicker the dielectric, or the greater the distance between the plates, the smaller the capacitance. If we were to combine capacitances so that the effective distance between the plates were increased, then the resulting capacitance would necessarily be less than before.

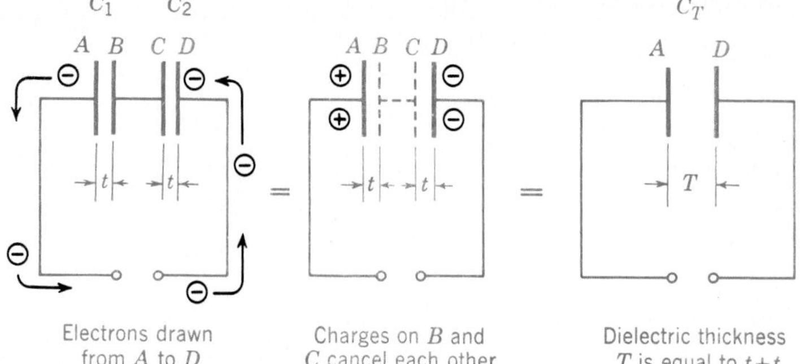


FIGURE 17-7
Capacitors in series increase the total dielectric thickness and decrease the total capacitance.

**What is the effect of placing capacitors in series?** In Fig. 17-7, capacitors $C_1$ and $C_2$ are connected in series. When a voltage is impressed on this combination, electrons are drawn from plate $A$ and deposited on plate $D$. The charges on plates $B$ and $C$ are equal and opposite and therefore neutralize each other. In this event, they may be considered to be eliminated and the combination replaced by a single capacitor whose dielectric thickness is equal to the sum of the dielectric thicknesses of the original capacitors. Thus, the effect of a series combination is to increase the dielectric thickness and therefore to decrease the capacitance.

**FORMULA**

$$\frac{1}{C_T} = \frac{1}{C_1} + \frac{1}{C_2} + \frac{1}{C_3} + \text{etc.} \tag{17-3}$$

All capacitances must be measured in the same units.

Notice the similarity between this formula and the formula for resistances in parallel given in Job 5-8 as formula (5-3). The similarity is continued through the formula for the total resistance in parallel of a number of equal resistors given in the same job as formula (5-4). For equal capacitors in *series*, we have

**FORMULA**

$$C_T = \frac{C}{N} \tag{17-4}$$

where $C_T$ = total capacitance
$C$ = capacitance of one of the equal capacitors
$N$ = number of equal capacitors

**Working voltage.** Within limits, the total voltage that may be applied across a group of capacitors in series is equal to the sum of the working voltages of the individual capacitors.

EXAMPLE 17-3. A 4-, a 5-, and a 10-$\mu$F capacitor are connected in series. Find the total capacitance.

SOLUTION. Review Job 5-4, Addition of Fractions.

$$\frac{1}{C_T} = \frac{1}{C_1} + \frac{1}{C_2} + \frac{1}{C_3} \qquad (17\text{-}3)$$

$$\frac{1}{C_T} = \frac{1}{4} + \frac{1}{5} + \frac{1}{10}$$

$$\frac{1}{C_T} = \frac{11}{20}$$

$$11 \times C_T = 20$$

$$C_T = \frac{20}{11} = 1.818\ \mu\text{F} \qquad \textit{Ans.}$$

EXAMPLE 17-4. A voltage-doubler circuit is shown in Fig. 17-8. Only $C_1$ is charged during the first half of the cycle. When the polarity reverses during the second part of the cycle, $C_2$ is charged to *twice* the peak voltage of the line because it is in series aiding with the line and $C_1$. What are the total capacitance and working voltage of the capacitor combination if $C_1$ and $C_2$ are both 100-μF 200-V capacitors?

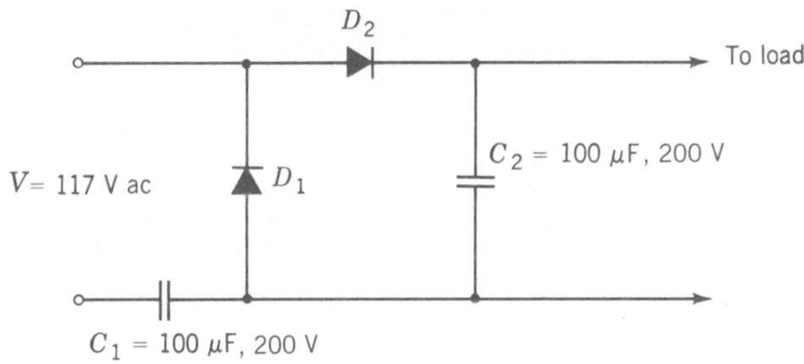


**FIGURE 17-8**
Voltage-doubler circuit.

SOLUTION

$$C_T = \frac{C}{N} = \frac{100}{2} = 50\ \mu\text{F} \qquad \textit{Ans.}$$

$$\text{Working voltage} = 200 + 200 = 400\ \text{V} \qquad \textit{Ans.}$$

EXAMPLE 17-5. The frequency which beats against the incoming frequency in the superheterodyne receiver is produced by an oscillator circuit. The Colpitts oscillator shown in Fig. 17-9 controls the frequency of oscillation by means of the variable inductance. What is the total capacity of the series combination of $C_1$ and $C_2$ if $C_1 = 0.0005\ \mu\text{F}$ and $C_2 = 0.000\ 25\ \mu\text{F}$?

SOLUTION. When only two capacitors are in series, we can use a formula very similar to the formula for finding the total resistance of two resistors in parallel.

**FORMULA**

$$C_T = \frac{C_1 \times C_2}{C_1 + C_2} \qquad \boxed{17\text{-}5}$$

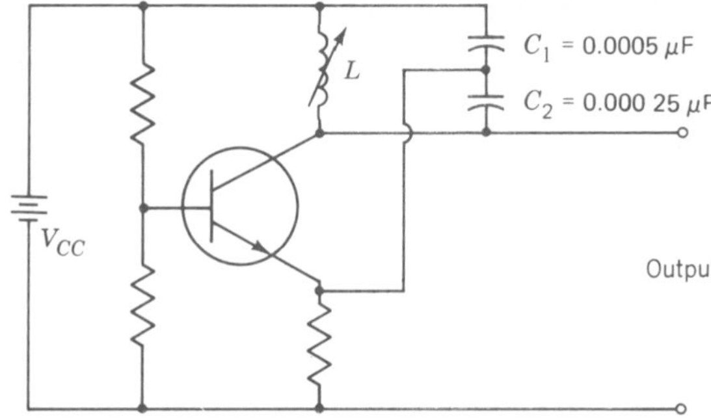


**FIGURE 17-9**
Colpitts oscillator.

In this formula, all the measurements must be in the same units. Since it is easier to use whole numbers than decimals in the formula, all measurements are changed to picofarads.

$$C_1 = 0.0005\ \mu\text{F} = 0.0005 \times 10^6 = 500\ \text{pF}$$
$$C_2 = 0.000\ 25\ \mu\text{F} = 0.000\ 25 \times 10^6 = 250\ \text{pF}$$
$$C_T = \frac{500 \times 250}{500 + 250} = \frac{125{,}000}{750} = 167\ \text{pF} \qquad \textit{Ans.}$$

**PROBLEMS**

1. Capacitors of 3, 4, and 6 μF are connected in series. What is their total capacitance?
2. What are the total capacitance and working voltage of a voltage doubler similar to that shown in Fig. 17-8 if the circuit uses two 40-μF 175-V capacitors?
3. Find the total capacitance of the series capacitors in a Colpitts oscillator similar to that shown in Fig. 17-9 if $C_1 = 0.0004$ μF and $C_2 = 0.000\ 02$ μF.
4. Find the total capacitance of the series capacitors in a Colpitts oscillator if (*a*) $C_1 = 0.01$ μF and $C_2 = 0.001$ μF, (*b*) $C_1 = 0.0015$ μF and $C_2 = 800$ pF.
5. What is the range of total capacitances available in an oscillator circuit which uses a tuning capacitor with a 35- to 350-pF range in series with a padder capacitor set at 300 pF?
6. In some vibrator power supplies operating from a storage battery, a pair of "buffer" capacitors are placed across the secondary of the transformer to reduce the voltage peaks. What are the total capacitance and working voltage of a pair of 0.0075-μF 800-V buffer capacitors in series?
7. Find the total capacitance of a 6-, a 10-, and a 15-μF capacitor in series.
8. What is the total capacitance of a 0.000 000 02-F, a 0.04-μF, and a 60,000-pF capacitor in series?
9. A series combination of a 4- and a 12-μF capacitor is connected in parallel with a 5-μF capacitor. Find the total capacitance of the combination.

10. A 30-pF capacitor is connected in series with a parallel combination of a 30- and a 60-pF capacitor. Find the total capacitance of the combination.

(See Instructor's Manual for Test 17-1)

## JOB 17-4 REACTANCE OF A CAPACITOR

We learned in Job 17-1 that as a capacitor is charged, electrons are drawn from one plate and deposited on the other. As more and more electrons accumulate on the second plate, they begin to act as an opposing voltage which attempts to stop the flow of electrons just as a resistor would do. This opposing effect is called the *reactance* of the capacitor and is measured in ohms.

### WHAT FACTORS DETERMINE THE REACTANCE?

1. The size of the capacitor is one factor. The larger the capacitor, the greater the number of electrons that may be accumulated on its plates. However, because the plate area is large, the electrons do not accumulate in one spot but spread out over the entire area of the plate and do not impede the flow of new electrons onto the plate. Therefore, a large capacitor offers a small reactance. If the capacitance were small, as in a capacitor with a small plate area, the electrons could not spread out and would attempt to stop the flow of electrons coming to the plate. Therefore, a small capacitor offers a large reactance. The reactance is therefore *inversely* proportional to the capacitance.

2. If an ac voltage is impressed across the capacitor, electrons are accumulated first on one plate and then on the other. If the frequency of the changes in polarity is low, the time available to accumulate electrons will be large. This means that a large number of electrons will be able to accumulate, which will result in a large opposing effect, or a large reactance. If the frequency is high, the time available to accumulate electrons will be small. This means that there will be only a few electrons on the plates, which will result in only a small opposing effect, or a small reactance. The reactance is therefore *inversely* proportional to the frequency.

3. A special constant of proportionality is necessary to change the current-reducing effect of the electron accumulation into ohms of reactance. This number is $2\pi$. The formula for capacitive reactance is

$$X_C = \frac{1}{2\pi f C}$$

with $C$ measured in farads. If the capacitance is measured in microfarads,

$$X_C = \frac{1}{2 \times 3.14 \times f \times (C/1{,}000{,}000)} = \frac{1{,}000{,}000}{6.28 \times f \times C}$$

and since

$$\frac{1{,}000{,}000}{6.28} = 159{,}000 \text{ (approx)}$$

**FORMULA**

$$X_C = \frac{159{,}000}{f \times C} \qquad \boxed{17\text{-}6}$$

where $X_C$ = capacitive reactance, Ω
$f$ = frequency, Hz
$C$ = capacitance, μF

*Note:* The larger the capacitance, the smaller the reactance. The larger the frequency, the smaller the reactance.

EXAMPLE 17-6. The antenna circuit of a two-transistor AM receiver is shown in Fig. 17-10. Find the capacitive reactance of the tuning capacitor to a frequency of 500 kHz when it is set at 300 pF.

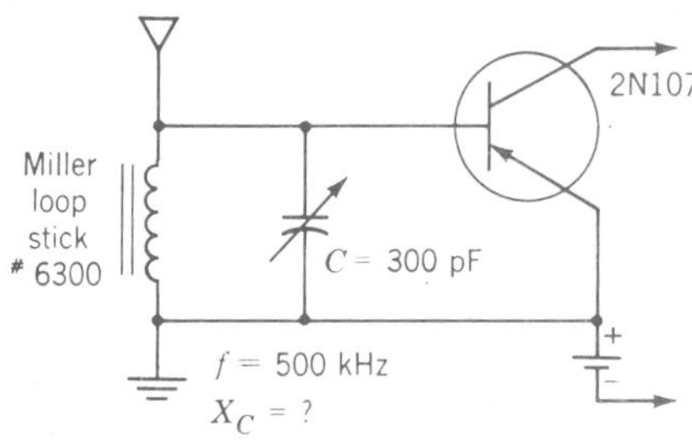


**FIGURE 17-10**
Find the capacitive reactance of the tuning capacitor.

SOLUTION

1. Change the measurements to the required units.

$$f = 500 \text{ kHz} = 500 \times 10^3 \text{ Hz}$$
$$C = 300 \text{ pF} = 300 \times 10^{-6} \ \mu\text{F}$$

2. Find the capacitive reactance.

$$X_C = \frac{159{,}000}{f \times C} \qquad (17\text{-}6)$$

$$X_C = \frac{159 \times \cancel{10^3}}{500 \times \cancel{10^3} \times 300 \times 10^{-6}}$$

$$X_C = \frac{159}{15 \times 10^4 \times 10^{-6}}$$

$$X_C = 10.6 \times 10^2 = 1060 \ \Omega \qquad \textit{Ans.}$$

EXAMPLE 17-7. A capacitor is formed whenever two metal pieces are separated by a dielectric. A capacitor formed by two wires of a circuit or two turns of a coil produces a *distributed capacitance.* This is extremely undesirable, because even a very small distributed capacitance can transfer energy from one circuit to another at radio frequencies, since the reactance at radio frequencies is very small. This unwanted transfer of energy represents wasted power. What is the current lost through a distributed capacitance of 100 pF formed by two parallel wires, one of which carries a 1000-kHz current? The difference in potential between the wires is 1 V.

SOLUTION. The diagram of the circuit is shown in Fig. 17-11.

1. Change the measurements to the required units.

$$1000\ \text{kHz} = 10^3 \times 10^3 = 10^6\ \text{Hz}$$
$$10\ \text{pF} = 10 \times 10^{-6} = 10^{-5}\ \mu\text{F}$$

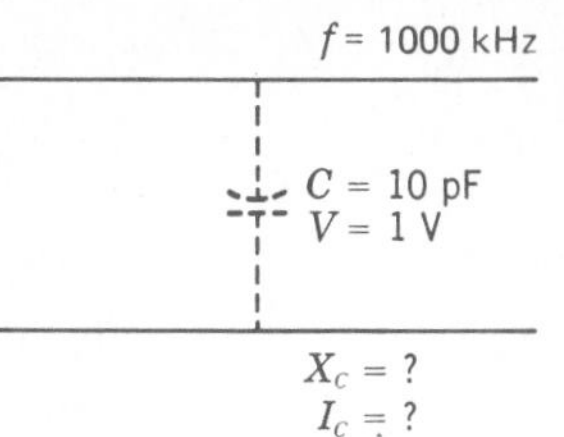


FIGURE 17-11

2. Find the capacitive reactance.

$$X_C = \frac{159{,}000}{f \times C} = \frac{159{,}000}{10^6 \times 10^{-5}} = \frac{159{,}000}{10} = 15{,}900\ \Omega$$

3. Find the current. Since the only opposition to the flow of current around a capacitor is its reactance, the formula for Ohm's law, $V = IR$, may be rewritten as

$$V_C = I_C \times X_C \qquad \boxed{17\text{-}7}$$
$$1 = I_C \times 15{,}900$$
$$I_C = \frac{1}{15{,}900} = 0.000\ 062\ \text{A} = 0.062\ \text{mA} \qquad \textit{Ans.}$$

EXAMPLE 17-8. Find the size of filter capacitor necessary to provide a capacitive reactance of 159 Ω at a frequency of 60 Hz.

SOLUTION. The diagram for the circuit is shown in Fig. 17-12.

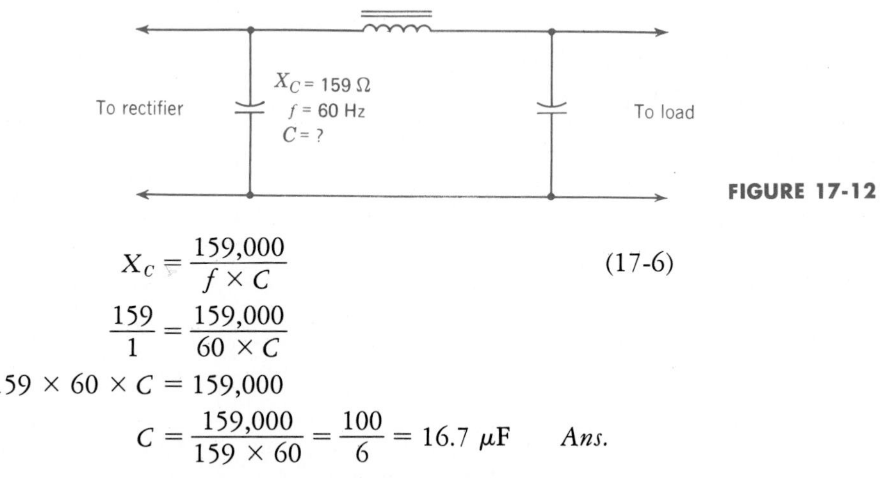


FIGURE 17-12

$$X_C = \frac{159{,}000}{f \times C} \qquad (17\text{-}6)$$
$$\frac{159}{1} = \frac{159{,}000}{60 \times C}$$
$$159 \times 60 \times C = 159{,}000$$
$$C = \frac{159{,}000}{159 \times 60} = \frac{100}{6} = 16.7\ \mu\text{F} \qquad \textit{Ans.}$$

Use the commercially available 16-μF 500-V capacitor.

SELF-TEST 17-9. A capacitance of 30 pF draws 20 mA when connected across a 106-V source. Find the frequency of the ac voltage.

SOLUTION. Given: $C = 30$ pF Find: $f = ?$

$I =$ ______ | 20 mA

$V = 106$ V

1. Change the measurements to the required units.

30 pF = 30 × ______ μF | $10^{-6}$

$$20 \text{ mA} = 20 \times 10^{-3} = \_\_\_\_\_\_ \text{ A}$$ 0.02

2. Find the capacitive reactance.

$$V_C = I_C \times X_C \qquad (17\text{-}7)$$
$$106 = 0.02 \times X_C$$
$$X_C = \frac{106}{0.02} = \_\_\_\_\_\_ \ \Omega$$ 5300

3. Find the frequency.

$$X_C = \frac{159{,}000}{f \times C} \qquad (17\text{-}6)$$
$$5300 = \frac{159 \times 10^3}{f \times 30 \times 10^{-6}}$$
$$f = \frac{159 \times 10^3}{30 \times 10^{-6} \times 53 \times 10^2}$$
$$f = \frac{159 \times 10^3}{159 \times ?}$$ $10^{-3}$
$$f = 1 \times 10^3 \times \_\_\_\_\_\_ = 1 \times 10^?$$ $10^3$ 6
$$f = 1 \ \_\_\_\_\_\_ \quad \textit{Ans.}$$ MHz

## PROBLEMS

1. What is the reactance of a 0.0003-$\mu$F capacitor at (*a*) 30 kHz, (*b*) 100 kHz, and (*c*) 800 kHz?
2. What is the reactance of an oscillator capacitor of 0.0004 $\mu$F to a frequency of 456 kHz?
3. A 10-$\mu$F capacitor in the emitter circuit of a 2N322 transistor used in a 12-V audio amplifier produces a voltage drop of 3 V at 1 kHz. Find the current passed by the capacitor.
4. What is the reactance of a 20-$\mu$F coupling capacitor to an audio frequency of 1 kHz? What current will flow if the voltage across the capacitor is 2.38 V as shown in Fig. 17-13?

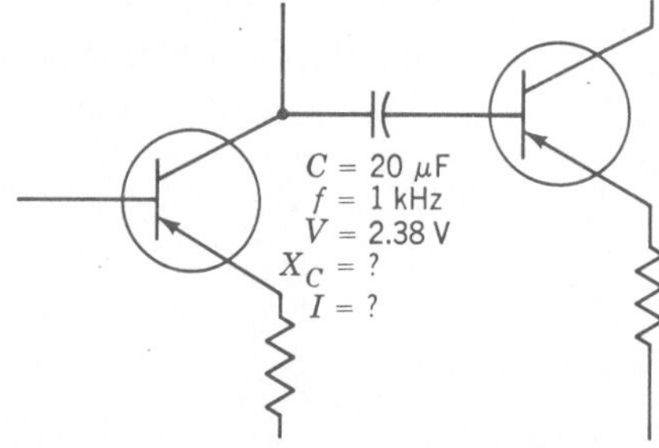


**FIGURE 17-13**

5. A 0.000 35-$\mu$F tuning capacitor is in parallel with a 0.000 05-$\mu$F trimmer capacitor. Find the total capacitance of the combination and its reactance to a frequency of 100 kHz.
6. A capacitor draws 4 A when connected across a 120-V 60-Hz line.

What will be the current drawn if both the capacitance and frequency are doubled?

7. What should be the capacitance of a thyratron control circuit if the reactance must be 1590 Ω at 60 Hz?
8. A capacitor in a telephone circuit has a capacitance of 2 μF. What current flows through it when an emf of 15 V at 795 Hz is impressed across it?
9. A pocket radio uses a 10-μF coupling capacitor in the base circuit of a 2N35 NPN transistor. Find its reactance to a frequency of (*a*) 1 kHz, (*b*) 5 kHz, and (*c*) 20 kHz.
10. The emitter resistor for a 2N1265 transistor is bypassed with a 5-μF capacitor. What is the reactance of this capacitor to a 1.5-kHz frequency? If the voltage across the capacitor is 10.6 V, what current will flow?
11. Find the capacitive reactance between two wires if the stray capacitance between them is 8 pF and one wire carries a radio frequency of 1500 kHz.
12. A 2N218 transistor acting as an FM sound detector has its emitter resistor bypassed with a 4-μF capacitor. What is the reactance of the capacitor to the 500-Hz AF current?
13. A capacitance of 1.1 μF draws 0.05 A when connected across a 120-V line. Find the frequency of the ac voltage.
14. An antenna tuning capacitor has a capacitance of 250 pF. What current flows through it when 41.8 V are impressed across it at 7.6 MHz?
15. A 2N190 PNP transistor used as an audio amplifier has its emitter stabilizing resistor bypassed with a 50-μF capacitor. Find the reactance of the capacitor to a frequency of (*a*) 100 Hz, and (*b*) 5 kHz.
16. In the Admiral transistorized television receiver model NA1-2B, the first sound IF transistor 2SC460 has its emitter resistor bypassed with a 0.01-μF capacitor. What is its reactance to a frequency of 21.25 MHz?
17. Find the bypass capacitor required for a 2N109 audio output transistor if it is to have a reactance of 795 Ω at 10 kHz.
18. A "leading" current of 5 A is to be obtained from a 220-V 60-Hz line by means of a capacitor. Find the required capacitance.
19. A static capacitor capable of passing 40 A at 240 V and 60 Hz is added to a line to correct the power factor. What must be its capacitance?

## JOB 17-5 IMPEDANCE OF A CAPACITOR

In the last job, the opposition to the flow of current offered by a capacitor was considered to be its reactance. Actually, however, it is impossible to obtain a circuit which contains only reactance. The plates of the capacitor and its connecting leads all have some resistance. The *impedance* is the total opposition to the flow of current and is equal to the *phasor (vector) sum* of the resistance and the reactance. The symbol for the impedance is $Z$. The complete explanation of impedance and the derivation of the formula are given in Job 18-4.

**FORMULA** $$Z = \sqrt{R^2 + X_C^2} \qquad \boxed{17\text{-}8}$$

where $Z$ = impedance, Ω
$R$ = resistance, Ω
$X_C$ = capacitive reactance, Ω

EXAMPLE 17-10. Find the impedance offered by a 10-μF filter capacitor to a 60-Hz frequency if its resistance is 200 Ω.

SOLUTION. Given: $C = 10$ μF Find: $Z = ?$
$f = 60$ Hz
$R = 200$ Ω

1. Find the capacitive reactance.

$$X_C = \frac{159{,}000}{f \times C} = \frac{159{,}000}{60 \times 10} = \frac{1590}{6} = 265\ \Omega \quad \textit{Ans.} \qquad (17\text{-}6)$$

2. Find the impedance.

$$Z = \sqrt{R^2 + X_C^2} = \sqrt{200^2 + 265^2} = \sqrt{40{,}000 + 70{,}225}$$
$$= \sqrt{110{,}225} = 332\ \Omega \quad \textit{Ans.}$$

## PROBLEMS

1. Find the impedance of a capacitor if its reactance is 40 Ω and its resistance is 9 Ω.
2. Find the impedance of a capacitor if its reactance is 24 Ω and its resistance is 7 Ω.
3. Using Fig. 17-14, find the impedance of a coupling-capacitor circuit to an audio frequency of 1 kHz if the capacitance is 0.01 μF and the resistance of the circuit is 3 kΩ.

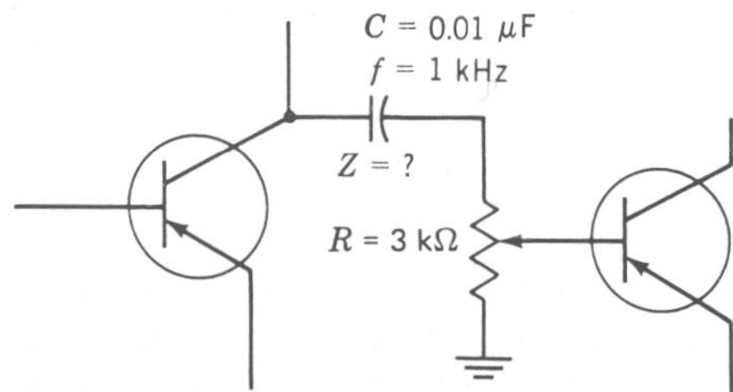


FIGURE 17-14

4. Find the impedance of the tone-control circuit shown in Fig. 17-15.
5. What is the total impedance of the circuit in Fig. 17-15 if $f = 5$ kHz, $C = 0.01\ \mu F$, and $R = 10{,}000\ \Omega$?

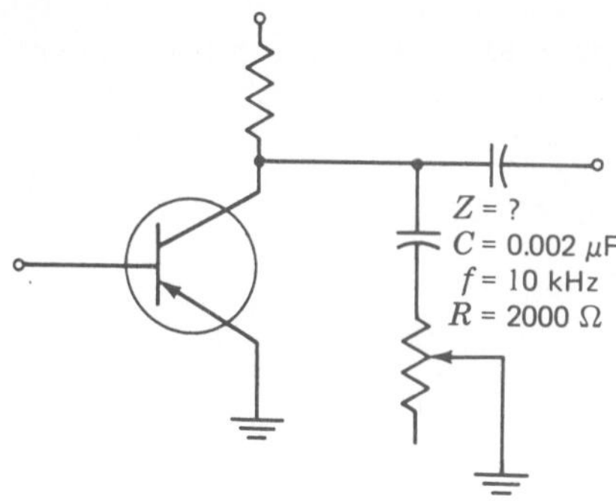


FIGURE 17-15

6. A 3000-Ω resistor is in series with a 0.02-μF capacitance. Find the impedance at (*a*) 500 kHz, (*b*) 5 kHz, and (*c*) 500 Hz.

(See Instructor's Manual for Test 17-2, Impedance)

## JOB 17-6 MEASUREMENT OF CAPACITY

**Voltmeter-ammeter method.** The circuit used to measure the capacitance of a capacitor is shown in Fig. 17-16. The voltmeter measures the voltage across the capacitor, and the ammeter measures the current in the circuit. Since the resistance of the capacitor is so very small when compared with its reactance, we can neglect it in this situation.

EXAMPLE 17-11. Find the capacitance of the capacitor shown in Fig. 17-16.

SOLUTION

1. Find the capacitive reactance.

$$V_C = I_C \times X_C \qquad (17\text{-}7)$$

$$110 = 0.02 \times X_C$$

$$X_C = \frac{110}{0.02} = 5500\ \Omega \qquad \textit{Ans.}$$

2. Find the capacitance.

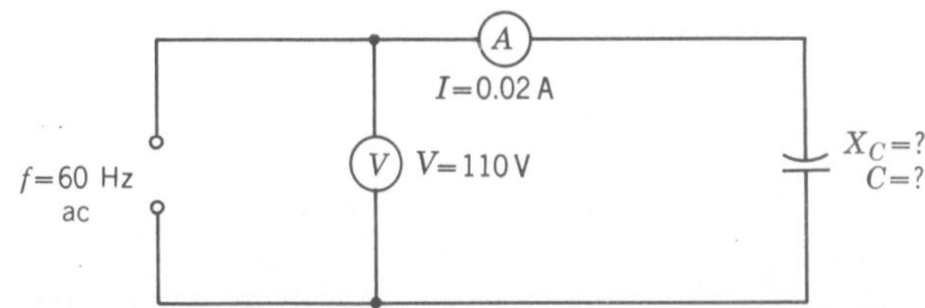


FIGURE 17-16
Circuit for measuring capacitance by the voltmeter-ammeter method.

$$X_C = \frac{159{,}000}{f \times C} \qquad (17\text{-}6)$$

$$\frac{5500}{1} = \frac{159{,}000}{60 \times C}$$

$$5500 \times 60 \times C = 159{,}000$$

$$C = \frac{159{,}000}{330{,}000} = 0.482\ \mu\text{F} \qquad \textit{Ans.}$$

## PROBLEMS

Find the capacitance of each capacitor if the measurements obtained by the voltmeter-ammeter method are given below.

| PROBLEM | $V$ | $I$ | $f$ |
|---|---|---|---|
| 1 | 110 V | 0.11 A | 60 Hz |
| 2 | 110 V | 0.5 A | 60 Hz |
| 3 | 18 V | 20 mA | 60 Hz |
| 4 | 50 V | 10 mA | 25 Hz |
| 5 | 318 mV | 20 mA | 1 kHz |

## JOB 17-7 REVIEW OF CAPACITANCE

A capacitor is made of two metallic plates separated by an ________ material, or dielectric. | insulating

A capacitor is used to store up an electric ________. | charge

The capacitance of a capacitor is a measure of its ability to store up ________. | electrons

The capacitance (increases/decreases) if the *plate area increases* or the *dielectric thickness* ________. | increases | decreases

The working voltage is the ________ voltage that may be placed across a capacitor before it breaks down. The working voltage in parallel is the working voltage of the (strongest/weakest) capacitor. The working voltage in series is the ________ of the working voltages of the series capacitors. | largest | weakest | sum

A *farad* is the capacitance of a capacitor which can hold 1 coulomb of electricity on its plates under a pressure of 1 ________. | V

The *reactance* of a capacitor is the opposition offered by the capacitor to the passage of an ________ current. The reactance decreases as the frequency or capacitance (increases/decreases) | alternating | increases

Distributed, or "stray," capacitance is the capacity formed when two wires run close together or when any two metal parts are separated by a thin ________ material. | insulating

### FORMULAS

*Capacitance:* $$C = \frac{Q}{V} \qquad \boxed{17\text{-}1}$$

where $C$ = capacitance, F
$Q$ = charge, coulombs
$V$ = voltage, V

*Capacitors in parallel:* $C_T = C_1 + C_2 + C_3$ **17-2**

where $C_T$ = total capacitance
$C_1, C_2, C_3$ = individual capacitances, all measured in the same units

*Capacitors in series:* $\frac{1}{C_T} = \frac{1}{C_1} + \frac{1}{C_2} + \frac{1}{C_3}$ **17-3**

where $C_T$ = total capacitance
$C_1, C_2, C_3$ = individual capacitances, all measured in the same units

*Equal capacitors in series:* $C_T = \frac{C}{N}$ **17-4**

where $C_T$ = total capacitance
$C$ = capacitance of one of the equal capacitors
$N$ = number of equal capacitors

*Two capacitors in series:* $C_T = \frac{C_1 \times C_2}{C_1 + C_2}$ **17-5**

where $C_T$ = total capacitance
$C_1, C_2$ = individual capacitances, all measured in the same units

*Reactance of a capacitor:* $X_C = \frac{159{,}000}{f \times C}$ **17-6**

where $X_C$ = reactance, Ω
$f$ = frequency, Hz
$C$ = capacitance, μF

*Impedance of a capacitor:* $Z = \sqrt{R^2 + X_C^2}$ **17-8**

where $Z$ = impedance, Ω
$R$ = resistance, Ω
$X_C$ = reactance, Ω

## PROBLEMS

1. Find the total capacitance in parallel of a 0.0035-μF, a 0.000 000 04-F, and a 6200-pF capacitor.
2. What amount of capacitance must be added in parallel to a 0.000 35-μF capacitor to obtain a total capacitance of 0.001 15 μF?
3. Find the total capacitance in series of a 3-, a 6-, and an 8-μF capacitor.
4. What are the total capacitance and working voltage of two 25-μF 180-V capacitors used in a voltage multiplier?

5. What is the total capacitance of the capacitors in a Colpitts oscillator similar to Fig. 17-9 if $C_1 = 0.0004\ \mu F$ and $C_2 = 0.000\ 35\ \mu F$?
6. An oscillator circuit contains a 0.000 35- and a 0.000 25-$\mu$F capacitor in series. Find the total capacitance.

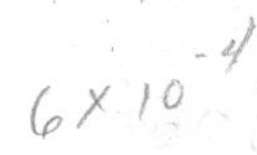

7. Part of the first stage of a hi-fi preamplifier is shown in Fig. 17-17. Find the capacitive reactance of $C_E$ at (*a*) 30 Hz, and (*b*) 15,000 Hz.

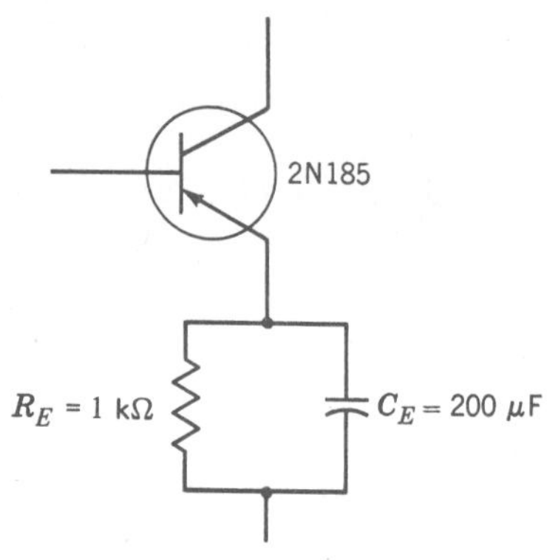


**FIGURE 17-17**

8. Find the reactance of a 0.0003-$\mu$F tuning capacitor to a frequency of 1350 kHz.
9. Find the impedance of a capacitor if its reactance is 40 Ω and its resistance is 20 Ω.
10. Find the impedance of a capacitive circuit to a 2-kHz audio frequency if the resistance is 2000 Ω and the capacitance is 0.02 $\mu$F.
11. What cathode bypass capacitor is needed to provide a reactance of 1000 Ω at an audio frequency of 500 Hz?
12. In finding the capacitance of a capacitor by the voltmeter-ammeter method, the current was 0.004 A and the voltage was 110 V at 60 Hz. Find the capacitance.

(See Instructor's Manual for Test 17-3, Capacitance)

# SERIES AC CIRCUITS

# 18

## JOB 18-1 SIMPLE SERIES AC CIRCUITS

Within certain limits (to be explained in Job 18-3), the general rules for solving dc series circuits are also applicable to the solution of ac series circuits.

1. The current in each part is equal to the current in every other part and equal to the total current.
2. The total voltage is equal to the sum of the voltages across all parts of the circuit.

**Series circuits containing only resistance.** As we have learned, an ac voltage consists of a number of different instantaneous voltages, each instant of time giving rise to a different value of voltage. If this voltage is impressed across a resistor, each instantaneous voltage will cause an instantaneous current to flow at that time. The increasing and decreasing voltages and currents are shown in Fig. 18-1. The current that flows as a result of an ac voltage will be an alternating current with the same frequency as the ac voltage. The voltage and the current both start at zero and rise to their maximum values, reaching them at the same instant. The voltage and current continue to rise and fall in step with each other

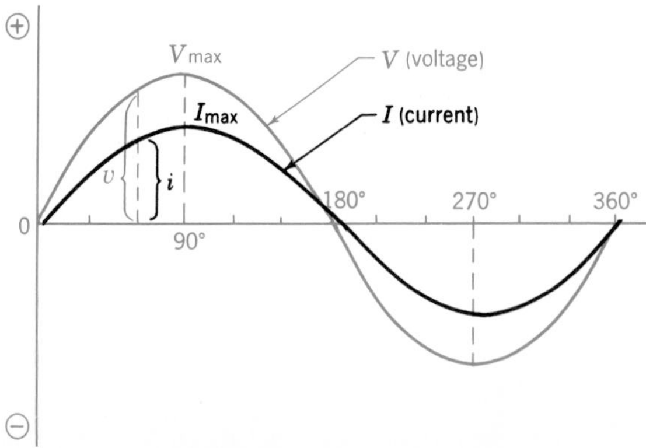


**FIGURE 18-1**
The voltage and current are in phase in a purely resistive circuit.

throughout the entire cycle. We say that the voltage and current are "in phase" or "in step" with each other. It is difficult to draw these curves whenever we wish to indicate this or any other condition, and so we shall use a method in which the voltage and current are indicated by straight lines which are drawn to a definite length and in a definite direction.

**Vectors.** A *scalar quantity* is one which has size and is independent of direction. Examples of scalar quantities are length and time.

A *vector quantity* is one which has size *and* direction. Examples of vector quantities are force and acceleration. Vectors are used whenever we wish to show the amount and *direction* of a quantity. For example, *both* the amount and direction must be indicated to describe adequately a 10-lb force which acts straight up. If 1 in represents 10 lb, then Fig. 18-2*a* describes this force. However, if the force were acting straight *down,* it would be described as shown in Fig. 18-2*b*. We shall use vectors to show the amount and *time* at which a voltage or current is acting. Vectors which are drawn in the *same direction* will indicate that they are happening at the *same time,* or are "in phase." Vectors which are drawn in *different directions* will indicate that they are happening at *different times* or are "out of phase." In electricity, since different directions really represent *time* expressed as a phase relationship, an electrical vector is called a *phasor.*

In a circuit containing only resistance, we have seen that the voltage and current occur at the *same time,* or are in phase. To indicate this condition by means of phasors, all that is necessary is to draw the phasors for the voltage and the current in the *same direction.* The value of each is indicated by the *length* of the phasor.

Since the current is constant in all the parts of a series circuit, the current vector is usually drawn on a horizontal line and is used as a reference line for the other vectors (phasors) in the same diagram. Draw a line *AC* from left to right as in Fig. 18-3 to represent the current phasor. Place an arrowhead on the phasor at *C* pointing to the right. Point *A* is the "tail" of the phasor, and point *C* is the "head" of the phasor. Since the voltage occurs at the *same time* as the current, the voltage phasor must be drawn in the *same direction* as the current phasor. Draw a line starting from the original point *A* to the right to point *V* with an arrow pointing to the right. This voltage phasor is drawn larger than the current phasor because the voltage is larger than the current.

**FIGURE 18-2**
A vector indicates both amount and direction.

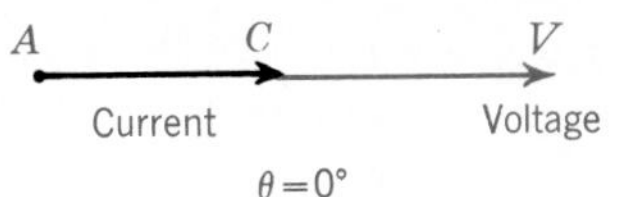


**FIGURE 18-3**
The voltage and current phasors are "in phase" in a circuit containing only resistance.

When phasors are drawn in the same direction, the angle between the phasors is 0°. This angle is called the phase angle and is denoted by the Greek letter theta ($\theta$).

EXAMPLE 18-1. A 22-Ω electric iron is operated from a 110-V 60-Hz line. Draw a phasor diagram, and find the current and power used by the iron.

SOLUTION. Given: $V = 110$ V, $R = 22\ \Omega$ Find: $I = ?$, $P = ?$

1. Draw the phasor diagram. Since the iron may be assumed to be made of a purely resistive element, the phasor diagram will be the same as that shown in Fig. 18-3. It is not drawn to scale.
2. In a purely resistive circuit, Ohm's law may be used.

$$V = IR \tag{2-1}$$

$$110 = I \times 22$$

$$I = \frac{110}{22} = 5 \text{ A} \qquad \textit{Ans.}$$

3. Find the power.

$$P = I \times V \tag{7-1}$$

$$P = 5 \times 110 = 550 \text{ W} \qquad \textit{Ans.}$$

## PROBLEMS

Draw a phasor diagram for each problem (not to scale).

1. What is the hot resistance of a tungsten lamp if it draws 2 A from a 120-V ac line?
2. Find the current drawn by a 50-Ω toaster from a 120-V ac line.
3. What is the voltage needed to operate a 600-W neon sign whose resistance is 20 Ω?
4. An electric soldering iron draws 0.8 A from a 120-V 60-Hz line. What is its resistance? How much power will it consume?
5. What ac voltage is required to force 0.02 A through an 8000-Ω radio resistor? What is the power used?
6. Find the current and power drawn from a 110-V 60-Hz line by a tungsten lamp whose hot resistance is 275 Ω.
7. What is the current drawn by a 200-W incandescent lamp from a 110-V 60-Hz line? What is the hot resistance of the lamp?
8. Find the power used by a 24-Ω soldering iron which draws 5 A.

9. Find the voltage needed to operate a 500-W electric percolator if it draws 4.5 A. What is its resistance?

**Series circuits containing only inductance.** When an ac voltage is impressed across a coil, it will produce an alternating current. The changing current will produce changing lines of force around the turns of the coil. The changing lines of force will cut actoss the wires forming the coil and induce an emf in the coil. This emf is a *back emf* which acts to oppose the original voltage. This opposition, called the *inductive reactance,* will reduce the current below that which would flow if there were no "cuttings" or back emf. This reactance does more than just reduce the current. It also prevents the current from appearing at the same time as the voltage. The current will be pushed back in *time* as well as in amount. We say that the current "lags" behind the voltage which produces it. In a perfect coil—one which has only inductance and zero resistance—the current will lag behind the voltage by an amount of time equal to the time required for $^1/_4$ cycle. It is easier to discuss this "time lag" in terms of the number of electrical degrees for $^1/_4$ cycle than in units of time. We say that the current lags the voltage by 90°, since $^1/_4$ cycle equals $^1/_4 \times 360° = 90°$.

The current that flows as a result of an ac voltage across a coil will be an alternating current with the same frequency as the ac voltage. The difference between this and the resistive circuit is that in the inductive circuit the current does *not* rise and fall in step with the voltage. The current remains forever 90 electrical degrees *behind* the voltage as shown in Fig. 18-4. The current lags behind the voltage by 90°, or the voltage "leads" the current by 90°. In an inductive circuit, the voltage and current are out of phase by 90°. The phase angle $\theta$ in an inductive circuit is 90°.

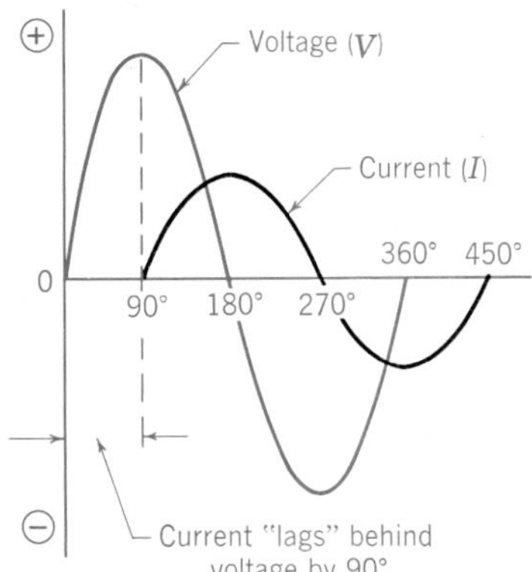


**FIGURE 18-4**
The voltage and current are *out of phase* in an ac circuit containing only inductance.

**Phasor diagram.** Since the voltage and current are out of phase by 90°, the phasors must be drawn in two different directions 90° apart. In addition to this, we must show which phasor is the "leading" phasor and which is the "lagging" phasor. To show quantities occurring "before" or "after" another, we shall use the numbers on a clock. Let us consider the hour hand of a clock as a phasor. A phasor pointing to 3 o'clock occurs after a phasor pointing to 12 o'clock, or noon. Similarly, a phasor pointing to 3 o'clock occurs before a phasor pointing to 6 o'clock.

We can now proceed to draw a phasor diagram for an inductive circuit. Since the current in a series circuit is constant, the current is used as the reference line upon which to draw the phasor diagram. Draw a line with an arrow pointing to the right as shown in Fig. 18-5. This will represent the current $I$ in the circuit. Since the voltage leads the current by 90°, we shall be forced to draw the voltage phasor in such a way so as to be 90° before the current phasor. If the current phasor already points to 3 o'clock, the voltage phasor must point to 12 o'clock in order to lead the current phasor by 90°.

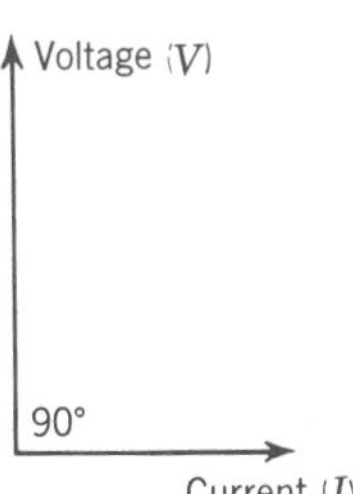


**FIGURE 18-5**
The voltage phasor leads the current phasor by 90° in a purely inductive circuit.

The amount of current that will flow in a "pure" inductance is found by Ohm's law. However, since a pure inductance contains zero resistance, the $R$ is replaced by $X_L$. Ohm's law for a pure inductance will then be

$$V_L = I_L \times X_L \qquad (16\text{-}2)$$

**Power.** In a purely resistive circuit, the voltage and current occur at the same time, or are in phase. The power is equal to the product of $V$ and $I$ and is expressed in *voltamperes* (VA) or kilovoltamperes (kVA). However, as we have just seen, the voltage and current do not always occur at the same time. In this event, only a portion of the current will occur at the same time as the voltage. In those ac circuits in which the voltage and current are *not* in phase, the useful or actual power is measured in watts and is equal to the voltage multiplied by only that portion of the current in phase with it. The amount of this in-phase current may be obtained from Fig. 18-6. The current in an ac circuit may be considered

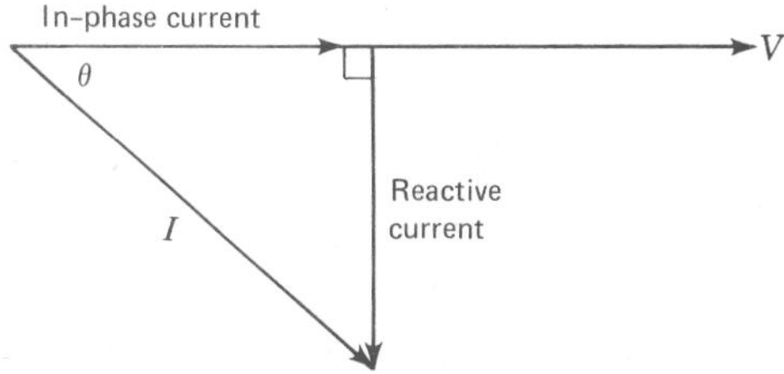


**FIGURE 18-6**
An alternating current is made up of two components.

to be made up of two components; a component in phase with the voltage and a component out of phase by 90°. By trigonometry, the in-phase component equals the current $I$ multiplied by the cosine of the phase angle.

$$\text{In-phase current} = I \times \cos\theta \qquad \boxed{18\text{-}1}$$

The general formula for power in an ac circuit is

$$W = V \times \text{in-phase current} \qquad \boxed{18\text{-}2}$$

By substitution, we obtain the

**FORMULA** $$W = V \times I \times \cos\theta \qquad \boxed{18\text{-}3}$$

where $W$ = effective or true power, W
$V$ = voltage, V
$I$ = current, A
$\theta$ = phase angle between voltage and current

The multiplier $\cos\theta$ is called the *power factor* (PF) of the circuit. $W$ is the true watts and $VI$ is the apparent watts, or voltamperes. Substituting PF for $\cos\theta$ in formula (18-3) gives

$$W = V \times I \times \text{PF} \qquad \boxed{18\text{-}4}$$

Since the PF is always less than 1 ($\cos 0° =$ a maximum of 1, decreasing to $\cos 90° = 0$), the true power will always be *less* than the apparent power $VI$.

Solving formula (18-4) for PF, we obtain

$$\text{PF} = \frac{W}{V \times I} \qquad \boxed{18\text{-}5}$$

or $$\text{PF} = \frac{\text{true power}}{\text{apparent power}} \qquad \boxed{18\text{-}6}$$

The power factor may be expressed as a decimal or as a percent. For example, a PF of 0.8 may be written as 80 percent. In this sense, the PF describes the *portion* of the voltampere input which is actually effective in operating the device. Thus, an 80 percent PF means that the device uses only 80 percent of the voltampere input in order to operate. The higher the PF, the greater is the efficiency of the device in using the voltampere input.

Low-power-factor, inefficient circuits occur when the phase angle between the current and voltage is large. This angle increases as the total reactance in the circuit increases. Practically pure resistive circuits (incandescent lamps) have high power factors of 95 to 100 percent. The power factors of circuits containing large induction motors (inductive reactance) range between 85 and 90 percent. Fractional-horsepower induction motors have power factors between 60 and 75 percent. Low power factors just mean wasted power, and should be increased whenever possible. We can increase a low PF caused by a lagging inductive reactance by adding an *opposite, leading* capacitive reactance, and vice versa. This will be discussed in greater detail in Chap. 21.

In a pure inductance, since the voltage leads the current by 90°, the power will be

$$W = V \times I \times \cos 90°$$
$$W = V \times I \times 0$$
$$W = 0 \text{ W}$$

Thus, the average power used by a pure inductance is zero. Actually, the inductance uses power to build up its magnetic field during one quarter of a cycle, but it delivers an equal amount of power back to the source while the field is collapsing during the second quarter of its cycle. The net result is that zero power is used by the inductance. A perfect inductance may be considered to be just like a perfect flywheel, which accumulates power during one revolution and delivers an equal amount of power back to the engine during its second revolution.

From the definition of the effective ac ampere (Job 15-4), power equals the effective current squared, multiplied by the resistance of the circuit, or

$$W = I^2R$$

This is true for any single-phase ac circuit and is independent of the phase relation between the current and the voltage. Thus, in a pure inductance, in which $R = 0$,

$$W = I^2R = I^2 \times 0 = 0 \text{ W} \qquad \textit{Check}$$

EXAMPLE 18-2. A 10-H filter choke coil is connected across a 120-V 60-Hz ac line. Assuming that the coil has zero resistance, draw the vector diagram and find (*a*) the current, (*b*) the PF, and (*c*) the effective power drawn.

SOLUTION. Given: $L = 10$ H Find: $I = ?$
$V = 120$ V $W = ?$
$f = 60$ Hz

Draw the phasor diagram. See Fig. 18-5.

*a.*

1. Find the inductive reactance.

$$X_L = 6.28fL = 6.28 \times 60 \times 10 = 3768\ \Omega \qquad (16\text{-}1)$$

2. Find the current.

$$V_L = I_L \times X_L \qquad (16\text{-}2)$$

$$120 = I_L \times 3768$$

$$I_L = \frac{120}{3768} = 0.0318 \text{ A} \qquad \textit{Ans.}$$

*b.*

$$\text{PF} = \cos\theta = \cos 90° = 0 \qquad \textit{Ans.}$$

*c.* Find the effective power.

$$W = V \times I \cos\theta \qquad (18\text{-}3)$$

$$W = 120 \times 0.0318 \times \cos 90°$$

$$W = 120 \times 0.0318 \times 0 = 0 \text{ W} \qquad \textit{Ans.}$$

*d. Check:*

$$W = I^2R$$
$$W = (0.0318)^2 \times 0 = 0 \text{ W} \qquad \textit{Check}$$

## PROBLEMS

Draw a phasor diagram for each problem (not to scale).

1. Find the current sent through a 0.03-H coil by a voltage of 188.4 V at a frequency of 1 kHz.
2. Find the current and effective power drawn by a 200-mH coil which is connected to a 31.4-V source at a frequency of 1000 Hz.
3. What voltage is needed to force 0.08 A through an inductance of 0.5 H at a frequency of 100 Hz?
4. What voltage at 10 kHz is necessary to send a current of 20 mA through an inductance of 50 mH? Find (*a*) the PF, and (*b*) the effective power consumed.
5. What must be the reactance of a filter choke in order for it to pass 80 mA of current when the voltage is 240 V? What must be the inductance of the choke if the frequency is 60 Hz?

**Series circuits containing only capacitance.** When an alternating voltage is impressed across a capacitor, the capacitor will be alternately charged and discharged. While it is charging, the flow of electrons to the plate of the capacitor is largest at the instant the charge is begun. This is so because there are no electrons already on the plate to exert an opposing force. As more and more electrons accumulate on the plate of the capacitor, they exert a greater and greater force which tends to stop the flow of electrons to the plate. When the capacitor is fully charged to the voltage of the source, the flow of current falls to zero, since the back pressure is equal to the pressure of the charging source. Notice that the maximum current occurs when the voltage is zero, and a zero current flows when the voltage is a maximum. If the capacitor is continually charged and discharged by a source of alternating voltage, the relation between current and voltage will be that shown in Fig. 18-7. This indicates that in a capacitive circuit the current *leads* the voltage or the voltage *lags* behind the current. This condition does not exist until the circuit is in operation for a few seconds, as it is obviously impossible for the current to start at any value other than zero. However, once the circuit is in operation, the phase relations are adjusted so that the current will *lead* the voltage by 1/4 cycle or 90 electrical degrees. As with an inductance, this applies only to a "perfect" capacitor—one in which there is no resistance due to the resistance of the capacitor plates or leads.

The current that flows as a result of an ac voltage across a capacitor will be an alternating current with the same frequency as the ac voltage. The current remains forever 90 electrical degrees *ahead* of the voltage. The current *leads* the voltage or the voltage *lags* behind the current by 90°. The phase angle $\theta$ in a capacitive circuit is 90°.

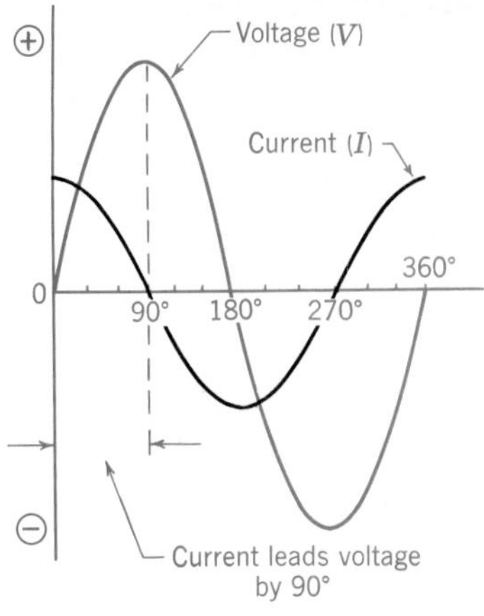


**FIGURE 18-7**
The voltage and current are *out of phase* in an ac circuit containing only capacitance.

**Phasor diagram.** To express this leading current by phasors, the phasors must be drawn 90° apart. Since the current in a series circuit is constant, the current is used as the reference line upon which to draw the phasor diagram. Draw a line with an arrow pointing to the right as shown in Fig. 18-8. This will represent the current $I$ in the circuit. Since the voltage *lags* the current by 90°, we shall be forced to draw the voltage phasor in such a way so as to be 90° *after* the current. Since the current phasor already points to 3 o'clock, the voltage phasor must point to 6 o'clock in order to *lag* the current vector by 90°.

The amount of current in a "pure" capacitance is found by Ohm's law. However, since a pure capacitance contains zero resistance, $R$ is replaced by $X_C$. Ohm's law for a pure capacitance will then be

$$V_C = I_C \times X_C \qquad (17\text{-}7)$$

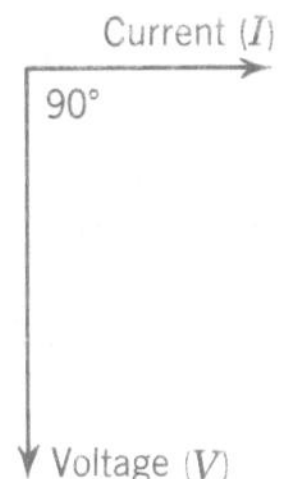


**FIGURE 18-8**
The voltage phasor lags the current phasor by 90° in a purely capacitive circuit.

**Power.** Since the voltage and current in a pure capacitive circuit are 90° out of phase, the power used is equal to zero. This fact is obtained by substituting in formula (18-3) for ac power.

$$W = V \times I \times \cos\theta \qquad (18\text{-}3)$$
$$W = V \times I \times \cos 90°$$
$$W = V \times I \times 0 = 0\text{ W}$$

EXAMPLE 18-3. A 10-$\mu$F coupling capacitor in a transistorized record player passes 300 mA at a frequency of 0.4 kHz. Find (*a*) the voltage drop across the capacitor, (*b*) the power factor, and (*c*) the effective power consumed.

SOLUTION. Given: $C = 10\ \mu\text{F}$ *Find:* $V = ?$

$I = 300\text{ mA} = 0.3\text{ A}$

$f = 0.4\text{ kHz} = 400\text{ Hz}$

*a.* Find the reactance of the capacitor.

$$X_C = \frac{159{,}000}{f \times C} = \frac{159{,}000}{400 \times 10} = \frac{159}{4} = 40\ \Omega\ \text{(approx)} \quad (17\text{-}6)$$

Find the voltage drop.

$$V_C = I_C \times X_C \quad (17\text{-}7)$$
$$V_C = 0.3 \times 40 = 12\ \text{V} \quad \textit{Ans.}$$

*b.* $$\text{PF} = \cos\theta = \cos 90° = 0 \quad \textit{Ans.}$$

*c.* Find the effective power.

$$W = V \times I \times \cos\theta \quad (18\text{-}3)$$
$$W = 12 \times 0.3 \times \cos 90°$$
$$W = 12 \times 0.3 \times 0 = 0\ \text{W} \quad \textit{Ans.}$$

*d.* $$W = I^2R = (0.3)^2 \times 0 = 0\ \text{W} \quad \textit{Check}$$

## PROBLEMS

1. A voltage of 9 V at a frequency of 10 kHz is impressed across a 4-μF capacitor. Find (*a*) the current, (*b*) the PF, and (*c*) the effective power used.
2. What current will flow through a 0.000 015-F capacitor if the voltage across it is 10.6 V at a frequency of 100 Hz?
3. An absorption-type wave trap in a television receiver is tuned to 27.25 MHz. What is the reactance of the 47-pF capacitor in the trap?
4. What is the reactance of a 0.06-μF coupling capacitor to an audio frequency of 2 kHz? What current will flow if the voltage across the capacitor is 6 V?
5. What is the reactance of a 0.02-μF coupling capacitor to a frequency of 200 kHz? What current will flow if the voltage across the capacitor is 4 V?
6. A 4-μF bypass capacitor passes 200 mA at a frequency of 1 kHz. Find (*a*) the voltage drop across the capacitor and (*b*) the effective power consumed.
7. The potential difference between two wires having a distributed capacity of 20 pF is 2 V. Find the flow of current between the wires if one of them carries a 1000-kHz current.
8. Find the voltage across a 10-μF filter capacitor if it passes 1 A at a frequency of 60 Hz.
9. What must be the reactance of a capacitor in order for it to pass 1 A of current when the voltage is 100 V? What is the capacitance of the capacitor if the frequency is 1000 kHz?
10. The plates of a tuning capacitor are set to provide a capacitance of 200 pF. If the capacitor passes 350 mA at 3 MHz, what is the voltage drop across it?

## JOB 18-2 THE PYTHAGOREAN THEOREM

In our next job we shall find the total voltage across a series ac circuit by adding the voltages *even though the voltages do not appear at the same time!* The formulas for this total voltage and all our work in ac power are applications of the *pythagorean theorem.*

This basic mathematical law about a *right triangle* was discovered by a Greek scholar named Pythagoras about 2500 years ago. In Fig. 18-9, squares are drawn on each of the three sides of the right triangle. The length of the sides of each square is equal to the length of the side of the triangle on which it is drawn. Pythagoras discovered that the *sum of the areas of the squares on the two legs of a right triangle is exactly equal to the area of the square erected on the hypotenuse.* This is true for *any right triangle.* Thus,

$$\text{Area I} + \text{area II} = \text{area III}$$
$$9 \quad + \quad 16 \quad = \quad 25$$

Since the area of a square is equal to a side times itself,

$$\text{Area I} = a \times a = a^2$$
$$\text{Area II} = b \times b = b^2$$
$$\text{Area III} = h \times h = h^2$$

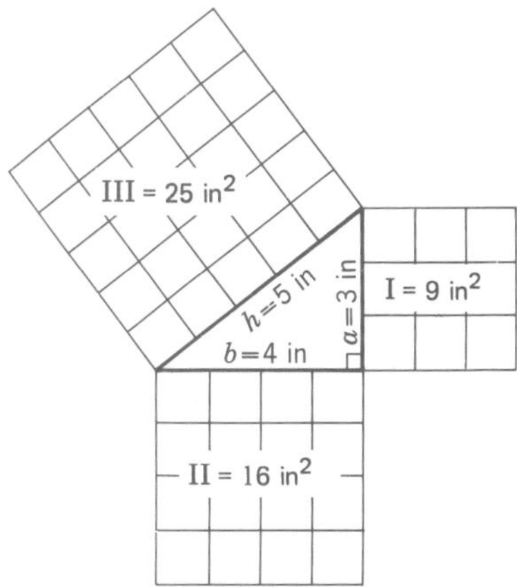


**FIGURE 18-9** Pictorial representation of the pythagorean theorem: $a^2 + b^2 = h^2$.

The theorem may be stated as the following rule.

**RULE** The sum of the squares of the legs of a right triangle is equal to the square of the hypotenuse.

**FORMULA** $a^2 + b^2 = h^2$ **18-7**

where $a, b$ = legs of a right triangle
$h$ = hypotenuse of a right triangle

The formula may be solved for $h$ by taking the square root of both sides.

$$\sqrt{a^2 + b^2} = \sqrt{h^2}$$

$$\sqrt{a^2 + b^2} = h \qquad \boxed{18\text{-}7a}$$

EXAMPLE 18-4. Find the hypotenuse of a right triangle whose altitude is 5 in and whose base is 12 in.

SOLUTION. The diagram is shown in Fig. 18-10.

$$h = \sqrt{a^2 + b^2} \qquad (18\text{-}7a)$$

$$h = \sqrt{5^2 + 12^2}$$

$$h = \sqrt{25 + 144}$$

$$h = \sqrt{169} = 13 \text{ in} \qquad \textit{Ans.}$$

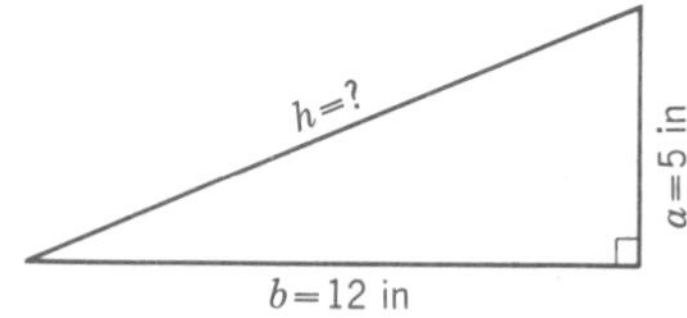


FIGURE 18-10

EXAMPLE 18-5. Find the altitude $a$ of a right triangle whose hypotenuse $h$ is 17 in and whose base $b$ is 15 in.

SOLUTION. Given: $h = 17$ in Find: $a = ?$
$b = 15$ in

$$a^2 + b^2 = h^2$$

$$a^2 + 15^2 = 17^2$$

$$a^2 + 225 = 289$$

$$a^2 = 289 - 225$$

$$a^2 = 64$$

$$a = \sqrt{64} = 8 \text{ in} \qquad \textit{Ans.}$$

SELF-TEST 18-6. An electrician's tool box measures 16 in × 12 in × 10 in. What is the length of the largest extension bit holder that may be placed in the tool box?

SOLUTION. The tool box is shown in Fig. 18-11.

1. Find the length of the diagonal $AC$. Triangle $ABC$ is a right triangle with the right angle at point ________. | $B$

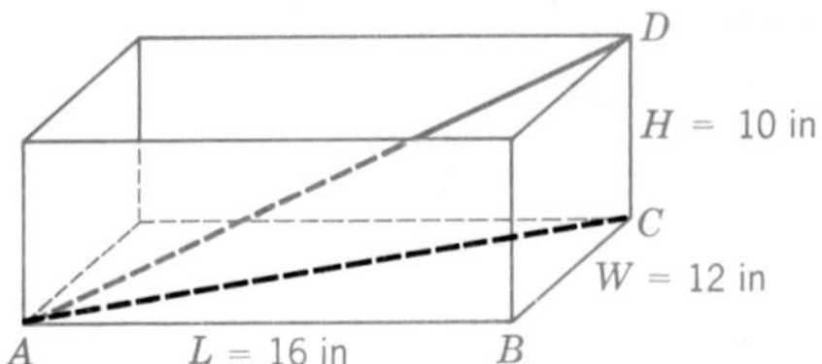


FIGURE 18-11
*AD* is the longest dimension in the tool box.

$$h = \sqrt{a^2 + b^2} \qquad (18\text{-}7a)$$
$$AC = \sqrt{16^2 + (\_\_\_\_\_)^2}$$
$$AC = \sqrt{256 + \_\_\_\_\_}$$
$$AC = \sqrt{400} = \_\_\_\_\_ \text{ in}$$

12
144
20

2. Find the length of the bit holder *AD* using triangle ______, in which the right angle is at point ______.

ACD
C

$$h = \sqrt{a^2 + b^2} \qquad (18\text{-}17a)$$
$$AD = \sqrt{10^2 + (\_\_\_\_\_)^2}$$
$$AD = \sqrt{100 + 400} = \sqrt{\_\_\_\_\_}$$
$$AD = \_\_\_\_\_ \text{ in} \quad \textit{Ans.}$$

20
500
22.3

**PROBLEMS**

Find the unknown side in each of the following right triangles:

| PROBLEM | $a$ | $b$ | $h$ |
|---|---|---|---|
| 1 | 6 | 8 | ? |
| 2 | 7 | 24 | ? |
| 3 | ? | 63 | 65 |
| 4 | 33 | ? | 65 |
| 5 | 14 | 22.5 | ? |
| 6 | 17.5 | 6 | ? |
| 7 | ? | 20 | 20.5 |
| 8 | 6.5 | ? | 42.5 |

9. A guy wire stretches from the top of an 40-m-high pole to a stake in the ground 60 m from the foot of the pole. Find the length of the wire.
10. A 10-ft ladder is placed against a wall with the foot of the ladder 6 ft from the base of the wall. How high above the ground will the ladder reach?
11. A doorway measures 3 by 7 ft. What is the diameter of the largest circular table that will pass through the doorway?
12. A conduit must be bent to provide a rise of 6 ft in a horizontal distance of 4 ft 6 in. What is the length of conduit between the bends?
13. Find the length of electrical conduit needed to include the 5-ft offset as shown in Fig. 18-12.

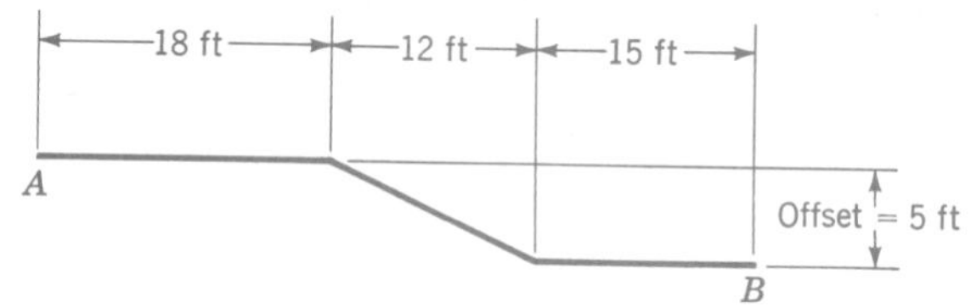


FIGURE 18-12
Find the length of conduit from $A$ to $B$.

14. Find the distance across the corners of a square nut measuring $3^1/_4$ in on a side.
15. Find the distance $X$ in Fig. 18-13.

(See Instructor's Manual for Test 18-1)

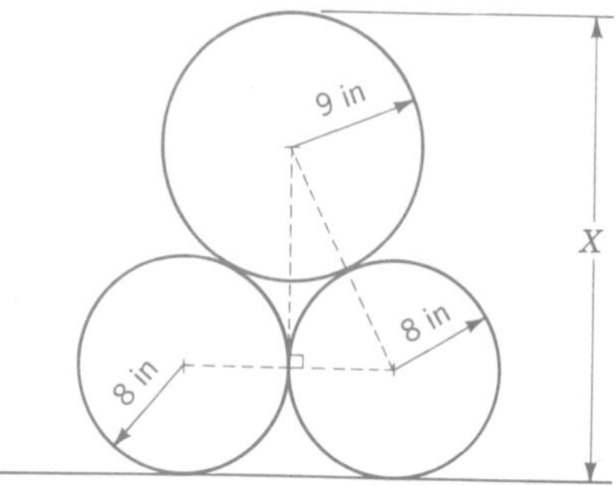


FIGURE 18-13

## JOB 18-3 RESISTANCE AND INDUCTANCE IN SERIES

Figure 18-14 shows a 100-Ω resistor connected in series with an inductance whose reactance is 100 Ω at a frequency of 60 Hz. A series current of 0.85 A produces a voltage drop of 85 V across both the resistor and the inductance. In a series circuit, the total voltage is ordinarily found by adding the voltages across all the parts of the circuit. This rule was used in dc circuits, but can we use it for ac circuits? Since the current in a series circuit remains unchanged throughout the circuit, we may draw the waveforms for the currents and voltages across each part of the circuit on the same drawing. Figure 18-15 shows the relationship of each voltage to the unchanging current. The voltage across

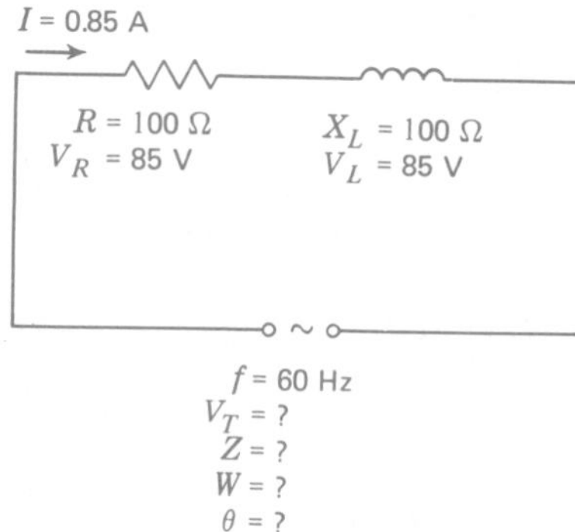


FIGURE 18-14
An ac series circuit containing resistance and inductance.

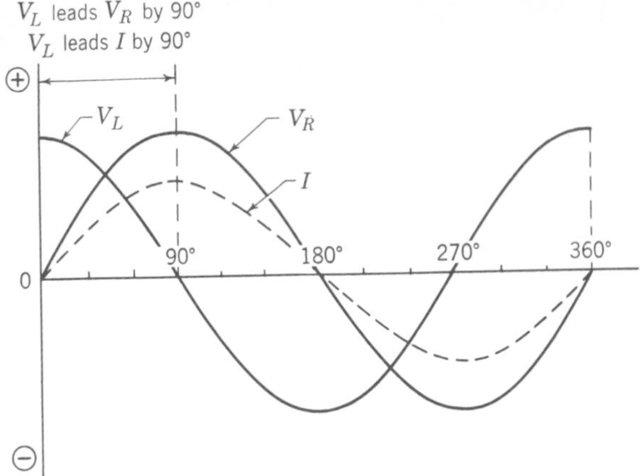


**FIGURE 18-15**
Voltages and currents in a series ac circuit containing resistance and inductance.

the resistor $V_R$ is *in phase* with the current $I$; that is, $V_R$ and $I$ reach their maximum and minimum values *at the same time.* The voltage across the inductance $V_L$ *leads* the current $I$ by 90°; that is, $V_L$ reaches its maximum 90° *before* $I$ reaches its maximum. Also, $V_L$ *leads* $V_R$ by 90°; that is, $V_L$ reaches its maximum 90° *before* $V_R$ reaches its maximum.

To obtain the total voltage, we must add the voltages across all parts of the circuit. But how are we going to add voltages that do not happen at the same time? The only way to do this is to add the *instantaneous* voltages that *do* occur at the same time. Thus, in Fig. 18-16, the instantaneous values of $v_R$ and $v_L$ are added for different instants of time.

At 0°, $v_R = 0$ and $v_L$ is a maximum. Therefore,

$$V_T = v_R + v_L$$
$$V_T = 0 + v_L$$
$$V_T = v_L \qquad \text{(point } a\text{)}$$

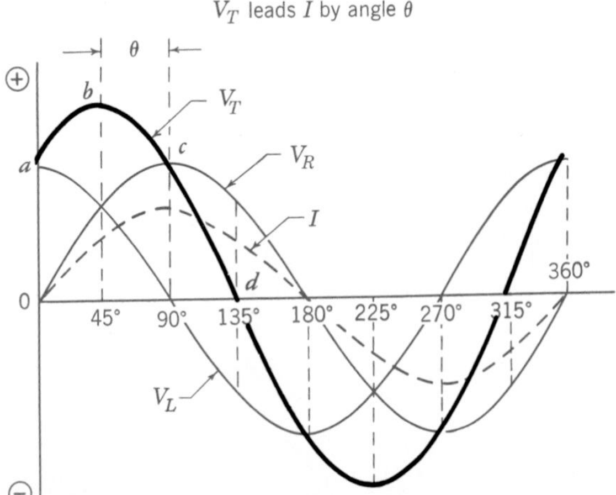


**FIGURE 18-16**
The total voltage in a series ac circuit is obtained by adding the instantaneous voltages, instant by instant.

At 45°, $v_R = v_L$ and

$$V_T = v_R + v_L$$

or $\quad V_T =$ twice the value of either (point *b*)

At 90°, $v_L = 0$ and $v_R$ is a maximum. Therefore,

$$V_T = v_R + v_L$$
$$V_T = v_R + 0$$
$$V_T = v_R \quad \text{(point } c\text{)}$$

At 135°, $v_R = v_L$, but they are of opposite polarity. Therefore,

$$V_T = v_R + (-v_L)$$
$$V_T = 0 \quad \text{(point } d\text{)}$$

By continuing in this manner, adding the voltages instant by instant, the waveform for the total voltage is obtained as shown in Fig. 18-16. Notice that the maximum value of $V_T$ (at 45°) is *still leading* the current maximum. In this particular problem, since $R$ and $X_L$ are equal, the angle of lead is equal to 45°. For other values of $R$ and $X_L$, angle $\theta$ will change.

Apparently, then, although we *do* add the voltages to get the total voltage, the addition is not just a simple arithmetic addition. We can see this more clearly if we draw the voltages and currents as phasors on the same unchanging current base as shown in Fig. 18-17*a*. This phasor diagram shows exactly the same relationships that were shown in Fig. 18-15 in wave form. Since the current in a series circuit is constant, the current is used as the reference line upon which to draw the phasor diagram. The voltage across the resistor $V_R$ is still in phase with the current $I$, and the voltage across the inductance $V_L$ still leads the current $I$ by 90°.

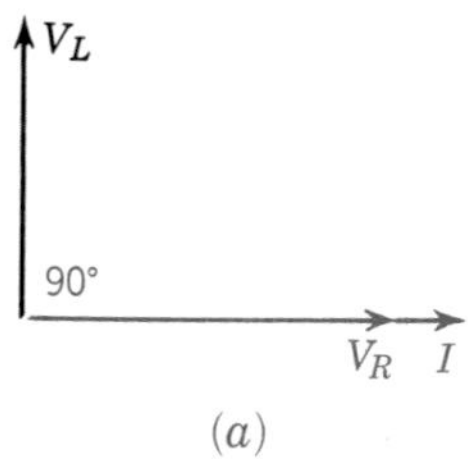


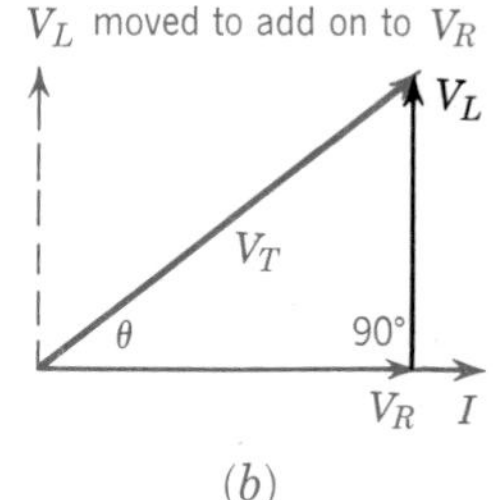


**FIGURE 18-17**
(*a*) Phasor diagram for a series ac circuit containing resistance and inductance. (*b*) $V_T$ represents the phasor sum of $V_R$ and $V_L$.

## ADDITION OF PHASORS

**RULE** Phasors are added by placing the tail of each phasor on the head of the preceding phasor and drawing it in its original direction and length.

**RULE** The sum of the phasors is the distance from the origin of the phasors to the head of the final phasor.

In Fig. 18-17*b*, the total voltage $V_T$ is obtained by adding the phasor $V_L$ to the phasor $V_R$. Place the tail of $V_L$ on the head of $V_R$ and draw it in its original direction and length. The distance from the origin of the phasors to the head of the final phasor is the *sum* of the phasors. In this instance, the phasor $V_T$ represents the sum of the phasors $V_R$ and $V_L$. The phase angle $\theta$ is the angle by which the total voltage leads the current. In this problem, the angle is 45° and is the same angle shown on the waveform diagram of Fig. 18-16.

By applying the pythagorean theorem to Fig. 18-17*b*, we obtain the

**FORMULA**

$$V_T^2 = V_R^2 + V_L^2 \tag{18-8}$$

**Total impedance.** To find the impedance of the series ac circuit, we must add $R$ and $X_L$ vectorially as was done with the voltages. This is shown in Fig. 18-18. The voltages in Fig. 18-18*a* are replaced by their Ohm's-law values in Fig. 18-18*b*. Now, by dropping out the common factor of the current $I$, we obtain the *impedance triangle* of Fig. 18-18*c*.

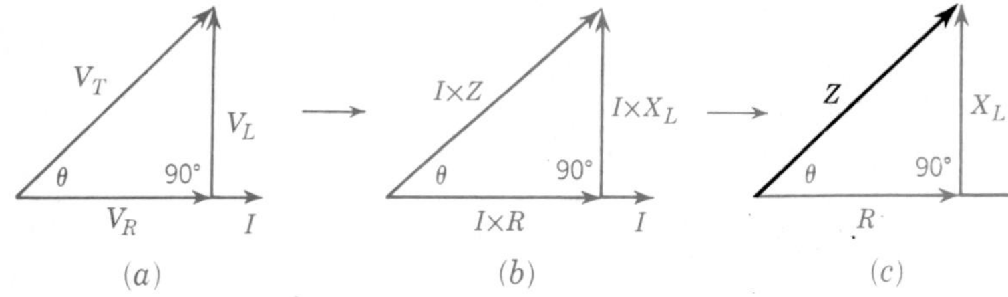


**FIGURE 18-18**
$Z$ represents the phasor sum of $R$ and $X_L$.

Applying the pythagorean theorem to Fig. 18-18*c*, we obtain the

**FORMULAS**

$$Z^2 = R^2 + X_L^2 \tag{18-9}$$

and

$$Z = \sqrt{R^2 + X_L^2} \tag{18-10}$$

And, by trigonometry,

$$\cos \theta = \frac{a}{h}$$

or

$$\cos \theta = \frac{R}{Z} \tag{18-11}$$

or

$$\text{PF} = \frac{R}{Z} \tag{18-12}$$

The effective power is still given by

$$W = V \times I \times \cos\theta \qquad (18\text{-}3)$$

or

$$W = V \times I \times \text{PF} \qquad (18\text{-}4)$$

or

$$W = I^2R \qquad (7\text{-}4)$$

EXAMPLE 18-7. In Fig. 18-14, find (*a*) the total voltage, (*b*) the impedance, (*c*) the phase angle, (*d*) the power factor, and (*e*) the power.

SOLUTION

*a.*

$$V_T^2 = V_R^2 + V_L^2 = 85^2 + 85^2 \qquad (18\text{-}8)$$
$$= 7225 + 7225$$
$$= 14{,}450$$
$$V_T = \sqrt{14{,}450} = 120 \text{ V} \qquad \textit{Ans.}$$

*b.*

$$Z = \sqrt{R^2 + X_L^2} \qquad (18\text{-}10)$$
$$Z = \sqrt{100^2 + 100^2} = \sqrt{10^4 + 10^4} = \sqrt{2 \times 10^4}$$
$$Z = 1.41 \times 10^2 = 141\ \Omega \qquad \textit{Ans.}$$

$$\cos\theta = \frac{R}{Z} = \frac{100}{141} = 0.709 \qquad (18\text{-}11)$$
$$\theta = 45° \qquad \textit{Ans.}$$

*d.*

$$\text{PF} = \frac{R}{Z} = 0.709 = 70.9\%\ \textit{lagging}^* \qquad (18\text{-}12)$$

*e.*

$$W = V \times I \times \text{PF} \qquad (18\text{-}4)$$
$$= 120 \times 0.85 \times 0.709 = 72.3 \text{ W} \qquad \textit{Ans.}$$

*f.*

$$W = I^2R = (0.85)^2 \times 100 = 72.3 \text{ W} \qquad \textit{Check}$$

In this series circuit, the total current of 0.85 A *lags* the total voltage of 120 V by 45°.

EXAMPLE 18-8. An inductance of 0.17 H and a resistance of 50 Ω are connected in series across a 110-V 60-Hz line. Find (*a*) the inductive reactance, (*b*) the impedance, (*c*) the total current, (*d*) the voltage drop across the resistor and the coil, (*e*) the phase angle, (*f*) the power factor, and (*g*) the power used.

SOLUTION. Given: $L = 0.17$ H, $R = 50\ \Omega$, $V_T = 110$ V, $f = 60$ Hz

Find: $X_L = ?$, $Z = ?$, $I_T = ?$, $V_R = ?$, $V_L = ?$, $\theta = ?$, $\text{PF} = ?$, $W = ?$

*In this book, a leading or lagging PF will always refer to a *current* which will lead or lag the voltage.

*a.* $$X_L = 6.28fL = 6.28 \times 60 \times 0.17 = 64\ \Omega \qquad \textit{Ans.}$$

*b.* $$Z = \sqrt{R^2 + X_L^2} = \sqrt{50^2 + 64^2} \qquad (18\text{-}10)$$
$$= \sqrt{2500 + 4096}$$
$$Z = \sqrt{6596} = 81\ \Omega \qquad \textit{Ans.}$$

*c.* Since the total opposition is the impedance $Z$, formula (4-7) becomes

$$V_T = I_T \times Z \qquad \boxed{18\text{-}13}$$
$$110 = I_T \times 81$$
$$I_T = \frac{110}{81} = 1.36\text{ A} \qquad \textit{Ans.}$$

*d.* Since

$$I_T = I_R = I_L = 1.36\text{ A} \qquad (4\text{-}1)$$

$$V_R = I_R \times R_R \qquad\qquad V_L = I_L \times X_L$$
$$V_R = 1.36 \times 50 = 68\text{ V} \qquad V_L = 1.36 \times 64 = 87\text{ V}$$

*e.* $$\cos\theta = \frac{R}{Z} = \frac{50}{81} = 0.6173 \qquad (18\text{-}8)$$
$$\theta = 52° \qquad \textit{Ans.}$$

*f.* $$\text{PF} = \frac{R}{Z} = 0.6173 = 61.7\%\ \textit{lagging} \qquad \textit{Ans.} \qquad (18\text{-}12)$$

*g.* $$W = V \times I \times \text{PF} = 110 \times 1.36 \times 0.617 = 92.5\text{ W} \qquad \textit{Ans.}$$
(18-4)

*h.* $$W = I^2R = (1.36)^2 \times 50 = 92.5\text{ W} \qquad \textit{Check}$$

In the series circuit, the total current of 1.36 A lags the total voltage of 110 V by 52°.

SELF-TEST 18-9. Figure 18-19 may be used to represent the plate circuit of an amplifier with transformer coupling. Find the value of $V_p$ when the frequency of the applied voltage is (*a*) 100 Hz and (*b*) 10 Hz.

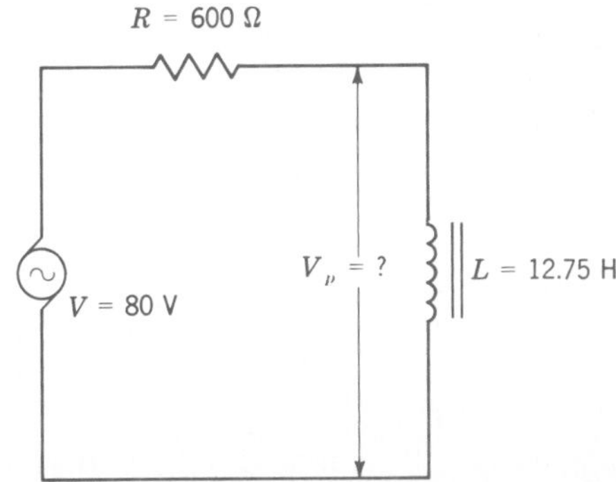


**FIGURE 18-19**
Representation of the plate circuit of a transformer-coupled amplifier.

SOLUTION.

*a.* At 100 Hz,

$$X_L = 6.28fL \qquad (16\text{-}1)$$
$$X_L = 6.28 \times 100 \times 12.75 = \_\_\_\_\_\_ \ \Omega$$ 8000

The factor that determines whether we should include the 600-Ω resistance in the calculation for $Z$ is called the ______ of the circuit. (See page 461.) $Q$

$$Q = \frac{X_L}{R} = \frac{8000}{600} = \_\_\_\_\_\_ \qquad (16\text{-}5)$$ 13.3

Since $Q$ is greater than 5, we ___(should/should not)___ include the resistance in the calculation for $Z$. Therefore, $Z$ = ______ Ω. should not 8000

Find the current.

$$V = IZ \qquad (16\text{-}6)$$
$$80 = I \times 8000$$
$$I = \_\_\_\_\_\_ \text{ A}$$ 0.01

Find $V_p$.

$$V_p = I \times X_L$$
$$= 0.01 \times 8000 = \_\_\_\_\_\_ \text{ V} \quad Ans.$$ 80

*b.* At 10 Hz,

$$X_L = 6.28 \times \_\_\_\_\_\_ \times 12.75 = \_\_\_\_\_\_ \ \Omega$$ 10 800

The $Q$ of the circuit is now 800/______, or 600

$$Q = \_\_\_\_\_\_$$ 1.33

Since $Q = 1.33$, the resistance ___(should/should not)___ be included in the calculation for $Z$. should

$$Z = \sqrt{R^2 + X_L^2} = \sqrt{(600)^2 + (800)^2}$$
$$= \sqrt{36 \times 10^4 + \_\_\_\_\_\_ \times 10^4}$$ 64
$$= \sqrt{(36 + 64) \times 10^4}$$
$$Z = \sqrt{100 \times 10^4} = \_\_\_\_\_\_ \ \Omega$$ 1000

Find the current.

$$V = IZ \qquad (16\text{-}6)$$
$$80 = I \times 1000$$
$$I = \_\_\_\_\_\_ \text{ A}$$ 0.08

Find $V_p$.

$$V_p = I \times X_L$$
$$= 0.08 \times \_\_\_\_\_\_ = \_\_\_\_\_\_ \text{ V} \quad Ans.$$ 800 64

**PROBLEMS**

1. A resistance of 5 Ω is in series with a coil whose inductive reactance

is 12 Ω. If the total voltage is 104 V, find (*a*) the impedance, (*b*) the total current, (*c*) the voltage drop across each part, (*d*) the phase angle, (*e*) the power factor, and (*f*) the power.

2. A 112-V 60-Hz ac voltage is applied across a series circuit of a 50-Ω resistor and a 100-Ω inductive reactance. Find (*a*) the impedance, (*b*) the total current, (*c*) the voltage drop across each part, (*d*) the phase angle, (*e*) the power factor, and (*f*) the power.
3. A 66-V 220-Hz ac voltage is applied across a series circuit of a 20-Ω resistor and a 0.05-H coil. Find the total current and the phase angle.
4. A tuning coil has an inductance of 48 μH and a resistance of 20 Ω. Find its impedance to a frequency of 100 kHz.
5. A fluorescent-lamp ballast has an inductance of 0.4 H and a resistance of 80 Ω. If the supply frequency is 60 Hz, find (*a*) the reactance of the ballast, (*b*) the impedance of the ballast, and (*c*) the voltage across the ballast when 0.5 A flows through it.
6. A transformer-coupled amplifier contains a 25,000-Ω resistance and a 20-H coil. At what frequency will the voltage across the coil equal that across the resistance? What current will flow if the impressed voltage is 35 V?
7. Part of the oscillator circuit for a continuous-wave transmitter is shown in Fig. 18-20. Find the current from point *A* to point *B* if the voltage drop is 70 V.

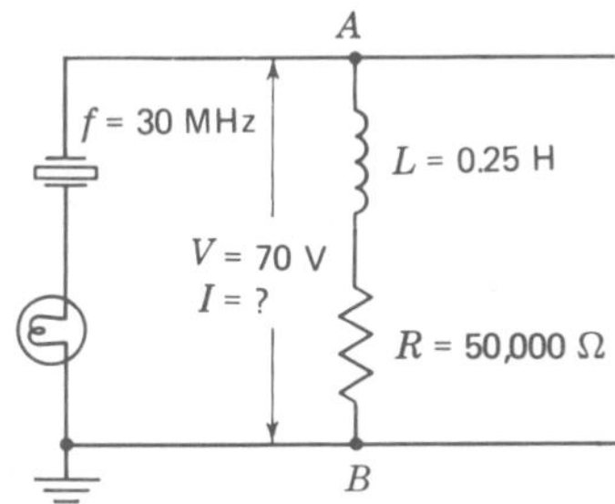


**FIGURE 18-20**
Portion of the oscillator circuit for a continuous-wave transmitter.

8. A 20-Ω resistor is in series with a 0.03-H dimmer coil. If the 230-V 60-Hz ac voltage is applied to the circuit, find (*a*) the current and (*b*) the power used.
9. A lightning-protector circuit contains a 63.7-mH coil in series with a 7-Ω resistor. What current will flow when it is tested with a 110-V 60-Hz ac voltage?
10. The coil of a telephone relay has a resistance of 500 Ω and an inductance of 0.32 H. When operated at a frequency of 500 Hz, find (*a*) the reactance of the coil, (*b*) the impedance of the coil, and (*c*) the voltage that must be impressed across the coil in order to operate the relay at its rated current of 5 mA.
11. The output voltage of an audio oscillator is 37.7 V at 3000 Hz. It is applied to a series circuit of 200 Ω resistance and 200 mH induc-

tance. Find the impedance of the circuit and the current that flows.

12. A filter choke coil is connected in series with a 400-Ω resistor. When the voltage across the circuit is 120 V, the current is 0.12 A. Find the inductance of the coil if the frequency is 60 Hz. *Hint:* See Job 16-5.
13. To measure the inductance of an audio choke, a 2000-Ω resistor is connected in series with the choke. A 110-V 60-Hz voltage is impressed across the circuit, and the current is measured at 10 mA. Find the inductance of the coil.
14. A 40-V emf at 1000 Hz is impressed across a loudspeaker of 5000 Ω resistance and 1.5 H inductance. Find the current and power drawn.
15. Find the inductive reactance of a single-phase motor if the line voltage is 220 V, the line current is 20 A, and the resistance of the motor coils is 8 Ω. What is the angle of lag?

(See Instructor's Manual for Test 18-2, *R* and *L* in Series)

## JOB 18-4 RESISTANCE AND CAPACITANCE IN SERIES

Figure 18-21 shows a 5-Ω resistor connected in series with a capacitance whose reactance is 12 Ω at a frequency of 60 Hz. A series current of 1 A produces a voltage drop of 5 V across the resistor and 12 V across the capacitor.

The total voltage across the circuit can be found by adding the voltage drops across each part. But, just as in the last job, since the voltages are *not* in phase, they must be added vectorially. The phase relations in a capacitive circuit are shown in Fig. 18-22*a*. The constant series current is used as the reference line. The voltage across the resistor $V_R$ is in phase with the current $I$; the voltage across the capacitor $V_C$ *lags* the current $I$ by 90°; $V_C$ *lags* $V_R$ by 90°.

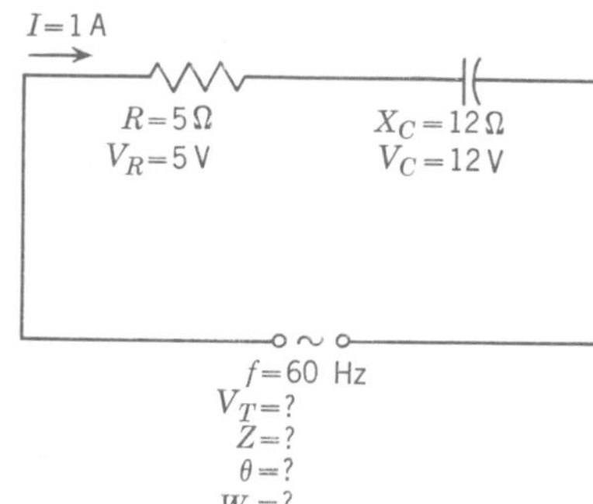


**FIGURE 18-21** Ac series circuit containing resistance and capacitance.

**Addition of the phasors.** In Fig. 18-22*b*, the total voltage $V_T$ is obtained by adding the phasor $V_C$ to the phasor $V_R$. Place the tail of $V_C$ on the head of $V_R$, and draw it in its original direction and length. The distance from the origin of phasors to the head of the final phasor is the

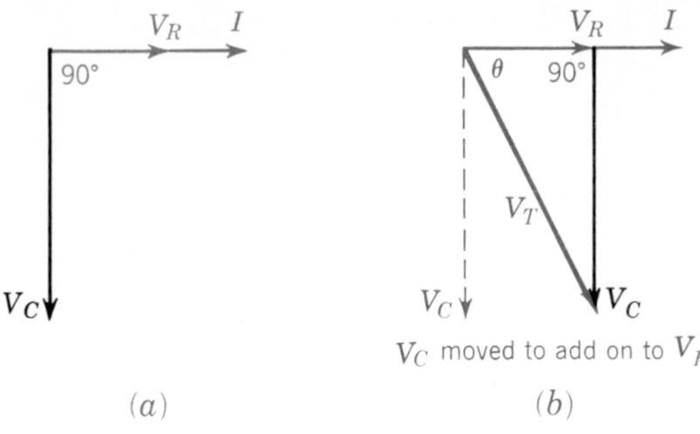


**FIGURE 18-22** (a) Phasor diagram for a series ac circuit containing resistance and inductance. (b) $V_T$ represents the phasor sum of $V_R$ and $V_C$.

sum of the phasors. In this instance, the phasor $V_T$ represents the sum of the phasors $V_R$ and $V_C$. The phase angle $\theta$ is the angle by which the total voltage *lags* behind the current. By applying the pythagorean theorem to Fig. 18-22*b*, we obtain the

**FORMULA** $$V_T^2 = V_R^2 + V_C^2 \qquad \boxed{18\text{-}14}$$

**Total impedance.** To find the impedance of the series ac circuit, we must add $R$ and $X_C$ vectorially as was done with the voltages. This is shown in Fig. 18-23. The voltages in Fig. 18-23*a* are replaced by their Ohm's-law values in Fig. 18-23*b*. Now, by dropping out the common factor of the current $I$, we obtain the *impedance triangle* of Fig. 18-23*c*.

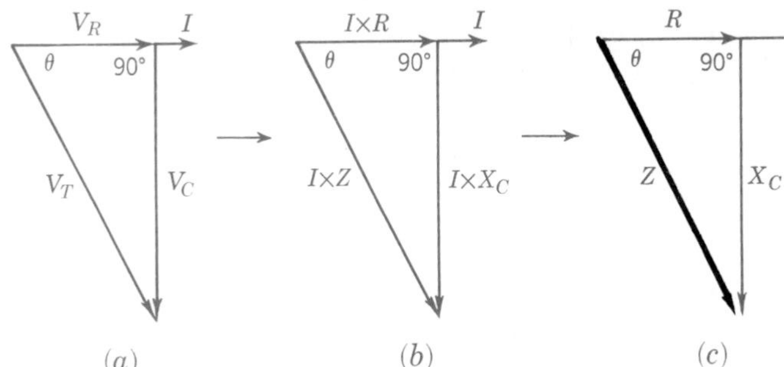


**FIGURE 18-23** $Z$ represents the phasor sum of $R$ and $X_C$.

Applying the pythagorean theorem to Fig. 18-23*c*, we obtain the

**FORMULAS**

$$Z^2 = R^2 + X_C^2 \qquad \boxed{18\text{-}15}$$

$$Z = \sqrt{R^2 + X_C^2} \qquad \boxed{18\text{-}16}$$

And, by trigonometry,

$$\cos\theta = \frac{a}{h}$$

or

$$\cos\theta = \frac{R}{Z} \qquad (18\text{-}11)$$

The power is still given by

$$W = V \times I \times \cos\theta \tag{18-3}$$

or

$$W = V \times I \times \text{PF} \tag{18-4}$$

or

$$W = I^2R \tag{7-4}$$

EXAMPLE 18-10. In Fig. 18-21, find (*a*) the total voltage, (*b*) the impedance, (*c*) the phase angle, (*d*) the power factor, and (*e*) the power.

SOLUTION

*a.*

$$V_T^2 = V_R^2 + V_C^2 = 5^2 + 12^2 \tag{18-14}$$

$$= 25 + 144$$

$$V_T^2 = 169$$

$$V_T = \sqrt{169} = 13 \text{ V} \quad \textit{Ans.}$$

*b.*

$$Z = \sqrt{R^2 + X_C^2} \tag{18-16}$$

$$= \sqrt{5^2 + 12^2}$$

$$= \sqrt{25 + 144}$$

$$Z = \sqrt{169} = 13\ \Omega \quad \textit{Ans.}$$

*c.*

$$\cos\theta = \frac{R}{Z} = \frac{5}{13} = 0.3846 \tag{18-11}$$

$$\theta = 67° \text{ (approx)} \quad \textit{Ans.}$$

*d.*

$$\text{PF} = \frac{R}{Z} = 0.3846 = 38.5\%\ \textit{leading} \quad \textit{Ans.} \tag{18-12}$$

*e.*

$$W = V \times I \times \text{PF} \tag{18-4}$$

$$W = 13 \times 1 \times 0.385 = 5 \text{ W} \quad \textit{Ans.}$$

*f.*

$$W = I^2R = (1)^2 \times 5 = 5 \text{ W} \quad \textit{Check}$$

In this series circuit, the total current of 1 A *leads* the total voltage of 13 V by 67°.

EXAMPLE 18-11. A capacitance of 4 $\mu$F and a resistance of 30 $\Omega$ are connected in series across a 100-V 1-kHz ac source. Find (*a*) the capacitive reactance, (*b*) the impedance, (*c*) the total current, (*d*) the voltage drop across the resistor and the capacitor, (*e*) the phase angle, (*f*) the power factor, and (*g*) the power.

SOLUTION. Given: $C = 4\ \mu\text{F}$ Find: $X_C = ?$

$R = 30\ \Omega$ $Z = ?$

$V_T = 100$ V $I_T = ?$

$f = 1 \text{ kHz} = 1000 \text{ Hz}$ $V_R = ?$

$V_C = ?$

$\theta = ?$

$\text{PF} = ?$

$W = ?$

*a.* $$X_C = \frac{159{,}000}{f \times C} = \frac{159{,}000}{1000 \times 4} = \frac{159}{4} = 40\ \Omega \text{ (approx)}$$

*b.* $$Z = \sqrt{R^2 + X_C^2} \qquad (18\text{-}16)$$
$$= \sqrt{30^2 + 40^2}$$
$$= \sqrt{900 + 1600} = \sqrt{2500} = 50\ \Omega \qquad \textit{Ans.}$$

*c.* $$V_T = I_T \times Z \qquad (18\text{-}13)$$
$$100 = I_T \times 50$$
$$I_T = \frac{100}{50} = 2\text{ A} \qquad \textit{Ans.}$$

*d.* Since $$I_T = I_R = I_C = 2\text{ A} \qquad (4\text{-}1)$$

$V_R = I_R \times R_R$ $\qquad V_C = I_C \times X_C$

$V_R = 2 \times 30 = 60$ V *Ans.* $\qquad V_C = 2 \times 40 = 80$ V *Ans.*

*e.* $$\cos\theta = \frac{R}{Z} = \frac{30}{50} = 0.6000 \qquad (18\text{-}11)$$
$$\theta = 53^\circ \text{ (approx)} \qquad \textit{Ans.}$$

*f.* $$\text{PF} = \frac{R}{Z} = 0.600 = 60\%\ \textit{leading} \qquad \textit{Ans.} \qquad (18\text{-}12)$$

*g.* $$W = V \times I \times \text{PF} = 100 \times 2 \times 0.6 = 120\text{ W} \quad \textit{Ans.}\ (18\text{-}4)$$

*h.* $$W = I^2R = (2)^2 \times 30 = 120\text{ W} \qquad \textit{Check}$$

In this series circuit, the total current of 2 A leads the total voltage of 100 V by 53°.

## PROBLEMS

1. A 119-V 60-Hz ac voltage is applied across a series circuit of an 8-Ω resistor and a capacitor whose reactance is 15 Ω. Find (*a*) the impedance, (*b*) the total current, (*c*) the voltage drop across each part, (*d*) the phase angle, (*e*) the power factor, and (*f*) the power.
2. A 134-V 60-Hz ac voltage is applied across a series circuit of a 30-Ω resistor and a 60-Ω capacitive reactance. Find (*a*) the impedance, (*b*) the total current, (*c*) the voltage drop across each part, (*d*) the phase angle, (*e*) the power factor, and (*f*) the power.
3. A 113-V 100-Hz ac voltage is applied across a series circuit of a 100-Ω resistor and a 15.9-μF capacitor. Find (*a*) the impedance, (*b*) the total current, (*c*) the voltage drop across each part, (*d*) the phase angle, (*e*) the power factor, and (*f*) the power.
4. Find the current and angle of lead for a series circuit of a 10-Ω resistor and an 8-μF capacitor if the applied voltage is 110 V at 60 Hz.
5. In the volume-control circuit shown in Fig. 18-24, find the impedance of the control unit to frequencies of (*a*) 1000 Hz and (*b*) 10 kHz.

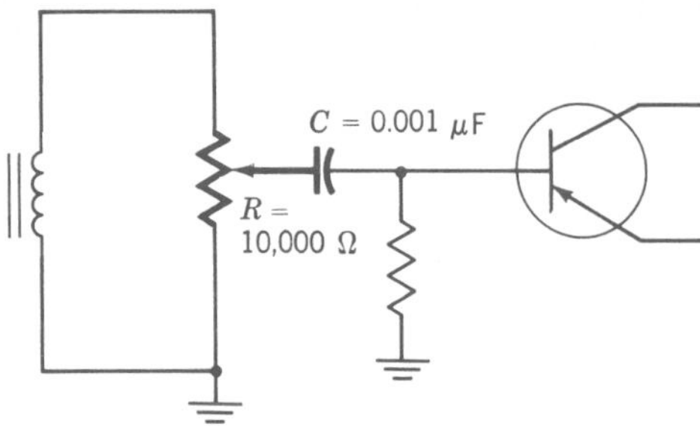


FIGURE 18-24

6. In the resistance-coupled stage shown in Fig. 18-25, the voltage drop between points $A$ and $B$ is 1.414 V. If the frequency of the current between these points is 10 kHz, find the voltage drop across the 1000-Ω resistor.

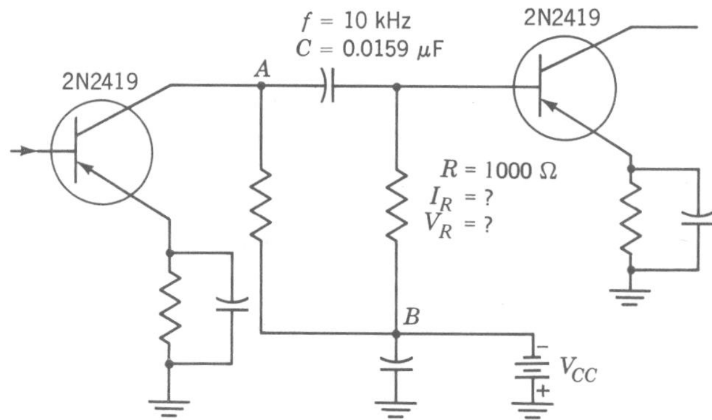


**FIGURE 18-25**
Resistance-coupled amplifier stage.

7. When the value of an unknown capacitor was calculated by the impedance method, a 300-Ω resistor was placed in series with the capacitor. A 110-V, 60-Hz ac voltage caused a current of 0.22 A to flow. Find ($a$) the impedance of the circuit, ($b$) the reactance of the capacitor, and ($c$) the capacitance of the capacitor.
8. A 10,000-Ω resistor and a capacitor are placed in series across a 60-Hz line. If the voltage across the resistor is 50 V, and across the capacitor 100 V, find ($a$) the current in the resistor, ($b$) the current in the capacitor, ($c$) the reactance of the capacitor by Ohm's law, and ($d$) the capacitance of the capacitor.
9. A 120-V source is connected to a resistance of 50 Ω in series with a capacitance of 10 μF. What frequency will permit a current of 0.8 A to flow?
10. A circuit consisting of a 40-μF capacitance in series with a rheostat is connected across a 138-V 60-Hz line. What must be the value of the resistance in order to permit a current of 2 A to flow?

## JOB 18-5 RESISTANCE, INDUCTANCE, AND CAPACITANCE IN SERIES

Figure 18-26 shows a 16-Ω resistor, an inductive reactance of 80 Ω, and a

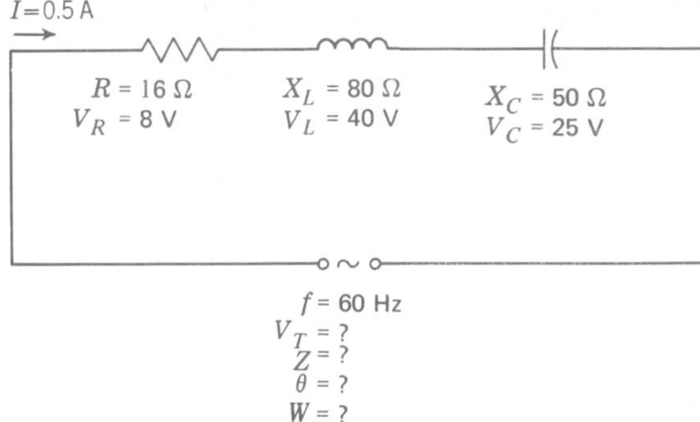


**FIGURE 18-26**
Ac series circuit containing resistance, inductance, and capacitance.

capacitive reactance of 50 Ω connected in series across a frequency of 60 Hz. A series current of 0.5 A produces a voltage drop of 8 V across the resistor, 40 V across the inductance, and 25 V across the capacitance.

The total voltage across the entire circuit may be found by adding the voltage drops across each part. However, since the voltages are *not* in phase, they must be added vectorially. The phase relations in the circuit are shown in Fig. 18-27*a*. The constant series current is used as the reference line. The voltage across the resistor $V_R$ is in phase with the current $I$. The voltage across the inductance $V_L$ *leads* the current $I$ by 90°. The voltage across the capacitance $V_C$ *lags* the current $I$ by 90°. Since $V_L$ and $V_C$ are exactly 180° out of phase and act in exactly *opposite* directions, the voltage $V_C$ is denoted by a minus sign.

**Addition of phasors.** When there are three phasors, it is best to add only two at a time. To add the phasor $V_C$ to $V_L$, place the tail of $V_C$ on the head of $V_L$ and draw it in its original direction and length, which will be straight *down* as shown in Fig. 18-27*b*. Since these phasors act in opposite directions, their *sum* is actually the *difference* between the phasors as indicated by $V_d$. After this addition, the phasor diagram looks like Fig. 18-28*a*. Notice that the effect of the capacitor has disappeared. The 40 V of coil voltage have exactly balanced the 25 V of capacitive voltage and have left an excess of 15 V of coil voltage. This 15 V of coil voltage must now be added vectorially to the 8 V of resistance voltage. In Fig. 18-28*b*, the total voltage $V_T$ is obtained by adding

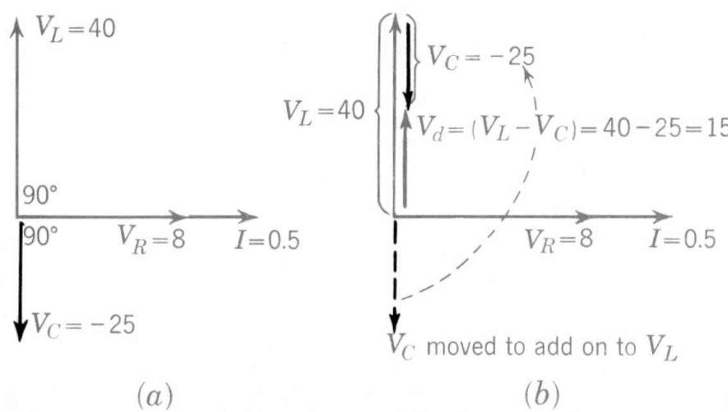


**FIGURE 18-27**
(*a*) Phasor diagram for a series ac circuit containing resistance, inductance, and capacitance. (*b*) $V_d$ represents the phasor sum of $V_L$ and $V_C$.

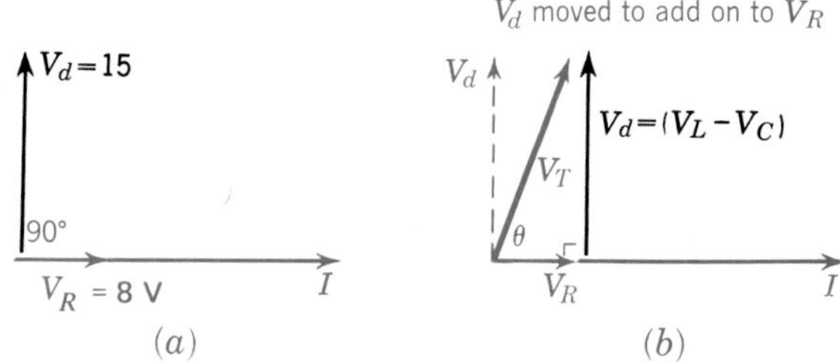


**FIGURE 18-28** (a) $V_d$ represents the result of adding $V_C$ to $V_L$. (b) $V_T$ represents the phasor sum of $V_R + V_L + V_C$.

the phasor $V_d$ to the phasor $V_R$. Place the tail of $V_d$ on the head of $V_R$, and draw it in the proper direction. If $V_C$ is larger than $V_L$, the phasor $V_d$ will have a *downward* direction. The distance from the origin of the phasors to the head of the final phasor is the sum of the phasors. In this instance, the phasor $V_T$ represents the sum of the phasors $V_R + V_L + V_C$. The phase angle $\theta$ is the angle by which the total voltage will lead the current. If $V_C$ were larger than $V_L$, the total voltage would *lag* behind the current by this angle. By applying the pythagorean theorem to Fig. 18-28*b*, we obtain the

**FORMULA** $$V_T^2 = V_R^2 + (V_L - V_C)^2 \qquad \boxed{18\text{-}17}$$

**Total impedance.** To find the impedance of the series ac circuit, we must add $R$, $X_L$, and $X_C$ vectorially as was done with the voltages. This is shown in Fig. 18-29, resulting in the impedance triangle of Fig. 18-29*c*.

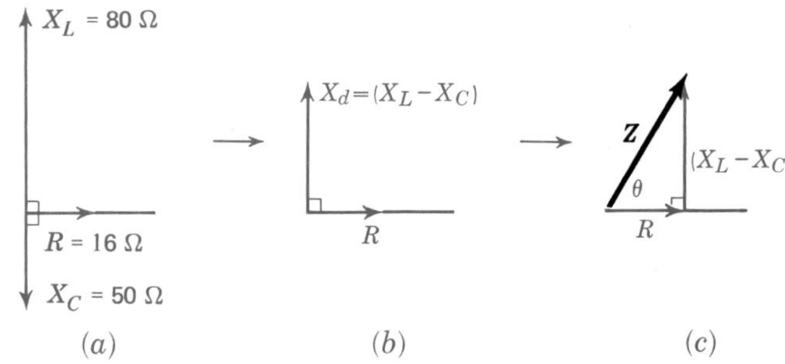


**FIGURE 18-29** Z represents the phasor sum of $R + X_L + X_C$.

Applying the pythagorean theorem to Fig. 18-29*c*, we obtain the

**FORMULAS**

$$Z^2 = R^2 + (X_L - X_C)^2 \qquad \boxed{18\text{-}18}$$

$$Z = \sqrt{R^2 + (X_L - X_C)^2} \qquad \boxed{18\text{-}19}$$

And, by trigonometry,

$$\cos\theta = \frac{a}{h}$$

or

$$\cos\theta = \frac{R}{Z} \qquad (18\text{-}11)$$

The power is still given by

$$W = V \times I \times \cos\theta \qquad (18\text{-}3)$$

or

$$W = V \times I \times \text{PF} \qquad (18\text{-}4)$$

or

$$W = I^2R \qquad (7\text{-}4)$$

EXAMPLE 18-12. In Fig. 18-26, find (*a*) the total voltage, (*b*) the impedance, (*c*) the phase angle, (*d*) the power factor, and (*e*) the power.

SOLUTION

*a.*

$$V_T^2 = V_R^2 + (V_L - V_C)^2 \qquad (18\text{-}17)$$
$$V_T^2 = 8^2 + (40 - 25)^2$$
$$V_T^2 = 8^2 + 15^2 = 64 + 225 = 289$$
$$V_T = \sqrt{289} = 17 \text{ V} \qquad \textit{Ans.}$$

*b.*

$$Z = \sqrt{R^2 + (X_L - X_C)^2} \qquad (18\text{-}19)$$
$$Z = \sqrt{16^2 + (80 - 50)^2}$$
$$Z = \sqrt{16^2 + 30^2} = \sqrt{256 + 900} = \sqrt{1156}$$
$$Z = 34\ \Omega \qquad \textit{Ans.}$$

*c.*

$$\cos\theta = \frac{R}{Z} = \frac{16}{34} = 0.4706 \qquad (18\text{-}11)$$
$$\theta = 62^\circ \text{ (approx)} \qquad \textit{Ans.}$$

*d.*

$$\text{PF} = \frac{R}{Z} = 0.4706 = 47.1\%\ \textit{lagging} \qquad \textit{Ans.} \qquad (18\text{-}12)$$

*e.*

$$W = V \times I \times \text{PF} = 17 \times 0.5 \times 0.471 = 4 \text{ W} \quad \textit{Ans.}\ (18\text{-}4)$$

*f.*

$$W = I^2R = (0.5)^2 \times 16 = 4 \text{ W} \qquad \textit{Check}$$

In this series circuit, the total current of 0.5 A *lags* the total voltage of 17 V by 62°.

EXAMPLE 18-13. An 18-Ω resistor, a 4-μF capacitor, and a 2.5-mH inductance are connected in series across a 60-V 1-kHz ac source. Find (*a*) the capacitive reactance, (*b*) the inductive reactance, (*c*) the impedance, (*d*) the total current, (*e*) the voltage drop across each part, (*f*) the phase angle, (*g*) the power factor, and (*h*) the power.

SOLUTION. Given: $R = 18\ \Omega$ Find: $X_C = ?$

$C = 4\ \mu\text{F}$ $X_L = ?$

$L = 2.5 \text{ mH} = 0.0025 \text{ H}$ $Z = ?$

$V_T = 60 \text{ V}$ $I_T = ?$

$f = 1 \text{ kHz} = 1000 \text{ Hz}$ $V_R = ?$

$V_L = ?$

$$V_C = ?$$
$$\theta = ?$$
$$\text{PF} = ?$$
$$W = ?$$

*a.* $$X_C = \frac{159{,}000}{f \times C} = \frac{159{,}000}{1000 \times 4} = \frac{159}{4} = 40\ \Omega \qquad \textit{Ans.}$$

*b.* $$X_L = 6.28fL = 6.28 \times 1000 \times 0.0025 \qquad (16\text{-}1)$$
$$= 6.28 \times 2.5 = 16\ \Omega \qquad \textit{Ans.}$$

*c.* $$Z = \sqrt{R^2 + (X_C - X_L)^2} \qquad (18\text{-}19)$$

(Notice that $X_C$ is written first because $X_C$ is larger than $X_L$.)

$$Z = \sqrt{18^2 + (40 - 16)^2} = \sqrt{18^2 + 24^2}$$
$$= \sqrt{324 + 576}$$
$$Z = \sqrt{900} = 30\ \Omega \qquad \textit{Ans.}$$

*d.* $$V_T = I_T \times Z \qquad (18\text{-}13)$$
$$60 = I_T \times 30$$
$$I_T = \frac{60}{30} = 2\ \text{A} \qquad \textit{Ans.}$$

*e.* Since $$I_T = I_R = I_L = I_C = 2\ \text{A} \qquad (4\text{-}1)$$

$$V_R = I_R \times R_R \qquad\qquad V_C = I_C \times X_C$$
$$V_R = 2 \times 18 = 36\ \text{V} \quad \textit{Ans.} \qquad V_C = 2 \times 40 = 80\ \text{V} \quad \textit{Ans.}$$

$$V_L = I_L \times X_L$$
$$V_L = 2 \times 16 = 32\ \text{V} \qquad \textit{Ans.}$$

*f.* $$\cos\theta = \frac{R}{Z} = \frac{18}{30} = 0.6000 \qquad (18\text{-}11)$$
$$\theta = 53^\circ\ \text{(approx)} \qquad \textit{Ans.}$$

*g.* $$\text{PF} = \frac{R}{Z} = 0.6000 = 60\%\ \textit{leading} \qquad \textit{Ans.} \qquad (18\text{-}12)$$

*h.* $$W = V \times I \times \text{PF} = 60 \times 2 \times 0.6 = 72\ \text{W} \qquad \textit{Ans.}\ (18\text{-}4)$$

*i.* $$W = I^2R = (2)^2 \times 18 = 72\ \text{W} \qquad \textit{Check}$$

In this series circuit, since the capacitive reactance is larger than the inductive reactance, the total current of 2 A *leads* the total voltage of 60 V by 53°.

## IMPEDANCES IN SERIES

The total impedance of a number of devices in series must include *all* the resistances and *all* the reactances. Since all the resistances are in phase with each other,

$$R_T = R_1 + R_2 + R_3 \qquad (4\text{-}3)$$

Since all the inductive reactances are in phase with each other,

$$X_{L_T} = X_{L_1} + X_{L_2} + X_{L_3} \qquad \boxed{18\text{-}20}$$

Since all the capacitive reactances are in phase with each other,

$$X_{C_T} = X_{C_1} + X_{C_2} + X_{C_3} \qquad \boxed{18\text{-}21}$$

EXAMPLE 18-14. In Fig. 18-30, find (*a*) the impedance, (*b*) the current, (*c*) the phase angle, (*d*) the PF, and (*e*) the power.

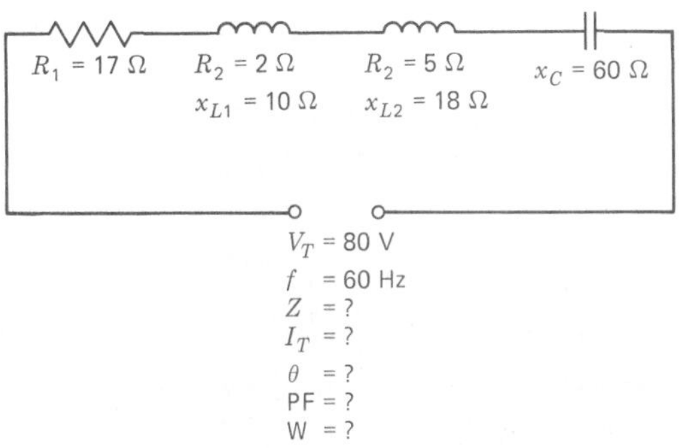

$R_1 = 17\ \Omega$
$R_2 = 2\ \Omega$, $X_{L1} = 10\ \Omega$
$R_2 = 5\ \Omega$, $X_{L2} = 18\ \Omega$
$X_C = 60\ \Omega$
$V_T = 80$ V
$f = 60$ Hz
$Z = ?$
$I_T = ?$
$\theta = ?$
PF = ?
W = ?

FIGURE 18-30

*a.*

$$R_T = R_1 + R_2 + R_3 = 17 + 2 + 5 = 24\ \Omega \qquad (4\text{-}3)$$

$$X_{L_T} = X_{L_1} + X_{L_2} = 10 + 18 = 28\ \Omega \qquad (18\text{-}20)$$

$$Z = \sqrt{R^2 + (X_C - X_L)^2} \qquad (18\text{-}19)$$

$$= \sqrt{24^2 + (60 - 28)^2}$$

$$= \sqrt{24^2 + 32^2} = \sqrt{576 + 1024}$$

$$= \sqrt{1600}$$

$$Z = 40\ \Omega \qquad \textit{Ans.}$$

*b.*

$$V_T = I_T \times Z$$

$$80 = I_T \times 40$$

$$I_T = \frac{80}{40} = 2\ \text{A} \qquad \textit{Ans.}$$

*c.*

$$\cos\theta = \frac{R_T}{Z} = \frac{24}{40} = 0.6000$$

$$\theta = 53^\circ\ \text{(approx)} \qquad \textit{Ans.}$$

*d.* $$\text{PF} = \frac{R_T}{Z} = 0.6000 = 60\%\ \textit{leading} \qquad \textit{Ans.}$$

*e.* $$W = V \times I \times \text{PF} = 80 \times 2 \times 0.6 = 96\text{ W} \qquad \textit{Ans.}$$

*f.* $$W = I^2R_T = (2)^2 \times 24 = 96\text{ W} \qquad \textit{Check}$$

In this series circuit, the total current of 2 A *leads* the total voltage of 80 V by 53°.

EXAMPLE 18-15. A rectifier delivers 200 V at 120 Hz to a filter circuit consisting of a 30-H filter choke coil and a 20-$\mu$F capacitor connected as shown in Fig. 18-31. How much 120-Hz voltage appears across the capacitor which feeds into the plate supply? Has this filter succeeded in removing the ac component from the plate supply?

SOLUTION. The circuit diagram is shown in Fig. 18-31.

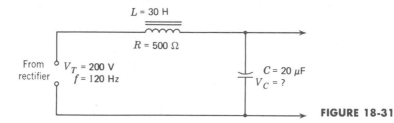

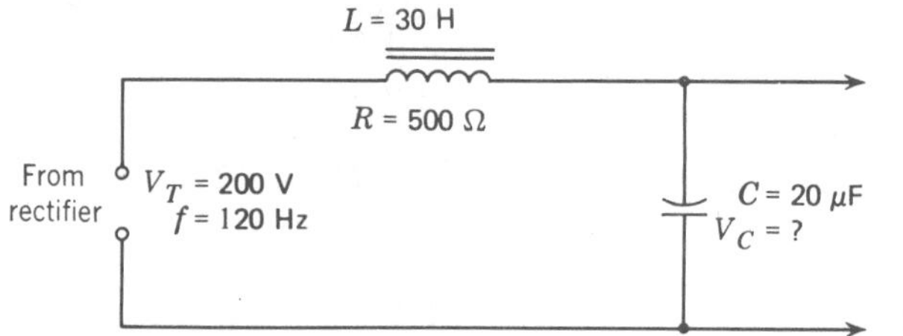


FIGURE 18-31

1. Find the reactance of the coil.

$$X_L = 6.28fL \qquad (16\text{-}1)$$
$$X_L = 6.28 \times 120 \times 30$$
$$X_L = 22{,}600\ \Omega \qquad \textit{Ans.}$$

2. Find the reactance of the capacitor.

$$X_C = \frac{159{,}000}{f \times C} = \frac{159{,}000}{120 \times 20} = \frac{1590}{24} = 66\ \Omega \qquad \textit{Ans.} \qquad (17\text{-}6)$$

3. Find the total reactance effect $X_d$ so that we may compare it with $R$. If $X_d/R$ is greater than 5, we shall be able to neglect $R$ and greatly simplify our calculations.

$$X_d = X_L - X_C$$
$$X_d = 22{,}600 - 66 = 22{,}534\ \Omega \qquad \textit{Ans.}$$

4. Find the $Q$ of the circuit.

$$Q = \frac{X_d}{R} = \frac{22{,}534}{500} = 45+ \qquad \textit{Ans.}$$

Therefore, since $Q$ is greater than 5, we shall neglect the resistance of 500 Ω, and

$$Z = X_d = 22{,}534\ \Omega \qquad \textit{Ans.}$$

5. Find the series current.

$$V_T = I_T \times Z \qquad (18\text{-}13)$$
$$200 = I_T \times 22{,}534$$
$$I_T = \frac{200}{22{,}534} = 0.0089\ \text{A} \qquad \textit{Ans.}$$

6. Find the voltage across the capacitor.

$$V_C = I_C \times X_C \qquad (17\text{-}7)$$
$$V_C = 0.0089 \times 66 = 0.59\ \text{V} \qquad \textit{Ans.}$$

Therefore this is a satisfactory filter, since only 0.59 V out of the total of 200 V of 120-Hz alternating current can get through to the plate supply.

SELF-TEST 18-16. The output voltage of an audio oscillator is 50 V at 3 kHz. It is applied to a series circuit of a 300-Ω resistor, a coil of 531-mH inductance and 75-Ω resistance, a 125-Ω resistor, and a 2650-pF capacitor. Find (*a*) the impedance, (*b*) the total current, and (*c*) the phase angle.

SOLUTION. Given: $R_1 = 300\ \Omega$ Find: $Z = ?$

Coil $\begin{cases} L = 531\ \text{mH} \\ R_2 = 75\ \Omega \end{cases}$ $I_T = ?$ $\theta = ?$

$R_3 = 125\ \Omega$

$C = 2650\ \text{pF}$

$V_T = 50\ \text{V}$

$f = 3\ \text{kHz}$

*a*. In order to find the impedance, we must know the values of $R_T$, $X_L$, and ________. — $X_C$

$R_T = R_1 + R_2 + R_3 = 300 + 75 +$ ________ (4-3) — 125

$R_T =$ ________ Ω — 500

$X_L = 6.28fL = 6.28 \times 3 \times 10^3 \times 531 \times$ ________ (16-1) — $10^{-3}$

$X_L =$ ________ Ω *Ans.* — 10,000

$$X_C = \frac{159{,}000}{f \times C} = \frac{159 \times 10^3}{3 \times 10^3 \times 2650 \times ?} \qquad (17\text{-}6)$$ — $10^{-6}$

$$= \frac{53 \times 10^6}{2650}$$

$X_C = 0.02 \times 10^6 =$ ________ Ω — 20,000

$X_d = X_C - X_L = 20{,}000 - 10{,}000 =$ ________ Ω — 10,000

The $Q$ of the circuit $= X_d/R = 10{,}000/500 =$ ________. Therefore, $R_T =$ 500 Ω __(may/may not)__ be neglected. — 20; may

$Z = X_d =$ ________ Ω *Ans.* — 10,000

*b.*

$$V_T = I_T \times Z$$

$$I_T = \frac{?}{10{,}000} = 0.005 \text{ A} \qquad \textit{Ans.}$$ 50

*c.*

$$\cos\theta = \frac{R}{Z} = \frac{?}{10{,}000} \qquad (18\text{-}8)$$ 500

$$\cos\theta = \_\_\_\_\_\_$$ 0.0500

$$\theta = \_\_\_\_\_\_ \qquad \textit{Ans.}$$ 87°

In this series circuit, since the capacitive reactance is larger than the inductive reactance, the total voltage of 50 V *lags* behind the total current of 0.005 A by 87°.

## PROBLEMS

1. A 16-Ω resistor, an 83-Ω inductive reactance, and a 20-Ω capacitive reactance are in series. A 130-V 60-Hz emf is impressed on the circuit. Find (*a*) the impedance, (*b*) the series current, (*c*) the voltage drops across all the parts, (*d*) the phase angle, (*e*) the power factor, and (*f*) the power.
2. A coil of 2.07-mH inductance, a 0.3-μF capacitor, and a 36-Ω resistor are connected in series across a 127.5-V 10-kHz ac source. Find (*a*) the impedance, (*b*) the total current, (*c*) the phase angle, (*d*) the power factor, and (*e*) the power.
3. A 125-V 100-Hz power supply is connected across a 4000-Ω resistor, a 0.5-μF capacitor, and a 10-H coil connected in series. Find (*a*) the individual reactances, (*b*) the impedance, (*c*) the total current, (*d*) the phase angle, (*e*) the power factor, and (*f*) the power.
4. The antenna circuit of a radio receiver consists of a 0.2-mH inductance and a 0.0001-μF capacitance. If the resistance of the antenna is small enough to be considered to be zero, what is the impedance of the antenna to a 1200-kHz signal? If this frequency induces a voltage of 100 μV in the antenna, what current will flow?
5. In a circuit similar to Fig. 18-31, $V_T = 250$ V, $f = 120$ Hz, $L = 25$ H, $R = 400$ Ω, and $C = 25$ μF. What amount of the 120-Hz voltage will appear across the capacitor?
6. A wave trap to eliminate a 13-kHz frequency is made of a 30-mH inductance of 40 Ω resistance and a 0.005-μF capacitor in series. What is the impedance of the circuit?
7. A 10.8-V 100-kHz emf is applied across a series circuit of a 6-Ω resistance, a 0.5-mH coil, and a 5000-pF capacitance. What is the total current?
8. A 300-Ω 100-μH resistor is in series with a capacitance of 2 μF. Find the impedance of the circuit at (*a*) 500 Hz, (*b*) 5 kHz, and (*c*) 500 kHz
9. A 5-H coil and a 1.67-μF capacitor are in series with an adjustable resistor. What must be the value of the resistance in order to draw 0.3 A from a 120-V 60-Hz line?

10. A 2.8-mH inductance and a 9-$\mu$F capacitance are in series with a 50-$\Omega$ resistor. At what frequency will the inductive reactance equal the capacitive reactance? What is the impedance of the circuit? What current will be drawn at this frequency from a 10-V source?
11. A 4-$\Omega$ resistor and an unknown inductance are connected across a 110-V 60-Hz ac line. Find the value of the inductance if the current flowing is 20 A.
12. A resistance and a capacitor in series take 120 W at a power factor of 54.5 percent (leading) from a 110-V 60-Hz line. Find (*a*) the current, (*b*) the impedance, (*c*) the resistance, and (*d*) the value of the capacitance.

(See Instructor's Manual for Test 18-3)

## JOB 18-6 SERIES RESONANCE

EXAMPLE 18-17. A 30-H coil, a 250-$\Omega$ resistor, and a variable capacitor are connected in series across a 110-V 60-Hz line. When the capacitor is adjusted to 0.2344 $\mu$F, find the impedance of the circuit.

SOLUTION. Given: $L = 30$ H Find: $Z = ?$

$R = 250\ \Omega$

$C = 0.2344\ \mu$F

$V = 110$ V

$f = 60$ Hz

1. Find the inductive reactance.

$$X_L = 6.28fL = 6.28 \times 60 \times 30 = 11{,}305\ \Omega \qquad \textit{Ans.}$$

2. Find the capacitive reactance.

$$X_C = \frac{159{,}000}{f \times C} = \frac{159{,}000}{60 \times 0.2344} = \frac{15{,}900}{1.4064} = 11{,}305\ \Omega \qquad \textit{Ans.}$$

3. Find the impedance.

$$Z = \sqrt{R^2 + (X_C - X_L)^2} = \sqrt{250^2 + (11{,}305 - 11{,}305)^2}$$
$$= \sqrt{250^2 + 0^2}$$
$$= \sqrt{250^2}$$
$$Z = 250\ \Omega \qquad \textit{Ans.}$$

Notice that for these particular values of $L$ and $C$, the inductive reactance and the capacitive reactance are exactly equal. Since their actions are directly opposed to each other, the total effect of both is equal to zero and the impedance of the circuit is equal to just the resistance of the circuit. This condition is called *resonance.* Series resonance is the condition of *smallest* circuit resistance. At resonance, since the reactance effect is zero, the *largest* amount of current will flow.

Any change in the value of either $L$ or $C$ would give a *different* value of $X_L$ or $X_C$, whose sum would no longer be zero. Under these conditions, since the reactance effect is *larger* than zero, the current that flows is *smaller* than the flow at resonance. In addition, since the values of the reactances depend on the frequency, any change in the frequency results in *different* values of reactances whose sum again would *not* be zero. Apparently, for any combination of $L$ and $C$ in series, there is only *one* frequency for which $X_L$ can equal $X_C$. This frequency is called the *resonant frequency*.

At this frequency, a large current will flow in the series circuit, since the reactance is zero at the resonant frequency. At any other frequency, since the sum of $X_L$ and $X_C$ is *not* equal to zero, the impedance will be *larger* and the current will be *smaller*.

**Why is the resonant frequency important?** The antenna of a radio receiver is receiving signals from many stations at the same time. Each station broadcasts at a different frequency. Each different frequency induces a signal voltage in the antenna so that at any one time there may be many different signal voltages in the same antenna. How shall we separate one of these signals from all the rest?

We can separate one frequency from the rest if we can find an $L$ and $C$ combination which is *resonant* to that same frequency which we are trying to separate. Only this frequency will encounter a *zero impedance* and therefore will produce a *large* current. All other frequencies will encounter large impedances, and the currents at these frequencies will be practically zero. The tuner of a receiver is a series circuit of an inductance and a capacitance which can be made resonant to different frequencies by changing the values of the capacitance. Thus it will select and pass on to the amplifying system only one frequency at a time.

There are many other uses for the series resonant circuit. In the superheterodyne receiver, the oscillator must deliver a definite frequency to the mixer tube. Values of $L$ and $C$ are chosen to make the combination resonant to that particular frequency. Bandpass filters and acceptance circuits are other common applications of the series resonant circuit.

**Calculating the resonant frequency.** At resonance, the inductive reactance is equal to the capacitive reactance. Write the equation.

$$X_L = X_C \tag{18-22}$$

Substitute reactances.

$$\frac{6.28fL}{1} = \frac{159{,}000}{f \times C}$$

Cross-multiply.

$$f^2 \times 6.28 \times L \times C = 159{,}000$$

Solve for $f$.

$$f^2 = \frac{159{,}000}{6.28 \times L \times C}$$

Divide.

$$f^2 = \frac{25{,}318}{L \times C}$$

Take the square root of both sides.

$$f = \sqrt{\frac{25{,}318}{L \times C}}$$

**FORMULA**

$$f = \frac{159}{\sqrt{L \times C}} \qquad \boxed{18\text{-}23}$$

where $f$ = frequency, Hz; $L$ = inductance, H; $C$ = capacitance, $\mu$F
or $f$ = frequency, kHz; $L$ = inductance, $\mu$H; $C$ = capacitance, $\mu$F

If we use the units of measurement commonly used in radio, electronics, and television work, the formula retains the same form *but* both the frequency and the inductance are expressed in different units as shown above.

EXAMPLE 18-18. Calculate the resonant frequency of a tuning circuit if the inductance is 300 $\mu$H and the capacitor is set at a capacity of 300 pF.

SOLUTION. The diagram of the circuit is shown in Fig. 18-32. If kilohertz are desired, the 300 pF must be changed to microfarads.

$$300 \text{ pF} = 300 \times 10^{-6}\ \mu\text{F}$$

$$f = \frac{159}{\sqrt{L \times C}} \qquad (18\text{-}23)$$

$$= \frac{159}{\sqrt{300 \times 300 \times 10^{-6}}} = \frac{159}{\sqrt{9 \times 10^{-2}}}$$

$$f = \frac{159}{0.3} = 530 \text{ kHz} \qquad \textit{Ans.}$$

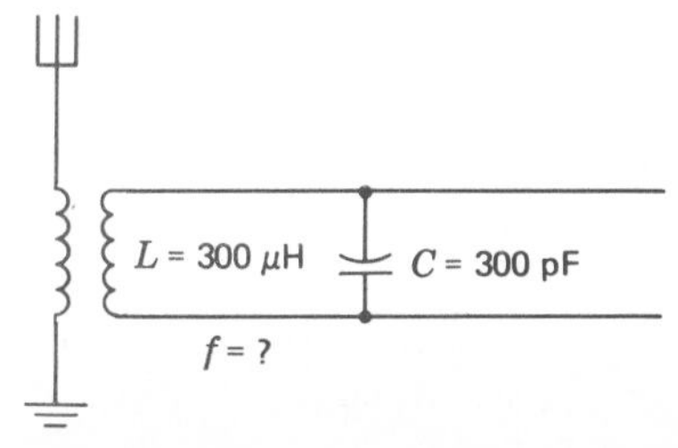


FIGURE 18-32

## PROBLEMS

1. A transmitting antenna has a capacitance of 200 pF, a resistance of 50 Ω, and an inductance of 200 μH. Find (*a*) the resonant frequency and (*b*) the impedance of the antenna.
2. Find the resonant frequency of a tuning circuit similar to that shown in Fig. 18-32 if $L = 250$ μH and $C = 40$ pF.
3. A series circuit consists of a 12-Ω resistance, an inductance of 0.04 H, and a capacitance of 0.16 μF. Find (*a*) the resonant frequency, (*b*) the reactance of the inductance, (*c*) the reactance of the capacitance, (*d*) the impedance, (*e*) the current at resonance if the impressed voltage is 6 V, and (*f*) the voltage drop across the capacitance.
4. What is the resonant frequency of a series circuit if the inductance is 270 μH and the capacitance is 0.003 μF?
5. What is the resonant frequency of the Hartley-type oscillator shown in Fig. 18-33 if the coil has an inductance of 40 μH and the capacitance is set at 160 pF?

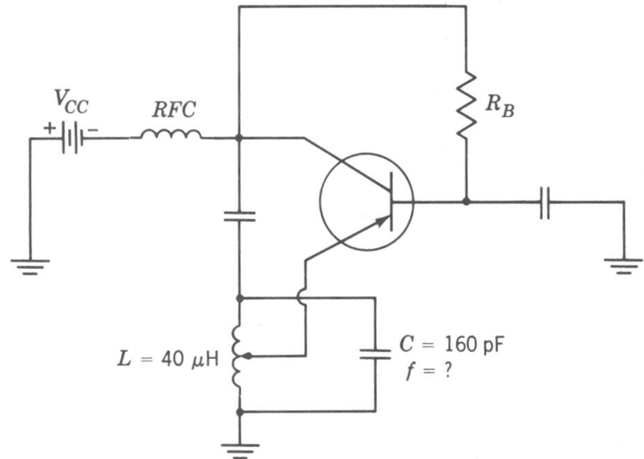


**FIGURE 18-33**
Hartley oscillator.

6. Find the resonant frequency of the series resonant section of the bandpass filter shown in Fig. 18-34.

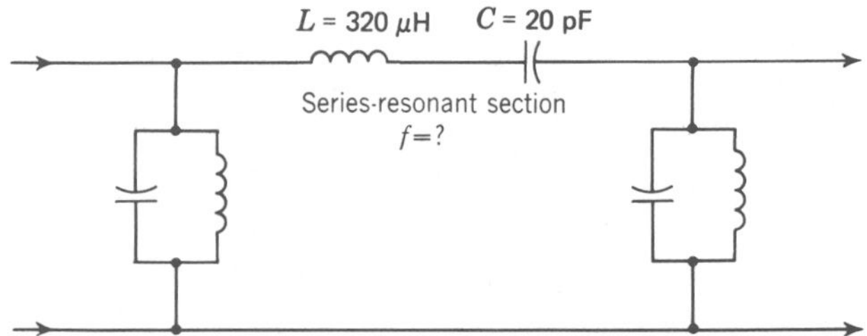


**FIGURE 18-34**
Bandpass filter.

7. Find the resonant frequency of the series resonant section of the wave trap or band-elimination filter shown in Fig. 18-35.

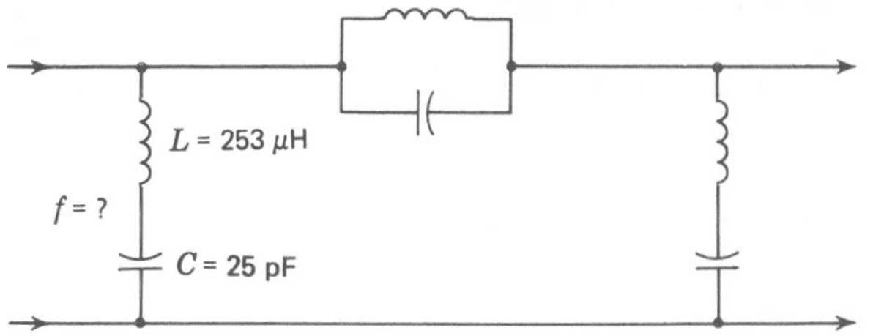


**FIGURE 18-35**
Band-elimination filter.

8. A 3-mH coil and a 40-pF capacitor are connected as shown in Fig. 18-36 to form the secondary side of an IF transformer. What is its resonant frequency? Explain why the secondary is a series tuned circuit while the primary is a parallel circuit.

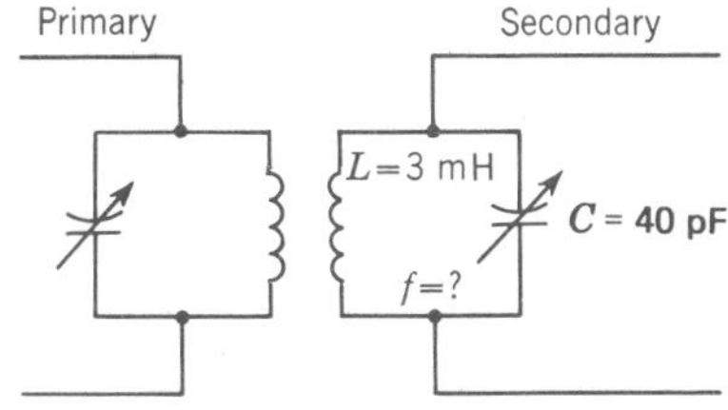


**FIGURE 18-36**
The secondary of the IF transformer is a series resonant circuit.

9. Find the resonant frequency of an antenna circuit if the inductance is 50 $\mu$H and the capacitance is 0.0002 $\mu$F.
10. Find the resonant frequency of a series circuit containing a 300-$\mu$H coil and a 365-pF capacitor.

## JOB 18-7 FINDING THE INDUCTANCE OR CAPACITANCE NEEDED TO MAKE A SERIES RESONANT CIRCUIT

Formula (18-23) may be transformed to find formulas which may be used to find the inductance or capacitance needed to form a series resonant circuit at a given frequency.

**FORMULAS**

$$L = \frac{25{,}300}{f^2 \times C} \qquad \boxed{18\text{-}24}$$

$$C = \frac{25{,}300}{f^2 \times L} \qquad \boxed{18\text{-}25}$$

where $L$ = inductance, $\mu$H
$C$ = capacitance, $\mu$F
$f$ = frequency, kHz

EXAMPLE 18-19. What value of inductance must be placed in series with a 253-pF tuning capacitor in order to provide resonance for a 500-kHz signal?

SOLUTION

Given: $C = 253 \text{ pF} = 253 \times 10^{-6}\ \mu\text{F}$ Find: $L = ?$

$f = 500 \text{ kHz}$

$$L = \frac{25{,}300}{f^2 \times C} = \frac{25{,}300}{(500)^2 \times 253 \times 10^{-6}} \qquad (18\text{-}24)$$

$$= \frac{25{,}300}{25 \times 10^4 \times 253 \times 10^{-6}}$$

$$= \frac{100 \times 10^2}{25}$$

$$L = 400\ \mu\text{H} \qquad \textit{Ans.}$$

EXAMPLE 18-20. An inductance of 40 $\mu$H is in series with a capacitor in a Hartley-type oscillator circuit. Find the value of the capacitance needed to produce resonance to a frequency of 5000 kHz.

SOLUTION. Given: $L = 40\ \mu\text{H}$ Find: $C = ?$

$f = 5000 \text{ kHz}$

$$C = \frac{25{,}300}{f^2 \times L} = \frac{25{,}300}{(5000)^2 \times 40} \qquad (18\text{-}25)$$

$$= \frac{25{,}300}{25 \times 10^6 \times 40}$$

$$= \frac{25{,}300}{10^9}$$

$$= 25{,}300 \times 10^{-9}\ \mu\text{F}$$

$$= 25{,}300 \times 10^{-9} \times 10^6 \text{ pF}$$

$$C = 25.3 \text{ pF} \qquad \textit{Ans.}$$

SELF-TEST 18-21. A series circuit has a resistance of 30 Ω, an inductance of 0.382 H, and a capacitance of 0.2 $\mu$F Find (*a*) the impedance of the circuit to a frequency of 500 Hz, (*b*) the capacitance that must be added in parallel with the 0.2-$\mu$F capacitor to produce resonance at this frequency, and (*c*) the impedance of the circuit at resonance.

SOLUTION. Given: $R = 30\ \Omega$ Find: $Z$ at 500 Hz

$L = 0.382$ H — $C$ to be added

$C = 0.2\ \mu$F — $Z$ at resonance

$f = 500$ Hz

*a.* In order to find $Z$, we must know $X_L$ and ________. | $X_C$

$$X_L = 6.28fL = 6.28 \times 500 \times 0.382$$

$X_L =$ ________ Ω | 1200

$$X_C = \frac{159{,}000}{f \times C} = \frac{159{,}000}{500 \times 0.2} = \text{\_\_\_\_\_\_\_\_}\ \Omega$$ | 1590

$$X_d = X_C - X_L = 1590 - 1200$$

$X_d =$ ______ Ω | 390

$Q = \frac{X_d}{R} = \frac{390}{30} = 13$

and we ___(should/should not)___ include $R$ in our calculations for $Z$. | should not

$Z = X_d =$ ______ Ω *Ans.* | 390

*b.* Find the $C$ that will produce resonance at 500 Hz.

$$C = \frac{25{,}300}{f^2 \times L} \qquad (18\text{-}25)$$

Using the units to obtain microfarads,

$$C = \frac{25{,}300}{(0.5)^2 \times 0.382 \times 10^6}$$

$$= \frac{25{,}300}{? \times 10^6}$$ | 0.0955

$$= \frac{25.3 \times 10^3}{95.5 \times ?}$$ | $10^3$

$C =$ ______ μF *Ans.* | 0.265

Since 0.265 μF are needed for resonance, and we have only 0.2 μF, we must add 0.265 − 0.2 = ______ μF. *Ans.* | 0.065

*c.* At resonance, $X_d$ will equal ______ Ω, and the only resistance in the circuit will be the original resistance of ______ Ω. Therefore, $Z$ = ______ Ω. *Ans.* | zero 30 30

## PROBLEMS

1. What value of inductance must be connected in series with a 0.0003-μF capacitor in order that the circuit be resonant to a frequency of 1000 kHz?
2. What value of capacitance must be connected in series with a 50-μH coil in order that the circuit be resonant to a frequency of 2000 kHz?
3. What value of inductance will produce resonance to 50 Hz if it is placed in series with a 20-μF capacitor?
4. What value of capacitance must be used in series with a 30-μH inductance in order to produce an oscillator frequency of 6000 kHz?
5. What value of capacitance must be added in series with a solenoid of 0.2-H inductance in order to be resonant to 60 Hz?
6. What capacity is necessary in series with a 100-μH coil to produce a wave trap for a 1200-kHz signal?
7. What is the inductance of the secondary winding of an RF transformer if it is in series with a 0.000 35-μF capacitor and is resonant to a frequency of 1000 kHz?
8. What is the capacity of an antenna circuit whose inductance is 50 μH if it is resonant to 1500 kHz?

9. A 0.000 04-μF capacitance is in series with the secondary of an RF transformer. What must be the inductance of the coil if the secondary is to be resonant to 500 kHz?
10. What must be the minimum and maximum values of the capacitor needed to produce resonance with a 240-μH coil to frequencies between 500 and 1500 kHz?

## JOB 18-8 REVIEW OF SERIES AC CIRCUITS

The voltages and current in a series ac circuit are not usually in phase with each other. Alternating-current voltages may not be added arithmetically but only by means of ______ addition. phasor

Since the ______ is constant in a series circuit, the ______ is used as the reference line upon which to draw the phasor diagram. current current

A phasor is a straight line drawn with a definite length and in a definite direction. In ac electricity, the direction of the phasor indicates the ______ at which the voltage or current occurs in relation to another voltage or current. time

Phasors drawn in the ______ direction are in phase. same

Phasors drawn in ______ directions are out of phase. different

Phasors are added by placing the tail of one phasor on to the ______ of another and drawing the phasor with its original length and direction. head

The sum of the phasors is the phasor drawn from the ______ of the phasors to the ______ of the last phasor. origin head

In a purely resistive circuit:

The voltage is __(in/out of)__ phase with the current. in

$$V_T = V_1 + V_2 + V_3 \quad \boxed{4\text{-}2}$$

$$V_T = I_T \times R_T \quad \boxed{4\text{-}7}$$

$$R_T = R_1 + R_2 + R_3 \quad \boxed{4\text{-}3}$$

$$P = V \times I \quad \boxed{6\text{-}1}$$

In a purely inductive circuit:

The voltage __(leads/lags)__ the current by 90°. leads

$$V_T = V_1 + V_2 + V_3 \quad \boxed{4\text{-}2}$$

$$V_L = I_L \times \_\_\_\_\_\_ \quad \boxed{16\text{-}2} \qquad X_L$$

$$X_{L_T} = X_{L_1} + X_{L_2} + X_{L_3} \quad \boxed{18\text{-}20}$$

$$W = V \times I \times \_\_\_\_\_\_ \quad \boxed{18\text{-}3} \qquad \cos\theta$$

$$W = V \times I \times \_\_\_\_\_\_ \quad \boxed{18\text{-}4} \qquad \text{PF}$$

In a purely capacitive circuit:

The voltage __(leads/lags)__ the current by 90°. lags

$$V_T = V_1 + V_2 + V_3 \quad \boxed{4\text{-}2}$$

$$V_C = I_C \times \_\_\_\_\_\_ \quad \boxed{17\text{-}7} \qquad X_C$$

| | | |
|---|---|---|
| $X_{C_T} = X_{C_1} + X_{C_2} + X_{C_3}$ | 18-21 | |
| $W = V \times I \times \cos\theta$ | 18-3 | |
| $W =$ ______ $\times$ ______ $\times$ PF | 18-4 | $V$ $I$ |

In an ac series circuit of resistance and inductance:
The total voltage ___(leads/lags)___ the current by some angle $\theta$. — leads

| | | |
|---|---|---|
| $V_T^2 = V_R^2 +$ ______ | 18-8 | $V_L^2$ |
| $Z =$ ______ | 18-10 | $\sqrt{R^2 + X_L^2}$ |
| $V_T = I_T \times$ ______ | 18-13 | $Z$ |
| $\cos\theta = \dfrac{R}{?}$ | 18-11 | $Z$ |
| $\text{PF} = \dfrac{?}{Z}$ | 18-12 | $R$ |
| $W = V \times I \times$ ______ | 18-4 | PF |

In an ac series circuit of resistance and capacitance:
The total voltage ___(leads/lags)___ the current by some angle $\theta$. — lags

| | | |
|---|---|---|
| $V_T^2 = V_R^2 + V_C^2$ | 18-14 | |
| $Z =$ ______ | 18-16 | $\sqrt{R^2 + X_C^2}$ |
| $V_T = I_T \times$ ______ | 18-13 | $Z$ |
| $\cos\theta = \dfrac{?}{Z}$ | 18-11 | $R$ |
| $\text{PF} = \dfrac{?}{?}$ | 18-12 | $R$ $Z$ |
| $W = V \times I \times$ ______ | 18-4 | PF |

In an ac series circuit of resistance, inductance, and capacitance:
The total voltage will lead or lag the current, depending on the values of $X_L$ and $X_C$. If $X_L$ is larger than $X_C$, the voltage will ___(lead/lag)___ the current. If $X_L$ is smaller than $X_C$, the voltage will ___(lead/lag)___ behind the current. The angle of lead or lag is given by the angle $\theta$. — lead, lag

| | | |
|---|---|---|
| $V_T^2 = V_R^2 + (V_L - V_C)^2$ | 18-17 | |
| $Z = \sqrt{R^2 + \text{\_\_\_\_\_\_}}$ | 18-19 | $(X_L - X_C)^2$ |
| $V_T = I_T \times$ ______ | 18-13 | $Z$ |
| $\cos\theta = \dfrac{R}{Z}$ | 18-11 | |
| $\text{PF} = \dfrac{R}{Z}$ | 18-12 | |
| $W = V \times I \times \cos\theta$ | 18-4 | |

Without regard for the PF of a circuit, the effective power is the power drawn by the *resistive* elements of the circuit, the formula is

$$W = \underline{\qquad} \qquad \boxed{7\text{-}4}$$ $I^2R$

Series resonance is the condition at which the inductive reactance is exactly equal to the capacitive reactance. For any given combination of coil and capacitor, there is only one frequency at which this situation can occur. This frequency is called the *resonant frequency.*

$$f = \frac{159}{\sqrt{L \times C}} \qquad \boxed{18\text{-}23}$$

where $f$ = frequency, measured in kHz
$L$ = inductance, measured in ______ μH
$C$ = capacitance, measured in ______ μF

The inductance or capacitance needed to make a circuit resonant to a given frequency is given by the formulas

$$L = \frac{25{,}300}{f^2 \times ?} \qquad \boxed{18\text{-}24}$$ $C$

$$C = \frac{25{,}300}{? \times L} \qquad \boxed{18\text{-}25}$$ $f^2$

where all units are measured in the units given for formula (18-23).

## PROBLEMS

1. What is the voltage drop across a 3000-Ω resistor carrying 60 mA of current at a frequency of 60 Hz?
2. Find the current and power drawn by a 20-mH coil from a 125.6-V 10-kHz source.
3. A filter choke passes 60 mA of current when it is connected across a 120-V 60-Hz line. What is its inductance?
4. Find the voltage drop across a 0.05-μF capacitor if it passes 50 mA at a frequency of 1 kHz.
5. A 50-V emf at 1-kHz frequency is impressed across a 1000-Ω resistor in series with a 1-H coil. Find (*a*) the impedance, (*b*) the total current, (*c*) the voltage drop across each part, (*d*) the phase angle, (*e*) the power factor, and (*f*) the power.
6. A 120-V 1-kHz ac voltage is applied across a series circuit of a 200-Ω resistor and a 1.6-μF capacitor. Find (*a*) the impedance, (*b*) the total current, (*c*) the voltage drop across each part, (*d*) the phase angle, (*e*) the power factor, and (*f*) the power.
7. An antenna circuit consists of a 10-Ω resistance, a 0.5-mH inductance, and a 50-pF capacitance. Find its impedance to a frequency of (*a*) 1000 kHz and (*b*) 500 kHz.
8. A 1000-Ω 100-μH coil is in series with a capacitance of 5 μF. Find the impedance at (*a*) 500 Hz, (*b*) 5 kHz, and (*c*) 500 kHz.

9. Find the frequency of operation of the Colpitts oscillator shown in Fig. 18-37. Note that $C_1$ and $C_2$ are in series.

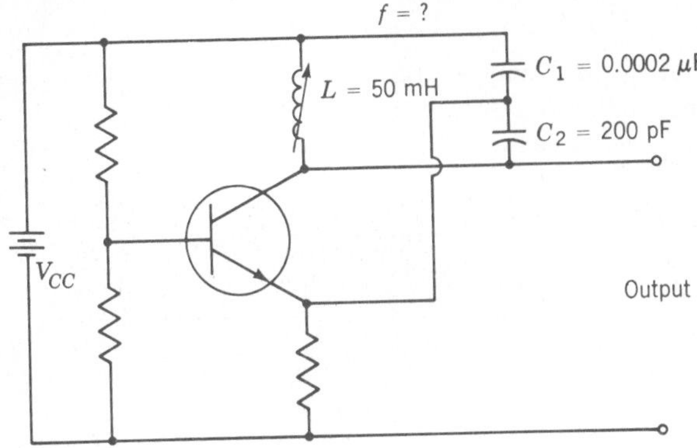


**FIGURE 18-37**
Colpitts oscillator.

10. What capacitance is needed in series with a 0.5-mH coil in order to produce resonance to 10 kHz?

(See Instructor's Manual for Test 18-4, Series AC Circuits)

# PARALLEL AC CIRCUITS

## JOB 19-1 SIMPLE PARALLEL AC CIRCUITS

The general rules for solving dc parallel circuits are also applicable to the solution of ac parallel circuits.

1. The voltages across all branches are equal to each other and to the total voltage.

$$V_T = V_1 = V_2 = V_3 \tag{5-1}$$

2. The total current is equal to the sum of all the branch currents.

$$I_T = I_1 + I_2 + I_3 \tag{5-2}$$

**Parallel circuits containing only resistance.** We can add the branch currents as indicated by formula (5-2) only if the branch currents are in phase. If they are out of phase, they may be added *only* by phasor addition. Let us draw the phasor diagrams for each branch of the circuit shown in Fig. 19-3 for Example 19-1 to discover the phase relationships in this type of circuit.

**Phasor diagrams.** Since the voltage in a parallel circuit is constant, the *voltage* is used as the reference line upon which to draw the phasors. In Fig. 19-1*a,* the current $I_1$ is drawn in the same direction as the voltage because the current in the purely resistive iron is in phase with the voltage. In Fig. 19-1*b,* the smaller current through the lamp $I_2$ is also drawn in the same direction as the voltage because the current through a purely resistive lamp is in phase with the voltage. Now let us draw both sets of phasors on the same voltage base as shown in Fig. 19-2*a.* This diagram indicates that the current in one resistance is in phase with the

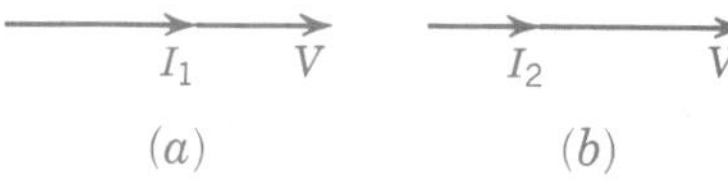


**FIGURE 19-1**
Currents in purely resistive parallel branches are in phase with the voltage.

current in the other resistance, since they are both drawn in the same direction. To find the total current, it is necessary only to add the two current phasors. This is done, as with any phasor quantities, by adding the tail of phasor $I_2$ on to the head of phasor $I_1$ as shown in Fig. 19-2*b*. The total current is then the distance from the origin of the phasors to the head of the last phasor. Since the two currents are in phase, the total current may be found by the direct arithmetical addition of the currents, using formula (5-2).

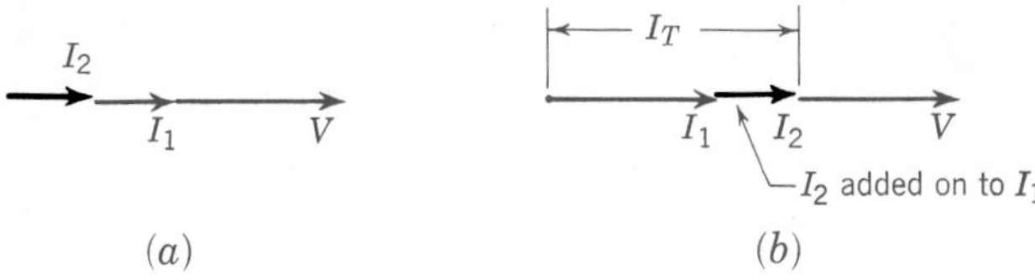


**FIGURE 19-2**
(*a*) Resistive branch currents are in phase with each other. (*b*) $I_T$ represents the phasor sum of $I_1$ and $I_2$.

EXAMPLE 19-1. A 20-Ω electric iron and a 100-Ω lamp are connected in parallel across a 120-V 60-Hz ac line. Find (*a*) the total current, (*b*) the total resistance, (*c*) the total power drawn by the circuit, and (*d*) the PF.

SOLUTION. The diagram for the circuit is shown in Fig. 19-3.

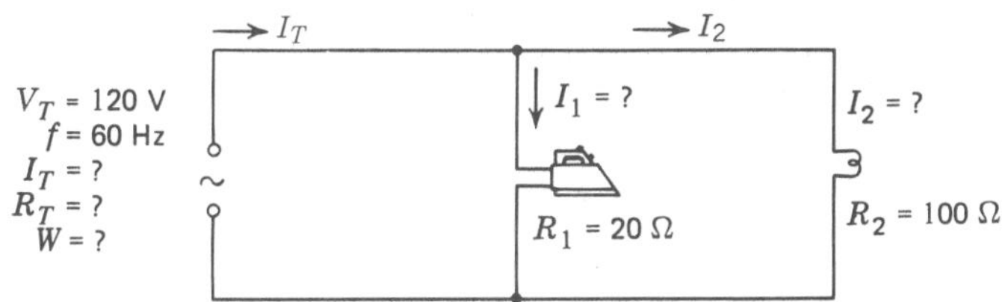


**FIGURE 19-3**

*a.* Find the branch currents. Since

$$V_T = V_1 = V_2 = V_3 = 120 \text{ V} \qquad (5\text{-}1)$$

$$V_1 = I_1 \times R_1 \qquad V_2 = I_2 \times R_2$$

$$120 = I_1 \times 20 \qquad 120 = I_2 \times 100$$

$$I_1 = \frac{120}{20} = 6 \text{ A} \qquad I_2 = \frac{120}{100} = 1.2 \text{ A}$$

Find the total current.

$$I_T = I_1 + I_2 = 6 + 1.2 = 7.2 \text{ A} \quad \textit{Ans.} \qquad (5\text{-}2)$$

*b.*

$$V_T = I_T \times R_T \qquad (4\text{-}7)$$

$$120 = 7.2 \times R_T$$

$$R_T = \frac{120}{7.2} = 16.7\ \Omega \quad \textit{Ans.}$$

*c.* In a purely resistive set of branch circuits, the total current is in phase with the total voltage. The phase angle is therefore equal to 0°.

$$W = V \times I \times \cos\theta = 120 \times 7.2 \times \cos 0° \qquad (18\text{-}3)$$

$$W = 120 \times 7.2 \times 1 = 864 \text{ W} \qquad \textit{Ans.}$$

*d.* $$\text{PF} = \cos\theta = \cos 0° = 1 \qquad \textit{Ans.}$$

**Parallel circuits containing only inductance.** This type of circuit is illustrated in Fig. 19-6 for Example 19-2 below.

**Phasor diagrams.** Since the voltage in a parallel circuit is constant, the *voltage* is used as the reference line upon which to draw the phasors. In Fig. 19-4*a*, the current $I_{L_1}$ is drawn lagging the voltage by 90°. In Fig. 19-4*b*, the current $I_{L_2}$ is also drawn lagging the voltage by 90°, since the current in *any* inductance lags the voltage by 90°. Now let us draw both sets of phasors on the same voltage base as shown in Fig. 19-5*a*. This diagram indicates that the current in one coil is in phase with the current in the second coil, since they are both drawn in the same direction. To find the total current, it is necessary only to add the two current phasors. This is done, as with any phasor quantities, by adding the tail of phasor $I_{L_2}$ on to the head of phasor $I_{L_1}$ as shown in Fig. 19-5*b*. The total current is then the distance from the origin of the phasors to the head of the last phasor. Since the two currents are in phase, the total current may be found by the direct arithmetical addition of the currents, using formula (5-2). The difference between this circuit and the purely resistive circuit lies in the fact that the total current *lags* behind the total voltage by 90°. The phase angle $\theta = 90°$.

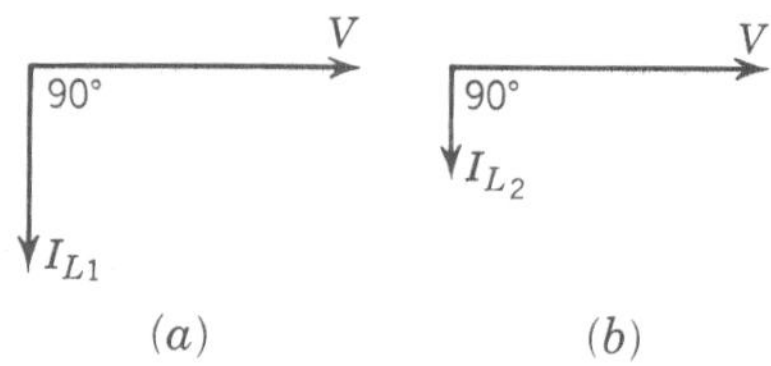


**FIGURE 19-4**
Currents in purely inductive parallel branches lag the voltage by 90°.

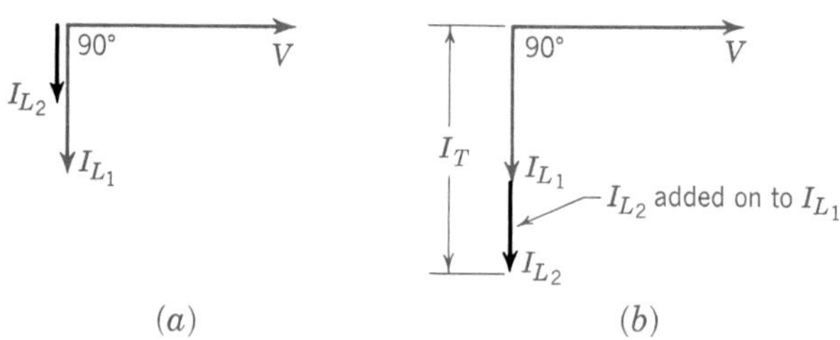


**FIGURE 19-5**
(*a*) Inductive branch currents are in phase with each other. (*b*) $I_T$ represents the phasor sum of $I_{L_1}$ and $I_{L_2}$.

EXAMPLE 19-2. Two coils of 20 and 30 Ω reactance, respectively, are connected in parallel across a 120-V 60-Hz ac line. Find (*a*) the total current, (*b*) the impedance, (*c*) the power factor, and (*d*) the power drawn by the circuit.

SOLUTION. The diagram for the circuit is shown in Fig. 19-6.

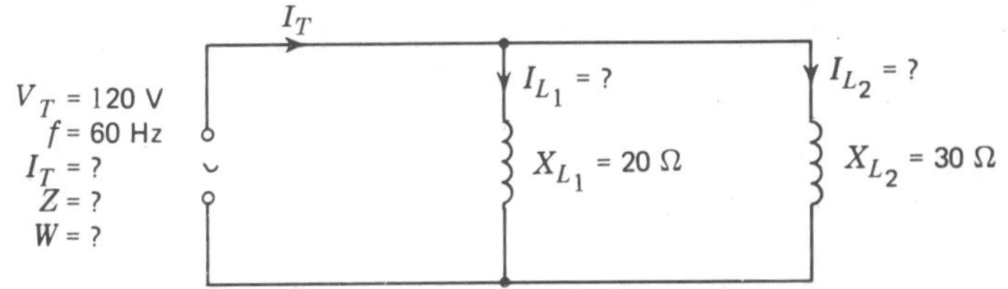


FIGURE 19-6

*a.* Find the branch currents. Since

$$V_T = V_1 = V_2 = 120 \text{ V} \tag{5-1}$$

$$V_1 = I_1 \times X_1 \qquad V_2 = I_2 \times X_2$$
$$120 = I_1 \times 20 \qquad 120 = I_2 \times 30$$
$$I_1 = \frac{120}{20} = 6 \text{ A} \qquad I_2 = \frac{120}{30} = 4 \text{ A}$$

Find the total current.

$$I_T = I_1 + I_2 = 6 + 4 = 10 \text{ A} \quad \textit{Ans.} \tag{5-2}$$

*b.*

$$V_T = I_T \times Z \tag{18-13}$$
$$120 = 10 \times Z$$
$$Z = \frac{120}{10} = 12\ \Omega \quad \textit{Ans.}$$

*c.* In a purely inductive set of branch circuits, the total current lags behind the voltage by 90°. The phase angle is therefore equal to 90°.

$$\text{PF} = \cos\theta = \cos 90° = 0 \quad \text{Ans.}$$

*d.*

$$W = V \times I \times \cos\theta = 120 \times 10 \times \cos 90° \tag{18-3}$$
$$W = 120 \times 10 \times 0 = 0 \text{ W} \quad \textit{Ans.}$$
$$W = I^2R = (10)^2 \times 0 = 0 \text{ W} \quad \textit{Check}$$

**Parallel circuits containing only capacitance.** This type of circuit is illustrated in Fig. 19-9 for Example 19-3 on page 551.

**Phasor diagrams.** Since the voltage in a parallel circuit is constant, the *voltage* is used as the reference line upon which to draw the phasors. In Fig. 19-7*a*, the current $I_{C_1}$ is drawn *leading* the voltage by 90°. In Fig. 19-7*b*, the current $I_{C_2}$ is also drawn *leading* the voltage by 90°, since the

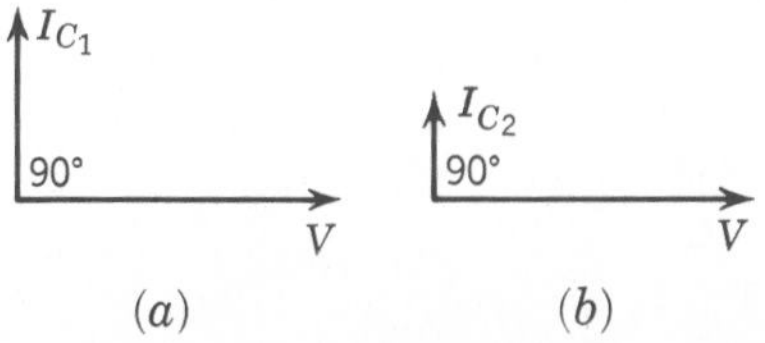

$I_{C_1}$ 90° $V$ (*a*) $I_{C_2}$ 90° $V$ (*b*)

FIGURE 19-7 Currents in purely capacitive parallel branches lead the voltage by 90°.

current in *any* capacitance leads the voltage by 90°. Now let us draw both sets of phasors on the same voltage base, as shown in Fig. 19-8*a*. This diagram indicates that the current in one capacitor is in phase with the current in the second capacitor, since they are both drawn in the same direction. To find the total current, it is necessary only to add the two current phasors. This is done, as with any phasor quantities, by adding the tail of the phasor $I_{C_2}$ on to the head of phasor $I_{C_1}$ as shown in Fig. 19-8*b*. The total current is then the distance from the origin of the phasors to the head of the last phasor. Since the two currents are in phase, the total current may be found by the direct arithmetical addition of the currents, using formula (4-2). The difference between this and the other two circuits lies in the fact that the total current *leads* the total voltage by 90°. The phase angle $\theta = 90°$.

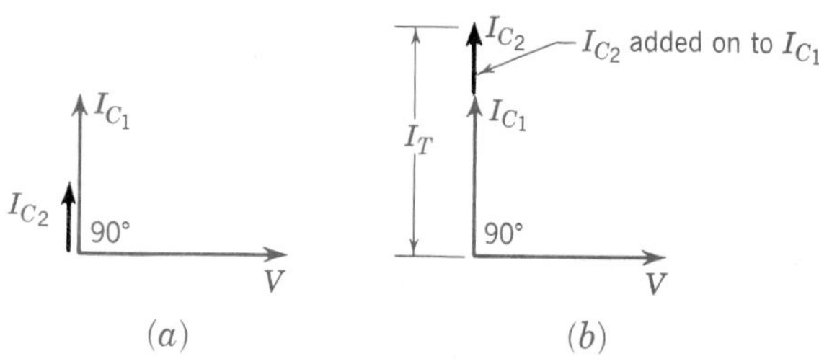


**FIGURE 19-8**
(a) Capcitive branch currents are in phase with each other. (b) $I_T$ represents the phasor sum of $I_{C_1}$ and $I_{C_2}$.

EXAMPLE 19-3. Two capacitors of 30 and 40 Ω reactance, respectively, are connected across a 120-V 60-Hz ac line. Find (*a*) the total current, (*b*) the impedance, (*c*) the power factor, and (*d*) the power drawn by the circuit.

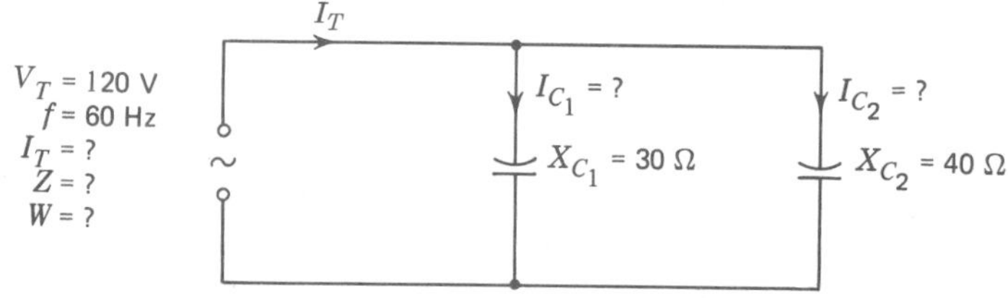


**FIGURE 19-9**

SOLUTION. The diagram for the circuit is shown in Fig. 19-9.

*a*. Find the branch currents. Since

$$V_T = V_1 = V_2 = 120 \text{ V} \qquad (5\text{-}1)$$

$$V_1 = I_1 \times X_{C_1} \qquad V_2 = I_2 \times X_{C_2}$$

$$120 = I_1 \times 30 \qquad 120 = I_2 \times 40$$

$$I_1 = \frac{120}{30} = 4 \text{ A} \qquad I_2 = \frac{120}{40} = 3 \text{ A}$$

Find the total current.

$$I_T = I_1 + I_2 = 4 + 3 = 7 \text{ A} \qquad \textit{Ans.} \qquad (5\text{-}2)$$

*b.*

$$V_T = I_T \times Z \qquad (18\text{-}13)$$

$$120 = 7 \times Z$$

$$Z = \frac{120}{7} = 17.1\ \Omega \qquad \textit{Ans.}$$

*c.* In a purely capacitive set of branch circuits, the total current leads the voltage by 90°. The phase angle is therefore equal to 90°.

$$\text{PF} = \cos\theta = \cos 90° = 0 \qquad \textit{Ans.}$$

*d.*

$$W = V \times I \times \cos\theta = 120 \times 7 \times \cos 90° \qquad (18\text{-}3)$$

$$W = 120 \times 7 \times 0 = 0 \text{ W} \qquad \textit{Ans.}$$

$$W = I^2R = (7)^2 \times 0 = 0 \text{ W} \qquad \textit{Check}$$

SUMMARY. In a parallel ac circuit,

1. The voltage across any branch is equal to the total voltage.
2. The current in a resistor is ___(in/out of)___ phase with the voltage. — in
3. The current in an inductance ___(leads/lags)___ the voltage by ______°. — lags 90
4. The current in a capacitance ___(leads/lags)___ the voltage by ______°. — leads 90
5. Currents that are in phase with each other are added ___(vectorially/arithmetically)___ using formula (5-2). — arithmetically
6. Ohm's law, $V_T = I_T \times Z$ [formula (18-13)], may be used for total values.
7. The power depends on the angle of lead or lag as shown by formula (18-3), $W = V \times I \times$ ______. — $\cos\theta$

## PROBLEMS

1. A 10-Ω electric heater and a 50-Ω incandescent lamp are placed in parallel across a 120-V 60-Hz ac line. Find (*a*) the total current, (*b*) the total resistance, and (*c*) the power drawn.
2. Two toy-train solenoids used in semaphore signals have inductive reactances of 24 and 48 Ω, respectively. They are connected in parallel across the 12-V winding of the power transformer. Find (*a*) the total current, (*b*) the impedance, (*c*) the power factor, and (*d*) the power drawn.
3. Two capacitors of 100 and 200 Ω capacitive reactance, respectively, are connected in parallel across a 100-V 60-Hz ac line. Find (*a*) the total current, (*b*) the impedance, (*c*) the power factor, and (*d*) the power drawn.
4. A 40-Ω soldering iron and a 100-Ω incandescent lamp are connected in parallel across a 110-V 60-Hz ac line. Find (*a*) the total current, (*b*) the total resistance, (*c*) the power factor, and (*d*) the power drawn.
5. Find the total current, impedance, and power used by the circuit shown in Fig. 19-10.

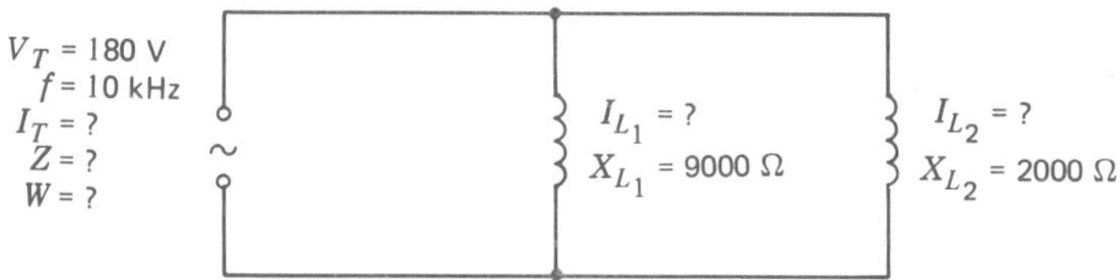


FIGURE 19-10

6. A mechanic replaced a leaky coupling capacitor with a parallel combination of two capacitors as shown in Fig. 19-11. If the voltage drop across the capacitors is 0.5 V at a frequency of 1 kHz, find the current in each capacitor, the total current, and the impedance of the combination.

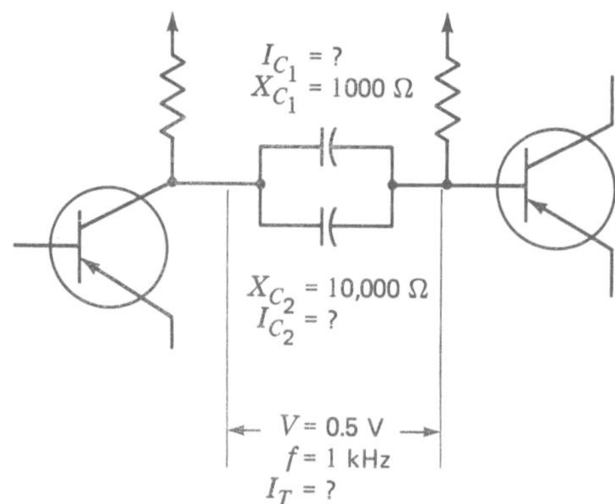


FIGURE 19-11

7. Two pure inductances of 2 and 5 H inductance are connected in parallel across a 120-V 60-Hz line. Find (*a*) the current in each inductance, (*b*) the total current, (*c*) the impedance, (*d*) the power factor, and (*e*) the power drawn.
8. Two pure capacitances of 0.159 and 0.04 μF are connected in parallel across a 50-V 1-kHz line. Find (*a*) the current in each capacitance, (*b*) the total current, (*c*) the impedance, (*d*) the power factor, and (*e*) the power drawn.

## JOB 19-2 RESISTANCE AND INDUCTANCE IN PARALLEL

Figure 19-13 shows a 24-Ω resistance and a 30-Ω inductive reactance in parallel across a 12-V 60-Hz ac source. The total current drawn by the circuit can be found by adding the currents in each branch. However, if they are not in phase, they must be added vectorially. The phase relations in the circuit are shown in Fig. 19-12*a*. The constant voltage is used as the reference line upon which to draw the phasor diagram. The current in the resistance is in phase with the voltage, and the current in the inductance lags the voltage by 90°.

**Addition of the phasors.** In Fig. 19-12*b*, the total current $I_T$ is ob-

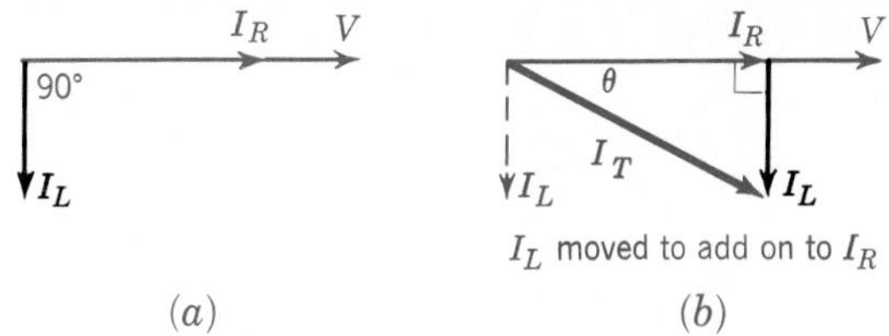


**FIGURE 19-12**
(a) Phasor diagram for a parallel ac circuit containing resistance and inductance. (b) $I_T$ represents the phasor sum of $I_R$ and $I_L$.

tained by adding the phasor $I_L$ to the phasor $I_R$. Place the tail of $I_L$ on the head of $I_R$, and draw it in its original direction and length. The distance from the origin of phasors to the head of the final phasor is the sum of the phasors. In this instance, the phasor $I_T$ represents the sum of the phasors $I_R$ and $I_L$. The phase angle $\theta$ is the angle by which the total current *lags* behind the total voltage. By applying the pythagorean theorem to Fig. 19-12*b*, we obtain the

**FORMULA**

$$I_T^2 = I_R^2 + I_L^2 \qquad \boxed{19\text{-}1}$$

By Ohm's law,

$$V_T = I_T \times Z \qquad (18\text{-}13)$$

Note that the impedance $Z$ is *not* found by the phasor addition of $R$ and $X$. The formula $Z^2 = R^2 + X^2$ applies *only* to series ac circuits. In parallel circuits, the *currents* are added vectorially as shown by formula (19-1) above. The impedance $Z$ is found by applying formula (18-13).

By trigonometry,

$$\cos\theta = \frac{a}{h}$$

**FORMULAS**

$$\cos\theta = \frac{I_R}{I_T} \qquad \boxed{19\text{-}2}$$

or

$$\text{PF} = \frac{I_R}{I_T} \qquad \boxed{19\text{-}3}$$

The power is still given by

$$W = V \times I \times \cos\theta \qquad (18\text{-}3)$$

or

$$W = V \times I \times \text{PF} \qquad (18\text{-}4)$$

or

$$W = I_R^2 \times R \qquad (7\text{-}4)$$

EXAMPLE 19-4. A toy electric-train semaphore is made of a 24-Ω lamp in parallel with a solenoid coil of 30 Ω inductive reactance. If it operates from the 12-V winding of the 60-Hz power transformer, find (*a*) the total current, (*b*) the impedance, (*c*) the phase angle, (*d*) the power factor, and (*e*) the power drawn.

SOLUTION. The diagram of the circuit is shown in Fig. 19-13.

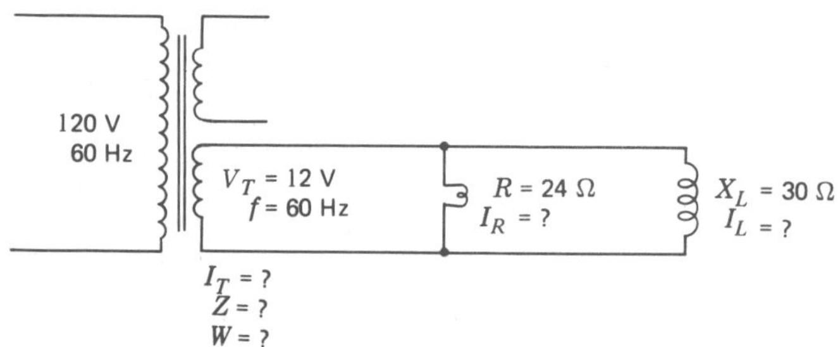


**FIGURE 19-13**
Ac parallel circuit containing resistance and inductance.

*a.* Find the branch currents. Since

$$V_T = V_R = V_L = 12 \text{ V} \qquad (5\text{-}1)$$

$$V_R = I_R \times R \qquad V_L = I_L \times X_L$$

$$12 = I_R \times 24 \qquad 12 = I_L \times 30$$

$$I_R = \frac{12}{24} = 0.5 \text{ A} \qquad I_L = \frac{12}{30} = 0.4 \text{ A}$$

Find the total current.

$$I_T^2 = I_R^2 + I_L^2 = 0.5^2 + 0.4^2 \qquad (19\text{-}1)$$

$$= 0.25 + 0.16$$

$$I_T^2 = 0.41$$

$$I_T = \sqrt{0.41} = 0.64 \text{ A} \qquad \textit{Ans.}$$

*b.*

$$V_T = I_T \times Z \qquad (18\text{-}13)$$

$$12 = 0.64 \times Z$$

$$Z = \frac{12}{0.64} = 18.75\ \Omega \qquad \textit{Ans.}$$

*c.*

$$\cos\theta = \frac{I_R}{I_T} = \frac{0.5}{0.64} = 0.7812 \qquad (19\text{-}2)$$

$$\theta = 39° \qquad \textit{Ans.}$$

*d.*

$$\text{PF} = \frac{I_R}{I_T} = 0.7812 = 78.1\%\ \textit{lagging} \qquad \textit{Ans.} \qquad (19\text{-}3)$$

*e.*

$$W = V \times I \times \cos\theta = 12 \times 0.64 \times \cos 39° \qquad (18\text{-}3)$$

$$= 12 \times 0.64 \times 0.781 = 6 \text{ W} \qquad \textit{Ans.}$$

$$W = I_R^2 \times R = (0.5)^2 \times 24 = 6 \text{ W} \qquad \textit{Check}$$

In the parallel circuit, the total current of 0.64 A *lags* the total voltage of 12 V by 39°.

EXAMPLE 19-5. The purpose of the "high-pass" circuit shown in Fig. 19-14 is to permit high frequencies to pass on to the load but to prevent the passage of low frequencies. Find the effectiveness of the circuit by calculating (*a*) the branch currents, (*b*) the total current, and (*c*) the percent of the total current in the resistor for (1) a 1-kHz audio frequency and (2) a 1000-kHz radio frequency.

SOLUTION. The diagram for the circuit is shown in Fig. 19-14.

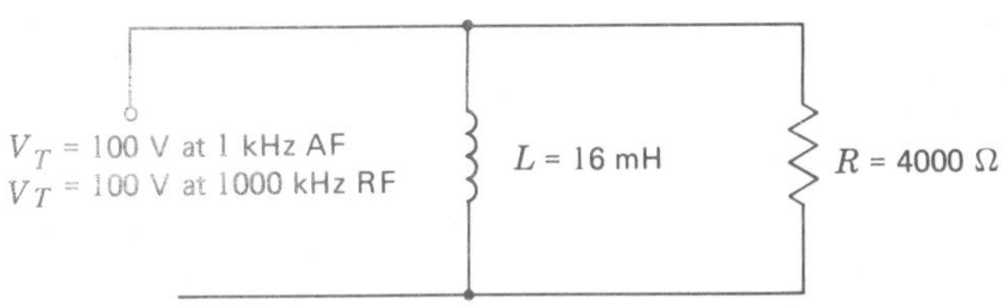


**FIGURE 19-14**
High-pass filter.

1. For the 1-kHz audio frequency,

*a.*
$$X_L = 6.28\,fL = 6.28 \times 10^3 \times 16 \times 10^{-3} \quad (16\text{-}1)$$
$$X_L = 6.28 \times 16 = 100\ \Omega \quad \textit{Ans.}$$

Find the branch currents. Since

$$V_T = V_R = V_L = 100 \text{ V} \quad (5\text{-}1)$$

$$V_R = I_R \times R \qquad V_L = I_L \times X_L$$
$$100 = I_R \times 4000 \qquad 100 = I_L \times 100$$
$$I_R = \frac{100}{4000} = 0.025 \text{ A} \qquad I_L = \frac{100}{100} = 1 \text{ A} \quad \textit{Ans.}$$

*b.* Find the total current.

$$I_T^2 = I_R^2 + I_L^2 = 0.025^2 + 1^2 \quad (19\text{-}1)$$
$$= 0.000\,625 + 1$$
$$I_T^2 = 1.000\,625$$
$$I_T = \sqrt{1.000\,625} = 1 \text{ A} \quad \textit{Ans.}$$

*c.* Find the percent of the total current passing through the resistor.

$$\text{Percent} = \frac{I_R}{I_T} \times 100 = \frac{0.025}{1} \times 100 = 2.5\% \quad \textit{Ans.}$$

That is, 2.5 percent of the 1-kHz audio frequency passes through the resistor.

2. For the 1000-kHz radio frequency: Since 1000 kHz is 1000 times as large as the audio frequency of 1 kHz, the $X_L$ at 1000 kHz will be equal to 1000 times the $X_L$ at 1 kHz. Therefore,

$$X_L = 1000 \times 100 = 100{,}000\ \Omega$$

Find the branch currents. Since

*a.*
$$V_T = V_R = V_L = 100 \text{ V} \quad (5\text{-}1)$$

$$V_R = I_R \times R \qquad V_L = I_L \times X_L$$
$$100 = I_R \times 4000 \qquad 100 = I_L \times 100{,}000$$
$$I_R = \frac{100}{4000} = 0.025 \text{ A} \qquad I_L = \frac{100}{100{,}000} = 0.001 \text{ A} \quad \textit{Ans.}$$

*b.* Find the total current.

$$I_T^2 = I_R^2 + I_L^2 = 0.025^2 + 0.001^2 \qquad (19\text{-}1)$$
$$= (25 \times 10^{-3})^2 + (1 \times 10^{-3})^2$$
$$= 625 \times 10^{-6} + 1 \times 10^{-6}$$
$$I_T^2 = 626 \times 10^{-6}$$
$$I_T = \sqrt{626 \times 10^{-6}} = 25 \times 10^{-3}$$
$$I_T = 0.025 \text{ A} \qquad \textit{Ans.}$$

*c.* Find the percent of the total current passing through the resistor.

$$\text{Percent} = \frac{I_R}{I_T} \times 100 = \frac{0.025}{0.025} \times 100 = 100\% \qquad \textit{Ans.}$$

That is, practically 100 percent of the 1000-kHz radio frequency passes through the resistor.

The circuit is evidently a good high-pass circuit, since it passes practically 100 percent of the high radio frequency and only 2.5 percent of the low audio frequency. The majority of the low audio frequency finds an easy path through the coil ($I_L$ for the 1-kHz audio frequency equals the total current of 1 A).

## PROBLEMS

1. A 24-Ω resistor and a 10-Ω inductive reactance are in parallel across a 120-V, 60-Hz ac line. Find (*a*) the total current, (*b*) the impedance, (*c*) the phase angle, (*d*) the power factor, and (*e*) the power drawn.
2. Repeat Prob. 1 for a 1000-Ω resistor and a 100-Ω inductive reactance.
3. A 50-Ω resistor and a 0.2-H coil are in parallel across a 100-V, 100-Hz ac line. Find (*a*) the total current, (*b*) the impedance, (*c*) the phase angle, (*d*) the power factor, and (*e*) the power drawn.
4. In Fig. 19-15*a*, find the percent of the total AF current that passes

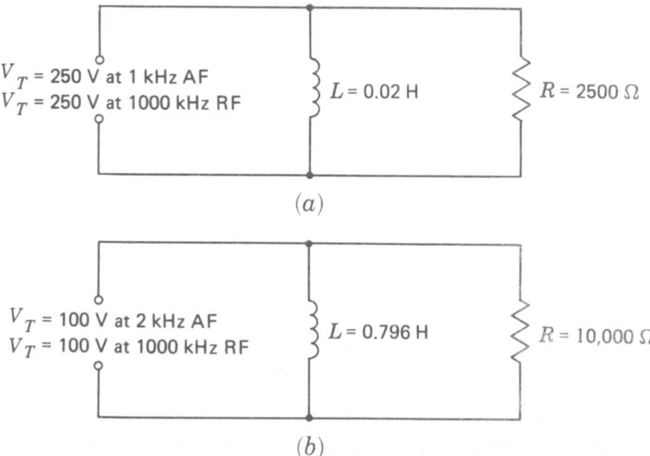


FIGURE 19-15

through the resistor. Find the percent of the total RF current that passes through the resistor. On the basis of these answers, may the circuit be classified as a high-pass circuit?

5. Repeat Prob. 4 for the circuit shown in Fig. 19-15*b*.

(See Instructor's Manual for Test 19-1, R and L in Parallel)

## JOB 19-3 RESISTANCE AND CAPACITANCE IN PARALLEL

Figure 19-17 shows a 20-Ω resistance and a 15-Ω capacitive reactance in parallel across a 60-V 60-Hz ac source. The total current drawn by the circuit can be found by adding the currents in each branch. However, if they are not in phase, they must be added vectorially. The phase relations in the circuit are shown in Fig. 19-16*a*. The constant voltage is used as the reference line upon which to draw the phasor diagram. The current in the resistance $I_R$ is in phase with the voltage, and the current in the capacitor $I_C$ *leads* the voltage by 90°.

**Addition of the phasors.** In Fig. 19-16*b*, the total current $I_T$ is obtained by adding the phasor $I_C$ to the phasor $I_R$. Place the tail of $I_C$ on the head of $I_R$, and draw it in its original direction and length. The

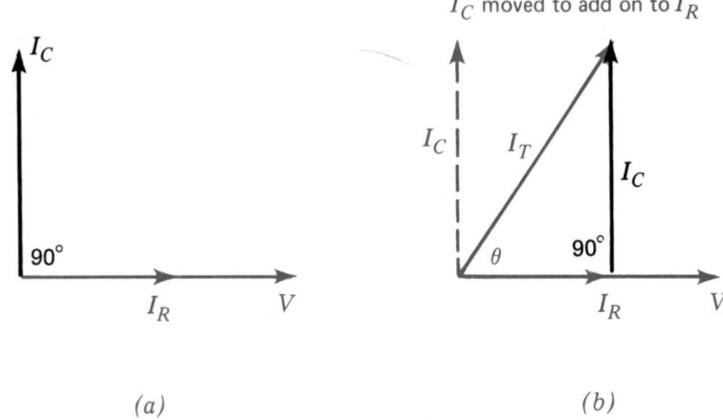


**FIGURE 19-16** (*a*) Phasor diagram for a parallel ac circuit containing resistance and capacitance. (*b*) $I_T$ represents the phasor sum of $I_R$ and $I_C$.

distance from the origin of the phasors to the head of the final phasor is the sum of the phasors. In this instance, the phasor $I_T$ represents the sum of the phasors $I_R$ and $I_C$. The angle $\theta$ is the angle by which the total current *leads* the total voltage. By applying the pythagorean theorem to Fig. 19-16*b*, we obtain the

**FORMULA** $$I_T^2 = I_R^2 + I_C^2 \qquad \boxed{19\text{-}4}$$

By Ohm's law, $$V_T = I_T \times Z \qquad (18\text{-}13)$$

By trigonometry, $$\cos\theta = \frac{a}{h}$$

$$\cos\theta = \frac{I_R}{I_T} \qquad (19\text{-}2)$$

or $$\mathrm{PF} = \frac{I_R}{I_T} \tag{19-3}$$

The power is still given by

$$W = V \times I \times \cos\theta \tag{18-3}$$

or $$W = V \times I \times \mathrm{PF} \tag{18-4}$$

or $$W = I^2 \times R \tag{7-4}$$

EXAMPLE 19-6. A 20-Ω resistor and a capacitor of 15 Ω capacitive reactance at 60 Hz are connected in parallel across a 60-V 60-Hz ac source. Find (*a*) the total current, (*b*) the impedance, (*c*) the phase angle, (*d*) the power factor, and (*e*) the power drawn by the circuit.

SOLUTION. The diagram for the circuit is shown in Fig. 19-17.

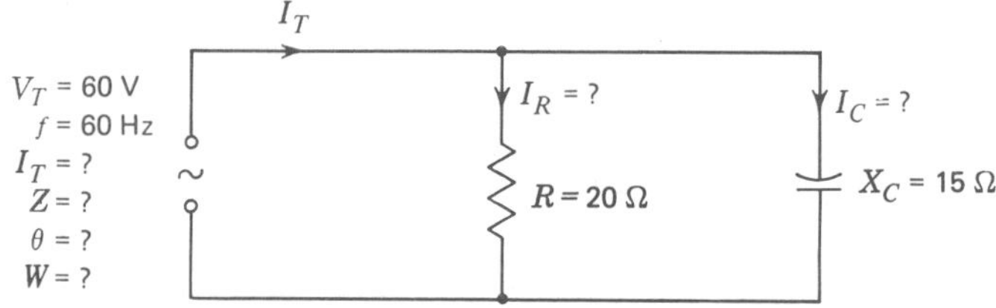


**FIGURE 19-17**
Ac parallel circuit containing resistance and capacitance.

*a.* Find the branch currents. Since

$$V_T = V_R = V_C = 60 \text{ V} \tag{5-1}$$

$$V_R = I_R \times R \qquad V_C = I_C \times X_C$$
$$60 = I_R \times 20 \qquad 60 = I_C \times 15$$
$$I_R = \frac{60}{20} = 3 \text{ A} \qquad I_C = \frac{60}{15} = 4 \text{ A}$$

Find the total current.

$$I_T^2 = I_R^2 + I_C^2 = 3^2 + 4^2 \tag{19-4}$$
$$= 9 + 16$$
$$I_T^2 = 25$$
$$I_T = \sqrt{25} = 5 \text{ A} \qquad \textit{Ans.}$$

*b.* $$V_T = I_T \times Z \tag{18-13}$$
$$60 = 5 \times Z$$
$$Z = \frac{60}{5} = 12\ \Omega \qquad \textit{Ans.}$$

*c.* $$\cos\theta = \frac{I_R}{I_T} = \frac{3}{5} = 0.6000 \tag{19-2}$$
$$\theta = 53° \qquad \textit{Ans.}$$

*d.* $$\mathrm{PF} = \frac{I_R}{I_T} = 0.6000 = 60\% \textit{ leading} \qquad \textit{Ans.} \tag{19-3}$$

*e.*
$$W = V \times I \times \cos\theta = 60 \times 5 \times \cos 53° \qquad (18\text{-}3)$$
$$= 60 \times 5 \times 0.6 = 180 \text{ W} \qquad \textit{Ans.}$$
$$W = I_R^2 \times R = (3)^2 \times 20 = 180 \text{ W} \qquad \textit{Check}$$

In the parallel circuit, the total current of 5 A leads the total voltage of 60 V by 53°.

EXAMPLE 19-7. The purpose of the "low-pass" circuit shown in Fig. 19-18 is to permit low frequencies to pass on to the load but to prevent the passage of high frequencies. Find the effectiveness of the circuit by calculating the percent of the total current in the resistor for (1) a 1-kHz audio frequency and (2) a 1000-kHz radio frequency.

SOLUTION. The diagram for the circuit is shown in Fig. 19-18.

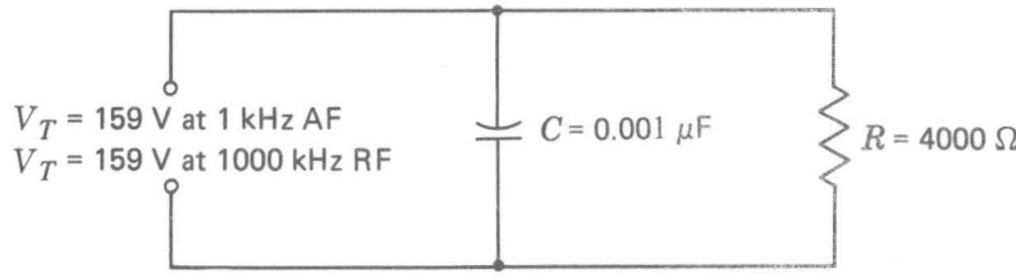


**FIGURE 19-18**
Low-pass filter.

1. For the 1-kHz audio frequency,

*a.*
$$X_C = \frac{159{,}000}{f \times C} = \frac{159{,}000}{10^3 \times 1 \times 10^{-3}} \qquad (17\text{-}6)$$
$$X_C = \frac{159{,}000}{1} = 159{,}000\ \Omega \qquad \textit{Ans.}$$

Find the branch currents. Since

$$V_T = V_R = V_C = 159 \text{ V} \qquad (5\text{-}1)$$

$$V_R = I_R \times R \qquad V_C = I_C \times X_C$$
$$159 = I_R \times 4000 \qquad 159 = I_C \times 159{,}000$$
$$I_R = \frac{159}{4000} = 0.04 \text{ A} \qquad I_C = \frac{159}{159{,}000} = 0.001 \text{ A} \qquad \textit{Ans.}$$

*b.* Find the total current.

$$I_T^2 = I_R^2 + I_C^2 = 0.04^2 + 0.001^2 \qquad (19\text{-}4)$$
$$= 0.0016 + 0.000\,001$$
$$I_T^2 = 0.001\,601$$
$$I_T = \sqrt{0.001\,601} = 0.04 \text{ A} \qquad \textit{Ans.}$$

*c.* Find the percent of the total current in the resistor.

$$\text{Percent} = \frac{I_R}{I_T} \times 100 = \frac{0.04}{0.04} \times 100 = 100\% \qquad \textit{Ans.}$$

That is, practically 100 percent of the 1-kHz audio frequency passes through the resistor.

2. For the 1000-kHz radio frequency: Since 1000 kHz is 1000 times as large as the audio frequency of 1 kHz, the $X_C$ at 1000 kHz will be equal to 1/1000 of the $X_C$ at 1 kHz. Therefore,

*a.* $$X_C = \frac{159{,}000}{1000} = 159\ \Omega \qquad \textit{Ans.}$$

Find the branch currents. Since

$$V_T = V_R = V_C = 159\text{ V} \qquad (5\text{-}1)$$

$$V_R = I_R \times R \qquad\qquad V_C = I_C \times X_C$$

$$159 = I_R \times 4000 \qquad\qquad 159 = I_C \times 159$$

$$I_R = \frac{159}{4000} = 0.04\text{ A} \qquad\qquad I_C = \frac{159}{159} = 1\text{ A} \qquad \textit{Ans.}$$

*b.* Find the total current.

$$I_T^2 = I_R^2 + I_C^2 = 0.04^2 + 1^2 \qquad (19\text{-}4)$$

$$= 0.0016 + 1$$

$$I_T^2 = 1.0016$$

$$I_T = \sqrt{1.0016} = 1.001\text{ A} \qquad \textit{Ans.}$$

*c.* Find the percent of the total current in the resistor.

$$\text{Percent} = \frac{I_R}{I_T} \times 100 = \frac{0.04}{1.001} \times 100 = 4\% \qquad \textit{Ans.}$$

That is, only 4 percent of the 1000-kHz current passes through the resistor. The circuit is evidently a good low-pass circuit. Practically all the low 1-kHz current goes through the resistor, but very little of the high 1000-kHz current gets through. The majority of the 1000-kHz RF current finds an easy path through the low reactance of the capacitor at this high frequency.

## PROBLEMS

1. An 8-Ω resistor and a 15-Ω capacitive reactance are in parallel across a 120-V 60-Hz ac line. Find (*a*) the total current, (*b*) the impedance, (*c*) the phase angle, (*d*) the power factor, and (*e*) the power drawn by the circuit.
2. A 26.5-Ω resistor and a 3-μF capacitor are in parallel across a 106-V 1-kHz ac source. Find (*a*) the total current, (*b*) the impedance, (*c*) the phase angle, (*d*) the power factor, and (*e*) the power drawn.
3. Repeat Prob. 1 for a 120-Ω resistor and a 60-Ω capacitive reactance.
4. An 8000-Ω emitter resistor in parallel with a 4-μF capacitor in a 500-Hz circuit is shown in Fig. 19-19. If the voltage drop across the combination is 8 V, find (*a*) the total current and (*b*) the impedance of the combination.
5. A 40-Ω resistor is bypassed with a 5-μF capacitor. Find the total current flowing if the voltage across the capacitor is 10.6 V at 1.5 kHz.
6. In Fig. 19-20*a*, find the percent of the total AF current that passes

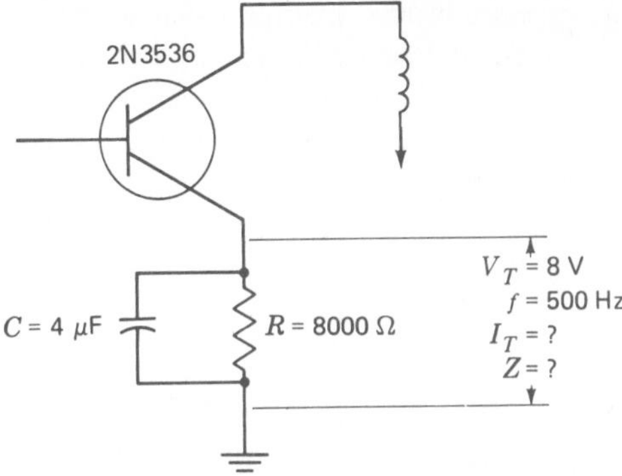


FIGURE 19-19

through the resistor. Find the percent of the total RF current that passes through the resistor. On the basis of these answers, may the circuit be classified as a low-pass circuit?

7. Repeat Prob. 6 for the circuit shown in Fig. 19-20*b*.

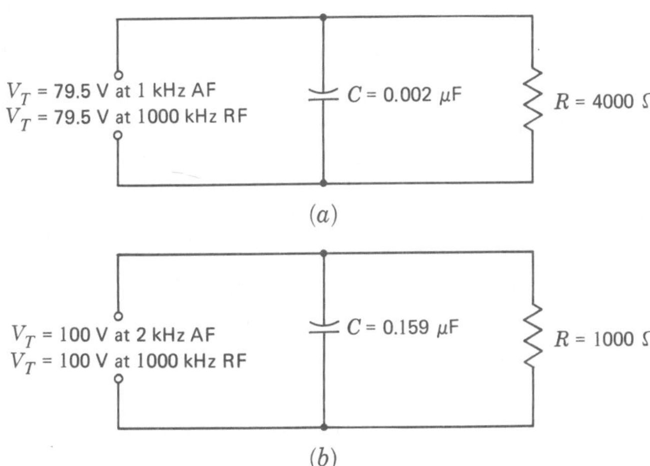


FIGURE 19-20

8. In a grid-leak circuit, $C = 250$ pF and $R = 1$ MΩ. If a 5-kHz AF signal causes a voltage drop of 0.6 V, find (*a*) the total current and (*b*) the impedance of the combination.
9. The grid-to-cathode capacity of 10 pF which is formed in the grid circuit of an amplifier is in parallel with the 1-MΩ grid resistor. If the signal voltage is 10 V at 318 kHz, find the current that will flow.

(See Instructor's Manual for Test 19-2, R and C in Parallel)

## JOB 19-4 RESISTANCE, INDUCTANCE, AND CAPACITANCE IN PARALLEL. THE EQUIVALENT SERIES CIRCUIT

Figure 19-23 shows a 30-Ω resistor, a 40-Ω inductive reactance, and a 60-Ω capacitive reactance connected in parallel across a 120-V 60-Hz ac line. $I_R = 4$ A, $I_L = 3$ A, and $I_C = 2$ A. The total current drawn by the circuit may be found by adding the currents in each branch. However, since the currents are not in phase, they must be added vectorially. The phase relations in the circuit are shown in Fig. 19-21*a*. The constant voltage is used as the reference line upon which to draw the phasor diagram. The current in the resistor $I_R$ is in phase with the voltage. The current in the capacitor $I_C$ *leads* the voltage by 90°. The current in the inductance $I_L$ *lags* the voltage by 90°. Since $I_L$ and $I_C$ are exactly 180° out of phase and act in exactly *opposite* directions, the current $I_C$ is denoted by a minus sign.

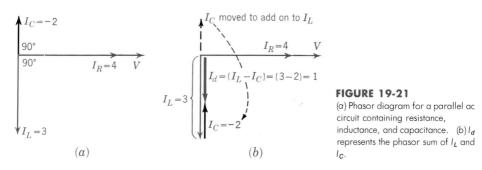

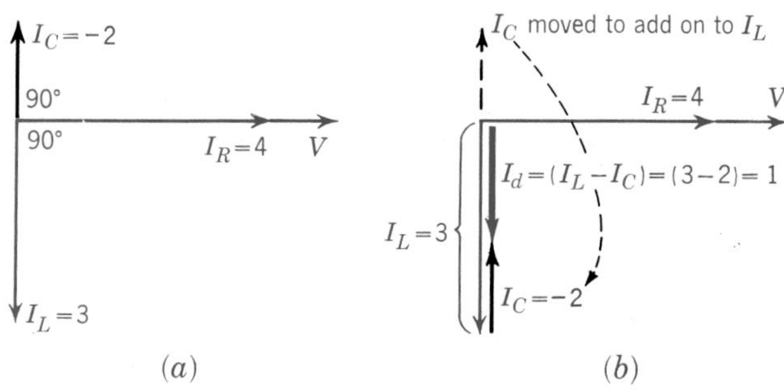


**FIGURE 19-21**
(a) Phasor diagram for a parallel ac circuit containing resistance, inductance, and capacitance. (b) $I_d$ represents the phasor sum of $I_L$ and $I_C$.

**Addition of phasors.** When there are three phasors, it is best to add only two at a time. To add the phasor $I_C$ to $I_L$, place the tail of $I_C$ on the head of $I_L$ and draw it in its original length and direction, which will be straight *up* as shown in Fig. 19-21*b*. Since these phasors are acting in *opposite* directions, their *sum* is actually the *difference* between the phasors as indicated by $I_d$. That is, the addition of $I_C$ to $I_L$ is really found by $I_d = I_L - I_C$. After this addition, the phasor diagram looks like Fig. 19-22*a*. Notice that the effect of the capacitor current has disap-

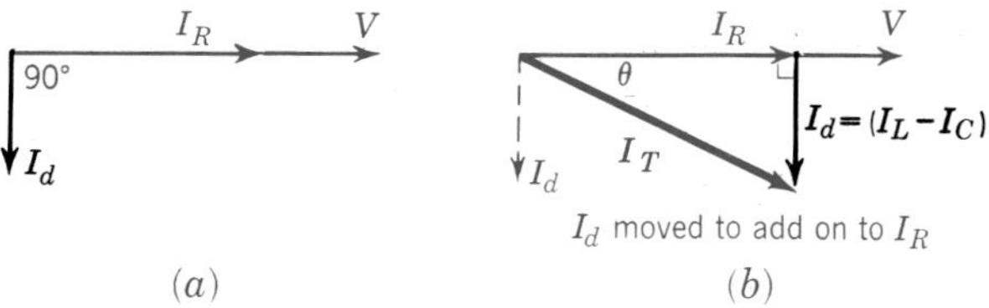


**FIGURE 19-22**
(a) $I_d$ represents the result of adding $I_C$ to $I_L$. (b) $I_T$ represents the phasor sum of $I_R + I_C + I_L$.

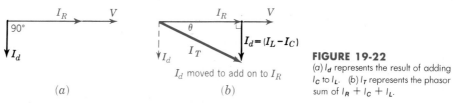

peared. The 3 A of coil current has exactly balanced the 2 A of capacitor current and has left an excess of 1 A of coil current. This 1 A of coil current must now be added to the 4 A of resistance current. In Fig. 19-22*b*, the total current $I_T$ is obtained by adding the phasor $I_d$ to the phasor $I_R$. Place the tail of $I_d$ on the head of $I_R$, and draw it in the proper direction. If $I_C$ were larger than $I_L$, the phasor $I_d$ would have an *upward* direction. The distance from the origin of the phasors to the head of the last phasor is the sum of the phasors. In this instance, the phasor $I_T$ represents the sum of the phasors $I_R + I_L + I_C$. The angle $\theta$ is the angle by which the total current *lags* the voltage. If $I_C$ were larger than $I_L$, the total current would *lead* the voltage by this angle. By applying the pythagorean theorem to Fig. 19-22*b*, we obtain the

**FORMULA** $I_T^2 = I_R^2 + (I_L - I_C)^2$ 19-5

By Ohm's law, $V_T = I_T \times Z$ (18-13)

By trigonometry, $\cos\theta = \frac{a}{h}$

$$\cos\theta = \frac{I_R}{I_T} \quad (19\text{-}2)$$

or $\text{PF} = \frac{I_R}{I_T}$ (19-3)

The power is still given by

$$W = V \times I \times \cos\theta \quad (18\text{-}3)$$

or $W = V \times I \times \text{PF}$ (18-4)

or $W = I_R^2 \times R$ (7-4)

EXAMPLE 19-8. A 30-Ω resistor, a 40-Ω inductive reactance, and a 60-Ω capacitive reactance are connected in parallel across a 120-V 60-Hz ac line. Find (*a*) the total current, (*b*) the impedance, (*c*) the phase angle, (*d*) the power factor, and (*e*) the power drawn by the circuit.

SOLUTION. The diagram for the circuit is shown in Fig. 19-23.

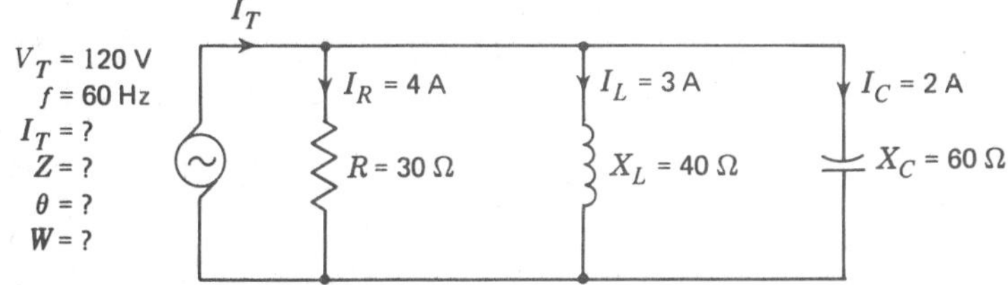


**FIGURE 19-23** Ac parallel circuit containing resistance, inductance, and capacitance.

*a.* Find the branch currents. Since

$$V_T = V_1 = V_2 = 120 \text{ V} \quad (5\text{-}1)$$

| | | |
|---|---|---|
| $V_R = I_R \times R$ | $V_L = I_L \times X_L$ | $V_C = I_C \times X_C$ |
| $120 = I_R \times 30$ | $120 = I_L \times 40$ | $120 = I_C \times 60$ |
| $I_R = \frac{120}{30} = 4$ A | $I_L = \frac{120}{40} = 3$ A | $I_C = \frac{120}{60} = 2$ A *Ans.* |

Find the total current.

$$I_T^2 = I_R^2 + (I_L - I_C)^2 \quad (19\text{-}5)$$
$$= 4^2 + (3 - 2)^2 = 4^2 + 1^2$$
$$= 16 + 1$$
$$I_T^2 = 17$$
$$I_T = \sqrt{17} = 4.12 \text{ A} \quad \textit{Ans.}$$

*b.*
$$V_T = I_T \times Z \qquad (18\text{-}13)$$
$$120 = 4.12 \times Z$$
$$Z = \frac{120}{4.12} = 29.1\ \Omega \qquad \textit{Ans.}$$

*c.*
$$\cos\theta = \frac{I_R}{I_T} = \frac{4}{4.12} = 0.9708 \qquad (19\text{-}2)$$
$$\theta = 14° \qquad \textit{Ans.}$$

*d.*
$$\text{PF} = \frac{I_R}{I_T} = 0.9708 = 97.1\%\ \textit{lagging} \qquad \textit{Ans.} \qquad (19\text{-}3)$$

*e.*
$$W = V \times I \times \cos\theta = 120 \times 4.12 \times \cos 14° \qquad (18\text{-}3)$$
$$= 120 \times 4.12 \times 0.971 = 480\text{ W} \qquad \textit{Ans.}$$
$$W = I_R^2 \times R = (4)^2 \times 30 = 480\text{ W} \qquad \textit{Check}$$

In the parallel circuit, the total current of 4.12 A lags the total voltage by 14°.

## THE EQUIVALENT SERIES CIRCUIT

EXAMPLE 19-9. In the circuit shown in Fig. 19-24, find (*a*) the impedance, (*b*) the phase angle, and (*c*) the equivalent series circuit.

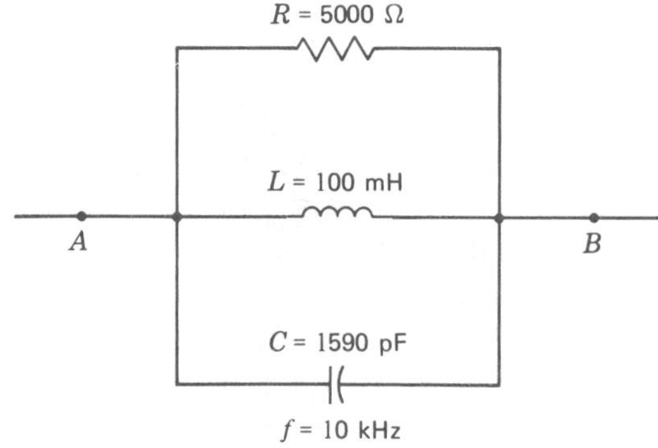


FIGURE 19-24

SOLUTION

*a.* Find the reactances.

$$X_L = 6.28fL = 6.28 \times 10 \times 10^3 \times 100 \times 10^{-3}$$
$$X_L = 6280\ \Omega$$

$$X_C = \frac{159{,}000}{f \times C} = \frac{159{,}000}{10 \times 10^3 \times 1590 \times 10^{-6}} = 10{,}000\ \Omega$$

Find the branch currents. If the voltage is unknown, it may be assumed to be any value. For ease in calculation, it should be equal to, or greater than, the largest impedance. Assume $V_{AB} = 10{,}000$ V. Since

$$V_{AB} = V_R = V_L = V_C$$

$$V_R = I_R \times R \qquad V_L = I_L \times X_L$$
$$10{,}000 = I_R \times 5000 \qquad 10{,}000 = I_L \times 6280$$
$$I_R = 2 \text{ A} \qquad I_L = 1.59 \text{ A}$$
$$V_C = I_C \times X_C$$
$$10{,}000 = I_C \times 10{,}000$$
$$I_C = 1 \text{ A}$$

Find the total current.

$$I_T^2 = I_R^2 + (I_L - I_C)^2 \tag{19-5}$$
$$I_T^2 = 2^2 + (1.59 - 1)^2$$
$$= 4 + 0.348$$
$$I_T^2 = 4.348$$
$$I_T = \sqrt{4.348} = 2.08 \text{ A}$$

Find the impedance.

$$V_T = I_T \times Z \tag{18-13}$$
$$10{,}000 = 2.08 \times Z$$
$$Z = 4800 \ \Omega \qquad \textit{Ans.}$$

*b.* Find the phase angle.

$$\cos \theta = \frac{I_R}{I_T} = \frac{2}{2.08} \tag{19-2}$$
$$\cos \theta = 0.9615$$
$$\theta = 16^\circ \textit{ lagging} \qquad \textit{Ans.}$$

*c.* Find the equivalent series circuit.

The impedance from $A$ to $B$ is 4800 Ω with the current lagging the voltage by 16°. If the circuit between $A$ and $B$ had been a *series circuit,* then the 4800 Ω of impedance would have been formed from an impedance triangle as shown in Fig. 19-25*b*, containing resistance and inductance only because of the *lagging* current of 16°.

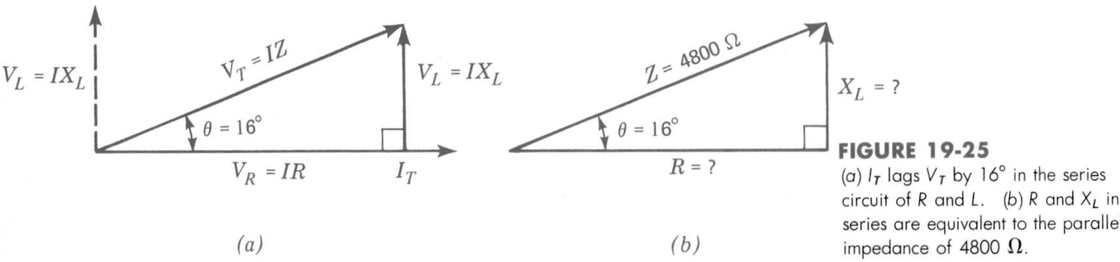


**FIGURE 19-25**
(a) $I_T$ lags $V_T$ by 16° in the series circuit of $R$ and $L$. (b) $R$ and $X_L$ in series are equivalent to the parallel impedance of 4800 Ω.

1. Find the equivalent series resistance. In Fig. 19-25*b*,

$$\cos 16^\circ = \frac{R}{4800} \tag{14-3}$$

$$R = 4800 \times \cos 16^\circ$$

$$R = 4800 \times 0.9615 = 4615\ \Omega \qquad \textit{Ans.}$$

2. Find the equivalent inductance. In Fig. 19-25*b*,

$$\sin 16^\circ = \frac{X_L}{4800}$$

$$X_L = 4800 \times \sin 16^\circ$$

$$X_L = 4800 \times 0.2756 = 1323\ \Omega$$

Also,

$$X_L = 6.28fL \qquad (16\text{-}1)$$

$$1323 = 6.28 \times 10 \times 10^3 \times L$$

$$L = 0.021\ \text{H} = 21\ \text{mH} \qquad \textit{Ans.}$$

Therefore, the equivalent series circuit consists of a resistance of 4615 Ω in series with an inductance of 21 mH. This circuit would produce the same load as the original parallel circuit.

## PROBLEMS

1. Find the series equivalent circuit for the circuit shown in Fig. 19-17 for Example 19-6.
2. Find the series equivalent circuit for the circuit shown in Fig. 19-13 for Example 19-4.
3. Find the series equivalent circuit for the circuit shown in Fig. 19-23 for Example 19-8.
4. A 24-Ω resistor, a 6-Ω inductive reactance, and a 15-Ω capacitive reactance are in parallel across a 120-V 60-Hz line. Find (*a*) the total current, (*b*) the impedance, (*c*) the phase angle, (*d*) the power factor, (*e*) the power drawn by the circuit, and (*f*) the series equivalent circuit.
5. Repeat Prob. 4 for a 30-Ω resistor, a 60-Ω inductive reactance, and a 40-Ω capacitive reactance across the same line in parallel.
6. A 50-Ω resistor, a 0.02-H coil, and a 3-μF capacitor are connected in parallel across a 100-V 1-kHz ac source. Find (*a*) the reactance of the coil and capacitor, (*b*) the current drawn by each branch, (*c*) the total current, (*d*) the impedance, (*e*) the phase angle, (*f*) the power factor, (*g*) the power drawn by the circuit, and (*h*) the series equivalent circuit.
7. Repeat Prob. 6 for a 2200-Ω resistor, a 20-H coil, and a 0.8-μF capacitor in parallel across a 220-V 60-Hz ac line.
8. In a circuit similar to that shown in Fig. 19-24, $R = 1590\ \Omega$, $L = 0.16$ H, and $C = 0.1\ \mu$F. Find the series equivalent circuit.

(See Instructor's Manual for Test 19-3)

## JOB 19-5 RESOLUTION OF PHASORS

Up to this point we have been considering only circuits which contained only "pure" inductances and capacitances. Actually, such pure components do not exist. There is always some resistance in every coil or capacitor. This resistance must be taken into account whenever the resistance is an appreciable value as compared with the reactance of the component. In addition, most motor loads may be considered to be a series combination of resistance and inductance or resistance and capacitance. For example, an *induction motor* may be considered to be a series combination of resistance and inductance in which the current *lags* behind the impressed voltage. A *synchronous motor* may be considered to be a series combination of resistance and capacitance in which the current *leads* the impressed voltage. The amount of lead or lag depends on the relative amounts of resistance in series with the inductance or capacitance.

These leading and lagging currents will not be out of phase by exactly 90° but may be out of phase by *any* angle. For example, the current in branch $A$ of a parallel circuit may *lead* the total voltage by 30° and the current in branch $B$ may *lag* the total voltage by 50°. The phasor diagram for this condition is shown in Fig. 19-26.

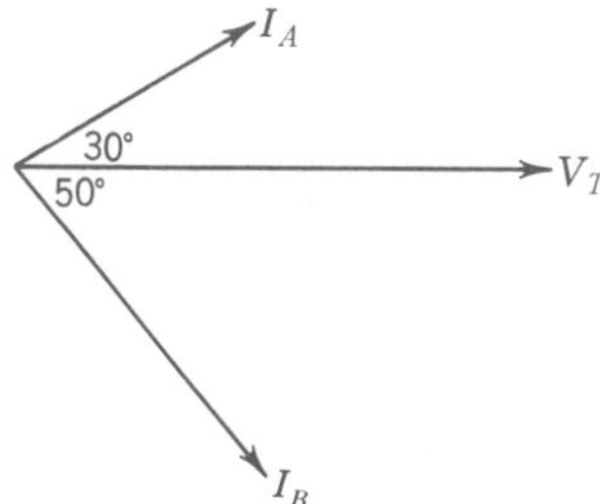


**FIGURE 19-26**
Leading and lagging branch currents in a parallel circuit.

The total current will still be the *phasor* sum of $I_A$ and $I_B$. However, the pythagorean theorem may not be used because the angle between the phasors is no longer 90°. The phasor addition may be accomplished if we first resolve each phasor into its component parts which *are* 90° out of phase. Consider the phasor $V$ in Fig. 19-27.

By the pythagorean theorem, $V$ is the phasor sum of $V_x$ and $V_y$,

$$V^2 = V_x^2 + V_y^2 \tag{18-4}$$

$$V^2 = 3^2 + 4^2 = 9 + 16 = 25$$

$$V = \sqrt{25} = 5$$

When the process is reversed, a phasor $V$ may be *resolved* into its two components whose phasor sum will be equal to the original phasor. These components are always at right angles to each other.

$V_x$ is called the *horizontal* or $x$ component of $V$.
$V_y$ is called the *vertical* or $y$ component of $V$.

The value of each component depends on the value of the total phasor and on the angle $\theta$ between the phasor and the $X$ axis. The relationships between the components and the phasor are determined by the basic

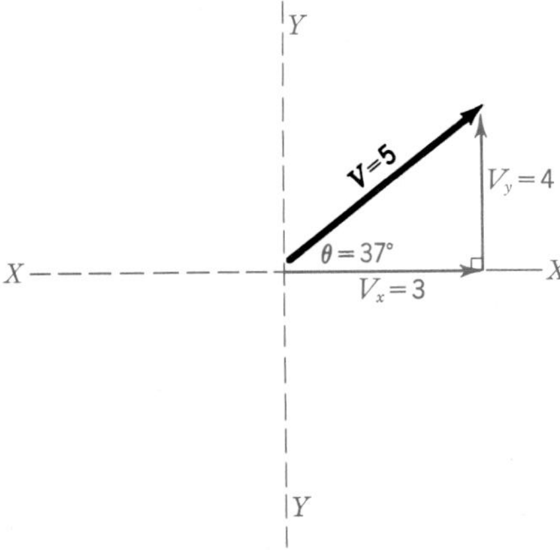


**FIGURE 19-27**
A phasor (or vector) $V$ may be resolved into its horizontal component $V_x$ and its vertical component $V_y$.

definitions of the sine and cosine of angle $\theta$. For example, in Fig. 19-27,

$$\sin\theta = \frac{V_y}{V} \quad \text{and} \quad \cos\theta = \frac{V_x}{V}$$

Cross-multiplying each equation yields the

**FORMULAS**

$$V_y = V\sin\theta \tag{19-6}$$

$$V_x = V\cos\theta \tag{19-7}$$

The rectangular components $V_x$ and $V_y$ of the phasor $V$ in Fig. 19-27 may now be found.

$$V_y = V\sin\theta \qquad V_x = V\cos\theta$$
$$V_y = 5 \times \sin 37° \qquad V_x = 5 \times \cos 37°$$
$$V_y = 5 \times 0.6018 = 3 \qquad V_x = 5 \times 0.7986 = 4$$

EXAMPLE 19-10. A 10-lb force acts up and to the right at an angle of 30° with the horizontal. Find its horizontal and vertical components.

SOLUTION. The diagram for the problem is shown in Fig. 19-28.

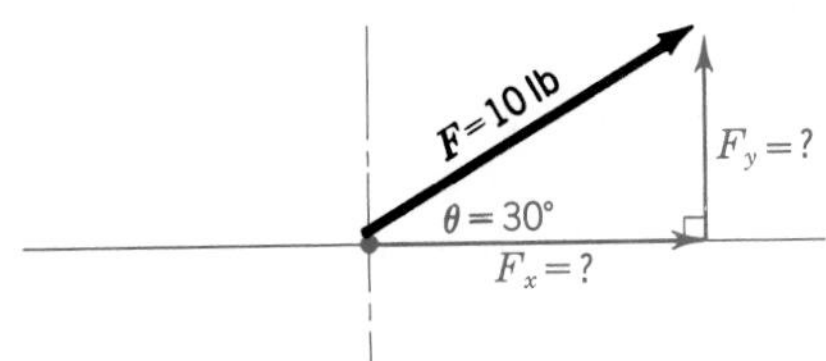


**FIGURE 19-28**

$$F_y = F\sin\theta \qquad (19\text{-}6) \qquad F_x = F\cos\theta \qquad (19\text{-}7)$$
$$F_y = 10\sin 30° \qquad F_x = 10\cos 30°$$
$$F_y = 10(0.5000) = 5\ \text{lb} \quad \textit{Ans.} \qquad F_x = 10(0.8660) = 8.66\ \text{lb} \quad \textit{Ans.}$$

EXAMPLE 19-11. A sled is being pulled with a force of 50 lb which is exerted through a rope held at an angle of 20° with the ground as shown in Fig. 19-29. How much of this force is useful in moving the sled horizontally?

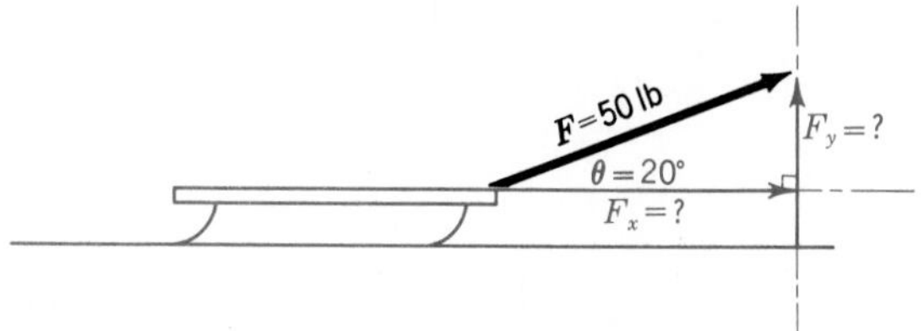


FIGURE 19-29

SOLUTION. Only the horizontal component $F_x$ is useful in moving the sled horizontally.

$$F_x = F \cos \theta \qquad (19\text{-}7)$$
$$F_x = 50 \cos 20°$$
$$F_x = 50(0.9397) = 47 \text{ lb} \qquad \textit{Ans.}$$

EXAMPLE 19-12. A window pole is used to pull down a window. If the force exerted through the pole is 30 lb at an angle of 72° with the horizontal, find the useful vertical component.

SOLUTION. The diagram for the problem is shown in Fig. 19-30.

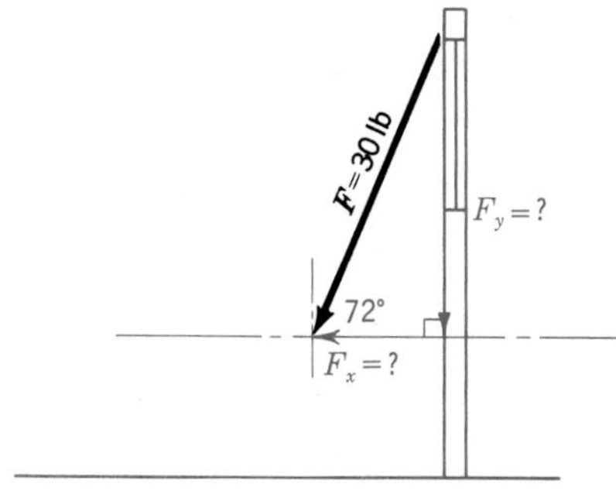


FIGURE 19-30
Only the vertical component of the force is effective in opening the window.

$$F_y = F \sin \theta \qquad (19\text{-}5)$$
$$F_y = 30 \sin 72°$$
$$F_y = 30(0.9511) = -28.5 \text{ lb} \qquad \textit{Ans.}$$

*Note:* The minus sign in the answer is used to indicate that the force is acting *downward.* The direction in which a component acts is indicated by either a plus (+) or a minus (−) sign. These signs are the same as those used to locate points on a graph.

$V_x$ components acting to the *right* are +.
$V_x$ components acting to the *left* are −.
$V_y$ components acting *upward* are +.
$V_y$ components acting *downward* are −.

EXAMPLE 19-13. A current of 20 A in one branch of an ac circuit leads the total voltage by 40°. Find its "in-phase" current and its "reactive" current.

SOLUTION. The phasor diagram is shown in Fig. 19-31*a*.

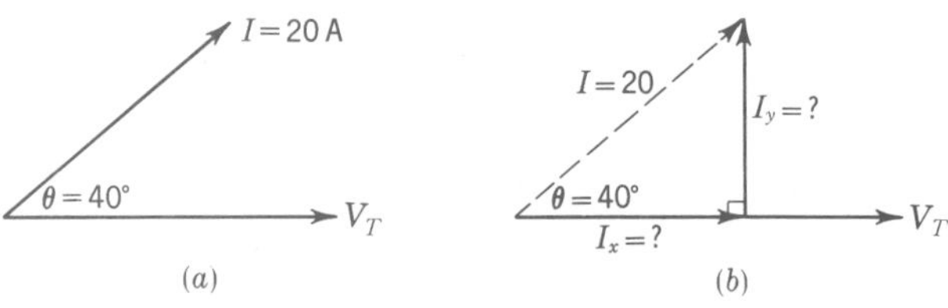


**FIGURE 19-31**
(*a*) Phasor diagram for a leading current. (*b*) The current $I$ is resolved into its components $I_x$ and $I_y$.

1. Resolve the current into its components $I_x$ and $I_y$ as shown in Fig. 19-31*b*. $I_x$ is the in-phase current, since it acts in the same direction as the total voltage $V_T$. $I_y$ is the reactive current, since it leads the total voltage $V_T$ by 90°.

2. Find each component.

$$I_x = I \cos \theta \qquad (19\text{-}7) \qquad I_y = I \sin \theta \qquad (19\text{-}6)$$
$$I_x = 20 \cos 40° \qquad I_y = 20 \sin 40°$$
$$I_x = 20(0.766) = 15.32 \text{ A} \qquad I_y = 20(0.6428) = 12.86 \text{ A} \qquad \textit{Ans.}$$

EXAMPLE 19-14. A current of 10 A supplying an induction motor lags the voltage by 53°. Find the in-phase and reactive currents.

SOLUTION. The phasor diagram is shown in Fig. 19-32*a*.

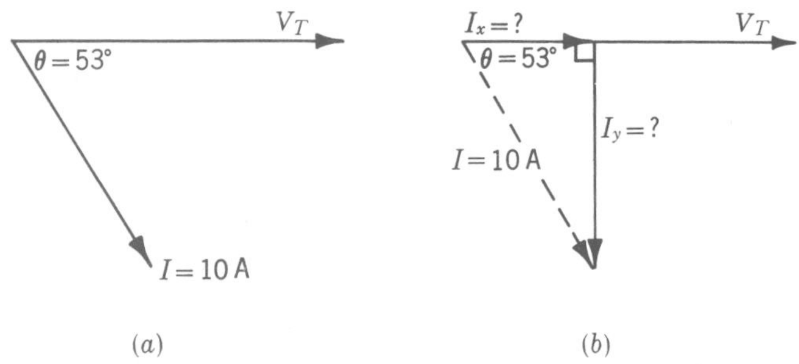


**FIGURE 19-32**
(*a*) Phasor diagram for a lagging current. (*b*) The current $I$ is resolved into its components $I_x$ and $I_y$.

1. Resolve the current into its in-phase component $I_x$ and its reactive component $I_y$ as shown in Fig. 19-32*b*.
2. Find each component.

$$I_x = I \cos \theta \qquad (19\text{-}7) \qquad I_y = I \sin \theta \qquad (19\text{-}6)$$
$$I_x = 10 \cos 53° \qquad I_y = 10 \sin 53°$$
$$I_x = 10(0.6018) = 6 \text{ A} \qquad I_y = 10(0.7986) = -8 \text{ A} \qquad \textit{Ans.}$$

## PROBLEMS

Find the in-phase current $I_x$ and the reactive current $I_y$ for each of the following currents.

| PROBLEM | $I$, A | $\theta$ |
|---|---|---|
| 1 | 20 | 30° leading |
| 2 | 10 | 35° leading |
| 3 | 20 | 60° lagging |
| 4 | 10 | 55° lagging |
| 5 | 26 | 42° leading |
| 6 | 1.8 | 23° leading |
| 7 | 0.1 | 50° lagging |
| 8 | 4.5 | 19° lagging |
| 9 | 14 | 25° leading |
| 10 | 31 | 28° lagging |

## JOB 19-6 PARALLEL-SERIES AC CIRCUITS

The current in any branch of an operating ac circuit is never exactly 90° out of phase with the voltage. There is always some resistance in series with the capacitance or inductance which reduces the phase angle from 90° to almost any angle. These "out-of-phase" currents must be added by phasor addition to get the total current.

**Addition of "out-of-phase" currents.** Let us try to find the total current in a parallel circuit if the current in branch $A$ leads the voltage by 30° and the current in branch $B$ leads the voltage by 60°.

SOLUTION. The phasor diagram for the conditions stated is shown in Fig. 19-33*a*.

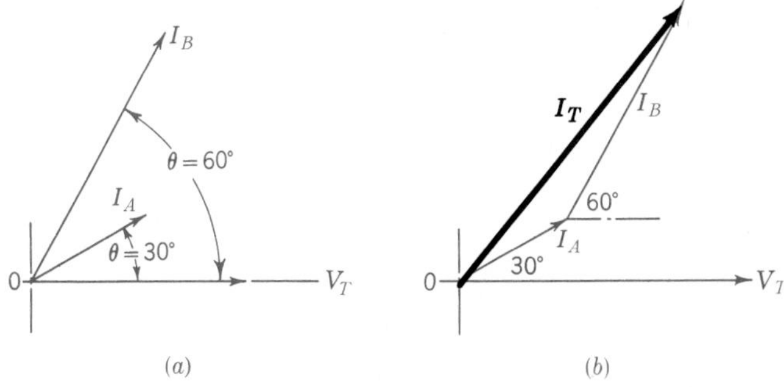


**FIGURE 19-33**
(a) Phasor diagram showing $I_A$ leading the voltage by 30° and $I_B$ leading the voltage by 60°. (b) $I_T$ represents the phasor sum of $I_A$ and $I_B$.

The total current is obtained by adding current $I_B$ to current $I_A$ *vectorially*. Place the tail of $I_B$ on to the head of $I_A$ and draw it in its original direction and length as shown in Fig. 19-33*b*. The distance from the origin of the phasors to the head of the final phasor is the sum of the phasors, or $I_T$.

Figure 19-33*b* may be redrawn as shown in Fig. 19-34 to show the $x$ and $y$ components of each current.

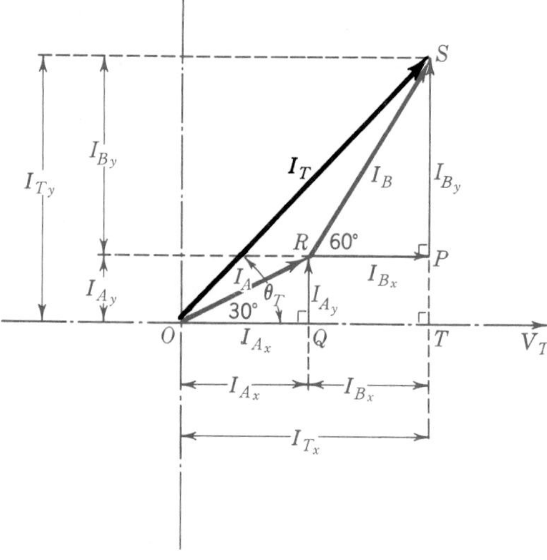


**FIGURE 19-34**
$I_A$ and $I_B$ have been resolved into their in-phase and reactive components. Their algebraic sums $I_{T_x}$ and $I_{T_y}$ are the components of the total current $I_T$.

$I_{A_x}$ and $I_{B_x}$ are the in-phase components of $I_A$ and $I_B$.
$I_{A_y}$ and $I_{B_y}$ are the reactive components of $I_A$ and $I_B$.
$I_{T_x}$ is the algebraic sum of all the $x$ components.
$I_{T_y}$ is the algebraic sum of all the $y$ components.

**FORMULAS**

$$I_{T_x} = I_{A_x} + I_{B_x} \qquad \boxed{19\text{-}8}$$

$$I_{T_y} = I_{A_y} + I_{B_y} \qquad \boxed{19\text{-}9}$$

By applying the pythagorean theorem to triangle $OTS$ in Fig. 19-34, we obtain the

**FORMULA**

$$I_T^2 = (I_{T_x})^2 + (I_{T_y})^2 \qquad \boxed{19\text{-}10}$$

and by trigonometry,

$$\cos \theta_T = \frac{I_{T_x}}{I_T} \qquad \boxed{19\text{-}11}$$

or

$$\text{PF} = \frac{I_{T_x}}{I_T} \qquad \boxed{19\text{-}12}$$

In actual practice, the resolution triangles $ORQ$ and $RSP$ are drawn on the same constant-voltage base and added as shown in the following example. This method eliminates the problem of complicated phasor diagrams resulting from combinations of leading and lagging phasors.

EXAMPLE 19-15. In a parallel circuit, a current of 10 A in branch $A$ leads the total voltage by 30°. A current of 20 A in branch $B$ leads the total voltage by 37°. Find (*a*) the total current and (*b*) the angle by which the total current leads the total voltage.

SOLUTION

*a.* 1. Draw the phasor diagram for the branch currents on the same voltage base as shown in Fig. 19-35*a*.

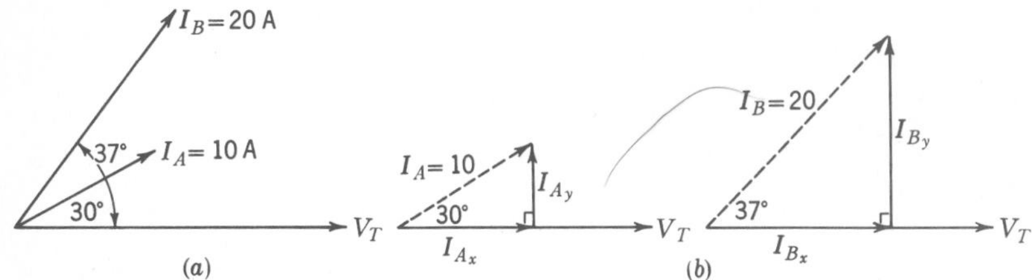


**FIGURE 19-35**
(*a*) Two leading currents in a parallel circuit. (*b*) Resolution of each current into its in-phase and reactive components.

2. Resolve the current in each branch into its components as shown in Fig. 19-35*b*. Calculate the components.

For branch *A*:

$$I_{A_x} = I_A \cos \theta \qquad (19\text{-}7)$$
$$= 10 \cos 30° = 10(0.866) = 8.66 \text{ A}$$
$$I_{A_y} = I_A \sin \theta \qquad (19\text{-}6)$$
$$= 10 \sin 30° = 10(0.5000) = 5 \text{ A}$$

For branch *B*:

$$I_{B_x} = I_B \cos \theta \qquad (19\text{-}7)$$
$$= 20 \cos 37° = 20(0.7986) = 16 \text{ A}$$
$$I_{B_y} = I_B \sin \theta \qquad (19\text{-}6)$$
$$= 20 \sin 37° = 20(0.6018) = 12 \text{ A}$$

3. Draw all the components on the same voltage base as shown in Fig. 19-36*a*. The total in-phase current $I_{T_x}$ and the total reactive current $I_{T_y}$ may now be found.

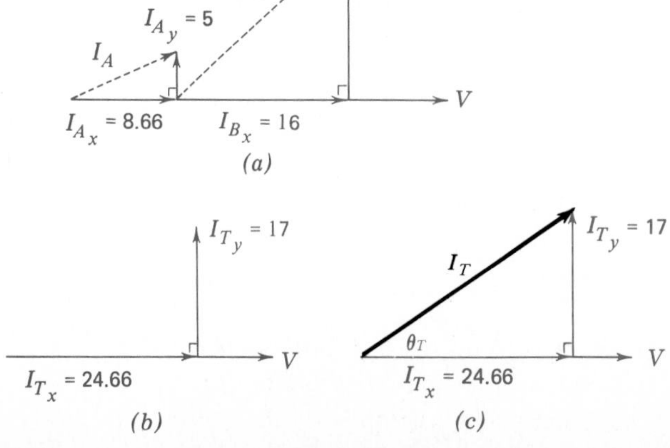


**FIGURE 19-36**
(*a*) The components of all the branch currents drawn on the same voltage base.
(*b*) $I_{T_x} = I_{A_x} + I_{B_x}$ and $I_{T_y} = I_{A_y} + I_{B_y}$. (*c*) $I_T$ represents the phasor sum of $I_{T_x}$ and $I_{T_y}$.

$$I_{T_x} = I_{A_x} + I_{B_x} \qquad (19\text{-}8)$$
$$= 8.66 + 16 = 24.66 \text{ A}$$
$$I_{T_y} = I_{A_y} + I_{B_y} \qquad (19\text{-}9)$$
$$= 5 + 12 = 17 \text{ A}$$

4. Draw these phasors as shown in Fig. 19-36*b*.
5. Draw the phasor diagram for the total current by adding $I_{T_x}$ and $I_{T_y}$ vectorially as shown in Fig. 19-36*c*.

Find $I_T$.

$$(I_T)^2 = (I_{T_x})^2 + (I_{T_y})^2 \qquad (19\text{-}10)$$
$$= (24.66)^2 + (17)^2$$
$$= 608 + 289 = 897$$
$$I_T = \sqrt{897} = 30 \text{ A } \textit{leading} \qquad \textit{Ans.}$$

*b.* Find the phase angle.

$$\cos \theta_T = \frac{I_{T_x}}{I_T} \qquad (19\text{-}11)$$
$$\cos \theta_T = \frac{24.66}{30} = 0.822$$
$$\theta_T = 35° \textit{ leading} \qquad \textit{Ans.}$$

EXAMPLE 19-16. An induction motor of 5 Ω impedance draws a current lagging by 26°. It is in parallel with a synchronous motor of 12 Ω impedance which draws a current leading by 37°. If the applied voltage is 120 V at 60 Hz, find (*a*) the current drawn by each motor, (*b*) the total current drawn, (*c*) the impedance, (*d*) the phase angle, (*e*) the power factor, and (*f*) the power drawn by the circuit.

SOLUTION. The diagram for the circuit is shown in Fig. 19-37.

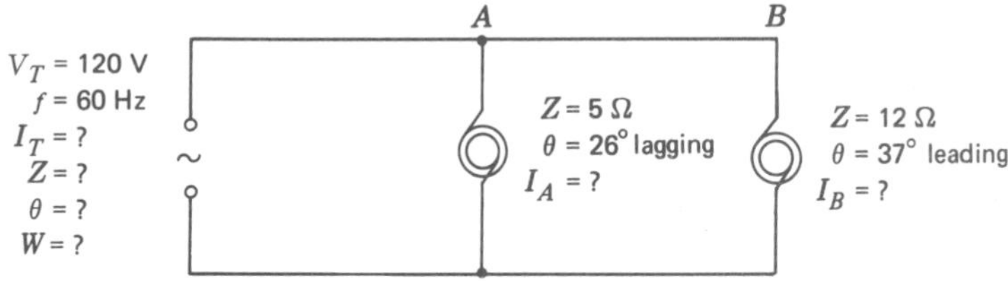


FIGURE 19-37

*a.* Find the current in each branch. Since

$$V_T = V_A = V_B = 120 \text{ V} \qquad (5\text{-}1)$$

$$V_A = I_A \times Z_A \qquad V_B = I_B \times Z_B$$
$$120 = I_A \times 5 \qquad 120 = I_B \times 12$$
$$I_A = \frac{120}{5} = 24 \text{ A } \textit{lagging} \qquad I_B = \frac{120}{12} = 10 \text{ A } \textit{leading} \quad \textit{Ans.}$$

*b.* The total current is found by adding the currents in the two branches. However, they are out of phase and must be added vectorially.

1. Draw the phasor diagram for the branch currents on the same voltage base as shown in Fig. 19-38*a*.

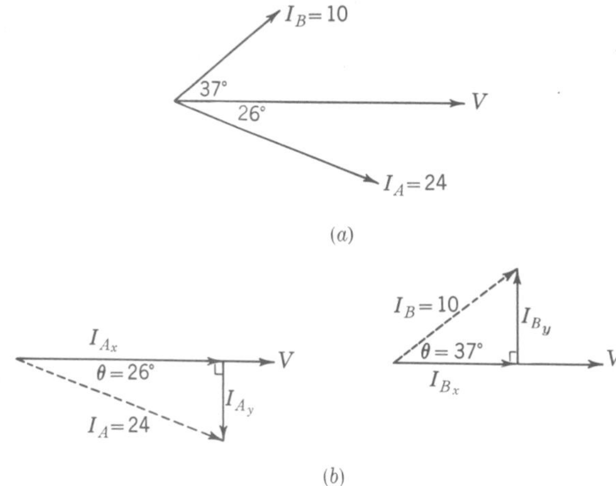


**FIGURE 19-38** (*a*) Leading and lagging currents in a parallel circuit. (*b*) Resolution of each current into its in-phase and reactive components.

2. Resolve the current in each branch into its components as shown in Fig. 19-38*b*. Calculate the components.

For branch *A*:

$$I_{A_x} = I_A \cos\theta \tag{19-7}$$
$$= 24 \cos 26° = 24(0.9) = 21.6 \text{ A}$$
$$I_{A_y} = I_A \sin\theta \tag{19-6}$$
$$= 24 \sin 26° = 24(0.44) = -10.56 \text{ A}$$

The minus sign in $I_{A_y}$ indicates a lagging reactive current.

For branch *B*:

$$I_{B_x} = I_B \cos\theta \tag{19-7}$$
$$= 10 \cos 37° = 10(0.8) = 8 \text{ A}$$
$$I_{B_y} = I_B \sin\theta \tag{19-6}$$
$$= 10 \sin 37° = 10(0.6) = 6 \text{ A}$$

3. Draw all the components on the same voltage base as shown in Fig. 19-39*a*. The total in-phase current $I_{T_x}$ and the total reactive current $I_{T_y}$ may now be found.

$$I_{T_x} = I_{A_x} + I_{B_x} \tag{19-8}$$
$$= 21.6 + 8 = 29.6 \text{ A}$$
$$I_{T_y} = I_{A_y} + I_{B_y} \tag{19-9}$$
$$= -10.56 + 6 = -4.56 \text{ A}$$

The minus sign indicates a lagging reactive current.

4. Draw the phasors as shown in Fig. 19-39*b*.

5. Draw the phasor diagram for the total current by adding $I_{T_x}$ and $I_{T_y}$ vectorially as shown in Fig. 19-39*c*. Find $I_T$.

$$(I_T)^2 = (I_{T_x})^2 + (I_{T_y})^2 \tag{19-10}$$

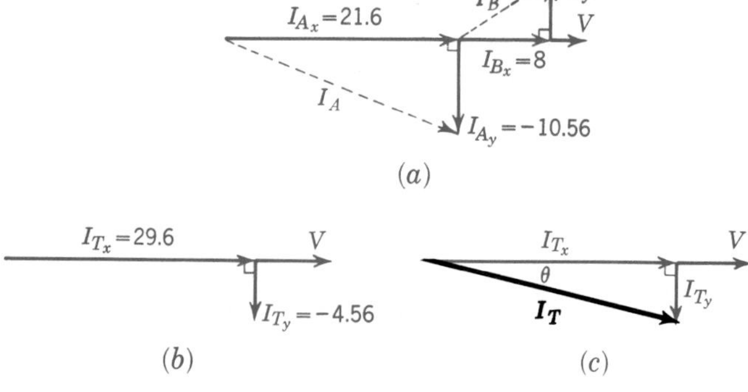


**FIGURE 19-39**
(a) The components of all the branch currents drawn on the same voltage base.
(b) $I_{T_x} = I_{A_x} + I_{B_x}$ and $I_{T_y} = I_{A_y} + I_{B_y}$. (c) $I_T$ represents the phasor sum of $I_{T_x}$ and $I_{T_y}$.

$$(I_T)^2 = (29.6)^2 + (-4.56)^2$$
$$= 876.2 + 20.8 = 897$$
$$I_T = \sqrt{897} = 30 \text{ A } \textit{lagging} \qquad \textit{Ans.}$$

*c.* Find the impedance of the circuit.

$$V_T = I_T \times Z \qquad (18\text{-}13)$$
$$120 = 30 \times Z$$
$$Z = \frac{120}{30} = 4\ \Omega \qquad \textit{Ans.}$$

*d.* Find the phase angle.

$$\cos\theta_T = \frac{I_{T_x}}{I_T} \qquad (19\text{-}11)$$
$$\cos\theta_T = \frac{29.6}{30} = 0.986$$
$$\theta_T = 10^\circ \textit{ lagging} \qquad \textit{Ans.}$$

*e.*
$$\text{PF} = \cos\theta_T = 98.6\% \qquad \textit{Ans.}$$

*f.* Find the power consumed.

$$W = V \times I \times \cos\theta = 120 \times 30 \times \cos 10^\circ \qquad (18\text{-}3)$$
$$W = 120 \times 30 \times 0.986 = 3546 \text{ W} \qquad \textit{Ans.}$$

In this parallel circuit, the total current of 30 A *lags* the total voltage of 120 V by 10°.

From this last example we can determine the procedure to follow in solving parallel-series ac circuits:

1. Find (*a*) the reactance, (*b*) the impedance, (*c*) the current, and (*d*) the phase angle for each branch of the parallel circuit.
2. Draw the phasor diagram for the branch currents on the same voltage base.
3. Resolve the current in each branch into its components.

$$I_y = I \times \sin\theta \qquad (19\text{-}6)$$
$$I_x = I \times \cos\theta \qquad (19\text{-}7)$$

*Note: y* components of lagging currents are negative (−). *y* components of leading currents are positive (+).

4. Find the total in-phase current $I_{T_x}$ and the total reactive current $I_{T_y}$.

$$I_{T_x} = I_{A_x} + I_{B_x} \tag{19-8}$$
$$I_{T_y} = I_{A_y} + I_{B_y} \tag{19-9}$$

5. Find the total current.

$$I_T^2 = (I_{T_x})^2 + (I_{T_y})^2 \tag{19-10}$$

6. Find the impedance.

$$V_T = I_T \times Z \tag{18-13}$$

7. Find the phase angle.

$$\cos \theta_T = \frac{I_{T_x}}{I_T} \tag{19-11}$$

8. Find the power factor.

$$\text{PF} = \frac{I_{T_x}}{I_T} \tag{19-12}$$

9. Find the total power.

$$W = V \times I \times \cos \theta \tag{18-3}$$

EXAMPLE 19-17. An induction motor of 6 Ω resistance and 8 Ω inductive reactance is in parallel with a synchronous motor of 8 Ω resistance and 15 Ω capacitive reactance and a third parallel branch of 15 Ω resistance. Find (*a*) the total current drawn from a 150-V 60-Hz source, (*b*) the total impedance, (*c*) the phase angle, (*d*) the power factor, and (*e*) the power drawn by the circuit.

SOLUTION. The diagram for the circuit is shown in Fig. 19-40.

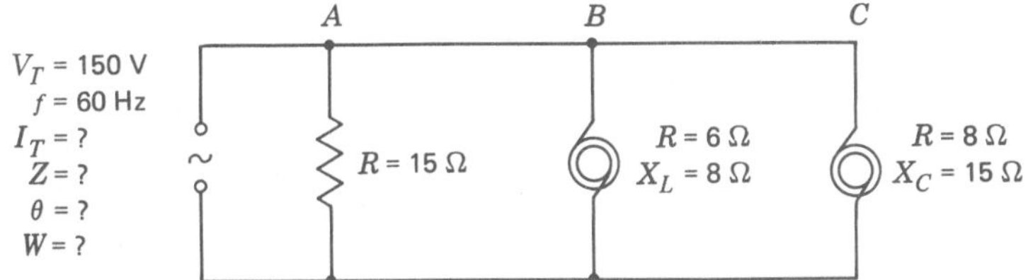


FIGURE 19-40

*a*. 1. Find the series impedance, current, and phase angle for each branch.

For branch *A*:

$$Z = R = 15\ \Omega$$
$$I_A = \frac{V}{R} = \frac{150}{15} = 10\text{ A} \tag{2-1}$$
$$\cos \theta = \frac{R}{Z} = \frac{15}{15} = 1 \tag{18-11}$$

Therefore, $\theta = 0°$

The current is *in phase* with the voltage.

For branch *B*:

$$Z = \sqrt{R^2 + X_L^2} = \sqrt{6^2 + 8^2} = \sqrt{36 + 64} = \sqrt{100} = 10\ \Omega$$

$$I_B = \frac{V}{Z} = \frac{150}{10} = 15\text{ A} \qquad (16\text{-}6)$$

$$\cos\theta = \frac{R}{Z} = \frac{6}{10} = 0.6000 \qquad (18\text{-}11)$$

Therefore, $\theta = 53°$

The current *lags* the voltage by 53°.

For branch $C$:

$$Z = \sqrt{R^2 + X_C^2} = \sqrt{8^2 + 15^2} = \sqrt{64 + 225} = \sqrt{289} = 17\ \Omega$$

$$I_C = \frac{V}{Z} = \frac{150}{17} = 8.8\text{ A} \qquad (17\text{-}7)$$

$$\cos\theta = \frac{R}{Z} = \frac{8}{17} = 0.4706 \qquad (18\text{-}11)$$

Therefore, $\theta = 62°$

The current *leads* the voltage by 62°.

2. Draw the phasor diagram for the branch currents on the same voltage base as shown in Fig. 19-41*a*.

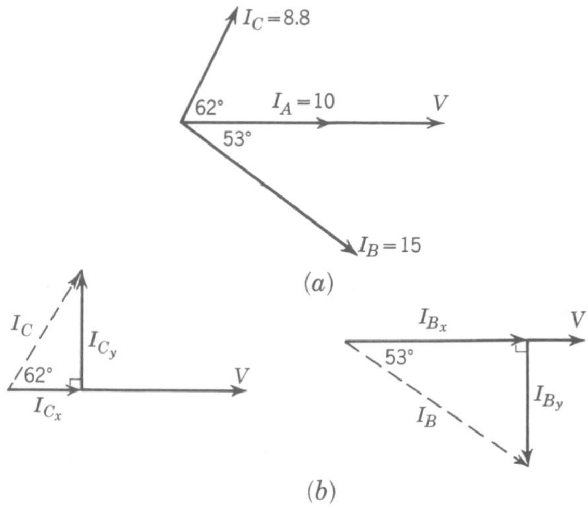


**FIGURE 19-41**
(*a*) Leading and lagging currents in a parallel circuit. (*b*) Resolution of each current into its in-phase and reactive components.

3. Resolve the current in each branch into its components as shown in Fig. 19-41*b*. Calculate the components.

For branch $A$:

$$I_{A_x} = I_A \times \cos\theta = 10 \times \cos 0° = 10 \times 1 = 10\text{ A} \qquad (19\text{-}7)$$
$$I_{A_y} = I_A \times \sin\theta = 10 \times \sin 0° = 10 \times 0 = 0\text{ A} \qquad (19\text{-}6)$$

For branch $B$:

$$I_{B_x} = I_B \times \cos\theta = 15 \times \cos 53° = 15 \times 0.6 = 9\text{ A} \qquad (19\text{-}7)$$
$$I_{B_y} = I_B \times \sin\theta = 15 \times \sin 53° = 15 \times 0.8 = -12\text{ A} \qquad (19\text{-}6)$$

*Note:* The minus sign in $I_{B_y}$ indicates a lagging reactive component.

For branch $C$:

$$I_{C_x} = I_C \times \cos\theta = 8.8 \times \cos 62° = 8.8 \times 0.470 = 4.14 \text{ A} \quad (19\text{-}7)$$
$$I_{C_y} = I_C \times \sin\theta = 8.8 \times \sin 62° = 8.8 \times .88 = 7.74 \text{ A} \quad (19\text{-}6)$$

4. Draw all the components on the same voltage base as shown in Fig. 19-42*a*. The total in-phase current $I_{T_x}$ and the total reactive current $I_{T_y}$ may now be found.

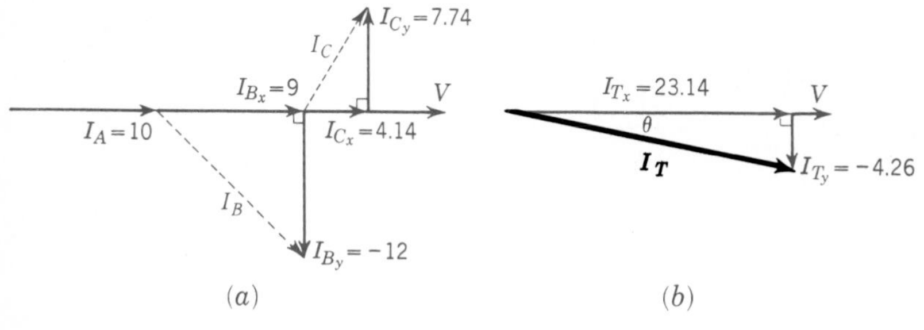


**FIGURE 19-42**
(*a*) The components of all the branch currents drawn on the same voltage base. (*b*) $I_T$ represents the phasor sum of $I_{T_x}$ and $I_{T_y}$.

$$I_{T_x} = I_{A_x} + I_{B_x} + I_{C_x} \quad (19\text{-}8)$$
$$= 10 + 9 + 4.14 = 23.14 \text{ A}$$
$$I_{T_y} = I_{A_y} + I_{B_y} + I_{C_y} \quad (19\text{-}9)$$
$$= 0 + (-12) + 7.74 = -4.26 \text{ A}$$

*Note:* The minus sign in $I_{T_y}$ indicates a lagging reactive component.

5. Draw the phasor diagram for the total current by adding $I_{T_x}$ and $I_{T_y}$ vectorially as shown in Fig. 19-42*b*. Notice that $I_{T_y}$ is drawn downward because $I_{T_y}$ is negative. Find $I_T$.

$$I_T^2 = (I_{T_x})^2 + (I_{T_y})^2 \quad (19\text{-}10)$$
$$= (23.14)^2 + (-4.26)^2$$
$$= 535.5 + 18.1 = 553.6$$
$$I_T = \sqrt{533.6} = 23.5 \text{ A } \textit{lagging} \qquad \textit{Ans.}$$

*b.* Find the impedance.

$$V_T = I_T \times Z \quad (18\text{-}13)$$
$$150 = 23.5 \times Z$$
$$Z = \frac{150}{23.5} = 6.38\ \Omega \qquad \textit{Ans.}$$

*c.* Find the phase angle.

$$\cos\theta_T = \frac{I_{T_x}}{I_T} \quad (19\text{-}11)$$
$$\cos\theta_T = \frac{23.14}{23.5} = 0.9847$$

Therefore, $\theta_T = 10°$ *lagging* *Ans.*

*d.* $$\text{PF} = \frac{I_{T_x}}{I_T} = 0.9847 = 98.5\%\ \textit{lagging} \qquad \textit{Ans.}$$

*e.* Find the power drawn by the circuit.

$$W = V \times I \times \cos\theta = 150 \times 23.5 \times \cos 10° \qquad (18\text{-}3)$$

$$W = 150 \times 23.5 \times 0.985 = 3472 \text{ W} \qquad \textit{Ans.}$$

*Check:*

$$W_A = I_A^2 \times R_A = (10)^2 \times 15 = 1500 \text{ W}$$
$$W_B = I_B^2 \times R_B = (15)^2 \times 6 = 1350 \text{ W}$$
$$W_C = I_C^2 \times R_C = (8.8)^2 \times 8 = \underline{\ 620 \text{ W}}$$
$$W_T = W_A + W_B + W_C = 3470 \text{ W} \qquad \textit{Check}$$

The total current of 23.5 A *lags* the total voltage of 150 V by 10°.

EXAMPLE 19-18. For the circuit shown in Fig. 19-43, find (*a*) the total current, (*b*) the total impedance, (*c*) the phase angle, (*d*) the power factor, and (*e*) the power drawn by the circuit.

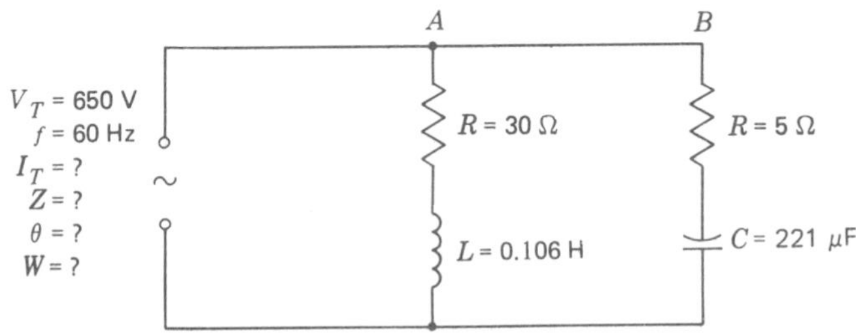


FIGURE 19-43

SOLUTION

*a.* 1. Find the reactance, series impedance, current, and phase angle for each branch.

For branch *A*:

$$X_L = 6.28fL = 6.28 \times 60 \times 0.106 = 40\ \Omega \qquad (16\text{-}1)$$

$$Z = \sqrt{R^2 + X_L^2} = \sqrt{30^2 + 40^2} \qquad (16\text{-}4)$$
$$= \sqrt{900 + 1600}$$
$$Z = \sqrt{2500} = 50\ \Omega$$

$$I_A = \frac{V}{Z} = \frac{650}{50} = 13 \text{ A} \qquad (16\text{-}6)$$

$$\cos\theta = \frac{R}{Z} = \frac{30}{50} = 0.6000 \qquad (18\text{-}11)$$

Therefore, $\theta = 53°$

The current of 13 A lags the total voltage by 53°.

For branch *B*:

$$X_C = \frac{159{,}000}{f \times C} = \frac{159{,}000}{60 \times 221} = 12\ \Omega \qquad (17\text{-}6)$$

$$Z = \sqrt{R^2 + X_C^2} = \sqrt{5^2 + 12^2} \qquad (17\text{-}8)$$
$$= \sqrt{25 + 144}$$
$$Z = \sqrt{169} = 13\ \Omega$$

$$I_B = \frac{V}{Z} = \frac{650}{13} = 50 \text{ A} \qquad (17\text{-}7)$$

$$\cos\theta = \frac{R}{Z} = \frac{5}{13} = 0.3846 \qquad (18\text{-}11)$$

Therefore, $\theta = 67°$

The current of 50 A leads the total voltage by 67°.

2. Draw the phasor diagram for the branch currents on the same voltage base as shown in Fig. 19-44*a*.

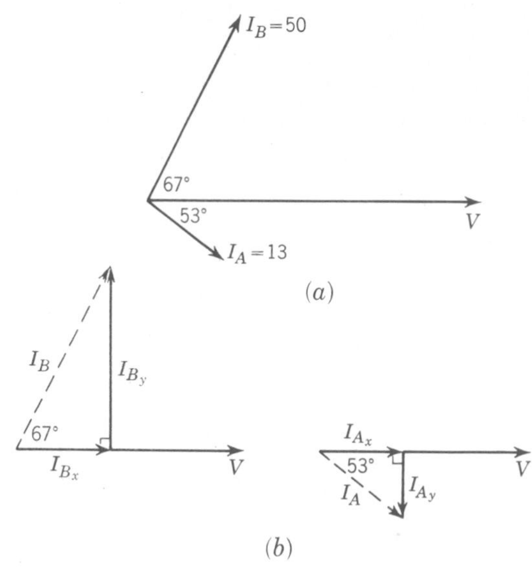


**FIGURE 19-44**
(*a*) Leading and lagging currents in a parallel circuit. (*b*) Resolution of each current into its in-phase and reactive components.

3. Resolve the current in each branch into its components as shown in Fig. 19-44*b*. Calculate the components.

For branch *A*:

$$I_{A_x} = I_A \times \cos\theta = 13 \times \cos 53° = 13 \times 0.6 = 7.8 \text{ A} \qquad (19\text{-}7)$$

$$I_{A_y} = I_A \times \sin\theta = 13 \times \sin 53° = 13 \times 0.8 = -10.4 \text{ A} \qquad (19\text{-}6)$$

For branch *B*:

$$I_{B_x} = I_B \times \cos\theta = 50 \times \cos 67° = 50 \times 0.385 = 19.25 \text{ A} \qquad (19\text{-}7)$$

$$I_{B_y} = I_B \times \sin\theta = 50 \times \sin 67° = 50 \times 0.92 = 46 \text{ A} \qquad (19\text{-}6)$$

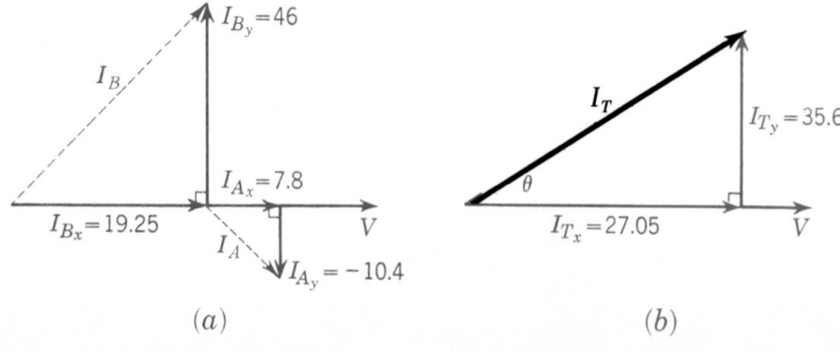


**FIGURE 19-45**
(*a*) The components of all the branch currents drawn on the same voltage base. (*b*) $I_T$ represents the phasor sum of $I_{T_x}$ and $I_{T_y}$.

4. Draw all the components on the same voltage base as shown in Fig. 19-45*a*. The total in-phase current $I_{T_x}$ and the total reactive current $I_{T_y}$ may now be found.

$$I_{T_x} = I_{A_x} + I_{B_x} = 7.8 + 19.25 = 27.05 \text{ A} \qquad (19\text{-}8)$$
$$I_{T_y} = I_{A_y} + I_{B_y} = -10.4 + 46 = 35.6 \text{ A} \qquad (19\text{-}9)$$

5. Draw the phasor diagram for the total current by adding $I_{T_x}$ and $I_{T_y}$ vectorially as shown in Fig. 19-45*b*. Notice that $I_{T_y}$ is drawn *upward* because $I_{T_y}$ is positive. Find $I_T$.

$$I_T^2 = (I_{T_x})^2 + (I_{T_y})^2 \qquad (19\text{-}10)$$
$$= (27.05)^2 + (35.6)^2$$
$$= 732 + 1267 = 1996$$
$$I_T = \sqrt{1999} = 44.7 \text{ A } \textit{leading} \qquad \textit{Ans.}$$

*b.* Find the impedance.

$$V_T = I_T \times Z \qquad (18\text{-}13)$$
$$650 = 44.7 \times Z$$
$$Z = \frac{650}{44.7} = 14.6\ \Omega \qquad \textit{Ans.}$$

*c.* Find the phase angle.

$$\cos\theta_T = \frac{I_{T_x}}{I_T} \qquad (19\text{-}11)$$
$$\cos\theta_T = \frac{27.05}{44.7} = 0.605$$

Therefore, $\theta_T = 53°$ *leading* *Ans.*

*d.* $$\text{PF} = \frac{I_{T_x}}{I_T} = 0.605 = 60.5\% \textit{ leading}$$

*e.* Find the power drawn by the circuit.

$$W = V \times I \times \cos\theta = 650 \times 44.7 \times \cos 53° \qquad (18\text{-}3)$$
$$W = 650 \times 44.7 \times 0.605 = 17{,}540 \text{ W} \qquad \textit{Ans.}$$

*Check:* $W_A = I_A^2 \times R_A = (13)^2 \times 30 = 5{,}070$ W
$W_B = I_B^2 \times R_B = (50)^2 \times 5 = \underline{12{,}500 \text{ W}}$
$W_T = W_A + W_B = 17{,}570$ W *Check*

The small error is due to the approximate value of $\theta = 53°$ obtained from the trigonometric tables.

The total current of 44.6 A *leads* the total voltage of 650 V by 53°.

**PROBLEMS**

Find (*a*) the total current, (*b*) the impedance, (*c*) the phase angle, and (*d*) the power drawn by each circuit shown in Fig. 19-46.

(See Instructor's Manual for Test 19-4)

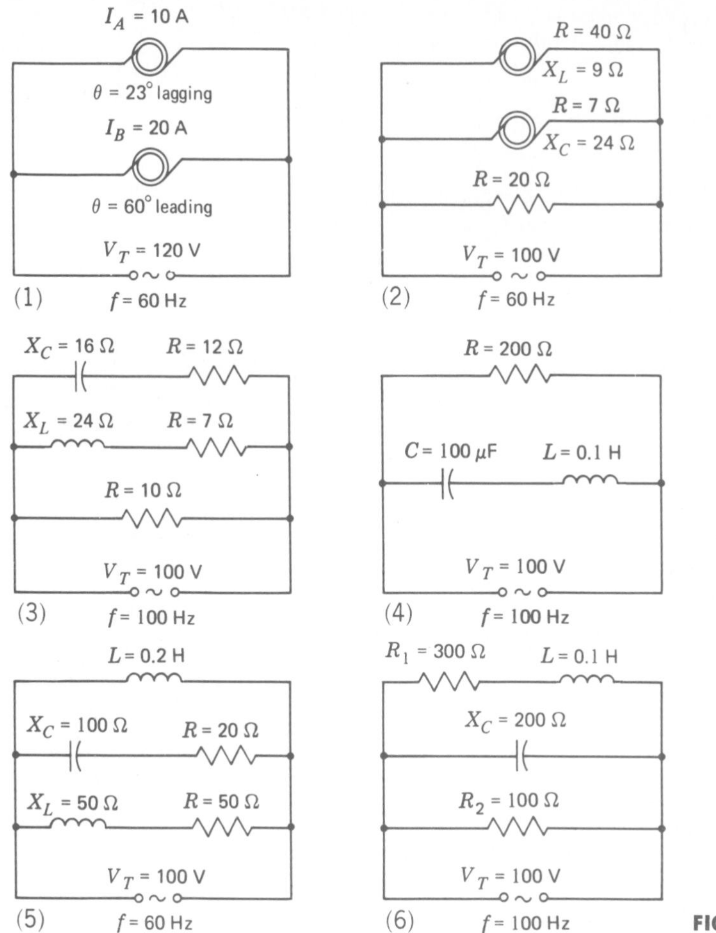


FIGURE 19-46

## JOB 19-7 SERIES-PARALLEL AC CIRCUITS

EXAMPLE 19-19. Solve the circuit shown in Fig. 19-47 for (*a*) the equivalent series impedance, (*b*) the total current, (*c*) the phase angle, (*d*) the power factor, and (*e*) the power drawn by the circuit.

SOLUTION

1. The parallel branches *A*, *B*, and *C* are solved in the same way in which we solved the branches in Example 19-18.

   *a.* Find the reactance, series impedance, current, and phase angle for each branch. Since the voltage across the section from *D* to *E* is unknown, a voltage may be assumed.

   *b.* Draw the phasor diagram for the branch currents on the same voltage base.

   *c.* Resolve the current in each branch into its in-phase and reactive components.

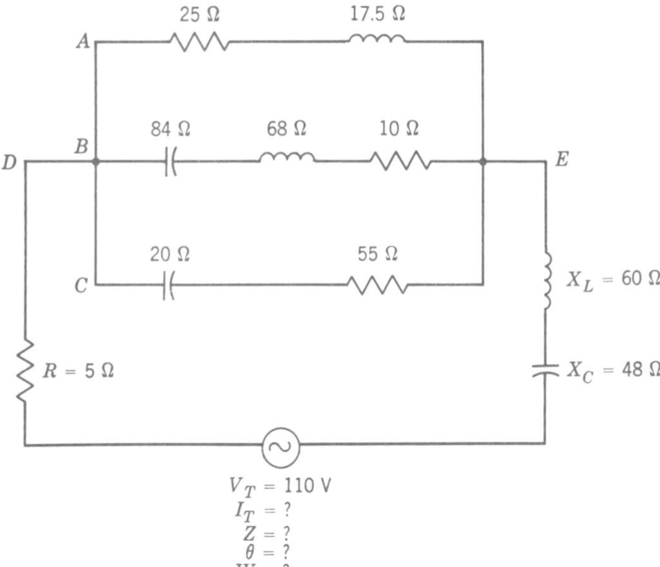


**FIGURE 19-47**
Series-parallel ac circuit.

*d.* Draw all the components on the same voltage base. Find the total in-phase current $I_{T_x}$ and the total reactive current $I_{T_y}$.

*e.* Find the total current for the parallel branches by adding $I_{T_x}$ and $I_{T_y}$ vectorially.

*f.* Find the parallel impedance.

*g.* Find the phase angle for the parallel circuit.

2. The impedance found in step *f* above must now be resolved into its equivalent series resistance and reactance.

3. These are now combined with the other series resistances and reactances to get the total series impedance.

4. The phase angle, total current, and power are found as in any series circuit.

The solution is as follows:

1. *a.* Find the series impedance, current, and phase angle for each branch. Assume 100 V across *DE*.

For branch *A*:

$$Z = \sqrt{R^2 + X_L^2} = \sqrt{(25)^2 + (17.5)^2}$$
$$Z = \sqrt{625 + 306} = 30.5\ \Omega$$
$$I_A = \frac{V}{Z} = \frac{100}{30.5} = 3.28\ \text{A}$$
$$\cos\theta = \frac{R}{Z} = \frac{25}{30.5} = 0.8197$$
$$\theta = 35^\circ\ \textit{lagging}$$

For branch *B*:

$$Z = \sqrt{R^2 + (X_C - X_L)^2} = \sqrt{(10)^2 + (84 - 68)^2}$$
$$= \sqrt{(10)^2 + (16)^2}$$
$$Z = \sqrt{100 + 256} = 18.9\ \Omega$$

$$I_B = \frac{V}{Z} = \frac{100}{18.9} = 5.30 \text{ A}$$

$$\cos\theta = \frac{R}{Z} = \frac{10}{18.9} = 0.5291$$

$$\theta = 58° \textit{ leading}$$

For branch $C$:

$$Z = \sqrt{R^2 + X_C^2} = \sqrt{(55)^2 + (20)^2}$$

$$Z = \sqrt{3025 + 400} = 58.5\ \Omega$$

$$I_C = \frac{V}{Z} = \frac{100}{58.5} = 1.71 \text{ A}$$

$$\cos\theta = \frac{R}{Z} = \frac{55}{58.5} = 0.9400$$

$$\theta = 20° \textit{ leading}$$

*b.* Draw the phasor diagram for the branch currents on the same voltage base as shown in Fig. 19-48*a*.

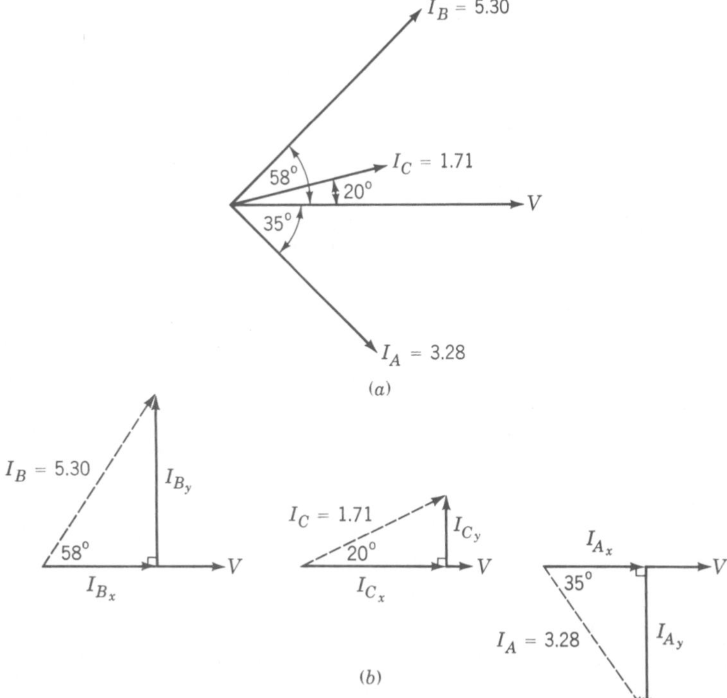


**FIGURE 19-48**
(*a*) Leading and lagging currents in a parallel circuit. (*b*) Resolution of each current into its in-phase and reactive components.

*c.* Resolve the current in each branch into its components as shown in Fig. 19-48*b*. Calculate the components.

For branch $A$:

$$I_{A_x} = I_A \times \cos\theta = 3.28 \times \cos 35° = 3.28 \times 0.819 = 2.68 \text{ A}$$

$$I_{A_y} = I_A \times \sin\theta = 3.28 \times \sin 35° = 3.28 \times 0.574 = -1.88 \text{ A}$$

For branch $B$:

$$I_{B_x} = I_B \times \cos\theta = 5.30 \times \cos 58° = 5.30 \times 0.53 = 2.82 \text{ A}$$
$$I_{B_y} = I_B \times \sin\theta = 5.30 \times \sin 58° = 5.30 \times 0.848 = 4.49 \text{ A}$$

For branch $C$:

$$I_{C_x} = I_C \times \cos\theta = 1.71 \times \cos 20° = 1.71 \times 0.94 = 1.61 \text{ A}$$
$$I_{C_y} = I_C \times \sin\theta = 1.71 \times \sin 20° = 1.71 \times 0.342 = 0.59 \text{ A}$$

*d.* Draw all the components on the same voltage base as shown in Fig. 19-49*a*. The total in-phase current $I_{T_x}$ and the total reactive current $I_{T_y}$ may now be found.

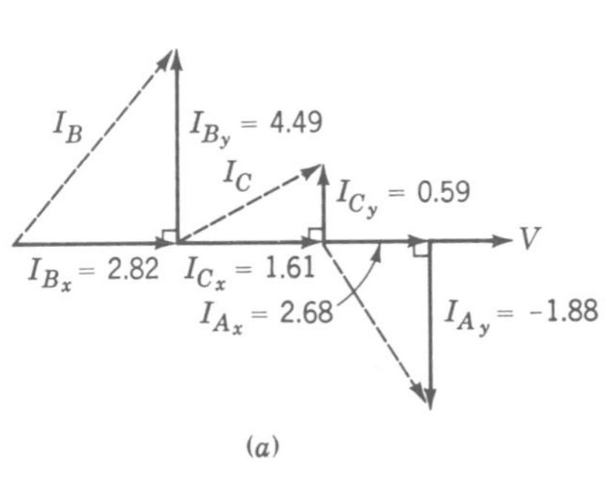


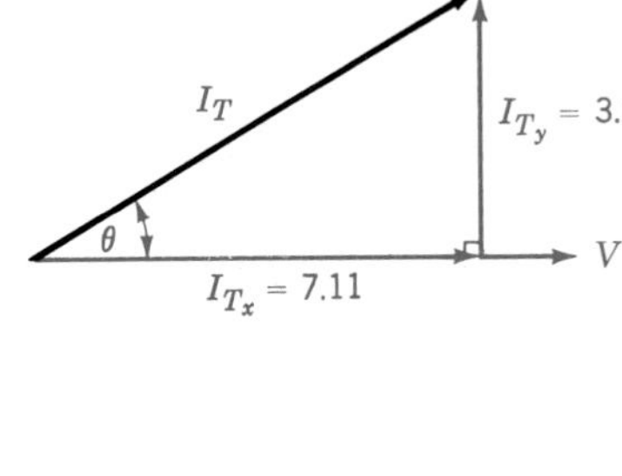


**FIGURE 19-49**
(*a*) The components of all the branch currents drawn on the same voltage base. (*b*) $I_T$ represents the phasor sum of $I_{T_x}$ and $I_{T_y}$.

$$I_{T_x} = I_{A_x} + I_{B_x} + I_{C_x} = 2.68 + 2.82 + 1.61 = 7.11 \text{ A}$$
$$I_{T_y} = I_{B_y} + I_{C_y} - I_{A_y} = 4.49 + 0.59 - 1.88 = 3.20 \text{ A}$$

*e.* Draw the phasor diagram for the total current by adding $I_{T_x}$ and $I_{T_y}$ vectorially as shown in Fig. 19-49*b*. Notice that $I_{T_y}$ is drawn *upward* because the total of all the $y$ components is positive. Find $I_T$.

$$\begin{aligned} I_T^2 &= (I_{T_x})^2 + (I_{T_y})^2 \\ &= (7.11)^2 + (3.20)^2 \\ &= 50.6 + 10.2 \\ &= 60.8 \\ I_T &= \sqrt{60.8} = 7.8 \text{ A} = I_{DE} \end{aligned}$$

*f.* Find the parallel impedance.

$$Z_{DE} = \frac{V_{DE}}{I_{DE}} = \frac{100}{7.8} = 12.8\ \Omega$$

*g.* Find the phase angle for the parallel circuit between $D$ and $E$.

$$\cos\theta = \frac{I_{T_x}}{I_T} = \frac{7.11}{7.8} = 0.9115$$
$$\theta = 24° \textit{ leading}$$

The total parallel impedance = 12.8 Ω at 24° *leading.*

2. The equivalent series impedance is made up of the resistive and reactive components into which this parallel impedance must be resolved.

$$Z_x = R = Z \times \cos\theta$$
$$R = 12.8 \times \cos 24° = 12.8 \times 0.914$$
$$R = 11.7\ \Omega \qquad \textit{Ans.}$$
$$Z_y = X_C = Z \times \sin\theta \qquad (X_C \text{ because of the } \textit{leading} \text{ current})$$
$$X_C = 12.8 \times \sin 24° = 12.8 \times 0.407$$
$$X_C = 5.26\ \Omega \qquad \textit{Ans.}$$

3. Find the total series impedance. The components of $Z_{DE}$ are now combined with the other parts of the series circuit as shown in Fig. 19-50.

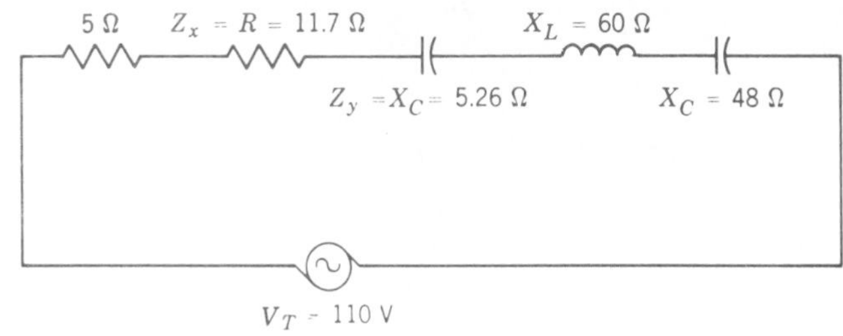


**FIGURE 19-50** The components of $Z_{DE}$ are parts of the equivalent series circuit.

$$R_T = 5 + 11.7 = 16.7\ \Omega$$
$$X_C = 5.26 + 48 = 53.26\ \Omega$$
$$X_L = 60\ \Omega$$
$$Z = \sqrt{R^2 + (X_L - X_C)^2}$$
$$= \sqrt{(16.7)^2 + (60.0 - 53.26)^2}$$
$$= \sqrt{(16.7)^2 + (6.74)^2}$$
$$= \sqrt{279 + 45.5} = \sqrt{324.5}$$
$$Z = 18\ \Omega \qquad \textit{Ans.}$$

4. Find the total series current.

$$I_T = \frac{V_T}{Z} = \frac{110}{18} = 6.11\ \text{A} \qquad \textit{Ans.}$$

5. Find the phase angle.

$$\cos\theta = \frac{R}{Z} = \frac{16.7}{18} = 0.926$$
$$\theta = 22° \textit{ lagging}$$

6. Find the power factor.

$$\text{PF} = \frac{R}{Z} = 0.926 = 92.6\% \textit{ lagging}$$

7. Find the power.

$$W = V \times I \times \cos\theta = 110 \times 6.11 \times \cos 22°$$
$$= 110 \times 6.11 \times 0.926 = 622\ \text{W} \qquad \textit{Ans.}$$

*Check:* $W = I_R^2 \times \text{R} = (6.11)^2 \times 16.7 = 623\ \text{W}$ *Check*

The total series current of 6.11 A lags the total voltage of 110 V by 22°.

## PROBLEMS

Solve each of the circuits shown in Fig. 19-51 for (*a*) the equivalent series impedance, (*b*) the total current, (*c*) the phase angle, (*d*) the power factor, and (*e*) the power drawn by the circuit.

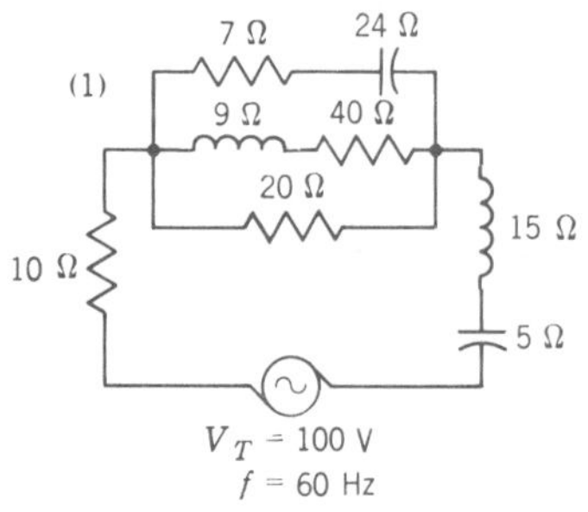


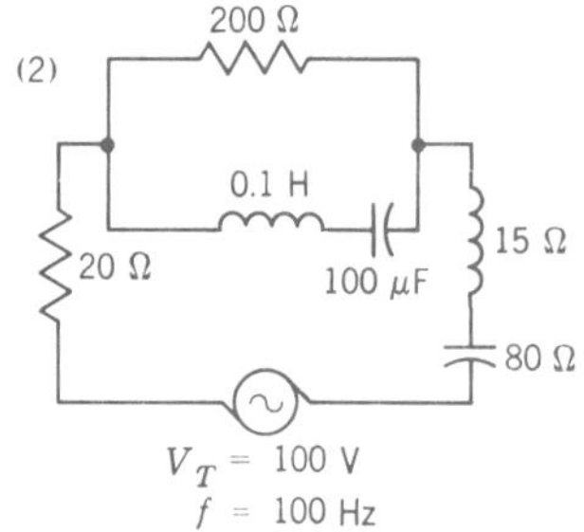


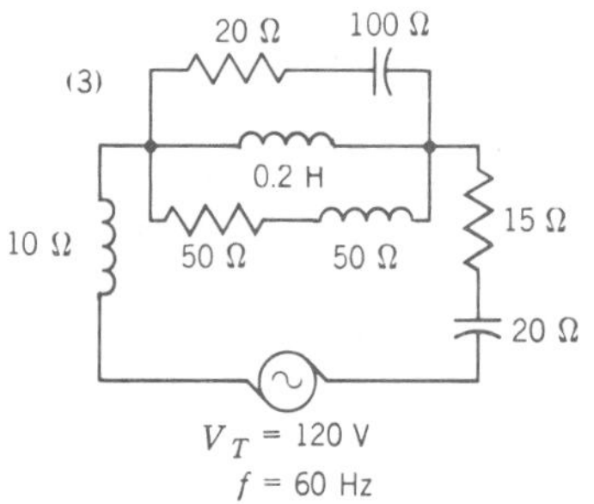


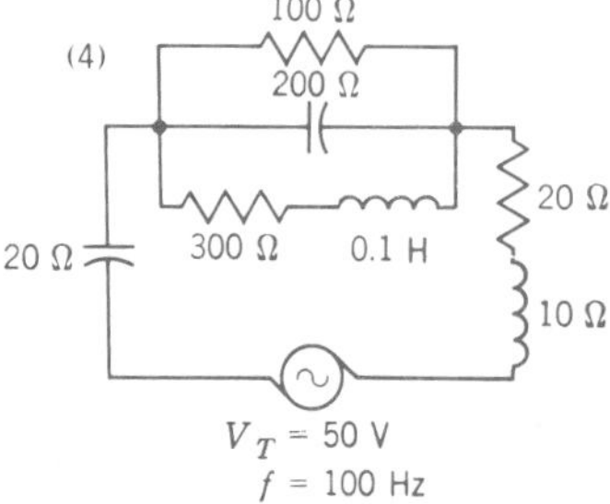


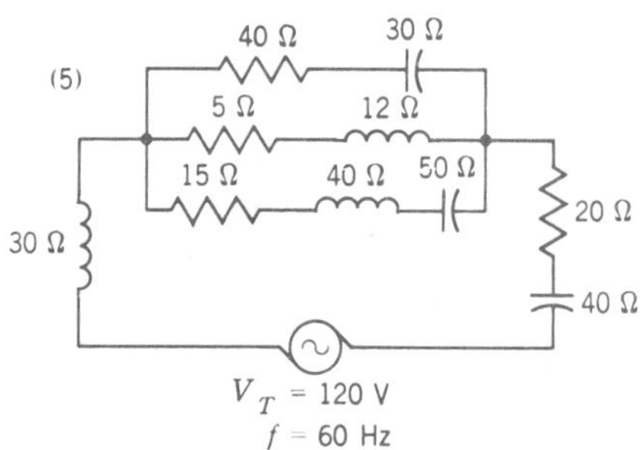


FIGURE 19-51

## JOB 19-8 PARALLEL RESONANCE

In a series circuit, it is possible to adjust the values of a coil and a capacitor so that their reactances will be equal for a definite frequency. The frequency at which the inductive and capacitive reactances are equal is the resonant frequency. Since the actions of the reactances are directly opposed to each other, the total reactance is zero and the impedance of the circuit becomes just the resistance of the circuit. Under these conditions, the current in the circuit will be a maximum.

Now let us arrange the coil and capacitor in *parallel* as shown in Fig. 19-52.

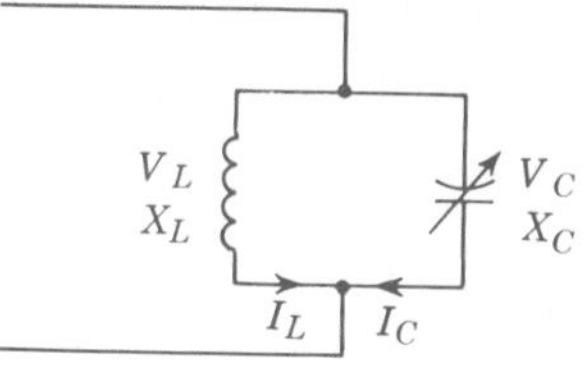


**FIGURE 19-52**
Parallel-resonant circuit.

At the resonant frequency,

$$V_L = V_C \qquad (5\text{-}1)$$

$$X_L = X_C \qquad (18\text{-}22)$$

By dividing Eq. (5-1) by Eq. (18-22), we obtain

$$\frac{V_L}{X_L} = \frac{V_C}{X_C}$$

or

$$I_L = I_C \qquad \boxed{19\text{-}13}$$

Since the currents are exactly opposed to each other, the total current is equal to $I_L - I_C$ or practically zero. At the resonant frequency, then, since the total current is very nearly equal to zero, the impedance of the circuit to currents at that frequency will be very large. At any other frequency, since $X_L$ is not equal to $X_C$, the currents will no longer be equal. The sum of the unequal currents may be quite large, which indicates a low impedance. To summarize, a parallel resonant circuit will offer a very large impedance to currents at the resonant frequency and a low impedance to currents at all other frequencies.

**Uses.** Just as a series resonant circuit is able to *accept* currents at the resonant frequency and reject all others, a parallel resonant circuit is able to *reject* currents at the resonant frequency and accept all others. This makes it possible to reject or "trap" a wave of a definite frequency in antenna and filter circuits. It is also a convenient method for obtaining the high impedance required in the primary of coupling transformers.

**Resonant frequency.** The formula for the resonant frequency of a parallel circuit is the same as that for a series circuit.

$$f = \frac{159}{\sqrt{L \times C}} \qquad \boxed{18\text{-}23}$$

where $f$ = frequency, kHz
$L$ = inductance, $\mu$H
$C$ = capacitance, $\mu$F

EXAMPLE 19-20. A 200-$\mu$H coil and a 50-pF capacitor are connected in parallel to form a "wave trap" in an antenna. What is the resonant frequency that the circuit will reject?

SOLUTION. Given: $L = 200\ \mu\text{H}$ Find: $f = ?$

$C = 50\ \text{pF} = 50 \times 10^{-6}\ \mu\text{F}$

$$f = \frac{159}{\sqrt{L \times C}} \qquad (18\text{-}23)$$

$$= \frac{159}{\sqrt{200 \times 50 \times 10^{-6}}}$$

$$= \frac{159}{\sqrt{10^4 \times 10^{-6}}} = \frac{159}{\sqrt{10^{-2}}} = \frac{159}{10^{-1}}$$

$$f = 159 \times 10 = 1590\ \text{kHz} \qquad \textit{Ans.}$$

## FINDING THE INDUCTANCE OR CAPACITANCE NEEDED TO PRODUCE RESONANCE

EXAMPLE 19-21. A 0.1-mH coil and a variable capacitor are connected in parallel to form the primary of an IF transformer as shown in Fig. 19-53. If the circuit is to be resonant to 456 kHz, what must be the value of the capacitor?

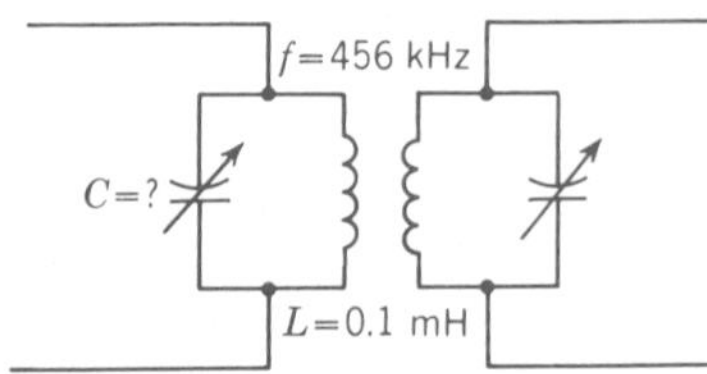


FIGURE 19-53

SOLUTION

1. Change 0.1 mH to 100 $\mu$H.
2. Find the required capacitance.

$$C = \frac{25{,}300}{f^2 \times L} \qquad (18\text{-}25)$$

$$= \frac{25{,}300}{(456)^2 \times 100}$$

$$C = \frac{25{,}300}{207{,}900 \times 100} = 0.0012\ \mu\text{F} \qquad \textit{Ans.}$$

**PROBLEMS**

1. Find the resonant frequency of a wave trap using a 45-pF capacitor and a 20-$\mu$H inductance.
2. Find the resonant frequency of a band-stop filter made of a 160-$\mu$H coil and a 40-pF capacitor in parallel.
3. The inductance of a parallel resonant circuit used as a wave trap in

an antenna circuit is 100 $\mu$H. What must be the value of the parallel capacitance in order to reject an 800-kHz wave?

4. The capacitor of a high-impedance primary of a transformer tuned to 460 kHz is 100 pF. What is the value of the inductance?
5. The tank circuit of an oscillator contains a coil of 320 $\mu$H. What is the value of the capacitance at the resonant frequency of 1000 kHz?
6. The tank circuit of an impedance-coupled AF amplifier circuit uses an inductance of 10 H. Find the value of the capacitance necessary to produce resonance at (*a*) 500 Hz and (*b*) 1 kHz.
7. What is the inductance of the coil in a 23.4-MHz trap of a video IF amplifier which uses a capacitor of 50 pF?
8. An IF coil in a superheterodyne receiver resonates at a frequency of 455 kHz. Find the inductance of the coil if the capacitor is valued at 50 pF.
9. A wave trap in the plate circuit of an IF stage is to resonate at 27.25 MHz. Find the capacity needed to produce resonance with a coil valued at 0.85 $\mu$H.

## JOB 19-9 REVIEW OF PARALLEL AC CIRCUITS

In a purely resistive circuit:

$$V_T = V_1 = V_2 = V_3$$ 5-1

$$V_T = I_T \times R_T$$ 4-7

$$I_T = I_1 + I_2 + I_3$$ 5-2

$$W = V \times I \times \underline{\qquad}$$ 18-3 — cos $\theta$

The total current is __(in/out of)__ phase with the total voltage. — in

In a purely inductive circuit:

$$V_T = V_1 = V_2 = V_3$$ 5-1

$$V_T = I_T \times \underline{\qquad}$$ 18-13 — $Z$

$$I_T = I_1 + I_2 + I_3$$ 5-2

$$W = V \times I \times \cos\theta$$ 18-3

The total current __(leads/lags)__ the total voltage by 90°. — lags

In a purely capacitive circuit:

$$V_T = V_1 = V_2 = V_3$$ 5-1

$$V_T = I_T \times Z$$ 18-13

$$I_T = I_1 + I_2 + I_3$$ 5-2

$$W = V \times I \times \cos\theta$$ 18-3

The total current __(leads/lags)__ the total voltage by 90°. — leads

In an ac parallel circuit of resistance and inductance:

$$I_T^2 = I_R^2 + \_\_\_\_\_\_ \qquad \boxed{19\text{-}1}$$ $I_L^2$

$$V_T = I_T \times Z \qquad \boxed{18\text{-}13}$$

$$\cos\theta = \frac{I_R}{?} \qquad \boxed{19\text{-}2}$$ $I_T$

$$\text{PF} = \frac{?}{I_T} \qquad \boxed{19\text{-}3}$$ $I_R$

$$W = V \times I \times \_\_\_\_\_\_ \qquad \boxed{18\text{-}3}$$ $\cos\theta$

or

$$W = V \times I \times \_\_\_\_\_\_ \qquad \boxed{18\text{-}4}$$ PF

The total current (leads/lags) the total voltage by angle $\theta$. — lags

In an ac parallel circuit of resistance and capacitance:

$$I_T^2 = I_R^2 + \_\_\_\_\_\_ \qquad \boxed{19\text{-}4}$$ $I_C^2$

$$V_T = I_T \times \_\_\_\_\_\_ \qquad \boxed{18\text{-}13}$$ $Z$

$$\cos\theta = \frac{?}{I_T} \qquad \boxed{19\text{-}2}$$ $I_R$

$$\text{PF} = \frac{I_R}{?} \qquad \boxed{19\text{-}3}$$ $I_T$

$$W = V \times I \times \cos\theta \qquad \boxed{18\text{-}3}$$

or

$$W = \_\_\_\_\_\_ \qquad \boxed{18\text{-}4}$$ $V \times I \times \text{PF}$

The total current (leads/lags) the total voltage by angle $\theta$. — leads

In an ac parallel circuit of resistance, inductance, and capacitance:

$$I_T^2 = I_R^2 + (I_L - ?)^2 \qquad \boxed{19\text{-}5}$$ $I_C$

$$V_T = I_T \times Z \qquad \boxed{18\text{-}13}$$

$$\cos\theta = \frac{I_R}{I_T} \qquad \boxed{19\text{-}2}$$

$$\text{PF} = \frac{?}{?} \qquad \boxed{19\text{-}3}$$ $I_R$, $I_T$

$$W = V \times I \times \cos\theta \qquad \boxed{18\text{-}3}$$

$$W = V \times I \times \text{PF} \qquad \boxed{18\text{-}4}$$

The total current will lead or lag the total voltage, depending on the values of $I_L$ and $I_C$. If $I_L$ is larger than $I_C$, the current will (lead/lag) the voltage. If $I_L$ is smaller than $I_C$, the current will (lead/lag) the voltage. The angle of lead or lag is given by the angle $\theta$. — lag; lead

## PARALLEL-SERIES AC CIRCUITS

1. Find the reactance, impedance, current, and phase angle for each branch of the parallel circuit.
2. Draw the phasor diagram for the branch currents on the same voltage base.

3. Resolve the current in each branch into its components.

$$I_y = I \times \_\_\_\_\_\_$$ 19-6 | sin $\theta$

$$I_x = I \times \_\_\_\_\_\_$$ 19-7 | cos $\theta$

($y$ components of lagging currents are negative.)

4. Find the total in-phase and total reactive currents.

$$I_{T_x} = I_{A_x} + I_{B_x}$$ 19-8

$$I_{T_y} = I_{A_y} + \_\_\_\_\_\_$$ 19-9 | $I_{B_y}$

5. Find the total current.

$$I_T^2 = (I_{T_x})^2 + \_\_\_\_\_\_$$ 19-10 | $(I_{T_y})^2$

6. Find the impedance.

$$V_T = I_T \times Z$$ 18-13

7. Find the phase angle.

$$\cos \theta_T = \frac{?}{I_T}$$ 19-11 | $I_{T_x}$

8. Find the power factor.

$$\text{PF} = \frac{I_{T_x}}{?}$$ 19-12 | $I_T$

9. Find the total power.

$$W = V \times I \times \cos \theta$$ 18-3

or

$$W = V \times I \times \text{PF}$$ 18-4

## SERIES-PARALLEL AC CIRCUITS

1. For the parallel branches: Repeat steps 1 through 7 for parallel-series circuits.
2. Resolve the parallel impedance found in step 6 above into its equivalent series resistance and reactance.
3. Combine these with the other resistances and reactances to get the total series impedance.
4. The total current, phase angle, and power are found as in any series circuit.

**Parallel resonance.** A parallel resonant circuit will offer a very large impedance to currents at the resonant frequency and a low impedance to currents at all other frequencies.

$$f = \frac{159}{\sqrt{L \times C}}$$ 18-23

where $L$ = inductance, measured in ______ | μH
$C$ = capacitance, measured in ______ | μF
$f$ = frequency, measured in ______ | kHz

## FINDING THE INDUCTANCE OR CAPACITANCE NEEDED TO PRODUCE A RESONANT CIRCUIT

$$L = \frac{25{,}300}{f^2 \times ?} \qquad \boxed{18\text{-}24}$$

C

$$C = \frac{25{,}300}{f^2 \times L} \qquad \boxed{18\text{-}25}$$

where $L$, $C$, and $f$ are measured in the same units as called for in formula (18-23).

## PROBLEMS

1. Two capacitors of 500 and 750 Ω reactance, respectively, are connected in parallel across a 25-V 25-Hz ac source. Find (*a*) the total current, (*b*) the impedance, and (*c*) the power drawn by the circuit.
2. A 0.01-H coil and a 5000-Ω resistor are connected in parallel to form a filter circuit. Find the percent of the total current passing through the resistor for (*a*) a 1-kHz AF frequency and (*b*) a 1000-kHz RF frequency. (*c*) Is this filter a high-pass or a low-pass filter?
3. A 200-Ω resistor and a 0.1-H coil are connected in parallel across a 100-V 1-kHz ac source. Find (*a*) the current in each branch, (*b*) the total current, (*c*) the impedance of the circuit, (*d*) the phase angle, (*e*) the power factor, and (*f*) the power drawn by the circuit.
4. A 500-Ω resistor in an emitter circuit similar to Fig. 19-19 is bypassed with a 5-μF capacitor. If a 1-kHz frequency causes a voltage drop of 10 V across the resistor, find (*a*) the current in each branch, (*b*) the total current, and (*c*) the impedance of the combination.
5. A 3000-Ω resistor, a 1200-Ω inductive reactance, and an 800-Ω capacitive reactance are connected in parallel across a 240-V line. Find (*a*) the total current, (*b*) the impedance, (*c*) the phase angle, (*d*) the power factor, and (*e*) the power drawn by the circuit.
6. A 0.5-μF capacitor, a 1-H coil, and a 2000-Ω resistor are connected in parallel across a 220-V 60-Hz ac line. Find (*a*) the total current, (*b*) the impedance, (*c*) the phase angle, (*d*) the power factor, and (*e*) the power drawn by the circuit.
7. In a circuit similar to that used for Prob. 1 in Fig. 19-46, $I_A = 8$ A with $\theta = 30°$ lagging, $I_B = 15$ A with $\theta = 60°$ leading, and $V_T = 120$ V at 60 Hz. Find (*a*) the total current, (*b*) the impedance, (*c*) the phase angle, (*d*) the power factor, and (*e*) the power drawn by the circuit.
8. In a circuit similar to that used for Prob. 6 in Fig. 19-46, $R_1 = 100$ Ω, $L = 0.2$ H, $X_C = 500$ Ω, $R_2 = 200$ Ω, $V_T = 100$ V, and $f = 100$ Hz. Find (*a*) the total current, (*b*) the impedance, (*c*) the phase angle, (*d*) the power factor, and (*e*) the power drawn by the circuit.
9. A 0.001-μF capacitor and a coil are connected in parallel to form the primary of an IF transformer similar to that shown in Fig.

19-53. What must be the inductance of the coil in order for the circuit to be resonant to a frequency of 460 kHz?

10. The collector circuit of the 2SC563 mixer transistor in the Emerson model 984 television receiver contains a 1-$\mu$H coil in parallel with a capacitance and resonant to 21.25 MHz. What is the capacitance?

11. Solve the circuit shown in Fig. 19-54 for (*a*) the equivalent series impedance, (*b*) the total current, (*c*) the phase angle, and (*d*) the power drawn by the circuit.

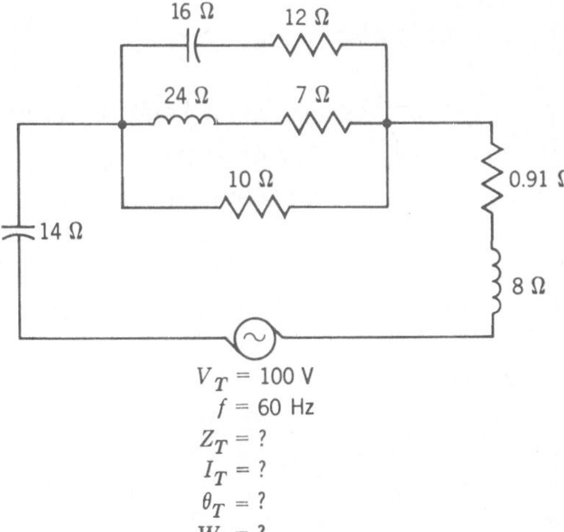


FIGURE 19-54

12. A capacitor, a 24-Ω resistor, and an inductive reactance of 20 Ω are connected in parallel across a 120-V 60-Hz supply. The total line current is read as 6.4 A lagging. Find (*a*) the total reactive current, (*b*) the capacitor current, and (*c*) the total power factor.

(See Instructor's Manual for Test 19-5, Parallel AC circuits)

# ALTERNATING–CURRENT POWER

# 20

## JOB 20-1 POWER AND POWER FACTOR

The power in any electric circuit is obtained by multiplying the voltage by the current flowing at that time. In a dc circuit, the unchanging voltage $V$ is multiplied by the unchanging current $I$ to give the power $P$. The formula for this was given in Job 7-1 as $P = V \times I$. In an ac circuit, the voltage and current are constantly changing. The power at any instant of time is obtained by multiplying the instantaneous voltage by the instantaneous current.

$$P_i = v \times i \tag{20-1}$$

**Power in a resistive circuit.** In a purely resistive circuit, each instantaneous current occurs at the same time as the instantaneous voltage which produced it. Since the worth of all the instantaneous values is the effective value, the power in a resistive circuit is found by multiplying the effective voltage by the effective current.

$$P = V \times I \tag{7-1}$$

**Power in a reactive circuit.** In an inductive circuit, the current will lag behind the voltage by some angle $\theta$ as shown in Fig. 20-1*a*. The power in this circuit will *not* be equal to the product of the voltage and the current, since they do not act at the same time. The actual power is equal to the voltage multiplied by *only that portion of the line current which is in phase with the voltage.* In Fig. 20-1*b*, $OE$ represents the line voltage. $OA$ represents the total line current as measured by an ammeter. This total line current may be resolved into its two component parts: $OB$, a component in phase with the voltage (the effective current), and $BA$, a component 90° out of phase with the voltage (the reactive current). By multiplying each of the current phasors of Fig. 20-1*b* by the line voltage, we can obtain a phasor diagram for the power in an *inductive* circuit as shown in Fig. 20-1*c*.

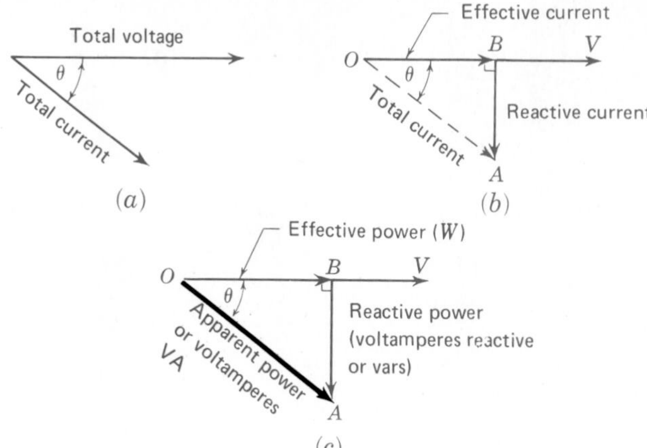


**FIGURE 20-1**
(a) Current lags behind voltage in an inductive circuit. (b) Resolution of the total current into its effective and reactive components. (c) Phasor diagram of the power.

The phasor diagram of Fig. 20-1*c* indicates that the apparent power *VA* is made of two component parts—the effective power *W* and the reactive power vars. The apparent power may be likened to the power delivered to a flywheel. The *portion* of the apparent power which is delivered to the shaft to operate some device is similar to the effective power. The *portion* of the apparent power which is delivered by the flywheel *back* to the engine to keep it running is similar to the reactive power. The reactive power does no work itself. In an inductive electric circuit it represents the power stored in the magnetic field (similar to the flywheel) and then returned to the line as the field collapses. This power merely moves back and forth between the coil and the line. In Fig. 20-1*c*, *OA* represents the *apparent power* as measured by a voltmeter and an ammeter. It is measured in voltamperes, *not* watts. *OB* represents the *effective,* or *true, power* and is measured by a wattmeter in watts. This is the power that does the work. *AB* represents the reactive, or *var,* power that does *no* work. It is measured in voltamperes reactive (var).

**FORMULAS**

$$VA = V \times I \tag{20-2}$$

$$\cos \theta = \frac{W}{VA} \tag{20-3}$$

By cross multiplication,

$$W = VA \times \cos \theta \tag{20-4}$$

Solving for *VA*,

$$VA = \frac{W}{\cos \theta} \tag{20-5}$$

Also, since $\sin \theta = \dfrac{\text{var power}}{VA}$

$$\text{var power} = VA \times \sin\theta \qquad \boxed{20\text{-}6}$$

where $VA$ = apparent power, VA
$V$ = voltage, V
$I$ = current, A
$W$ = true power, W
$\theta$ = phase angle of the circuit, degrees
var power = voltamperes reactive, var

**Power factor.** The ratio of the effective power as read by a wattmeter $W$ to the apparent power $VA$ is called the *power factor* (PF).

**FORMULA**

$$\text{PF} = \frac{W}{VA} \qquad \boxed{20\text{-}7}$$

By comparing formulas (20-3) and (20-7), we can see that the power factor is equal to the cosine of angle $\theta$.

**FORMULA**

$$\text{PF} = \cos\theta \qquad \boxed{20\text{-}8}$$

By substituting PF for cos $\theta$ in formulas (20-4) and (20-5), we obtain

**FORMULAS**

$$W = VA \times \text{PF} \qquad \boxed{20\text{-}9}$$

$$VA = \frac{W}{\text{PF}} \qquad \boxed{20\text{-}10}$$

The power factor may be expressed as a decimal or as a percent. For example, a PF of 0.8 may be written as 80 percent. In this sense, the PF describes the *portion* of the voltampere input which is actually effective in operating the device. Thus, an 80 percent PF means that the device uses only 80 percent of the voltampere input in order to operate, and 20 percent is wasted as reactive power. This is uneconomical. Generator capacity, copper size, fusing, etc., must be provided for the apparent power although part of it does not work. Low power factors may be improved by the methods discussed in Job 20-4. Circuits with

1. Resistance only will have a high PF with $I$ in phase with $V$
2. High inductive reactance will have a low PF and a lagging current
3. High capacitive reactance will have a low PF and a leading current
4. Equal amounts of inductive and capacitive reactance will have a high PF with $I$ in phase with $V$

EXAMPLE 20-1. Find the power factor of a washing-machine motor if it draws 5 A and 440 W from a 110-V 60-Hz line.

SOLUTION. Given: $I = 5$ A Find: PF = ?
$V = 110$ V
$W = 440$ W
$f = 60$ Hz

$$\text{PF} = \frac{W}{VA} = \frac{440}{110 \times 5} = \frac{440}{550} = 0.8 = 80\% \qquad \textit{Ans.} \ (20\text{-}7)$$

EXAMPLE 20-2. A single-phase capacitor-type motor operates at a power factor of 70 percent leading. The line voltage is 110 V, and the line current is 10 A. Find (*a*) the apparent power and (*b*) the true power taken in watts.

SOLUTION. Given: PF = 70% = 0.70 Find: $VA = ?$
$I = 10$ A $W = ?$
$V = 110$ V

*a.* $$VA = V \times I = 10 \times 110 = 1100 \text{ VA} \qquad \textit{Ans.} \qquad (20\text{-}2)$$

*b.* $$W = VA \times \text{PF} = 1100 \times 0.70 = 770 \text{ W} \qquad \textit{Ans.} \quad (20\text{-}9)$$

EXAMPLE 20-3. A 5-kVA single-phase ac generator operates at a power factor of 90 percent. Find (*a*) the full-load power supplied and (*b*) the full-load current if the terminal voltage is 120 V.

SOLUTION. Given: $VA = 5000$ VA Find: $W = ?$
PF = 90% = 0.9 $I = ?$
$V = 120$ V

*a.* $$W = VA \times \text{PF} \qquad (20\text{-}9)$$
$$= 5000 \times 0.9 = 4500 \text{ W} \qquad \textit{Ans.}$$

*b.* $$P = I \times V \qquad (7\text{-}1)$$
$$4500 = I \times 120$$
$$I = \frac{4500}{120} = 37.5 \text{ A} \qquad \textit{Ans.}$$

EXAMPLE 20-4. An impedance coil of 3 Ω resistance and 4 Ω inductive reactance is connected across a 24-V 60-Hz ac source. Find (*a*) the PF, (*b*) the current drawn, and (*c*) the effective power consumed by the coil.

SOLUTION. Given: $R = 3$ Ω Find: PF = ?
$X_L = 4$ Ω $I = ?$
$V = 24$ V $W = ?$
$f = 60$ Hz

*a.* Find the impedance.

$$Z = \sqrt{R^2 + X_L^2} = \sqrt{3^2 + 4^2} = \sqrt{9 + 16} = \sqrt{25} = 5\ \Omega$$

Find the PF.

$$PF = \frac{R}{Z} = \frac{3}{5} = 0.60 = 60\% \qquad Ans. \qquad (18\text{-}12)$$

*b.* Find the current drawn.

$$V_T = I_T \times Z \qquad (18\text{-}13)$$

$$24 = I_T \times 5$$

$$I_T = \frac{24}{5} = 4.8 \text{ A} \qquad Ans.$$

*c.* Find the effective power.

$$W = V \times I \times PF \qquad (18\text{-}4)$$

$$= 24 \times 4.8 \times 0.6 = 69 \text{ W} \qquad Ans.$$

## PROBLEMS

1. Find the power factor of a refrigerator motor if it draws 288 W and 3 A from a 120-V 60-Hz line.
2. A single-phase circuit has unity power factor. Find the true power in watts if the applied voltage is 120 V and the line current is 20 A.
3. Find the effective power used by a capacitor-type jigsaw motor operating at a power factor of 75 percent if it draws 4 A at 120 V.
4. The lights and motors in a shop draw 16 kW of power. The power factor of the entire load is 80 percent. Find the voltamperes of power delivered to the shop.
5. A circuit consumes 4 kW. If the line current is 40 A and the voltage is 110 V, find (*a*) the apparent power in voltamperes, and (*b*) the power factor of the circuit.
6. A capacitance of 5 Ω resistance and 12 Ω reactance is connected across a 117-V 60-Hz ac line. Find the power factor and the effective power.
7. A motor operating at 90 percent PF draws 270 W from a 120-V line. Find the current drawn.
8. Find the PF of a motor in an air-conditioning unit if it draws 500 W and 5 A from a 120-V line.
9. A motor operates on a 220-V line at a power factor of 0.8 lagging. If it consumes 8 kW, find the motor line current.
10. A certain load draws 90 kW and 110 kVA from an ac supply. At what power factor does it operate?
11. An industrial load draws 15 A from a 230-V line at a PF of 85 percent. Find the voltamperes of apparent power and the effective power taken from the line.
12. A 40-V emf at 1 kHz is impressed across a loudspeaker of 5000 Ω resistance and 1.5 H inductance. Find (*a*) the impedance, (*b*) the PF, (*c*) the current drawn, and (*d*) the effective power drawn by the speaker.

13. A 10-hp motor operates at an efficiency of 80 percent and a PF of 90 percent. Find the voltamperes of apparent power delivered to the motor. *Hint:* Find the watt input to the motor by the efficiency formula (11-2).
14. An inductive load operating at a phase angle of 53° draws 1200 W from a 120-V line. Find the current taken.
15. Find the current drawn from a 230-V line by a motor if it uses 6 kW of power at a PF of 0.65.
16. A 120-V 2-hp motor operates at an efficiency of 80 percent. Find the power input to the motor. If the current drawn by the motor is 20 A, find the PF.
17. A motor draws 600 W and 1000 VA from a 120-V 60-Hz ac line. Find the reactive voltamperes of the motor.
18. A single-phase alternator delivers 80 kW at 80 percent power factor. Find the kilovoltampere output of the alternator.
19. A wattmeter connected in a 120-V inductive circuit reads 720 W. Find the line current if the power factor of the circuit is 60 percent.
20. A single-phase motor draws 25 A when connected to a 220-V ac supply. If the power factor of the motor is 80 percent lagging, find (*a*) the voltamperes supplied, (*b*) the power input, and (*c*) the reactive voltamperes.

## JOB 20-2 TOTAL POWER DRAWN BY COMBINATIONS OF REACTIVE LOADS

As we learned in Job 7-2, the total power in a circuit may be found by adding the power taken by the individual parts. If the power drawn by one branch is not in phase with the power drawn by another branch, however, the addition of the power must be made by *phasor addition.* This will be done, as in Job 19-5, by resolving the power into its components and adding the components.

**Resolving the apparent power into its components.** In Fig. 20-1*c*,

$$\sin \theta = \frac{\text{var power}}{VA} \qquad \text{and} \qquad \cos \theta = \frac{W}{VA}$$

Cross-multiplying each equation yields the

**FORMULAS**

$$\text{var power} = VA \times \sin \theta \qquad \boxed{20\text{-}4}$$

$$W = VA \times \cos \theta \qquad \boxed{20\text{-}6}$$

or

$$W = VA \times \text{PF} \qquad \boxed{20\text{-}9}$$

Also, using the tangent function, we obtain the

**FORMULA** $$\tan \theta = \frac{\text{var power}}{W}$$ **20-11**

Applying the pythagorean theorem, we obtain the

**FORMULA** $$(VA)^2 = W^2 + (\text{var } P)^2$$ **20-12**

## Combinations of devices of different power factors

EXAMPLE 20-5. Find (*a*) the total effective power, (*b*) the total apparent power, (*c*) the total PF, and (*d*) the total current of the combination shown in Fig. 20-2.

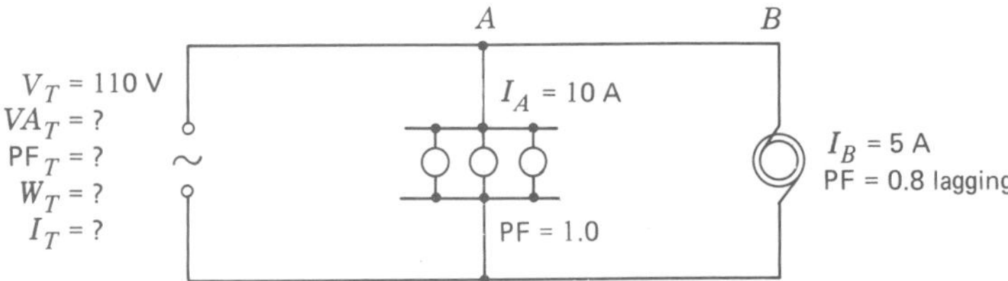


FIGURE 20-2

SOLUTION. When the power in one branch is not in phase with the power in another branch, the apparent powers must be resolved into their components and the components added as shown below.

*a.* Find the total effective power.

1. Find the apparent power $VA$, the effective power $W$, the phase angle $\theta$, and the var power (var $P$) for each load.

For branch *A*: $VA_A = I_A \times V_A = 110 \times 10 = 1100 \text{ VA}$ (20-2)

$W_A = VA_A \times \text{PF}_A = 1100 \times 1 = 1100 \text{ W}$ (20-9)

Since $\text{PF}_A = \cos\theta = 1.0 \qquad \theta = 0°$ (20-8)

$$\text{var } P_A = VA_A \times \sin\theta \quad (20\text{-}6)$$
$$= 1100 \times \sin 0°$$
$$= 1100 \times 0$$
$$\text{var } P_A = 0 \text{ var}$$

For branch *B*: $VA_B = I_B \times V_B = 5 \times 110 = 550 \text{ VA}$ (20-2)

$W_B = VA_B \times \text{PF}_B = 550 \times 0.8 = 440 \text{ W}$ (20-9)

Since $\text{PF}_B = \cos\theta = 0.8 \qquad \theta = 37° \textit{ lagging}$ (20-8)

$$\text{var } P_B = VA_B \times \sin\theta \quad (20\text{-}6)$$
$$= 550 \times \sin 37°$$
$$= 550 \times 0.6$$
$$\text{var } P_B = -330 \text{ var}$$

2. Draw the phasor diagram for the apparent powers on the same voltage base as shown in Fig. 20-3*a*.

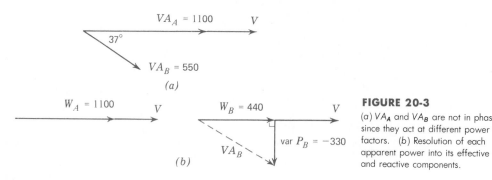

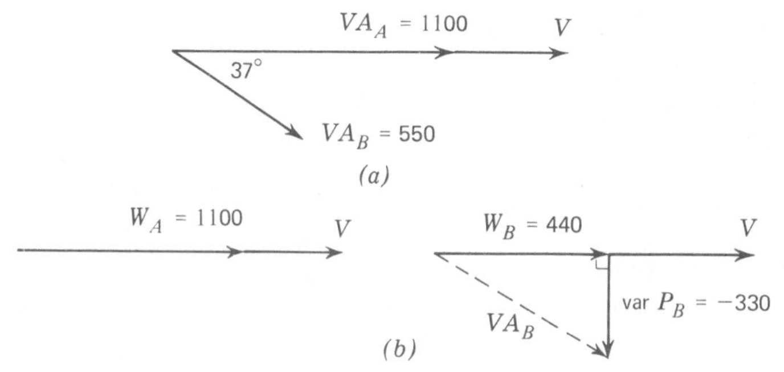


**FIGURE 20-3** (*a*) $VA_A$ and $VA_B$ are not in phase, since they act at different power factors. (*b*) Resolution of each apparent power into its effective and reactive components.

3. Resolve the power in each branch into its components as shown in Fig. 20-3*b*.

4. Draw all the components on the same voltage base as shown in Fig. 20-4*a*. The total effective power $W_T$ and the total reactive power var $P_T$ may now be found by the

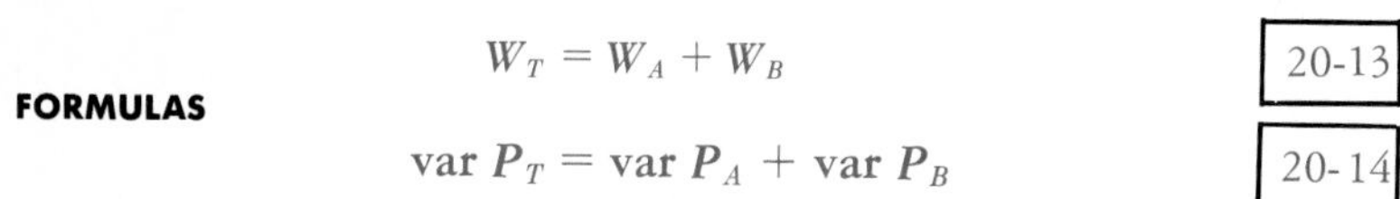

**FORMULAS**

$$W_T = W_A + W_B \qquad \boxed{20\text{-}13}$$

$$\text{var } P_T = \text{var } P_A + \text{var } P_B \qquad \boxed{20\text{-}14}$$

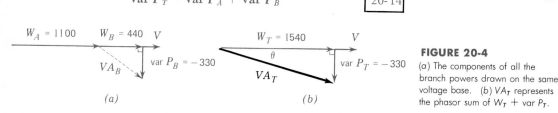

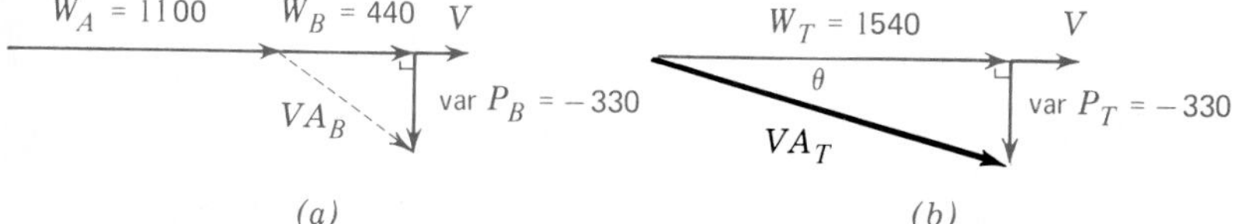


**FIGURE 20-4** (*a*) The components of all the branch powers drawn on the same voltage base. (*b*) $VA_T$ represents the phasor sum of $W_T$ + var $P_T$.

$$W_T = W_A + W_B = 1100 + 440 = 1540 \text{ W} \qquad (20\text{-}13)$$

$$\text{var } P_T = \text{var } P_A + \text{var } P_B = 0 + (-330) = -330 \text{ var} \qquad (20\text{-}14)$$

*b*. Find the total apparent power. Draw the phasor diagram for the total power by adding $W_T$ and var $P_T$ vectorially as shown in Fig. 20-4*b*. Notice that var $P_T$ is drawn *downward* because var $P_T$ is negative.

Applying the pythagorean theorem, we obtain the

**FORMULAS**

$$VA_T^2 = W_T^2 + \text{var } P_T^2 \qquad \boxed{20\text{-}15}$$

$$\text{PF}_T = \frac{W_T}{VA_T} \qquad \boxed{20\text{-}16}$$

Find $VA_T$.

$$VA_T^2 = W_T^2 + \text{var } P_T^2 \qquad (20\text{-}15)$$

$$= (1540)^2 + (-330)^2$$

$$VA_T^2 = 237 \times 10^4 + 11 \times 10^4$$
$$= 248 \times 10^4$$
$$VA_T = \sqrt{248 \times 10^4} = 1575 \text{ VA} \qquad \textit{Ans.}$$

*c.* Find the total PF.

$$\text{PF}_T = \frac{W_T}{VA_T} = \frac{1540}{1575} = 0.977 = 97.7\% \textit{ lagging} \qquad \textit{Ans.} \quad (20\text{-}16)$$

*d.* Find the total current.

$$VA_T = V_T \times I_T \qquad (20\text{-}2)$$
$$1575 = 110 \times I_T$$
$$I_T = \frac{1575}{110} = 14.3 \text{ A} \qquad \textit{Ans.}$$

ALTERNATIVE SOLUTION. The phase angle and the apparent power may be found by applying the tangent formula to the right triangle shown in Fig. 20-4*b*. This solution is simpler, since it does not involve the labor of squaring numbers and finding the square root of the sum.

**FORMULAS**

$$\tan \theta_T = \frac{\text{var } P_T}{W_T} \qquad \boxed{20\text{-}17}$$

$$VA_T = \frac{W_T}{\text{PF}_T} \qquad \boxed{20\text{-}18}$$

*a.* Find the phase angle in Fig. 20-4*b*.

$$\tan \theta_T = \frac{\text{var } P_T}{W_T} = \frac{330}{1540} = 0.214 \qquad \text{therefore} \qquad \theta = 12° \quad (20\text{-}17)$$
$$\text{PF} = \cos \theta = \cos 12° = 0.978 = 97.8\% \textit{ lagging} \qquad \textit{Ans.} \quad (20\text{-}8)$$

*b.* Find the total apparent power.

$$VA_T = \frac{W_T}{\text{PF}_T} = \frac{1540}{\cos 12°} = \frac{1540}{0.978} = 1574 \text{ VA} \qquad \textit{Ans.} \quad (20\text{-}18)$$

EXAMPLE 20-7. Find (*a*) the total effective power, (*b*) the total power factor, and (*c*) the total apparent power for the circuit shown in Fig. 20-5.

SOLUTION

*a.* 1. Find the effective power and the var power for each load.

For load *A*:

$$\text{PF} = \cos \theta = 0.500 \qquad \text{therefore} \qquad \theta = 60° \qquad (20\text{-}8)$$
$$W_A = VA_A \times \text{PF}_A = 20 \times 0.5 = 10 \text{ kW} \qquad (20\text{-}9)$$
$$\text{var } P_A = VA_A \times \sin \theta = 20 \times \sin 60° = 20 \times 0.866 = -17.32 \text{ kvar}$$

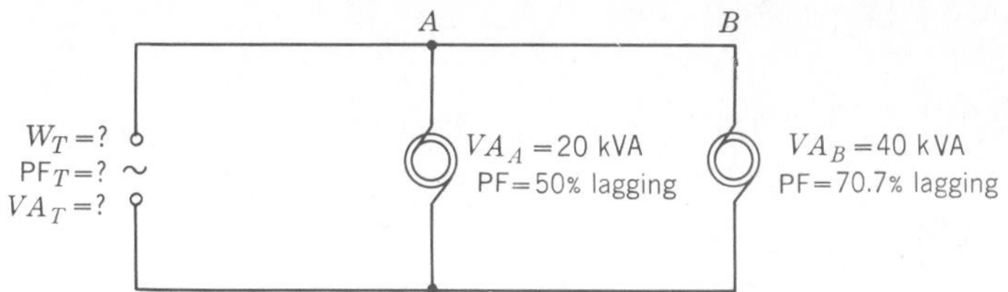


FIGURE 20-5

For load $B$:

$$\text{PF} = \cos\theta = 0.707 \qquad \text{therefore} \qquad \theta = 45° \tag{20-8}$$

$$W_B = VA_B \times \text{PF}_B = 40 \times 0.707 = 28.28 \text{ kW} \tag{20-9}$$

$$\text{var } P_B = VA_B \times \sin\theta = 40 \times \sin 45° = 40 \times 0.707 = -28.28 \text{ kvar}$$

2. Draw the phasor diagram for the apparent powers on the same voltage base as shown in Fig. 20-6*a*.

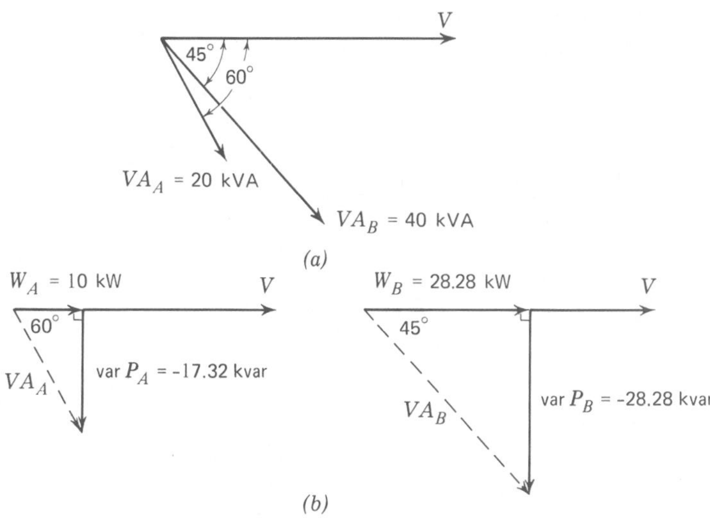


FIGURE 20-6
(*a*) $VA_A$ and $VA_B$ are not in phase, since they act at different power factors. (*b*) Resolution of each apparent power into its effective and reactive components.

3. Resolve the power in each branch into its components as shown in Fig. 20-6*b*.

4. Draw all the components on the same voltage base as shown in Fig. 20-7*a*.

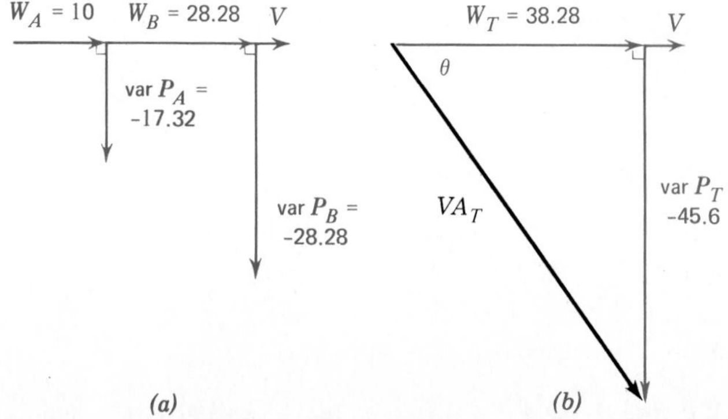


FIGURE 20-7
(*a*) The components of all the branch powers drawn on the same voltage base. (*b*) $VA_T$ represents the phasor sum of $W_T$ + var $P_T$.

The total effective power $W_T$ and the total reactive power var $P_T$ may now be found.

$$W_T = W_A + W_B = 10 + 28.28 = 38.28 \text{ kW} \quad \textit{Ans.} \quad (20\text{-}13)$$
$$\text{var } P_T = \text{var } P_A + \text{var } P_B = (-17.32) + (-28.28) = -45.6 \text{ kvar}$$

*b.* Find the total power factor by adding $W_T$ and var $P_T$ vectorially as shown in Fig. 20-7*b*.

$$\tan \theta_T = \frac{\text{var } P_T}{W_T} = \frac{-45.6}{38.28} = -1.191 \quad \text{therefore} \quad \theta = 50° \quad (20\text{-}17)$$
$$\text{PF} = \cos \theta = \cos 50° = 0.643 = 64.3\% \textit{ lagging} \quad \textit{Ans.} \quad (20\text{-}8)$$

*c.* Find the total apparent power.

$$VA_T = \frac{W_T}{\text{PF}_T} = \frac{38.28}{\cos 50°} = \frac{38.28}{0.643} = 59.5 \text{ kVA} \quad \textit{Ans.} \quad (20\text{-}18)$$

EXAMPLE 20-8. An inductive load taking 10 A and 2000 W from a 220-V line is in parallel with a motor taking 1400 W at a PF of 50 percent lagging. Find (*a*) the total effective power, (*b*) the total PF, (*c*) the total apparent power, and (*d*) the total current drawn by the circuit.

SOLUTION. The diagram for the circuit is shown in Fig. 20-8.

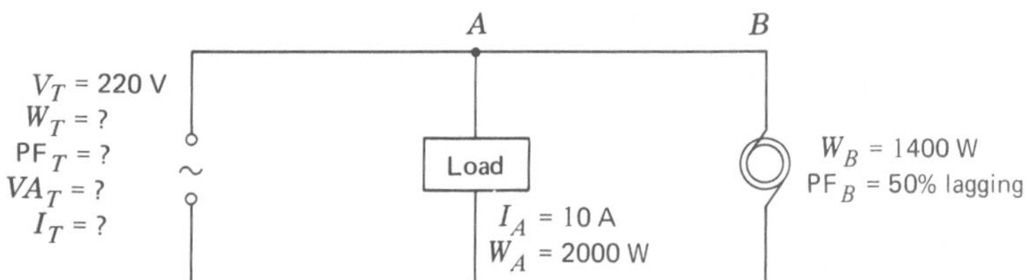


FIGURE 20-8

*a.* 1. Find the apparent power, the effective power, the phase angle, and the var power for each load.

For load $A$:

$$VA_A = I_A \times V_A = 10 \times 220 = 2200 \text{ VA} \quad (20\text{-}2)$$
$$W_A = 2000 \text{ W}$$
$$\cos \theta = \frac{W}{VA} = \frac{2000}{2200} = 0.909 \quad \text{therefore} \quad \theta = 25° \textit{ lagging} \quad (20\text{-}3)$$
$$\text{var } P_A = VA_A \times \sin \theta = 2200 \times \sin 25° = 2200 \times 0.423 = -931 \text{ var}$$

For load $B$:

$$W_B = 1400 \text{ W}$$
$$VA_B = \frac{W}{\text{PF}} = \frac{1400}{0.5} = 2800 \text{ VA} \quad (20\text{-}10)$$
$$\text{PF}_B = \cos \theta = 0.500 \quad \text{so} \quad \theta = 60° \quad (20\text{-}8)$$
$$\text{var } P_B = VA_B \times \sin \theta \quad (20\text{-}6)$$
$$= 2800 \times \sin 60°$$
$$= 2800 \times 0.866$$
$$\text{var } P_B = -2425 \text{ var}$$

2. Find the total effective and reactive power.

$$W_T = W_A + W_B = 2000 + 1400 = 3400 \text{ W} \qquad \textit{Ans.}$$
$$\text{var } P_T = \text{var } P_A + \text{var } P_B = (-931) + (-2425) = -3356 \text{ var}$$

*b.* Find the phase angle and the PF.

$$\tan \theta_T = \frac{\text{var } P_T}{W_T} = \frac{3356}{3400} = 0.987 \qquad \text{therefore} \qquad \theta = 45° \quad (20\text{-}17)$$
$$\text{PF} = \cos \theta = \cos 45° = 0.707 = 70.7\% \textit{ lagging} \quad \textit{Ans.} \qquad (20\text{-}8)$$

*c.* Find the total apparent power.

$$VA_T = \frac{W_T}{\text{PF}_T} = \frac{3400}{0.707} = 4809 \text{ VA} \qquad \textit{Ans.} \qquad (20\text{-}18)$$

*d.* Find the total current.

$$I_T = \frac{VA_T}{V_T} = \frac{4809}{220} = 21.8 \text{ A} \qquad \textit{Ans.} \qquad (20\text{-}2)$$

SUMMARY

## Total power drawn by combinations of reactive loads

1. Find the apparent power for each load using any of the following formulas.

$$VA = V \times I \qquad (20\text{-}2)$$

or

$$VA = \frac{W}{\text{PF}} \qquad (20\text{-}10)$$

or

$$VA = \frac{W}{\cos \theta} \qquad (20\text{-}5)$$

2. Find the effective and reactive power for each load.

$$W = VA \times \text{PF} \qquad (20\text{-}9)$$

or

$$W = VA \times \cos \theta \qquad (20\text{-}4)$$

and

$$\text{var } P = VA \times \sin \theta \qquad (20\text{-}6)$$

*Note:* If the PF or the phase angle is not given, it may be found by either of the following formulas:

$$\text{PF} = \frac{W}{VA} \qquad (20\text{-}7)$$

$$\cos \theta = \frac{W}{VA} \qquad (20\text{-}3)$$

3. Find the total effective power.

$$W_T = W_A + W_B \qquad (20\text{-}13)$$

4. Find the total reactive power.

$$\text{var } P_T = \text{var } P_A + \text{var } P_B \qquad (20\text{-}14)$$

5. Find the phase angle $\theta$.

$$\tan \theta_T = \frac{\text{var } P_T}{W_T} \qquad (20\text{-}17)$$

6. Find the total PF.

$$\text{PF} = \cos \theta \qquad (20\text{-}8)$$

7. Find the total apparent power.

$$VA_T = \frac{W_T}{\text{PF}_T} \qquad (20\text{-}18)$$

8. Find the total current drawn, using any of the following formulas.

$$VA_T = V_T \times I_T \qquad (20\text{-}2)$$

$$W_T = V_T \times I_T \times \cos \theta_T \qquad \boxed{20\text{-}19}$$

$$W_T = V_T \times I_T \times \text{PF}_T \qquad \boxed{20\text{-}20}$$

## PROBLEMS

1. A refrigerator motor drawing 6 A at 80 percent PF lagging is in parallel with a washing-machine motor drawing 8 A at 80 percent PF lagging from a 110-V line. Find (*a*) the total effective power, (*b*) the total PF, (*c*) the total apparent power, and (*d*) the total current drawn.
2. Motor *A* draws 10 A and 800 W from a 110-V 60-Hz ac line. Motor *B* draws 6 A and 480 W from the same line in parallel. Find (*a*) the PF of each motor, (*b*) the total effective power, (*c*) the total PF, (*d*) the total apparent power, and (*e*) the total current drawn.
3. A purely resistive lamp load (PF = 1) drawing 8 A from a 110-V 60-Hz line is in parallel with an induction motor taking 10 A at 70 percent PF. Find (*a*) the total effective power, (*b*) the total PF, (*c*) the total apparent power, and (*d*) the total current drawn.
4. A lamp bank (PF = 1) drawing 1200 W from a 110-V 60-Hz line is in parallel with an induction motor taking 8 A at a power factor of 90 percent. Find (*a*) the total effective power, (*b*) the total PF, (*c*) the total apparent power, and (*d*) the total current.
5. Find (*a*) the total effective power, (*b*) the total PF, and (*c*) the total apparent power for the circuit shown in Fig. 20-9.

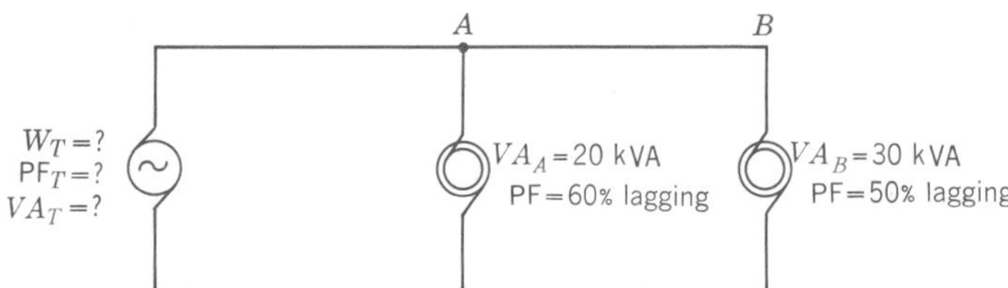


FIGURE 20-9

6. Repeat Prob. 4 for a 2-kW lamp load in parallel with a motor

operating at a PF of 60 percent lagging and drawing 3 kW from a 110-V 60-Hz line.

7. Repeat Prob. 4 for the following circuit: A motor drawing 5 kW at 80 percent PF lagging is in parallel with a second motor drawing 8 kW at 70 percent PF lagging from a 220-V line.
8. Repeat Prob. 4 for the circuit shown in Fig. 20-10.

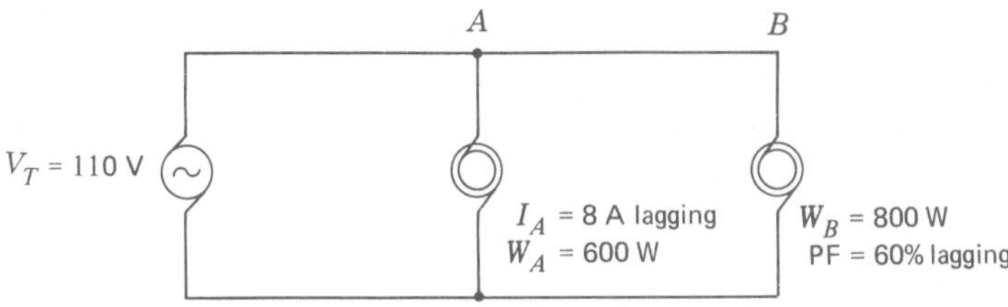


FIGURE 20-10

9. A 2-kW lamp load, a 70-kVA motor operating at a PF of 60 percent lagging, and a 40-kVA motor operating at a PF of 70 percent lagging are connected in parallel. Find (*a*) the total effective power, (*b*) the total PF, and (*c*) the total apparent power.
10. Repeat Prob. 4 for the circuit shown in Fig. 20-11.

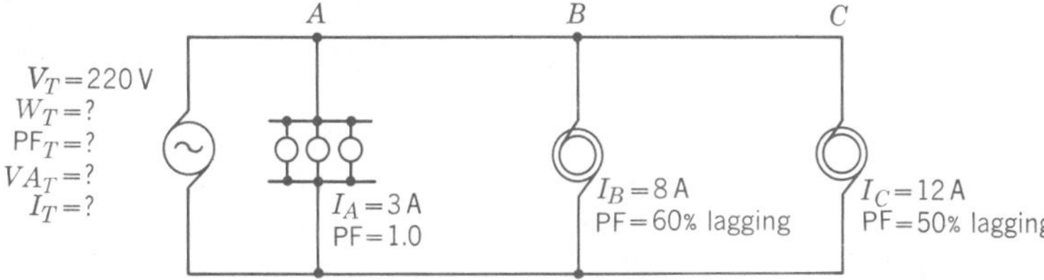


FIGURE 20-11

(See Instructor's Manual for Test 20-1)

## JOB 20-3 POWER DRAWN BY COMBINATIONS OF RESISTIVE, INDUCTIVE, AND CAPACITIVE LOADS

**Power in capacitive loads.** When the current is out of phase with the voltage, only that portion of the current which is in phase with the voltage is useful in producing usable power.

In an inductive circuit, the current lags behind the voltage, but in a capacitive circuit, the current leads the voltage. In such a circuit, the phasor diagrams will be very similar to those of an inductive circuit, except that now there will be a *leading* current and a *leading* power. All the formulas developed for the inductive circuits in the last job will also apply to a capacitive circuit. The only difference is that the reactive power will *lead* in a capacitive circuit.

EXAMPLE 20-9. A lamp bank drawing 1 kW of power is in parallel with a synchronous motor drawing 2 kW at a leading PF of 80 percent from a 220-V 60-Hz ac line. Find (*a*) the total effective power, (*b*) the total PF, (*c*) the total apparent power, and (*d*) the total current drawn.

SOLUTION. The diagram of the circuit is shown in Fig. 20-12.

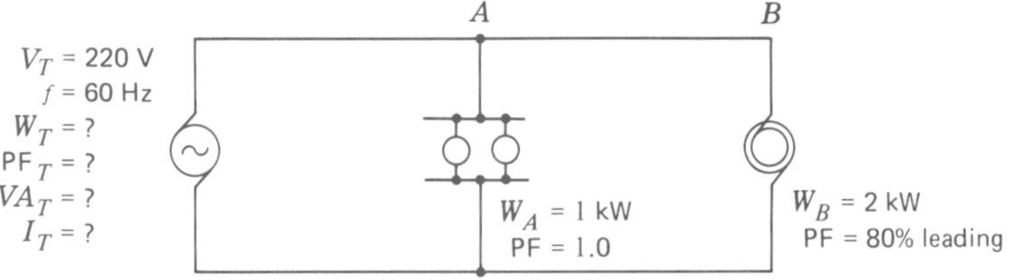


FIGURE 20-12

*a.* 1. Find the apparent power, the effective power, the phase angle, and the var power for each load.

For load $A$:

$$W_A = 1 \text{ kW} = 1000 \text{ W} \qquad \text{(given)}$$

$$VA_A = \frac{W_A}{\text{PF}} = \frac{1000}{1} = 1000 \text{ VA} \qquad (20\text{-}10)$$

$$\text{PF} = \cos\theta = 1.0 \qquad \text{therefore} \qquad \theta = 0° \qquad (20\text{-}8)$$

$$\text{var } P_A = VA_A \times \sin\theta = 1000 \times \sin 0° = 1000 \times 0 = 0 \text{ var} \qquad (20\text{-}6)$$

For load $B$:

$$W_B = 2 \text{ kW} = 2000 \text{ W} \qquad \text{(given)}$$

$$VA_B = \frac{W_A}{\text{PF}} = \frac{2000}{0.8} = 2500 \text{ VA} \qquad (20\text{-}10)$$

$$\text{PF} = \cos\theta = 0.8 \qquad \text{therefore} \qquad \theta = 37° \textit{ leading} \qquad (20\text{-}8)$$

$$\text{var } P_B = VA_B \times \sin\theta = 2500 \times \sin 37° = 2500 \times 0.6 = 1500 \text{ var}$$

2. Find the total effective and reactive power.

$$W_T = W_A + W_B = 1000 + 2000 = 3000 \text{ W}$$

$$\text{var } P_T = \text{var } P_A + \text{var } P_B = 0 + 1500 = 1500 \text{ var} \qquad (20\text{-}14)$$

*b.* Find the phase angle and the PF.

$$\tan\theta_T = \frac{\text{var } P_T}{W_T} = \frac{1500}{3000} = 0.5 \qquad \text{therefore} \qquad \theta = 27° \quad (20\text{-}17)$$

$$\text{PF}_T = \cos\theta = \cos 27° = 89.1\% \textit{ leading} \qquad (20\text{-}8)$$

*c.* Find the total apparent power.

$$VA_T = \frac{W_T}{\text{PF}_T} = \frac{3000}{0.891} = 3367 \text{ VA} \qquad \textit{Ans.} \qquad (20\text{-}18)$$

*d.* Find the total current drawn.

$$I_T = \frac{VA_T}{V_T} = \frac{3367}{220} = 15.3 \text{ A} \qquad \textit{Ans.} \qquad (20\text{-}2)$$

EXAMPLE 20-10. A 10-kVA induction motor operating at 85 percent lagging PF and a 5-kVA synchronous motor operating at 68.2 percent leading PF are connected in parallel across a 220-V 60-Hz ac line. Find (*a*) the total effective power, (*b*) the total PF, (*c*) the total apparent power, and (*d*) the total current drawn.

SOLUTION. The diagram for the circuit is shown in Fig. 20-13.

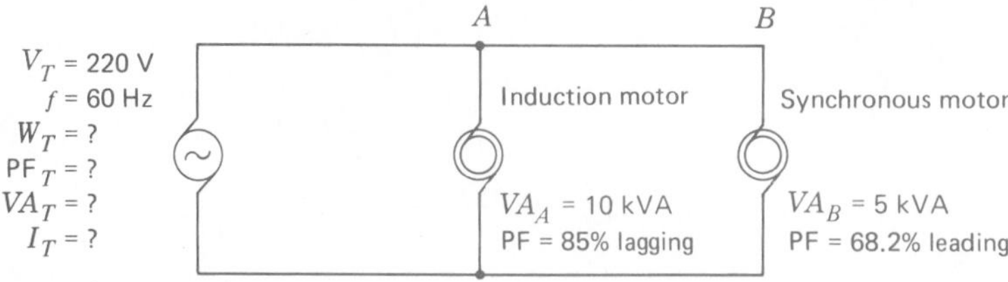


**FIGURE 20-13**

*a*. 1. Find the apparent power, the effective power, the phase angle, and the var power for each load.

For the induction motor *A*:

$$VA_A = 10 \text{ kVA} \qquad \text{(given)}$$
$$W_A = VA_A \times \text{PF}_A = 10 \times 0.85 = 8.5 \text{ kW} \qquad (20\text{-}9)$$
$$\text{PF}_A = \cos\theta = 0.85 \qquad \text{therefore} \qquad \theta = 32° \qquad (20\text{-}8)$$
$$\text{var } P_A = VA_A \times \sin\theta = 10 \times \sin 32° = 10 \times 0.53 \qquad (20\text{-}6)$$
$$\text{var } P_A = -5.3 \text{ kvar } \textit{lagging}$$

For the synchronous motor *B*:

$$VA_B = 5 \text{ kVA} \qquad \text{(given)}$$
$$W_B = VA_B \times \text{PF}_B = 5 \times 0.682 = 3.41 \text{ kW} \qquad (20\text{-}9)$$
$$\text{PF}_A = \cos\theta = 0.682 \qquad \text{therefore} \qquad \theta = 47° \qquad (20\text{-}8)$$
$$\text{var } P_B = VA_B \times \sin\theta = 5 \times \sin 47° = 5 \times 0.73 = 3.65 \text{ kvar } \textit{leading}$$

2. Draw the phasor diagram for the apparent powers on the same voltage base as shown in Fig. 20-14*a*.

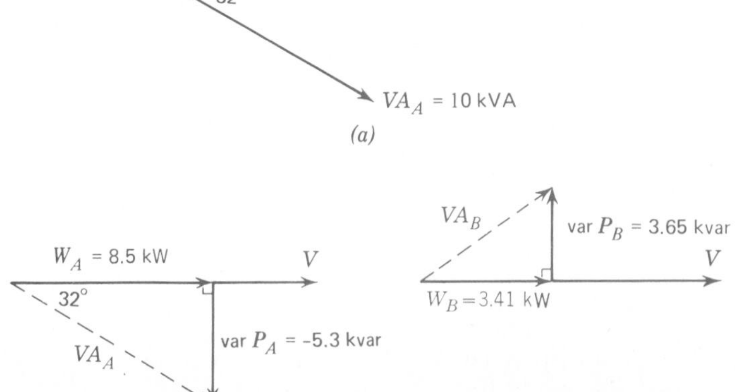


**FIGURE 20-14**
(*a*) Leading and lagging apparent powers. (*b*) Resolution of each power into its effective and reactive components.

3. Resolve the power in each branch into its components as shown in Fig. 20-14*b*.

4. Draw all the components on the same voltage base as shown in Fig. 20-15*a*. The total effective power $W_T$ and the total reactive power var $P_T$ may now be found.

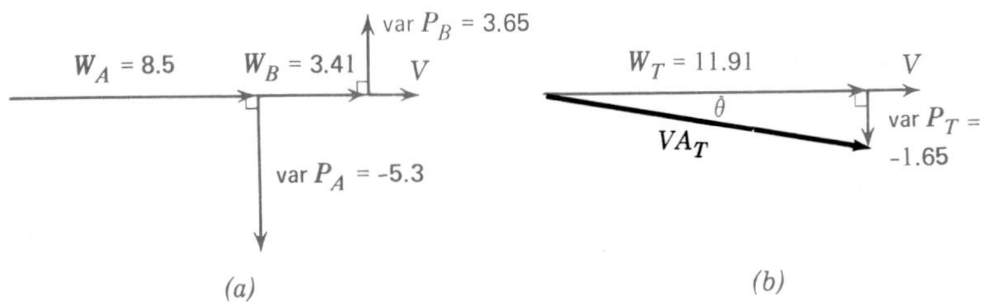


**FIGURE 20-15**
(*a*) The components of all the branch powers drawn on the same voltage base. (*b*) $VA_T$ represents the phasor sum of $W_T$ + var $P_T$.

$$W_T = W_A + W_B = 8.5 + 3.41 = 11.91 \text{ kW} \qquad \textit{Ans.} \qquad (20\text{-}13)$$

$$\text{var } P_T = \text{var } P_A + \text{var } P_B = (-5.3) + 3.65 = -1.65 \text{ kvar } \textit{lagging}$$

*b*. Find the total PF by adding $W_T$ and var $P_T$ vectorially as shown in Fig. 20-15*b*.

$$\tan \theta_T = \frac{\text{var } P_T}{W_T} = \frac{-1.65}{11.91} = -0.1385 \quad \text{so} \quad \theta = 8° \quad (20\text{-}17)$$

$$\text{PF} = \cos \theta = \cos 8° = 0.99 = 99\% \textit{ lagging} \qquad (20\text{-}8)$$

*c*. Find the total apparent power.

$$VA_T = \frac{W_T}{\text{PF}_T} = \frac{11.91}{0.99} = 12.03 \text{ kVA} = 12{,}030 \text{ VA} \qquad \textit{Ans.}$$

*d*. Find the total current drawn.

$$I_T = \frac{VA_T}{V_T} = \frac{12{,}030}{220} = 54.7 \text{ A } \textit{lagging} \qquad \textit{Ans.} \qquad (20\text{-}2)$$

EXAMPLE 20-11. A synchronous motor drawing 10 A at 60 percent leading PF from a 110-V line is in parallel with an induction motor drawing 1 kW at 80 percent PF lagging. Find (*a*) the total effective power, (*b*) the total PF, (*c*) the total apparent power, and (*d*) the total current drawn.

SOLUTION. The diagram for the circuit is shown in Fig. 20-16.

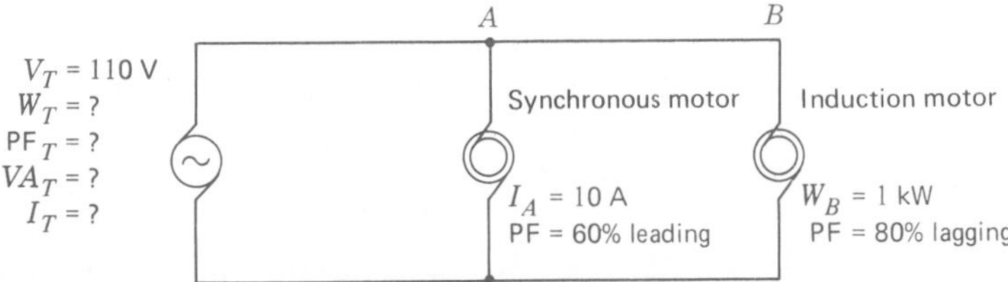


**FIGURE 20-16**

*a*. 1. Find the apparent power, the effective power, the phase angle, and the var power for each load.

For the synchronous motor *A*:

$$VA_A = I_A \times V_A = 10 \times 110 = 1100 \text{ VA} \qquad (20\text{-}2)$$

$$W_A = VA_A \times \text{PF}_A = 1100 \times 0.6 = 660 \text{ W} \qquad (20\text{-}9)$$

$$\text{PF}_A = \cos\theta = 0.600 \quad \text{therefore} \quad \theta = 53° \qquad (20\text{-}8)$$

$$\text{var } P_A = VA_A \times \sin\theta = 1100 \times \sin 53° = 1100 \times 0.8 \qquad (20\text{-}6)$$

$$\text{var } P_A = 880 \text{ var } \textit{leading}$$

For the induction motor *B*:

$$W_B = 1 \text{ kW} = 1000 \text{ W} \quad \text{(given)}$$

$$VA_B = \frac{W_B}{\text{PF}_B} = \frac{1000}{0.8} = 1250 \text{ VA} \qquad (20\text{-}10)$$

$$\text{PF}_B = \cos\theta = 0.800 \quad \text{therefore} \quad \theta = 37° \qquad (20\text{-}8)$$

$$\text{var } P_B = VA_B \times \sin\theta = 1250 \times \sin 37° = 1250 \times 0.6 \qquad (20\text{-}6)$$

$$\text{var } P_B = -750 \text{ var } \textit{lagging}$$

2. Find the total effective and reactive power.

$$W_T = W_A + W_B = 660 + 1000 = 1660 \text{ W} \quad \textit{Ans.}$$

$$\text{var } P_T = \text{var } P_A + \text{var } P_B = (+880) + (-750) = 130 \text{ var } \textit{leading}$$

*b*. Find the phase angle and the PF.

$$\tan\theta_T = \frac{\text{var } P_T}{W_T} = \frac{130}{1660} = 0.078 \quad \text{so} \quad \theta = 4° \qquad (20\text{-}17)$$

$$\text{PF}_T = \cos\theta = \cos 4° = 0.998 = 99.8\% \textit{ leading} \quad \textit{Ans.}$$

*c*. Find the total apparent power.

$$VA_T = \frac{W_T}{\text{PF}_T} = \frac{1660}{0.998} = 1663 \text{ VA} \quad \textit{Ans.} \qquad (20\text{-}18)$$

*d*. Find the total current drawn.

$$I_T = \frac{VA_T}{V_T} = \frac{1663}{110} = 15.1 \text{ A } \textit{leading} \quad \textit{Ans.} \qquad (20\text{-}2)$$

## PROBLEMS

1. An induction motor drawing 400 VA at 80 percent lagging PF is in parallel with a synchronous motor drawing 700 VA at 90 percent leading PF. Find (*a*) the total effective power, (*b*) the total PF, and (*c*) the total apparent power.
2. Repeat Prob. 1 for a circuit containing an induction motor that draws 100 kVA at a PF of 85 percent in parallel with a capacitive load that draws 80 kVA at a PF of 70 percent.
3. A capacitive load drawing 20 A at 70 percent PF from a 120-V 60-Hz line is in parallel with an induction motor drawing 4 kW at

70 percent PF. Find (*a*) the total effective power, (*b*) the total PF, (*c*) the total apparent power, and (*d*) the total current drawn.

4. Repeat Prob. 3 for a circuit containing a 60 percent PF induction motor drawing 1 kW in parallel with a snychronous motor rated at 120 V, 15 A, and 70 percent PF.
5. Find the total PF of a parallel combination of a 5-kW lamp load, a 10-kW inductive load (PF = 80 percent), and an 8-kW capacitive load (PF = 90 percent).
6. Find (*a*) the total effective power, (*b*) the total PF, (*c*) the total apparent power, and (*d*) the total current drawn by the circuit shown in Fig. 20-17.

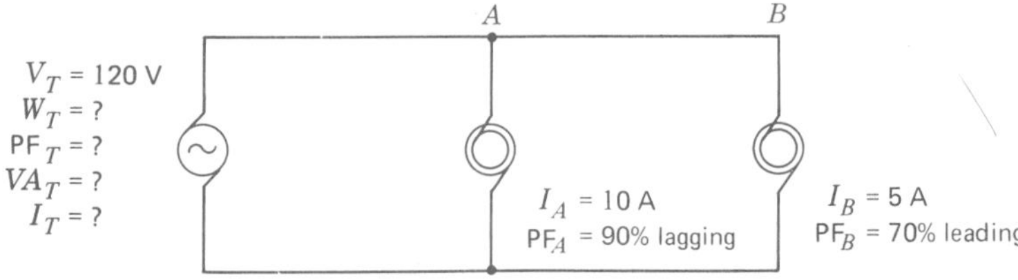


**FIGURE 20-17**

7. Repeat Prob. 6 for the circuit shown in Fig. 20-18.

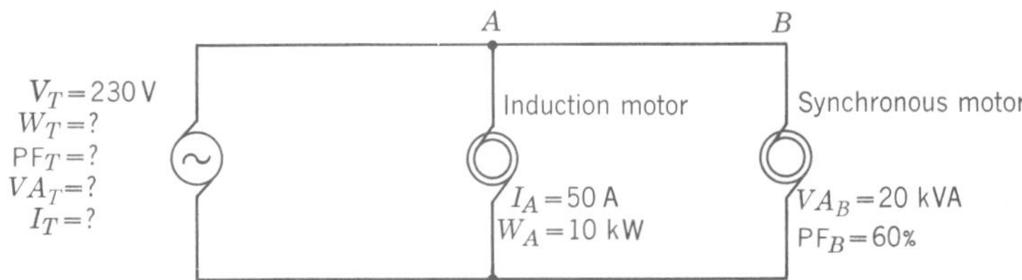


**FIGURE 20-18**

8. An induction motor draws 20 A at a PF of 80 percent from a 120-V line. What is the total PF of the circuit if (*a*) a lamp load drawing 15 A is connected in parallel and (*b*) a capacitive load drawing 5 A at a PF of 60 percent leading is connected in parallel?
9. Repeat Prob. 6 for a circuit containing an induction motor taking 4 A at 80 percent PF lagging, a synchronous motor taking 8 A at 50 percent PF leading, and a lamp load taking 6 A at 100 percent PF, all connected in parallel across a 120-V line.
10. Repeat Prob. 6 for a circuit containing a 100-W inductive load at a PF of 80 percent, a 600-W capacitive load at a PF of 60 percent, and an induction motor drawing 1 kW at a PF of 90 percent, all connected in parallel across a 120-V line.

(See Instructor's Manual for Test 20-2, AC Power)

## JOB 20-4 POWER-FACTOR CORRECTION

Consider the three circuits shown in Fig. 20-19. Each circuit draws the same effective power, but at decreasing PFs.

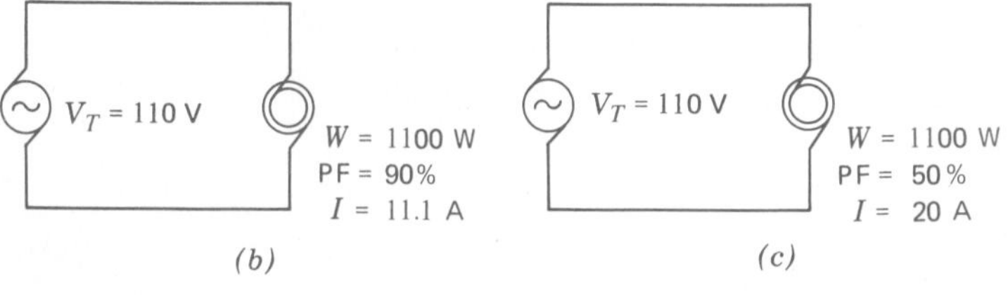

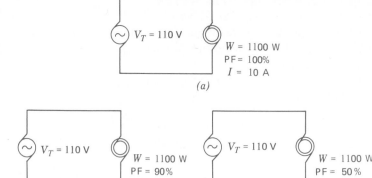

**FIGURE 20-19** In order to produce a constant effective power, the current must increase as the power factor decreases.

For the circuit of Fig. 20-19*a*:

$$W = V \times I \times \text{PF} \qquad (18\text{-}4)$$

$$1100 = 110 \times I \times 1$$

$$I = \frac{1100}{110} = 10 \text{ A}$$

For the circuit of Fig. 20-19*b*:

$$W = V \times I \times \text{PF} \qquad (18\text{-}4)$$

$$1100 = 110 \times I \times 0.9$$

$$1100 = 99 \times I$$

$$I = \frac{1100}{99} = 11.1 \text{ A}$$

For the circuit of Fig. 20-19*c*:

$$W = V \times I \times \text{PF} \qquad (18\text{-}4)$$

$$1100 = 110 \times I \times 0.5$$

$$1100 = 55 \times I$$

$$I = \frac{1100}{55} = 20 \text{ A}$$

By an investigation of the amount of current drawn in each of the circuits, we can see that as the PF decreases, more and more current must be supplied in order to produce the same effective power. Now, regardless of the current that is drawn to provide this 1100 W of effective power, the consumer pays for only 1100 W. Therefore, if the power company is forced to send 20 instead of 10 A to provide 1100 W of power, it must provide heavier wires to carry the larger current. This is expensive. In addition, since heating losses depend on the *square* of the current, these losses will be higher at low PF than at high PF. For these reasons, the power company demands an extra premium payment if the PF falls below a certain value for a particular installation.

A low PF is generally due to the large consumption of power by underloaded induction motors which take a *lagging* current. This will produce a large, wasteful, *lagging* reactive power. In order to correct this low power factor, an equal amount of *leading* reactive power should be connected into the circuit. To do this, synchronous motors, which operate at unity or leading power factors are placed in parallel with the inductive load. The leading reactive kilovars supplied by the synchronous motor compensate for the lagging reactive kilovars and result in an improvement of the total power factor. Any improvement in the PF will release supply capacity, increase efficiency, and generally improve the operating characteristics of the system.

Capacitors, which take a leading current, may also be used for this purpose. They are rated in kilovars instead of microfarads. Standard ratings are either 15 or 25 kvar. Although it is desirable to raise the power factor to 100 percent, any standard capacitor which raises the power factor to 90 to 95 percent is considered acceptable.

EXAMPLE 20-12. An induction motor draws 1300 W and 1500 VA from an ac line. (*a*) What is its power factor? (*b*) What is the value of the reactive voltamperes, or vars, drawn by the motor? (*c*) How many vars of leading reactive power are needed to raise the power factor to unity?

SOLUTION

*a.*

$$\text{PF} = \frac{W}{VA} = \frac{1300}{1500} = 0.8667 = 86.7\% \qquad \textit{Ans.} \qquad (20\text{-}7)$$

*b.* 1.

$$\text{Since } \cos\theta = \text{PF} \qquad (20\text{-}8)$$

$$\cos\theta = 0.8667$$

$$\theta = 30°$$

2.

$$\text{var power} = VA \times \sin\theta \qquad (20\text{-}6)$$

$$= 1500 \times \sin 30°$$

$$= 1500 \times 0.5 = 750 \text{ var} \qquad \textit{Ans.}$$

*c.* The leading var power should equal the lagging var power, or 750 var. *Ans.*

EXAMPLE 20-13. An induction motor takes 15 kVA at 220 V and 80 percent lagging PF. What must be the PF of a 10-kVA synchronous motor connected in parallel in order to raise the total PF to 100 percent, or unity?

SOLUTION

Given: $VA = 15$ kVA, PF = 80% *lagging*, $V = 220$ V } induction motor; $VA = 10$ kVA synchronous motor

Find: To get a PF = 1, PF of synchronous motor = ?

1. Find the var $P$ of the induction motor.

Since

$$\text{PF} = \cos\theta = 0.8 \qquad \theta = 37^\circ \tag{20-8}$$

$$\text{var } P = VA \times \sin\theta \tag{20-6}$$

$$\text{var } P = 15 \times \sin 37^\circ = 15 \times 0.6 = -9 \text{ kvar } \textit{lagging}$$

The phasor diagram for the induction motor is shown in Fig. 20-20*a*.

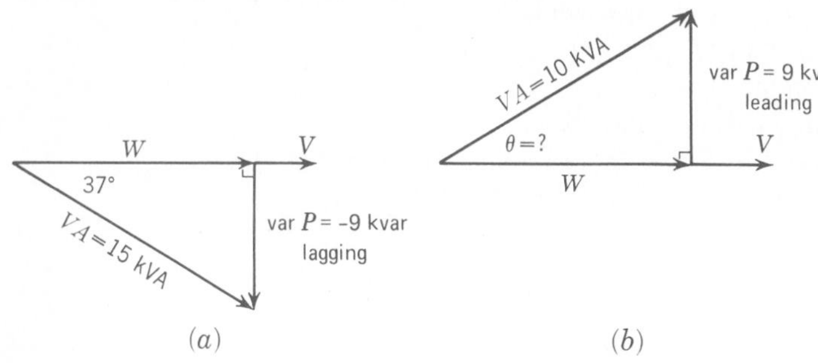


**FIGURE 20-20**
(*a*) Phasor diagram for the induction motor. (*b*) Phasor diagram for the synchronous motor.

2. To adjust the PF to unity means that $\cos\theta = 1.0$. At unity PF, if $\cos\theta = 1.0$, $\theta$ will equal 0°. When the phase angle is 0°,

$$\text{var } P = VA \times \sin\theta \tag{20-6}$$

$$\text{var } P = VA \times \sin 0^\circ$$

$$\text{var } P = VA \times 0 = 0 \text{ var}$$

Thus, to adjust a circuit to unity PF, all that is required is to make the total reactive power equal to 0 var. Since we already have 9 kvar *lagging* in the circuit, we must bring this down to zero by adding 9 kvar *leading*. This *leading* reactive power must come from the synchronous motor. The phasor diagram for the synchronous motor must be as shown in Fig. 20-20*b*. We can find the phase angle for the synchronous motor from this diagram.

3.

$$\sin\theta = \frac{\text{var } P}{VA} \tag{20-21}$$

**FORMULAS**

$$\sin\theta = \frac{\text{var } P \text{ of induction motor}}{VA \text{ of synchronous motor}} \tag{20-22}$$

$$\sin\theta = \frac{9}{10} = 0.9000$$

$$\theta = 64^\circ$$

4. Find the PF of the synchronous motor.

$$\text{PF} = \cos\theta = \cos 64^\circ \tag{20-8}$$

$$= 0.438$$

$$\text{PF} = 43.8\% \qquad \textit{Ans.}$$

EXAMPLE 20-14. A 220-V 50-A induction motor draws 10 kW of power. An 8-kVA synchronous motor is placed in parallel with it in order to adjust the PF to unity. What must be the PF of the synchronous motor?

SOLUTION

Given: $\left.\begin{array}{r} V = 220 \text{ V} \\ I = 50 \text{ A} \\ W = 10 \text{ kW} \end{array}\right\}$ induction motor

$VA = 8$ kVA synchronous motor

Find: To get a PF = 1.0, PF of synchronous motor = ?

1. Find the var $P$ of the induction motor.

$$VA = I \times V = 50 \times 220 = 11{,}000 \text{ VA} = 11 \text{ kVA}$$

Since

$$\cos \theta = \frac{W}{VA} \qquad (20\text{-}3)$$

$$\cos \theta = \frac{10}{11} = 0.909$$

$$\theta = 25°$$

$$\text{var } P = VA \times \sin \theta = 11 \times \sin 25° \qquad (20\text{-}6)$$

$$\text{var } P = 11 \times 0.423 = -4.65 \text{ kvar } \textit{lagging}$$

2. Draw the phasor diagram for the induction motor as shown in Fig. 20-21*a*.

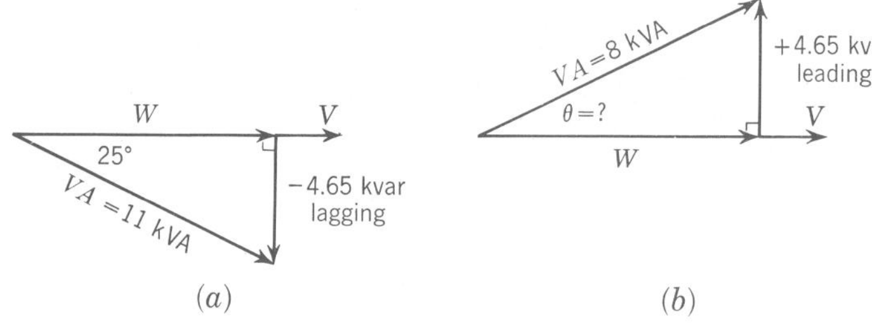


**FIGURE 20-21**
(*a*) Phasor diagram for the induction motor. (*b*) Phasor diagram for the synchronous motor.

To adjust the PF of the circuit to unity, the phasor diagram for the synchronous motor must indicate a *leading* power as shown in Fig. 20-21*b*.

3. Find the phase angle for the synchronous motor.

$$\sin \theta = \frac{\text{var } P}{VA} = \frac{4.65}{8} = 0.581$$

$$\theta = 36°$$

4. Find the PF of the synchronous motor.

$$\text{PF} = \cos \theta = \cos 36° = 0.809 = 80.9\% \qquad \textit{Ans.} \qquad (20\text{-}8)$$

*Note:* If only the effective power $W$ of the synchronous motor is given, find angle $\theta$ for the synchronous motor by using the

**FORMULA** $$\tan\theta = \frac{\text{lagging var } P}{W \text{ of synchronous motor}}$$ 20-23

Then $$\text{PF} = \cos\theta \qquad (20\text{-}8)$$

EXAMPLE 20-15. An induction motor takes 64 kW at 80 percent PF from an ac line. Find (*a*) the apparent power, (*b*) the phase angle, (*c*) the lagging var power, (*d*) the number of standard 15- or 25-kvar capacitors needed to improve the PF without producing a leading PF, and (*e*) the new PF.

SOLUTION. Given: $W = 64$ kW Find: $VA = ?$
PF = 80% *lagging* Phase angle = ?
Lagging var power = ?
$C = ?$
New PF = ?

*a.* Find the apparent power of the induction motor.

$$VA = \frac{W}{\text{PF}} = \frac{64\text{ kW}}{0.8} = 80\text{ kVA} \quad \textit{Ans.} \qquad (20\text{-}10)$$

*b.* $$\text{PF} = \cos\theta = 0.800 \qquad (20\text{-}8)$$
$$\theta = 37° \quad \textit{Ans.}$$

*c.* Find the var $P$ of the induction motor.

$$\text{var } P = VA \times \sin\theta = 80 \times \sin 37° \qquad (20\text{-}6)$$
$$= 80 \times 0.6 = -48\text{ kvar} \quad \textit{Ans.}$$

*d.* Two 25-kvar capacitors would produce a leading PF, and so we use one 15-kvar and one 25-kvar capacitor to balance the lagging 48 kvar.
*e.* This will still leave 8 kvar of reactive power in the circuit (48 − 40 = 8).

$$\sin\theta = \frac{\text{var } P}{VA} = \frac{8}{80} = 0.1000 \qquad (20\text{-}21)$$
$$\theta = 6°$$
$$\text{New PF} = \cos\theta = \cos 6° = 0.9945 \qquad (20\text{-}8)$$
$$\text{PF} = 99.5\% \quad \textit{Ans.}$$

## POWER-FACTOR CORRECTION WITH CAPACITORS

Figure 20-22*a* shows the phasor diagram for a circuit with a given power factor ($\theta_1$), and Fig. 20-22*b* shows the phasor diagram for the circuit with an improved power factor ($\theta_2$). In Fig. 20-22*a*,

$$\tan\theta_1 = \frac{\text{kvar}_1}{\text{kW}}$$

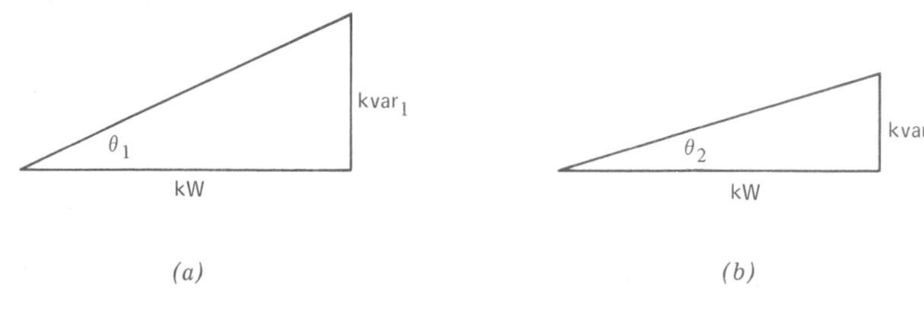


**FIGURE 20-22**
(*a*) Phasor diagram for the original circuit. (*b*) Phasor diagram for the circuit with improved PF.

or $$\text{kvar}_1 = \text{kW} \times \tan \theta_1$$

In Fig. 20-22*b*, $$\tan \theta_2 = \frac{\text{kvar}_2}{\text{kW}}$$

or $$\text{kvar}_2 = \text{kW} \times \tan \theta_2$$

The *difference* between these kvar values will be the capacitor kvar ($C$ kvar) needed to improve the power factor from the first situation ($\theta_1$) to the improved second situation ($\theta_2$).

$$\begin{aligned} C \text{ kvar} &= \text{kvar}_1 - \text{kvar}_2 \\ &= \text{kW} \times \tan \theta_1 - \text{kW} \times \tan \theta_2 \\ C \text{ kvar} &= \text{kW} (\tan \theta_1 - \tan \theta_2) \end{aligned} \quad \boxed{20\text{-}24}$$

EXAMPLE 20-16. A plant has a load of 300 kW with an average PF of 80 percent. How much capacitor kvar is required to raise the power factor to 90 percent? How many standard 25-$C$ kvar capacitors would be needed?

SOLUTION. Given: $W = 300$ kW, PF = 80%, Required PF = 90%. Find: $C$ kilovars needed.

1. Find the phase angles.

$$\cos \theta_1 = \text{PF} = 80\% = 0.8000 \qquad (20\text{-}8)$$
$$\theta_1 = 37°$$
$$\cos \theta_2 = \text{PF} = 90\% = 0.9000 \qquad (20\text{-}8)$$
$$\theta_2 = 26°$$

2. Find the $C$ kilovars needed.

$$\begin{aligned} C \text{ kvar} &= \text{kW}(\tan \theta_1 - \tan \theta_2) \qquad (20\text{-}24) \\ &= 300(\tan 37° - \tan 26°) \\ &= 300(0.7536 - 0.4877) \\ &= 300(0.2659) = 79.77 \text{ kvar} \end{aligned}$$

Use three standard 25-$C$ kvar capacitors. *Ans.*

EXAMPLE 20-17. Find the system capacity that has been released by the improvement in the power factor of Example 20-16.

SOLUTION

1. Find the original $VA$.

$$VA = \frac{W}{\text{PF}} = \frac{300}{0.8} = 375 \text{ kVA} \tag{20-10}$$

2. Find the improved $VA$.

$$VA = \frac{W}{\text{PF}} = \frac{300}{0.9} = 333 \text{ kVA}$$

3. Find the released capacity. At 80 percent PF, the circuit required 375 kVA. At 90 percent PF, the circuit only requires 333 kVA.

Released capacity = 375 − 333 = 42 kVA at 80% *Ans.*

EXAMPLE 20-18. Examples 20-16 and 20-17 may both be solved quickly (although not so accurately) by the use of the graphs shown in Fig. 20-23.

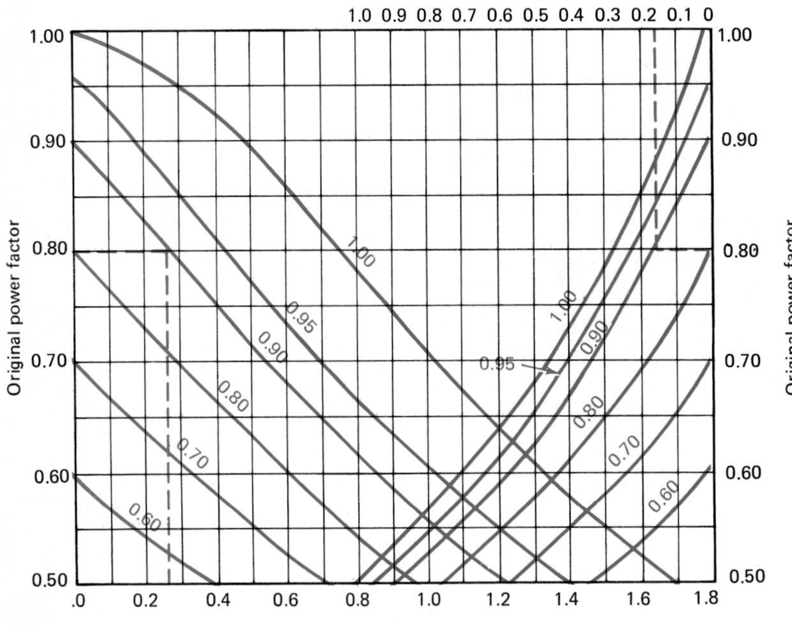


**FIGURE 20-23**
Composite graph to determine (*a*) capacitor kvar required per kW load to improve PF, (*b*) system capacity released (kVA per kW load) for improved power factors.

Find the $C$ kilovars for Example 20-16.

1. Locate the original power factor 0.80 at the *left* of the graph.
2. Read horizontally to the curve marked 0.90 (the new PF).
3. Read down to find 0.27 $C$ kvar per kilowatt load.
4. Multiply this 0.27 by the 300 kW to obtain 81 $C$ kvar. *Ans.*

Use the nearest standard value of 80 $C$ kvar.

Find the system capacity released for Example 20-16.

1. Locate the original power factor 0.80 at the *right* side of the graph.
2. Read horizontally to the left to the curve marked 0.90 (the new PF).
3. Read *up* to find 0.15 kVA released per kilowatt load.
4. Multiply this 0.15 by the 300 kW to obtain 45 kVA released.

This means that an additional 45 kVA at 80 percent PF load may be added to the existing circuits as a result of the improved PF.

SELF-TEST 20-19. A circuit operates with a load of 10 kW at a power factor of 85 percent. How much capacitor kilovar is required to raise the power factor to unity? Find the amount of released capacity.

SOLUTION

1. Find the phase angles.

$$\cos\theta_1 = \text{PF} = 85\% = \underline{\qquad} \quad (20\text{-}8)$$ 0.85

$$\theta_1 = \underline{\qquad}^\circ$$ 32

$$\cos\theta_2 = \text{PF} = \underline{\qquad}$$ 1.000

$$\theta_2 = \underline{\qquad}^\circ$$ 0

2. Find the $C$ kilovars needed.

$$C \text{ kvar} = \underline{\qquad}(\tan\theta_1 - \tan\theta_2)$$ kW

$$= 10(\tan 32^\circ - \tan 0^\circ)$$

$$= 10(\underline{\qquad} - \underline{\qquad})$$ 0.6249 0.000

$$= 10(\underline{\qquad}) = \underline{\qquad}\ C \text{ kvar} \quad \textit{Ans.}$$ 0.6249 6.25

3. Find the amount of released capacity. Find the original $VA$.

$$VA = \frac{?}{?} \quad (20\text{-}10)$$ $\frac{W}{\text{PF}}$

$$= \frac{10}{0.85} = \underline{\qquad} \text{ kVA}$$ 11.8

Find the improved $VA$.

$$VA = \frac{10}{1} = 10 \text{ kVA}$$

Find the released capacity.

$$11.8 - 10 = \underline{\qquad} \text{ kVA} \quad \textit{Ans.}$$ 1.8

SELF-TEST 20-20. Check Self-Test 20-19 using the graph of Fig. 20-23.

SOLUTION. Find the $C$ kilovars needed.

1. Locate the original power factor 0.85 at the ___(left/right)___ side of the graph. left
2. Read horizontally to reach the curve marked ______. 1.00
3. Read ___(up/down)___ to read ______ $C$ kvar per kilowatt load. down 0.63

| | | |
|---|---|---|
| 4. Multiply this 0.63 by ______ kW to get ______ $C$ kvar. | 10 | 6.3 |
| Find the amount of increased capacity. | | |
| 1. Locate the original power factor 0.85 at the ___(left/right)___ side of the graph. | right | |
| 2. Read horizontally to the left to reach the curve marked ______. | 1.00 | |
| 3. Read ___(up/down)___ to read ______ released kVA per kilowatt load. | up | 0.18 |
| 4. Multiply this 0.18 by ______ to get ______ kVA. | 10 | 1.8 |

EXAMPLE 20-21. When operating at full load, an induction motor draws 800 W and 4 A from a 220-V 60-Hz line. Find (*a*) the PF of the motor, (*b*) the lagging reactive power, (*c*) the apparent power drawn by a capacitor in order to raise the PF to unity, (*d*) the current drawn by this capacitor, (*e*) the reactance of the capacitor, and (*f*) the capacitance of this capacitor.

SOLUTION. Given: $W = 800$ W, $I = 4$ A, $V = 220$ V, $f = 60$ Hz

Find: PF = ?, var $P$ = ?, $VA$ of capacitor = ?, $I_C$ = ?, $X_C$ = ?, $C$ = ?

*a.* Find the PF of the motor.

1.
$$VA = V \times I = 220 \times 4 = 880 \text{ VA} \qquad (20\text{-}2)$$

2.
$$\cos\theta = \frac{W}{VA} = \frac{800}{880} = 0.909 \quad \textit{Ans.} \qquad (20\text{-}3)$$
$$\theta = 25°$$

*b.* Find the lagging reactive power.

$$\text{var } P = VA \times \sin\theta = 880 \times \sin 25° \qquad (20\text{-}6)$$
$$\text{var } P = 880 \times 0.423 = -372 \text{ var } \textit{lagging} \quad \textit{Ans.}$$

*c.* In order to raise the PF to unity, the capacitor must provide 372 var of reactive power leading. Since the apparent power in a capacitor is equal to the var $P$, 372 VA of capacitor power will exactly balance the lagging var $P$.

*d.* Find the current drawn by this capacitor.

$$VA = I \times V \qquad (20\text{-}2)$$
$$372 = I \times 220$$
$$I = \frac{372}{220} = 1.69 \text{ A} \quad \textit{Ans.}$$

*e.* Find the reactance of the capacitor.

$$V_C = I_C \times X_C \qquad (17\text{-}7)$$
$$220 = 1.69 \times X_C$$
$$X_C = \frac{220}{1.69} = 130\ \Omega \qquad \textit{Ans.}$$

*f.* Find the capacitance of the capacitor.

$$X_C = \frac{159{,}000}{f \times C} \qquad (17\text{-}6)$$
$$131.2 = \frac{159{,}000}{60 \times C}$$
$$131.2 \times 60 \times C = 159{,}000$$
$$C = \frac{159{,}000}{7872} = 20.2\ \mu\text{F} \qquad \textit{Ans.}$$

**PROBLEMS**

1. A 440-V line delivers 15 kVA to a load at 75 percent PF lagging. To what PF should a 10-kVA synchronous motor be adjusted in order to raise the PF to unity when connected in parallel?
2. A 220-V line delivers 10 kVA to a load at 80 percent PF lagging. What must be the PF of an 8-kW synchronous motor in parallel in order to raise the PF to unity?
3. A 220-V 20-A induction motor draws 3 kW of power. A 4-kVA synchronous motor is placed in parallel to adjust the PF to unity. What must be the PF of the synchronous motor?
4. A bank of motors draws 20 kW at 75 percent PF lagging from a 440-V 60-Hz line. What must be the capacity of a static capacitor connected across the motor terminals if it is to raise the total PF to 1.0?
5. A motor draws 1500 W and 7.5 A from a 220-V 60-Hz line. What must be the capacity of a capacitor in parallel which will raise the total PF to unity?
6. A 4-hp motor operates at an efficiency of 85 percent and a PF of 80 percent lagging when connected across a 120-V 60-Hz line. Find the capacity of the capacitor needed to raise the total PF to 100 percent. *Hint:* Find the kilowatt input to the motor, and proceed as before.
7. An inductive load draws 5 kW at 60 percent PF from a 220-V 60-Hz line. Find the kilovoltampere rating of the capacitor needed to raise the total PF to 100 percent.
8. *a.* A 30-kW motor operates at 80 percent PF lagging. In parallel with it is a 50-kW motor which operates at 90 percent PF lagging. Find (1) the total effective power, (2) the total PF, and (3) the total apparent power.

   *b.* Find the PF adjustment which must be made on a 20-kW synchronous motor in parallel with the motors in (*a*) in order to raise the PF of the circuit to unity.

9. An industrial plant has a load of 200 kW with an average power factor of 70 percent. How much capacitor kilovar is required to raise the power factor to 90 percent? How many standard 25-*C* kvar capacitors would be needed? Find the system capacity that would be released by this improvement in the power factor. Check the results using the graph shown in Fig. 20-23.
10. A circuit takes 150 kVA at an 80 percent lagging power factor. (*a*) Find the *C* kilovars needed to raise the power factor to 95 percent. (*b*) Find the system capacity released. (*c*) What values are obtained for (*a*) and (*b*) using the graph in Fig. 20-23?
11. One load at a factory takes 50 kVA at 50 percent power factor lagging, and another load connected to the same line takes 100 kVA at 86.6 percent power factor lagging. Find the total (*a*) effective power, (*b*) reactive power, (*c*) power factor, (*d*) apparent power, and (*e*) number and size of standard *C* kilovar capacitors needed to raise the power factor to 90 percent.
12. A group of induction motors takes 100 kVA at 84 percent power factor lagging. To improve the power factor, a synchronous motor taking 60 kVA at 70.7 percent power factor leading was connected in the same line. Find the total (*a*) effective power, (*b*) reactive power, (*c*) new power factor, and (*d*) apparent power.

## JOB 20-5 REVIEW OF AC POWER

The apparent power *VA* is the product of the voltage and the current used by a circuit as measured by ac meters.

The effective power *W* is the actual power used by the circuit as measured by a wattmeter.

The power factor (PF) of a circuit is the ratio of the effective power to the apparent power. It may be expressed as a decimal or as a percent. The PF is also equal to the cosine of the phase angle of the circuit.

$$\text{PF} = \cos\theta \qquad \boxed{20\text{-}8}$$

The relations among *VA*, *W*, and PF are given by

$$\text{PF} = \frac{W}{VA} \qquad \boxed{20\text{-}7} \qquad \text{or} \qquad \cos\theta = \frac{W}{VA} \qquad \boxed{20\text{-}3}$$

$$W = VA \times \text{PF} \qquad \boxed{20\text{-}9} \qquad \text{or} \qquad W = VA \times \cos\theta \qquad \boxed{20\text{-}4}$$

$$VA = \frac{W}{\text{PF}} \qquad \boxed{20\text{-}10} \qquad \text{or} \qquad VA = \frac{W}{\cos\theta} \qquad \boxed{20\text{-}5}$$

The apparent power may be resolved into its components—the effective power *W* and the reactive power *var*—by the following formulas:

$$W = VA \times \cos\theta \qquad \boxed{20\text{-}4}$$

$$\text{var } P = VA \times \sin\theta \qquad \boxed{20\text{-}6}$$

The procedure for solving problems involving resistive, inductive, and capacitive loads is as follows:

1. For each load, find

   *a.* The apparent power, using the formula

$$VA = V \times I \qquad \boxed{20\text{-}2}$$

or

$$VA = \frac{W}{\text{PF}} \qquad \boxed{20\text{-}10}$$

or

$$VA = \frac{W}{\cos\theta} \qquad \boxed{20\text{-}5}$$

or

$$(VA)^2 = W^2 + (\text{var } P)^2 \qquad \boxed{20\text{-}12}$$

   *b.* The effective power, using the formula

$$W = VA \times \text{PF} \qquad \boxed{20\text{-}9}$$

or

$$W = VA \times \cos\theta \qquad \boxed{20\text{-}4}$$

   *c.* The phase angle, using the formula

$$\cos\theta = \frac{W}{VA} \qquad \boxed{20\text{-}3}$$

or

$$\cos\theta = \text{PF} \qquad \boxed{20\text{-}8}$$

or

$$\tan\theta = \frac{\text{var } P}{W} \qquad \boxed{20\text{-}11}$$

   *d.* The var power.

$$\text{var } P = VA \times \sin\theta \qquad \boxed{20\text{-}6}$$

2. Find the total effective power.

$$W_T = W_A + W_B \qquad \boxed{20\text{-}13}$$

3. Find the total reactive power.

$$\text{var } P_T = \text{var } P_A + \text{var } P_B \qquad \boxed{20\text{-}14}$$

4. Find the phase angle for the entire circuit by adding $W_T$ and var $P_T$ vectorially. In the triangle thus formed,

$$\tan\theta_T = \frac{\text{var } P_T}{W_T} \qquad \boxed{20\text{-}17}$$

5. Find the total PF.

$$\text{PF}_T = \cos\theta_T \qquad \boxed{20\text{-}8}$$

or
$$\text{PF}_T = \frac{W_T}{VA_T} \qquad \boxed{20\text{-}16}$$

6. Find the total apparent power.

$$VA_T = \frac{W_T}{\text{PF}_T} \qquad \boxed{20\text{-}18}$$

or
$$VA_T^2 = W_T^2 + \text{var } P_T^2 \qquad \boxed{20\text{-}15}$$

7. Find the total current drawn, using the formula

$$VA_T = I_T \times V_T \qquad \boxed{20\text{-}2}$$

or
$$W_T = V_T \times I_T \times \cos \theta_T \qquad \boxed{20\text{-}19}$$

or
$$W_T = V_T \times I_T \times \text{PF}_T \qquad \boxed{20\text{-}20}$$

## POWER-FACTOR CORRECTION

1. By synchronous motors:
   *a.* Find the var $P$ of the induction motor given.
   *b.* For correction to unity PF, the leading var $P$ of the synchronous motor must equal the lagging var $P$ of the induction motor. Therefore, for the synchronous motor,

$$\sin \theta = \frac{\text{var } P \text{ of induction motor}}{VA \text{ of synchronous motor}} \qquad \boxed{20\text{-}22}$$

or
$$\tan \theta = \frac{\text{lagging var } P}{W \text{ of synchronous motor}} \qquad \boxed{20\text{-}23}$$

   from either of which the value of angle $\theta$ may be obtained.
   *c.* The PF of the synchronous motor is

$$\text{PF} = \cos \theta \qquad \boxed{20\text{-}8}$$

2. By capacitors rated in $C$ kilovars:
   *a.* (1) Find the kilovar $P$ of the induction motor load.
   (2) Balance this reactive power using standard 15- or 25-$C$ kvar capacitors. It is not necessary to achieve 100 percent PF; 90 to 95 percent is considered good practice.
   *b.* Use the formula $C \text{ kvar} = \text{kW}(\tan \theta_1 - \tan \theta_2)$
3. By capacitors rated in microfarads:
   *a.* Find the var $P$ of the induction motor given.
   *b.* Set this var $P$ equal to the apparent power $VA$ drawn by the capacitor, since the apparent power of a capacitor is equal to its var $P$.
   *c.* Solve for $I$ in the formula

$$VA = I \times V \qquad \boxed{20\text{-}2}$$

   *d.* Find the reactance of the capacitor.

$$V_C = I_C \times X_C \qquad \boxed{17\text{-}7}$$

   *e.* Find the capacitance at the given frequency.

$$X_C = \frac{159{,}000}{f \times C} \qquad \boxed{17\text{-}6}$$

## SYSTEM CAPACITY RELEASED BY IMPROVED POWER FACTOR

1. Find the original voltamperes by the formula

$$VA = \frac{W}{PF_1} \qquad \boxed{20\text{-}10}$$

2. Find the new voltamperes with the improved PF.

$$VA = \frac{W}{PF_2} \qquad \boxed{20\text{-}10}$$

3. Subtract. This will be the number of voltamperes newly available at the original PF.

### PROBLEMS

1. A coil of 7 Ω resistance and 24 Ω reactance is connected across a 120-V 60-Hz line. Find the PF and the current drawn.
2. A 2-hp motor operates at a PF of 80 percent and an efficiency of 90 percent. Find (*a*) the watt input to the motor and (*b*) the apparent power delivered to the motor.
3. A motor drawing 10 A at 90 percent PF lagging is in parallel with another motor drawing 8 A at 90 percent PF lagging from a 120-V line. Find (*a*) the total effective power, (*b*) the total PF, (*c*) the total apparent power, and (*d*) the total current drawn.
4. A purely resistive lamp load drawing 20 A from a 220-V 60-Hz line is in parallel with a capacitive motor taking 8 A at 34.2 percent PF. Find (*a*) $W_T$, (*b*) $\mathrm{PF}_T$, (*c*) $VA_T$, and (*d*) $I_T$.
5. A 1-kW lamp load, a 20-kVA induction motor operating at a PF of 70 percent, and a 15-kVA motor operating at a PF of 50 percent lagging are connected in parallel. Find (*a*) $W_T$, (*b*) $\mathrm{PF}_T$, and (*c*) $VA_T$.
6. An induction motor drawing 250 VA at 70 percent PF is in parallel with a synchronous motor drawing 400 VA at 60 percent PF leading. Find (*a*) the total effective power, (*b*) the total PF, and (*c*) the total apparent power.
7. An induction motor drawing 10 A at 80 percent PF is in parallel with a capacitive load drawing 1 kW at 90 percent PF from a 117-V 60-Hz line. Find (*a*) the total effective power, (*b*) the total PF, (*c*) the total apparent power, and (*d*) the total current drawn.
8. A lamp bank drawing 10 A, a synchronous motor drawing 8 A at 75 percent PF leading, and an induction motor drawing 5 A at 80 percent PF are all connected in parallel across a 230-V 60-Hz line. Find (*a*) the total apparent power and (*b*) the total current drawn.
9. A 120-V 20-A induction motor draws 2 kW of power. A 1.5-kVA synchronous motor is placed in parallel with it to adjust the PF to unity. What is the PF of the synchronous motor?
10. An induction motor draws 1 kW from a 110-V 60-Hz line at a PF of 70 percent. Find (*a*) the apparent power, (*b*) the lagging var

power, and (*c*) the capacitance needed in parallel to raise the total PF to 1.0.

11. An induction motor draws 1200 W and 1500 VA from an ac line. (*a*) What is its power factor? (*b*) What is the value of the reactive voltamperes drawn by the motor? (*c*) How many vars of leading reactive power are needed to raise the power factor to unity?
12. An induction motor takes 20 kVA at 220 V and 80 percent PF. What must be the PF of a 15-kVA synchronous motor connected in parallel in order to raise the total PF to 100 percent?
13. An inductive load takes 70 kW at 85 percent PF from an ac line. Find (*a*) the apparent power, (*b*) the phase angle, (*c*) the lagging var power, (*d*) the number of standard 15- or 25-kvar capacitors needed to improve the PF without producing a leading PF, and (*e*) the new power factor.
14. A plant has a load of 250 kW with an average PF of 85 percent. How much capacitor kilovar is required to raise the power factor to 90 percent? Find the system capacity that has been released.
15. A circuit operates with a load of 30 kW at a power factor of 80 percent. How much capacitor kilovar is required to raise the power factor to unity? Find the system capacity that has been released.

(See Instructor's Manual for Test 20-3, AC Power)

# THREE–PHASE SYSTEMS

# 21

## JOB 21-1 THE WYE CONNECTION

A three-phase circuit is simply a combination of three single-phase circuits. This combination is desirable for many reasons. In general, three-phase equipment is more economical than single-phase equipment. For any particular physical size, the horsepower ratings of three-phase motors and the kilovoltampere ratings of three-phase generators are larger than those of single-phase motors and generators. Three-phase motors are more efficient than single-phase types because three-phase power is more uniform, and does not pulsate as does single-phase power. Also, the distribution of three-phase power may be accomplished with only three-fourths the line copper required by a single-phase system.

**Generation of three-phase voltages.** In Fig. 21-1*a*, three sets of coils are mounted 120° apart on an axis which is free to turn. The coils, with their six leads, are represented in Fig. 21-1*b*. The coil end at which the voltage is into the wire is called the *start* end (1, 3, 5), and the end at which the voltage is out of the wire is called the *finish* end (2, 4, 6). At the instant shown, coil $X$ is traveling parallel to the lines of force, and the emf induced in it is zero. As the coils rotate, the voltage induced in coil $X$ will be that shown as $V_X$ in Fig. 21-1*c*. After a rotation of 120°, the emf in coil $Y$ will be zero and starting a new cycle, shown as $V_Y$ in Fig. 21-1*c*. After another 120°, coil $Z$ will reach this neutral position, and will begin the cycle shown as $V_Z$ in Fig. 21-1*c*. If we place these three waves on the same electrical time base, we get the three-phase system shown in Fig. 21-2*b*. Actually, this figure represents three separate waves, each 120° apart. Each of these phases may be used to operate any single-phase device. The vector diagram for these three voltages is shown in Fig. 21-2*c*.

The three coils, with their six leads, may be connected inside the alternator, which results in the more practical three final leads. This interconnection of the single-phase windings of generators, motors,

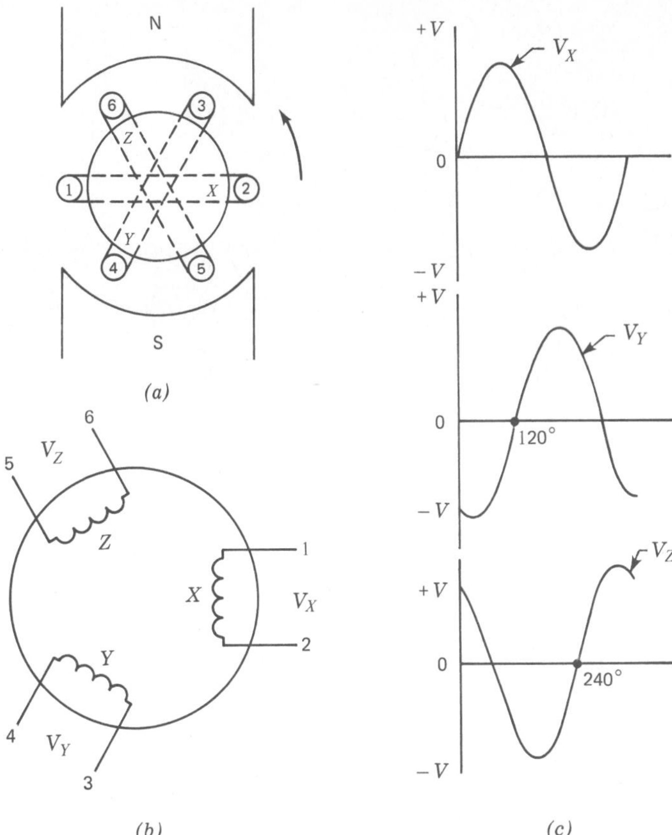


**FIGURE 21-1**
(*a*) Three sets of coils mounted 120° apart. (*b*) Diagrammatic representation of the coils. (*c*) Waveforms produced by the individual coils.

transformers, etc. can be done by either of two standard methods—the *wye* connection or the *delta* connection. Most generators are wye-connected, but loads may be either wye-connected or delta-connected.

**The wye connection.** In this connection, the start ends (1, 3, 5) are connected together inside the machine as shown in Fig. 21-3*a,* and the finish ends (2, 4, 6) are brought out as the terminals. Any loads would be connected to these terminals. The connection is shown in diagram form in Fig. 21-3*b.* Since this connection looks like a star, it is often called the *star* connection. The voltage between any two terminals may be described as $V_{XY}$, $V_{XZ}$, or $V_{YZ}$.

**Line voltages in wye.** Let us find the voltage between terminals $X$ and $Y$ in Fig. 21-3*b*. In this balanced system, $V_X = V_Y$ or simply the phase voltage $V_p$. The line voltage $V_{XY}$ is *not* simply twice the phase voltage, because the phase voltages are 120° out of phase and must be added vectorially. Let us trace through the circuit from terminal $X$ to terminal $Y$ as shown in Fig. 21-4*a*. According to Kirchhoff's laws, if the direction

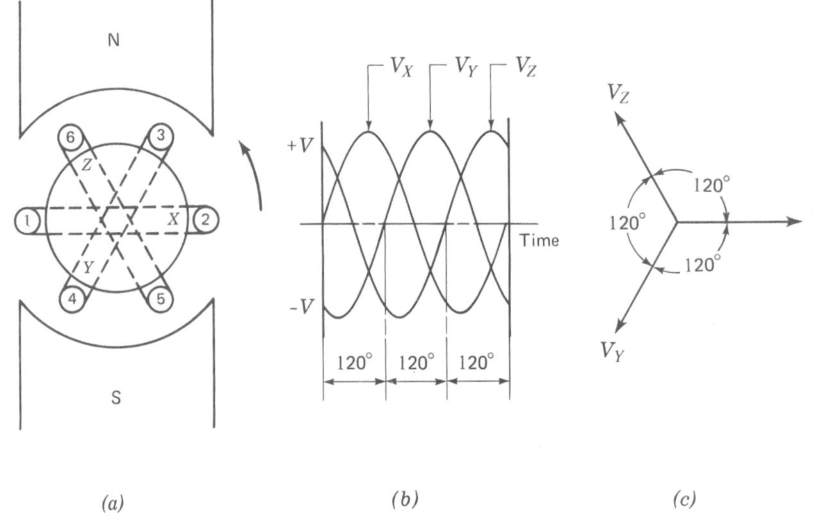


**FIGURE 21-2**
(a) Three separate coils mounted 120° apart. (b) Waveforms produced by these coils. (c) Vector diagram of the voltages.

of the path traced through a voltage source is the *same* as that of the arrow, the sign of the voltage is *plus*; if the direction of the trace is *opposite* to the arrow, the sign of the voltage is *minus*. Since our trace through $V_X$ indicates an opposing, or negative voltage, the vector diagram must be redrawn as shown in Fig. 21-4*b*.

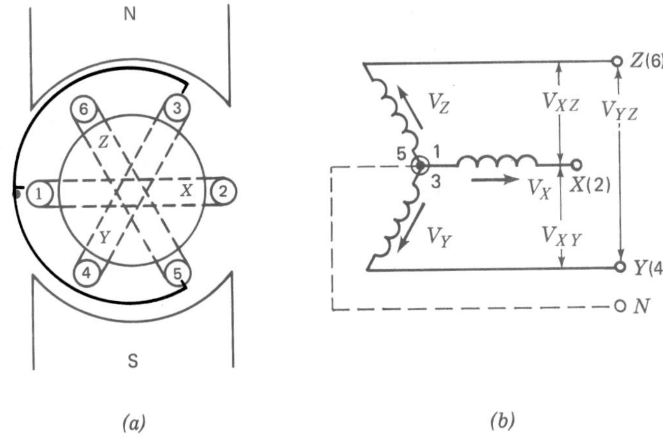


**FIGURE 21-3**
(a) In the wye connection, the start ends (1, 3, 5) are connected inside the machine; the finish ends (2, 4, 6) form the terminals. (b) Diagram form of the wye or *star* connection. A common lead N may or may not be brought out from the connection at point O.

Add vector $V_Y$ to vector $V_X$ by placing the tail of $V_Y$ on the head of $V_X$ and drawing it in its original size and direction as shown in Fig. 21-5*a*. The total voltage ($V_{\text{line}}$ or $V_L$) will be the resultant vector $OY$. Since the phase voltages $V_X$ and $V_Y$ are equal, each vector is described as simply $V_p$. Resolve the vector $XY$ into its components as shown in Fig. 21-5*b*.

$$XC = V_p \cos 60° = 0.5V_p$$
$$CY = V_p \sin 60° = 0.866V_p$$

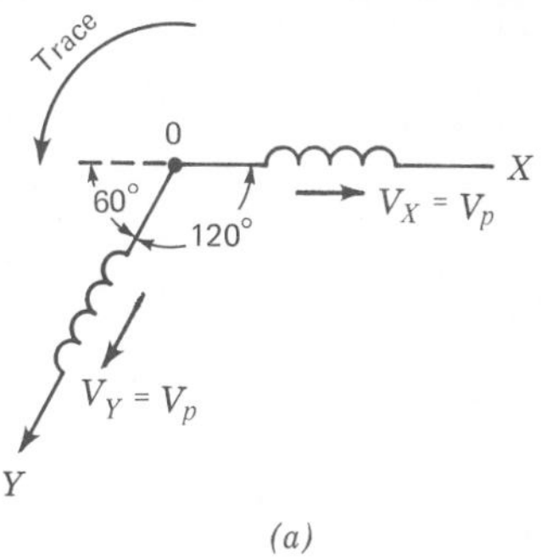


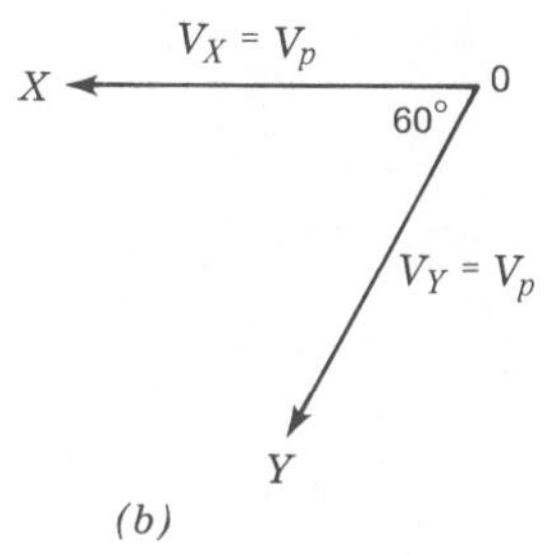


**FIGURE 21-4**
(a) The trace from X to Y indicates that the vector OX ($V_X$) opposes the vector OY ($V_Y$). (b) True vector diagram with OX ($V_X$) reversed to show a negative voltage.

Adding the in-phase components will give the vector diagram shown in Fig. 21-5*c*. ($OC = OX + CX = 1V_p + 0.5V_p = 1.5V_p$.) In $\triangle\ OCY$,

$$V_L^2 = (1.5V_p)^2 + (0.866V_p)^2$$
$$V_L = \sqrt{2.25V_p^2 + 0.75V_p^2}$$
$$= \sqrt{3V_p^2}$$
$$V_L = \sqrt{3}\ V_p$$

This relationship is true for any pair of line-to-line voltages.

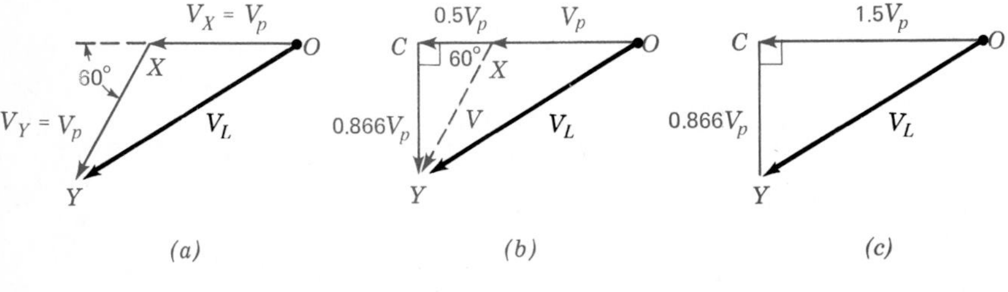


**FIGURE 21-5**
(a) OY is the vector sum of OX and XY. (b) Resolution of vector XY into its components. (c) Vector triangle resulting from the addition of the in-phase components; $V_L$ is the vector sum of the phase voltages.

**FORMULA**

$$V_L = \sqrt{3}\ V_p \qquad \boxed{21\text{-}1}$$

where $V_L$ = line voltage
$V_p$ = phase voltage

The complete vector diagram for the three phase voltages and the three line voltages is shown in Fig. 21-6. Note that (1) the line voltages are equal and out of phase with each other by 120°. (2) Each phase voltage is in phase with its own phase current. (3) The phase current (which is the line current) is 30° out of phase with the line voltage at unity power factor.

**Line currents in wye.** The current in any line wire *X, Y,* or *Z* flows from the neutral point *0* through the generator coils to the lines. This means that the current $I_L$ in any line wire is equal to the current in the phase ($I_p$) to which it is connected.

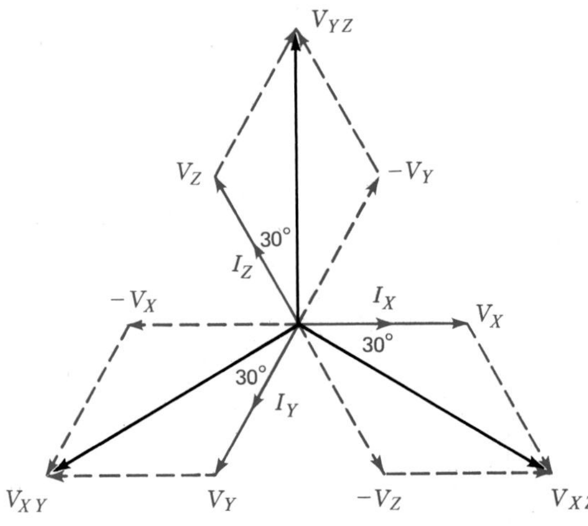


**FIGURE 21-6**
Vector diagram showing the relations between the phase and line voltages at 100 percent PF.

**FORMULA** $$I_L = I_p \tag{21-2}$$

**Three-phase power.** The power developed in each phase $W_p$ of a wye-connected three-phase circuit is the same as the power in a single-phase circuit.

$$W_p = V_p I_p \cos \theta \tag{21-3}$$

where $\theta$ is the angle between the phase current and the phase voltage. The power developed in all three phases is

$$W_T = 3W_p = 3V_p I_p \cos \theta \tag{21-4}$$

Since it is more practical to measure line voltages and line currents than phase values, let us convert phase values to line values.

In a wye connection,

$$I_p = I_L \quad \text{and} \quad V_p = \frac{V_L}{\sqrt{3}}$$

Substituting these values in formula (21-4) gives

$$W_T = 3\frac{V_L}{\sqrt{3}} \times I_L \times \cos \theta$$

or

**FORMULA** $$W_T = \sqrt{3} V_L I_L \cos \theta \tag{21-5}$$

Also, since $VA = W/\cos \theta$, (20-5)

$$VA = \frac{\sqrt{3} V_L I_L \overset{1}{\cancel{\cos \theta}}}{\underset{1}{\cancel{\cos \theta}}}$$

or

**FORMULA** $$VA = \sqrt{3}V_L I_L \tag{21-6}$$

Expressed in kilounits,

$$\text{kW} = \frac{\sqrt{3}V_L I_L \cos\theta}{1000} \tag{21-7}$$

$$\text{kVA} = \frac{\sqrt{3}V_L I_L}{1000} \tag{21-8}$$

The system power factor of a balanced three-phase, wye-connected system, which is also the coil power factor, is still given by the formula PF $= \cos\theta = W/VA$, where $\theta$ is the angle between the *phase* current and the *phase* voltage. This may be restated as the

**FORMULA** $$\text{PF} = \frac{W_T}{\sqrt{3}V_L I_L} \tag{21-9}$$

*Note:* A system is *balanced* when each of its three elements is equal to the others.

EXAMPLE 21-1. If a three-phase, wye-connected generator delivers 120 V per phase, what is the line voltage between any two phases?

SOLUTION

$$V_L = \sqrt{3}V_p \qquad (21\text{-}1)$$

$$V_L = 1.732 \times 120 = 208 \text{ V} \qquad \textit{Ans.}$$

EXAMPLE 21-2. A three-phase, wye-connected ac generator has a terminal voltage of 480 V and delivers a full-load current of 300 A per terminal at a power factor of 80 percent. Find (*a*) the full-load current per phase, (*b*) the phase voltage, (*c*) the kilovoltampere rating, and (*d*) the full-load power in kilowatts.

SOLUTION

*a.* In a wye-connected system, $I_p = I_L$: (21-2)

$$I_p = 300 \text{ A} \qquad \textit{Ans.}$$

*b.* $$V_p = \frac{V_L}{\sqrt{3}} = \frac{480}{1.732} = 277 \text{ V} \qquad \textit{Ans.} \qquad (21\text{-}1)$$

*c.* $$\text{kVA} = \frac{\sqrt{3}V_L I_L}{1000} \qquad (21\text{-}8)$$

$$= \frac{1.732 \times 480 \times 300}{1000} = 249.4 \text{ kVA} \qquad \textit{Ans.}$$

*d.*
$$\text{kW} = \frac{\sqrt{3}V_L I_L \cos\theta}{1000} \tag{21-7}$$
$$= \frac{1.732 \times 480 \times 300 \times 0.8}{1000}$$
$$= 199.5 \text{ kW} \quad \textit{Ans.}$$

Alternative method

$$\text{kW} = \text{kVA} \times \text{PF} \tag{20-9}$$
$$= 249.4 \times 0.8 = 199.5 \text{ kW} \quad \textit{Ans.}$$

EXAMPLE 21-3. A balanced load of three 10-Ω resistors (PF = 100 percent) is connected across a 480-V three-phase line as shown in Fig. 21-7*a*. With the neutral wire omitted, the circuit is seen to be wye-connected, as shown in Fig. 21-7*b*. Find (*a*) the phase voltage for each resistor, (*b*) the current in each resistor, (*c*) the line current, (*d*) the power taken by each resistor, (*e*) the total three-phase power, and (*f*) the total kilovoltampere input to the load.

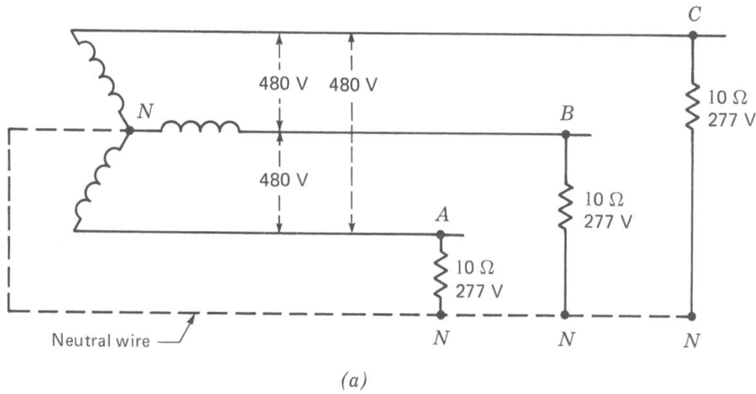


(*a*)

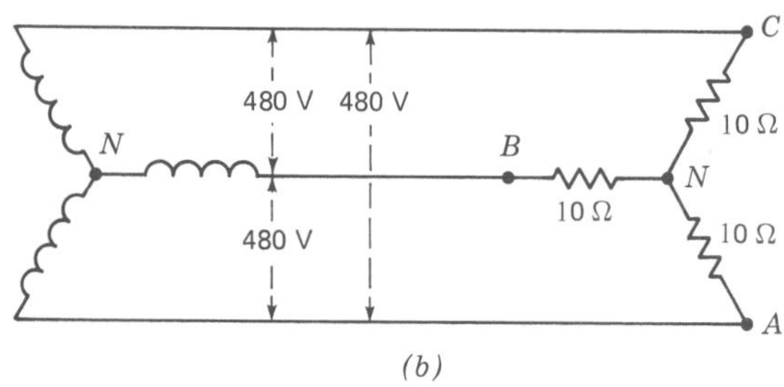


(*b*)

FIGURE 21-7

SOLUTION

*a.* The phase voltage across each resistor is

$$V_p = \frac{V_L}{\sqrt{3}} = \frac{480}{1.732} = 277 \text{ V} \qquad \textit{Ans.} \qquad (21\text{-}1)$$

*b.* The current in each resistor with no reactance present (PF = 100 percent) is

$$I_p = \frac{V_p}{Z_p} = \frac{V_p}{R_p} = \frac{277}{10} = 27.7 \text{ A} \qquad \textit{Ans.}$$

*c.* In a wye-connected circuit, the line and phase currents are equal.

$$I_L = I_p = 27.7 \text{ A} \qquad \textit{Ans.} \qquad (21\text{-}2)$$

*d.* The power drawn by each resistor in each phase is

$$\begin{aligned} W_p &= V_p I_p \cos\theta \qquad (21\text{-}3) \\ &= 277 \times 27.7 \times 1 = 7673 \text{ W} \qquad \textit{Ans.} \end{aligned}$$

*e.* The total power may be found in three different ways.

$$\begin{aligned} W_T &= 3W_p & W_T &= 3V_p I_p \cos\theta \\ &= 3 \times 7673 & &= 3 \times 277 \times 27.7 \times 1 \\ &= 23{,}020 \text{ W} & &= 23{,}020 \text{ W} \end{aligned}$$

$$\begin{aligned} W_T &= \sqrt{3} V_L I_L \cos\theta \\ &= 1.732 \times 480 \times 27.7 \\ &= 23{,}030 \text{ W} \qquad \textit{Ans.} \end{aligned}$$

*f.* The kilovoltampere input is

$$\frac{\sqrt{3} V_L I_L}{1000} = \frac{1.732 \times 480 \times 27.7}{1000} = 23.03 \text{ kVA} \qquad \textit{Ans.}$$

The total power $W_T = 23.03$ kW is equal to the kilovoltamperes because the power factor is 100 percent. The phase angle is 0°.

EXAMPLE 21-4. Each phase of a balanced wye-connected three-phase load consists of a resistor of 12 Ω in series with a pure inductive reactance of 5 Ω. The load is supplied with a 480-V 60-Hz three-phase line. Find (*a*) the impedance per phase, (*b*) the power factor, (*c*) the current in each branch of the load, (*d*) the line current, (*e*) the total power, and (*f*) the kilovoltamperes input. (See Fig. 21-8.)

SOLUTION

*a.*

$$\begin{aligned} Z_p &= \sqrt{R^2 + X_L^2} \qquad (16\text{-}4) \\ &= \sqrt{12^2 + 5^2} = \sqrt{169} = 13 \ \Omega \qquad \textit{Ans.} \end{aligned}$$

*b.*

$$\text{PF} = \cos\theta = \frac{R_p}{Z_p} = \frac{12}{13} = 0.923 = 92.3\% \qquad \textit{Ans.}$$

*c.* Find the phase voltage.

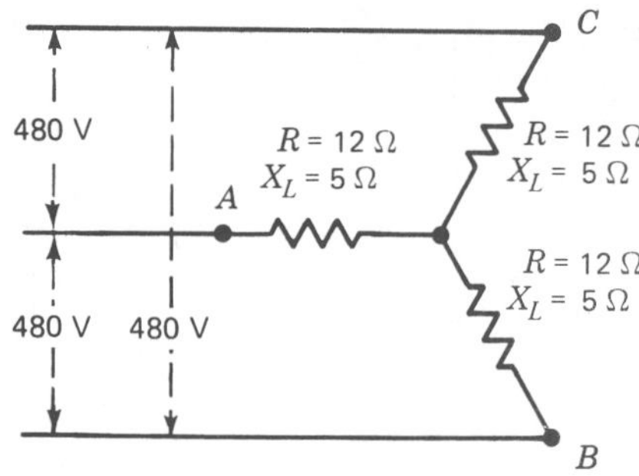

FIGURE 21-8

$$V_p = \frac{V_L}{\sqrt{3}} = \frac{480}{1.732} = 277 \text{ V} \qquad (21\text{-}1)$$

Find the phase current.

$$I_p = \frac{V_p}{Z_p} = \frac{277}{13} = 21.3 \text{ A} \qquad Ans.$$

*d.*

$$I_L = I_p = 21.3 \text{ A} \qquad Ans. \qquad (21\text{-}2)$$

*e.*

$$W_T = \sqrt{3}\, V_L I_L \cos\theta \qquad (21\text{-}5)$$
$$= 1.732 \times 480 \times 21.3 \times 0.923$$
$$= 16{,}344 \text{ W} = 16.34 \text{ kW} \qquad Ans.$$

*f.*

$$\text{kVA} = \frac{\sqrt{3}\, V_L I_L}{1000} = \frac{1.732 \times 480 \times 21.3}{1000} = 17.71 \text{ kVA} \qquad Ans.$$

(21-8)

EXAMPLE 21-5. A 220-V three-phase induction motor draws a full-load current of 35 A per terminal when wye-connected. Find (*a*) the full-load kilovoltamperes input and (*b*) the kilowatt input if the full-load power factor is 80 percent.

SOLUTION

*a.*

$$\text{kVA} = \frac{\sqrt{3}\, V_L I_L}{1000} = \frac{1.732 \times 220 \times 35}{1000} = 13.34 \text{ kVA} \qquad Ans.$$

(21-8)

*b.*

$$\text{kW} = \text{kVA} \times \text{PF} \qquad (20\text{-}9)$$
$$= 13.34 \times 0.8 = 10.67 \text{ kW} \qquad Ans.$$

or

$$\text{kW} = \frac{\sqrt{3}\, V_L I_L \cos\theta}{1000} \qquad (21\text{-}7)$$
$$= \frac{1.732 \times 220 \times 35 \times 0.8}{1000} = 10.67 \text{ kW} \qquad Ans.$$

## PROBLEMS

1. The phase voltage of a wye-connected generator is 1200 V. What is the line voltage?

2. If the terminal voltage of a wye-connected generator is 4160 V, what is the voltage generated in each of the phases?
3. A wye-connected generator has a phase current of 40 A. Find the current in the line wires.
4. A three-phase motor takes a current of 30 A per terminal. If the line voltage is 440 V and the power factor is 80 percent, find the power used by the motor.
5. A wye-connected generator delivers a phase current of 20 A at a phase voltage of 265 V and a power factor of 85 percent. Find (*a*) the generator terminal voltage, (*b*) the power developed per phase, and (*c*) the total three-phase power.
6. A three-phase load draws 50 kW from a 440-V line. If the line current is 80 A, find (*a*) the kilovoltamperes, and (*b*) the power factor of the load.
7. A three-phase, 208-V 6-pole 60-Hz induction motor draws 25 A and 7.93 kW from a line while driving a 10-hp load. Find (*a*) the total kilovoltamperes, and (*b*) the power factor of the motor.
8. A certain three-phase, wye-connected alternator is rated at 1000 kVA and 2300 V at unity power factor. Find (*a*) the current rating per terminal, (*b*) the rated phase current, and (*c*) the rated coil voltage.
9. A three-phase, 60-Hz 6600-V wye-connected alternator delivers 4000 kW and 425 A at 6600 V. Find (*a*) the kilovoltamperes delivered, (*b*) the power factor, and (*c*) the coil voltage.
10. A three-phase, 60-Hz wye-connected synchronous motor draws 150 A and 500 kW from a 2300-V three-phase supply. Find (*a*) the kilovoltamperes, (*b*) the power factor, and (*c*) the phase voltage.
11. Each phase winding of a three-phase, wye-connected alternator is rated at 6930 V and 361 A. If the alternator is designed to operate at full load with a lagging power factor of 80 percent, find (*a*) the line voltage, (*b*) the line current, (*c*) the kilovoltampere rating at full load, and (*d*) the full-load power in kilowatts.
12. A three-phase, wye-connected alternator is rated at 12,500 kVA, 10,000 V, and 60 Hz. Find (*a*) the full-load kilowatt output of the generator at a power factor of 80 percent lagging, (*b*) the full-load line current of the generator, and (*c*) the voltage rating of each of the three windings.
13. Three resistive loads of 10 Ω each are connected to a three-phase 120-V 60-Hz supply. If the loads are connected in wye, find (*a*) the current in each resistor, (*b*) the current in the line, and (*c*) the power used by the system.
14. A wye-connected load of three 10-Ω resistors is connected to a 220-V three-phase supply. At 100 percent power factor, find (*a*) the voltage applied across each resistor, (*b*) the current in each phase, (*c*) the line current, and (*d*) the total power used.
15. Each coil of a three-phase 60-Hz wye-connected alternator generates 277 V. The alternator supplies a load consisting of three

wye-connected impedances, each with 12-Ω resistance and 5-Ω inductive reactance. Find (*a*) the line voltage, (*b*) the impedance of each resistor, (*c*) the current through each load impedance, (*d*) the line current, (*e*) the power factor, and (*f*) the kilowatt load being supplied.

16. A three-phase, wye-connected load takes 80 kW of power at a power factor of 80 percent lagging from a three-phase 440-V 60-Hz line. Find (*a*) the phase voltage, (*b*) the total kilovoltamperes, (*c*) the line current, (*d*) the phase current, and (*e*) the reactive kilovoltamperes.
17. Each coil of the armature of a three-phase wye-connected alternator is rated at 50 kVA, 2400 V, 60 Hz. If the power factor of the system is 85 percent, find (*a*) the line voltage, (*b*) the full-load kilowatt output of the alternator, (*c*) the full-load line current, and (*d*) the full-load coil current.
18. A three-phase 60-Hz wye-connected turboalternator is supplying 14,000 kW at 13,200 V and 80 percent power factor to a load. Find (*a*) the kilovoltampere output, (*b*) the line current, (*c*) the coil current, and (*d*) the coil voltage.
19. Three loads, each with a series resistance of 86.6 Ω and an inductive reactance of 50 Ω, are wye-connected to a 240-V three-phase power supply. Find (*a*) the impedance of each phase, (*b*) the voltage of each phase, (*c*) the line current, (*d*) the power factor, (*e*) the three-phase power, and (*f*) the voltamperes.
20. Each phase of a balanced wye-connected, three-phase load consists of a resistor of 24 Ω in series with a pure inductive reactance of 10 Ω. If the load is supplied from a 440-V 60-Hz three-phase line, find (*a*) the impedance per phase, (*b*) the power factor, (*c*) the current in each branch of the load, (*d*) the line current, (*e*) the total power, and (*f*) the kilovoltampere input.

## JOB 21-2 THE DELTA CONNECTION

The coils of a three-phase alternator may also be connected as shown in Fig. 21-9*a*. The finish end of coil *X* (point 2) is connected to the start end of coil *Y* (point 3); then the finish end of coil *Y* (point 4) is connected to the start end of coil *Z* (point 5); finally the finish end of coil *Z* (point 6) is connected to the start end of coil *X* (point 1).

The junction of points 2 and 3 is brought out as line 1.
The junction of points 4 and 5 is brought out as line 2.
The junction of points 1 and 6 is brought out as line 3.

This is shown in diagram form in Fig. 21-9*b*. Since the configuration looks like the Greek letter delta (Δ), this connection is known as the *delta* connection. In this figure, notice that each pair of line leads is connected directly across a coil winding. For example, the voltage generated in phase *Y* is also the voltage between lines 1 and 2. Thus,

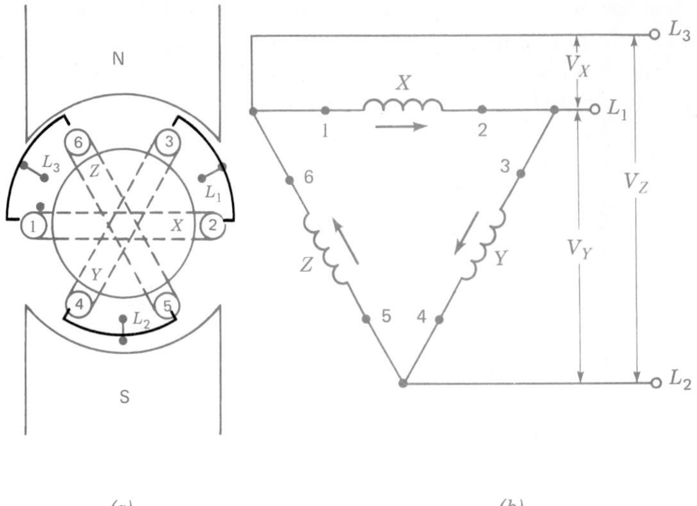


**FIGURE 21-9**
(*a*) Coils connected in delta. (*b*) Standard delta configuration.

**FORMULA** $V_L = V_p$ 21-10

**Currents in delta-connected generators.** The current in any line wire is the vector sum of the currents in the two coils to which it is attached. That is, for example, $I_{XY}$ is equal to the vector sum of $I_X$ and $I_Y$ in Fig. 21-10*a*. Let us assume that the coils are loaded so that equal currents flow in the three coils, and that these currents are in phase with their respective voltages as shown in Fig. 21-10*b*.

According to Kirchhoff's law, the total current leaving a junction must equal the total current entering the junction. At point *A* in Fig. 21-10*a*,

$$I_Y + I_{XY} = I_X$$

or

$$I_{XY} = I_X - I_Y$$

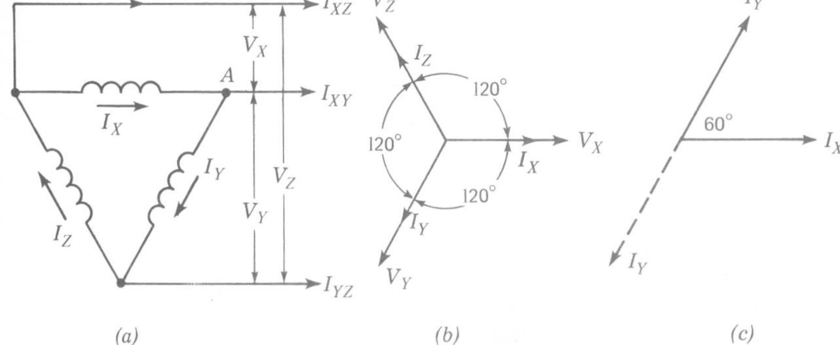


**FIGURE 21-10**
(*a*) Coils in delta. (*b*) Vector diagram of the phase currents. (*c*) Vector diagram for the addition of opposing currents.

This means that the vector total will be obtained by *reversing* the direction of $I_Y$ and adding *this* vector to $I_X$ as shown in Fig. 21-10*c*. To add the vectors, place the tail of $I_Y$ on the head of $I_X$ and draw it in its original size and direction as shown in Fig. 21-11*a*. The total ($I_L$) will be the vector from *0* to the tip of $I_Y$. Resolve the vector $I_Y$ into its components as shown in Fig. 21-11*b*. Since $I_X = I_Y =$ the phase current $I_p$, we may drop the individual subscripts and use just the letter $I_p$ to describe the current vectors.

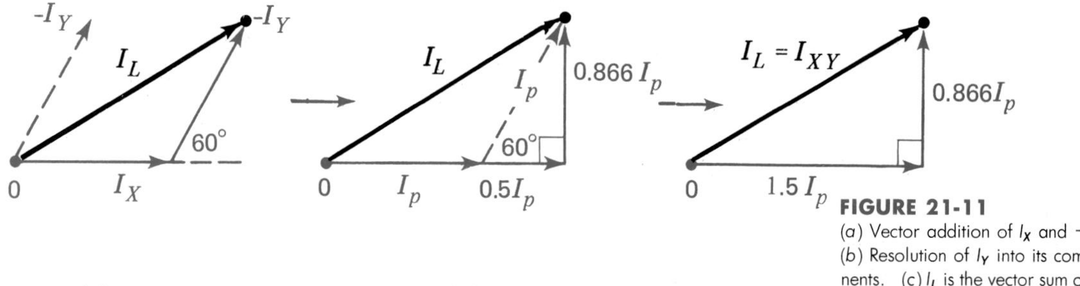


**FIGURE 21-11**
(*a*) Vector addition of $I_X$ and $-I_Y$. (*b*) Resolution of $I_Y$ into its components. (c) $I_L$ is the vector sum of the phase currents.

$$\text{Horizontal component} = I_p \cos 60° = 0.5I_p$$
$$\text{Vertical component} = I_p \sin 60° = 0.866I_p$$

Adding the in-phase components will give the vector diagram shown in Fig. 21-11*c*.

$$I_L^2 = (1.5I_p)^2 + (0.866I_p)^2$$
$$I_L = \sqrt{2.25I_p^2 + 0.75I_p^2}$$
$$= \sqrt{3I_p^2}$$
$$I_L = \sqrt{3}I_p$$

A similar situation exists for the currents in the other line wires. Thus, for a balanced load,

**FORMULA** $$I_L = \sqrt{3}I_p \qquad \boxed{21\text{-}11}$$

where $I_L$ = line current
$I_p$ = phase current

**Power in delta.** The power developed in each phase $W_p$ of a delta-connected, three-phase circuit is the same as the power in a single-phase circuit.

$$W_p = V_pI_p \cos \theta \qquad (21\text{-}3)$$

where $\theta$ is the angle between the phase current and the phase voltage.

The power developed in all three phases is

$$W_T = 3W_p = 3V_pI_p \cos\theta \qquad (21\text{-}4)$$

Since it is more practical to measure line voltages and line currents than phase values, let us convert phase values to line values.

In a delta connection,

$$V_p = V_L \qquad \text{and} \qquad I_p = \frac{I_L}{\sqrt{3}}$$

Substituting these values in formula (21-4) gives

$$W_T = 3\,V_L \times \frac{I_L}{\sqrt{3}} \cos\theta$$

or

**FORMULA** $$W_T = \sqrt{3}V_L I_L \cos\theta \qquad \boxed{21\text{-}5}$$

Also, since $VA = \dfrac{W}{\cos\theta}$, (20-5)

$$VA = \frac{\sqrt{3}V_L I_L \overset{1}{\cancel{\cos\theta}}}{\underset{1}{\cancel{\cos\theta}}}$$

or

**FORMULA** $$VA = \sqrt{3}V_L I_L \qquad \boxed{21\text{-}6}$$

Expressed in kilo units,

$$\text{kW} = \frac{\sqrt{3}V_L I_L \cos\theta}{1000} \qquad \boxed{21\text{-}7}$$

$$\text{kVA} = \frac{\sqrt{3}V_L I_L}{1000} \qquad \boxed{21\text{-}8}$$

$$\text{PF} = \frac{W_T}{\sqrt{3}V_L I_L} \qquad \boxed{21\text{-}9}$$

It should be noted that these formulas for power are identical for the wye and delta systems. This should be true, because the relations in a three-phase line are the same whether the power originates in a wye-connected or a delta-connected generator.

EXAMPLE 21-6. If the current through each phase of a delta-connected generator is 30 A, find the current in the line wires.

SOLUTION

$$I_L = \sqrt{3}I_p \qquad (21\text{-}11)$$
$$I_L = 1.732 \times 30 = 52 \text{ A} \qquad Ans.$$

EXAMPLE 21-7. Each phase of a three-phase, delta-connected generator supplies a full-load current of 50 A at a voltage of 220 V and at a power factor of 80 percent lagging. Find (*a*) the line voltage, (*b*) the line current, (*c*) the three-phase kilovoltamperes, and (*d*) the three-phase power in kilowatts.

SOLUTION

*a.* In a delta-connected system,

$$V_L = V_p \qquad (21\text{-}10)$$
$$V_L = 220 \text{ V} \qquad Ans.$$

*b.* $$I_L = \sqrt{3}I_p = 1.732 \times 50 = 86.6 \text{ A} \qquad Ans. \qquad (21\text{-}11)$$

*c.* $$\text{kVA} = \frac{\sqrt{3}V_L I_L}{1000} = \frac{1.732 \times 220 \times 86.6}{1000} = 33 \text{ kVA} \qquad Ans. \qquad (21\text{-}8)$$

*d.* $$\text{kW} = \text{kVA} \times \text{PF} = 33 \times 0.8 = 26.4 \text{ kW} \qquad Ans. \qquad (20\text{-}9)$$

EXAMPLE 21-8. A balanced load of three 10-Ω resistors (PF = 1) is connected across a 480-V three-phase line as shown in Fig. 21-12*a*. This is a delta connection, as shown in Fig. 21-12*b*. Find (*a*) the phase voltage across each resistor, (*b*) the current in each resistor, (*c*) the line current, (*d*) the three-phase power in kilowatts, and (*e*) the total kilovoltamperes input.

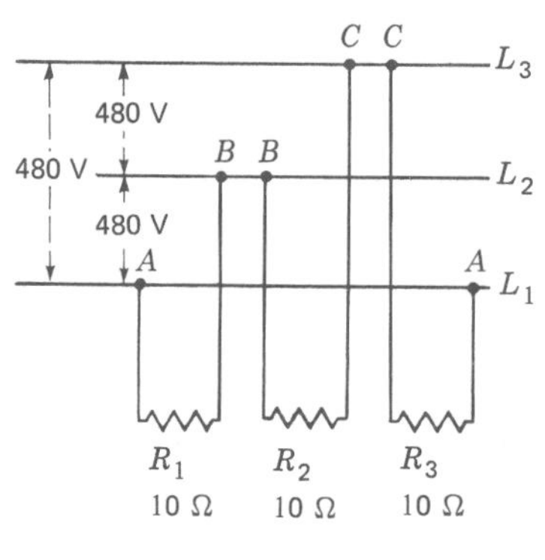


(*a*)

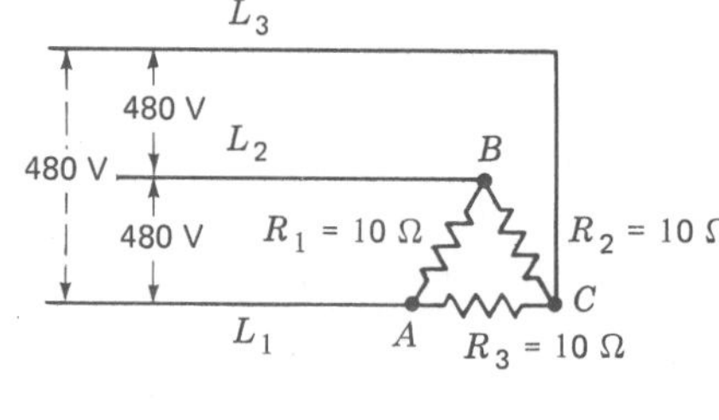


(*b*)

**FIGURE 21-12**

SOLUTION

*a.* $$V_p = V_L = 480 \text{ V} \qquad Ans. \qquad (21\text{-}10)$$

*b.* The current in each resistor with zero reactance (PF = 1) is

$$I_p = \frac{V_p}{Z_p} = \frac{V_p}{R_p} = \frac{480}{10} = 48 \text{ A} \qquad \textit{Ans.}$$

*c.* In a delta-connected circuit, the line current is

$$I_L = \sqrt{3}I_p = 1.732 \times 48 = 83.1 \text{ A} \qquad \textit{Ans.} \qquad (21\text{-}11)$$

*d.* $$\text{kW} = \frac{\sqrt{3}V_L I_L \cos\theta}{1000} = \frac{1.732 \times 480 \times 83.1 \times 1}{1000} = 69.1 \text{ kW} \qquad (21\text{-}7)$$

*e.* $$\text{kVA input} = \frac{\sqrt{3}V_L I_L}{1000} = \frac{1.732 \times 480 \times 83.1}{1000} = 69.1 \text{ kVA} \qquad (21\text{-}8)$$

or $$\text{kVA} = \frac{\text{kW}}{\text{PF}} = \frac{69.1}{1} = 69.1 \text{ kVA} \qquad \textit{Ans.} \qquad (20\text{-}10)$$

EXAMPLE 21-9. Each phase of a balanced delta-connected load consists of a resistor of 56 Ω in series with a pure inductive reactance of 33 Ω. The load is supplied with a 220-V 60-Hz three-phase line. Find (*a*) the impedance per phase, (*b*) the power factor, (*c*) the phase voltage, (*d*) the phase current, (*e*) the line current, (*f*) the total power consumed, and (*g*) the kilovoltampere input.

SOLUTION

*a.* $$Z_p = \sqrt{R^2 + X_L^2}$$
$$= \sqrt{56^2 + 33^2} = \sqrt{3136 + 1089} = \sqrt{4225} = 65\ \Omega \qquad \textit{Ans.}$$

*b.* $$\text{PF} = \cos\theta = \frac{R_p}{Z_p} = \frac{56}{65} = 0.862 \qquad \textit{Ans.}$$

*c.* $$V_p = V_L = 220 \text{ V} \qquad \textit{Ans.} \qquad (21\text{-}10)$$

*d.* $$I_p = \frac{V_p}{Z_p} = \frac{220}{65} = 3.38 \text{ A} \qquad \textit{Ans.}$$

*e.* $$I_L = \sqrt{3}I_p = 1.732 \times 3.38 = 5.85 \text{ A} \qquad \textit{Ans.} \qquad (21\text{-}11)$$

*f.* $$\text{kW} = \frac{\sqrt{3}V_L I_L \cos\theta}{1000} = \frac{1.732 \times 220 \times 5.85 \times 0.862}{1000} \qquad (21\text{-}7)$$
$$= 1.92 \text{ kW} \qquad \textit{Ans.}$$

*g.* $$\text{kVA} = \frac{\text{kW}}{\text{PF}} = \frac{1.92}{0.862} = 2.23 \text{ kVA} \qquad \textit{Ans.} \qquad (20\text{-}10)$$

EXAMPLE 21-10. A 220-V delta-connected, three-phase induction motor draws a full-load current of 35 A per terminal. Find (*a*) the phase voltage, (*b*) the phase current, (*c*) the full-load kilovoltampere input, and (*d*) the kilowatt input if the full-load power factor of the motor is 75 percent.

SOLUTION

*a.* $$V_p = V_L = 220 \text{ V} \qquad Ans. \qquad (21\text{-}10)$$

*b.* $$I_p = \frac{I_L}{\sqrt{3}} = \frac{35}{1.732} = 20.2 \text{ A} \qquad Ans. \qquad (21\text{-}11)$$

*c.* $$\text{kVA} = \frac{\sqrt{3}V_L I_L}{1000} = \frac{1.732 \times 220 \times 35}{1000} = 13.3 \text{ kVA} \qquad Ans. \qquad (21\text{-}8)$$

*d.* $$\text{kW} = \text{kVA} \times \text{PF} = 13.3 \times 0.75 = 9.98 \text{ kW} \qquad Ans. \qquad (20\text{-}9)$$

EXAMPLE 21-11. A three-phase induction motor draws 35 kW from a 220-V main. If the line current is 100 A, find (*a*) the power factor of the motor and (*b*) the reactive kilovoltampere.

SOLUTION

*a.* $$\text{PF} = \frac{\text{W}}{\text{VA}} = \frac{\text{kW} \times 1000}{\sqrt{3}V_L I_L} \qquad (20\text{-}7)$$

$$= \frac{35 \times 1000}{1.732 \times 220 \times 100} = 0.919 \qquad Ans.$$

*b.* $$\text{PF} = \cos\theta = 0.919$$

$$\theta = 24°$$

$$\tan\theta = \frac{\text{var}}{\text{W}} \qquad (20\text{-}11)$$

or $$\text{var} = \text{W} \tan\theta$$

or $$\text{kvar} = \text{kW} \tan\theta = 35 \times \tan 24°$$

$$= 35 \times 0.4452 = 15.6 \text{ kvar} \qquad Ans.$$

*Alternative method*

$$\text{kVA} = \frac{\sqrt{3}V_L I_L}{1000} = \frac{1.732 \times 220 \times 100}{1000} = 38.1 \text{ kVA} \qquad (21\text{-}8)$$

$$\text{kvar} = \text{kVA} \times \sin\theta \qquad (20\text{-}6)$$

$$= 38.1 \times \sin 24° = 38.1 \times 0.4067 = 15.5 \text{ kvar} \qquad Ans.$$

*Note:* The slight difference in the values is due to the use of the approximate phase angle of 24°. Greater accuracy would be obtained by the use of decimal trigonometric tables.

**PROBLEMS**

1. The phase current in a delta-connected generator is 60 A. Find the line current.

2. If the line current in a delta-connected generator is 50 A, find the coil current.
3. If the voltage per phase of a delta-connected generator is 440 V, find the line voltage.
4. Three equal resistive heating elements in delta are supplied by a three-phase 220-V line. If 8.66 A flows through each line wire, find the resistance of each element.
5. Six 24-Ω resistors are connected in two groups: three in wye, and three in delta. If the two groups in parallel are supplied by a three-phase 240-V line, find (*a*) the current in each wye-connected resistor, and (*b*) the current in each delta-connected resistor.
6. Each phase of a three-phase, delta-connected generator supplies a full-load current of 20 A at a voltage of 120 V and at a power factor of 80 percent lagging. Find (*a*) the line voltage, (*b*) the line current, (*c*) the total kilovoltamperes, and (*d*) the total kilowatts.
7. A heating load consists of three 24-Ω resistive heating elements connected in delta. If the load is supplied by a 240-V three-phase line, find (*a*) the voltage across each heating element, (*b*) the current in each heater, (*c*) the line current, (*d*) the total power in kilowatts, and (*e*) the total kilovoltampere input.
8. Each of the three parts of a balanced delta-connected load consists of a resistance of 16 Ω in series with a pure inductive reactance of 12 Ω. The load is supplied from a 220-V three-phase line. Find (*a*) the impedance per phase, (*b*) the power factor, (*c*) the phase voltage, (*d*) the phase current, (*e*) the line current, (*f*) the total kilowatts, and (*g*) the total kilovoltamperes.
9. A three-phase induction motor draws 50 kW from a 440-V line. If the line current is 80 A, find (*a*) the power factor of the motor, (*b*) the phase angle, (*c*) the kilovoltampere input, and (*d*) the reactive kilovars.
10. A delta-connected load draws 38.4 kW from a three-phase supply. If the line current is 69.3 A and the power factor is 80 percent, find (*a*) the line voltage, (*b*) the voltage across each phase, (*c*) the current per phase, (*d*) the impedance per phase, and (*e*) the resistance and reactance of each phase.
11. Each coil of a three-phase generator has a current-carrying capacity of 50 A and a voltage rating of 480 V. Find the kilovoltampere rating of the alternator when it is (*a*) connected in delta and (*b*) connected in wye.
12. A three-phase, delta-connected alternator is rated at 75 kVA, 240 V, and 60 Hz. Find (*a*) the full-load line current, (*b*) the full-load line voltage, and (*c*) the kilovoltamperes, current, and voltage of one phase of the alternator.

## JOB 21-3 REVIEW OF THREE-PHASE CONNECTIONS

For either connection,

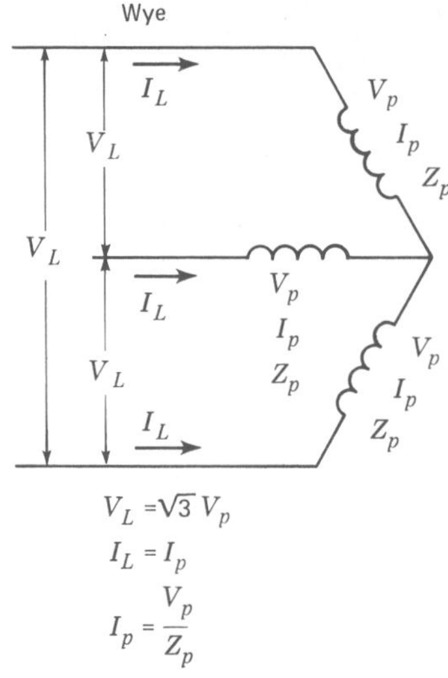


Delta

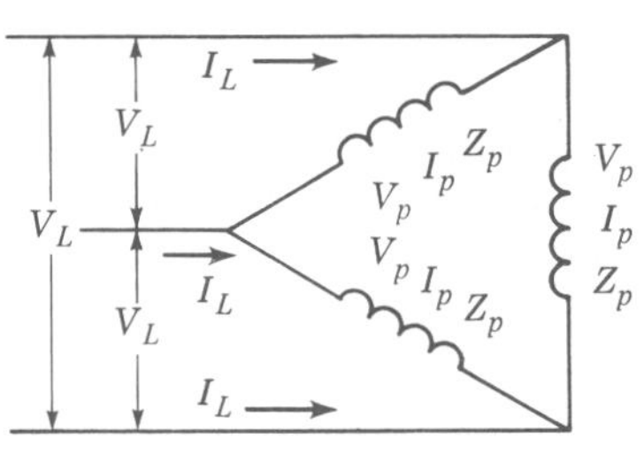


$$V_L = V_p$$
$$I_L = \sqrt{3}\, I_p$$
$$I_p = \frac{V_p}{Z_p}$$

**FIGURE 21-13**
Formulas for the wye and delta connections.

$$Z \text{ per phase} = \frac{V_p}{I_p}$$
$$VA \text{ per phase} = V_p I_p$$
$$\text{Three-phase } VA = VA_T = 3V_p I_p$$
$$= \sqrt{3} V_L I_L$$
$$\text{Three-phase kVA} = \frac{\sqrt{3} V_L I_L}{1000}$$
$$\text{Power factor} = \cos\theta$$
$$\theta = \text{angle between } V_p \text{ and } I_p$$
$$\cos\theta = \frac{R_p}{Z_p}$$
$$\text{Power per phase} = W = V_p I_p \cos\theta$$
$$\text{Three-phase power} = W_T = 3V_p I_p \cos\theta$$
$$W_T = \sqrt{3} V_L I_L \cos\theta$$
$$\text{Three-phase kW} = \frac{\sqrt{3} V_L I_L \cos\theta}{1000}$$
$$\text{PF} = \frac{W_T}{VA_T}$$
$$= \frac{W_T}{\sqrt{3} V_L I_L}$$
$$\text{kW} = \text{kVA} \times \text{PF}$$
$$\text{kvar} = \text{kVA} \times \sin\theta$$
$$\text{kvar} = \text{kW} \times \tan\theta$$

EXAMPLE 21-12. A 60-kVA wye-connected alternator rated at 120 V per coil is connected to a balanced load consisting of three 3-Ω resistors connected in delta. Find (*a*) the current per phase in the load, (*b*) the

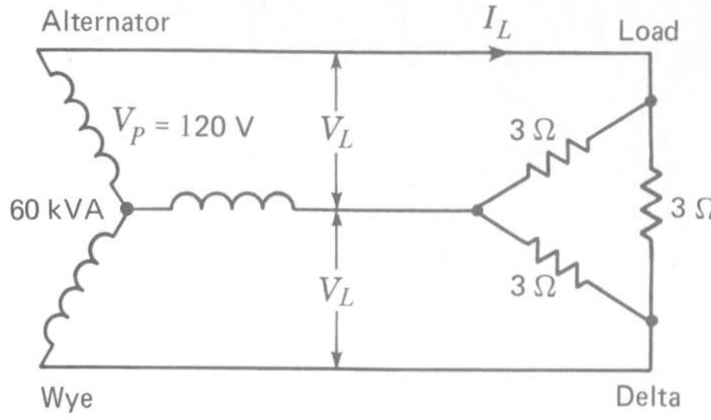


FIGURE 21-14

current per coil in the alternator due to the load, and (*c*) the kilovoltamperes delivered by the alternator to the load.

SOLUTION

*a.* Find the line voltage of the alternator. In wye,

$$V_L = \sqrt{3}V_p = 1.732 \times 120 = 208 \text{ V} \qquad (21\text{-}1)$$

Find the phase voltage in the load. In delta,

$$V_p = V_L = 208 \text{ V} \qquad (21\text{-}10)$$

Find the phase current in the load.

$$I_p = \frac{V_p}{Z_p} = \frac{V_p}{R_p} = \frac{208}{3} = 69.3 \text{ A} \qquad \textit{Ans.}$$

*b.* Find the line current to the load. In delta,

$$\begin{aligned} I_L &= \sqrt{3}I_p \qquad (21\text{-}11) \\ &= 1.732 \times 69.3 = 120 \text{ A} \end{aligned}$$

Find the current in the alternator coil to create this line current. In wye,

$$I_p = I_L = 120 \text{ A} \qquad \textit{Ans.} \qquad (21\text{-}2)$$

*c.*

$$\begin{aligned} \text{kVA} &= \frac{\sqrt{3}V_L I_L}{1000} \qquad (21\text{-}8) \\ &= \frac{1.732 \times 208 \times 120}{1000} = 43.2 \text{ kVA} \qquad \textit{Ans.} \end{aligned}$$

EXAMPLE 21-13. Each phase winding of a wye-connected generator is rated at 20 kVA and 120 V. Find (*a*) the line voltage, (*b*) the line current at full load, (*c*) the kilowatt output at full load and 80 percent power factor, and (*d*) the reactive kilovoltamperes at full load.

SOLUTION

*a.*

$$V_L = \sqrt{3}V_p = 1.732 \times 120 = 208 \text{ V} \qquad \textit{Ans.} \qquad (21\text{-}1)$$

*b.*

$$\begin{aligned} VA &= V_p I_p \\ 20{,}000 &= 120 \times I_p \end{aligned}$$

$$I_p = \frac{20{,}000}{120} = 166.7 \text{ A}$$

In wye,
$$I_L = I_p = 166.7 \text{ A} \qquad Ans. \qquad (21\text{-}2)$$

*c.*
$$\text{kW} = \frac{\sqrt{3}\,V_L I_L \cos\theta}{1000} \qquad (21\text{-}7)$$
$$= \frac{1.732 \times 208 \times 166.7 \times 0.8}{1000}$$
$$\text{kW} = 48 \text{ kW} \qquad Ans.$$

*d.* If the PF = 0.8, then

$$\cos\theta = 0.8$$

and
$$\theta = 37°$$
$$\tan\theta = \frac{\text{kvar}}{\text{kW}}$$

or
$$\text{kvar} = \text{kW} \times \tan\theta$$
$$= 48 \times \tan 37° = 48 \times 0.7536 = 36.2 \text{ kvar} \qquad Ans.$$

*Check:* If each coil is rated at 20 kVA, then the total kilovoltamperes equal 3 × 20 = 60 kVA.

$$\text{kvar} = \text{kVA} \times \sin\theta = 60 \times \sin 37°$$
$$= 60 \times 0.602 = 36.1 \text{ kvar} \qquad Check$$

EXAMPLE 21-14. A three-phase 220-V feeder supplies two motors as shown in Fig. 21-15. (*a*) Find the kilovoltamperes, kilowatts, and lagging var power for the induction motor. (*b*) Find the kilovoltamperes, kilowatts, and leading var power for the synchronous motor. (*c*) Find the total load supplied to the two motors. (*d*) Find the total var power drawn by the two motors. (*e*) Find the total power factor for the two loads. (*f*) Find the line current drawn by the two motors.

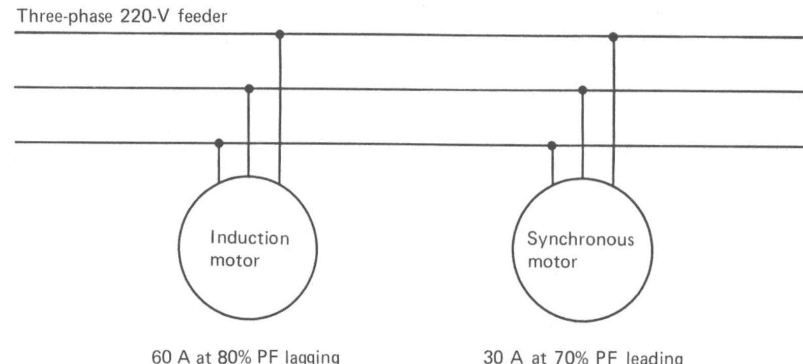


**FIGURE 21-15**

SOLUTION

*a.* For the induction motor:

$$\text{kVA} = \frac{\sqrt{3}V_L I_L}{1000} = \frac{1.732 \times 220 \times 60}{1000} = 22.86 \text{ kVA} \qquad \textit{Ans.}$$

$$\text{kW} = \frac{\sqrt{3}V_L I_L \cos\theta}{1000} = \frac{1.732 \times 220 \times 60 \times 0.8}{1000} = 18.29 \text{ kW} \qquad \textit{Ans.}$$

If the PF = 80 percent, then

$$\cos\theta = 0.8$$

and

$$\theta = 37°$$

$$\text{kvar} = \text{kVA} \times \sin\theta = 22.86 \times \sin 37°$$
$$= 22.86 \times 0.6 = 13.72 \text{ kvar } \textit{lagging} \qquad \textit{Ans.}$$

*b.* For the synchronous motor:

$$\text{kVA} = \frac{\sqrt{3}V_L I_L}{1000} = \frac{1.732 \times 220 \times 30}{1000} = 11.43 \text{ kVA} \qquad \textit{Ans.}$$

$$\text{kW} = \frac{\sqrt{3}V_L I_L \cos\theta}{1000} = \frac{1.732 \times 220 \times 30 \times 0.7}{1000} = 8.00 \text{ kW} \qquad \textit{Ans.}$$

If the PF = 70 percent, then

$$\cos\theta = 0.7$$

and

$$\theta = 45°$$

$$\text{kvar} = \text{kVA} \times \sin\theta = 11.43 \times \sin 45°$$
$$= 11.43 \times 0.7 = 8.00 \text{ kvar } \textit{leading}$$

*c.* Since the kilowatts drawn by the two motors are in phase,

$$\text{Total kW} = W_1 + W_2 = 18.29 + 8.00 = 26.29 \text{ kW} \qquad \textit{Ans.}$$

*d.* since the reactive powers are 180° out of phase and oppose each other,

$$\text{Total kvar} = \text{kvar}_1 - \text{kvar}_2$$
$$= 13.72 - 8.00 = 5.72 \text{ kvar } \textit{lagging} \qquad \textit{Ans.}$$

*e.* The vector diagram for the combination is shown in Fig. 21-16.

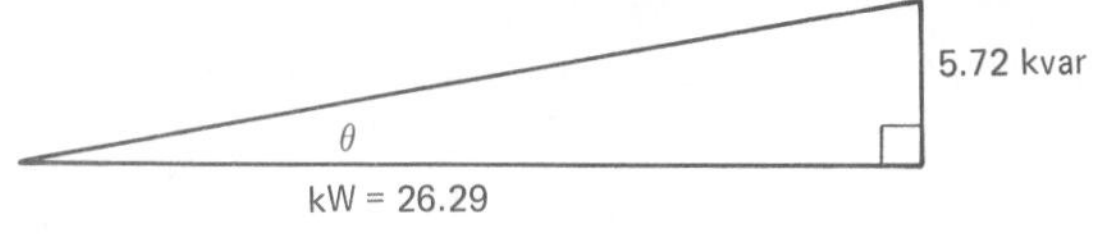


FIGURE 21-16

$$\tan\theta = \frac{\text{kvar}}{\text{kW}}$$

$$\tan\theta = \frac{5.72}{26.29} = 0.2176$$

$$\theta = 12°$$

$$\text{PF} = \cos\theta = \cos 12° = 0.978 \qquad \textit{Ans.}$$

*f.*
$$W_T = \sqrt{3} V_L I_L \cos\theta$$
$$26{,}290 = 1.732 \times 220 \times I_L \times 0.978$$
$$I_L = 70.5 \text{ A} \qquad \textit{Ans.}$$

## PROBLEMS

1. Three coils are connected in delta across a 240-V three-phase supply. If the line current is 30 A, find (*a*) the coil current, (*b*) the coil voltage, and (*c*) the total kilovoltamperes.
2. A 120-V three-phase feeder delivers power to twenty-four 120-V 60-W lamps. If this load is balanced in three groups of eight lamps each, find the current through each line wire.
3. A 220-V three-phase feeder delivers 90 A at 80 percent power factor to a distribution panel. Find the kilowatts delivered.
4. The coils of a three-phase 60-Hz generator are wye-connected. Each coil is rated at 4800 VA and 120 V. Find (*a*) the line voltage, (*b*) the coil current, (*c*) the full-load line current, and (*d*) the full-load kilovoltampere rating of the generator.
5. Each phase of a three-phase alternator is rated at 42 kVA and 120 V. It is wye-connected and supplying power at its rated output to a load at 80 percent power factor. Find (*a*) the line current, (*b*) the line voltage, and (*c*) the total power.
6. If the alternator of Prob. 5 is now reconnected delta, and each phase delivers full load, find (*a*) the line current, (*b*) the line voltage, and (*c*) the total power.
7. If each phase of a three-phase 440-V line carries a load of 40 kW, find the total power delivered.
8. A three-phase, wye-connected generator is rated at 650 kVA, 1300 V, and 60 Hz. Find (*a*) the kilowatt output at a power factor of 80 percent, (*b*) the full-load line current, (*c*) the full-load current rating of each coil, and (*d*) the voltage across each coil.
9. A three-phase, delta-connected generator is rated at 650 kVA, 1300 V, and 60 Hz. Find (*a*) the kilowatt output at a power factor of 80 percent, (*b*) the full-load coil current, (*c*) the full-load line current, and (*d*) the voltage rating of each coil.
10. Each phase of a balanced delta-connected load consists of a resistor of 24 Ω in series with a pure inductive reactance of 7 Ω. The load is supplied from a 220-V 60-Hz three-phase line. Find (*a*) the impedance per phase, (*b*) the power factor, (*c*) the phase voltage, (*d*) the phase current, (*e*) the line current, (*f*) the total power, and (*g*) the kilovoltampere input.
11. A three-phase, delta-connected alternator is rated at 15 kVA and 208 V. It is supplying a balanced wye-connected load which has an impedance of 20 Ω per phase and a power factor of 80 percent lagging. Find (*a*) the line current for the given load, (*b*) the total power delivered to the load in kilowatts, (*c*) the current in one coil

of the alternator due to the load, and (*d*) the resistance and the inductive reactance of the load per phase.

12. A delta-connected motor with an efficiency of 87 percent is operated at the rated output of 20 hp (1 hp = 746 W) on a three-phase line. If its phase power factor is 85 percent lagging and the phase voltage is 208 V, find (*a*) the phase current of the motor, (*b*) the line current, and (*c*) the line voltage.
13. An inductive load of 55 kVA at 74 percent power factor lagging is being operated from a 480-V three-phase line. In order to raise the power factor, a synchronous motor drawing 16 kW at 80 percent leading power factor is added in parallel. (*a*) Find the kilowatts and kilovars for the induction motor. (*b*) Find the kilovoltamperes and kilovars for the synchronous motor. (*c*) Find the total kilowatts and kilovars drawn by the combination of motors. (*d*) Find the power factor of the combination. (*e*) Find the total line current drawn.
14. A three-phase, three-wire 1200-V feeder supplies a plant which has a 250-kVA load operating at 75 percent lagging power factor. A 1200-V 100-kVA synchronous motor operating at a power factor of 80 percent leading is added in parallel to raise the power factor. Find (*a*) the kilowatts and kilovars of the inductive load, (*b*) the kilowatts and kilovars of the synchronous motor, (*c*) the total kilowatts and total kilovars of the combination, and (*d*) the new power factor.

(See Instructor's Manual for Test 21-1, Three-Phase Circuits)

# TRANSFORMER CONNECTIONS FOR THREE-PHASE CIRCUITS

# 22

## JOB 22-1 THE DELTA CONNECTION

Electric energy may be transformed in the three-phase system using the same devices that were used to transform single-phase voltages. Three-phase power may be supplied through the use of three separate single-phase transformers, with four possible combinations of their primaries and secondaries. These connections are (1) primaries in delta and secondaries in delta, (2) primaries in wye and secondaries in wye, (3) primaries in delta and secondaries in wye, and (4) primaries in wye and secondaries in delta.

Figure 22-1*a* shows three separate single-phase transformers with their high-side or primary terminals labeled $H_1$ and $H_2$. The primary windings are connected in series, with the end of one winding connected to the beginning of the next winding. Each connection, labeled *A, B,* and *C,* would be connected to the lines of the three-phase source. The low-side or secondary terminals are labeled $L_1$ and $L_2$. These are also connected in series, with the end of one winding connected to the beginning of the next. Each connection, labeled $A'$, $B'$, and $C'$, would become a line to the load. The connections of the primary coils are shown in diagram form in Fig. 22-1*b,* and the connections of the secondary coils in Fig. 22-1*c.* This type of connection is known as a *closed-delta* connection.

The delta-connected primary may be considered to be a load on the three-phase line $ABC$. The normal delta-connected voltage relation $V_L = V_p$ will apply in this situation. Also, as shown in Fig. 22-1*b,* each primary winding is connected directly across two line leads, confirming the relationship. The current relationships are the same as for any delta circuit, namely $I_L = \sqrt{3}I_p$.

The delta-connected secondary may be considered to be a generator supplying the load lines $A'B'C'$. The secondary line voltage $V_L = V_p$, and $I_L = \sqrt{3}I_p$.

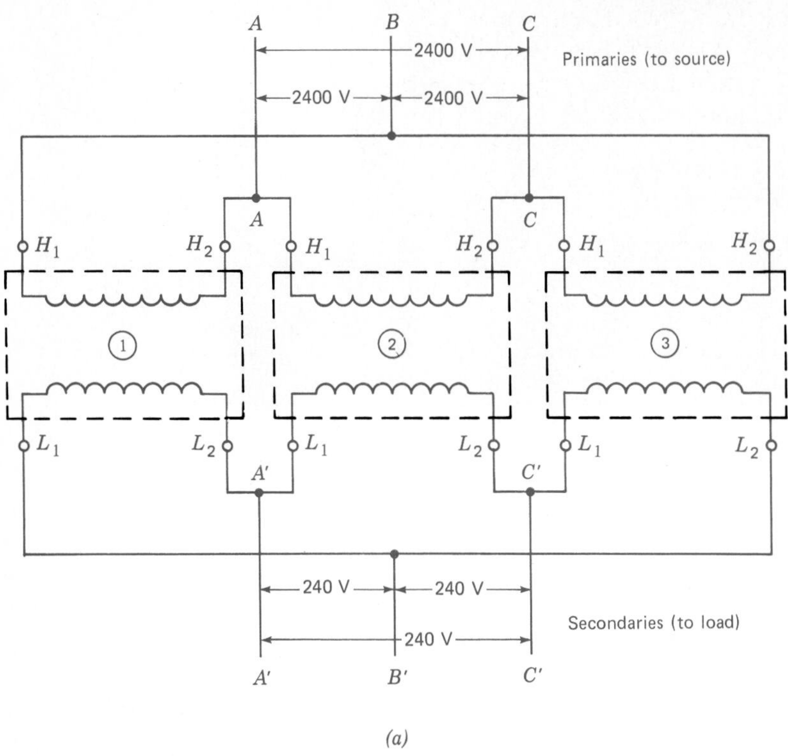


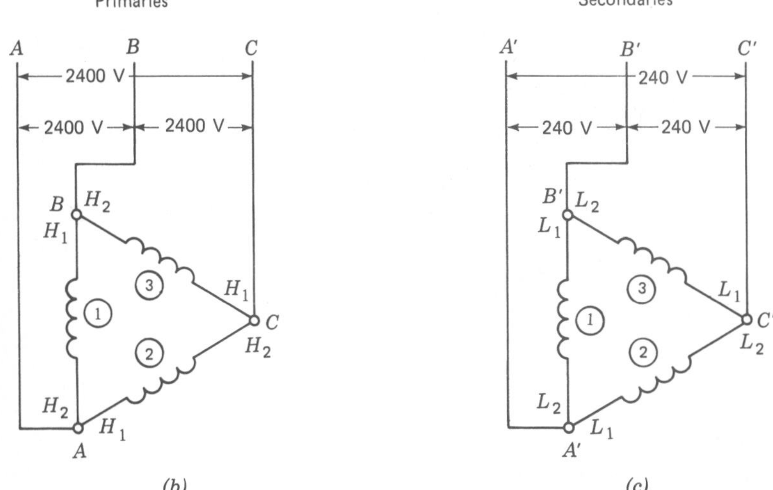


**FIGURE 22-1**
(*a*) Delta-delta connection for a three-phase 10:1 transformer. (*b*) The connections of the primary coils in diagram form. (*c*) The connections of the secondary coils in diagram form.

EXAMPLE 22-1. Each of the three delta-delta connected transformers shown in Fig. 22-1 is rated at 40 kVA and 2400 V, with a step-down voltage ratio of 10:1. The secondary leads each carry 200 A to a three-phase load with a power factor of 80 percent lagging. Find (*a*) the voltage across each primary coil, (*b*) the voltage across each secondary coil, (*c*) the secondary line voltage $A'B'C'$, (*d*) the kilovoltampere load on the transformer bank, (*e*) the kilowatt load on the transformer bank, (*f*) the current in each secondary winding, (*g*) the current in each primary winding, (*h*) the current in each primary line *A*, *B*, and *C*, (*i*) the kilovoltampere load on each transformer, and (*j*) the kilovoltampere input to the transformer bank.

SOLUTION

*a.* In delta, the primary coil voltage is

$$V_1 = V_p = V_L = 2400 \text{ V} \qquad \textit{Ans.}$$

*b.* In Fig. 22-2, the secondary coil voltage is $V_2$.

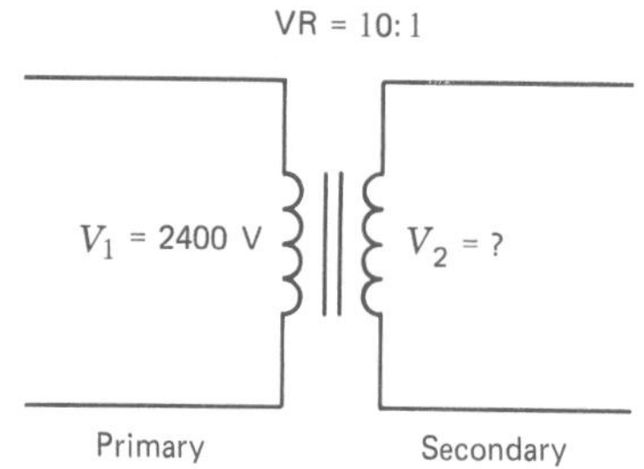


FIGURE 22-2

$$\text{VR} = \frac{10}{1} = \frac{V_1}{V_2}$$

$$\frac{10}{1} = \frac{2400}{V_2}$$

$$V_2 = 240 \text{ V} \qquad \textit{Ans.}$$

*c.* The secondary line voltage is $V_L$. In delta,

$$V_L = V_p = V_2 = 240 \text{ V} \qquad \textit{Ans.}$$

*d.* $$\text{kVA} = \frac{\sqrt{3}\,V_L I_L}{1000} = \frac{1.732 \times 240 \times 200}{1000} = 83.1 \text{ kVA} \qquad \textit{Ans.}$$

*e.* $$\text{kW} = \frac{\sqrt{3}\,V_L I_L \cos\theta}{1000} = \frac{1.732 \times 240 \times 200 \times 0.8}{1000}$$

$$= 66.5 \text{ kW} \qquad \textit{Ans.}$$

or $\text{kW} = \text{kVA} \times \text{PF} = 83.1 \times 0.8 = 66.5 \text{ kW}$ *Ans.*

*f.* In delta, the secondary coil current is

$$I_2 = I_p = \frac{I_L}{\sqrt{3}} = \frac{200}{1.732}$$

$$I_2 = 115.5 \text{ A} \qquad \textit{Ans.}$$

*g.* In Fig. 22-3, the current in the primary winding is $I_1$.

$$\frac{I_1}{I_2} = \frac{V_2}{V_1}$$

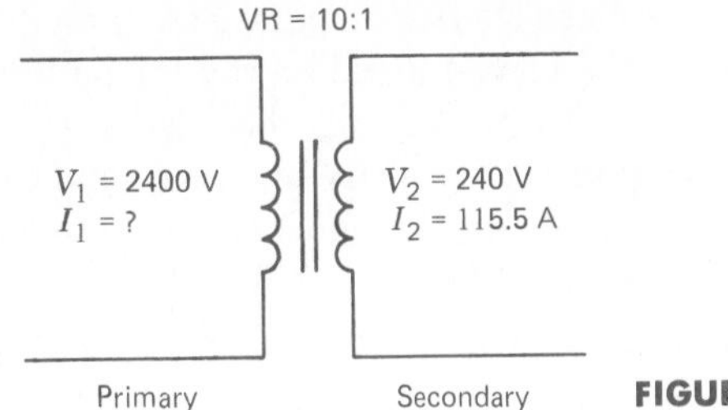


FIGURE 22-3

$$\frac{I_1}{115.5} = \frac{240}{2400}$$

$$I_1 = I_p = \frac{115.5 \times 240}{2400} = 11.6 \text{ A} \qquad \textit{Ans.}$$

*h.* The primary line current is

$$I_L = \sqrt{3} I_p = 1.732 \times 11.6 = 20 \text{ A} \qquad \textit{Ans.}$$

*i.*
$$\text{kVA per transformer} = \frac{V_p \times I_p}{1000} = \frac{240 \times 115.5}{1000}$$
$$= 27.7 \text{ kVA} \qquad \textit{Ans.}$$

or
$$\text{kVA per transformer} = \frac{\text{total kVA}}{3} = \frac{83.1}{3}$$
$$= 27.7 \text{ kVA} \qquad \textit{Ans.}$$

*j.*
$$\text{kVA input} = \frac{\sqrt{3} V_L I_L}{1000} = \frac{1.732 \times 2400 \times 20}{1000}$$
$$= 83.1 \text{ kVA} \qquad \textit{Ans.}$$

*Check:* Since we are assuming no losses, the kilovoltampere input should equal the kilovoltampere output from step (*d*) above, or 83.1 kVA. *Check*

## JOB 22-2 THE WYE CONNECTION

Figure 22-4*a* is a schematic diagram of three single-phase transformers with both their primary and secondary windings connected in wye. The $H_2$ leads of the windings are connected together. The $H_1$ leads are brought out as the line wires *A, B,* and *C,* which will be connected to a three-phase source. Similarly, the $L_2$ leads of the secondary windings are connected together. The $L_1$ leads are brought out as the line wires $A', B'$, and $C'$, which will be connected to a three-phase load. As in any wye connection, the current in each transformer winding equals the line current ($I_p = I_L$), and the voltage impressed on each winding equals the line voltage divided by $\sqrt{3}$ ($V_p = V_L \div \sqrt{3}$).

The total voltampere load transmitted by a balanced three-phase line equals $\sqrt{3}\, V_L I_L$. The voltampere capacity of each transformer should equal one-third of this total voltampere load, or

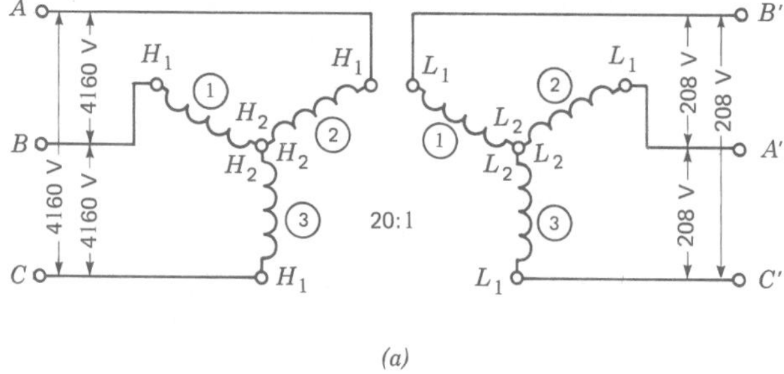


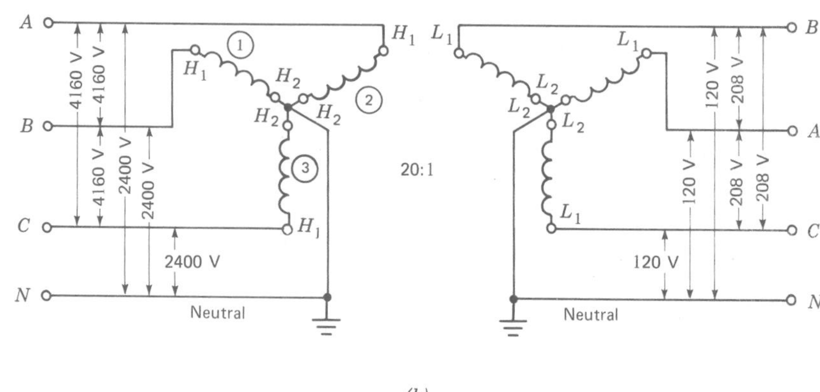


**FIGURE 22-4**
(*a*) Wye-wye connected transformer. (*b*) Wye-wye with neutral wire.

$$\text{VA per transformer} = \frac{\sqrt{3}V_L I_L}{3} = \frac{1.732 V_L I_L}{3} = 0.58 V_L I_L$$

A neutral wire is usually used with this connection, as shown in Fig. 22-4*b*, in order to avoid voltage instabilities on the secondary side.

In Fig. 22-4*b*, if the primary line voltage $V_L = 4160$ V, the primary coil voltage $V_p = V_L/\sqrt{3} = 4160/1.732 = 2400$ V.

If the voltage ratio is 20:1, then the secondary coil voltage will be $V_p = 2400 \div 20 = 120$ V

The secondary line voltage $V_L = \sqrt{3}V_p$ or $1.732 \times 120 = 208$ V.

The voltage between the line wire and the neutral will equal the secondary coil voltage of 120 V. This means that a three-phase, four-wire wye system is able to supply (1) a 208-V three-phase industrial power source and (2) a 120-V source for any lighting load.

## JOB 22-3 THE DELTA-WYE CONNECTION

Figure 22-5 describes a typical delta-wye connection. The primaries of the single-phase transformers are connected in series to form a delta connection across a 2400-V three-phase supply. The secondaries are connected in wye with all $L_2$ leads connected to the grounded neutral wire.

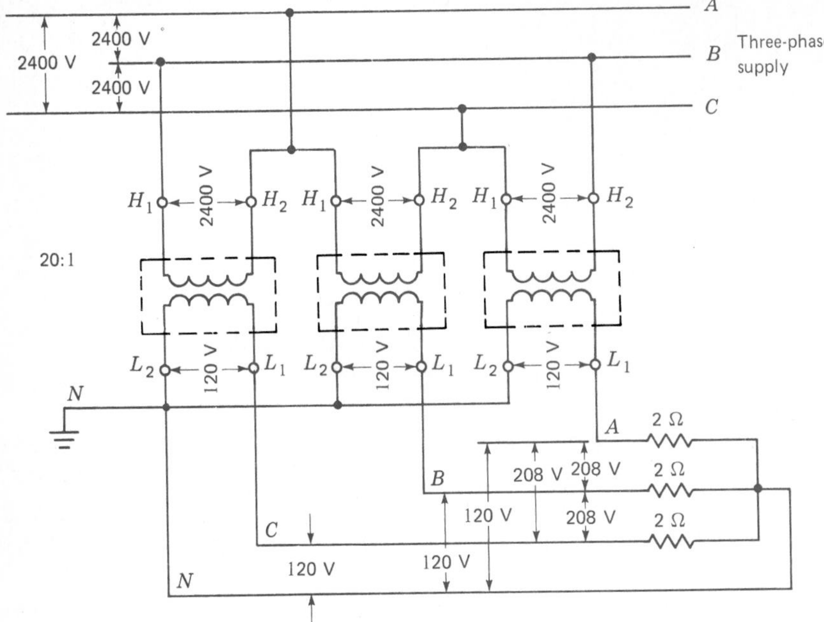


**FIGURE 22-5**
Delta-wye transformer bank.

EXAMPLE 22-2. Each 20:1 step-down transformer shown in Fig. 22-5 is rated at 15 kVA and 2400 V. The load consists of three 2-Ω resistors connected in wye to the three-phase, four-wire secondary. Find (*a*) the primary coil voltage, (*b*) the secondary coil voltage, (*c*) the secondary line voltage, (*d*) the voltage across each load resistor, (*e*) the current in each resistor, (*f*) the current in each secondary coil, (*g*) the secondary line current, (*h*) the kilovoltampere load on the transformer bank, (*i*) the kilowatt load on the bank, (*j*) the primary coil current, (*k*) the primary line current, and (*l*) the kilovoltampere capacity of the transformer bank.

SOLUTION

*a.* In delta,

$$V_p = V_L = 2400 \text{ V}$$

*b.* At a voltage ratio of 20:1, the secondary coil voltage is

$$2400 \div 20 = 120 \text{ V} \qquad \textit{Ans.}$$

*c.* In wye, $V_L = \sqrt{3}V_p$.

$$V_L = 1.732 \times 120 = 208 \text{ V} \qquad \textit{Ans.}$$

*d.* Each load resistor is connected between a line and the neutral wire, or $V_R = 120$ V. *Ans.*

*e.* $$I_R = \frac{V_R}{R_R} = \frac{120}{2} = 60\ \text{A} \qquad \textit{Ans.}$$

*f.* Since the secondary coil is in series with the resistor,

$$I_{\text{coil}} = 60\ \text{A} \qquad \textit{Ans.}$$

*g.* In wye,

$$I_L = I_{\text{coil}} = 60\ \text{A} \qquad \textit{Ans.}$$

*h.* $$\text{kVA load} = \frac{\sqrt{3}V_L I_L}{1000} = \frac{1.732 \times 208 \times 60}{1000} = 21.6\ \text{kVA} \qquad \textit{Ans.}$$

*i.* kW = kVA (since the PF = 1 for a resistor) = 21.6 kW *Ans.*

*j.* Working with coil values and a VR = 20:1,

$$\text{VR} = \frac{V_1}{V_2} = \frac{I_2}{I_1}$$

$$\frac{20}{1} = \frac{60}{I_1}$$

$$I_1 = I_p = \frac{60}{20} = 3\ \text{A} \qquad \textit{Ans.}$$

*k.* In delta,

$$I_L = \sqrt{3}I_{\text{coil}} = 1.732 \times 3 = 5.2\ \text{A} \qquad \textit{Ans.}$$

*l.* kVA capacity of transformer bank = 3 × kVA per transformer
= 3 × 15 = 45 kVA *Ans.*

*Check:*

$$\text{kVA input} = \frac{\sqrt{3}V_L I_L}{1000} = \frac{1.732 \times 2400 \times 5.2}{1000} = 21.6\ \text{kVA}$$

$$\text{kVA output} = \frac{\sqrt{3}V_L I_L}{1000} = \frac{1.732 \times 208 \times 60}{1000} = 21.6\ \text{kVA} \qquad \textit{Check}$$

(Since we are assuming no losses, the output should equal the input.)

## JOB 22-4 THE WYE-DELTA CONNECTION

Figure 22-6 shows three single-phase transformers with wye-connected primaries and delta-connected secondaries. This arrangement is generally used to transform high voltages into lower ones. With the primaries in wye, the line voltage is 1.732 times the phase voltage, and the line current will be proportionally smaller. This will reduce the line losses and improve the efficiency of the transmission line.

EXAMPLE 22-3. In Fig. 22-6, the delta-connected secondary supplies 200 A to a motor-bank load. Find (*a*) the secondary coil current, (*b*) the primary coil current, (*c*) the primary line current, (*d*) the kilovoltampere load on the transformer bank, and (*e*) the kilovoltampere input to the transformer bank.

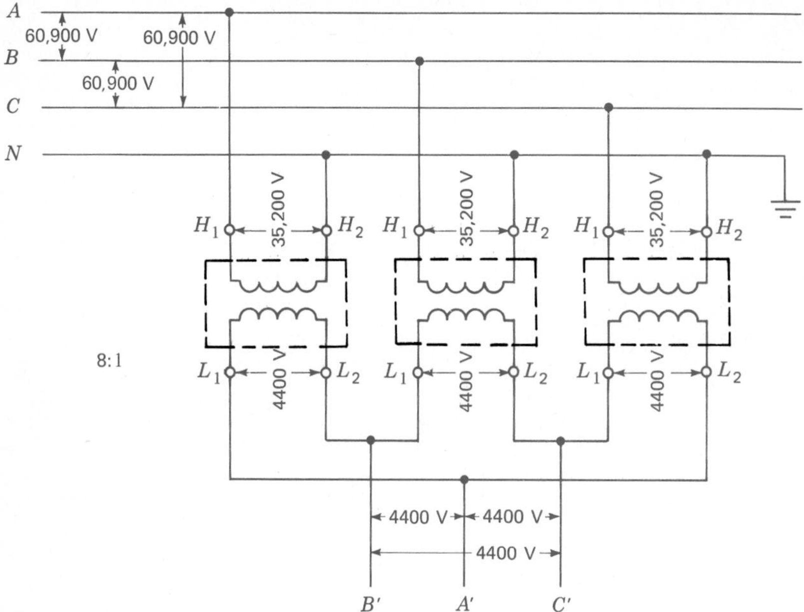


**FIGURE 22-6**
Wye-delta transformer bank.

SOLUTION

*a.* In delta,

$$I_{\text{coil}} = \frac{I_L}{\sqrt{3}} = \frac{200}{1.732} = 115.5 \text{ A} \qquad \textit{Ans.}$$

*b.* Working with coil values, for a voltage ratio of 8:1,

$$\text{VR} = \frac{8}{1} = \frac{V_1}{V_2} = \frac{I_2}{I_1}$$

$$\frac{8}{1} = \frac{115.5}{I_1}$$

$$I_1 = I_p = \frac{115.5}{8} = 14.44 \text{ A} \qquad \textit{Ans.}$$

*c.* In wye, the line current equals the coil current.

$$I_L = I_p = 14.44 \text{ A} \qquad \textit{Ans.}$$

*d.* $$\text{kVA} = \frac{\sqrt{3}V_L I_L}{1000} = \frac{1.732 \times 4400 \times 200}{1000} = 1524 \text{ kVA} \qquad \textit{Ans.}$$

*e.* $$\text{kVA input} = \frac{\sqrt{3}V_L I_L}{1000} = \frac{1.732 \times 60{,}900 \times 14.44}{1000}$$

$$= 1523 \text{ kVA} \qquad \textit{Ans.}$$

(Since we are assuming no losses, the kilovoltampere input should equal the kilovoltampere output.) *Check*

## JOB 22-5 THE OPEN-DELTA CONNECTION

Figure 22-7*a* shows three balanced delta-connected transformer secondaries supplying a 150-kW load at 250 V. If one of the transformers should be damaged, it may be cut out completely, as shown in Fig. 22-7*b*. This connection is known as the open-delta connection. In the closed-delta circuit of Fig. 22-7*a*,

$$\text{Rating per transformer} = \frac{150}{3} = 50 \text{ kW per transformer}$$

$$\text{Phase current} = \frac{\text{watts}}{\text{volts}} = \frac{50{,}000}{250} = 200 \text{ A}$$

$$\text{Line current } I_L = 1.732 \times 200 = 346 \text{ A}$$

In the open-delta circuit of Fig. 22-7*b*, the line current $I_L$ is now equal to the phase current, and $I_L = 200$ A. At unity power factor,

$$\text{Total kilowatts delivered} = \frac{\sqrt{3}V_L I_L \cos\theta}{1000} = \frac{1.732 \times 250 \times 200 \times 1}{1000}$$
$$= 86.6 \text{ kW}$$

This means that the open-delta circuit delivers only 86.6/150 of the original closed-delta circuit. Since 86.6/150 = 0.58, the kilovoltampere capacity of the open-delta connection equals 58 percent of the capacity of the closed-delta connection.

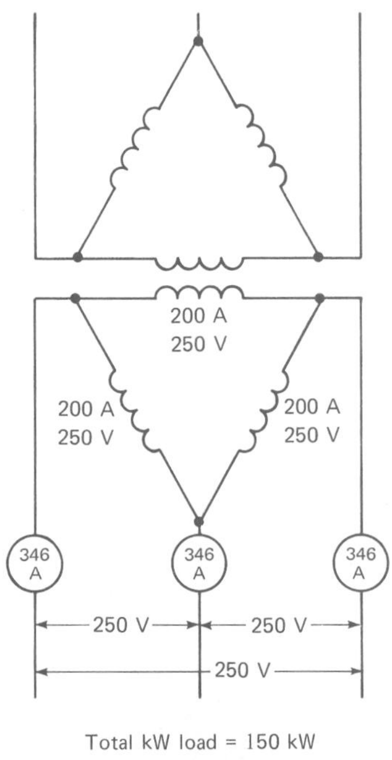


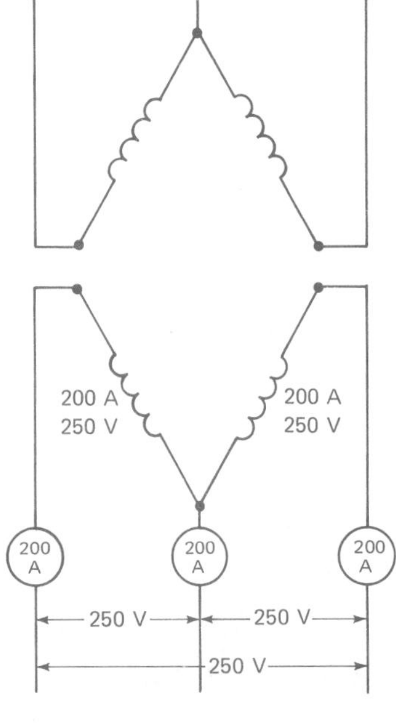


**FIGURE 22-7**
(*a*) Closed-delta connection. (*b*) Open-delta connection.

There are some situations in which the original installation may be an open-delta connection, owing to low demand for power. It may be easily converted to a regular delta-delta connection by the addition of a third transformer when more capacity is needed. The capacity of a two-transformer open-delta connection is *not* twice as much as the capacity of a single transformer. In our example above, two 50-kW transformers (total kilowatts = 100) can carry only 86.6 kW, or only 86.6 percent of the capacity of the two transformers.

EXAMPLE 22-4. Three 40-kVA 2400-V 10:1 transformers were connected delta-delta. One became defective and was cut out. The remaining two were connected open-delta. (*a*) Find the kilovoltampere capacity of the open-delta bank. (*b*) Find the kilowatt output when the bank is loaded to capacity at 70 percent power factor lagging. (*c*) Find the line current for this load. (*d*) Check the kilovoltampere capacity of the open-delta connection.

SOLUTION

*a.* Total KVA in closed-delta = 3 × 40 = 120 kVA

Open-delta KVA = 0.58 × 120 = 69.6 kVA *Ans.*

*b.* kW output = kVA × PF = 69.6 × 0.7 = 48.7 kW *Ans.*

*c.*

$$\text{Line voltage} = \frac{2400}{10} = 240 \text{ V}$$

$$\text{kVA} = \frac{\sqrt{3}V_L I_L}{1000}$$

$$69.6 = \frac{1.732 \times 240 \times I_L}{1000}$$

$$I_L = 167.4 \text{ A} \qquad \textit{Ans.}$$

*d.* Open-delta capacity = 86.6% of two-transformer capacity

= 0.866 × (2 × 40) = 69.3 kVA *Check*

## PROBLEMS

1. Three 20:1 single-phase transformers are delta-delta connected to a 2400-V three-phase line. (*a*) Find the secondary line voltage. (*b*) Find the secondary line voltage if the connection is made wye-wye.
2. A 50:1 step-down delta-wye transformer bank is supplied from a 13,800-V three-phase system. Find (*a*) the primary coil voltage, (*b*) the secondary coil voltage, (*c*) the voltage to the neutral on the secondary, and (*d*) the secondary line voltage.
3. A transformer bank consists of three 100-kVA single-phase transformers, each rated at 4800/240 V. The bank is supplied from a 4800-V three-phase, three-wire line and delivers 240 V to a

three-phase, three-wire secondary system. (*a*) Describe the type of connection. (*b*) If a balanced load draws 150 kW at 75 percent lagging power factor from the bank, find the secondary line current.

4. A motor draws 50 kW at 80 percent power factor from a 240-V three-phase line. If the motor is supplied from a 4160/240-V wye-delta step-down transformer bank, find (*a*) the secondary line current, (*b*) the secondary coil current, (*c*) the primary coil voltage, and (*d*) the primary line current.
5. A three-phase 10:1 transformer bank has its windings connected wye-delta. The bank is fed from a 4160/2400-V three-phase line and delivers 50 A per terminal on the secondary side. Find (*a*) the secondary line voltage, (*b*) the total kilovoltampere output, (*c*) the secondary coil current, (*d*) the primary line current, and (*e*) the total kilovoltampere input.
6. A delta-wye transformer bank is used to transform 2000 kVA at 90 percent power factor from 13,800 V to 4000 V. If the bank supplies a 4000-V three-phase, four-wire distribution system, find (*a*) the voltage ratio of each transformer, (*b*) the primary coil current, (*c*) the secondary coil current, and (*d*) the voltage to the neutral on the secondary.
7. A three-phase 60-Hz transformer bank is connected delta-wye and steps up the primary line voltage of 13,800 V to the secondary line voltage of 119,500 V. The bank is made of three single-phase transformers, each rated at 25,000 kVA. Find the following for the rated kilovoltampere load: (*a*) the primary line current, (*b*) the primary coil current, (*c*) the transformer secondary coil voltage, (*d*) the primary coil voltage, (*e*) the voltage ratio of each transformer, and (*f*) the secondary line current.
8. A 480-V three-phase alternator is delta-connected and feeds a transformer rated at 100 kVA, 480 V, and 60 Hz. The transformer is connected delta-wye and feeds a three-phase, three-wire 208-V distribution line. If a heater unit with 20 Ω in each section is connected in wye across the secondary, find (*a*) the secondary line current and (*b*) the primary line current.
9. Each of the three single-phase transformers in a bank connected wye-delta is rated at 2000 kVA. The bank is supplied from a three-phase, four-wire 60-Hz 60,900-V line. If the voltage ratio is 8:1, find (*a*) the primary coil voltage, (*b*) the secondary coil voltage, (*c*) the total kilovoltampere rating, (*d*) the primary line current, (*e*) the primary coil current, (*f*) the secondary coil current, (*g*) the secondary line current, and (*h*) the kilovoltampere output.
10. A factory is serviced from a 2400-V three-phase, three-wire line. Three 50-kVA single-phase transformers, each rated at 2400/240 V, are used to supply the factory's 240-V three-phase three-wire system. If the load on the system is 100 kW at 80 percent power factor lagging, find (*a*) the primary line voltage, (*b*) the primary winding voltage, (*c*) how the transformer primaries must be con-

nected, (*d*) the secondary winding voltage, (*e*) the secondary line voltage, (*f*) how the transformer secondaries must be connected, (*g*) the secondary line current, (*h*) the secondary coil current, (*i*) the voltage ratio, (*j*) the primary coil current, (*k*) the primary line current, (*l*) the total kilovoltampere input, (*m*) the total kilovoltampere output, and (*n*) the total kilovoltampere rating.

11. An 18,000-kVA 6600-V, 60-Hz three-phase, wye-connected alternator supplies its rated load at rated voltage to a transmission line at 90 percent power factor lagging. A delta-wye transformer bank then steps up the voltage to 33,000 V for transmission. At the end of the line, the voltage is stepped down to 13,800 V with a delta-delta transformer bank. Find (*a*) the coil voltage of the alternator, (*b*) the coil current of the alternator, (*c*) the voltage and current of the primaries of the step-up transformer, (*d*) the voltage and current of the secondaries of the step-up transformer, (*e*) the voltage and current of the primaries of the step-down transformer, (*f*) the voltage and current of the secondaries of the step-down transformer, and (*g*) the kilovoltampere rating of each transformer coil.
12. The power generated by a three-phase, 20,000-kVA 60-Hz wye-connected alternator at 6600 V is transformed by a delta-wye-connected transformer bank for transmission at 66,000 V. The voltage at the end of the transmission line is then stepped down by a delta-delta bank to 13,200 V. Find (*a*) the voltage and current of the primaries of the step-up transformer, (*b*) the voltage and current of the secondaries of the step-up transformer, (*c*) the voltage and current of the primaries of the step-down transformer, (*d*) the voltage and current of the secondaries of the step-down transformer, and (*e*) the kilovoltampere rating of each transformer coil.
13. A substation receives 345 kW of three-phase power at 13,800 V. This voltage is stepped down to 2300 V by two open-delta-connected transformers. If the load power factor is 100 percent, find (*a*) the minimum rating of each transformer, (*b*) the voltage and current rating of each primary winding, (*c*) the voltage and current rating of each secondary coil, and (*d*) the power factor at which each winding operates.

(See Instructor's Manual for Test 22-1, Three-Phase Transformer Connections)

# MATHEMATICS FOR LOGIC CONTROLS

## JOB 23-1 INTRODUCTION

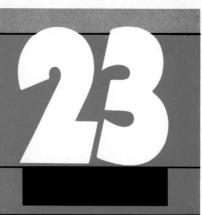

The very rapidly expanding field of microelectronics is causing major changes in many fields of electricity and electronics. Sophisticated control systems are being added to such entities as elevators, energy systems, and home lighting and power control, not to mention the rapidly growing area of robotics and other manufacturing and industrial controls. Hardly anyone is not being affected by the new forms of binary signals and controls. The intent of this chapter is to introduce the basic concepts of the mathematics of logic that are used in these modern control systems.

The field of logic and binary arithmetic is at such a primitive stage that we still have the advantage of starting with very basic concepts and learning and applying from that point. At first you may feel that this is a drastically new form of mathematics, but you will find out that it is actually very similar to all the mathematics that has preceded this chapter. There are rules, just like algebra, and if these rules are followed, you have fewer chances to make errors.

What we are learning is a means to analyze logical problems. This is not like the previous mathematics in which numbers represent the size of components or the value of a voltage.

In modern controls we find two common types of components used to build control circuits. One type is composed of switches and relays, while the other contains modern solid-state electronics. The former could be called "switching circuits" and the latter, "digital logic circuits."

In switching circuits the primary emphasis is on the continuity or "connectedness" of the circuit. The analysis of these circuits more closely resembles the thinking of an electrician who wires a system. The electrician follows the lines on a wiring diagram to determine where it comes from and where it goes. If the line is continuous from a switch to a motor, the electrician knows that the switch controls the motor.

The second class of circuits, digital logic, is usually voltage-level-dependent. This means that the primary interest is on various control signals and what their voltage is. There is less physical resemblance between these circuits and the "connected" switching circuits. Both types of circuit are used in modern electronics, sometimes in the same system. We shall study the fundamentals of both and see how they differ and how they are the same.

### BINARY CONCEPTS

In all the previous mathematics that has been studied in this book, we have become accustomed to the use of voltage, current, and resistance in problems. Each of these quantities has been used in equations and could have different values depending on the problem. These quantities are called *variables*. Resistors, capacitors, and inductors come in many sizes. This means that there must be many numbers so that we can tell just what size they are. Because we place no limits on their size, we need a number system that can handle very large and very small values.

The fundamental characteristic of the modern mathematics is that there are only two states for each variable that we use. This goes to the other extreme of complexity. Instead of having many numbers for a variable in this new type of mathematics, we require only two numbers. This is called *binary mathematics*.

We can select any two binary conditions. Some common ones that are used are yes-no, true-false, connected–not connected, closed–open, high-voltage–low-voltage, up–down, and heads–tails. Rather than develop a yes–no or an up–down system of mathematics, we use numbers to represent each of these conditions; thus false = 0 and true = 1. We can then develop a single type of binary mathematics and apply it to any of the above sets of conditions. All we have to do is substitute the descriptive terms we want for the numbers 0 and 1.

SELF-TEST 23-1

1. In the field of industrial controls two classes of logic circuit are ______ and ______. | switching circuits; digital logic circuits
2. Switching circuits emphasize ______ or ______. | continuity connectedness
3. Digital circuits are usually dependent on ______. | voltage
4. ______ systems have only two states or values. | Binary

## JOB 23-2 SIMPLE SWITCHES

Consider a simple switch and see how it can be analyzed as a binary device. Figure 23-1 shows the circuit that we want to analyze. A battery is connected through a switch to a light. We can consider this problem in various ways. We could say the light is on or off, or the battery is turned on or turned off, or the circuit is open or closed, or the lamp is connected or not connected. These are all binary types of description of the system operation that we use in our common language or thought. They in no way confuse us. We shall now use these common concepts in a binary form of mathematical analysis.

In this problem there are two variables, the switch and the lamp. Call

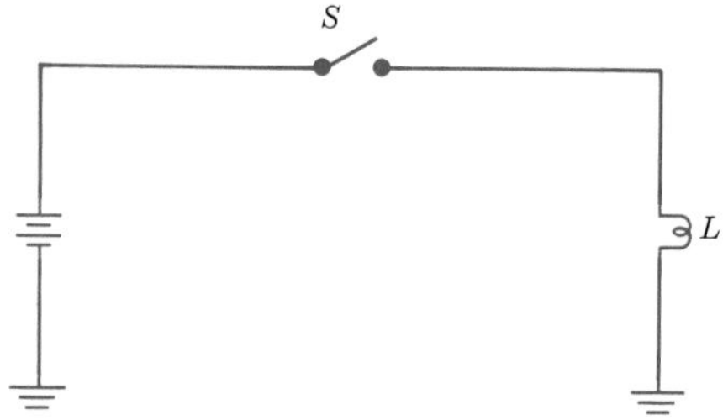


**FIGURE 23-1**
The basic switch-lamp circuit.

the variable associated with the switch $S$ and that with the lamp $L$. We will say that if the switch is closed, then $S = 1$, and if the lamp is on, then $L = 1$. We thus have:

$$S = \begin{array}{l} 0\text{—switch open} \\ 1\text{—switch closed} \end{array}$$

$$L = \begin{array}{l} 0\text{—lamp off} \\ 1\text{—lamp on} \end{array}$$

**RULE** In a switching circuit an open switch is represented by a 0 and a closed switch is represented by a 1.

In the example above the output of the circuit is the lamp. If the lamp is on, there is current flowing through it and the variable is given the value 1.

**RULE** In a switching circuit an output that has current flowing through it has the logic value 1. If no current flows through it, it has the logic value 0.

We can see from the circuit that if $S = 1$, then $L = 1$. In words, this means that if the switch is closed ($S = 1$), the lamp is on ($L = 1$). We also see that if $S = 0$, then $L = 0$. In terms of our variables in this problem, the relationship can be stated mathematically as $S = L$.

Care must be taken in how the relationships are defined. The relationship $S = L$ is not true if we invert either one of the definitions for $L$ or $S$. For example, if we decide that $S = 1$ means that the switch is open rather than closed, $S$ is *not* equal to $L$ as above. There are some cases in which people have chosen to use this different definition of $S$. Always be sure that the definitions are the same as those that are being used here.

## THE AND FUNCTION

The first of the basic logic functions is the AND function. The word "and" is used in our common language, and we have no problem with it.

There is no change at all when we go to the arithmetic of logic. Figure 23-2 shows a circuit with two single-pole–single-throw switches in series. Anyone with a basic understanding of circuits knows that both $S_1$ *and* $S_2$ must be closed so that the lamp will light. This is all that there is to the AND function. It is a mathematical tool that helps us to describe and work with the concept of AND.

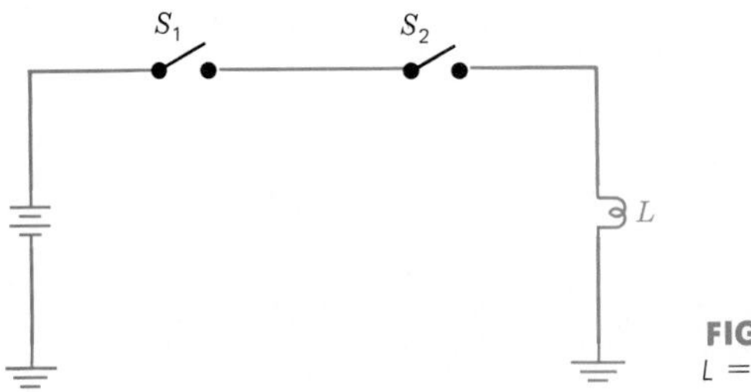


**FIGURE 23-2**
$L = S_1$ AND $S_2$.

Using the above conventions for the values of $S$ and $L$, we can now tabulate the values as shown in table 23-1.

**TABLE 23-1**
**THE AND FUNCTION**

| $S_1$ | $S_2$ | $L$ |
|---|---|---|
| 0 | 0 | 0 |
| 0 | 1 | 0 |
| 1 | 0 | 0 |
| 1 | 1 | 1 |

Table 23-1 merely states that the light will be on ($L = 1$) only if both switches are closed ($S_1 = 1$ AND $S_2 = 1$). The light will be off ($L = 0$) in any other combination of switch positions.

The mathematical symbol for the AND function is a dot $\cdot$. The equation relating $L$ to $S_1$ and $S_2$ is written as:

$$L = S_1 \cdot S_2$$

This dot symbol is sometimes used to represent multiplication in arithmetic. Multiplication helps us to remember the above AND table (Table 23-1); the value of $L$ is the product of the variables $S_1$ and $S_2$. This is just like the rules of algebra that we are accustomed to; 1 times 1 is 1, 1 times 0 is 0.

Now that we have an introduction to the AND function, we shall study some of its properties. It is obvious from the circuit diagram that it does not make any difference whether $S_1$ comes before $S_2$ or after $S_2$. In either position the operation of the switches and the lamp are exactly the same. This is also true for the mathematical equation:

$$L = S_1 \cdot S_2 = S_2 \cdot S_1$$

**RULE** The AND function is independent of the order in which the variables are listed.

Another important property of the AND function can be seen from Fig. 23-2. If switch $S_1$ is open, it is irrelevant whether switch $S_2$ is open or closed, the lamp will not light. In other words, if $S_1 = 0$, the AND of $S_1$ and $S_2$ is 0. As an equation, this would be written as:

$$S \cdot 0 = 0$$

this gives the following rule:

**RULE** The AND function of "0" and any variable is 0.

Another important rule for the AND function can be seen from Fig. 23-2 also. If switch $S_2$ is closed, the lamp depends only on $S_1$. If $S_1$ is closed, the lamp is on. If $S_1$ is open, the lamp is off; that is, $L = S_1$. This can be written as the equation:

$$S_1 \cdot 1 = S_1$$

This gives the following rule:

**RULE** The AND function of any variable and a 1 is only that variable.

Another similar property of the AND function can be seen from Fig. 23-2. Assume that $S_1$ is a double-pole single-throw switch and that the second pole $S_1$ is switch $S_2$ as in Fig. 23-2. This would give the logic function $S_1 \cdot S_1$. Obviously, the second section of the switch serves no real purpose as far as the logical operation is concerned. The equation for the lamp is $L = S_1$. The equation for this is:

$$S_1 \cdot S_1 = S_1$$

We therefore have another rule for the AND function.

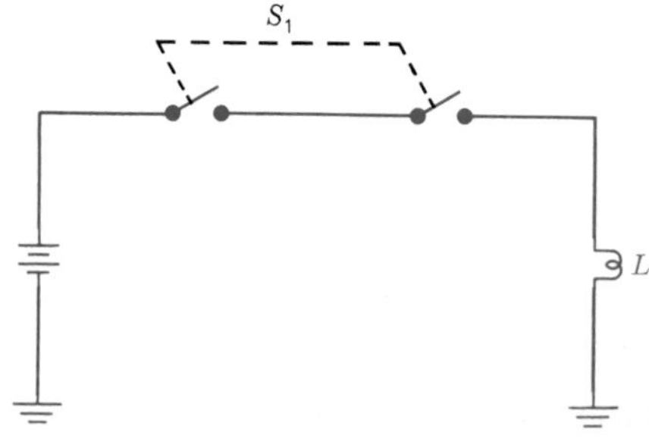


**FIGURE 23-3**
$L=S_1$ AND $S_1$

S

**RULE** The AND function of any variable and itself is only that variable.

SELF-TEST 23-2

1. When analyzing a switching circuit, the convention we use is that the logic value for a closed switch is ______ and an open switch is ______. | 1 0
2. When an output is active or on, we say that it has a logic value ______. | 1
3. When an output is inactive or off, we say that the logic value is ______. | 0
4. For a logical AND function we (may/may not) take the variables in any order that we please. | may
5. The AND function is (dependent/independent) of the order of the variables. | independent
6. 0 AND any variable is ______. | 0
7. 1 AND any variable is ______. | that variable

**PROBLEMS**

1. For the following problems (*a*, *b*, and *c*), what is the value of $L$ for the given values of $S_1$ and $S_2$ if $L = S_1 \cdot S_2$?
   *a*. $S_1 = 0$ AND $S_2 = 1$.
   *b*. $S_2 = 1$ AND $S_1 = 1$.
   *c*. $S_2 = 0$ AND $S_1 = 0$.
2. Give a circuit diagram for the logic function $S_1 \cdot S_2 \cdot S_3 = 1$.
3. What is the equation for the circuit in Prob. 2 if $S_3$ is permanently closed?
4. What is the equation for the circuit in Prob. 2 if $S_3$ is permanently open (broken)?

## JOB 23-3 TRUTH TABLES

Table 23-1 described the AND function exactly. It listed every possible combination of the input or independent variables ($S_1$ and $S_2$). Independent variables are those variables such as switches that we can set to any value as we please or as the system pleases. The dependent variables are those whose values are determined by other variables in the system. For our problem, the dependent variable is $L$. Tables of the form shown above are often used to describe any logic function. For this reason they are given the special name "truth table." Note the following facts about a truth table:

1. It has every possible combination of the input variables.
2. The combinations of input variables are listed in an ordered manner on the left of the table.
3. The output or independent variables are listed on the right adjacent to the combination of input variables that cause the output.

How many input combinations are there? Remember that each input can have either of two values (0 or 1). This means that $S_1$ can have two values and that for each of these values, $S_2$ can have two values; therefore, there are two times two or four possible input combinations. This is shown in Table 23-2.

**TABLE 23-2**
**COMBINATIONS OF TWO VARIABLES**

| $S_1$ | $S_2$ |
|---|---|
| 0 | 0 |
| | 1 |
| 1 | 0 |
| | 1 |

Suppose that we had a third variable, $S_3$. For each of the two possible values for $S_3$, the other two variables could have any of their four possible combinations as shown in Table 23-3. The number of combinations is again doubled.

**TABLE 23-3**
**COMBINATIONS OF THREE VARIABLES**

| $S_3$ | $S_2$ | $S_1$ |
|---|---|---|
| | 0 | 0 |
| 0 | 0 | 1 |
| | 1 | 0 |
| | 1 | 1 |
| | 0 | 0 |
| 1 | 0 | 1 |
| | 1 | 0 |
| | 1 | 1 |

The general rule is that each additional independent variable doubles the number of combinations; in other words, the total number of input combinations is $2^N$, or two times two $N$ times. Table 23-4 below shows the eight combinations for three variables.

**TABLE 23-4**
**COMBINATIONS OF THREE VARIABLES**

| No. | $S_1$ | $S_2$ | $S_3$ |
|---|---|---|---|
| 0 | 0 | 0 | 0 |
| 1 | 0 | 0 | 1 |
| 2 | 0 | 1 | 0 |
| 3 | 0 | 1 | 1 |
| 4 | 1 | 0 | 0 |
| 5 | 1 | 0 | 1 |
| 6 | 1 | 1 | 0 |
| 7 | 1 | 1 | 1 |

This leads to an important rule for the number of different states that a logic circuit can have.

| **RULE** | The number of different combinations of input variables is $2^N$, where $N$ is the number of input variables. |
|---|---|

The practicality of a truth table soon comes to an abrupt halt. For three variables there are 8 combinations, for 4 variables there are 16, for 5 variables 32, and so on. These numbers may be all right for computers, but people start to tire with 5, 6, or more variables. This still leaves us with a lot of useful applications and at least a very practical concept for problem solution.

What is the order in which the values of the independent variables are listed? It is important to have some orderly manner to list the different combinations so that none will be overlooked and so that it is easy to find any particular combination that is sought. The following is the accepted order in which to list the independent variables.

Start with the rightmost variable and give it the values 0 and then 1 while all those variables to the left are given the value 0. You then give the next variable to the left the value 1 and then repeat the combinations for those to the right. This process continues to the left until all variables have been used.

For those of you who are familiar with the above number scheme, it is called the *binary number system*. The combination of $S_1$, $S_2$, and $S_3$ corresponds to the decimal number listed in the number (leftmost) column.

The digit on the right is called the least significant digit (LSD). Like the rightmost number on the gasoline pump, it is the one that changes value the most rapidly. The digit on the left is the most significant digit (MSD). It changes the least rapidly.

| **RULE** | On a truth table the input variables are arranged in the binary number sequence. |
|---|---|

SELF-TEST 23-3

| Question | Answer |
|---|---|
| 1. Variables that can be set to any desired value are called ______. | independent variables |
| 2. Variables that are determined by other variables are called ______. | dependent variables |
| 3. The total number of input combinations for N input variables is ______. | $2^N$ |
| 4. The order in which the variables are listed in a truth table is called the ______ ______ system. | binary number |
| 5. The abbreviation MSD means ______ ______ ______. | most significant digit |
| 6. The abbreviation LSD means ______ ______ ______. | least significant digit |
| 7. The (MSD/LSD) changes the most rapidly. | LSD |
| 8. The (MSD/LSD) changes the least rapidly. | MSD |

## PROBLEMS

1. How many possible combinations are there for three single-pole switches?
2. List the combinations for two switches $S_1$ and $S_2$.
3. If a system has 16 combinations of inputs, how many switches does it have?
4. A circuit has a limit switch, an on/off switch, and an over temperature switch. How many combinations of control switches can it have?

(See Instructor's Manual for Test 23-1, Elementary Functions.)

## JOB 23-4 THE OR FUNCTIONS

We are now ready to look at another logic function called the OR function. Sometimes this function is called the INCLUSIVE OR function. The reasons for this will become apparent later. Here again mathematics is nothing more than an orderly expression of concepts that are familiar to us. Refer to the circuit in Fig. 23-4. If $S_1$ is on the front door and $S_2$ is on the back door, the light will go on if either $S_1$ OR $S_2$ is closed. In other words, the light will go on if *either* the front door OR the back door is opened.

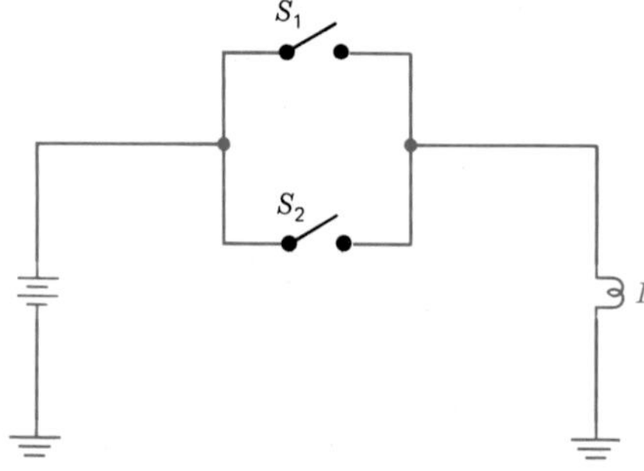


**FIGURE 23-4**
$L = S_1$ OR $S_2$

The truth table for the OR function is given in Table 23-5.

**TABLE 23-5**
**TRUTH TABLE FOR THE OR FUNCTION**

| $S_1$ | $S_2$ | $L$ |
|---|---|---|
| 0 | 0 | 0 |
| 0 | 1 | 1 |
| 1 | 0 | 1 |
| 1 | 1 | 1 |

The mathematical symbol for the OR function is +. The rules for the OR function are similar to the rules of addition, and the + sign helps us to remember this. We have $0 + 0 = 0$ and $1 + 0 = 1$. The only unusual value is the entry $1 + 1 = 1$. In this case we resort to the fact that this is a mathematical function of logic and not numbers. A 1 means that in some manner the value desired is present. Logically speaking, in Table 23-5, the 1 under $S_1$ means that we have $S_1$, or that $S_1$ is closed. Similarly, the 1 under $S_2$ means that $S_2$ is closed; therefore, the 1 under $L$ should mean that we have the lamp (the lamp is on). According to this reasoning, $1 + 1 = 1$.

Another property of the OR function is that, as with the AND function, it does not make any difference which order we put the variables in. As can be seen from Fig. 23-4, interchanging $S_1$ and $S_2$ gives the same circuit performance. This means that we can write:

$$L = S_1 + S_2 = S_2 + S_1$$

This also is the same as algebra.

**RULE** The OR function is independent of the order in which the variables are taken.

A very important property of the OR function can be seen from Fig. 23-4. If switch $S_1$ is closed, the lamp will be on no matter what position $S_2$ is in; that is, $L$ is independent of $S_2$. This shows the desired property of the OR function; namely:

$$S_2 + 1 = 1$$

This gives the following rule:

**RULE** Any variable ORed with a 1 is a 1.

Another property of the OR function is that if switch $S_2$ is open, the lamp will depend on $S_1$ only. This means that:

$$L = S_1 + 0 = S_1$$

Another rule of the OR function is:

**RULE** The OR function of any variable and a 0 is that variable only.

We have now reached a point where the precision and clarity of mathematics helps to remove the confusion of everyday language. No one had any problem with the above example. There is a different use of the word "or." Suppose that there are two substitute players that can be put into a game. The manager wants Kline or Gross to go into the game depending on what the other team does. We know by experience and common sense that Kline and Gross will not both go into the game. Only one—not both. We used the same word "or," but we had a slightly different meaning. This meaning is called the EXCLUSIVE OR. It says that one, but not both alternatives can be used. The truth table for the EXCLUSIVE OR is shown in Table 23-6.

**TABLE 23-6**
**TRUTH TABLE FOR THE EXCLUSIVE OR**

| $S_1$ | $S_2$ | $L$ |
|---|---|---|
| 0 | 0 | 0 |
| 0 | 1 | 1 |
| 1 | 0 | 1 |
| 1 | 1 | 0 |

Notice that this is exactly like the OR function except for the last line. In the last line when both $S_1$ and $S_2$ are true, $L$ is not true.

It should be expected that the symbol for the EXCLUSIVE OR function should be similar to that for the OR functions. The EXCLUSIVE OR symbol is $\oplus$.

It is now apparent why the OR function is sometimes called the INCLUSIVE OR as opposed to the EXCLUSIVE OR. The INCLUSIVE OR includes the situation in which you have both alternatives. The EXCLUSIVE OR function excludes that case.

### THE INVERT FUNCTION

The last basic function is the INVERT function. It is sometimes called the complement function. This is really the easiest of all of the functions. It merely inverts the variable, that is, it changes the value from a 1 to a 0 or from a 0 to a 1. If $S_1$ is true then the inverter makes it false and vice versa. This is a single-variable function, and its truth table looks like that shown in Table 23-7.

**TABLE 23-7**
**TRUTH TABLE FOR THE INVERT FUNCTION**

| $S_1$ | $L$ |
|---|---|
| 0 | 1 |
| 1 | 0 |

The mathematical symbol for inversion is a bar over the variable as $\overline{S}_1$. It is read "$S_1$ bar." It is also written $S_1'$. This is read "$S_1$ prime."

A warning is necessary about the notation of inverted quantities. If you have the term $(S_1 \cdot S_2)'$, it means that you first take $S_1$ AND $S_2$ and then take the complement of the result of that AND function. For example, if $S_1 = 1$ and $S_2 = 1$, then $S_1 S_2 = 1$. This gives $(1 \cdot 1)' = 1' = 0$.

The parentheses must be used because $(S_1 S_2)'$ is not equal to $S_1 S_2'$. If you let $S_1 = 0$ and $S_2 = 1$, then:

$$(S_1 \cdot S_2)' = (0 \cdot 1)' = 0' = 1$$

while

$$S_1 \cdot S_2' = 0 \cdot 1' = 0 \cdot 0 = 0$$

These two values are not the same.

Note also that it is not equal to $S_1' S_2'$. You can use truth tables to show that $(S_1 \cdot S_2)' = S_1' + S_2'$.

SELF-TEST 23-4

1. The two types of OR function are ______ and ______. — inclusive exclusive
2. The OR functions are (dependent/independent) of the order of the variables. — independent
3. Any variable OR 1 is ______. — 1
4. Any variable OR 0 is ______. — that variable
5. The OR function truth table is like addition except that 1 + 1 = ______. — 1
6. The EXCLUSIVE OR function is like the OR function except that 1 + 1 = ______. — 0

**PROBLEMS**

1. How many input combinations are there for a truth table with four independent variables?
2. Give the truth table for $S_1 S_2 + S_3$.
3. Construct a circuit for the equation in Prob. 2.
4. Construct a truth table for the function $(S_1 S_2)' + S_3$.
5. Show that the truth table for $(S_1 S_2)'$ is the same as the truth table for the function $S_1' + S_2'$.

## JOB 23-5 SUMMARY OF THE ELEMENTARY FUNCTIONS

The following functions have been introduced: AND, OR, EXCLUSIVE OR, and INVERT. The truth tables for these functions are given in Table 23-8.

**TABLE 23-8**
**TRUTH TABLES FOR THE ELEMENTARY FUNCTIONS**

| AND | | | INVERT | | OR | | | EXCLUSIVE OR | | |
|---|---|---|---|---|---|---|---|---|---|---|
| $S_1$ | $S_2$ | $L$ | $S_1$ | $L$ | $S_1$ | $S_2$ | $L$ | $S_1$ | $S_2$ | $L$ |
| 0 | 0 | 0 | 0 | 1 | 0 | 0 | 0 | 0 | 0 | 0 |
| 0 | 1 | 0 | 1 | 0 | 0 | 1 | 1 | 0 | 1 | 1 |
| 1 | 0 | 0 | | | 1 | 0 | 1 | 1 | 0 | 1 |
| 1 | 1 | 1 | | | 1 | 1 | 1 | 1 | 1 | 0 |

Rules of the AND function are as follows:

**RULE** The AND function is independent of the order in which the variables are listed.

**RULE** The AND function of 0 and any variable is 0.

**RULE** The AND function of any variable and a 1 is only that variable.

**RULE** The AND function of any variable and itself is only that variable.

Rules of the OR function are as follows:

**RULE** The OR function is independent of the order in which the variables are taken.

**RULE** Any variable ORed with a 1 is a 1.

**RULE** The OR function of any variable and a 0 is only that variable.

Facts about a truth table are as follows:

1. It has every possible combination of the input variables.
2. The combinations of input variables are listed in an ordered manner on the left of the table.
3. The output or independent variables are listed on the right adjacent to the combination of input variables that cause the output.

**RULE** The number of different combinations of input variables is $2^N$, where $N$ is the number of input variables.

**RULE** On a truth table the input variables are arranged in the binary number sequence.

## JOB 23-6 SYSTEMS OF ELEMENTARY FUNCTIONS

This section starts to build complex networks out of the elementary logic building blocks that have been learned in the earlier sections. It will show how logic equations for switching circuits can be obtained from the "paths" of the circuit. This means that you can take a diagram of a control circuit and trace all the paths with your finger and find the logic equation for the circuit.

The first step toward understanding the operation of a complex system is to break it into segments or blocks of elementary functions. Refer to Fig. 23-5. This circuit is the combination of the AND and the OR circuits that we analyzed previously. The two switches that are in parallel compose the function $S_2 + S_3$. The result of this composition is then in series with $S_1$, or in other words it is ANDED with $S_1$. This gives the equation for $L$ as:

$$L = S_1(S_2 + S_3)$$

In this equation the dot that represents AND was not used. By common usage it is not used when there is no confusion as to the meaning. This is just the same as the multiplication sign in arithmetic. It would not be wrong to write $S_1 \times (S_2 + S_3)$, but it is common to write $S_1(S_2 + S_3)$.

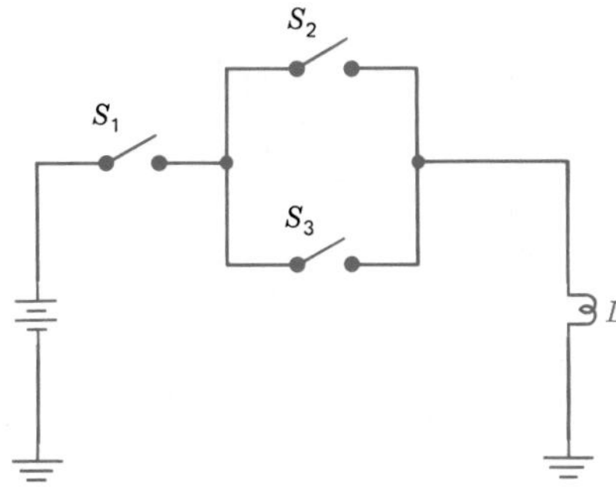


**FIGURE 23-5** $L = S_1(S_2 + S_3)$.

The equation may also be expanded in the same way that it could be in algebra, where we could write $S_1 \times (S_2 + S_3) = S_1 \times S_2 + S_1 \times S_3$. We can similarly write the equation for $L$ as:

$$L = S_1\,S_2 + S_1\,S_3$$

Again the dot symbol may or may not be used, depending only on whether there would be any confusion about the meaning.

When the conventions that have been chosen are used, there is a physical significance to the above equation for $L$. If you put your finger on the battery, the first term for $L$ represents a path that you can follow with your finger that goes from the battery through $S_1$ and $S_2$ and then through the lamp and back to the battery. In other words, the term $S_1 S_2$ represents a path from the battery through the network and the output and back again to the battery.

In a similar manner, the term $S_1 S_3$ is a second path through the network and the load and back to the battery. Again you can put your finger or a pencil on the battery, go through the network of switches, through the output (the lamp) and back to the battery.

In other words, we say that the lamp is on if $S_1$ is closed and $S_2$ is closed (path 1) OR if $S_1$ is closed and $S_3$ is closed (path 2). A path is a continuous line that goes from the power source through the logic elements and the output and back to the power source without intersecting itself. A path forms no loops.

When the equation for $L$ is expressed as above, it is in the form of the sum of paths (SOP). This could also mean the sum of products, as each term of the equation for $L$ is a set of products and the overall equation is the sum of these products. There is a slightly different definition of SOP in logic network theory; however, it is consistent with what we are saying.

Once the logic equation is obtained, it is easy to construct a truth table for the above circuit. In this case there are three input variables that can be set independently. These are $S_1$, $S_2$, and $S_3$. The output is the lamp $L$. In the truth table given in Table 23-9, the mathematical symbols have been included to emphasize the close resemblance between logic functions and conventional algebra.

**TABLE 23-9**
**TRUTH TABLE FOR $S_1(S_2 + S_3)$**

| $S_1$ | $\times$ | $(S_2$ | $+$ | $S_3)$ | $=$ | $L$ |
|---|---|---|---|---|---|---|
| 0 | $\times$ | (0 | + | 0) | | 0 |
| 0 | $\times$ | (0 | + | 1) | | 0 |
| 0 | $\times$ | (1 | + | 0) | | 0 |
| 0 | $\times$ | (1 | + | 1) | | 0 |
| 1 | $\times$ | (0 | + | 0) | | 0 |
| 1 | $\times$ | (0 | + | 1) | | 1 |
| 1 | $\times$ | (1 | + | 0) | | 1 |
| 1 | $\times$ | (1 | + | 1) | | 1 |

The only unusual entry is the last one. In conventional algebra the answer would be 2. There is no 2 in the binary system. All the other results are exactly the same as conventional algebra. It must be remembered that for the OR function $1 + 1$ is 1. When this is used, the conventional algebra can be used again.

It was shown that just as in algebra, there are alternate ways to write the equation. These alternative equations represent different sets of paths. Each equation also has a different physical representation.

Figure 23-6 shows the physical representation of the equation $L = S_1 S_2 + S_1 S_3$. If you make a truth table for this equation, you will find that it is exactly the same as the truth table for $L = S_1(S_2 + S_3)$. Check the operation of the system. Logically it is exactly the same. The difference is only in $S_1$. In the form of the equation for Fig. 23-6, $S_1$ is a double-pole single-throw switch.

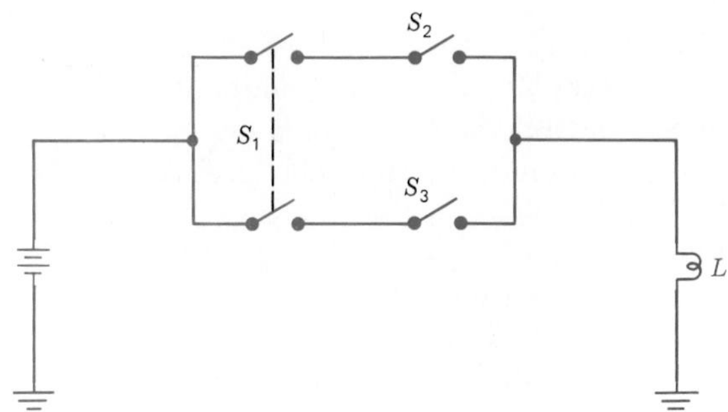


**FIGURE 23-6**
$L = S_1S_2 + S_1S_3$.

The result above can be very important. It shows that the fundamental intent of logic analysis is the solution of a logic problem, not a physical one. Solving a physical problem in the mathematics of logic can help you to realize new and different configurations of hardware to achieve the same result. This can be done by changing the form of the equation.

Sometimes with complicated circuits you may find that two circuits give the same logic equation. This means that they are identical in operation and differ only in form.

For another example, analyze the circuit shown in Fig. 23-7. The circuit can be broken into two sections. The one on the left is the function $S_1 + S_2$, while the one on the right is $S_3 + S_4$. The overall circuit is the AND function of these two sections. This gives the logic equation $L = (S_1 + S_2)(S_3 + S_4)$. This says that the lamp will light if $S_1$ or $S_2$ is closed AND $S_3$ or $S_4$ is closed.

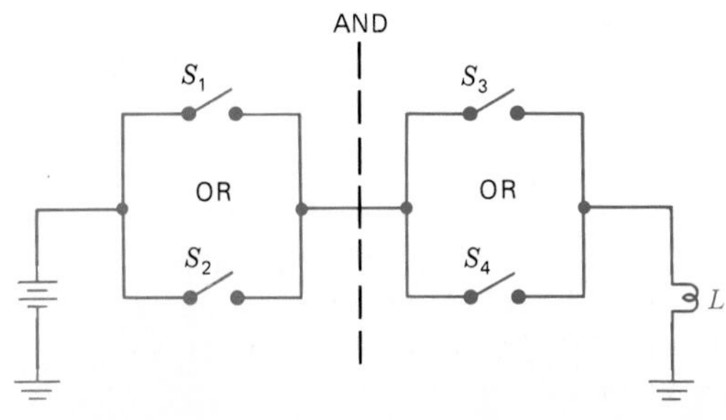


**FIGURE 23-7**
$L = (S_1 + S_2)(S_3 + S_4)$.

A word of caution is necessary here about the use of brackets in these logic equations. Many times people forget to use them and write the above equation as:

$$L = S_1 + S_2 S_3 + S_4$$

The problem with this is that there can be confusion about the terms in the equation. The above equation really has only three terms; $S_1$, $S_2 S_3$, and $S_4$. This is not the correct equation for $L$. The correct equation has four terms in it. When properly expanded the equation for $L$ is:

$$\begin{aligned} L &= (S_1 + S_2)(S_3 + S_4) \\ &= S_1(S_3 + S_4) + S_2(S_3 + S_4) \\ &= S_1 S_3 + S_1 S_4 + S_2 S_3 + S_2 S_4 \end{aligned}$$

The expansion of the above equation is done in exactly the same manner that algebra removes parentheses. Each term in parentheses on the left is multiplied by each term in parentheses on the right. In this equation $S_1$ is multiplied by $S_3$ and then $S_4$, the terms in the right-hand parentheses. The next term in the left-hand parentheses, $S_2$, is then multiplied by the terms in the right-hand parentheses (Table 23-10).

**TABLE 23-10**
**TRUTH TABLE FOR $L = (S_1 + S_2)(S_3 + S_4)$**

| $S_1$ | $S_2$ | $S_3$ | $S_4$ | $L$ | Path |
|---|---|---|---|---|---|
| 0 | 0 | 0 | 0 | 0 | |
| 0 | 0 | 0 | 1 | 0 | |
| 0 | 0 | 1 | 0 | 0 | |
| 0 | 0 | 1 | 1 | 0 | |
| 0 | 1 | 0 | 0 | 0 | |
| 0 | 1 | 0 | 1 | 1 | $S_2 S_4$ |
| 0 | 1 | 1 | 0 | 1 | $S_2 S_3$ |
| 0 | 1 | 1 | 1 | 1 | $S_2 S_3 + S_2 S_4$ |
| 1 | 0 | 0 | 0 | 0 | |
| 1 | 0 | 0 | 1 | 1 | $S_1 S_4$ |
| 1 | 0 | 1 | 0 | 1 | $S_1 S_3$ |
| 1 | 0 | 1 | 1 | 1 | $S_1 S_3 + S_1 S_4$ |
| 1 | 1 | 0 | 0 | 0 | |
| 1 | 1 | 0 | 1 | 1 | $S_1 S_4 + S_2 S_4$ |
| 1 | 1 | 1 | 0 | 1 | $S_1 S_3 + S_2 S_3$ |
| 1 | 1 | 1 | 1 | 1 | $S_1 S_3 + S_1 S_4 + S_2 S_3 + S_2 S_4$ |

Again it is obvious that each of the terms in the equation corresponds to a path in the network. The lamp would be lighted if any one of the terms in the equation was a 1. It would also light if any two of the paths were connected or if all four were. The truth table (Fig. 23-10) shows that the 1s in the $L$ column correspond to these situations.

An interesting point is that you cannot have a combination of three paths, only one, two, or four paths. Look at the circuit and see if you can reason why this is true.

SELF-TEST 23-5

1. The first step toward analysis of complex systems is to break the network into ______ or ______. | segments blocks
2. Each term of the logic equation for a switching circuit corresponds to a ______ of the circuit. | path
3. Logic equations may be ______ just as though they were algebraic equations. | expanded
4. Alternative forms of the logic equations correspond to ______ forms of the hardware. | alternative
5. Circuits that have equivalent logic equations perform ______ logic function. | the same
6. Each line of the truth table for which the output is a 1 corresponds to a ______ or a combination of ______ through the network and the output. | path paths

## PROBLEMS

1. Give a circuit for and list the paths of the equation

$$L = S_1 S_2 + S_1 S_3$$

2. Repeat Prob. 1 for the equation

$$L = S_1(S_2 + S_1)S_3$$

3. Repeat Prob. 1 for the equation

$$L = S_1 S_2 + (S_1 + S_3)$$

4. Determine whether there is any operational difference between the equation in Prob. 3 and the equation

$$L = S_1 + S_3$$

Compare the circuits and the truth tables for the two equations.

5. Repeat Prob. 1 for the equation

$$L = S_1(S_2 + S_1 + S_3)$$

(See Instructor's Manual for Test 23-2, Systems of Elementary Functions.)

## JOB 23-7 SWITCHING CIRCUIT INVERTER

The previous developments have been with single-pole or double-pole switches. A switch has been considered either open or closed. To implement the invert function for the switching circuit concepts, a single-pole–double-throw switch is used as shown in Fig. 23-8. The switch position that previously would have been considered open is now the complement position. To use the complement of a variable in a logic equation, a connection is made to the other pole of the switch. For

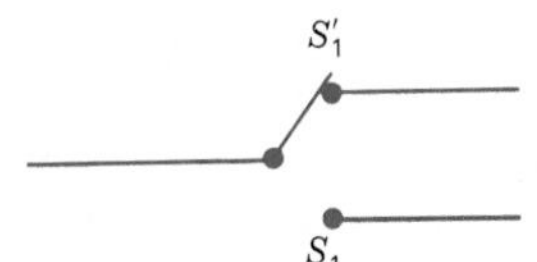


**FIGURE 23-8**
Switch arrangement for generation of a complement.

example, to implement the logic equation $L = S_1' S_2$, a circuit such as that shown in Fig. 23-9 would be used.

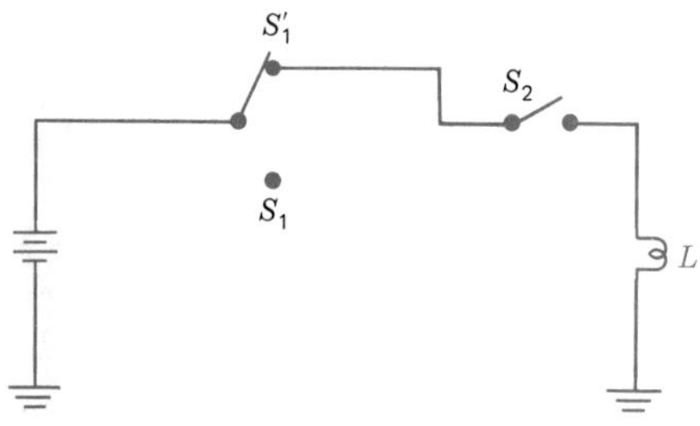


**FIGURE 23-9**
$L = S_1' S_2$.

An identity for the complement that is easy to see from a circuit point of view is $S + S' = 1$. Figure 23-10 shows the circuit for the equation $L = (S_1' + S_1)S_2$. The lamp in Fig. 23-10 will light no matter what position switch $S_1$ is in. The operation depends only on $S_2$. In logic arithmetic this is shown as:

$$L = (S_1 + S_1')S_2$$
$$= 1\,S_2 = S_2$$

from which the desired identity is obvious:

$$S + S' = 1$$

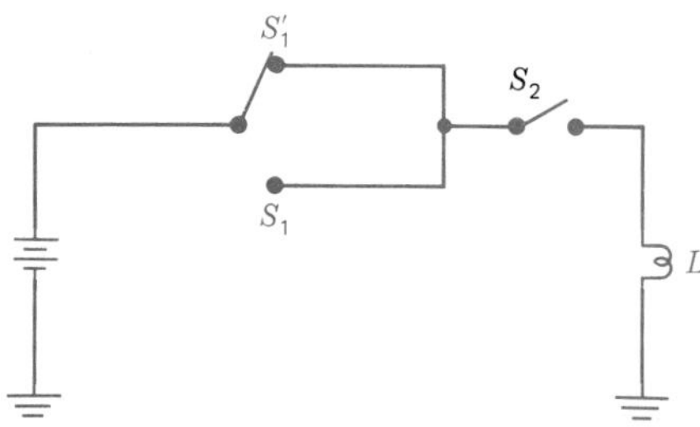


**FIGURE 23-10**
$L = (S_1' + S_1)\,S_2$.

**RULE** Whenever there is an OR combination of a variable and its complement, the result is always a 1.

## EQUATIONS USING THE COMPLEMENT

Logic is often used in such common locations that we never realize it is being used. The EXCLUSIVE OR can be written in logic equation form as:

$$L = S_1 S_2' + S_1' S_2$$

This can be derived from the paths that give the truth table for the EXCLUSIVE OR. Another way to reason it is to express the EXCLUSIVE OR in words. The lamp will light only if one switch is on and the other is off. The circuit for the above equation is shown in Fig. 23-11. As soon

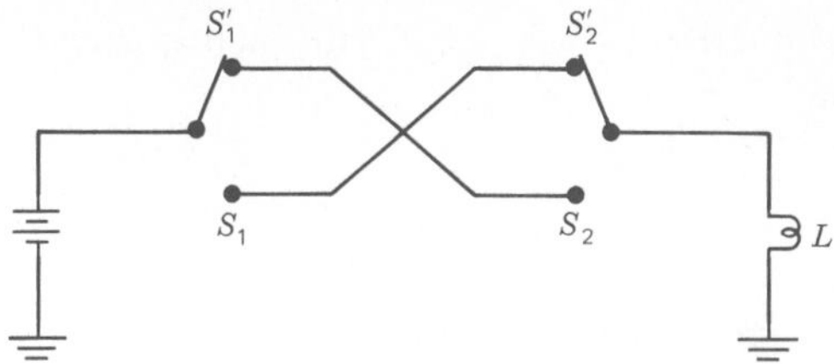


FIGURE 23-11
$L = S_1' S_2 + S_1 S_2'$.

as it is drawn out, you probably recognize this as the "three-way switch." The light can be turned on or off from the top of the stairs or from the bottom of the stairs.

By this time you should have a feeling for the close relationship between logic arithmetic and your everyday experiences. The purpose of this mathematics is to describe things that are familiar and common to you. The correctness of the fundamental rules and applications that we have seen is determined not by mysterious and obscure processes, but by what is common sense.

## PROBLEMS

1. Prepare a truth table for the function $L = S_1' (S_2 + S_3) + S_1$.
2. Make a truth table for the function $L_1 = S_1 S_2 + S_1' S_2'$.
3. Make a circuit diagram for the equation in Prob. 2.
4. Construct a truth table for the two-way switch shown in Fig. 23-10.
5. What is the relationship between the equation for $L$ in Prob. 4 and that for $L_1$ in Prob. 2? (the function $L_1$ is called an EXCLUSIVE NOR). Is there any logical difference in the operation of the two circuits? Does this make any difference to the user? Could he tell which circuit is being used?
6. Give a truth table for a "four-way switch"; that is, a light can be turned on or off from any of three possible locations. Start by assuming that for the switch positions $S_1'$, $S_2'$, $S_3'$ the lamp is on. For the truth table, let $S_1$ be the LSD and $S_3$ the MSD. (Hint: reason that from any given set of switch settings, if only one switch position is changed, the light will change state. This means that if the light was on it will go off and if it was off, it will go on.)

(See Instructor's Manual for Test 23-3, Switching Complements.)

## JOB 23-8 RELAYS IN LOGIC CIRCUITS

Many control circuits use conventional or solid-state relays to control higher-power circuits. The contacts on a relay are treated as any other switch contacts would be treated in a switching circuits problem. The primary difference is that the coil or actuator of the relay can be controlled by other logic variables; that is, the coil could be treated as the lamp was in the above problems. It would be considered as an output function and given a variable name such as $R_a$. The contacts that are associated with the relay are given the same name as the coil. This is

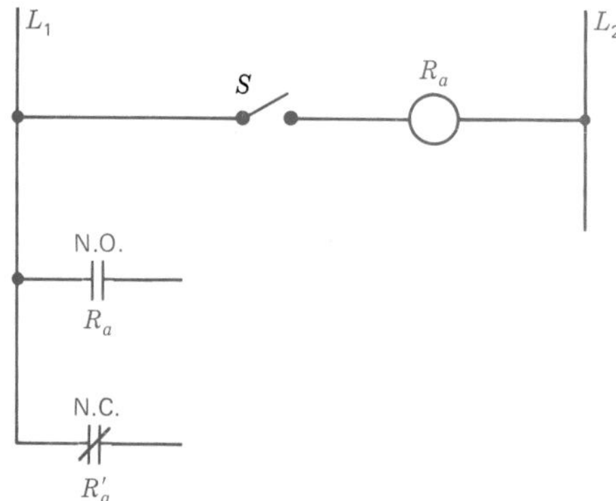


**FIGURE 23-12**
A switch-controlled relay showing relay-controlled contacts.

shown in Fig. 23-12, where the relay coil is shown actuated by switch $S$. Two sets of contacts are shown associated with relay $R_a$. One set is normally open (NO), and the other set is normally closed (NC). If the conventions that we have chosen are used, the normally open contacts would represent the variable $R_a$. When $R_a$ is turned on ($R_a = 1$), the contacts would be closed (= 1). The normally closed contacts, on the other hand, represent the complement of $R_a$. When the relay is not actuated ($R_a = 0$), the normally closed contacts would be closed ($R_a =$ 1). In relay control terminology, a relay that is energized is usually said to be picked up or picked. When the relay is turned off, it is said to be deenergized or dropped.

**RULE** The NO contacts of a relay represent the relay variable, and the NC contacts represent the complement of the relay variable.

Figure 23-12 shows only one set of contacts that are normally opened and one set that are normally closed. Relays are made with many combinations of contacts. The identification system for the more complicated devices would be similar to the above. The identification of the contacts is associated with the coil that actuates them. The separate contacts may be further marked, but by being able to associate contacts with coils the logic equations are easier to understand.

To further explain the relationships between word descriptions of operation, hardware, and logic equations, let's take an example of a simple motor control. The motor operates a garage door. Pushing a button turns the motor on and raises the door as long as the button is depressed. Releasing the button stops the door. To prevent the door from going off the track or the motor from becoming overstressed, there is a limit switch that stops the motor when the door is fully opeed. The door is very large, so in order to use a smaller switch, a relay is used to actuate the motor. This is a verbal description of the operation.

The above description could be reduced to a logical description of the operation. We want the relay to be energized if the button is pushed and the limit switch has not been opened. This means that we want an AND function in the operation. We are assuming that the limit switch is a normally closed type. This gives us a logic equation for the relay operation.

$$R_a = L_s S$$

Since $L_s$ is assumed NC under normal conditions, the value of the variable $L_s$ is 1. This makes the above equation

$$R_a = 1 S = S$$

The relay will actuate according to switch $S$ as long as the limit switch is closed. If switch $S$ is closed, the relay turns on. This is the desired operation.

The motor will actuate when the relay actuates; therefore

$$M = R_a$$

The logic equations say everything that the verbal description says but they are more compact and precise. The entire verbal description is exactly the same as the following set of logic equations:

$$R_a = L_s S$$
$$M = R_a$$

The above equations represent paths for the logic of the controls. The relay coil has a path $L_s S$ that goes through the limit switch, through the switch, and then through the coil of the relay. The motor is actuated by a path that goes through the NO set of contacts on the relay and then through the motor. These paths can be implemented as shown in Fig. 23-13. Notice how easy it was to take the equation, express it in paths, and then implement the logic.

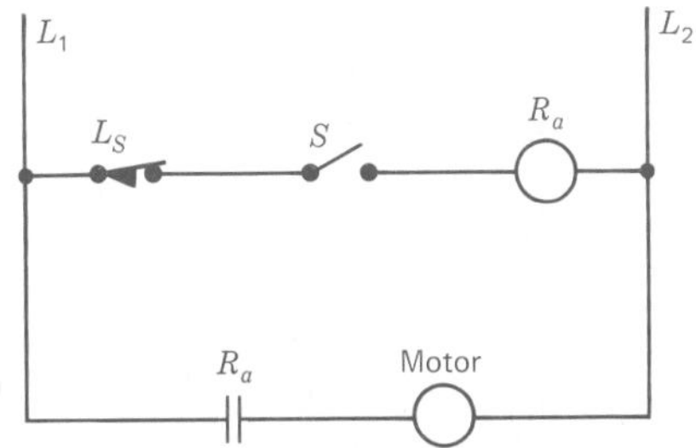


**FIGURE 23-13**
Diagram for relay-controlled motor.

Now modify the operation a little. We want the door to start to open as soon as the button is pushed, but when the button is released, we want the door to go to its fully opened position. What we are now saying logically is that we want the door to operate as before OR if it is once actuated to continue until the limit switch is reached. The logic equation is now:

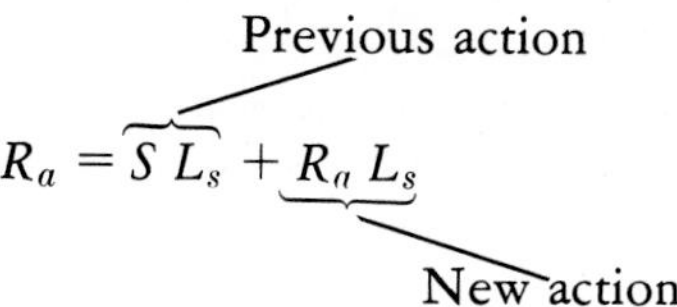


The second term is the new condition for operation. It says that if the relay is actuated and the limit switch is still closed, the relay will remain actuated. To add this new operation to the system, all that was necessary was to add a term to the equation.

The new equation could be implemented by adding a path that goes through the relay coil, through the limit switch, and then through a set of NO contacts on the relay. This looks as shown in Fig. 23-14. The limit switch is now a little more complicated than it was before. It is now a double-pole–single-throw switch. If the logic equation is rearranged a little, just as you would with an algebraic equation, we get the following form:

$$R_a = L_s\,(S + R_a)$$

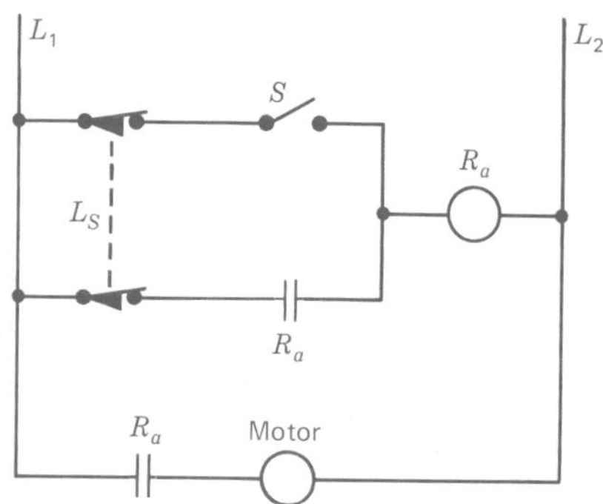


**FIGURE 23-14**
Alternate circuit for motor control.

In this form we have the combination of an AND and an OR circuit that has removed the necessity of the more complicated limit switch. The limit switch is now back to being a single-pole–single-throw switch. This gives the simplified configuration shown in Fig. 23-15.

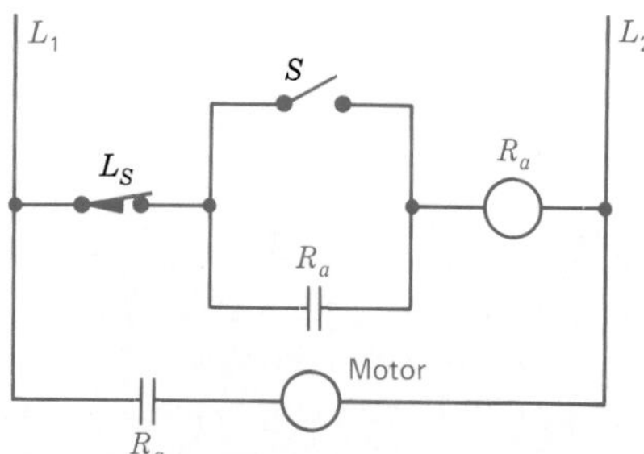


**FIGURE 23-15**
Alternate form for circuit shown in Fig. 23-14.

The operation of the circuit shown in Fig. 23-15 is called "latching" or "locking." This means that once the relay is energized, it keeps itself energized. This comes about because the state of the relay depends on itself; in other words, the relay variable is on both sides of the logic equation:

$$R_a = L_s(S + R_a)$$

Using the rules of the OR function, it can be seen that when the relay is not actuated ($R_a = 0$), the equation for $R_a$ is:

$$R_a = L_s(S + 0) = L_s\,S$$

In this condition or "state," the relay depends only on the limit switch and the control switch.

Once the relay becomes actuated ($R_a = 1$), the logic equation is:

$$R_a = L_s(S + 1) = L_s$$

In this state the relay depends only on the limit switch.

**RULE** If the state of a relay depends on its own state, that is, the relay variable is on both sides of the logic equation, the system has modes of operation.

This type of performance is very common in control circuits. Systems may have many "modes" of operation with the logic of operation changing depending on which "mode" the system is in. To the observer, the system looks like different systems, one in each mode of operation.

In the above example, "moding" or "latching" was desired. Sometimes in complicated systems this type of operation comes in by accident. A very unusual sequence of events may cause the system to get into the undesired mode, where unplanned operation can occur.

Another time that this can occur is when there is some type of equipment failure that puts the system into a form of operation other than what it was intended to be. Sometimes these faults can be very difficult to locate. One way is to look at the circuit diagram and make logic equations of the operation. After this is done, you can make different assumptions about parts that might be broken. Substitute 0s or 1s for variables according to the type of failure that you think they may have. Then look at the equations and see if they describe the type of operation that you are seeing. If this does not work, the problem is that the circuit is not like the diagram. Either something has failed in a manner that changes the circuit, or it was built incorrectly.

Now consider some additional problems of this type. While doing these problems try to improve your skills with the rules for the basic logic functions. Also try to learn the truth tables for these functions so that you don't have to keep referring to them. Another purpose of these examples is to help improve your skills in going directly from circuit diagrams to logic equations so that you can better understand the operation of the circuit.

The next example of a latching circuit is shown in Fig. 23-16. The circuit is supposed to be a self-latching relay that can be released with a release switch.

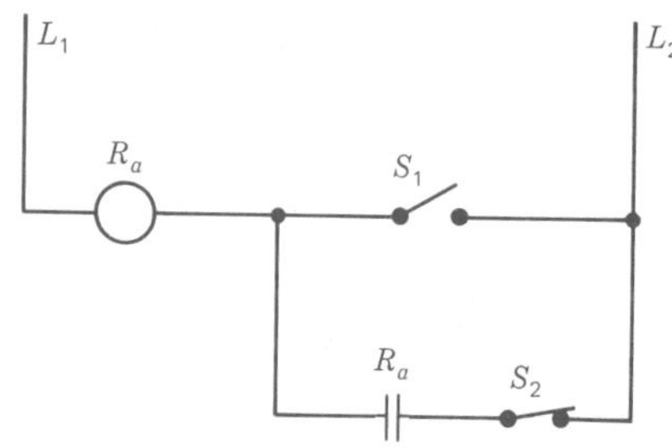


**FIGURE 23-16**
Self-latching circuit with release.

The contacts on the relay are drawn as NO contacts. This means that the function that the relay contacts represent is $R_a$. With this background on the intent of the circuit, determine the actual operation.

First list the paths that the network has. Start at the line $L_1$ and go through the relay coil. This now completes two important elements of the path requirement. We have gone from the source through the output. At the other side of the relay we can go directly through switch $S_1$ to line $L_2$. This completes a path. Therefore, the first path is just $S_1$.

At the output of the coil the other path that we could have taken goes through the relay contacts and then through the NC switch $S_2$ to ground. This path has the value $R_a S_2$. Since this is all the paths that the circuit has, we can now write the logic equation for the relay operation as:

$$R_a = S_1 + R_a S_2$$

With $R_a$ on both sides of the equation, the system has different modes of operation. Let's look at each of these modes. First let $R_a = 0$. This corresponds to the initial conditions before the relay is turned on. When this is done, then

$$R_a = S_1 + 0\, S_2$$
$$R_a = S_1$$

This means that when the relay is off ($R_a = 0$), only $S_1$ controls the operation of the circuit. As long as $A_1$ is open, the relay will remain off. When the switch closes, the relay will be energized. This will cause the system to go into the other mode of operation.

The second mode of operation is when the relay is energized and $R_a =$ 1, the equation becomes

$$R_a = S_1 + 1\, S_2$$
$$R_a = S_1 + S_2$$

Switch $S_2$ is a NC; therefore, under normal conditions the equation is

$$R_a = S_1 + 1 = 1$$

As long as $S_2$ is closed, it makes no difference what is done with $S_1$. The equation is independent of $S_1$; thus the relay remains energized or is "locked."

The relay can be "unlocked" and turned off if switch $S_2$ is opened. The switch $S_2$ then becomes ($S_2 = 0$), and the equation for the relay becomes

$$R_a = S_1 + 0 = S_1$$

If $S_1$ is open so that $S_1 = 0$, the relay will turn off. If $S_1$ is closed, the relay will not turn off. This may not be the type of operation that was desired. As the system is designed, the relay can be turned off only if $S_1$ and $S_2$ are both open. If $S_2$ is supposed to be an emergency switch, the operation is improper. Suppose that this were the door operation example that we discussed earlier and that $S_2$ were the limit switch. This could mean that when the operation limit is reach the limit switch could be defeated by someone closing $S_1$.

On the other hand, if this were a ventilation fan that must be kept on, then turning it off would require two switches to be thrown. For this type of application, the operation provides additional safety. The logic of the circuit may be an advantage to one application and a disadvantage to another. The design and analysis of these circuits are complicated by the fact that what is proper operation depends on human decisions.

With a little experience you can just look at a circuit diagram and write the logic equation without going through all the above discussion. In your mind you can see all the paths that control a device. This is faster than following them with your finger or a pencil. Notice how fast and easy it is and how it implies all the subtle aspects of the operation.

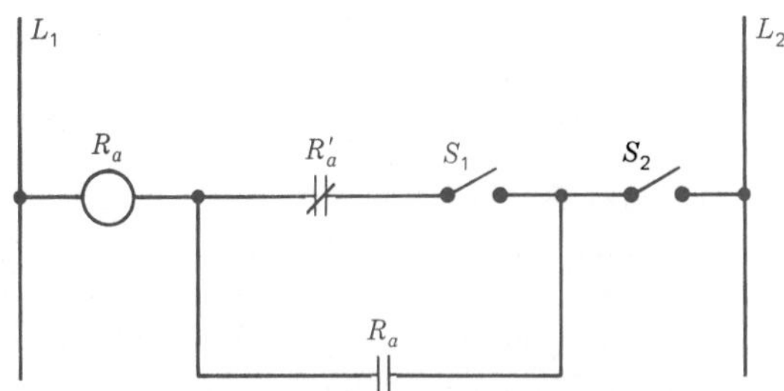


**FIGURE 23-17**
Self-latching circuit.

As another example look at the circuit shown in Fig. 23-17. This relay has two sets of contacts; one is NO and the other is NC. Find the logical operation of the system. First list the paths of the network:

Path 1—$R_a\ S_2$
Path 2—$R'_a\ S_1\ S_2$

The equation for the operation of the relay is obtained as the sum of the paths:

$$R_a = R_a S_2 + R'_a S_1 S_2$$

This equation can be simplified a little if the $S_2$ is factored out to give

$$R_a = S_2(R_a + R'_a S_1)$$

Again the system has two modes of operation. To better understand these modes of operation, first let $R_a = 0$. This corresponds to the start of operation when the relay is not operated. The logic equation is then:

$$R_a = S_2(0 + 1\ S_1) = S_2 S_1$$

This shows that both $S_1$ and $S_2$ have to be closed so that the relay will be activated. Once it is activated, $R_a = 1$ and the equation becomes:

$$R_a = S_2\ (1 + 0\ S_1) = S_2$$

This says that when the relay is actuated, $S_2$ has full control of its operation; that is, when the relay is actuated, only $S_2$ has to open so that the relay will be deenergized or "dropped."

SELF-TEST 23-6

1. The ______ contacts on a relay represent the relay variable and the ______ contacts represent the complement of the variable. — NO, NC
2. When a relay turns on and keeps itself on the relay is said to be ______ or ______. — latched, locked
3. When a relay is turned on it is said to be ______ or ______ or ______. — energized, picked up, picked
4. When a relay is turned off it is said to be ______ or ______. — dropped, deenergized
5. When a control system operates differently depending on the internal states of the relays, we say that the system has ______ of operation. — modes
6. Bimodal operation means a system has ______ modes of operation. — two
7. Bimodal operation of a relay is indicated when the logic equation for the relay has the relay variable on ______ sides of the equation. — both

## PROBLEMS

For the following problems, use the motor control circuit shown in Fig. 23-18.

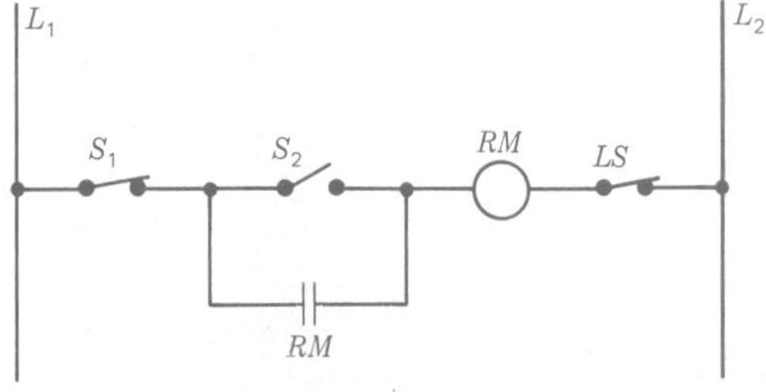


**FIGURE 23-18**
Control circuit for Probs. 1 to 5.

1. Show that for the motor control relay, the logic equation is given by:

$$R_m = S_1\, L_s(S_2 + R_m)$$

2. Assume that the motor is not operating and show that the control equation is:

$$R_m = S_1\, L_s\, S_2$$

3. Once the motor starts to operate, show that it can be stopped by either $S_1$ or $L_s$.
4. What would happen if the contacts broke off of relay $R_m$?
5. What would be the system operation if the contacts on relay $R_m$ burned shut? If $S_1$ were a NC push-button switch, how could you stop the motor?

(See Instructor's Manual for Test 23-4, Relays in Logic Circuits.)

## JOB 23-9 DIGITAL LOGIC CIRCUITS

Starting in this section a slightly different type of circuit is studied. All the basic concepts of logic that were learned previously will remain the same. The only difference is the relationship between the logic and the circuit form. The previous circuits were switching circuits, and the fundamental interest was whether there is a path for the current or not.

Digital logic circuits are usually voltage-level-operated. The most common of such devices is the transistor-transistor logic (TTL) family of devices. Many digital components that are not actually TTL are often designed to be "TTL-compatible." This means that the voltages are designed to be compatible as well as other aspects of power, supply voltages, and so on. First we shall look at the symbols that are used in this type of logic.

### SYMBOLS AND OTHER CONVENTIONS

It is necessary that certain conventions be established so that communications in this field can be easier. If you find circuit diagrams that are 15 or 20 years old, you may find many different styles of symbols. To eliminate this confusion, some standards or conventions have been established. Most present-day diagrams will conform to the symbols and conventions given below. As stated earlier, this is the very early phase of a new field of mathematics. Because of the rapid changes in the technology of integrated circuits, the techniques of binary and logic circuits are changing very rapidly. At the present time a totally new set of standard symbols is evolving. These symbols are coming from the International Electrotechnical Commission, the Institute of Electrical and Electronic Engineers, and the American National Standards Institute.

The new symbols mostly represent new development in the technology. The symbols given there are likely to continue in use for the type of problems that we are analyzing. The symbols for the elementary logic functions are shown in Fig. 23-19.

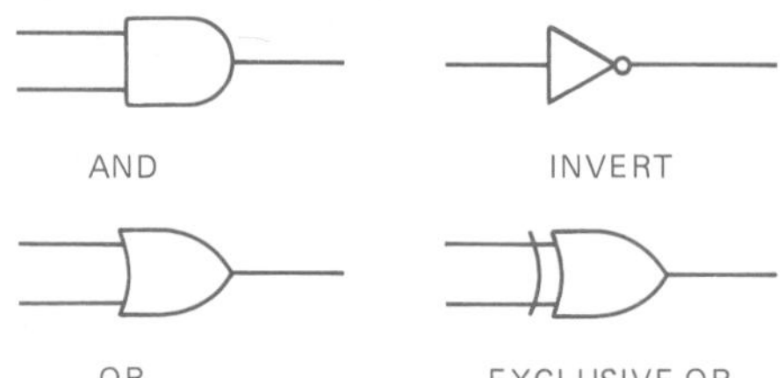


**FIGURE 23-19**
Elementary logic function symbols.

There is one subtle convention in the symbols given in Fig. 23-19. The symbol for the inverter is really two symbols. The inverter is considered to be an amplifier followed by an inverter. For this reason the symbol has a triangle which represents an amplifier and an open circle which represents inversion of logic level. When we later review more complicated logic functions, it will be found that the open circle is used frequently on both inputs and outputs to logic devices (Fig. 23-20).

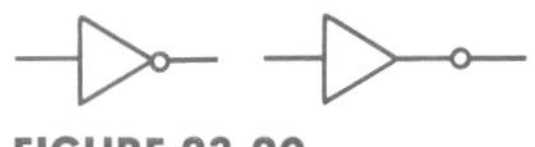

**FIGURE 23-20**
The inverter symbol.

The logic as we have used it in the above truth tables is called "positive logic." In positive logic a high or true logic value is represented by a 1 while a low or false value is represented by a 0. The opposite values can be used. When they are, it is called negative logic. For negative logic, a high or true value is represented by a 0 and a low or false value is represented by a 1.

When manufacturers specify a logic device, they generally state the logic convention that they are using. For example, they would say that a device is a "positive logic AND gate." This means that the device performs the logic function that we have described above as an AND function. In many newer control systems, particularly microprocessor-oriented systems, many control signals are negative logic. The logic performed on a negative logic signal by a positive logic device is not the same as it would be on a positive logic signal.

To better understand the characteristics of the digital logic devices, we shall look at the 54/74 families of TTL components a little more closely. The individual devices are generally numbered 5400, 74164, and so on. The leading 54/ or 74/ indicates that the device belongs to these families. The prefix 54- indicates a broader operating temperature range that is usually associated with military-type devices. The 7400 series is a commercial-grade device. Devices with comparable numbers other than the prefix are functionally interchangeable.

To conform to the standards, each device must give an output high voltage $V_{oh}$ that is greater than 2 V and an output low voltage $V_{ol}$ that is less than 0.4 V. At the input a device will treat any voltage less than 0.8 V as a low and any voltage greater than 2.4 V as a high. An additional characteristic is that if the input has nothing connected to it then the device will treat it as an input high. The inputs are said to "float" high.

It is apparent that the voltages on which these devices operate are much less than most relays and house wiring. It is also apparent that the single-pole–single-throw switch configuration that we used at first in the switching networks section will not work for the control of inputs to a TTL device. If the switch were closed, the input would be a high. If the switch were open, the input would have nothing attached to it, so the device would again assume a high. The input logic level could not be changed.

The single-pole–double-throw configuration that was used will not work, either. In the switching circuits the complement was indicated by switching the voltage from one line to another. It did not change the voltage level, as is required by the TTL devices.

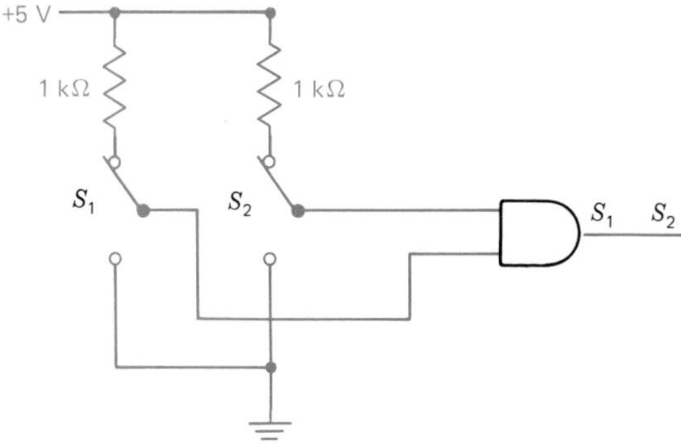


**FIGURE 23-21**
Input signals to digital logic using switches.

The manner in which a switch would be used in a digital logic circuit is shown in Fig. 23-21. To provide a high voltage to the input, the input is connected through a 1-kΩ resistor to the 5-V line. A low is provided by connecting the input to ground. If a low-voltage lamp such as a light-emitting diode (LED) is connected to the output of the AND gate, the logic operation is exactly the same as the switching circuit and function. Let $L = 1$ mean that the output is high and the lamp is on. An input high is represented by $S = 1$ and a low, by $S = 0$. The truth table would then be as shown in Table 23-11.

**TABLE 23-11**
**AND FUNCTION**

| $S_1$ | $S_2$ | $L$ |
|---|---|---|
| 0 | 0 | 0 |
| 0 | 1 | 0 |
| 1 | 0 | 0 |
| 1 | 1 | 1 |

There is no difference in the logical operation of the switching circuit AND function and the digital circuit AND function. There is a physical difference in the circuits. The digital circuit does not have the path configuration that the switching circuit had. This means only that different techniques are used when going back and forth between the circuits and the logic equations. We first consider how logic signals are traced through a logic circuit to determine the output for a given set of input conditions.

Assume the logic circuit shown in Fig. 23-22. Assume also that the input signals are $S_1 = 1$, $S_2 = 1$, and $S_3 = 0$. AND gate 1 has two 1s as inputs; therefore, the output is a 1 also. This means that the inputs to AND gate 2 are a 1 and a 0. The output from this gate is therefore 0; thus $L = 0$.

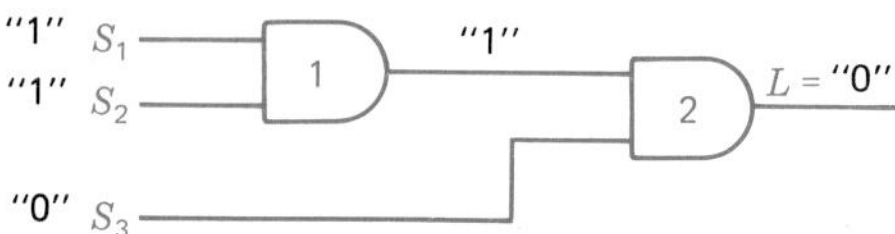


FIGURE 23-22

EXAMPLE 23-7. For practice in following logic signals through a digital circuit, show that if the inputs to the above circuit are all 1s, the output is $L = 1$. Show also that any other combination of inputs results in the output $L = 0$.

SOLUTION

1. Look at gate 1. The inputs are 1 AND 1. From the truth table for an AND function (Table 23-11), the ouput is also a 1. This output is an input to gate 2.
2. Now look at the inputs to gate 2. They also are two 1s. The output is the desired result, which is a 1.
3. To show that any other combination of inputs will give a zero, prepare a truth table for the circuit. Use as inputs $S_1$, $S_2$, and $S_3$. Use $L$ as the output.

For a second example, find the output of the circuit shown in Fig. 23-23 if the inputs are $S_1 = 1$, $S_2 = 1$, and $S_3 = 0$. The AND gate has the same inputs as in the previous example; therefore, the output is again 1. This 1 is now an input with $S_3$ to the EXCLUSIVE OR. Since only one input is a 1, the output of the EXCLUSIVE OR is 1. Thus $L = 1$ in this case.

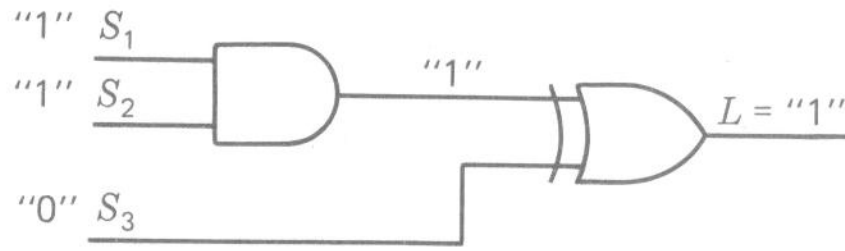


FIGURE 23-23

Now that we have a better understanding of the concepts of digital circuits, let's make a comparison of the two techniques we have studied: switching circuits and digital circuits. We shall implement some of the same logic functions that we used in the switching circuits sections, but this time in digital circuits.

First recall Fig. 23-5, where we generated the logic function $S_1(S_2 + S_3) = L$. Figure 23-24 illustrates the equivalent function in digital logic. The path structure that was in the switching circuit is not present. To determine the logic function performed by this circuit, it is necessary to

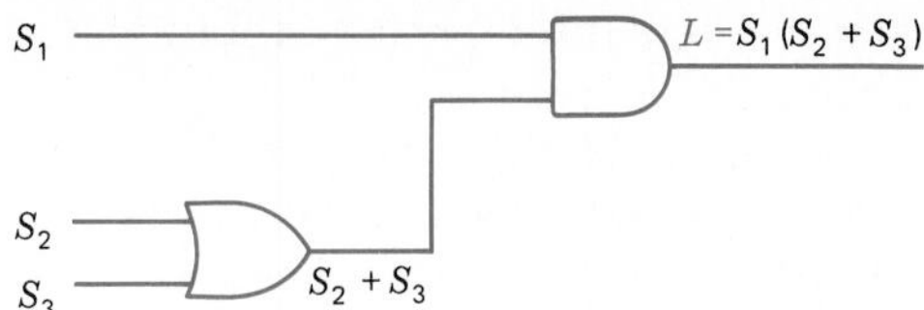


FIGURE 23-24

"trace" the signals through the circuit as shown in Fig. 23-24. The inputs to each gate are labeled with their variable names. The output is then labeled with the result of the function performed. This then becomes the input to the next gate. This process continues until the desired output is determined.

For one more example, construct the digital equivalent of Fig. 23-7. The first step is to generate the two OR functions and then to AND these together to achieve the desired result (Fig. 25-25).

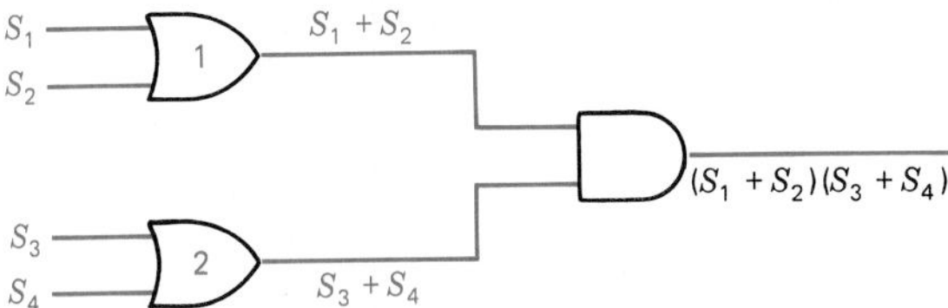


FIGURE 23-25

At this point you may ask why anyone would want to use such an apparently more complicated system. The reasons lie in cost, reliability, and size. Because of the development of the semiconductor fabrication industry, what appears to be very complicated logic can be made on very small "chips." These chips are not only small, but in large volume they become very inexpensive. Experience has shown that they are also very reliable and durable. Of course, they do not handle the power that switching circuits handle.

The result is that hybrid or combination circuits are likely to be around for quite a while. Switching circuits may be more cost-effective in units that handle higher power and are manufactured in small numbers. If the logic is complicated, it may be done digitally and then interfaced to switching circuits. This means that persons working at the technology level must be able to use both types of analysis.

We now do some problems that will help relate the two fields and to further develop skills in digital circuit analysis. Remember that once you are at the logic equation level, it makes no difference which system you are using. Truth tables, and so on are exactly the same for each type of circuit.

## PROBLEMS

1. Convert the circuit shown in Fig. 23-11 to digital logic.
2. Convert the circuit shown in Fig. 23-16 to digital logic.
3. Convert the circuit shown in Fig. 23-17 to digital logic.

4. Make a truth table for the circuit in Prob. 1.
5. Make a truth table for the circuit shown in Fig. 23-24.
6. Make a truth table for the circuit shown in Fig. 23-25.
7. In the circuit shown in Figure 23-25, what is the output if the inputs are $S_1 = 0, S_2 = 1$, and $S_3 = 1$? Do you have to know the value of $S_4$ to determine the output? Why?
8. Using Fig. 23-25, determine the output if $S_1 = 0$ and $S_2 = 0$. Do you have to know the values of $S_3$ and $S_4$? Why?

(See Instructor's Manual for Test 23-5, Digital Logic Circuits.)

# APPENDIX

## COLOR CODES

### FINDING THE VALUES OF FIXED RESISTORS

Instead of the number of ohms of resistance being stamped on carbon-type resistors, the resistors are colored according to a definite system approved by the EIA (Electronics Industries Association). Each color represents a number according to the plan in Table A-1.

TABLE A-1

| COLOR* | NUMBER | COLOR | NUMBER |
|---|---|---|---|
| Black | 0 | Green | 5 |
| Brown | 1 | Blue | 6 |
| Red | 2 | Violet | 7 |
| Orange | 3 | Gray | 8 |
| Yellow | 4 | White | 9 |

*Gold—multiply by 0.1; silver—multiply by 0.01.

The value of the resistor is obtained by reading the colors according to the following systems.

**The three-band system.** The first band represents the first number in the value. The second band represents the second number. The third band represents the number of zeros to be added after the first two numbers. If the third band is gold or silver, multiply the value indicated by the first two bands by 0.1 or 0.01, respectively, as indicated above.

EXAMPLE A-1. Find the resistance of a resistor marked red, violet, yellow as shown in Fig. A-1.

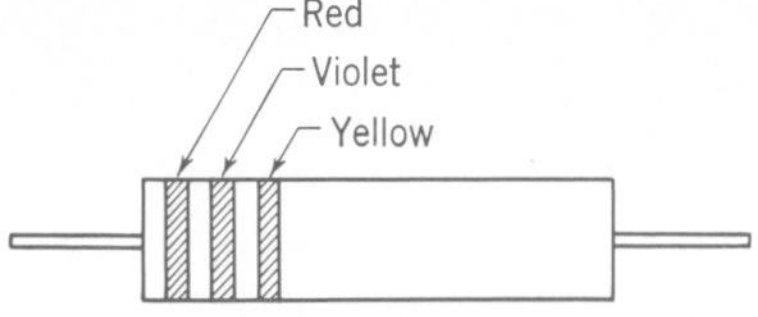


**FIGURE A-1**
The resistance value is indicated by three bands of color read in order from left to right.

SOLUTION

| First band | Second band | Third band |
|---|---|---|
| Red | Violet | Yellow |
| 2 | 7 | 0000 |

$$\text{Resistance} = 270{,}000\ \Omega \qquad \textit{Ans.}$$

EXAMPLE A-2. A resistor is marked yellow, orange, black. What is its resistance?

SOLUTION

| First band | Second band | Third band |
|---|---|---|
| Yellow | Orange | Black |
| 4 | 3 | No zeros |

$$\text{Resistance} = 43\ \Omega \qquad \textit{Ans.}$$

EXAMPLE A-3. A resistor is marked green, blue, gold. What is its resistance?

SOLUTION

| First band | Second band | Third band |
|---|---|---|
| Green | Blue | Gold |
| 5 | 6 | Multiply by 0.1 |

The first two bands indicate a value of 56 Ω. Therefore,

$$56 \times 0.1 = 5.6\ \Omega \qquad \textit{Ans.}$$

**The body-tip-dot system.** The colors must be read in the following order: body, tip, and dot. The body color represents the first number, the right-hand tip represents the second number, and the center dot represents the number of zeros to be added after the first two numbers. If the dot is gold or silver, multiply the value indicated by the first two numbers by 0.1 or 0.01, respectively.

EXAMPLE A-4. What is the resistance of the resistor shown in Fig. A-2?

SOLUTION

| Body | Tip | Dot |
|---|---|---|
| Violet | Green | Red |
| 7 | 5 | 00 |

$$\text{Resistance} = 7500\ \Omega \qquad \textit{Ans.}$$

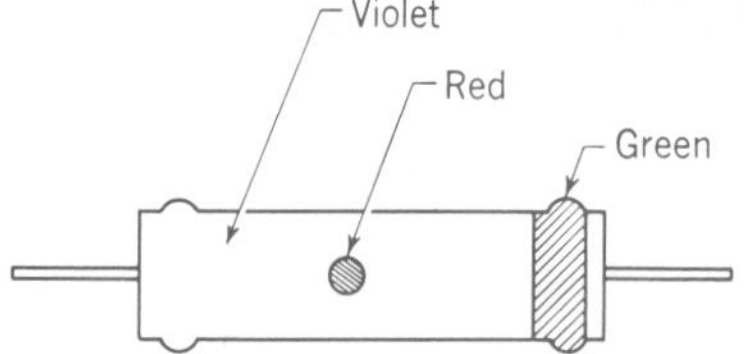


**FIGURE A-2**
The resistance value is indicated by the colors in the order: body, right-end tip, and center dot.

EXAMPLE A-5. A resistor has a brown body, a blue right end, and an orange dot in the center. What is its resistance?

SOLUTION

| Body | Tip | Dot |
|---|---|---|
| Brown | Blue | Orange |
| 1 | 6 | 000 |

$$\text{Resistance} = 16{,}000\ \Omega \quad \textit{Ans.}$$

EXAMPLE A-6. A resistor has a gray body, a red right end, and a silver dot in the center. What is its resistance?

SOLUTION

| Body | Tip | Dot |
|---|---|---|
| Gray | Red | Silver |
| 8 | 2 | Multiply by 0.01 |

The first two colors indicate a value of 82 Ω. Therefore,

$$82 \times 0.01 = 0.82\ \Omega \quad \textit{Ans.}$$

EXAMPLE A-7. A mechanic needs a 510,000-Ω resistor. What combination of colors in the body-tip-dot system is needed?

SOLUTION

The first digit is a 5, indicating green.
The second digit is a 1, indicating brown.
The four zeros that remain indicate yellow.

Therefore, the resistor will be color-coded as follows:

| Body | Tip | Dot | |
|---|---|---|---|
| Green | Brown | Yellow | *Ans.* |

EXAMPLE A-8. What color combination is needed to indicate a 6.8-Ω resistor in the three-band system?

SOLUTION

The first digit is a 6, indicating blue.
The second digit is an 8, indicating gray.

To obtain the number 6.8 from the number 68 it is necessary to multiply 68 by 0.1, which indicates gold. Therefore, the resistor will be color-coded as follows:

| First band | Second band | Third band | |
|---|---|---|---|
| Blue | Gray | Gold | *Ans.* |

**Tolerance markings.** A fourth band of color in the band system or a color on the left-hand side of the resistor in the body-tip-dot system is used to indicate how accurately the part is made to conform to the indicated markings. Gold means that the value is not more than 5 percent away from the indicated value. Silver indicates a tolerance of 10 percent, and black a tolerance of 20 percent. If the tolerance is not indicated by a color, it is assumed to be 20 percent.

## PROBLEMS

What value of resistance is indicated by each of the following color combinations?

1. Brown, black, yellow
2. Red, yellow, red
3. Gray, red, orange
4. Green, brown, red
5. Violet, green, black
6. Red, black, green
7. Yellow, violet, gold
8. Brown, green, yellow
9. Brown, red, gold
10. Brown, red, red
11. Brown, gray, green
12. Orange, white, silver
13. Blue, red, brown
14. Brown, gray, brown
15. White, brown, red
16. Orange, orange, gold
17. Yellow, violet, silver
18. Yellow, green, silver
19. Gray, red, black
20. Green, blue, gold

What color combination is needed to indicate each of the following resistances?

21. 240,000 Ω
22. 430,000 Ω
23. 51 Ω
24. 150 Ω
25. 10,000,000 Ω
26. 5.6 Ω
27. 3900 Ω
28. 0.47 Ω
29. 12 Ω
30. 750,000 Ω
31. 0.68 Ω
32. 1.2 Ω
33. 2.2 Ω
34. 100 Ω
35. 1.8 Ω
36. 1,000,000 Ω
37. 360 Ω
38. 6200 Ω
39. 1.5 Ω
40. 1 Ω

## PREFERRED VALUES OF COMPONENTS

In both of the conventions shown there are two digits and a multiplier. The first two digits have only certain values according to industry and military standards. The number that they can have depends on the tolerance that the resistor has. For tolerances of 5, 10, and 20 percent, the standard values are:

| TOLERANCE, % | | |
|---|---|---|
| 5 | 10 | 20 |
| 10 | 10 | 10 |
| 11 | | |
| 12 | 12 | |
| 13 | | |
| 15 | 15 | 15 |
| 16 | | |
| 18 | 18 | |
| 20 | | |
| 22 | 22 | 22 |
| 24 | | |
| 27 | 27 | |
| 30 | | |
| 33 | 33 | 33 |
| 36 | | |
| 39 | 39 | |
| 43 | | |
| 47 | 47 | 47 |
| 51 | | |
| 56 | 56 | |
| 62 | | |
| 68 | 68 | 68 |
| 75 | | |
| 82 | 82 | |
| 91 | | |

These preferred values are also used for some capacitors and zener diodes.

In the problems in the book no consideration was given to the standard resistor values. If a solution required a 65-Ω resistor, the answer stated a 65-Ω resistor. This is not a standard value, and it would have to be made specially for the application. According to the table, a choice would have to be made between 62 and 68 Ω; 65 Ω is 3 Ω greater than 62 and 3 Ω smaller than 68 Ω. There is no particular reason to select either 62 or 68 Ω over the other. Notice that 62 Ω is available only in 5 percent tolerance. The other choice, 68 Ω, can be obtained in 5, 10, or 20 percent tolerances. The higher-precision 5 percent resistor is usually more expensive and less readily available; therefore, you may decide to use 68-Ω 20 percent tolerance rather than 62-Ω 5 percent tolerance.

## PROBLEMS

For problems 21 through 40 of the appendix, select the 10 percent tolerance resistor that is closest to the value given.

# TRIGONOMETRIC FUNCTIONS OF DECIMAL DEGREES

## NATURAL TRIGONOMETRIC FUNCTIONS

| DEGREES | FUNCTION | 0.0° | 0.1° | 0.2° | 0.3° | 0.4° | 0.5° | 0.6° | 0.7° | 0.8° | 0.9° |
|---|---|---|---|---|---|---|---|---|---|---|---|
| 0 | sin | 0.0000 | 0.0017 | 0.0035 | 0.0052 | 0.0070 | 0.0087 | 0.0105 | 0.0122 | 0.0140 | 0.0157 |
| | cos | 1.0000 | 1.0000 | 1.0000 | 1.0000 | 1.0000 | 1.0000 | 0.9999 | 0.9999 | 0.9999 | 0.9999 |
| | tan | 0.0000 | 0.0017 | 0.0035 | 0.0052 | 0.0070 | 0.0087 | 0.0105 | 0.0122 | 0.0140 | 0.0157 |
| 1 | sin | 0.0175 | 0.0192 | 0.0209 | 0.0227 | 0.0244 | 0.0262 | 0.0279 | 0.0297 | 0.0314 | 0.0332 |
| | cos | 0.9998 | 0.9998 | 0.9998 | 0.9997 | 0.9997 | 0.9997 | 0.9996 | 0.9996 | 0.9995 | 0.9995 |
| | tan | 0.0175 | 0.0192 | 0.0209 | 0.0227 | 0.0224 | 0.0262 | 0.0279 | 0.0297 | 0.0314 | 0.0332 |
| 2 | sin | 0.0349 | 0.0366 | 0.0384 | 0.0401 | 0.0419 | 0.0436 | 0.0454 | 0.0471 | 0.0488 | 0.0506 |
| | cos | 0.9994 | 0.9993 | 0.9993 | 0.9992 | 0.9991 | 0.9990 | 0.9990 | 0.9989 | 0.9988 | 0.9987 |
| | tan | 0.0349 | 0.0367 | 0.0384 | 0.0402 | 0.0419 | 0.0437 | 0.0454 | 0.0472 | 0.0489 | 0.0507 |
| 3 | sin | 0.0523 | 0.0541 | 0.0558 | 0.0576 | 0.0593 | 0.0610 | 0.0628 | 0.0645 | 0.0663 | 0.0680 |
| | cos | 0.9986 | 0.9985 | 0.9984 | 0.9983 | 0.9982 | 0.9981 | 0.9980 | 0.9979 | 0.9978 | 0.9977 |
| | tan | 0.0524 | 0.0542 | 0.0559 | 0.0577 | 0.0594 | 0.0612 | 0.0629 | 0.0647 | 0.0664 | 0.0682 |
| 4 | sin | 0.0698 | 0.0715 | 0.0732 | 0.0750 | 0.0767 | 0.0785 | 0.0802 | 0.0819 | 0.0837 | 0.0854 |
| | cos | 0.9976 | 0.9974 | 0.9973 | 0.9972 | 0.9971 | 0.9969 | 0.9968 | 0.9966 | 0.9965 | 0.9963 |
| | tan | 0.0699 | 0.0717 | 0.0734 | 0.0752 | 0.0769 | 0.0787 | 0.0805 | 0.0822 | 0.0840 | 0.0857 |
| 5 | sin | 0.0872 | 0.0889 | 0.0906 | 0.0924 | 0.0941 | 0.0958 | 0.0976 | 0.0993 | 0.1011 | 0.1028 |
| | cos | 0.9962 | 0.9960 | 0.9959 | 0.9957 | 0.9956 | 0.9954 | 0.9952 | 0.9951 | 0.9949 | 0.9947 |
| | tan | 0.0875 | 0.0892 | 0.0910 | 0.0928 | 0.0945 | 0.0963 | 0.0981 | 0.0998 | 0.1016 | 0.1033 |
| 6 | sin | 0.1045 | 0.1063 | 0.1080 | 0.1097 | 0.1115 | 0.1132 | 0.1149 | 0.1167 | 0.1184 | 0.1201 |
| | cos | 0.9945 | 0.9943 | 0.9942 | 0.9940 | 0.9938 | 0.9936 | 0.9934 | 0.9932 | 0.9930 | 0.9928 |
| | tan | 0.1051 | 0.1069 | 0.1086 | 0.1104 | 0.1122 | 0.1139 | 0.1157 | 0.1175 | 0.1192 | 0.1210 |
| 7 | sin | 0.1219 | 0.1236 | 0.1253 | 0.1271 | 0.1288 | 0.1305 | 0.1323 | 0.1340 | 0.1357 | 0.1374 |
| | cos | 0.9925 | 0.9923 | 0.9921 | 0.9919 | 0.9917 | 0.9914 | 0.9912 | 0.9910 | 0.9907 | 0.9905 |
| | tan | 0.1228 | 0.1246 | 0.1263 | 0.1281 | 0.1299 | 0.1317 | 0.1334 | 0.1352 | 0.1370 | 0.1388 |
| 8 | sin | 0.1392 | 0.1409 | 0.1426 | 0.1444 | 0.1461 | 0.1478 | 0.1495 | 0.1513 | 0.1530 | 0.1547 |
| | cos | 0.9903 | 0.9900 | 0.9898 | 0.9895 | 0.9893 | 0.9890 | 0.9888 | 0.9885 | 0.9882 | 0.9880 |
| | tan | 0.1405 | 0.1423 | 0.1441 | 0.1459 | 0.1477 | 0.1495 | 0.1512 | 0.1530 | 0.1548 | 0.1566 |
| 9 | sin | 0.1564 | 0.1582 | 0.1599 | 0.1616 | 0.1633 | 0.1650 | 0.1668 | 0.1685 | 0.1702 | 0.1719 |
| | cos | 0.9877 | 0.9874 | 0.9871 | 0.9869 | 0.9866 | 0.9863 | 0.9860 | 0.9857 | 0.9854 | 0.9851 |
| | tan | 0.1584 | 0.1602 | 0.1620 | 0.1638 | 0.1655 | 0.1673 | 0.1691 | 0.1709 | 0.1727 | 0.1745 |
| 10 | sin | 0.1736 | 0.1754 | 0.1771 | 0.1778 | 0.1805 | 0.1822 | 0.1840 | 0.1857 | 0.1874 | 0.1891 |
| | cos | 0.9848 | 0.9845 | 0.9842 | 0.9839 | 0.9836 | 0.9833 | 0.9829 | 0.9826 | 0.9823 | 0.9820 |
| | tan | 0.1763 | 0.1781 | 0.1799 | 0.1817 | 0.1835 | 0.1853 | 0.1871 | 0.1890 | 0.1908 | 0.1926 |
| 11 | sin | 0.1908 | 0.1925 | 0.1942 | 0.1959 | 0.1977 | 0.1994 | 0.2011 | 0.2028 | 0.2045 | 0.2062 |
| | cos | 0.9816 | 0.9813 | 0.9810 | 0.9806 | 0.9803 | 0.9799 | 0.9796 | 0.9792 | 0.9789 | 0.9785 |
| | tan | 0.1944 | 0.1962 | 0.1980 | 0.1998 | 0.2016 | 0.2035 | 0.2053 | 0.2071 | 0.2089 | 0.2107 |
| 12 | sin | 0.2079 | 0.2096 | 0.2113 | 0.2130 | 0.2147 | 0.2164 | 0.2181 | 0.2198 | 0.2215 | 0.2232 |
| | cos | 0.9781 | 0.9778 | 0.9774 | 0.9770 | 0.9767 | 0.9763 | 0.9759 | 0.9755 | 0.9751 | 0.9748 |
| | tan | 0.2126 | 0.2144 | 0.2162 | 0.2180 | 0.2199 | 0.2217 | 0.2235 | 0.2254 | 0.2272 | 0.2290 |
| 13 | sin | 0.2250 | 0.2267 | 0.2284 | 0.2300 | 0.2318 | 0.2334 | 0.2351 | 0.2368 | 0.2385 | 0.2402 |
| | cos | 0.9744 | 0.9740 | 0.9736 | 0.9732 | 0.9728 | 0.9724 | 0.9720 | 0.9715 | 0.9711 | 0.9707 |
| | tan | 0.2309 | 0.2327 | 0.2345 | 0.2364 | 0.2382 | 0.2401 | 0.2419 | 0.2438 | 0.2456 | 0.2475 |
| 14 | sin | 0.2419 | 0.2436 | 0.2453 | 0.2470 | 0.2487 | 0.2504 | 0.2521 | 0.2538 | 0.2554 | 0.2571 |
| | cos | 0.9703 | 0.9699 | 0.9694 | 0.9690 | 0.9686 | 0.9681 | 0.9677 | 0.9673 | 0.9668 | 0.9664 |
| | tan | 0.2493 | 0.2512 | 0.2530 | 0.2549 | 0.2568 | 0.2586 | 0.2605 | 0.2623 | 0.2642 | 0.2661 |
| 15 | sin | 0.2588 | 0.2605 | 0.2622 | 0.2639 | 0.2656 | 0.2672 | 0.2689 | 0.2706 | 0.2723 | 0.2740 |
| | cos | 0.9659 | 0.9655 | 0.9650 | 0.9646 | 0.9641 | 0.9636 | 0.9632 | 0.9627 | 0.9622 | 0.9617 |
| | tan | 0.2679 | 0.2698 | 0.2717 | 0.2736 | 0.2754 | 0.2773 | 0.2792 | 0.2811 | 0.2830 | 0.2849 |
| 16 | sin | 0.2756 | 0.2773 | 0.2790 | 0.2807 | 0.2823 | 0.2840 | 0.2857 | 0.2874 | 0.2890 | 0.2907 |
| | cos | 0.9613 | 0.9608 | 0.9603 | 0.9598 | 0.9593 | 0.9588 | 0.9583 | 0.9578 | 0.9573 | 0.9568 |
| | tan | 0.2867 | 0.2886 | 0.2905 | 0.2924 | 0.2943 | 0.2962 | 0.2981 | 0.3000 | 0.3019 | 0.3038 |
| 17 | sin | 0.2924 | 0.2940 | 0.2957 | 0.2974 | 0.2990 | 0.3007 | 0.3024 | 0.3040 | 0.3057 | 0.3074 |
| | cos | 0.9563 | 0.9558 | 0.9553 | 0.9548 | 0.9542 | 0.9537 | 0.9532 | 0.9527 | 0.9521 | 0.9516 |
| | tan | 0.3057 | 0.3076 | 0.3096 | 0.3115 | 0.3134 | 0.3153 | 0.3172 | 0.3191 | 0.3211 | 0.3230 |
| 18 | sin | 0.3090 | 0.3107 | 0.3123 | 0.3140 | 0.3156 | 0.3173 | 0.3190 | 0.3206 | 0.3223 | 0.3239 |
| | cos | 0.9511 | 0.9505 | 0.9500 | 0.9494 | 0.9489 | 0.9483 | 0.9478 | 0.9472 | 0.9466 | 0.9461 |
| | tan | 0.3249 | 0.3269 | 0.3288 | 0.3307 | 0.3327 | 0.3346 | 0.3365 | 0.3385 | 0.3404 | 0.3424 |
| 19 | sin | 0.3256 | 0.3272 | 0.3289 | 0.3305 | 0.3322 | 0.3338 | 0.3355 | 0.3371 | 0.3387 | 0.3404 |
| | cos | 0.9455 | 0.9449 | 0.9444 | 0.9438 | 0.9432 | 0.9426 | 0.9421 | 0.9415 | 0.9409 | 0.9403 |
| | tan | 0.3443 | 0.3463 | 0.3482 | 0.3502 | 0.3522 | 0.3541 | 0.3561 | 0.3581 | 0.3600 | 0.3620 |
| DEGREES | FUNCTION | 0′ | 6′ | 12′ | 18′ | 24′ | 30′ | 36′ | 42′ | 48′ | 54′ |

## NATURAL TRIGONOMETRIC FUNCTIONS, CONTINUED

| DEGREES | FUNCTION | 0.0° | 0.1° | 0.2° | 0.3° | 0.4° | 0.5° | 0.6° | 0.7° | 0.8° | 0.9° |
|---|---|---|---|---|---|---|---|---|---|---|---|
| 20 | sin | 0.3420 | 0.3437 | 0.3453 | 0.3469 | 0.3486 | 0.3502 | 0.3518 | 0.3535 | 0.3551 | 0.3567 |
| | cos | 0.9397 | 0.9391 | 0.9385 | 0.9379 | 0.9373 | 0.9367 | 0.9361 | 0.9354 | 0.9348 | 0.9342 |
| | tan | 0.3640 | 0.3659 | 0.3679 | 0.3699 | 0.3719 | 0.3739 | 0.3759 | 0.3779 | 0.3799 | 0.3819 |
| 21 | sin | 0.3584 | 0.3600 | 0.3616 | 0.3633 | 0.3649 | 0.3665 | 0.3681 | 0.3697 | 0.3714 | 0.3730 |
| | cos | 0.9336 | 0.9330 | 0.9323 | 0.9317 | 0.9311 | 0.9304 | 0.9298 | 0.9291 | 0.9285 | 0.9278 |
| | tan | 0.3839 | 0.3859 | 0.3879 | 0.3899 | 0.3919 | 0.3939 | 0.3959 | 0.3979 | 0.4000 | 0.4020 |
| 22 | sin | 0.3746 | 0.3762 | 0.3778 | 0.3795 | 0.3811 | 0.3827 | 0.3843 | 0.3859 | 0.3875 | 0.3891 |
| | cos | 0.9272 | 0.9265 | 0.9259 | 0.9252 | 0.9245 | 0.9245 | 0.9239 | 0.9232 | 0.9225 | 0.9219 |
| | tan | 0.4040 | 0.4061 | 0.4081 | 0.4101 | 0.4122 | 0.4142 | 0.4263 | 0.4183 | 0.4204 | 0.4224 |
| 23 | sin | 0.3907 | 0.3923 | 0.3939 | 0.3955 | 0.3971 | 0.3987 | 0.4003 | 0.4019 | 0.4035 | 0.4051 |
| | cos | 0.9205 | 0.9198 | 0.9191 | 0.9184 | 0.9178 | 0.9171 | 0.9164 | 0.9157 | 0.9150 | 0.9143 |
| | tan | 0.4245 | 0.4265 | 0.4286 | 0.4307 | 0.4327 | 0.4348 | 0.4369 | 0.4390 | 0.4411 | 0.4431 |
| 24 | sin | 0.4067 | 0.4083 | 0.4099 | 0.4115 | 0.4131 | 0.4147 | 0.4163 | 0.4179 | 0.4195 | 0.4210 |
| | cos | 0.9135 | 0.9128 | 0.9121 | 0.9114 | 0.9107 | 0.9100 | 0.9092 | 0.9085 | 0.9078 | 0.9070 |
| | tan | 0.4452 | 0.4473 | 0.4494 | 0.4515 | 0.4536 | 0.4557 | 0.4578 | 0.4599 | 0.4621 | 0.4642 |
| 25 | sin | 0.4226 | 0.4242 | 0.4258 | 0.4274 | 0.4289 | 0.4305 | 0.4321 | 0.4337 | 0.4352 | 0.4368 |
| | cos | 0.9063 | 0.9056 | 0.9048 | 0.9041 | 0.9033 | 0.9026 | 0.9018 | 0.9011 | 0.9003 | 0.8996 |
| | tan | 0.4663 | 0.4684 | 0.4706 | 0.4727 | 0.4748 | 0.4770 | 0.4791 | 0.4813 | 0.4834 | 0.4856 |
| 26 | sin | 0.4384 | 0.4399 | 0.4415 | 0.4431 | 0.4446 | 0.4462 | 0.4478 | 0.4493 | 0.4509 | 0.4524 |
| | cos | 0.8988 | 0.8980 | 0.8973 | 0.8965 | 0.8957 | 0.8949 | 0.8942 | 0.8934 | 0.8926 | 0.8918 |
| | tan | 0.4877 | 0.4899 | 0.4921 | 0.4942 | 0.4964 | 0.4986 | 0.5008 | 0.5029 | 0.5051 | 0.5073 |
| 27 | sin | 0.4540 | 0.4555 | 0.4571 | 0.4586 | 0.4602 | 0.4617 | 0.4633 | 0.4648 | 0.4664 | 0.4679 |
| | cos | 0.8910 | 0.8902 | 0.8894 | 0.8886 | 0.8878 | 0.8870 | 0.8862 | 0.8854 | 0.8846 | 0.8838 |
| | tan | 0.5095 | 0.5117 | 0.5139 | 0.5161 | 0.5184 | 0.5206 | 0.5228 | 0.5250 | 0.5272 | 0.5295 |
| 28 | sin | 0.4695 | 0.4710 | 0.4726 | 0.4741 | 0.4756 | 0.4772 | 0.4787 | 0.4802 | 0.4818 | 0.4833 |
| | cos | 0.8829 | 0.8821 | 0.8813 | 0.8805 | 0.8796 | 0.8788 | 0.8780 | 0.8771 | 0.8763 | 0.8755 |
| | tan | 0.5317 | 0.5340 | 0.5362 | 0.5384 | 0.5407 | 0.5430 | 0.5452 | 0.5475 | 0.5498 | 0.5520 |
| 29 | sin | 0.4848 | 0.4863 | 0.4879 | 0.4894 | 0.4909 | 0.4924 | 0.4939 | 0.4955 | 0.4970 | 0.4985 |
| | cos | 0.8746 | 0.8738 | 0.8729 | 0.8721 | 0.8712 | 0.8704 | 0.8695 | 0.8686 | 0.8678 | 0.8669 |
| | tan | 0.5543 | 0.5566 | 0.5589 | 0.5612 | 0.5635 | 0.5658 | 0.5681 | 0.5704 | 0.5727 | 0.5750 |
| 30 | sin | 0.5000 | 0.5015 | 0.5030 | 0.5045 | 0.5060 | 0.5075 | 0.5090 | 0.5105 | 0.5120 | 0.5135 |
| | cos | 0.8660 | 0.8652 | 0.8643 | 0.8634 | 0.8625 | 0.8616 | 0.8607 | 0.8599 | 0.8590 | 0.8581 |
| | tan | 0.5774 | 0.5797 | 0.5820 | 0.5844 | 0.5967 | 0.5890 | 0.5914 | 0.5938 | 0.5961 | 0.5985 |
| 31 | sin | 0.5150 | 0.5165 | 0.5180 | 0.5195 | 0.5210 | 0.5225 | 0.5240 | 0.5255 | 0.5270 | 0.5284 |
| | cos | 0.8572 | 0.8563 | 0.8554 | 0.8545 | 0.8536 | 0.8526 | 0.8517 | 0.8508 | 0.8499 | 0.8490 |
| | tan | 0.6009 | 0.6032 | 0.6056 | 0.6080 | 0.6104 | 0.6128 | 0.6152 | 0.6176 | 0.6200 | 0.6224 |
| 32 | sin | 0.5299 | 0.5314 | 0.5329 | 0.5344 | 0.5358 | 0.5373 | 0.5388 | 0.5402 | 0.5417 | 0.5432 |
| | cos | 0.8480 | 0.8471 | 0.8462 | 0.8453 | 0.8443 | 0.8434 | 0.8425 | 0.8415 | 0.8406 | 0.8396 |
| | tan | 0.6249 | 0.6273 | 0.6297 | 0.6322 | 0.6346 | 0.6371 | 0.6395 | 0.6420 | 0.6445 | 0.6469 |
| 33 | sin | 0.5446 | 0.5461 | 0.5476 | 0.5490 | 0.5505 | 0.5519 | 0.5534 | 0.5548 | 0.5563 | 0.5577 |
| | cos | 0.8387 | 0.8377 | 0.8368 | 0.8358 | 0.8348 | 0.8339 | 0.8329 | 0.8320 | 0.8310 | 0.8300 |
| | tan | 0.6494 | 0.6519 | 0.6544 | 0.6569 | 0.6594 | 0.6619 | 0.6644 | 0.6669 | 0.6694 | 0.6720 |
| 34 | sin | 0.5592 | 0.5606 | 0.5621 | 0.5635 | 0.5650 | 0.5664 | 0.5678 | 0.5693 | 0.5707 | 0.5721 |
| | cos | 0.8290 | 0.8281 | 0.8271 | 0.8261 | 0.8251 | 0.8241 | 0.8231 | 0.8221 | 0.8211 | 0.8202 |
| | tan | 0.6745 | 0.6771 | 0.6796 | 0.6822 | 0.6847 | 0.6873 | 0.6899 | 0.6924 | 0.6950 | 0.6976 |
| 35 | sin | 0.5736 | 0.5750 | 0.5764 | 0.5779 | 0.5793 | 0.5807 | 0.5821 | 0.5835 | 0.5850 | 0.5864 |
| | cos | 0.8192 | 0.8181 | 0.8171 | 0.8161 | 0.8151 | 0.8141 | 0.8131 | 0.8121 | 0.8111 | 0.8100 |
| | tan | 0.7002 | 0.7028 | 0.7054 | 0.7080 | 0.7107 | 0.7133 | 0.7159 | 0.7186 | 0.7212 | 0.7239 |
| 36 | sin | 0.5878 | 0.5892 | 0.5906 | 0.5920 | 0.5934 | 0.5948 | 0.5962 | 0.5976 | 0.5990 | 0.6004 |
| | cos | 0.8090 | 0.8080 | 0.8070 | 0.8059 | 0.8049 | 0.8039 | 0.8028 | 0.8018 | 0.8007 | 0.7997 |
| | tan | 0.7265 | 0.7292 | 0.7319 | 0.7346 | 0.7373 | 0.7400 | 0.7427 | 0.7454 | 0.7481 | 0.7508 |
| 37 | sin | 0.6018 | 0.6032 | 0.6046 | 0.6060 | 0.6074 | 0.6088 | 0.6101 | 0.6115 | 0.6129 | 0.6143 |
| | cos | 0.7986 | 0.7976 | 0.7965 | 0.7955 | 0.7944 | 0.7934 | 0.7923 | 0.7912 | 0.7902 | 0.7891 |
| | tan | 0.7536 | 0.7563 | 0.7590 | 0.7618 | 0.7646 | 0.7673 | 0.7701 | 0.7729 | 0.7757 | 0.7785 |
| 38 | sin | 0.6157 | 0.6170 | 0.6184 | 0.6198 | 0.6211 | 0.6225 | 0.6239 | 0.6252 | 0.6266 | 0.6280 |
| | cos | 0.7880 | 0.7869 | 0.7859 | 0.7848 | 0.7837 | 0.7826 | 0.7815 | 0.7804 | 0.7793 | 0.7782 |
| | tan | 0.7813 | 0.7841 | 0.7869 | 0.7898 | 0.7926 | 0.7954 | 0.7983 | 0.8012 | 0.8040 | 0.8069 |
| 39 | sin | 0.6293 | 0.6307 | 0.6320 | 0.6334 | 0.6347 | 0.6361 | 0.6374 | 0.6388 | 0.6401 | 0.6414 |
| | cos | 0.7771 | 0.7760 | 0.7749 | 0.7738 | 0.7727 | 0.7716 | 0.7705 | 0.7694 | 0.7683 | 0.7672 |
| | tan | 0.8098 | 0.8127 | 0.8156 | 0.8185 | 0.8214 | 0.8243 | 0.8273 | 0.8302 | 0.8332 | 0.8361 |
| 40 | sin | 0.6428 | 0.6441 | 0.6455 | 0.6468 | 0.6481 | 0.6494 | 0.6508 | 0.6521 | 0.6534 | 0.6547 |
| | cos | 0.7660 | 0.76.49 | 0.7638 | 0.7627 | 0.7615 | 0.7604 | 0.7593 | 0.7581 | 0.7570 | 0.7559 |
| | tan | 0.8391 | 0.8421 | 0.8451 | 0.8481 | 0.8511 | 0.8541 | 0.8571 | 0.8601 | 0.8632 | 0.8662 |
| DEGREES | FUNCTION | 0′ | 6′ | 12′ | 18′ | 24′ | 30′ | 36′ | 42′ | 48′ | 54′ |

## NATURAL TRIGONOMETRIC FUNCTIONS, CONTINUED

| DEGREES | FUNCTION | 0.0° | 0.1° | 0.2° | 0.3° | 0.4° | 0.5° | 0.6° | 0.7° | 0.8° | 0.9° |
|---|---|---|---|---|---|---|---|---|---|---|---|
| | sin | 0.6561 | 0.6574 | 0.6587 | 0.6600 | 0.6613 | 0.6626 | 0.6639 | 0.6652 | 0.6665 | 0.6678 |
| 41 | cos | 0.7547 | 0.7536 | 0.7524 | 0.7513 | 0.7501 | 0.7490 | 0.7478 | 0.7466 | 0.7455 | 0.7443 |
| | tan | 0.8693 | 0.8724 | 0.8754 | 0.8785 | 0.8816 | 0.8847 | 0.8878 | 0.8910 | 0.8941 | 0.8972 |
| | sin | 0.6691 | 0.6704 | 0.6717 | 0.6730 | 0.6743 | 0.6756 | 0.6769 | 0.6782 | 0.6794 | 0.6807 |
| 42 | cos | 0.7431 | 0.7420 | 0.7408 | 0.7396 | 0.7385 | 0.7373 | 0.7361 | 0.7349 | 0.7337 | 0.7325 |
| | tan | 0.9004 | 0.9036 | 0.9067 | 0.9099 | 0.9131 | 0.9163 | 0.9195 | 0.9228 | 0.9260 | 0.9293 |
| | sin | 0.6820 | 0.6833 | 0.6845 | 0.6858 | 0.6871 | 0.6884 | 0.6896 | 0.6909 | 0.6921 | 0.6934 |
| 43 | cos | 0.7314 | 0.7302 | 0.7290 | 0.7278 | 0.7266 | 0.7254 | 0.7242 | 0.7230 | 0.7218 | 0.7206 |
| | tan | 0.9325 | 0.9358 | 0.9391 | 0.9424 | 0.9457 | 0.9490 | 0.9523 | 0.9556 | 0.9590 | 0.9623 |
| | sin | 0.6947 | 0.6959 | 0.6972 | 0.6984 | 0.6997 | 0.7009 | 0.7022 | 0.7034 | 0.7046 | 0.7059 |
| 44 | cos | 0.7193 | 0.7181 | 0.7169 | 0.7157 | 0.7145 | 0.7133 | 0.7120 | 0.7108 | 0.7096 | 0.7083 |
| | tan | 0.9657 | 0.9691 | 0.9725 | 0.9759 | 0.9793 | 0.9827 | 0.9861 | 0.9896 | 0.9930 | 0.9965 |
| | sin | 0.7071 | 0.7083 | 0.7096 | 0.7108 | 0.7120 | 0.7133 | 0.7145 | 0.7157 | 0.7169 | 0.7181 |
| 45 | cos | 0.7071 | 0.7059 | 0.7046 | 0.7034 | 0.7022 | 0.7009 | 0.6997 | 0.6984 | 0.6972 | 0.6959 |
| | tan | 1.0000 | 1.0035 | 1.0070 | 1.0105 | 1.0141 | 1.0176 | 1.0212 | 1.0247 | 1.0283 | 1.0319 |
| | sin | 0.7193 | 0.7206 | 0.7218 | 0.7230 | 0.7242 | 0.7254 | 0.7266 | 0.7278 | 0.7290 | 0.7302 |
| 46 | cos | 0.6947 | 0.6934 | 0.6921 | 0.6909 | 0.6896 | 0.6884 | 0.6871 | 0.6858 | 0.6845 | 0.6833 |
| | tan | 1.0355 | 1.0392 | 1.0428 | 1.0464 | 1.0501 | 1.0538 | 1.0575 | 1.0612 | 1.0649 | 1.0686 |
| | sin | 0.7314 | 0.7325 | 0.7337 | 0.7349 | 0.7361 | 0.7373 | 0.7385 | 0.7396 | 0.7408 | 0.7420 |
| 47 | cos | 0.6820 | 0.6807 | 0.6794 | 0.6782 | 0.6769 | 0.6756 | 0.6743 | 0.6730 | 0.6717 | 0.6704 |
| | tan | 1.0724 | 1.0761 | 1.0799 | 1.0837 | 1.0875 | 1.0913 | 1.0951 | 1.0990 | 1.1028 | 1.1067 |
| | sin | 0.7431 | 0.7443 | 0.7455 | 0.7466 | 0.7478 | 0.7490 | 0.7501 | 0.7513 | 0.7524 | 0.7536 |
| 48 | cos | 0.6691 | 0.6678 | 0.6665 | 0.6652 | 0.6639 | 0.6626 | 0.6613 | 0.6600 | 0.6587 | 0.6574 |
| | tan | 1.1106 | 1.1145 | 1.1184 | 1.1224 | 1.1263 | 1.1303 | 1.1343 | 1.1383 | 1.1423 | 1.1463 |
| | sin | 0.7547 | 0.7559 | 0.7570 | 0.7581 | 0.7593 | 0.7604 | 0.7615 | 0.7627 | 0.7638 | 0.7649 |
| 49 | cos | 0.6561 | 0.6547 | 0.6534 | 0.6521 | 0.6508 | 0.6494 | 0.6481 | 0.6468 | 0.6455 | 0.6441 |
| | tan | 1.1504 | 1.1544 | 1.1585 | 1.1626 | 1.1667 | 1.1708 | 1.1750 | 1.1792 | 1.1833 | 1.1875 |
| | sin | 0.7660 | 0.7672 | 0.7683 | 0.7694 | 0.7705 | 0.7716 | 0.7727 | 0.7738 | 0.7749 | 0.7760 |
| 50 | cos | 0.6428 | 0.6414 | 0.6401 | 0.6388 | 0.6374 | 0.6361 | 0.6347 | 0.6334 | 0.6320 | 0.6307 |
| | tan | 1.1918 | 1.1960 | 1.2002 | 1.2045 | 1.2088 | 1.2131 | 1.2174 | 1.2218 | 1.2261 | 1.2305 |
| | sin | 0.7771 | 0.7782 | 0.7793 | 0.7804 | 0.7815 | 0.7826 | 0.7837 | 0.7848 | 0.7859 | 0.7869 |
| 51 | cos | 0.6293 | 0.6280 | 0.6266 | 0.6252 | 0.6239 | 0.6225 | 0.6211 | 0.6198 | 0.6184 | 0.6170 |
| | tan | 1.2349 | 1.2393 | 1.2437 | 1.2482 | 1.2527 | 1.2572 | 1.2617 | 1.2662 | 1.2708 | 1.2753 |
| | sin | 0.7880 | 0.7891 | 0.7902 | 0.7912 | 0.7923 | 0.7934 | 0.7944 | 0.7955 | 0.7965 | 0.7976 |
| 52 | cos | 0.6157 | 0.6143 | 0.6129 | 0.6115 | 0.6101 | 0.6088 | 0.6074 | 0.6060 | 0.6046 | 0.6032 |
| | tan | 1.2799 | 1.2846 | 1.2892 | 1.2938 | 1.2985 | 1.3032 | 1.3079 | 1.3127 | 1.3175 | 1.3222 |
| | sin | 0.7986 | 0.7997 | 0.8007 | 0.8018 | 0.8028 | 0.8039 | 0.8049 | 0.8059 | 0.8070 | 0.8080 |
| 53 | cos | 0.6018 | 0.6004 | 0.5990 | 0.5976 | 0.5962 | 0.5948 | 0.5934 | 0.5920 | 0.5906 | 0.5892 |
| | tan | 1.3270 | 1.3319 | 1.3367 | 1.3416 | 1.3465 | 1.3514 | 1.3564 | 1.3613 | 1.3663 | 1.3713 |
| | sin | 0.8090 | 0.8100 | 0.8111 | 0.8121 | 0.8131 | 0.8141 | 0.8151 | 0.8161 | 0.8171 | 0.8181 |
| 54 | cos | 0.5878 | 0.5864 | 0.5850 | 0.5835 | 0.5821 | 0.5807 | 0.5793 | 0.5779 | 0.5764 | 0.5750 |
| | tan | 1.3764 | 1.3814 | 1.3865 | 1.3916 | 1.3968 | 1.4019 | 1.4071 | 1.4124 | 1.4176 | 1.4229 |
| | sin | 0.8192 | 0.8202 | 0.8211 | 0.8221 | 0.8231 | 0.8241 | 0.8251 | 0.8261 | 0.8271 | 0.8281 |
| 55 | cos | 0.5736 | 0.5721 | 0.5707 | 0.5693 | 0.5678 | 0.5664 | 0.5650 | 0.5635 | 0.5621 | 0.5606 |
| | tan | 1.4281 | 1.4335 | 1.4388 | 1.4442 | 1.4496 | 1.4550 | 1.4605 | 1.4659 | 1.4715 | 1.4770 |
| | sin | 0.8290 | 0.8300 | 0.8310 | 0.8320 | 0.8329 | 0.8339 | 0.8348 | 0.8358 | 0.8368 | 0.8377 |
| 56 | cos | 0.5592 | 0.5577 | 0.5563 | 0.5548 | 0.5534 | 0.5519 | 0.5505 | 0.5490 | 0.5476 | 0.5461 |
| | tan | 1.4826 | 1.4882 | 1.4938 | 1.4994 | 1.5051 | 1.5108 | 1.5166 | 1.5224 | 1.5282 | 1.5340 |
| | sin | 0.8387 | 0.8396 | 0.8406 | 0.8415 | 0.8425 | 0.8434 | 0.8443 | 0.8453 | 0.8462 | 0.8471 |
| 57 | cos | 0.5446 | 0.5432 | 0.5417 | 0.5402 | 0.5388 | 0.5373 | 0.5358 | 0.5344 | 0.5329 | 0.5314 |
| | tan | 1.5399 | 1.5458 | 1.5517 | 1.5577 | 1.5637 | 1.5697 | 1.5757 | 1.5818 | 1.5880 | 1.5941 |
| | sin | 0.8480 | 0.8490 | 0.8499 | 0.8508 | 0.8517 | 0.8526 | 0.8536 | 0.8545 | 0.8554 | 0.8563 |
| 58 | cos | 0.5299 | 0.5284 | 0.5270 | 0.5255 | 0.5240 | 0.5225 | 0.5210 | 0.5195 | 0.5180 | 0.5165 |
| | tan | 1.6003 | 1.6066 | 1.6128 | 1.6191 | 1.6255 | 1.6319 | 1.6383 | 1.6447 | 1.6512 | 1.6577 |
| | sin | 0.8572 | 0.8581 | 0.8590 | 0.8599 | 0.8607 | 0.8616 | 0.8625 | 0.8634 | 0.8643 | 0.8652 |
| 59 | cos | 0.5150 | 0.5135 | 0.5120 | 0.5105 | 0.5090 | 0.5075 | 0.5060 | 0.5045 | 0.5030 | 0.5015 |
| | tan | 1.6643 | 1.6709 | 1.6775 | 1.6842 | 1.6909 | 1.6977 | 1.7045 | 1.7113 | 1.7182 | 1.7251 |
| | sin | 0.8660 | 0.8669 | 0.8678 | 0.8686 | 0.8695 | 0.8704 | 0.8712 | 0.8721 | 0.8729 | 0.8738 |
| 60 | cos | 0.5000 | 0.4985 | 0.4970 | 0.4955 | 0.4939 | 0.4924 | 0.4909 | 0.4894 | 0.4879 | 0.4863 |
| | tan | 1.7321 | 1.7391 | 1.7461 | 1.7532 | 1.7603 | 1.7675 | 1.7747 | 1.7820 | 1.7893 | 1.7966 |
| | sin | 0.8746 | 0.8755 | 0.8763 | 0.8771 | 0.8780 | 0.8788 | 0.8796 | 0.8805 | 0.8813 | 0.8821 |
| 61 | cos | 0.4848 | 0.4833 | 0.4818 | 0.4802 | 0.4787 | 0.4772 | 0.4756 | 0.4741 | 0.4726 | 0.4710 |
| | tan | 1.8040 | 1.8115 | 1.8190 | 1.8265 | 1.8341 | 1.8418 | 1.8495 | 1.8572 | 1.8650 | 1.8728 |
| DEGREES | FUNCTION | 0′ | 6′ | 12′ | 18′ | 24′ | 30′ | 36′ | 42′ | 48′ | 54′ |

## NATURAL TRIGONOMETRIC FUNCTIONS, CONTINUED

| DEGREES | FUNCTION | 0.0° | 0.1° | 0.2° | 0.3° | 0.4° | 0.5° | 0.6° | 0.7° | 0.8° | 0.9° |
|---|---|---|---|---|---|---|---|---|---|---|---|
| 62 | sin | 0.8829 | 0.8838 | 0.8846 | 0.8854 | 0.8862 | 0.8870 | 0.8878 | 0.8886 | 0.8894 | 0.8902 |
| | cos | 0.4695 | 0.4679 | 0.4664 | 0.4648 | 0.4633 | 0.4617 | 0.4602 | 0.4586 | 0.4571 | 0.4555 |
| | tan | 1.8807 | 1.8887 | 1.8967 | 1.9047 | 1.9128 | 1.9210 | 1.9292 | 1.9375 | 1.9458 | 1.9542 |
| 63 | sin | 0.8910 | 0.8918 | 0.8926 | 0.8934 | 0.8942 | 0.8949 | 0.8957 | 0.8965 | 0.8973 | 0.8980 |
| | cos | 0.4540 | 0.4524 | 0.4509 | 0.4493 | 0.4478 | 0.4462 | 0.4446 | 0.4431 | 0.4415 | 0.4399 |
| | tan | 1.9626 | 1.9711 | 1.9797 | 1.9883 | 1.9970 | 2.0057 | 2.0145 | 2.0233 | 2.0323 | 2.0413 |
| 64 | sin | 0.8988 | 0.8996 | 0.9003 | 0.9011 | 0.9018 | 0.9026 | 0.9033 | 0.9041 | 0.9048 | 0.9056 |
| | cos | 0.4384 | 0.4368 | 0.4352 | 0.4337 | 0.4321 | 0.4305 | 0.4289 | 0.4274 | 0.4258 | 0.4242 |
| | tan | 2.0503 | 2.0594 | 2.0686 | 2.0778 | 2.0872 | 2.0965 | 2.1060 | 2.1155 | 2.1251 | 2.1348 |
| 65 | sin | 0.9063 | 0.9070 | 0.9078 | 0.9085 | 0.9092 | 0.9100 | 0.9107 | 0.9114 | 0.9121 | 0.9128 |
| | cos | 0.4226 | 0.4210 | 0.4195 | 0.4179 | 0.4163 | 0.4147 | 0.4131 | 0.4115 | 0.4099 | 0.4083 |
| | tan | 2.1445 | 2.1543 | 2.1642 | 2.1742 | 2.1842 | 2.1943 | 2.2045 | 2.2148 | 2.2251 | 2.2355 |
| 66 | sin | 0.9135 | 0.9143 | 0.9150 | 0.9157 | 0.9164 | 0.9171 | 0.9178 | 0.9184 | 0.9191 | 0.9198 |
| | cos | 0.4067 | 0.4051 | 0.4035 | 0.4019 | 0.4003 | 0.3987 | 0.3971 | 0.3955 | 0.3939 | 0.3923 |
| | tan | 2.2460 | 2.2566 | 2.2673 | 2.2781 | 2.2889 | 2.2998 | 2.3109 | 2.3220 | 2.3332 | 2.3445 |
| 67 | sin | 0.9205 | 0.9212 | 0.9219 | 0.9225 | 0.9232 | 0.9239 | 0.9245 | 0.9252 | 0.9259 | 0.9265 |
| | cos | 0.3907 | 0.3891 | 0.3875 | 0.3859 | 0.3843 | 0.3827 | 0.3811 | 0.3795 | 0.3778 | 0.3762 |
| | tan | 2.3559 | 2.3673 | 2.3789 | 2.3906 | 2.4023 | 2.4142 | 2.4262 | 2.4383 | 2.4504 | 2.4627 |
| 68 | sin | 0.9272 | 0.9278 | 0.9285 | 0.9291 | 0.9298 | 0.9304 | 0.9311 | 0.9317 | 0.9323 | 0.9330 |
| | cos | 0.3746 | 0.3730 | 0.3714 | 0.3697 | 0.3681 | 0.3665 | 0.3649 | 0.3633 | 0.3616 | 0.3600 |
| | tan | 2.4751 | 2.4876 | 2.5002 | 2.5129 | 2.5257 | 2.5386 | 2.5517 | 2.5649 | 2.5782 | 2.5916 |
| 69 | sin | 0.9336 | 0.9342 | 0.9348 | 0.9354 | 0.9361 | 0.9367 | 0.9373 | 0.9379 | 0.9385 | 0.9391 |
| | cos | 0.3584 | 0.3567 | 0.3551 | 0.3535 | 0.3518 | 0.3502 | 0.3486 | 0.3469 | 0.3453 | 0.3437 |
| | tan | 2.6051 | 2.6187 | 2.6325 | 2.6464 | 2.6605 | 2.6746 | 2.6889 | 2.7034 | 2.7179 | 2.7326 |
| 70 | sin | 0.9397 | 0.9403 | 0.9409 | 0.9415 | 0.9421 | 0.9426 | 0.9432 | 0.9438 | 0.9444 | 0.9449 |
| | cos | 0.3420 | 0.3404 | 0.3387 | 0.3371 | 0.3355 | 0.3338 | 0.3322 | 0.3305 | 0.3289 | 0.3272 |
| | tan | 2.7475 | 2.7625 | 2.7776 | 2.7929 | 2.8083 | 2.8239 | 2.8397 | 2.8556 | 2.8716 | 2.8878 |
| 71 | sin | 0.9455 | 0.9461 | 0.9466 | 0.9472 | 0.9478 | 0.9483 | 0.9489 | 0.9494 | 0.9500 | 0.9505 |
| | cos | 0.3256 | 0.3239 | 0.3223 | 0.3206 | 0.3190 | 0.3173 | 0.3156 | 0.3140 | 0.3123 | 0.3107 |
| | tan | 2.9042 | 2.9208 | 2.9375 | 2.9544 | 2.9714 | 2.9887 | 3.0061 | 3.0237 | 3.0415 | 3.0595 |
| 72 | sin | 0.9511 | 0.9516 | 0.9521 | 0.9527 | 0.9532 | 0.9537 | 0.9542 | 0.9548 | 0.9553 | 0.9558 |
| | cos | 0.3090 | 0.3074 | 0.3057 | 0.3040 | 0.3024 | 0.3007 | 0.2990 | 0.2974 | 0.2957 | 0.2940 |
| | tan | 3.0777 | 3.0961 | 3.1146 | 3.1334 | 3.1524 | 3.1716 | 3.1910 | 3.2106 | 3.2305 | 3.2506 |
| 73 | sin | 0.9563 | 0.9568 | 0.9573 | 0.9578 | 0.9583 | 0.9588 | 0.9593 | 0.9598 | 0.9603 | 0.9608 |
| | cos | 0.2924 | 0.2907 | 0.2890 | 0.2874 | 0.2857 | 0.2840 | 0.2823 | 0.2807 | 0.2790 | 0.2773 |
| | tan | 3.2709 | 3.2914 | 3.3122 | 3.3332 | 3.3544 | 3.3759 | 3.3977 | 3.4197 | 3.4420 | 3.4646 |
| 74 | sin | 0.9613 | 0.9617 | 0.9622 | 0.9627 | 0.9632 | 0.9636 | 0.9641 | 0.9646 | 0.9650 | 0.9655 |
| | cos | 0.2756 | 0.2740 | 0.2723 | 0.2706 | 0.2689 | 0.2672 | 0.2656 | 0.2639 | 0.2622 | 0.2605 |
| | tan | 3.4874 | 3.5105 | 3.5339 | 3.5576 | 3.5816 | 3.6059 | 3.6305 | 3.6554 | 3.6806 | 3.7062 |
| 75 | sin | 0.9659 | 0.9664 | 0.9668 | 0.9673 | 0.9677 | 0.9681 | 0.9686 | 0.9690 | 0.9694 | 0.9699 |
| | cos | 0.2588 | 0.2571 | 0.2554 | 0.2538 | 0.2521 | 0.2504 | 0.2487 | 0.2470 | 0.2453 | 0.2436 |
| | tan | 3.7321 | 3.7583 | 3.7848 | 3.8118 | 3.8391 | 3.8667 | 3.8947 | 3.9232 | 3.9520 | 3.9812 |
| 76 | sin | 0.9703 | 0.9707 | 0.9711 | 0.9715 | 0.9720 | 0.9724 | 0.9728 | 0.9732 | 0.9736 | 0.9740 |
| | cos | 0.2419 | 0.2402 | 0.2385 | 0.2368 | 0.2351 | 0.2334 | 0.2317 | 0.2300 | 0.2284 | 0.2267 |
| | tan | 4.0108 | 4.0408 | 4.0713 | 4.1022 | 4.1335 | 4.1653 | 4.1976 | 4.2303 | 4.2635 | 4.2972 |
| 77 | sin | 0.9744 | 0.9748 | 0.9751 | 0.9755 | 0.9759 | 0.9763 | 0.9767 | 0.9770 | 0.9774 | 0.9778 |
| | cos | 0.2250 | 0.2232 | 0.2215 | 0.2198 | 0.2181 | 0.2164 | 0.2147 | 0.2130 | 0.2113 | 0.2096 |
| | tan | 4.3315 | 4.3662 | 4.4015 | 4.4374 | 4.4737 | 4.5107 | 4.5483 | 4.5864 | 4.6252 | 4.6646 |
| 78 | sin | 0.9781 | 0.9785 | 0.9789 | 0.9792 | 0.9796 | 0.9799 | 0.9803 | 0.9806 | 0.9810 | 0.9813 |
| | cos | 0.2079 | 0.2062 | 0.2045 | 0.2028 | 0.2011 | 0.1994 | 0.1977 | 0.1959 | 0.1942 | 0.1925 |
| | tan | 4.7046 | 4.7453 | 4.7867 | 4.8288 | 4.8716 | 4.9152 | 4.9594 | 5.0045 | 5.0504 | 5.0970 |
| 79 | sin | 0.9816 | 0.9820 | 0.9823 | 0.9826 | 0.9829 | 0.9833 | 0.9836 | 0.9839 | 0.9842 | 0.9845 |
| | cos | 0.1908 | 0.1891 | 0.1874 | 0.1857 | 0.1840 | 0.1822 | 0.1805 | 0.1788 | 0.1771 | 0.1754 |
| | tan | 5.1446 | 5.1929 | 5.2422 | 5.2924 | 5.3435 | 5.3955 | 5.4486 | 5.5026 | 5.5578 | 5.6140 |
| 80 | sin | 0.9848 | 0.9851 | 0.9854 | 0.9857 | 0.9860 | 0.9863 | 0.9866 | 0.9869 | 0.9871 | 0.9874 |
| | cos | 0.1736 | 0.1719 | 0.1702 | 0.1685 | 0.1668 | 0.1650 | 0.1633 | 0.1616 | 0.1599 | 0.1582 |
| | tan | 5.6713 | 5.7297 | 5.7894 | 5.8502 | 5.9124 | 5.9758 | 6.0405 | 6.1066 | 6.1742 | 6.2432 |
| 81 | sin | 0.9877 | 0.9880 | 0.9882 | 0.9885 | 0.9888 | 0.9890 | 0.9893 | 0.9895 | 0.9898 | 0.9900 |
| | cos | 0.1564 | 0.1547 | 0.1530 | 0.1513 | 0.1495 | 0.1478 | 0.1461 | 0.1444 | 0.1426 | 0.1409 |
| | tan | 6.3138 | 6.3859 | 6.4596 | 6.5350 | 6.6122 | 6.6912 | 6.7720 | 6.8548 | 6.9395 | 7.0264 |
| 82 | sin | 0.9903 | 0.9905 | 0.9907 | 0.9910 | 0.9912 | 0.9914 | 0.9917 | 0.9919 | 0.9921 | 0.9923 |
| | cos | 0.1392 | 0.1374 | 0.1357 | 0.1340 | 0.1323 | 0.1305 | 0.1288 | 0.1271 | 0.1253 | 0.1236 |
| | tan | 7.1154 | 7.2066 | 7.3002 | 7.3962 | 7.4947 | 7.5958 | 7.6996 | 7.8062 | 7.9158 | 8.0285 |
| DEGREES | FUNCTION | 0′ | 6′ | 12′ | 18′ | 24′ | 30′ | 36′ | 42′ | 48′ | 54′ |

## NATURAL TRIGONOMETRIC FUNCTIONS, CONTINUED

| DEGREES | FUNCTION | 0.0° | 0.1° | 0.2° | 0.3° | 0.4° | 0.5° | 0.6° | 0.7° | 0.8° | 0.9° |
|---|---|---|---|---|---|---|---|---|---|---|---|
| 83 | sin | 0.9925 | 0.9928 | 0.9930 | 0.9932 | 0.9934 | 0.9936 | 0.9938 | 0.9940 | 0.9942 | 0.9943 |
| | cos | 0.1219 | 0.1201 | 0.1184 | 0.1167 | 0.1149 | 0.1132 | 0.1115 | 0.1097 | 0.1080 | 0.1063 |
| | tan | 8.1443 | 8.2636 | 8.3863 | 8.5126 | 8.6427 | 8.7769 | 8.9152 | 9.0579 | 9.2052 | 9.3572 |
| 84 | sin | 0.9945 | 0.9947 | 0.9949 | 0.9951 | 0.9952 | 0.9954 | 0.9956 | 0.9957 | 0.9959 | 0.9960 |
| | cos | 0.1045 | 0.1028 | 0.1011 | 0.0993 | 0.0976 | 0.0958 | 0.0941 | 0.0924 | 0.0906 | 0.0889 |
| | tan | 9.5144 | 9.6768 | 9.8448 | 10.02 | 10.20 | 10.39 | 10.58 | 10.78 | 10.99 | 11.20 |
| 85 | sin | 0.9962 | 0.9963 | 0.9965 | 0.9966 | 0.9968 | 0.9969 | 0.9971 | 0.9972 | 0.9973 | 0.9974 |
| | cos | 0.0872 | 0.0854 | 0.0837 | 0.0819 | 0.0802 | 0.0785 | 0.0767 | 0.0750 | 0.0732 | 0.0715 |
| | tan | 11.43 | 11.66 | 11.91 | 12.16 | 12.43 | 12.71 | 13.00 | 13.30 | 13.62 | 13.95 |
| 86 | sin | 0.9976 | 0.9977 | 0.9978 | 0.9979 | 0.9980 | 0.9981 | 0.9982 | 0.9983 | 0.9984 | 0.9985 |
| | cos | 0.0698 | 0.0680 | 0.0663 | 0.0645 | 0.0628 | 0.0610 | 0.0593 | 0.0576 | 0.0558 | 0.0541 |
| | tan | 14.30 | 14.67 | 15.06 | 15.46 | 15.89 | 16.35 | 16.83 | 17.34 | 17.89 | 18.46 |
| 87 | sin | 0.9986 | 0.9987 | 0.9988 | 0.9989 | 0.9990 | 0.9990 | 0.9991 | 0.9992 | 0.9993 | 0.9993 |
| | cos | 0.0523 | 0.0506 | 0.0488 | 0.0471 | 0.0454 | 0.0436 | 0.0419 | 0.0401 | 0.0384 | 0.0366 |
| | tan | 19.08 | 19.74 | 20.45 | 21.20 | 22.02 | 22.90 | 23.86 | 24.90 | 26.03 | 27.27 |
| 88 | sin | 0.9994 | 0.9995 | 0.9995 | 0.9996 | 0.9996 | 0.9997 | 0.9997 | 0.9997 | 0.9998 | 0.9998 |
| | cos | 0.0349 | 0.0332 | 0.0314 | 0.0297 | 0.0279 | 0.0262 | 0.0244 | 0.0227 | 0.0209 | 0.0192 |
| | tan | 28.64 | 30.14 | 31.82 | 33.69 | 35.80 | 38.19 | 40.92 | 44.07 | 47.74 | 52.08 |
| 89 | sin | 0.9998 | 0.9999 | 0.9999 | 0.9999 | 0.9999 | 1.000 | 1.000 | 1.000 | 1.000 | 1.000 |
| | cos | 0.0175 | 0.0157 | 0.0140 | 0.0122 | 0.0105 | 0.0087 | 0.0070 | 0.0052 | 0.0035 | 0.0017 |
| | tan | 57.29 | 63.66 | 71.62 | 81.85 | 95.49 | 114.6 | 143.2 | 191.0 | 286.5 | 573.0 |
| DEGREES | FUNCTION | 0′ | 6′ | 12′ | 18′ | 24′ | 30′ | 36′ | 42′ | 48′ | 54′ |

# BIBLIOGRAPHY


Calter, Paul: **Technical Mathematics,** Prentice-Hall, Inc., Englewood Cliffs, New Jersey, 1983.

Cooke, Nelson M., Herbert Adams, and Peter Dell: **Basic Mathematics for Electronics,** 5th ed., McGraw-Hill Book Company, New York, 1982.

Croft, Terrell, Clifford C. Carr, and John H. Watt: **American Electricians Handbook,** 9th ed., McGraw-Hill Book Company, New York, 1970.

Fowler, Richard J.: **Electricity Principles and Applications,** McGraw-Hill Book Company, New York, 1979.

Gilmore, Charles M.: **Instruments and Measurements,** McGraw-Hill Book Company, New York, 1980.

Hubert, Charles I.: **Electric Circuits AC/DC—An Integrated Approach,** McGraw-Hill Book Company, New York, 1982.

Leach, Donald P.: **Mathematics for Electronics,** Reston Publishing Company, Reston, Virginia, 1979.

Miller, Gary M.: **Modern Electricity and Electronics,** Prentice-Hall, Inc., Englewood Cliffs, New Jersey, 1981.

"National Electrical Code for Electric Wiring and Apparatus," NBFU Pamphlet No. 70, National Board of Fire Underwriters, New York, 1975.

New York Institute of Technology: "A Programmed Course in Basic Electricity," 2d ed., McGraw-Hill Book Company, New York, 1970.

Rexford, Kenneth B.: **Electrical Control for Machines,** Delmar Publishers, Inc., 1981.

Richter, H. P.: **Practical Electrical Wiring, Residential, Farm, and Industrial,** 10th ed., McGraw-Hill Book Company, New York, 1976.

Tokheim, Roger L.: **Digital Electronics,** McGraw-Hill Book Company, New York, 1979.

Tontsch, John W.: **Applied Electronic Math with Calculators,** Science Research Associates, Inc., Chicago, 1982.

Westinghouse Electric Corporation: "Westinghouse Electrical Maintenance Hints," Trafford, Pennsylvania, 1974.

# ANSWERS TO SELECTED PROBLEMS

Job 2-1: Page 12

1. $25^1/_2$ ft
2. $3^3/_4$ hp
3. 90 V
4. 7 Ω
5. \$2.81
6. 228 kWh
7. $37^1/_2$ h
8. $94^1/_2$ W
9. 36 lb
10. 20 ft 3 in
11. 8
12. 8
13. 66
14. $^1/_6$
15. $^1/_{50}$
16. $^1/_2$
17. $45^1/_3$
18. 16

Job 2-2: Page 15

1. $^1/_2$
3. $^1/_4$
5. $^3/_4$
7. $^1/_4$
9. $^1/_3$
11. $^3/_{100}$
13. $^3/_8$
15. $^7/_{16}$
17. $^1/_3$
19. $^3/_5$
21. $^2/_1$
23. $^{49}/_1$
25. $^2/_5$
27. $^3/_4$

Job 2-2: Page 17 (Top)

1. $^9/_4$
3. $^{20}/_9$
5. $^{13}/_{10}$
7. $^{25}/_6$
9. $^{37}/_8$
11. $^{10}/_3$

Job 2-2: Page 17 (Bottom)

1. $1^3/_4$
3. $4^1/_3$
5. $3^3/_4$
7. $1^3/_5$
9. $3^3/_{10}$
11. 10
13. $9^2/_3$
15. 16
17. $3^3/_5$
19. 12
21. 7
23. $1^7/_{10}$
25. $5^2/_9$
27. 12
29. $12^1/_4$

Job 2-2: Page 19

1. 4
3. $^3/_{10}$
5. $2^1/_2$
7. $^1/_{15}$
9. 6
11. $5^1/_4$ hp
13. $^1/_2$ hp
15. 3200 W
17. $^1/_3$
19. $^5/_7$

Job 2-2: Page 20

1. $^2/_3$
3. $9^1/_2$
5. $^1/_3$
7. 5
9. $^2/_3$
11. $4^1/_2$ V
13. $7^1/_8$ h
15. $67^1/_2$ ft
17. $4^1/_4$ hp

Job 2-3: Page 24

1. 55.8 V
3. 44 V
5. 117.6 V
7. 3.72 V
9. 18.75 V
11. 0.05 V
13. $V_1 = 150$ V
    $V_2 = 100$ V
    $V_3 = 50$ V

Job 2-4: Page 26

1. 5.55 A
2. 0.015 25 μF
3. 0.0378 in
4. 0.35 A
5. 2.6 V
6. 1.185 in
7. 65.75 V
8. \$0.23
9. 0.004 Ω
10. 12.55
11. 0.4 A
12. (*a*) Four and three tenths
    (*b*) Three hundred and fifty-nine thousandths
    (*c*) Forty-one hundredths
13. 0.015
14. (*a*) 0.25

(b) 0.571
(c) 0.625
(d) 0.0625
15. 6.24
16. 1.1, 0.30, 0.050, 0.0070
17. 0.05
18. 2.25

**Job 2-5: Page 29 (Top)**
1. 0.7
3. 0.114
5. 0.06
7. 0.018
9. 0.11
11. 0.013
13. 0.045
15. 0.6
17. 0.4, 0.16, 0.007
19. 0.8, 0.06, 0.040
21. 0.5, 0.18, 0.051
23. 0.4, 0.236, 0.1228
25. 0.19, 0.08, 0.004

**Job 2-5: Page 29 (Bottom)**
1. 2.3
3. 3.144
5. 2.025
7. 2.020
9. 1.002
11. 9.145

**Job 2-5: Page 32 (Top)**
1. 0.25
3. 0.625
5. 0.4
7. 0.15
9. 0.875
11. 0.444
13. 0.094
15. 0.625
17. 0.02
19. 0.192
21. 0.375 A

**Job 2-5: Page 32 (Bottom)**
1. 0.625
3. 0.5625
5. 2.25
7. 1.140625
9. 3/8
11. 29/64
13. 1/32
15. 13/32
17. 3/4
19. 9/64
21. 2 13/16
23. 1 27/32
25. The lamp

**Job 2-6: Page 36 (Top)**
1. 7.2
3. 3090
5. 3700
7. 0.6
9. 2700
11. 0.078
13. 15,400
15. 0.5
17. 80
19. 2340

**Job 2-6: Page 36 (Bottom)**
1. 6.5
3. 880
5. 0.06
7. 0.835
9. 0.6538
11. 0.000 45
13. 0.0085
15. 0.0286
17. 0.002
19. 0.0398

**Job 2-6: Page 39**
1. 350 mm
3. 24 cm
5. 360 dm
7. 500 cm
9. 2450 mm
11. 12.5 cm
13. 42 dam
15. 750 hm
17. 390 m
19. 20,000 m
21. 150 dm
23. 345.6 mm

**Job 2-7: Page 40**
1. 13.492
3. 22.832
5. 11.967
7. 45.46 V
9. 0.1019 in
11. 0.0315 A
13. 25.38 mm
15. $44.90

**Job 2-7: Page 41**
1. 0.23
3. 0.34
5. 7.56
7. 0.317
9. 2.92
11. 5.9
13. 2.39
15. 0.262
17. 2.62
19. 0.515
21. (a) 0.304
(b) 1.106
(c) 0.39
(d) 5.75
(e) 1.98
23. 0.1446 in
25. 0.43 mm
27. 0.000 15 μF

**Job 2-7: Page 42**
1. 0.41
3. 0.463
5. 30.31
7. 0.221
9. 0.0186
11. 4.77
13. 4.55 A
15. (a) 250 mA
(b) 25 mA
(c) 2500 mA
17. 41.850 in
19. 69.6 A·h

**Job 2-7: Page 44**
1. 13
3. 161
5. 4.7
7. 786
9. 2.32
11. 200 ft
13. 60.7
15. 40

**Job 2-7: Page 46**
1. $A = 3.09$ in, $B = 2.8475$ in
3. 0.082
5. (a) 735
(b) 81,700,000
(c) 3.5
(d) 0.04
(e) 0.0007
(f) 0.006 28
(g) 47.5
(h) 0.000 397
7. $43.31
9. 22.62 mm
11. 1.875 A
13. 163.9 W
15. No. 10

**Job 3-1: Page 48**
1. 0.6 A
2. 300 Ω
3. 111 V
4. 4 A
5. $\text{kW} = \dfrac{I \times V}{1000}$
6. 2 A
7. 25
8. 0.125 A
9. 11 A
10. 100 V

**Job 3-2: Page 52**
1. $P = IV$
3. $\text{Eff} = \dfrac{P_o}{P_i}$
5. $X_C = \dfrac{159{,}000}{fC}$
7. $P = 2L + 2W$
9. $I_T = \dfrac{V_T}{R_1 + R_2}$

**Job 3-2: Page 55**
1. 14
3. 20
5. 18
7. 6 1/4
9. 7
11. 135 ft$^2$
13. 10 A
15. 140.4 Ω
17. 20 Ω
19. 1/4 A
21. 72.6 Ω

**Job 3-3: Page 56**
1. −$3
3. +23°

5. +2 dB
7. −10 mph
9. −5 blocks
11. −4 V

Job 3-3: Page 58
1. +12
3. +6
5. −17
7. +7
9. −17
11. −14
13. +16
15. −52
17. +1.9
19. $+^1/_4$
21. $-^5/_6$
23. $-5^{11}/_{16}$
25. +3
27. +7
29. −25
31. +0.5
33. −2.9

Job 3-3: Page 59 (Multiplication)
1. +18
3. −8
5. +30
7. $+18R$
9. $-10R$
11. $+10R$
13. 0
15. −104
17. $-85R$
19. $-144R$
21. −6
23. $-2.4I$
25. $-3R$
27. −14.72
29. $-5.1R$

Job 3-3: Page 59 (Division)
1. +4
3. −4
5. +12
7. $-6R$
9. $-^1/_2$
11. $+^1/_2$
13. 0
15. −7
17. $-7^1/_2$
19. −3
21. $+3^1/_5$
23. +130
25. −2
27. −3.2
29. −62.5

Job 3-4: Page 60
1. 200
2. 360
3. 20 V
4. 65 Ω
5. 0.000 000 5 F

Job 3-5: Page 62
1. 3600 cmil
3. 1024 ft
5. 0.45 A
7. 0.032 V
9. 132.25 W

Job 3-6: Page 63 (Top)
1. 7.2
3. 3090
5. 3700
7. 0.6
9. 2700
11. 0.078
13. 15,400
15. 1
17. 80
19. 234,000

Job 3-6: Page 63 (Bottom)
1. 65
3. 880
5. 0.06
7. 0.835
9. 0.6538
11. 0.0045
13. 0.0085
15. 0.0286
17. 0.000 02

Job 3-6: Page 65 (Top)
1. 25
3. 0.15
5. 6.25
7. 0.0025
9. 0.007 54

Job 3-6: Page 65 (Bottom)
1. 1920
3. 850
5. 7200
7. 0.000 045
9. 88,000,000
11. 3000
13. 3,000,000
15. 0.0006

Job 3-6: Page 67 (Top)
1. $6 \times 10^3$
3. $1.5 \times 10^5$
5. $2.35 \times 10^5$
7. $4.96 \times 10^3$
9. $9.8 \times 10^2$
11. $1.25 \times 10$
13. $4.82 \times 10$
15. $8.8 \times 10^8$
17. $3.83 \times 10^4$
19. $1.75 \times 10^6$
21. $4.83 \times 10^5$

Job 3-6: Page 67 (Bottom)
1. $6 \times 10^{-3}$
3. $3.5 \times 10^{-3}$
5. $4.56 \times 10^{-1}$
7. 7.85
9. $9.65 \times 10^{-2}$
11. $5 \times 10^{-1}$
13. $8.15 \times 10^{-3}$
15. $7.25 \times 10$
17. $6 \times 10^{-1}$
19. $3.6 \times 10^{-2}$

Job 3-6: Page 69
1. 5
3. 6,400,000 or $6.4 \times 10^6$
5. 30,000 or $3 \times 10^4$
7. 120
9. 0.03
11. $5 \times 10^{-7}$
13. 960
15. 628,000 or $6.28 \times 10^5$

Job 3-6: Page 70
1. $10^5$
3. 120
5. 0.0002
7. $2 \times 10^{-7}$
9. 8
11. $5 \times 10^{-3}$
13. (*a*) 53 Ω
(*b*) 63 Ω
(*c*) 3.18 Ω

Job 3-7: Page 73
1. 0.225 A
3. 3,500,000 Ω
5. 550,000 Hz
7. 0.07 MΩ
9. 65 mA
11. 0.075 V
13. 0.006 A
15. 0.0039 A
17. 5000 pF
19. 1,000,000 Hz
21. 8 mV
23. 60 pF
25. 0.000 000 15 F
27. 8 kW
29. 4 μA
31. 500,000 MHz
33. 0.015 GV

Job 3-8: Page 75
1. 100 μV or 0.1 mV
3. 0.6 kW
5. 0.5 μA
7. 215 kΩ
9. 2625 pF
11. $4 \times 10^{-5}$
13. 14.1 V

Job 3-9: Page 80
1. 5
3. 6
5. 9
7. $13^2/_3$
9. 5.85
11. 40
13. 400
15. 200
17. 800
19. 390
21. $^1/_{16}$

Job 3-9: Page 81
1. 2 Ω
3. 13 Ω
5. $1^1/_2$ A

7. 20 A
9. 25 Ω
11. 0.133 Ω
13. 1200 Ω
15. 2.25 A
17. 3.67 A
19. 0.0015 Ω
21. 1.76 Ω
23. 0.002 A
25. 21.43 Ω

**Job 3-10: Page 85**

1. 110 V
3. Yes
5. 18 V
7. 0.0007 Ω
9. 10,000 Ω
11. 0.004 Ω
13. 10 A
15. 0.03 A
17. 0.0081 A

**Job 4-1: Page 91**

1. (*a*) 24 V
   (*b*) 0.2 A
   (*c*) 120 Ω
3. (*a*) 42.9 V
   (*b*) 0.3 A
   (*c*) 143 Ω

**Job 4-2: Page 94**

1. (*a*) 4 A
   (*b*) 74 V
   (*c*) 18.5 Ω
3. (*a*) Dash = 2.5 Ω, tail = 5 Ω
   (*b*) 7.5 Ω
   (*c*) 0.8 A
5. 0.0088 A
7. (*a*) 4.31 V
   (*b*) 7.7 V
   (*c*) 12.01 V

**Job 4-3: Page 98**

1. 550 V
3. 4.8 V
5. 0.012 A
7. (*a*) 0.8 A
   (*b*) 1.6 V
   (*c*) 2 Ω
9. 157.5 V

**Job 4-4: Page 100**

1. 54 V
3. 6 V
5. (*a*) 45, 75, 150 V
   (*b*) 270 V
   (*c*) 18,000 Ω
7. (*a*) 1.2 A
   (*b*) 132 V
   (*c*) 110 Ω
9. $V_T = 32$ V
   $I_T = 0.4$ A
   $I_1 = I_2 = I_3 = 0.4$ A
   $R_1 = 20\ \Omega$
   $R_2 = 25\ \Omega$
   $R_3 = 35\ \Omega$

**Job 4-5: Page 101**

1. 29
2. 26
3. 1.1 A
4. 17.4
5. 7
6. 45
7. 12
8. 3
9. 25 Ω
10. $103^1/_3$ Ω

**Job 4-6: Page 103**

1. 6
3. 70
5. 52
7. 25
9. 7.3
11. 28.2
13. $5^1/_4$
15. 6.1
17. 56.1
19. 120
21. 73
23. 0.000 08
25. 1.45

**Job 4-6: Page 105**

1. 17
3. 22
5. 5
7. 12.5
9. 79
11. $10^3/_4$
13. $15^1/_4$
15. 80
17. 0.099
19. $21.83
21. 172
23. 17.05
25. 109

**Job 4-6: Page 108**

1. 11
3. 6
5. 10
7. 12
9. 8
11. 1.54
13. 10
15. 11.5
17. 0.021
19. 11.48
21. $1^1/_2$
23. 180
25. 30
27. 76.5
29. 0.000 25

**Job 4-7: Page 110**

1. 25 Ω
3. 8 Ω
5. 84 Ω
7. 90 Ω
9. 202 V
11. $I_{max} = 11$ A
    $I_{min} = 2$ A
13. 0.025 Ω

**Job 4-8: Page 113**

1. $I_T = I_1 = I_3 = 2$ A
   $V_T = 110$ V
   $R_T = 55\ \Omega$
   $R_1 = 12\ \Omega$
   $V_2 = 60$ V
   $V_3 = 26$ V
3. 12 Ω
5. 28 Ω, 2.67 A, 74.8 V
7. (*a*) 2.2 A
   (*b*) 10 Ω
9. 30 Ω

**Job 5-1: Page 119**

1. (*a*) 5 A
   (*b*) $V_1 = V_2 = V_3 = 110$ V
   (*c*) 22 Ω
3. 4 A
5. 40 lamps; 30 lamps
7. 9.1 A; yes
9. 9.5 A, 1.26 Ω

**Job 5-2: Page 121**

1. (*a*) 18 V
   (*b*) $I_1 = 6$ A, $I_2 = 3$ A, $I_T = 9$ A
   (*c*) 2 Ω
3. 2 A, 6 A; $I_T = 12$ A
5. (*a*) $V_T = V_1 = V_2 = V_3 = 114$ V
   (*b*) $I_2 = 6$, $I_3 = 2$, $I_T = 20$ A
   (*c*) $R_1 = 9.5\ \Omega$
   $R_T = 5.7\ \Omega$
7. (*a*) 30 Ω for 3 lamps; 15 Ω for fourth lamp
   (*b*) 0.25 A
   (*c*) $1^1/_2$ V
   (*d*) 6 Ω
9. $I_T = 1.05$ A
   $V_T = 6.8$ V
   $R_T = 6.47\ \Omega$

**Job 5-3: Page 122**

1. $^{17}/_{24}$ hp
2. $^1/_6$
3. $^9/_{64}$ in
4. $^{15}/_{32}$ in
5. $22^7/_8$ lb
6. $^{19}/_{30}$
7. $1^{13}/_{16}$
8. $2^{11}/_{24}$
9. $1^{11}/_{18}$
10. $1^{27}/_{28}$
11. $2^1/_8$
12. $4^{15}/_{16}$
13. $4^1/_4$
14. $2^{11}/_{16}$
15. $^1/_{60}$

**Job 5-4: Page 127**

1. $1^3/_8$
3. $1^{11}/_{24}$
5. $^1/_6$

7. $^{2}/_{5}$
9. $^{1}/_{120}$
11. $^{57}/_{64}$
13. $^{3}/_{20}$
15. $1^{11}/_{32}$
17. $15^{7}/_{16}$
19. $^{15}/_{16}$
21. $^{11}/_{600}$ S
23. $29^{23}/_{24}$ ft

**Job 5-5: Page 131**

1. $^{1}/_{2}$
3. $^{3}/_{8}$
5. $^{5}/_{16}$
7. $^{1}/_{12}$
9. $^{1}/_{6}$
11. $^{1}/_{12}$
13. $^{1}/_{80}$
15. $^{1}/_{500}$
17. $4^{3}/_{4}$
19. $7^{1}/_{3}$
21. $5^{7}/_{16}$
23. $^{1}/_{300}$
25. $76^{3}/_{4}$ ft
27. $^{7}/_{32}$ in
29. $^{7}/_{8}$ A

**Job 5-6: Page 132**

1. 1800
2. $62^{1}/_{2}$
3. 2 A
4. 843 Ω
5. 6 V

**Job 5-7: Page 133**

1. 24
3. 12
5. 1
7. 2
9. 8
11. 6750
13. 54
15. 4000
17. 30
19. 93.97

**Job 5-7: Page 136**

1. 2
3. 4
5. $3^{1}/_{3}$
7. 15
9. 15
11. 1.2
13. 10
15. $3^{1}/_{2}$
17. 7.8
19. 315
21. 120 turns
23. $33^{1}/_{3}$
25. 60 Ω
27. 500 lb
29. 1.02 Ω

**Job 5-7: Page 138**

1. 6
3. 80
5. 51
7. 27
9. 25
11. 10
13. 1.44
15. $5^{1}/_{3}$
17. 0.1
19. 1.84

**Job 5-8: Page 140**

1. 1.5 Ω
3. 1.64 Ω
5. 1.9 Ω
7. 2.5 Ω
9. $R_A = 12.4\ \Omega$
   $R_B = 11.25\ \Omega$
   $R_C = 9.47\ \Omega$
11. 60 Ω
13. 1200 Ω
15. 20 Ω

**Job 5-8: Page 142 (Top)**

1. 50 Ω
3. 0.75 Ω
5. (*a*) 7.33 Ω
   (*b*) 15 A

**Job 5-8: Page 142 (Bottom)**

1. 20 Ω
3. 24 Ω
5. 7143 Ω
7. 3.53 kΩ
9. 28.6 Ω

**Job 5-9: Page 144**

1. 40 V
3. 4.8 V
5. 14.4 V
7. 18.5 V
9. 4.8 V

**Job 5-10: Page 148**

1. $I_1 = I_2 = 4$ A
   (The current in a circuit divides equally between equal parallel branches.)
3. $V_T = 6$ V
   $I_1 = I_2 = 4$ A
5. 9 A, 6 A, 4.5 A
7. $I_2 = 0.024$ A
   $I_1 = 0.006$ A
9. (*a*) 2.14 Ω
   (*b*) 28 V
   (*c*) $V_1 = V_2 = V_3 = 28$ V
   (*d*) $I_1 = 5.6$ A
   $I_2 = 4$ A
   $I_3 = 3.5$ A

**Job 5-11: Page 151**

1. (*a*) 5.75 A
   (*b*) 110 V
   (*c*) 19.1 Ω
3. $I_T = 47$ A
   $R_T = 9.36\ \Omega$
   $V_1 = V_2 = V_3 = 440$ V
   $I_2 = 22$ A
   $R_1 = 44\ \Omega$
   $R_3 = 29.3\ \Omega$
5. 26.7 Ω
7. (*a*) 10 Ω
   (*b*) 8 Ω
   (*d*) Series
   (*e*) 18 Ω
   (*f*) 20 A
9. 600 Ω
11. 85.7 Ω
13. (*a*) 2550 Ω
    (*b*) 102 V

**Job 6-2: Page 157**

1. $R_T = 24\ \Omega$
   $I_T = 5$ A
   $I_1 = I_2 = 3$ A
   $I_3 = 2$ A
   $V_1 = 30$ V
   $V_2 = 90$ V
   $V_3 = 120$ V
3. $R_T = 6\ \Omega$
   $I_T = 2$ A
5. (*a*) 12,000 Ω
   (*b*) 0.03 A
   (*c*) 360 V
   (*d*) 0.012 A
   (*e*) 0.018 A
   (*f*) 54 V
   (*g*) 306 V
7. 8 Ω

**Job 6-3: Page 168**

1. $I_T = 10$ A
   $I_1 = 10$ A
   $I_2 = 7.5$ A
   $I_3 = 2.5$ A
   $R_T = 13\ \Omega$
   $V_1 = 100$ V
3. 0.5 A
5. (*a*) 6640 Ω
   (*b*) 0.0015 A
   (*c*) 3 V
7. $R_T = 30\ \Omega$
9. (*a*) 60 kΩ
   (*b*) 90 kΩ
11. $R_T = 25\ \Omega$

**Job 6-4: Page 176**

1. $V_l = 2.4$ V
   $V_L = 114.6$ V
3. 0.25 Ω
5. 6.048 V, No. 14
7. 113.3 V

**Job 6-5: Page 180**

1. (*a*) $V_1 = 15.2$ V
   $V_2 = 4.8$ V
   (*b*) 101.8 V
   (*c*) 97 V
3. $V_A = 112$ V
   $V_B = 109.6$ V
5. $V_G = 126.4$ V
   $V_B = 111.1$ V
7. 125.8 V
9. $V_{M_1} = 109$ V
   $V_{M_2} = 110$ V

**Job 7-1: Page 185**

1. 480 W
3. 22 W

5. 374 W
7. 2.2 W
9. 2.3 W
11. 900 W
13. 1 W
15. 2508 W

**Job 7-2: Page 188**

1. 11.2 W
3. Yes. (0.135 W developed)
5. 58.8 W
7. 2246 W
9. $I_T = 1$ A
   $P_1 = 2$ W
   $P_2 = 0.64$ W
   $P_3 = 3.84$ W
   $P_4 = 1.92$ W
   $P_5 = 1.6$ W
   $P_6 = 3$ W
   $P_T = 13$ W

**Job 7-3: Page 189**

1. 5 A
3. 50 A
5. 1.2 W
7. 0.5 A
9. 5.45 A
11. 20 A

**Job 7-4: Page 190**

1. 720 W
3. 6.25 W, 13 W
5. 0.006 W
7. 3361 W
9. 661 W
11. (*a*) 60 Ω
    (*b*) 2 A
    (*c*) 240 W
13. (*a*) 100 Ω
    (*b*) 2 A
    (*c*) 400 W

**Job 7-5: Page 193**

1. $10^{10}$
3. $9 \times 10^6$
5. $10^{10}$
7. $64 \times 10^6$
9. $121 \times 10^4$
11. $625 \times 10^{-10}$
13. $2.25 \times 10^6$
15. $1.44 \times 10^{-6}$

**Job 7-6: Page 195**

1. 2.25 kW
3. 132.3 W
5. 6.5 W
7. 6.36 A; 16.2 W
9. 11.1 W; 28.9 W

**Job 7-7: Page 196**

1. 8
2. 13
3. 4.2
4. 59
5. 112
6. 7.6
7. 3.74
8. 23.8
9. 276.5
10. 5.48
11. 0.807
12. 932.6

**Job 7-8: Page 202**

1. 59
3. 3.9
5. 137
7. 6.32
9. 8.06

**Job 7-9: Page 204**

1. $10^4$
3. $4 \times 10^2 = 400$
5. $4 \times 10^2 = 400$
7. 0.005
9. 60
11. $6.32 \times 10^2 = 632$
13. 4
15. $2 \times 10^2 = 200$
17. $1.732 \times 10^3 = 1732$
19. $12 \times 10^3$
21. $1.5 \times 10^{-2}$
23. 250

**Job 7-10: Page 206**

1. 10 A
3. 2 A
5. 20 V
7. 144 Ω
9. 38.7 A
11. 50 mA

**Job 7-11: Page 208**

1. 660 W
3. 550 W
5. 43.2 W
7. 6.25 V
9. (*a*) $R_T = 16.8$ Ω
   (*b*) $P_T = 6.05$ W
11. 12.3 W
13. 90 Ω
15. $I_l = 20$ A
    $V_L = 110$ V
    $P_L = 2200$ W
    No. = 44 lamps
17. (*a*) 116.5 V
    (*b*) 112.5 W
    (*c*) 46.75 W
    (*d*) 757.25 W

**Job 8-1: Page 212**

1. $8x$
3. $6R$
5. $12x$
7. $5x$
9. $4.4R$
11. $3.5x$
13. $\frac{5}{6} T$
15. $9R$
17. $7.5x$
19. $4.5x$
21. $6.6R$
23. $2\frac{7}{8}x$
25. $3x$
27. $1.12x$

**Job 8-2: Page 214**

1. $10R + 7I$
3. $2x + 9$
5. $6R + 8$
7. $3R + 60$
9. $6x + 5y$
11. $0.9I + 1.9R$
13. $8I_1 + 10I_2$
15. $4x + 5y + 10$
17. $1.2I_2 + 2$
19. $2.4 + 3.4I$
21. $6x + 55$
23. $0.45R + 25$
25. $19I_3 + 3I_2$

**Job 8-3: Page 218**

1. 7
3. 5
5. 5
7. 5
9. 12
11. 5
13. $5\frac{1}{3}$
15. 200
17. 30
19. 70
21. 5
23. 0.8
25. 5
27. 2
29. $24\frac{1}{2}$
31. 7
33. 3
35. 6
37. 6
39. $\frac{1}{4}$
41. 0.7
43. $6\frac{2}{3}$
45. $10\frac{4}{5}$
47. 2
49. 40
51. 6 in, 3 in, 12 in
53. $A = 13, B = 0.4, C = 21.6$
55. $A = 32, B = 16, C = 8$
57. (*a*) $2R + 150 = 200$
    (*b*) $R = 25$ Ω
59. Motor = 4 A
    Lamp = 2 A
    Iron = 7 A
61. $V_1 = 30$ V
    $V_2 = 60$ V
    $V_3 = 30$ V

**Job 8-4: Page 221**

1. 10
3. 30
5. 16
7. 15

9. 10
11. 13.2
13. 3
15. 28
17. 12
19. 14 ft

Job 8-5: Page 223
1. $-2x + 7$
3. $-4I_1 - 2I_2$
5. $-2x - 4y$
7. $-3x - 7$
9. $-20 - 2x$

Job 8-6: Page 225
1. 2
3. 9
5. 4
7. 10
9. 5
11. 7
13. 1/3
15. 5
17. 4
19. 9
21. 7
23. 8
25. 5
27. 1/3
29. 2.1

Job 8-7: Page 226
1. −8
3. −5
5. −1/2
7. −8
9. +7
11. −4
13. −1
15. −5
17. −11
19. −3
21. −2
23. −7

Job 8-8: Page 228
1. $6 - 8x$
3. $-3x + 12$
5. $3x - 5$
7. $6I - 24$
9. $1 - 2R$
11. $29 - 3x$
13. $20 + 2I_1 + 2I_2$
15. $7R - 12$
17. $5 - 2R$
19. $3I_1 + I_2 - 24$

Job 8-9: Page 230
1. 1
3. −2
5. 2
7. −3
9. 3
11. −1
13. 3
15. −4
17. 7
19. 4
21. −21
23. 1/2
25. 3
27. 5 1/2
29. 1.1
31. 25
33. 48 Ω
35. 650 W

Job 8-10: Page 233
1. 2
3. 40
5. 30
7. 8
9. 7
11. 6
13. 300
15. 10
17. 2
19. 17 1/7
21. 108

Job 8-11: Page 239
1. $x = 3, y = 2$
3. $x = 5, y = 2$
5. $I_1 = 6, I_2 = 2$
7. $V = 7, R = 2$
9. $I = 7, V = -2$
11. $a = 1, b = -3$
13. $x = 5, y = 1$
15. $a = 6, b = 10$
17. $I_2 = 7, I_3 = -3$
19. $I_2 = 5, I_3 = 2$

Job 8-12: Page 243
1. $x = 5, y = 3$
3. $x = 5, y = 2$
5. $x = 2, y = 7$
7. $x = 5, y = 3$
9. $x = 1, y = 4$
11. $x = 1, y = 4$

Job 9-3: Page 254
1. (a) $I_T = 4$ A
(b) $V_2 = 20$ V
$V_4 = 12$ V
3. (a) $I_T = 2.4$ A
(b) $V_2 = 14.4$ V
$V_3 = 9.6$ V
$V_5 = 12$ V
5. (a) $I_T = 3$ A
(b) $V_2 = 12$ V
$V_4 = 15$ V
$V_6 = 6$ V
7. 8.43 A

Job 9-4: Page 258
1. $V_T = V_1 = V_2$
$= V_3 = 90$ V
$I_T = 15$ A
$I_2 = 4.5$ A
$I_3 = 1.5$ A
$R_2 = 20\ \Omega$
3. $V_T = V_1 = V_2$
$= V_3 = 300$ V
$I_T = 20$ A
$I_2 = 2.5$ A
$I_3 = 7.5$ A
$R_2 = 120\ \Omega$
5. $V_T = 108$ V
$R_3 = 6\ \Omega$
7. $I_1 = 1.95$ A
$I_2 = 0.65$ A
$V_T = 7.80$ V
9. Same as Prob. 7
11. (a) $I_{16} = 6$ A
$I_{48} = 2$ A
$I_{24} = 4$ A
(b) $V_G = 96$ V

Job 9-5: Page 272
Set No. 1
1. $x = 2$ A, $y = 3$ A,
$z = 5$ A
3. $x = 3$ A, $y = 2$ A,
$z = 1$ A
5. $v = 12$ A, $w = 3$ A,
$x = 5$ A, $y = 10$ A,
$z = 2$ A, $V_T = 104$ V
7. $x = 12$ A, $y = 4$ A,
$z = 8$ A

Job 9-5: Page 273
Set No. 2
1. $x = 4$ A, $y = 2$ A,
$z = 6$ A
3. $x = 1$ A, $y = 0.5$ A,
$z = 0.5$ A
5. $v = 6$ A, $w = 5$ A,
$x = 3$ A, $y = 8$ A,
$V_T = 170$ V, $z = 2$ A
7. $x = 7$ A, $y = 3$ A,
$z = 4$ A

Job 9-5: Page 274
Set No. 3
1. $x = 0.5$ A,
$y = 1.5$ A, $z = 2$ A
3. $x = 0.8$ A,
$y = 0.6$ A, $z = 0.2$ A
5. $v = 1$ A, $w = 2.5$ A,
$x = 0.5$ A, $y = 3$ A,
$z = 2$ A, $V_T = 26$ V
7. $x = 5.5$ A, $y = 2$ A,
$z = 3.5$ A

Job 9-6: Page 277
9. 223.5 W

Job 9-7: Page 283
1. $R_a = R_b = R_c$
$= 20\ \Omega$
3. $R_a = 6\ \Omega$
$R_b = 4\ \Omega$
$R_c = 2.4\ \Omega$
5. $R_{AD} = 3.67\ \Omega$
7. $R_{AD} = 10.95\ \Omega$
9. (a) $R_T = 5\ \Omega$
(b) $v = 5$ A,

$w = 5$ A,
$x = 4.5$ A,
$y = 5.5$ A,
$z = 0.5$ A
(*c*) $V_T = 50$ V
11. (*a*) $R_T = 12.67\ \Omega$
(*b*) $v = 4$ A,
$w = 6$ A,
$x = 7.87$ A,
$y = 2.13$ A,
$z = 1.87$ A
(*c*) $V_T = 126.7$ V

**Job 9-8: Page 294**
1. $I_L = 1$ A
$V_L = 10$ V
3. $I_L = 12$ A
$V_L = 4.32$ V
5. $I_L = 0.2$ A
$V_L = 4.6$ V
7. $I_5 = 3$ A
$V_5 = 18$ V
9. $I_L = 0.0025$ A
$V_L = 25$ V

**Job 10-1: Page 303**
1. 5.56
3. 3
5. 0.051
7. 0.747
9. 0.21

**Job 10-2: Page 304**
1. 15 mA
3. 40.4 mA
5. 70.7 A

**Job 10-3: Page 308**
1. (*a*) 290,000
(*b*) 168 V
3. (*a*) 2995
(*b*) 21.6 V
5. (*a*) 17,000
(*b*) 51,000
7. (*a*) 45,000
(*b*) 95,000
(*c*) 145,000

**Job 10-4: Page 310**
1. 0.182
3. 60.6 mA
5. 333 Ω/V
7. 50 μA
9. 1500
11. (*a*) 2250
(*b*) 12.8 V
13. 0.56

**Job 10-5: Page 314**
1. $R_1 = 50{,}000\ \Omega$
$R_2 = 7143\ \Omega$
$R_3 = 20{,}000\ \Omega$
3. $R_1 = 10{,}000\ \Omega$
$R_2 = 3333\ \Omega$
$R_3 = 1111\ \Omega$
5. $R_1 = 10{,}000\ \Omega$
$R_2 = 7143\ \Omega$
$R_3 = 12{,}500\ \Omega$

**Job 10-6: Page 319**
1. 1250 Ω
3. 11.1 Ω
5. $V_k = 33.3$ V
$V_x = 66.7$ V
7. (*a*) $V_1 = 25.6$ V
(*b*) $V_2 = 58.1$ V
(*c*) $V_3 = 116.3$ V
9. 130 V
11. 0.1 V

**Job 10-7: Page 322**
1. 843 Ω
3. 6.04 Ω
5. 1289 Ω
7. 465 Ω
9. 0.16 Ω

**Job 10-8: Page 326**
1. $z = 9$ A
3. $z = 3$ A
5. $I_5 = 2.25$ A

**Job 10-9: Page 332**
1. (*a*) $I_{RL} = 1.44$ mA
(*b*) $V_{RL} = 7.2$ V
(*c*) $V_C = -2.8$ V
3. (*a*) $I_C = 0.25$ mA
(*b*) $I_E = 0.25$ mA
(*c*) $V_{RL} = 10$ V
(*d*) $V_C = -3$ V
(*e*) $V_{RE} = 0.5$ V
(*f*) $V_{CE} = 2.5$ V

**Job 10-10: Page 336**
1. (*a*) $I_C = 1$ mA
(*b*) $V_{RL} = 1$ V
(*c*) $V_C = -9$ V
3. (*a*) $I_E = 3.67$ mA
(*b*) $V_E = -0.73$ V
(*c*) $I_C = 3.63$ mA
(*d*) $V_{RL} = 3.63$ V
(*e*) $V_C = -4.37$ V

**Job 10-11: Page 341**
1. $4 \times 10^{-6}$ A
3. $V_{BE} = 11.5$ mV

**Job 10-12: Page 346**
1. $R_1 = 12\ \Omega$,
$R_2 = 3.75\ \Omega$
3. $R_1 = 5\ \Omega$,
$R_2 = 5.45\ \Omega$
5. $R_1 = 36\ \Omega$,
$R_2 = 40\ \Omega$
7. $R_1 = 20\ \Omega$,
$R_2 = 400\ \Omega$
9. $R_1 = 75\ \Omega$,
$R_2 = 33.3\ \Omega$

**Job 10-12: Page 348**
1. $R_1 = R_3 = 20\ \Omega$
$R_2 = 80\ \Omega$
3. $R_1 = R_3 = 20\ \Omega$
$R_2 = 80\ \Omega$
5. $R_1 = R_3 = 7.14\ \Omega$
$R_2 = 3.43\ \Omega$
7. $R_1 = R_2 = 77.7\ \Omega$
$R_2 = 25.3\ \Omega$

**Job 10-13: Page 349**
1. $R_1 = 20$ Ω—10 W
$R_2 = 40$ Ω—5 W
$R_L = 40$ Ω—5 W
3. $R_1 = 10$ Ω—55 W
$R_2 = 25$ Ω—113 W
$R_3 = 20$ Ω—113 W
5. $R_1 = 3$ Ω, 7 W
$R_2 = 4$ Ω, 7 W
$R_3 = 12$ Ω, 2 W

**Job 11-1: Page 351**
1. 100 Ω
2. 0.765 lb
3. 17.25 A
4. No
5. 70.5 V
6. 90 percent
7. 93.3 percent
8. 57.6 W
9. 117.6 V
10. $81.25
11. $79.62
12. 25 percent
13. $3.33
14. 12.8 oz

**Job 11-2: Page 352**
1. 0.38
3. 0.06
5. 0.04
7. 0.036
9. 1.25
11. 0.167
13. 0.625
15. 0.125
17. 0.0225
19. 0.0425

**Job 11-2: Page 353**
1. 50 percent
3. 20 percent
5. 145 percent
7. 100 percent
9. $62\frac{1}{2}$ percent
11. 22.2 percent
13. 70 percent
15. $5\frac{1}{2}$ percent

**Job 11-2: Page 354**
1. 25 percent
3. 40 percent
5. $62\frac{1}{2}$ percent
7. 30 percent
9. 50 percent
11. 65 percent
13. $66\frac{2}{3}$ percent
15. 45.4 percent
17. $33\frac{1}{3}$ percent
19. 22.2 percent

**Job 11-2: Page 355**
1. 33
3. $7.50
5. $56.25
7. 17.25 A

9. 116.7 V
11. 5
13. 7.5 W
15. $19.18

## Job 11-2: Page 357

1. 25 percent
3. 40 percent
5. 2 percent
7. 150 percent
9. 33 1/3 percent
11. 12 1/2 percent
13. 40 percent
15. 2.25 percent
17. 20 percent
19. 5 percent

## Job 11-2: Page 358

1. 20
3. 40
5. 2000
7. $238.33
9. 15 A

## Job 11-2: Page 359

1. (*a*) 0.62
   (*b*) 0.03
   (*c*) 0.056
   (*d*) 0.008
   (*e*) 1.16
   (*f*) 0.045
   (*g*) 0.0625
3. (*a*) 75 percent
   (*b*) 60 percent
   (*c*) 42.9 percent
   (*d*) 30 percent
   (*e*) 23.1 percent
   (*f*) 16 2/3 percent
   (*g*) 46.1 percent
5. $90
7. 40 percent
9. $48.75
11. 20.8 percent
13. 1.5 percent
15. 116.4 V
17. $22.80, $307.80
19. $88.80

## Job 11-3: Page 361

1. 6500 W
3. 2.3 kW
5. 50 W
7. 3/4 kW
9. 1.69 kW
11. 0.094 kW
13. 5.33 hp
15. 13 1/3 hp

## Job 11-4: Page 363

1. 83 1/3 percent
3. 96 percent
5. 88.9 percent
7. 91.3 percent
9. 80.2 percent;
   19.8 percent

## Job 11-5: Page 365

1. 3.6 hp
3. 0.587 hp
5. 4508 W
7. 2647 W
9. 2200 W, 3.26 hp
11. 220 V

## Job 11-6: Page 367

1.

| $R_L$ | $I_L$ | $P_{RL}$ | $P_{P_{th}}$ | Eff, % |
|---|---|---|---|---|
| 6 | 3/4 A | 5.63 | 3.38 | 38 |
| 8 | 2/3 A | 4.44 | 3.56 | 45 |
| 10 | 3/5 A | 3.60 | 3.60 | 50 |
| 12 | 6/11 A | 2.98 | 3.57 | 55 |
| 14 | 1/2 A | 2.58 | 3.50 | 58 |

3. $R = 100\ \Omega$

## Job 11-7: Page 368

1. (*a*) 0.5 kW
   (*b*) 2/3 hp
   (*c*) 1500 W
   (*d*) 1125 W
   (*e*) 13.3 hp
   (*f*) 7.5 kW
3. 91.4 percent
5. 0.62 hp
7. 0.625 kW

## Job 12-1: Page 370

1. 2:5
2. 4:1
3. 9:1
4. 10:11
5. 3:4
6. 0.007 Ω
7. $23.75
8. 45 oz
9. 15 V
10. 104 Ω
11. $\dfrac{R_1}{R_2} = \dfrac{I_2}{I_1}$
12. 2 V

## Job 12-2: Page 373

1. 1:4
3. 2:3
5. 2:1
7. 4:1
9. (*a*) 3:2
   (*b*) 2:3
11. 1:4
13. 1:4
15. 83.3 percent
17. 171
19. 40
21. 10
23. 1:48
25. (*a*) 2 V
   (*b*) 1.7 percent
   (*c*) 98.3 percent

## Job 12-2: Page 377

1. 2750 lb
3. 4 1/2 h
5. 71.4 ft$^3$
7. 1000 Ω
9. 45 V

## Job 12-2: Page 380

1. 5 in
3. 66.7
5. 500
7. 0.08 A
9. 0.93 Ω
11. 0.25 A

## Job 12-3: Page 382

1. (*a*) 1:3
   (*b*) 1:6
   (*c*) 1:10
   (*d*) 5:8
   (*e*) 3:4
3. 36
5. 72
7. 0.6 A
9. 110 V

## Job 12-4: Page 384

1. No. 21, 0.0285 in
3. No. 2, 250 mil
5. No. 6, 162 mil
7. 64.08 mil, 0.0641 in
9. No. 18, 0.0403 in

## Job 12-4: Page 386

1. 0.010 in, 100 mil
3. 0.064 in, 64 mil
5. 32 mil, 1024 cmil
7. 0.0872 in, 7604 cmil
9. 0.173 in, 173 mil
11. 397,908 cmil

## Job 12-5: Page 388

1. 19.6 Ω
3. 12.5 Ω

## Job 12-6: Page 389

1. 3.68 Ω
3. 4.06 Ω
5. 1.205 Ω, 5 A
7. 32.82 Ω, 3.35 A
9. 33.7 Ω, 6.74 V

## Job 12-7: Page 391

1. 52.5 Ω
3. 508 Ω
5. 12.86 Ω

## Job 12-8: Page 393

1. 0.104 Ω
3. 0.0062 Ω
5. 4.46 Ω, 24.6 A,
   2.71 kW
7. (*a*) 0.145 A
   (*b*) 0.87 W
9. 0.2 V

## Job 12-9: Page 395

1. 1.7 Ω
3. 96.1 ft
5. 3 ft
7. 370 ft
9. Nichrome

**Job 12-10: Page 396**

1. 9.4 mil
3. 6.93 mil
5. 32.2 mil
7. No. 17

**Job 12-11: Page 398**

1. (*a*) 25 mil, No. 22
   (*b*) 102 mil, No. 10
   (*c*) 125 mil, No. 8
   (*d*) 128 mil, No. 8
3. 44.9 Ω
5. 1.4 Ω
7. 3.07 ft

**Job 13-1: Page 401**

1. 55 A
3. No. 6
5. 34.5 V
7. 0.000 25 Ω/ft
9. No. 3

**Job 13-2: Page 403**

1. No. 6
3. No. 4
5. No. 1

**Job 13-3: Page 407**

1. No. 4
3. No. 0
5. No. 3
7. No. 12
9. (*a*) 5250 W
   (*b*) 250 W
   (*c*) 44 A
   (*d*) 5.7 V
   (*e*) No. 1
11. 450 ft

**Job 13-4: Page 409**

1. 40 A
3. No. 3
5. No. 14
7. No. 10
9. No. 1
11. No. 14

**Job 14-1: Page 415**

1. sin = 0.2462
   cos = 0.9692
   tan = 0.2540
3. sin = 0.3243
   cos = 0.9459
   tan = 0.3428
5. sin = 0.8823
   cos = 0.4706
   tan = 1.8750

**Job 14-2: Page 418 (Top)**

1. 18°
3. 80°
5. 30°
7. 30°
9. 60°

**Job 14-2: Page 418 (Bottom)**

1. 15°
3. 58°
5. 65°
7. 16°
9. 46°

**Job 14-3: Page 420**

| | ∠A | ∠B |
|---|---|---|
| 1. | 53° | 37° |
| 3. | 30° | 60° |
| 5. | 55° | 35° |
| 7. | 51° | 39° |
| 9. | 11° | 79° |

11. 67°, 113°
13. 3°
15. 58°

**Job 14-4: Page 425**

1. 10 in, 30°
3. 89.4 ft, 40°
5. 151 Ω, 60°
7. 940.4 ft, 62°
9. 123 Ω, 75°
11. 35.1 m, 33 m
13. 8.5 cm
15. 9.1 m
17. $I_x$ = 12.69 A,
    $I_y$ = 5.92 A
19. 2128 Ω

**Job 14-5: Page 427**

1. 0.5878
3. 2.6051
5. 0.9063
7. 65°
9. 11°
11. 17°
13. 73°
15. 37°
17. 71°
19. 58°
21. 930 W
23. 38.5
25. 833 Ω
27. 2828 W
29. 236 V
31. 81°, 99°
33. 3°
35. 1.72 in
37. 23°
39. 233.3 Ω

**Job 15-1: Page 435**

1. (*a*) 0.54 A
   (*b*) 0.63 A
   (*c*) 0.66 A
   (*d*) 0.72 A
   (*e*) 0.79 A
   (*f*) 0.87 A
   (*g*) 0.99 A
5. Max eff = 80 percent; 50 hp

**Job 15-3: Page 447**

1. 10,000 meters
3. 126.4 V
5. 17°
7. 43.3 A
9. (*a*) 30°
   (*b*) 37°
   (*c*) 60°

**Job 15-4: Page 451**

| | MAX | EFF | INSTANT |
|---|---|---|---|
| 1. | — | 24.7 | 17.5 |
| 3. | 622 | — | 476 |
| 5. | — | 109.6 | 77.5 |
| 7. | 155.5 | — | 140.9 |
| 9. | 35 | 24.7 | — |
| 11. | — | 70.7 | 26° |

13. 17.3 V
15. (*a*) 10.6 A
    (*b*) 900 W
17. 285 W

**Job 15-5: Page 454**

1. 90 V
3. (*a*) 150 mV
   (*b*) 0.2 Hz

**Job 16-2: Page 460**

1. (*a*) 439.6 Ω
   (*b*) 8792 Ω
3. (*a*) 7536 Ω
   (*b*) 0.02 A
5. 37.7 Ω
7. 5 H
9. 62,800 Ω
11. 2864 Ω
13. 2.7 mH
15. 255 μH

**Job 16-3: Page 463**

1. 4
3. 31.4
5. (*a*) 10
   (*b*) 1000 Ω
7. (*a*) 9420 Ω
   (*b*) 10,665 Ω
   (*c*) 3.8 mA
9. (*a*) 314 Ω
   (*b*) 330 Ω
11. 0.7 mA

**Job 16-4: Page 466**

1. 3.18 H
3. 0.796 H
5. 0.51 H

**Job 16-5: Page 467**

1. (*a*) 125.6 Ω
   (*b*) 1256 Ω
   (*c*) 12,560 Ω
   (*d*) 125,600 Ω
3. 3140 Ω
5. 60
7. 30 Ω
9. 0.61 H
11. 0.25 H

**Job 16-6: Page 471**

1. 360 V
3. (*a*) 48:1
   (*b*) 48:1

5. 600 turns
7. 6000 V
9. 1.5 V
11. 300 turns
13. 240 V
15. (*a*) 2 turns
    (*b*) 5 1/4 turns
    *c*) 500 turns

**Job 16-7: Page 475**
1. 2 A
3. (*a*) 0.16 A
   (*b*) 19.2 W
5. (*a*) 0.136 A
   (*b*) 7.5 V
7. (*a*) 0.086 A
   (*b*) 9.45 W
9. (*a*) 5 A
   (*b*) 1500 V
11. 511 A

**Job 16-8: Page 477**
1. 87.5 percent
3. (*a*) 180 W
   (*b*) 168 W
   (*c*) 93.3 percent
5. 80 percent
7. (*a*) 10,417 W
   (*b*) 5.21 A
9. (*a*) 77 W
   (*b*) 88 V

**Job 16-9: Page 480**
1. 20:1
3. 4:1
5. 1:5
7. 12,100 Ω
9. 8:1
11. 39:1
13. 1:3

**Job 16-10: Page 482**
1. (*a*) 176 turns
   (*b*) 8:1
3. 1750 turns
5. (*a*) 100 turns
   (*b*) 22.7 turns
7. (*a*) 1.25 A
   (*b*) 150 W
9. (*a*) 0.063 A
   (*b*) 7.56 W
11. 93.3 percent
13. 12.5:1
15. (*a*) 240 V
    (*b*) 12 A
    (*c*) 40 mA
    (*d*) 2880 W

**Job 17-2: Page 488**
1. 373 pF
3. 5065 pF
5. 30,085 pF
7. 150 pF

**Job 17-3: Page 491**
1. 1.333 μF
3. 19 pF
5. 31 to 162 pF
7. 3 μF
9. 8 μF

**Job 17-4: Page 495**
1. (*a*) 17,667 Ω
   (*b*) 5300 Ω
   (*c*) 662 Ω
3. 0.19 A
5. 400 pF; 3975 Ω
7. 1.67 μF
9. (*a*) 15.9 Ω
   (*b*) 3.18 Ω
   (*c*) 0.795 Ω
11. 13,250 Ω
13. 60 Hz
15. (*a*) 31.8 Ω
    (*b*) 0.636 Ω
17. 0.02 μF
19. 442 μF

**Job 17-5: Page 497**
1. 41 Ω
3. 15,900 Ω (the resistance may be neglected)
5. 10,500 Ω

**Job 17-6: Page 499**
1. 2.65 μF
3. 2.94 μF
5. 10 μF

**Job 17-7: Page 500**
1. 0.0497 μF
3. 1.6 μF
5. 187 pF
7. (*a*) 26.5 Ω
   (*b*) 0.053 Ω
9. 44.7 Ω
11. 0.318 μF

**Job 18-1: Page 504**
1. 60 Ω
3. 110 V
5. 160 V, 3.2 W
7. 1.82 A, 60.4 Ω
9. 111 V, 24.7 Ω

**Job 18-1: Page 509**
1. 1 A
3. 25.12 V
5. 3000 Ω, 7.96 H

**Job 18-1: Page 511**
1. (*a*) 2.25 A
   (*b*) 0
   (*c*) 0 W
3. 124 Ω
5. 39.75 Ω, 0.1 A
7. 0.25 mA
9. 100 Ω, 0.001 59 μF

**Job 18-2: Page 514**
1. 10
3. 16
5. 26.5
7. 4.5
9. 72.11 m
11. 7.62 ft
13. 46 ft
15. 32 in

**Job 18-3: Page 521**
1. (*a*) 13 Ω
   (*b*) 8 A
   (*c*) $V_R = 40$ V
   $V_L = 96$ V
   (*d*) 67°
   (*e*) 39.1 percent
   (*f*) 321 W
3. 0.92 A, 74°
5. (*a*) 150 Ω
   (*b*) 170 Ω
   (*c*) 85 V
7. 1.5 μA
9. 4.5 A
11. 3770 Ω, 0.01 A
13. 28.6 H
15. $X_L = 7.5$ Ω; 43°

**Job 18-4: Page 526**
1. (*a*) 17 Ω
   (*b*) 7 A
   (*c*) $V_R = 56$ V
   $V_C = 105$ V
   (*c*) 62°
   (*e*) 47.1 percent
   (*f*) 392 W
3. (*a*) 141 Ω
   (*b*) 0.8 A
   (*c*) $V_R = 80$ V
   $V_C = 80$ V
   (*d*) 45°
   (*e*) 70.9 percent
   (*f*) 64 W
5. 159,000 Ω, 18,800 Ω
7. (*a*) 500 Ω
   (*b*) 400 Ω
   (*c*) 6.63 μF
9. 113 Hz

**Job 18-5: Page 535**
1. (*a*) 65 Ω
   (*b*) 2 A
   (*c*) $V_R = 32$ V
   $V_L = 166$ V
   $V_C = 40$ V
   (*d*) 76°
   (*e*) 24.6 percent
   (*f*) 64 W
3. (*a*) $X_L = 6280$ Ω
   $X_C = 3180$ Ω
   (*b*) 5060 Ω
   (*c*) 0.025 A
   (*d*) 37°
   (*e*) 79.1 percent
   (*f*) 2.47 W
5. 0.7 V
7. 1.5 A
9. 271 Ω
11. 0.01 H

**Job 18-6: Page 539**
1. (*a*) 795 kHz
   (*b*) 50 Ω

3. (*a*) 2 kHz
   (*b*) 502 Ω
   (*c*) 502 Ω
   (*d*) 12 Ω
   (*e*) 0.5 A
   (*f*) 251 V
5. 1988 kHz
7. 1988 kHz
9. 1590 kHz

**Job 18-7: Page 542**

1. 84.3 μH
3. 0.506 H
5. 35.1 μF
7. 72 μH
9. 2.53 mH

**Job 18-8: Page 545**

1. 180 V
3. 5.3 H
5. (*a*) 6360 Ω
   (*b*) 7.9 mA
   (*c*) $V_R$ = 7.9 V
   $V_L$ = 49.6 V
   (*d*) 81°
   (*e*) 15.7 percent
   (*f*) 0.06 W
7. (*a*) 41.2 Ω
   (*b*) 4790 Ω
9. 71.3 kHz

**Job 19-1: Page 552**

1. (*a*) 14.4 A
   (*b*) 8.33 Ω
   (*c*) 1728 W
3. (*a*) 1.5 A
   (*b*) 66.7 Ω
   (*c*) 0
   (*d*) 0 W
5. 0.11 A, 1636 Ω; 0 W
7. (*a*) 0.159 A, 0.064 A
   (*b*) 0.223 A
   (*c*) 538 Ω
   (*d*) 0
   (*e*) 0 W

**Job 19-2: Page 557**

1. (*a*) 13 A
   (*b*) 9.23 Ω
   (*c*) 67°
   (*d*) 38.5 percent
   (*e*) 600 W
3. (*a*) 2.15 A
   (*b*) 46.5 Ω
   (*c*) 22°
   (*d*) 93 percent
   (*e*) 200 W
5. 70 percent, 100 percent; Yes, but not a good filter, since it also passes low frequencies.

**Job 19-3: Page 561**

1. (*a*) 17 A
   (*b*) 7.06 Ω
   (*c*) 28°
   (*d*) 88.2 percent
   (*e*) 1799 W
3. (*a*) 2.23 A
   (*b*) 53.9 Ω
   (*c*) 63°
   (*d*) 44.8 percent
   (*e*) 120 W
5. 0.566 A
7. 44.8 percent AF, 0.05 percent RF; Yes, but not a good one.
9. 200 μA

**Job 19-4: Page 567**

1. $R$ = 7.2 Ω; $C$ = 276 μF
3. $R$ = 28.2 Ω; $L$ = 16 mH
5. (*a*) 4.12 A
   (*b*) 29.1 Ω
   (*c*) 14° leading
   (*d*) 97.1 percent
   (*e*) 480 W
   (*f*) $R$ = 28.2 Ω; $C$ = 376 μF
7. (*a*) $X_L$ = 7536 Ω
   $X_C$ = 3312 Ω
   (*b*) $I_L$ = 0.029 A
   $I_C$ = 0.066 A
   $I_R$ = 0.1 A
   (*c*) 0.106 A
   (*d*) 2075 Ω
   (*e*) 19° leading
   (*f*) 94.3 percent
   (*g*) 22 W
   (*h*) $R$ = 1963 Ω; $C$ = 3.9 μF

**Job 19-5: Page 571**

1. $I_x$ = 17.32 A
   $I_y$ = 10 A
3. $I_x$ = 10 A
   $I_y$ = −17.32 A
5. $I_x$ = 19.32 A
   $I_y$ = 17.4 A
7. $I_x$ = 64.3 mA
   $I_y$ = −76.6 mA
9. $I_x$ = 12.69 A
   $I_y$ = 5.92 A

**Job 19-6: Page 583**

1. (*a*) 23.4 A
   (*b*) 5.13 Ω
   (*c*) 35° leading
   (*d*) 2305 W
3. (*a*) 14.1 A
   (*b*) 7.09 Ω
   (*c*) 0°
   (*d*) 1410 W
5. (*a*) 1.8 A
   (*b*) 55.5 Ω
   (*c*) 48° lagging
   (*d*) 120 W

**Job 19-7: Page 589**

1. (*a*) $Z$ = 21.1 Ω
   (*b*) $I_T$ = 4.74 A
   (*c*) $\theta$ = 17° lagging
   (*d*) 95.6 percent
   (*d*) $P$ = 454 W
3. (*a*) $Z$ = 60.7 Ω
   (*b*) $L_T$ = 1.98 A
   (*c*) $\theta$ = 31° lagging
   (*d*) 85.9 percent
   (*e*) $P$ = 204 W
5. (*a*) $Z$ = 30.9 Ω
   (*b*) $I_T$ = 3.9 A
   (*c*) $\theta$ = 12° leading
   (*d*) 97.6 percent
   (*e*) $P$ = 457 W

**Job 19-8: Page 591**

1. 5300 kHz
3. 400 pF
5. 80 pF
7. 0.92 μH
9. 40.2 pF

**Job 19-9: Page 595**

1. (*a*) 0.083 A
   (*b*) 301 Ω
   (*c*) 0 W
3. (*a*) $I_R$ = 0.5 A
   $I_L$ = 0.16 A
   (*b*) $I_T$ = 0.52 A
   (*c*) 192 Ω
   (*d*) 16°
   (*e*) 96.2 percent
   (*f*) 50 W
5. (*a*) 128 mA
   (*b*) 1875 Ω
   (*c*) 51°
   (*d*) 62.5 percent
   (*e*) 19.2 W
7. (*a*) 17.1 A
   (*b*) 7 Ω
   (*c*) 31° leading
   (*d*) 85.6 percent
   (*e*) 1756 W
9. 120 μH
11. (*a*) 10 Ω
   (*b*) $I_T$ = 10 A
   (*c*) 37° leading
   (*d*) 800 W

**Job 20-1: Page 601**

1. 80 percent
3. 360 W
5. (*a*) 4400 VA
   (*b*) 90.9 percent
7. 2.5 A
9. 45.45 A
11. 3450 VA; 2933 W
13. 10,417 VA
15. 40 A
17. 800 var
19. 10 A

**Job 20-2: Page 609**

1. (*a*) 1232 W
   (*b*) 80 percent leading
   (*c*) 1540 VA
   (*d*) 14 A
3. (*a*) 1650 W
   (*b*) 89.9 percent
   (*c*) 1835 VA

(*d*) 16.7 A
5. (*a*) 27 kW
(*b*) 54.5 percent
(*c*) 49.5 kVA
7. (*a*) 13 kW
(*b*) 73.1 percent
(*c*) 17.8 kVA
(*d*) 80.9 A
9. (*a*) 72 kW
(*b*) 64.3 percent
(*c*) 112 kVA

**Job 20-3: Page 614**
1. (*a*) 950 W
(*b*) 99.8 percent
(*c*) 952 VA
3. (*a*) 5680 W
(*b*) 92.1 percent
(*c*) 6167 VA
(*d*) 51.4 A
5. 98.8 percent
7. (*a*) 22 kW
(*b*) 90.6 percent
(*c*) 24.28 kVA
(*d*) 105.6 A
9. (*a*) 1584 W
(*b*) 94.6 percent
(*c*) 1674 VA
(*d*) 13.95 A

**Job 20-4: Page 625**
1. 17.4 percent
3. 58.8 percent
5. 38 $\mu$F
7. 6.67 kVA
9. 109.6 kvar; Graph = 108; 4 standard 25-*C* kvar; 63.5 kVA released; Graph = 64 kVA
11. (*a*) 111.6 kW
(*b*) 93.3 kvar
(*c*) 76.6 percent
(*d*) 145.7 kVA
(*e*) (1) 25 *C* kvar plus (1) 15 *C* kvar

**Job 20-5: Page 629**
1. 28 percent, 4.8 A
3. (*a*) 1944 W
(*b*) 90 percent
(*c*) 2160 VA
(*d*) 18 A
5. (*a*) 22.5 kW
(*b*) 62.9 percent
(*c*) 35.8 kVA
7. (*a*) 1936 W
(*b*) 99.5 percent
(*c*) 1946 VA
(*d*) 16.6 A
9. 45.4 percent
11. (*a*) 80 percent
(*b*) 900 var
(*c*) 900 var
13. (*a*) 82.4 kVA
(*b*) 32°
(*c*) 43.7 kvar
(*d*) 1-15 *C* kvar, plus 1-25 *C* kvar
(*e*) 99.9 percent
15. 22.6 *C* kvar; 7.5 kVA released

**Job 21-1: Page 639**
1. 2078 V
3. 40 A
5. (*a*) 460 V
(*b*) 4505 W
(*c*) 13.52 kW
7. (*a*) 9.01 kVA
(*b*) 88 percent
9. (*a*) 4858 kVA
(*b*) 0.823
(*c*) 3810 V
11. (*a*) 12,000 V
(*b*) 361 A
(*c*) 7500 kVA
(*d*) 6000 kW
13. (*a*) 6.93 A
(*b*) 6.93 A
(*c*) 1440 W
15. (*a*) 480 V
(*b*) 13 Ω
(*c*) 21.3 A
(*d*) 21.3 A
(*e*) 0.923
(*f*) 16.34 kW
17. (*a*) 4157 V
(*b*) 127.4 kW
(*c*) 20.8 A
(*d*) 20.8 A
19. (*a*) 100 Ω
(*b*) 138.6 V
(*c*) 1.39 A
(*d*) 0.866
(*e*) 500 W
(*f*) 577 VA

**Job 21-2: Page 647**
1. 103.9 A
3. 440 V
5. (*a*) 5.77 A
(*b*) 10 A
7. (*a*) 240 V
(*b*) 10 A
(*c*) 17.3 A
(*d*) 7.2 kW
(*e*) 7.2 kVA
9. (*a*) 0.82
(*b*) 35°
(*c*) 61 kVA
(*d*) 35 kvar
11. (*a*) 72 kVA
(*b*) 72 kVA

**Job 21-3: Page 653**
1. (*a*) 17.3 A
(*b*) 240 V
(*c*) 12.5 kVA
3. 27.4 kW
5. (*a*) 350 A
(*b*) 208 V
(*c*) 100.8 kW
7. 120 kW
9. (*a*) 520 kW
(*b*) 166.7 A
(*c*) 288.7 A
(*d*) 1300 V
11. (*a*) 6 A
(*b*) 1.73 kW
(*c*) 3.46 A
(*d*) $R = 16\ \Omega$
$X_L = 12\ \Omega$
13. (*a*) 40.7 kW; 36.8 kvar
(*b*) 20.0 kVA; 12.0 kvar
(*c*) 56.7 kW; 24.8 kvar
(*d*) 0.914
(*e*) 74.62 A

**Job 22-5: Page 664**
1. (*a*) 120 V
(*b*) 120 V
3. (*a*) Delta-delta
(*b*) 481 A
5. (*a*) 240 V
(*b*) 20.8 kVA
(*c*) 28.9 A
(*d*) 2.89 A
(*e*) 20.8 kVA
7. (*a*) 3138 A
(*b*) 1812 A
(*c*) 69,000 V
(*d*) 13,800 V
(*e*) 1:5
(*f*) 362.4 A
9. (*a*) 35,160 V
(*b*) 4400 V
(*c*) 6000 kVA
(*d*) 56.9 A
(*e*) 56.9 A
(*f*) 455.2 A
(*g*) 788.4 A
(*h*) 6000 kVA
11. (*a*) 3810 V
(*b*) 1575 A
(*c*) 6600 V; 909 A
(*d*) 19,050 V; 314.9 A
(*e*) 33,000 V; 181.8 A
(*f*) 13,800 V; 434.7 A
(*g*) 18,000 kVA per bank, or 6000 kVA per coil
13. (*a*) 200 kVA
(*b*) 13,800 V; 14.4 A
(*c*) 2300 V; 86.4 A
(*d*) 86.3 percent

**Job 23-2: Page 672**
1. (*a*) 0
(*b*) 1
(*c*) 0
3. $L = S_1 \cdot S_3$

**Job 23-3: Page 675**
1. 8
3. 4

### Job 23-4: Page 678

1. 16
3.

5.

| $S_1$ | $S_2$ | $(S_1S_2)'$ | $S_1' + S_2'$ |
|---|---|---|---|
| 0 | 0 | 1 | 1 |
| 0 | 1 | 1 | 1 |
| 1 | 0 | 1 | 1 |
| 1 | 1 | 0 | 0 |

### Job 23-6: Page 684

1.

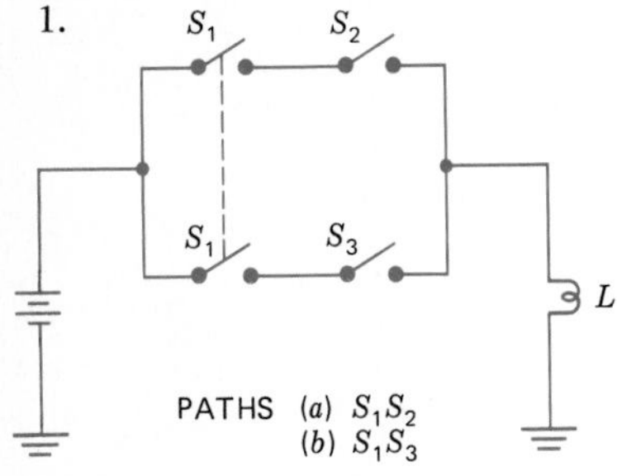


3.

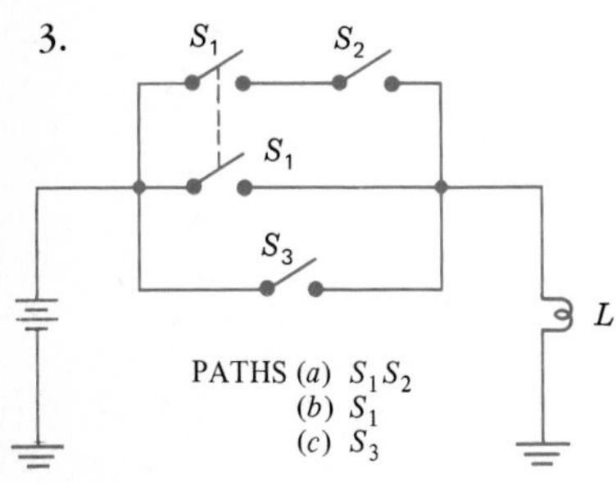


5.

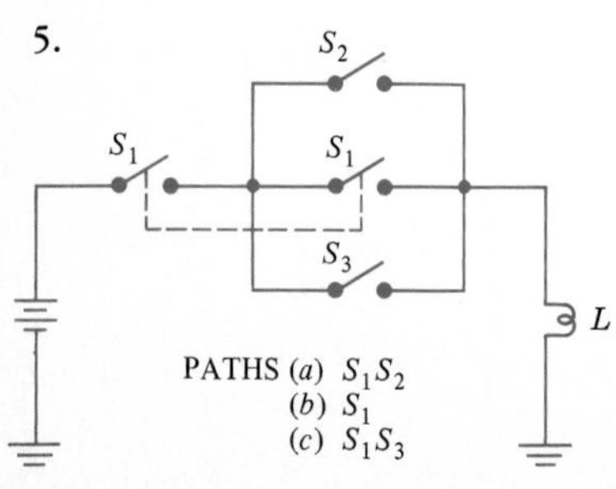


### Job 23-7: Page 686

1.

| $S_1'$ | $S_2$ | $S_3$ | $(S_2 + S_3)$ | $S_1$ | $L$ |
|---|---|---|---|---|---|
| 0 | 0 | 0 | 0 | 1 | 1 |
| 0 | 0 | 1 | 1 | 1 | 1 |
| 0 | 1 | 0 | 1 | 1 | 1 |
| 0 | 1 | 1 | 1 | 1 | 1 |
| 1 | 0 | 0 | 0 | 0 | 0 |
| 1 | 0 | 1 | 1 | 0 | 1 |
| 1 | 1 | 0 | 1 | 0 | 1 |
| 1 | 1 | 1 | 1 | 0 | 1 |

3.

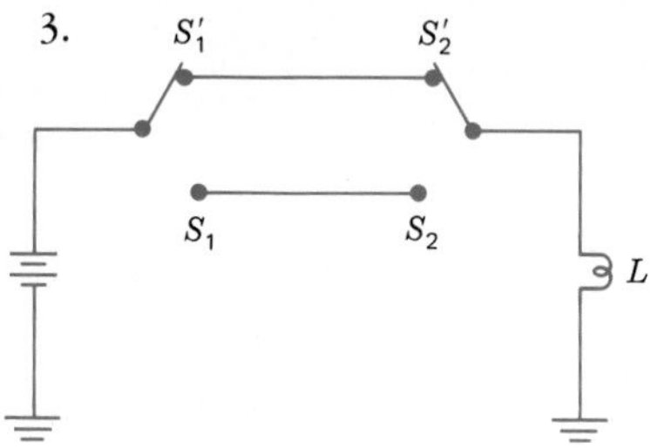


5. $L_1 = L_1'$
The user cannot tell the difference in the operation.

### Job 23-8: Page 693

1.

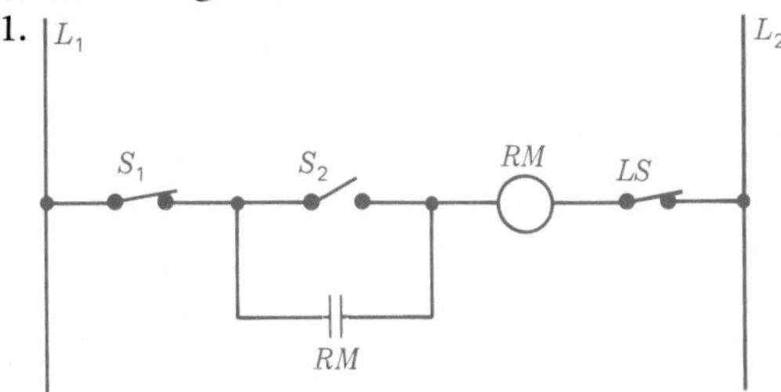


3. If $R_M = 1$, then $R_M = S_1L_S$
5. (*a*) If the relay burned shut, then $R_M = S_1L_S$.
(*b*) Either push $S_1$ or let the motor run to the limit switch.

### Job 23-9: Page 698

1.

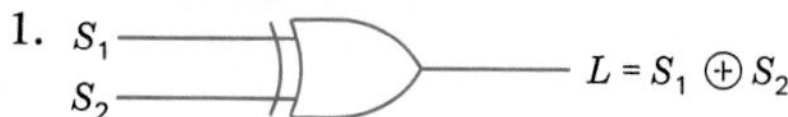

3.

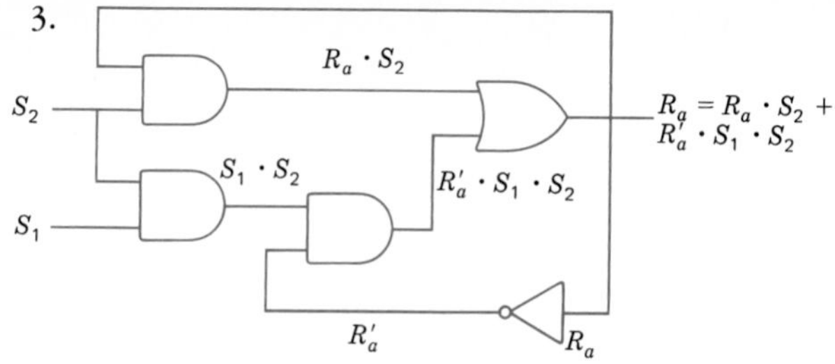


5. $L = S_1(S_2 + S_3)$

| $S_3$ | $S_2$ | $S_1$ | $S_2 + S_3$ | $L$ |
|---|---|---|---|---|
| 0 | 0 | 0 | 0 | 0 |
| 0 | 0 | 1 | 0 | 0 |
| 0 | 1 | 0 | 1 | 0 |
| 0 | 1 | 1 | 1 | 1 |
| 1 | 0 | 0 | 1 | 0 |
| 1 | 0 | 1 | 1 | 1 |
| 1 | 1 | 0 | 1 | 0 |
| 1 | 1 | 1 | 1 | 1 |

7. 1; No;
Since $S_3 = 1$, $S_3 + S_4 = 1$. The output is independent of $S_4$.

Appendix: Color Codes: Page 704

1. 100,000 Ω
3. 82,000 Ω
5. 75 Ω
7. 4.7 Ω
9. 1.2 Ω
11. 1,800,000 Ω
13. 620 Ω
15. 9100 Ω
17. 0.47 Ω
19. 82 Ω
21. Red, yellow, yellow
23. Green, brown, black
25. Brown, black, blue
27. Orange, white, red
29. Brown, red, black
31. Blue, gray, silver
33. Red, red, gold
35. Brown, gray, gold
37. Orange, blue, brown
39. Brown, green, gold

Appendix: Preferred Values: Page 705

21. 220,000 Ω
23. 47 Ω
25. 10,000,000 Ω
27. 3900 Ω
29. 12 Ω
31. 0.68 Ω
33. 2.2 Ω
35. 1.8 Ω
37. 33 Ω or 39 Ω
39. 1.5 Ω

# INDEX

# SYMBOLS AND ABBREVIATIONS

| TERM | SYMBOLIC QUANTITY | ABBREVI-ATION |
|---|---|---|
| Alternating current | | ac |
| Ampere (unit of current) | $I$ | A |
| Milliampere | | mA |
| Microampere | | μA |
| American Wire Gage | | AWG |
| Apparent power | $VA$ | VA |
| Angle | ∠ | |
| Area | $A$ | |
| Circular mils | | cmil |
| Square inches | | in² |
| Base | $B$ | |
| Candela | | cd |
| Capacitance | $C$ | |
| Collector | $C$ | |
| Constant | $K$ | |
| Continuous wave | | cw |
| Cosine | | cos |
| Coulomb | $C$ | C |
| Current | $I$ | |
| Average value | $I_{av}$ | |
| Change in current | $\Delta I$ | |
| Effective value | $I$ | |
| Instantaneous value | $i$ | |
| Maximum value | $I_{max}$ | |
| Cycles | | |
| Cycles per second | | Hz |
| Kilocycles per second | | kHz |
| Megacycles per second | | MHz |
| Decibel | | dB |
| Delta (Greek letter) | Δ | |
| Diameter | $D, d$ | diam |
| Degree | ° | deg |
| Diode | $D$ | |
| Direct current | | dc |
| Efficiency | | Eff |
| Electromotive force | $E$ | emf |
| Emiter | $E$ | |
| Energy | $W$ | |

| TERM | SYMBOLIC QUANTITY | ABBREVI-ATION |
|---|---|---|
| Farad (unit of capacitance) | $C$ | F |
| Microfarad | $C$ | μF |
| Picofarad | $C$ | pF |
| Foot | | ft |
| Frequency | $f$ | freq |
| Audio | | AF |
| Intermediate | | IF |
| High | | HF |
| Low | | LF |
| Radio | | RF |
| Resonant | | f |
| Grid | $G$ | |
| Ground | | gnd |
| Henry (unit of inductance) | $L$ | H |
| Millihenry | $L$ | mH |
| Microhenry | $L$ | μH |
| Hertz | | Hz |
| High pass | | h-p |
| Horsepower | | hp |
| Hour | $T$ | h |
| Impedance | $Z$ | |
| Inch | | in |
| Inductance | $L$ | |
| Kilo (1000) | | k |
| Low pass | | l-p |
| Maximum | | max |
| Mega (1,000,000) | | M |
| Micro (one-millionth) | | μ |
| Milli (one-thousandth) | | m |
| Minimum | | min |
| Ohm (unit of resistance) | $R$ | Ω |
| Pi (Greek letter) | $\pi$ | |
| Pico | | p |
| Phase angle (theta) | $\theta$ | |
| Power | $P$ | |
| Apparent power | $VA$ | VA |
| Effective power | $W$ | |
| Input power | $P_i$ | |